McGraw-Hill Ryerson Limited 

A Subsidiary of The ***McGraw·Hill*** *Companies*

COPIES OF THIS BOOK MAY BE OBTAINED BY CONTACTING:
McGraw-Hill Ryerson Ltd.

E-MAIL:
orders@mcgrawhill.ca

TOLL FREE FAX:
1-800-463-5885

TOLL FREE CALL:
1-800-565-5758

OR BY MAILING YOUR ORDER TO:
McGraw-Hill Ryerson
Order Department
300 Water Street
Whitby, Ontario
L1N 9B6

Please quote the ISBN and title when placing your order.

McGraw-Hill Ryerson Biology 12

Copyright © 2002, McGraw-Hill Ryerson Limited, a Subsidiary of The McGraw-Hill Companies. All rights reserved. No part of this publication may be reproduced or transmitted in any form or by any means, or stored in a data base or retrieval system, without the prior written permission of McGraw-Hill Ryerson Limited, or, in the case of photocopying or other reprographic copying, a licence from CANCOPY (Canadian Copyright Licensing Agency), One Yonge Street, Suite 1900, Toronto, Ontario, M5E 1E5.

Any request for photocopying, recording, or taping of this publication shall be directed in writing to CANCOPY.

The information and activities in this textbook have been carefully developed and reviewed by professionals to ensure safety and accuracy. However, the publishers shall not be liable for any damages resulting, in whole or in part, from the reader's use of the material. Although appropriate safety procedures are discussed in detail and highlighted throughout the textbook, safety of students remains the responsibility of the classroom teacher, the principal, and the school board/district.

0-07-088713-6 / 0-07-091674-8

http://www.mcgrawhill.ca

2 3 4 5 6 7 8 9 0 TRI 0 9 8 7 6 5 4 3 2

Printed and bound in Canada

Care has been taken to trace ownership of copyright material contained in this text. The publisher will gladly take any information that will enable them to rectify any reference or credit in subsequent printings. Please note that products shown in photographs in this textbook do not reflect an endorsement by the publisher of those specific brand names.

National Library of Canada Cataloguing in Publication Data

Main entry under title:

McGraw-Hill Ryerson biology 12

Includes index.

ISBN 0-07-088713-6

1. Biology. I. Blake, Leesa II. Title: Biology 12.
III. Title: Biology twelve. IV. Title: McGraw-Hill Ryerson biology twelve.

QH308.7.M32 2002 570 C2002-900243-5

The Biology 12 Team

SCIENCE PUBLISHER: Jane McNulty
PROJECT MANAGER: Alan Simpson
DEVELOPMENTAL EDITORS: Jonathan Bocknek, Anita Drabyk, Dan Kozlovic, Lesley McKarney, Frances Purslow, Neil Purslow, Tom Shields, Michael J. Webb
SENIOR SUPERVISING EDITOR: Linda Allison
PROJECT CO-ORDINATOR: Valerie Janicki
PROJECT ASSISTANTS: Melissa Nippard, Janie Reeson
COPY EDITOR: Linda Jenkins
PROOFREADER: Carol-Ann Freeman
PERMISSIONS EDITOR: Karen Shore, Pronk&Associates Inc.
SPECIAL FEATURES CO-ORDINATOR: Keith Owen Richards
PRODUCTION CO-ORDINATOR: Jennifer Wilkie
PRODUCTION SUPERVISOR: Yolanda Pigden
COVER DESIGN, INTERIOR DESIGN, AND ART DIRECTION: Pronk&Associates Inc.
ELECTRONIC PAGE MAKE-UP: Pronk&Associates Inc.
SET-UP PHOTOGRAPHY: Ian Crysler
SET-UP PHOTOGRAPHY CO-ORDINATOR: Shannon O'Rourke
TECHNICAL ILLUSTRATION: Pronk&Associates Inc., Imagineering Scientific, Brett Clayton, Bernadette Lau, Dave Mazierski, Jun Park, Theresa Sakno, Jane Whitney
COVER IMAGE: A. Syred/Photo Researchers Inc.

Acknowledgements

Producing a textbook of high quality is a true team effort, requiring the input and expertise of a very large number of people. The authors, consultants, project manager, and publisher of this book would like to convey our sincere thanks, first and foremost, to the reviewers listed below who provided critical analysis of our draft manuscript and who checked content for scientific accuracy. Their assistance was invaluable in helping us to develop a text that we hope you will find completely appropriate for your teaching and your students' learning. We extend particular thanks to Leesa Blake and Marshall Letcher, who wrote additional questions and unit material, provided additional activities, and supplied ongoing advice and support "beyond the call of duty." Our appreciation is extended to Catherine Little and D'Arcy Little, who co-authored the Biology Course Challenge. We thank the following writers who authored the Special Features in *Biology 12*: Karin Banerd, Kirsten Craven, Eric Grace, Denyse O'Leary, Trevor Tucker, and Elma Schemenauer. We would also like to thank Jonathan Bocknek for his insightful comments on Units 1 and 2; Tom Dickinson for his considerable assistance and advice in developing Unit 5; and to Zoltan and Margit Koritar, Meg O'Mahony, and Angela Yu for their advice and assistance in the preparation of visual material for Chapter 9. We acknowledge, with gratitude and respect, Trudy Rising, who initiated McGraw-Hill Ryerson's senior science program, and who worked tirelessly in support of the program, its authors, and its development team.

Finally, we thank a wonderfully co-operative design studio, Pronk&Associates Inc., and its talented staff, who collaborated with us closely to make this book come to life.

Pedagogical and Academic Reviewers

Dr. Sumihisa Aota
Nepean High School
Ottawa, Ontario

Marietta (Mars) Bloch
Toronto District School Board
Toronto, Ontario

Stuart Cumner
Crescent School
Willowdale, Ontario

Bryan Dixon
Head of Science
William Lyon Mackenzie
Collegiate Institute
Toronto, Ontario

Clayton Ellis
Turner Fenton
Secondary School
Brampton, Ontario

Bruce Evans
St. Joan of Arc Catholic
High School
Maple, Ontario

Keith Gibbons
Catholic Central High School
London, Ontario

Ed Hitchcock
Bayview Glen School
North York, Ontario

Terry Kilroy
Widdifield Secondary School
North Bay, Ontario

Lucy Kisway
Glendale Secondary School
Hamilton, Ontario

David Kondziolka
Cardinal Newman High School
Toronto, Ontario

James E.E. Kushny
Renfrew County District
School Board
Barry's Bay, Ontario

Kirby Lee
Dr. Norman Bethune
Collegiate Institute
Toronto, Ontario

Mieah Lee
Peoples Christian Academy
North York, Ontario

D'Arcy Little, M.D.
York Community Services and
University of Toronto
Toronto, Ontario

Philip Marsh
Rick Hansen Secondary School
Mississauga, Ontario

Donna Matovinovic
Edmonton Public Schools
Edmonton, Alberta

Matthew McKinlay
Head of Science
Sir John A. MacDonald
Collegiate Institute
Toronto, Ontario

Amish Parikh, M.D.
University of Toronto,
Faculty of Medicine
Toronto, Ontario

Dennis Pikulyk
Wexford Collegiate Institute
Scarborough, Ontario

Chris Schramek
John Paul II Catholic
Secondary School
London, Ontario

Sandy Searle
Western Canada High School
Calgary, Alberta

Charles Sims
Mackenzie High School
Deep River, Ontario

Gail de Souza
Marshall McLuhan Catholic
Secondary School
Toronto, Ontario

Accuracy Reviewers

Dr. Heather Addy
University of Calgary
Calgary, Alberta

Dr. Catherine Christie
Queen's University
Kingston, Ontario

Dr. David Ng
University of British Columbia
Vancouver, British Columbia

Dr. J.R. Pickavance
Memorial University
of Newfoundland
St. John's, Newfoundland
and Labrador

Dr. Safia Wasi
Education Consultant
Toronto, Ontario

Dr. Michael Webb
(Ph.D. Chemistry)
Michael J. Webb Consulting Inc.
Toronto, Ontario

Safety Reviewer

Dr. Margaret-Ann Armour
University of Alberta
Edmonton, Alberta

Contents

Safety in Your Biology Laboratory

Active involvement in science enhances learning. Thus, investigations are integrated throughout this textbook. Keep in mind at all times that working in a laboratory can involve some risks. *Therefore, become familiar with all facets of laboratory safety, especially for performing investigations safely.* To make the investigations and activities in *Biology 12* safe and enjoyable for you and others who share a common working environment,

- become familiar with and use the following safety rules and procedures,
- follow any special instructions from your teacher, and
- *always* read over the safety notes before beginning each and every lab activity.

Your teacher will tell you about any additional safety rules that are in place at your school.

General Rules

1. Read through all of the steps in the investigation before beginning. Be sure to read and understand the *Cautions* and safety symbols at the beginning of each Investigation or MiniLab.
2. Listen carefully to any special instructions your teacher provides. Get your teacher's approval before beginning any investigation that you have designed yourself.
3. Inform your teacher if you have any allergies, medical conditions, or physical problems (including a hearing impairment) that could affect your work in the laboratory.
4. Inform your teacher if you wear contact lenses. If possible, wear eyeglasses instead of contact lenses, but remember that eyeglasses are not a substitute for proper eye protection.
5. Never eat, drink, or taste any substances in the lab. Never pipette with your mouth. If you are asked to smell a substance, do not hold it directly under your nose. Keep the object at least 20 cm away, and waft the fumes toward your nostrils with your hand.
6. When you are directed to do so, wear safety goggles and protective equipment in the laboratory. Be sure you understand all safety labels on materials and pieces of equipment. Familiarize yourself with the safety symbols used in this textbook, and with the WHMIS symbols found in *Appendix 1*.

Safety Equipment and First Aid

7. Know the location and proper use of the nearest fire extinguisher, fire blanket, fire alarm, first-aid kit, and eye-wash station (if available).
8. Never use water to fight an electrical equipment fire. Severe electrical shock may result. Use a carbon dioxide or dry chemical fire extinguisher. Report any damaged equipment or frayed cords to your teacher.
9. Cuts, scratches, or any other injuries in the laboratory should receive immediate medical attention, no matter how minor they seem. If any part of your body comes in contact with a potentially dangerous substance, wash the area immediately and thoroughly with water.
10. If you get any material in your eyes, do not touch them. Wash your eyes immediately and continuously for 15 min, and make sure your teacher is informed. If you wear contact lenses, take your lenses out immediately if you get material in your eyes. Failing to do so may result in material being trapped behind the contact lenses. Flush your eyes with water for 15 min, as above.

Lab Precautions

11. Make sure your work area is clean, dry, and well-organized.
12. Wear heat-resistant safety gloves, and any other safety equipment that your teacher or the *Cautions* suggest, when heating any item. Be especially careful with a hot plate that may look as though it has cooled down. If you do receive a burn, apply cold water to the burned area immediately. Make sure your teacher is notified.
13. Make sure the work area, the area of the socket, and your hands are dry when touching electrical cords, plugs, sockets, or equipment such as hot plates and microscopes. Ensure the cords on your equipment are placed neatly where they

will not cause a tripping hazard. Turn OFF all electrical equipment before connecting to or disconnecting from a power supply. When unplugging electrical equipment, do not pull the cord — grasp the plug firmly at the socket and pull gently.

14. When using a scalpel or knife, cut away from yourself and others. Always keep the pointed end of any sharp objects directed away from yourself and others when carrying such objects.
15. When you are heating a test tube, always slant it so the mouth points away from you and others.

Safety for Animal Dissections

16. Ensure your work area is well ventilated.
17. Always wear appropriate protective equipment for your skin, clothing, and eyes. This will prevent preservatives from harming you in any way.
18. Make sure you are familiar with the proper use of all dissecting equipment. Whenever possible, use a probe or your gloved fingers to explore the specimen. Scalpels are not appropriate for this. They can damage the structures you are examining.
19. If your scalpel blade breaks, do not replace it yourself. Your teacher will do this for you.

Laboratory Clean-up

20. Wipe up all spills immediately, and always inform your teacher. Acid or base spills on clothing or skin should be diluted and rinsed with water. Small spills of acid solutions can be neutralized with sodium hydrogen carbonate (baking soda). Small spills of basic solutions can be neutralized with sodium hydrogen sulfate or citric acid.
21. Never use your hands to pick up broken glass. Use a broom and dustpan. Dispose of broken glass and solid substances in the proper containers, as directed by your teacher.
22. Dispose of all specimens, materials, chemicals, and other wastes as instructed by your teacher. Do not dispose of materials in a sink or drain unless directed to do so.
23. Clean equipment before putting it away, according to your teacher's instructions. Turn off the water and gas. Disconnect electrical devices. Wash your hands thoroughly after all laboratory investigations.

Working with Living Organisms

24. When in the field, be careful and observant at all times to avoid injury, such as tripping, being poked by branches, etc., or coming into contact with poisonous plants.
25. On a field trip, try not to disturb the area any more than is absolutely necessary. If you must move anything, do so carefully. If you are asked to remove plant material, do so gently. Take as little as possible.
26. In the classroom, remember to treat living organisms with respect. Make sure all living organisms receive humane treatment while they are in your care. If it is possible, return living organisms to their natural environment when your work is done.

NOTE: Some schools do not permit labs that involve bacteria. Your teacher will inform you of your school board's policy in this regard.

27. When working with micro-organisms, observe your results through the clear lid of the petri dish. Do *not* open the cover. Make sure that you do not touch your eyes, mouth, or any other part of your face.
28. When handling live bacterial cultures, always wear gloves and eye protection. Be careful not to spill the cultures. Wash your hands thoroughly with soap immediately after handling any bacterial culture.
29. Carefully clean and disinfect your work area after handling bacterial cultures and other living organisms.
30. Follow your teacher's instructions about disposal of your swabs, petri dishes containing your cultures, and any other disposable materials used in the lab.
31. Your teacher will autoclave cultures before discarding them, if an autoclave is available. If an autoclave is not available, the culture surface should be sprayed with a 10% solution of chlorine bleach. (Your school may have other disposal techniques.)

Here is a quick glimpse at the learning that lies before you in this course. Expand your knowledge and skills from your grade 11 biology course and experience biology in action.

How are bonds formed between atoms, and how do molecules interact? In Unit 1, you will study the chemical reactions that take place inside cells. You will discover how special proteins, called enzymes, help speed up chemical reactions. Where does the energy to fuel these reactions come from? Both plant and animal cells synthesize energy-rich molecules that are used in a variety of cellular processes. At the end of the unit, you will design an investigation to study enzyme activity.

Eating ice cream stimulates the secretion of insulin, one of the hormones that control blood sugar. In Unit 2, you will learn how the body responds to change and how it maintains a stable internal environment. You will see how the nervous system and the endocrine system regulate life processes and how the immune system fends off foreign invaders. You will study how hormones control growth, development, and reproduction. You will also conduct investigations into various metabolic disorders. At the end of Unit 2, you will complete a project on the metabolic disorder diabetes.

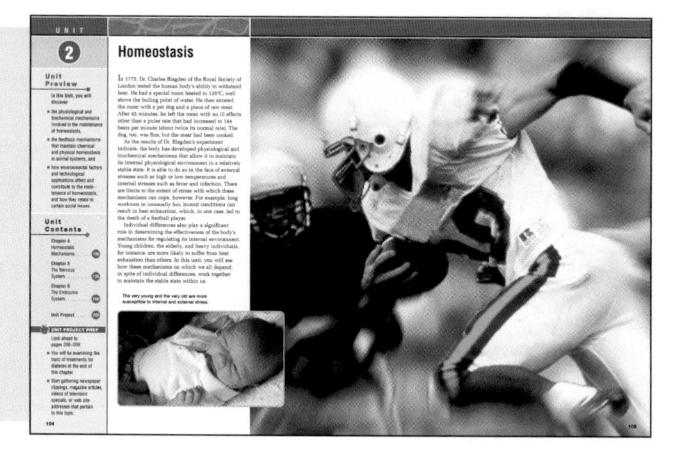

What do the colour of a newborn bear and the development of a new strain of rice have in common? The same genetic structures and processes are at work in each case. In Unit 3, you will discover how certain molecules provide coded instructions for life processes for generation after generation, while allowing new organisms to arise. You will see how scientists are learning to manipulate those instructions. At the end of the unit, you will research a disease that can develop when these instructions fail.

How has life on Earth changed over the millennia? What mechanisms give rise to new species? In Unit 4, you will evaluate how fossil evidence, observation of species over time, and genetic analysis have shaped our understanding of evolution. You will examine how genetic variation within a population allows changes to occur within a species. At the end of the unit, you will research and present a report on the topic of the common ancestry of living organisms.

UNIT 4

Evolution

UNIT 5

Population Dynamics

What effect do changing populations have on Earth? This question may be vital to the survival of life on our planet. In Unit 5, you will analyze the components required for population growth, examine the interrelationships of these components within ecosystems, and evaluate the carrying capacity of Earth with respect to the growth of populations. At the end of Unit 5, you will choose a country and analyze an environmental issue that affects that country's human population.

Once you have completed the five units of the textbook, you will have an opportunity to use your skills and knowledge in a Biology Course Challenge. You will research and analyze information related to a topic of your choice, and then communicate your work by presenting a paper or a poster in a simulated Biology Symposium. Through examination of your topic from different perspectives, you can demonstrate the relationship of the topic to science, technology, society, and the environment.

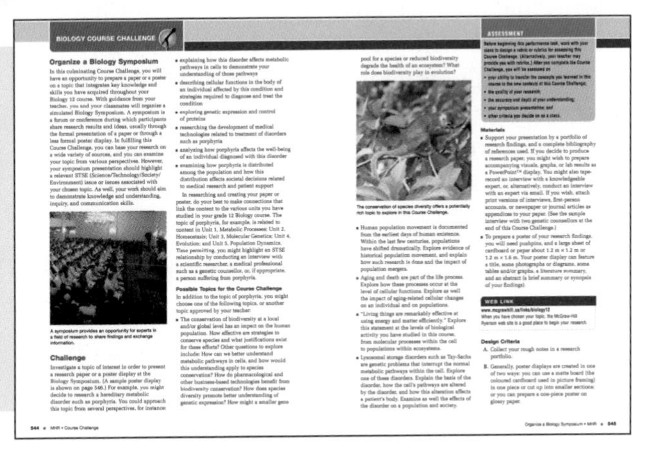
BIOLOGY COURSE CHALLENGE

Organize a Biology Symposium

COURSE CHALLENGE

The evolution of rattlesnake venom or the evolution of sickle cell disease are two potential topics that can be examined in the Course Challenge. Start making notes on the links that evolution (in general) has with metabolic processes, homeostasis, and molecular genetics. In the next unit, think about how the population dynamics might, in turn, be affected by changes in populations.

Watch for this feature, which appears in every unit of the textbook. It will prompt you to think about the upcoming Course Challenge and remind you to begin planning for it.

- Use the cues to trigger your thought processes and point you to a line of research.
- Keep a file of ideas and preliminary research for your Course Challenge.

Reviewing Essential Skills

The appendices at the back of your textbook can help you review information and skills that are essential to all scientific investigation. Safety symbols and WHMIS symbols are featured in Appendix 1. If you would like to review metric measurements, turn to Appendix 2, Units of Measurement. The illustration below surveys the range of sizes, from atoms to whole organisms, that you will encounter in this course.

The appendices at the back of the book include Significant Digits (Appendix 3), Cell Division (Appendix 4), Shapes of Selected Macromolecules (Appendix 5), Electronegativity (Appendix 6), and the Periodic Table and List of Elements (Appendix 7).

Achieving Excellence

The Achievement Chart on the facing page identifies four main categories that you may use to assess your own learning during the course. Your teacher will use these categories to assess your achievement of curriculum expectations for your biology course.

You may be asked to design or to review rubrics that are based on the Achievement Chart. These rubrics will provide specific criteria for assessing each of the four Achievement Chart categories.

The features outlined below are developed to provide both you and your teacher with a variety of ways to assess your achievement.

Section Review

- Each numbered section in each chapter is followed by a set of section review questions.
- These questions, indicated by the following symbols, are organized according to the categories given in the Achievement Chart.

  K/U Knowledge and Understanding

  I Inquiry　　C Communication

  MC Making Connections
- In general, Section Review questions appear in order from least to most challenging.

Chapter Review

- Each chapter ends with a set of features and questions to review and reinforce your understanding of the knowledge, concepts, skills, and issues presented in the chapter.
- Summary of Expectations is a list of the key concepts and themes presented in the chapter. Each concept includes the section number(s) within the chapter where it is discussed.
- Language of Biology is a list of key terminology that appeared in boldfaced type in the chapter.
- Review questions for the entire chapter are provided, organized according to the four categories given in the Achievement Chart.

Unit Review

- Following each unit, a review allows you to assess your understanding of the entire unit.

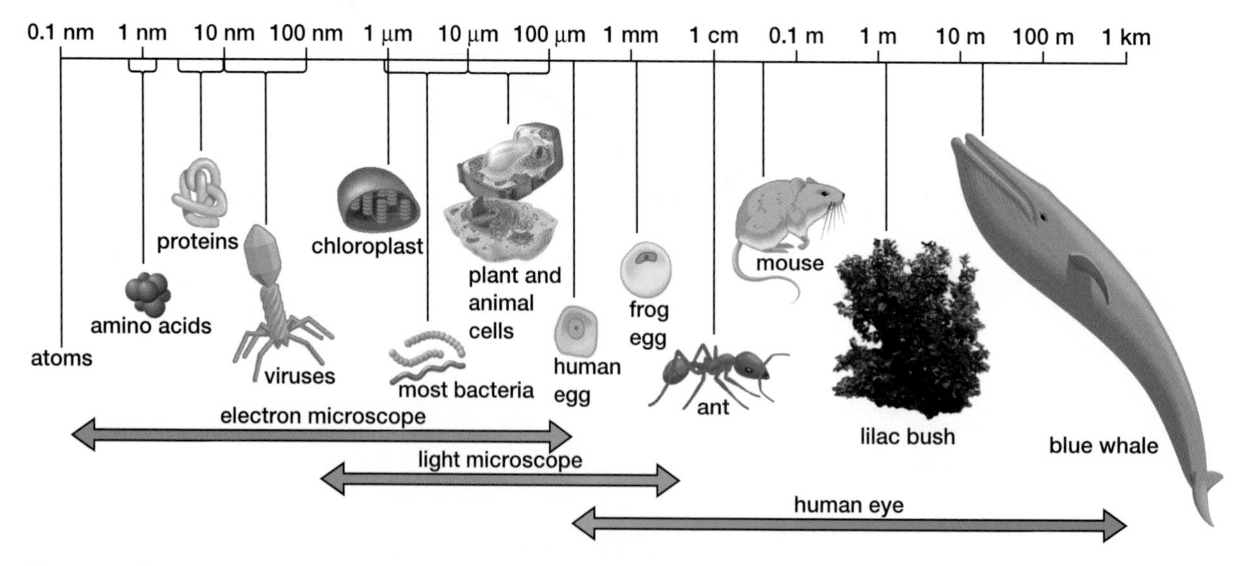

How large does an organism have to be before it can be seen by the unaided human eye?

- Questions are organized according to the four categories of the Achievement Chart. Within each category, questions are generally listed in order of increasing difficulty.

Other Features to Assist Your Understanding

A casual flip through the pages of your textbook will show you a range of features and items to stimulate your interest, aid your understanding, and practise your skills. The following represent an overview of these helpful features.

UNIT INVESTIGATION PREP

UNIT PROJECT PREP

UNIT ISSUE PREP

Unit Investigation/Project/Issue Prep: At the end of each unit, you will find a culminating performance task that your teacher may assign. This task involves either an issue to analyze, a project to undertake, or an investigation to design and perform. Throughout the units, one of these logos will remind you of the upcoming assignment, and provide helpful tips.

Probeware

Probeware: Use electronic probes for certain investigations and projects, if they are available in your school.

WEB LINK

Web Link: At the McGraw-Hill Biology 12 web site, you will find links to many useful, reputable educational institutions and organizations — excellent sources of material for culminating projects, for example.

CAUTION: The Use and Sharing of Computer Files
Be sure to have an up-to-date virus checker on your computer. The checker is critical because even a disk given to you by a close friend could contain a virus. Never open an e-mail file if you do not know the source of the message, since you might, inadvertently, infect your computer with a virus that could destroy files containing many hours of your work. Even if you do know the individual who sent you the message, some of the newer viruses are transmitted via address books of friends and peers. If you discover that you have a virus in your computer, notify your teacher and peers immediately.

Biology At Work

Biology at Work: Explore the range of biology careers and opportunities available to you.

Biology Magazine TECHNOLOGY • SOCIETY • ENVIRONMENT

Biology Magazine: Reflect on topical issues that touch your life and encourage you to make connections among science, technology, society, and the environment.

Canadians in Biology

Canadians in Biology: Meet Canadian scientists involved in important research and discoveries.

ELECTRONIC LEARNING PARTNER

Electronic Learning Partner: Refer to animations, simulations, and videoclips on your student e-book.

Achievement Chart

Knowledge and Understanding	Inquiry	Communication	Making Connections
■ Understanding of concepts, principles, laws and theories ■ Knowledge of facts and terms ■ Transfer of concepts to new contexts ■ Understanding of relationships between concepts	■ Application of skills and strategies of scientific inquiry ■ Application of technical skills and procedures ■ Use of tools, equipment, and materials	■ Communication of information and ideas ■ Use of scientific terminology, symbols, conventions, and standard (SI) units ■ Communication for different audiences and purposes ■ Use of various forms of communication ■ Use of information technology for scientific purposes	■ Understanding of connections among science, technology, society, and the environment ■ Analysis of social and economic issues involving science and technology ■ Assessment of impacts of science and technology on the environment ■ Proposing courses of practical action in relation to science-and-technology-based problems

UNIT

Metabolic Processes

Unit Preview

In this Unit, you will discover

- what molecules are necessary for metabolic functions in cells,
- which major reactions occur in cells,
- how thermodynamic principles maintain metabolic function,
- which processes are involved in cellular respiration and photosynthesis, and
- how knowledge of metabolic processes can contribute to technological development.

Unit Contents

UNIT INVESTIGATION

Look ahead to pages 98–99.

- You can start planning your investigation well in advance by organizing what you will need.
- As you work through the unit, watch for ideas and materials that will help you prepare your experimental design.

Like large emeralds encrusted with gold, thousands of chrysalides (cocoons) hang from milkweed plants in southern Ontario. Within each of these chrysalides, a monarch butterfly caterpillar will undergo a metamorphosis to become an adult butterfly. This process requires much energy to fuel the tremendous changes that occur in a caterpillar's physical appearance and abilities.

All organisms require energy to survive. Cells in a eukaryotic organism contain organelles, such as the mitochondria shown below, that transform the energy in food into energy that can be used for various cellular processes. Without mitochondria, organisms such as the monarch caterpillar would not be able to perform the metabolic processes they need for metamorphosis. Metabolic processes involve all the chemical reactions that take place in cells, as well as the chemical reactions that need energy to transport molecules and build the cellular structures necessary for all life processes.

In this unit, you will learn about the chemical reactions that form molecules and see how the laws of thermodynamics govern all reactions between molecules. You will discover how special proteins are essential to metabolic processes in the cell. You will explore the series of metabolic reactions that take place in cells and learn how energy is transformed and used in these reactions. Finally, you will explore how the study of cell biology relates to your life and lifestyle.

How do organisms obtain the energy they need for life processes?

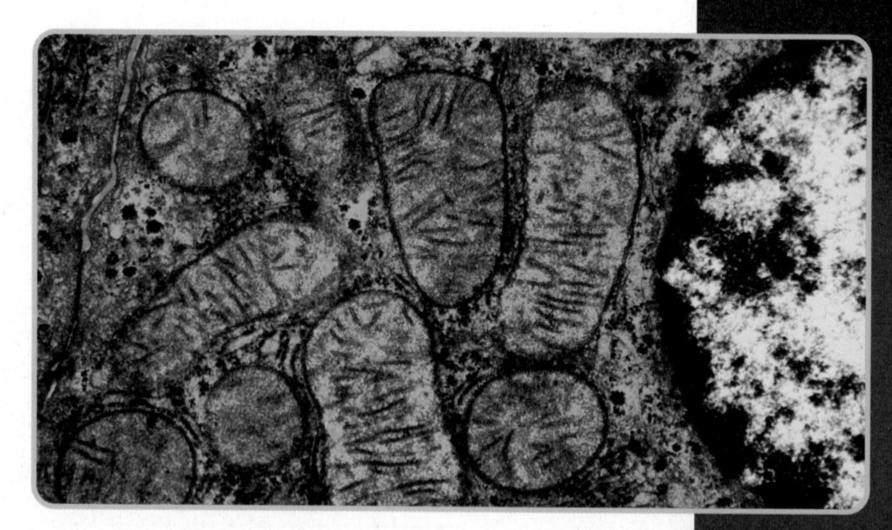

CHAPTER 1

The Chemistry of Life

Reflecting Questions

- How are bonds formed between atoms?
- What is ATP, and what reactions within the cell govern its formation?
- What reactions build and break down molecules?
- Why are large molecules essential for life?

Within each living cell, chemical reactions take place millions of times every second. These reactions involve the transport of material, the removal of wastes, and the formation of structures, such as the molecule of protein shown in the image on the facing page. How do these reactions proceed and where are they carried out? Questions such as these have been at the centre of many investigations into metabolic processes in unicellular and multicellular organisms. Knowing how metabolic reactions take place and understanding the intricacies of the key steps in those reactions have helped researchers further their knowledge of metabolic processes.

In the field of human biology, for example, radiologists have been using radioactive isotopes since the 1930s to treat and diagnose disease. The electron micrograph on the right shows healthy human thyroid tissue. Radioactive iodine is used as a marker to locate thyroid tissue that may be cancerous. Iodine can only be utilized in thyroid tissue. If the machine tracing the radioactive iodine shows that some of it remains in the patient's body, doctors and technicians know that cancerous tissue remains to be removed. How are substances, like iodine, that are part of the non-living world related to substances that are part of the living world? What chemical processes are involved and what molecules play key roles in these processes?

In this chapter, you will explore the basis for biochemical processes. Although biologists are not usually concerned with individual atoms, they must know how atoms combine to form the molecules that animals and plant cells need to perform daily body functions such as growth, maintenance, and repair. Metabolic processes involve the interactions of electrons and the formation or breaking of bonds between atoms to produce or break down molecules. Understanding the chemistry of metabolism is key to understanding its many reactions and the products of those reactions.

How do biologists use their knowledge of molecules and chemical reactions in order to understand metabolic processes?

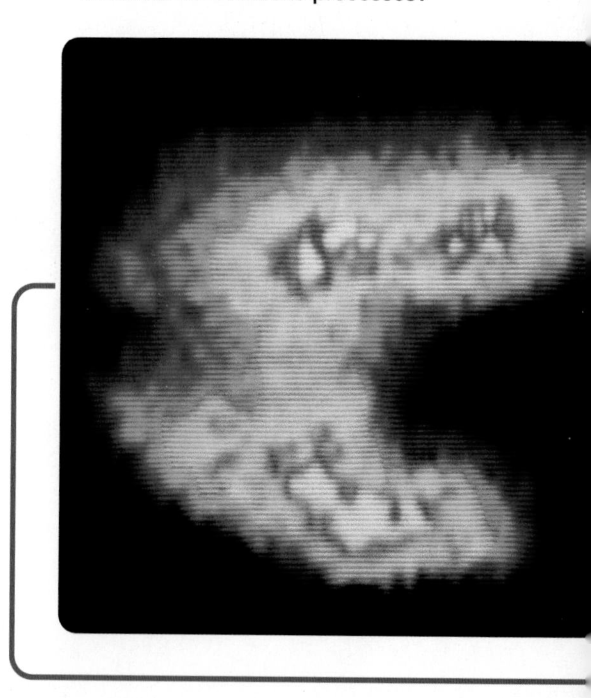

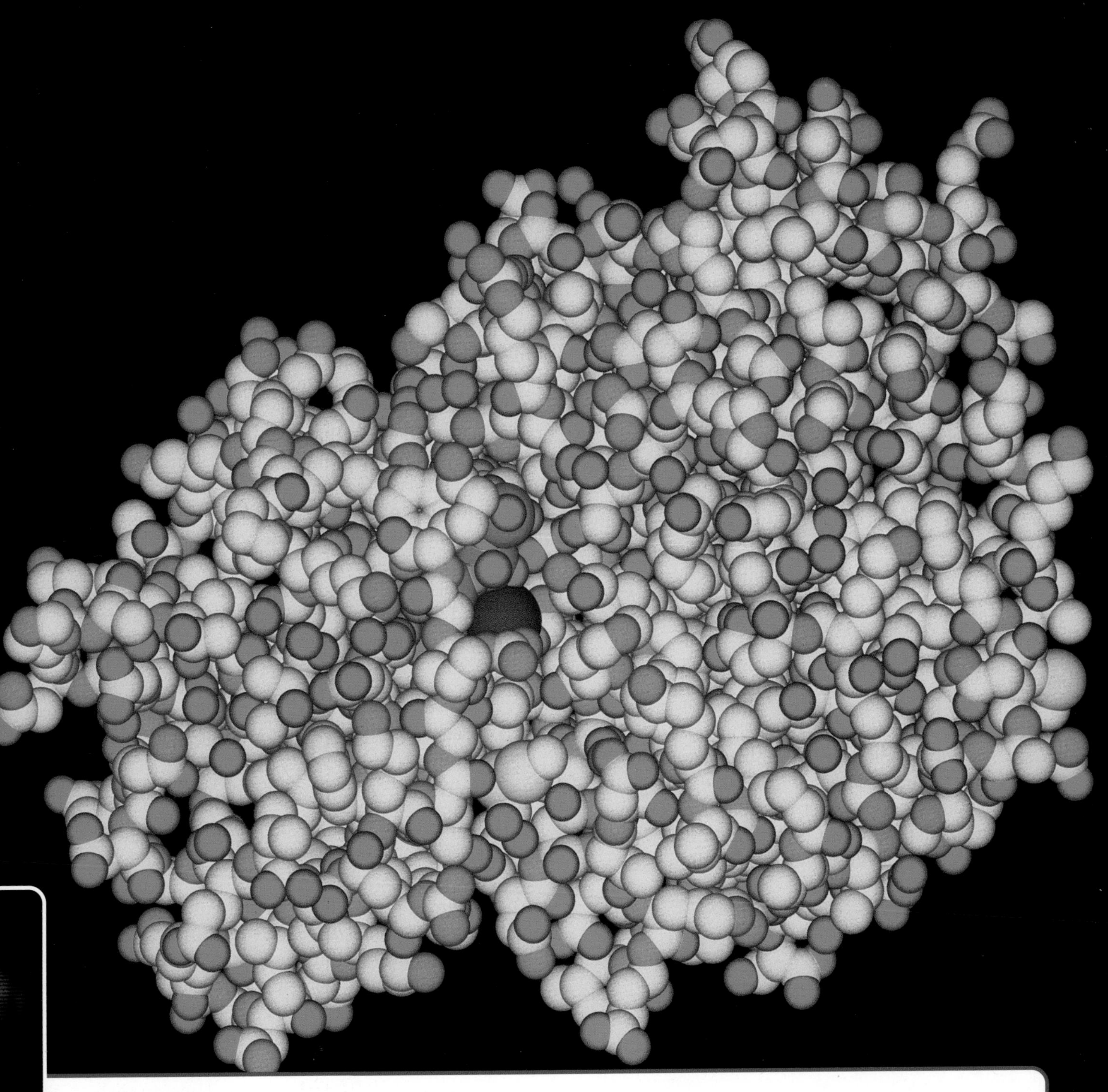

Chapter Contents

Atoms and Bonding

EXPECTATIONS

- **Explain the term electronegativity.**
- **Express the relationship between electronegativity, bonding, and the polarity of molecules.**

Living things are unique among all forms of matter. Unlike non-living things, all living things — from single-celled organisms such as the *Euglena* in Figure 1.1 to multicelled organisms such as whales and redwood trees — interact with and manipulate matter and energy. For example, all cells take in essential substances such as oxygen, water, and nutrients from their external environment. Inside cells, these substances undergo chemical reactions of several types. These reactions may be used to break down substances, synthesize others, and repair defective structures. Chemical reactions also provide energy for these life-sustaining activities, as well as others such as reproduction. Unneeded (or harmful) products of the reactions are eliminated as wastes.

Collectively, these processes — intake of substances, processing of substances, and elimination of wastes — are called metabolic processes, or **metabolism**. The substances involved in metabolism are molecules. The bonds that form between atoms define the structure and properties of these molecules. In this section, you will review several key ideas about atoms and bonding.

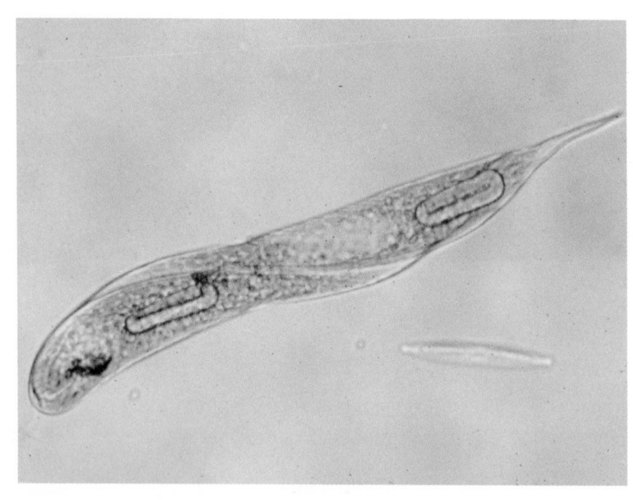

Figure 1.1 *Euglena*, a unicellular freshwater organism, carries out the same metabolic processes that your cells do.

Atoms and Elements

As you have learned in previous studies, all matter is formed of atoms. The **atom** is the smallest unit of matter involved in chemical reactions. Although tiny, atoms are complex structures, composed of even smaller subatomic particles. Most students of chemistry still study the model of the atom that Danish physicist Niels Bohr presented in the early twentieth century (see Figure 1.2). In this model, an atom consists of a small, dense core called a **nucleus**. It is composed of two kinds of subatomic particles — the positively (+) charged **protons** and the uncharged, or neutral, **neutrons**. Also in the Bohr model, negatively (–) charged **electrons** orbit the nucleus in one or more energy levels, or shells.

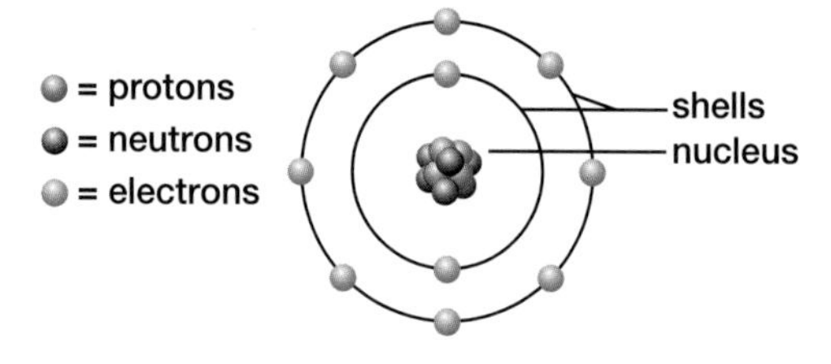

Figure 1.2 Niels Bohr's model of the neon atom

An **element** is a substance that cannot be broken down into simpler substances by chemical means. Substances such as calcium, oxygen, potassium, iron, and carbon are all elements. A few elements, such as helium, occur as single atoms. Several elements, such as hydrogen, nitrogen, and oxygen, occur as molecules made up of two atoms. Such molecules are called diatomic. Other elements such as phosphorus and sulfur occur as molecules made up of more than two atoms.

All atoms of an element have the same number of protons in their nuclei. This number, called the atomic number, is different for every element. The nuclei of carbon atoms, for example, each contain six protons. Because the nuclei of most atoms also contain neutrons, another important characteristic of an atom is its mass number. The mass number of an atom is the total number of protons and

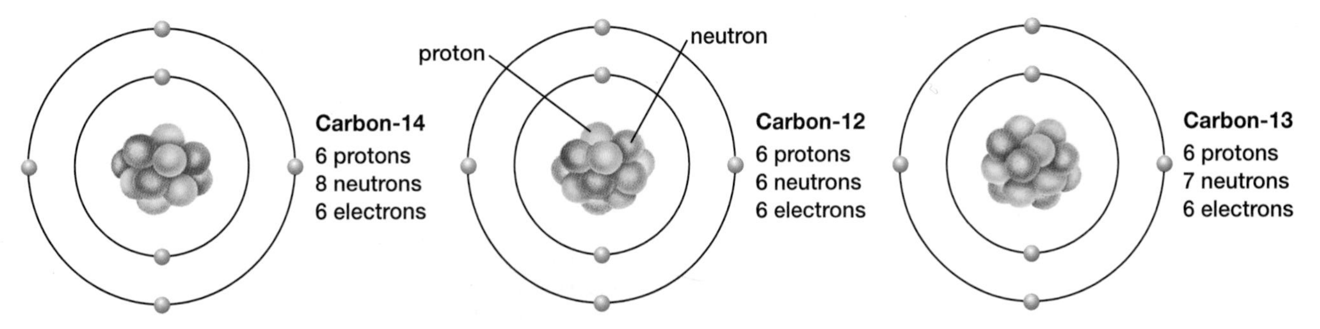

Figure 1.3 Carbon, one of the most important elements in living matter, has three naturally occurring isotopes. The nucleus of each isotope contains 6 protons, but the number of neutrons in the nucleus is 6, 7, or 8. In each isotope, 6 electrons exist outside the nucleus.

neutrons in its nucleus. Atoms of the same element that contain different numbers of neutrons are called **isotopes** of that element. Refer to Figure 1.3 to see the numbers of protons, neutrons, and electrons in three isotopes of carbon. Their names include the mass number of each isotope: carbon-12, carbon-13, and carbon-14.

Some isotopes are stable, whereas others are unstable and break down (decay). The unstable isotopes are known as radioactive isotopes. Carbon-12 and carbon-13 are both stable isotopes, whereas carbon-14 is unstable and decays. Many radioactive isotopes decay at known rates. The rate at which a radioactive isotope decays may be used scientifically. The decay of carbon-14 can be used by archeologists, in a process called radiocarbon dating, to find the ages of some objects up to about 50 000 years old.

Table 1.1
The most abundant elements in living organisms

Element	Symbol	Atomic mass (u)	Abundance (% by mass)
oxygen	O	16.0	62.0
carbon	C	12.0	20.0
hydrogen	H	1.0	10.0
nitrogen	N	14.0	3.3
calcium	Ca	40.1	2.5
phosphorus	P	31.0	1.0
sulfur	S	32.1	0.25
potassium	K	39.1	0.25
chlorine	Cl	35.5	0.2
sodium	Na	23.0	0.1
magnesium	Mg	24.3	0.07
iron	Fe	55.9	0.01

Table 1.1 shows the atomic masses of the elements that are most abundant in living organisms. Notice that, unlike atomic numbers and mass numbers, some atomic masses are not whole numbers. This is the case because the atomic mass of an element is the average mass of all the naturally occurring isotopes of that element. Chlorine, for example, naturally occurs as a mixture of two isotopes: chlorine-35 and chlorine-37. There are three chlorine-35 atoms for every chlorine-37 atom. Therefore, the average mass of chlorine atoms is closer to 35 than to 37. The atomic mass of chlorine is, in fact, 35.5 u (atomic mass units). Appendix 7 provides atomic masses for all the known elements.

Electron Energy

Biologists usually study the groups of atoms that make up molecules rather than atoms and subatomic particles themselves. All cells obtain the energy to function from chemical reactions that involve molecules. The actions of electrons are key to this process.

According to the Bohr model, electrons orbit the nucleus of an atom within energy levels, or shells. An electron in the first shell (nearest the nucleus) has the lowest amount of potential energy. Any electrons in the remaining shells have more potential energy. Each shell can hold a maximum number of electrons. The first shell, for example, can hold a maximum of two electrons, while the second shell can hold a maximum of eight. Refer to Figure 1.2, which shows that in a neon atom the first two shells are filled. In general, the maximum number of electrons that a shell can hold is given by the formula $2n^2$, where n is the number of the shell. For example, the third shell can hold a total of $2(3)^2 = 18$ electrons.

The chemical properties of atoms rely mostly on the number of electrons in the outermost, occupied shell of an atom in its lowest energy state. This shell is known as the **valence shell**. The electrons that occupy the valence shell of any atom are called **valence electrons**. The elements in the periodic table that are least reactive are the noble gases, such as neon, found in group 18(8A) (see Appendix 7). Atoms of the other elements in the periodic table are more reactive than the noble gases. These elements can form chemical bonds with each other. The MiniLab examines safety issues involving the use of chemicals and how they react with each other during chemical bonding.

Ionic and Covalent Bonds

Most atoms can form **chemical bonds** with other atoms. These bonds are the forces that hold the atoms together in the form of compounds. For example, two chlorine atoms can combine (chemically react) to form a diatomic molecule of the element chlorine (Cl_2). Atoms of sodium and chlorine can combine to form the ionic compound sodium chloride ($NaCl$).

There are two general types of chemical bonds. One type involves the sharing of electrons between atoms, and is known as a **covalent bond**. The other type involves the transferring of one or more electrons from one atom to another, and is called an **ionic bond**. How are these bonds formed?

MINI LAB

Using Chemicals Safely

Throughout this course, you will use solutions of chemicals such as hydrochloric acid and sodium hydroxide to help you isolate, identify, or investigate the properties of various substances. Understanding how to use these chemicals properly, and what to do in case of an emergency, is vital to ensure your safety as well as the safety of your classmates.

Obtain a copy of an MSDS (Materials Data Safety Sheet) for the following:

- hydrochloric acid
- sodium hydroxide
- Benedict's solution (reagent)
- Biuret solution
- Lugol's solution
- silver nitrate

Copy the chart below into your notebook, and use the MSDS information to complete it.

Analyze

1. When you use these chemicals for a laboratory activity, what personal safety supplies will you need? Explain your choices.
2. What concentration of substance are you likely to use in your laboratory activities?
3. (a) Substance A is a clear, colourless liquid. How could you safely determine the identity of this liquid and/or its components?
   (b) Substance B is a blue liquid. How could you safely determine the identity of this liquid and/or its components?
4. Write a short scenario that involves a materials spill for dilute hydrochloric acid. Explain how to respond safely to the spill.

	Hydrolochloric acid	Sodium hydroxide	Benedict's solution	Biuret's solution	Lugol's solution	Silver nitrate
Physical/Chemical Properties ■ Appearance? ■ Odour? ■ Solubility in Water? ■ Toxicity?						
Health Hazard Data ■ List three health hazards.						
Safety ■ Identify three safety precautions related to the health hazards listed above. ■ Spill Procedure? ■ Neutralizing Agent?						

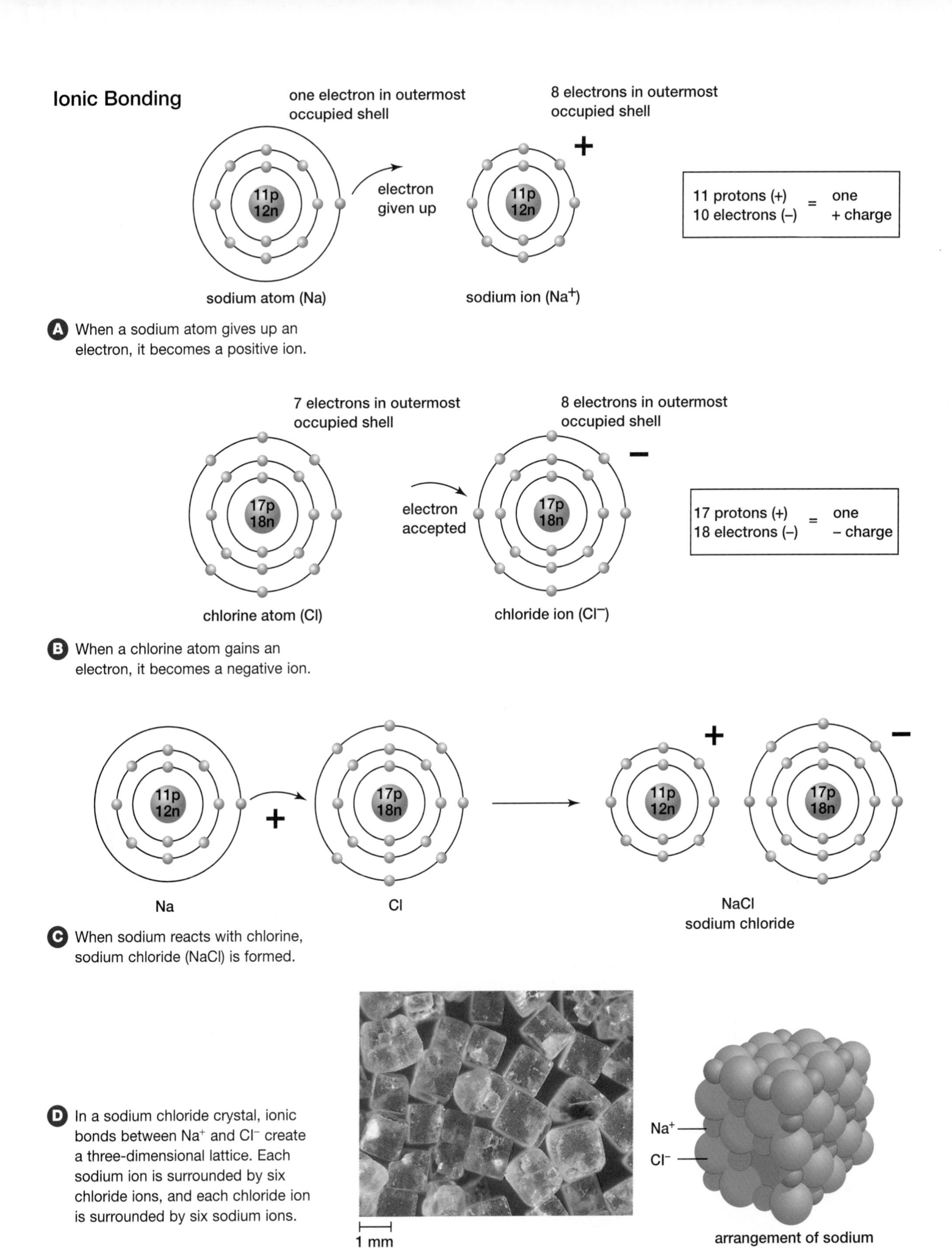

Figure 1.4 Reaction between sodium and chlorine to form ionic sodium chloride

Any atom has the same number of electrons and protons. Therefore, the atom has no charge and is said to be neutral. However, if an atom loses or gains electrons, that atom becomes an **ion**. If an atom loses electrons, the ion formed has more protons than electrons and therefore has a positive charge. A positively charged ion is called a *cation*. In contrast, if an atom gains one or more electrons the ion formed has a negative charge. A negatively charged ion is called an *anion*. When sodium (Na) and chlorine (Cl) atoms react, they form an ionic bond, as shown in Figure 1.4. The sodium atom gives up its only valence electron and becomes a sodium ion, with 11 protons and 10 electrons. This number of electrons is arranged in the same way that the 10 electrons are arranged in the neon atom. The chlorine atom gains an electron and becomes a chloride ion, with 17 protons and 18 electrons. This number of electrons is arranged in the same way as the 18 electrons in an argon (Ar) atom. Because the sodium ion is positively charged and the chloride ion is negatively charged, they attract each other to form an ionic bond.

The tendency of chlorine to gain electrons is characteristic of atoms with a few electrons less than a noble gas atom. For example, atoms of fluorine and oxygen also tend to gain electrons when they form ionic bonds. One way to understand which elements form ionic bonds when they react is to use the principle of electronegativity. **Electronegativity** is a measure of the relative abilities of bonding atoms to attract electrons. The Pauling scale is the most commonly used measure of electronegativities of atoms. Fluorine, the most

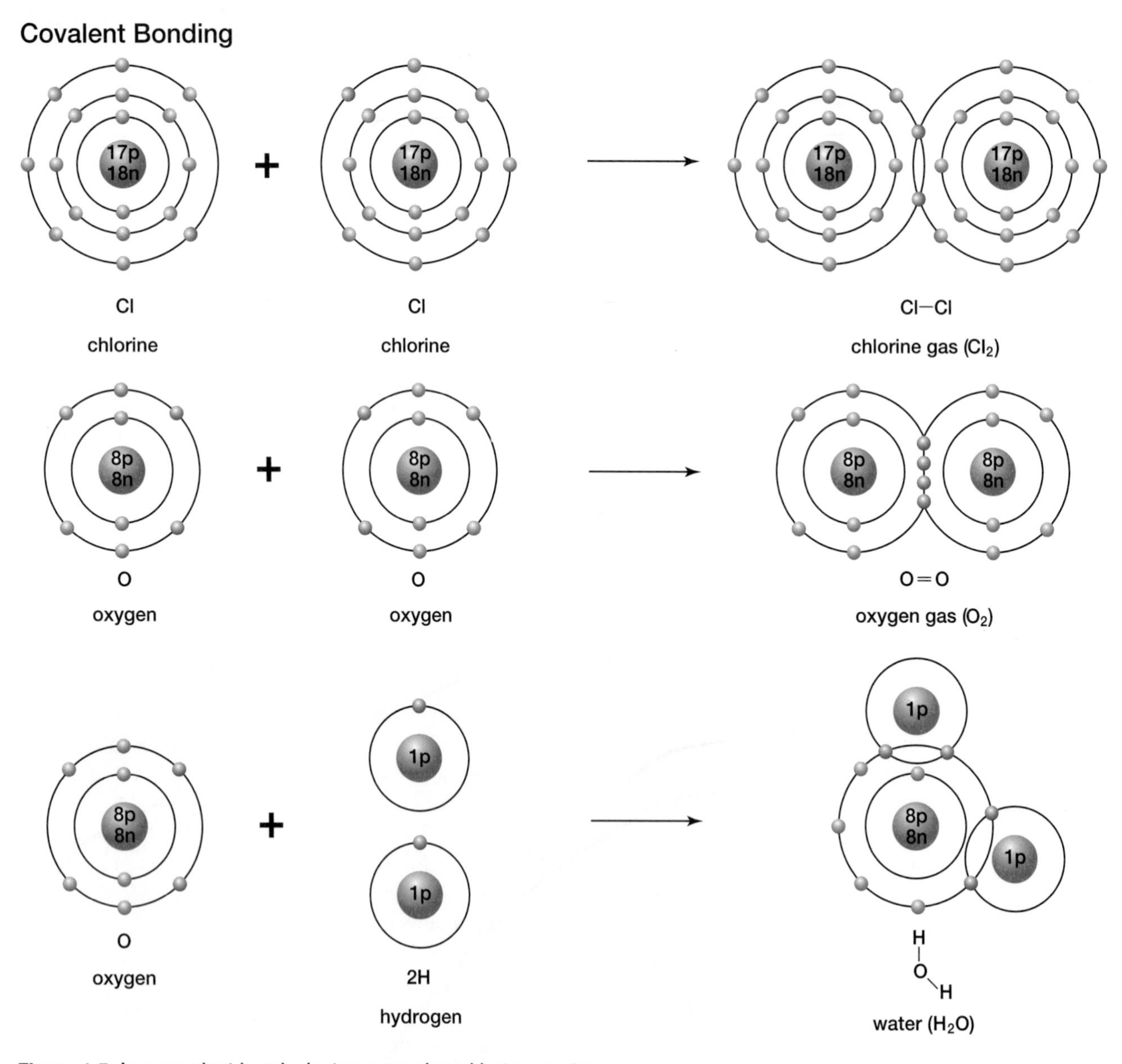

Figure 1.5 In a covalent bond, electrons are shared between atoms.

electronegative element, is found near the top right corner of the periodic table and has an electronegativity value of 4.0. Both cesium and francium, the least electronegative elements, are found near the bottom left corner of the periodic table and each has an electronegativity value of 0.7. Elements that are most likely to form ionic bonds, such as sodium and chlorine, are far apart in the periodic table and have a large difference in their electronegativity.

Elements that are close together in the periodic table have a small difference in their electronegativity. If two of these elements react to form a compound, their similar abilities to attract electrons results in the formation of a covalent bond, in which electrons are shared. In a covalent bond, atoms share two valence electrons. An example of this is the covalent bonding of two chlorine atoms, as shown in Figure 1.5, top. Double covalent bonds involve the sharing of two pairs of shared valence electrons. The two oxygen atoms in an oxygen molecule are joined by a double covalent bond, as shown in Figure 1.5, middle.

The shared electrons in covalent bonds belong exclusively to neither one nor the other atom. However, by sharing these valence electrons, both atoms appear to have the same number of valence electrons as a noble gas atom. In a covalent bond formed by two atoms of the same element, the electronegativity difference is zero. Therefore, the electrons in the bond are shared equally between the two atoms. This type of bond is described as **non-polar covalent**. Examples of non-polar covalent bonds are found in chlorine and carbon dioxide molecules.

A covalent bond is said to be **polar covalent** when the electronegativity difference between the atoms is not zero and the electrons are therefore shared unequally. In a water molecule (see Figure 1.5, bottom), oxygen is more electronegative than is hydrogen. The shared electrons spend more of their time near the oxygen nucleus than near the hydrogen nucleus. As a result, the oxygen atom gains a slight negative charge and the hydrogen atoms become slightly positively charged.

Chemists represent molecules formed through covalent bonds with various formulas, such as those in Figure 1.6. Electron-dot and structural formulas are simplified ways of showing what electrons are being shared.

Electron-Dot formula	Structural formula	Molecular formula
:Ö::C::Ö: carbon dioxide	O=C=O carbon dioxide	CO_2 carbon dioxide
H :N̈:H Ḧ ammonia	H–N(–H)–H ammonia	NH_3 ammonia
H :Ö:H water	H–O–H water	H_2O water

Figure 1.6 In an electron-dot formula, only the electrons in the valence shell are shown. In a structural formula, each line represents a pair of electrons shared by two atoms. A molecular formula shows only the number of each type of atom in a molecule.

ELECTRONIC LEARNING PARTNER

To learn more about ionic and covalent bonding, go to your Electronic Learning Partner now.

Hydrogen Bonds and the Properties of Water

Some molecules with polar covalent bonds are known as polar molecules. A **polar molecule** has an unequal distribution of charge as a result of its polar bonds and its shape. More information about polar molecules is provided in Appendix 6. Water is a common example of a polar molecule. In a water molecule, as shown in Figure 1.7, the slightly negative end of each bond can be labelled $\delta-$ and the slightly positive end can be labelled $\delta+$. These two ends, with slightly different charges, are sometimes referred to as "poles." Because a water molecule is polar, it can attract other water molecules, due to the attraction between negative poles and positive poles (see Figure 1.7). The attractions between water molecules are called **hydrogen bonds**. Hydrogen bonds can also be found between other molecules that contain hydrogen atoms bonded covalently to atoms of a much more electronegative element. Examples include ammonia (NH_3) and hydrogen fluoride (HF) in their liquid states.

A hydrogen bond is a force between molecules, not a chemical bond within a molecule. Hydrogen bonds are usually weaker than chemical bonds. For instance, a hydrogen bond may be only five percent the strength of a covalent bond, but it is sufficient to hold one water molecule to another in liquid water or ice. Under normal conditions, water molecules are attracted to each other in such a way that they are neither attracted too strongly (to form a solid) nor too weakly (allowing water to become a vapour). For this reason, under normal conditions on Earth, water exists as a liquid.

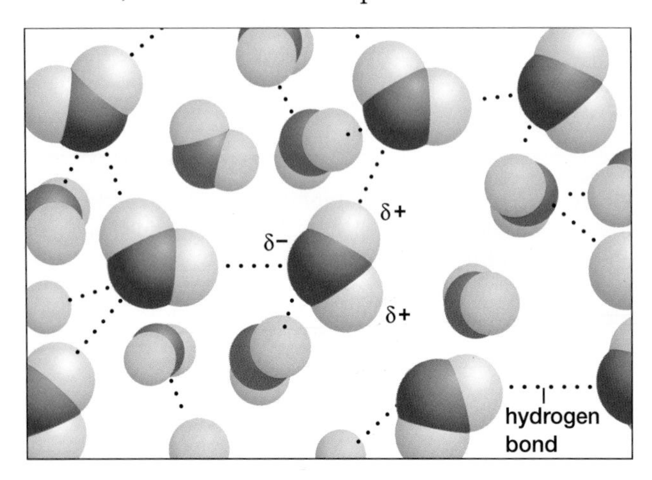

Figure 1.7 Hydrogen bonding (shown via dotted lines) between water molecules

Solubility of Substances in Water

All cells depend on liquid water. In fact, living organisms contain more molecules of water than any other substance; water comprises as much as 90 percent of a typical cell. Water is a perfect fluid environment through which other molecules can move and interact.

Sodium chloride (table salt), and many other ionic compounds or salts, dissolve readily in water. This occurs because the positively charged poles of the water molecule are attracted to the anions (chloride ions) in the salt. The negatively charged pole of the water molecule is similarly attracted to the cations (sodium ions) in the salt as shown in Figure 1.8. These two attractions pull the sodium ions and chloride ions away from each other. The salt is now dissociated, which means that the sodium ions and chloride ions have separated and have dissolved in the water.

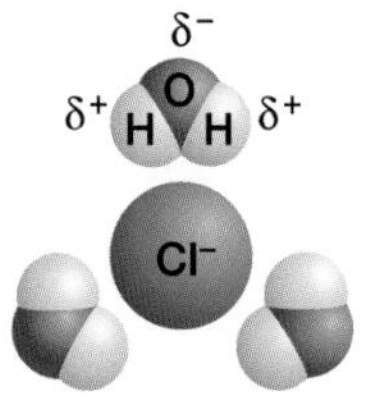

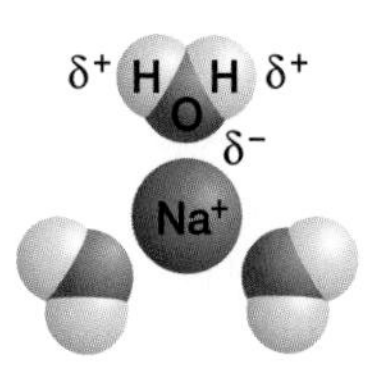

Figure 1.8 The salt NaCl dissolves in water because chloride ions and sodium ions attract water molecules.

Compounds that interact with water — for example, by dissolving in it — are called **hydrophilic**. In contrast, compounds that do not interact with water are called **hydrophobic**. Non-polar compounds are hydrophobic. They cannot form hydrogen bonds with water in the same way that ionic or polar compounds can. Therefore, hydrophobic molecules are insoluble in water. For example, when you place a drop of oil (a non-polar compound of carbon and hydrogen) into water, the oil does not mix with the water — they remain separate.

In this section, you have learned that the type of chemical bond that joins individual atoms together determines whether the resulting compound is ionic or covalent. Covalent molecules may be polar or non-polar, depending on the electronegativities of the bonded atoms and shape of the molecule. You have learned that hydrogen bonds form between molecules in water, which interacts very differently with hydrophobic and hydrophilic compounds. Hydrophobic interactions especially have a great effect on many biological molecules. For instance, many protein molecules have hydrophobic regions in portions of their structure. Interactions of these regions with water cause the molecules to adopt specific shapes. You will see examples of this in the next section, which reviews the four main kinds of molecules that make up all cells.

COURSE CHALLENGE

What you have learned about molecules and polarity may be useful in preparing your Biology Course Challenge. How could you use this knowledge to help you prepare for your science symposium?

SECTION REVIEW

1. K/U Is this atom chemically unreactive or chemically reactive? Explain your answer in terms of valence electrons.

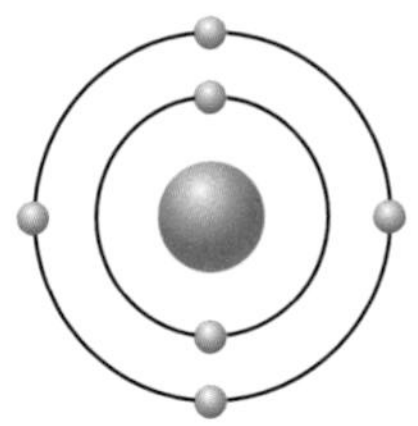

2. K/U Use examples of ionic, covalent, and polar covalent bonds to help explain how they are different.

3. K/U Describe a polar covalent bond using the term "electronegativity."

4. K/U How is a hydrogen bond different from other bonds described in this chapter?

5. C Explain why hydrophobic and hydrophilic compounds behave differently in water.

6. I If you pour a few drops of oil into water, you will notice that the oil forms a thin layer at the surface of the water almost immediately. Based on your knowledge of hydrophobic and hydrophilic compounds and bonding, form a hypothesis as to why oil and water interact in this way. How would you test your hypothesis?

7. MC A graduated cylinder with 10 mL of dilute sodium hydroxide solution is accidentally spilled onto your work surface. List the steps you will follow in response to this spill.

8. MC If the spill in question 7 came in contact with human skin, how would your list of steps change?

UNIT INVESTIGATION PREP

The reactivity of atoms is a concept central to metabolic processes. How might this knowledge help you plan an experiment to study a metabolic process?

Biological Macromolecules and Their Subunits

EXPECTATIONS

- **Identify functional groups and explain their effect on the properties of molecules.**
- **Review the types of macromolecules.**
- **Explain the meaning of the term isomer, with examples.**

The atoms of four elements make up roughly 99 percent of the mass of most cells: hydrogen, nitrogen, carbon, and oxygen. With only a few exceptions, molecules that contain carbon atoms are called **organic** compounds. There are millions of different organic compounds. Nearly all organic compounds contain hydrogen as well as carbon, and most of these also include oxygen. Pure carbon and carbon compounds that lack hydrogen — such as carbon dioxide and calcium carbonate — are considered *inorganic*. Inorganic compounds are, nevertheless, integral components of living systems. See Figure 1.9. For example, water — an inorganic compound — provides a medium in which various substances may be dissolved and transported within and between cells.

Figure 1.9 In what ways do living and non-living systems, and organic and inorganic compounds interact?

The Central Atom: Carbon

The diversity of life relies greatly upon the versatility of carbon. Recall that a carbon atom in its most stable state has two occupied energy levels, the second of which contains four valence electrons. This means that, in covalent molecules, a carbon atom can form bonds with as many as four other atoms. In biological systems, these atoms are mainly hydrogen, oxygen, nitrogen, phosphorus, sulfur, and — importantly — carbon itself. Carbon's ability to bond covalently with other carbon atoms enables carbon to form a variety of geometrical structures, including straight chains, branched chains, and rings. Figure 1.10 shows the shapes of several simple organic molecules that contain only carbon and hydrogen atoms. These molecules, called hydrocarbons, comprise the fossil fuels that serve as the main fuel source for much of the world's industrial activities. Hydrocarbons are themselves not components of living systems. However,

Name	Molecular formula	Structural formula
methane	CH_4	H H—C—H H
ethane	C_2H_6	H H H—C—C—H H H
benzene	C_6H_6	H H—C C—H C C H—C C—H H

Figure 1.10 Carbon atoms can bond in several ways — from the simple tetrahedral structure of methane, to the short chain of carbon atoms in ethane, to the ring of carbon atoms in benzene.

substantial portions of many biological molecules consist of bonded chains of carbon and hydrogen.

Molecular Isomers

Because carbon can form so many compounds with so many elements, it is common to encounter several organic compounds with the same molecular formula but different structures. Such compounds are known as **isomers**. For example, two isomers of glucose, a six-carbon sugar, are fructose and galactose. Glucose, fructose, and galactose all have the same molecular formula ($C_6H_{12}O_6$). However, they differ in their molecular structures, as shown in Figure 1.11.

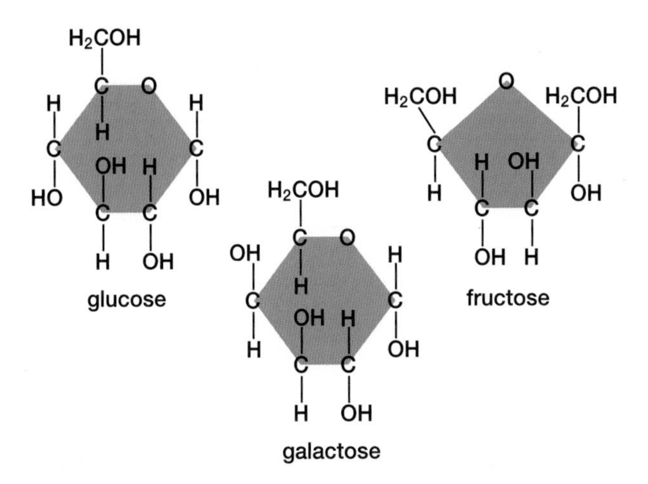

Figure 1.11 The different ways in which the same atoms are arranged in glucose, galactose, and fructose make them isomers.

There are two main types of isomers. **Structural isomers** are two or more compounds with the same atoms bonded differently. Glucose and fructose, for example, are structural isomers. Notice that a glucose molecule contains a ring of five carbon atoms and an oxygen atom, whereas a fructose molecule contains a ring of four carbon atoms and an oxygen atom. Because their structures are different, glucose and fructose have different properties, and cells metabolize them differently.

Stereoisomers are two or more compounds with their atoms bonded in the same way, but with atoms arranged differently in space. Stereoisomers may be geometrical or optical. Geometrical isomers can have very different physical properties (such as different melting points), but they tend to have the same chemical properties. Glucose and galactose are examples of geometrical isomers.

Optical isomers, shown in Figure 1.12, are non-superimposable mirror images of each other. They usually have similar chemical and physical properties, but enzymes or proteins on the cell membrane can distinguish between them. Usually, one optical isomer is biologically active and the other biologically inactive. In some cases however, this is not always true. For example, sometimes one optical isomer of a drug is not as effective as the other or can even cause complications. In the early 1960s, many pregnant women were prescribed a drug called thalidomide for morning sickness. Thalidomide is a mixture of two optical isomers; one produced the desired effect, but the other caused major birth defects. As the thalidomide example demonstrates, organisms can be very sensitive to minute variations in molecular geometry.

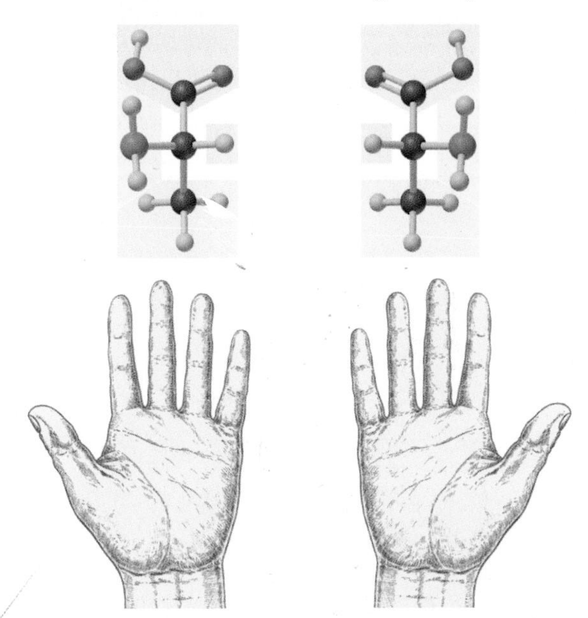

Figure 1.12 Optical isomers, such as these molecules of the amino acid, alanine, are identical in their structures except that they are non-superimposable mirror images of each other. Your left and right hands are also non-superimposable images of each other.

The Functional Groups

Chemical reactions involve breaking or forming chemical bonds. These processes can transform simple molecules such as glucose into complex molecules such as starch or cellulose. Many of these complex molecules contain groups of atoms with characteristic chemical properties. These groups of atoms, known as **functional groups**, include hydroxyl, carbonyl, carboxyl, amino, sulfhydryl, and phosphate groups, as shown in Figure 1.13 on page 16. Many compounds have more than one functional group in their structure.

These functional groups are hydrophilic. Except for the phosphate group, they are polar and so they increase the solubility in water of the organic molecules to which they are attached. Each functional group also has capabilities to change the chemical properties of the organic molecules to which it bonds. For example, if a hydrogen atom in ethane is replaced by a sulfhydryl group, the result is ethanethiol, also known as ethyl mercaptan. While ethanethiol in small amounts stabilizes protein structures, it is also a dangerous neurotoxin and respiratory toxin.

Each functional group has a specific role in cell metabolism. Phosphates are essential to the metabolic processes of photosynthesis and cellular respiration. For example, the transfer of a phosphate group from ATP (adenosine triphosphate) begins the very important process of glycolysis — the first step in cellular respiration. You will discover more about this process in Chapter 3.

While amino and phosphate groups contribute to energy transactions in the cell, the sulfhydryl (–SH) group is essential to protein stabilization. Amino acids with –SH groups form bonds called disulfide bridges (S–S bonds) that help protein molecules to take on and maintain a specific shape.

Functional group	Formula	Name of compound	Structural example
hydroxyl	—OH	alcohols	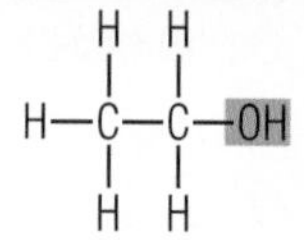 CH_3—CH_2—**OH** ethanol
carbonyl	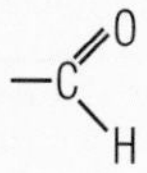 —C(=O)H	aldehydes	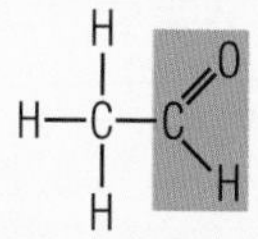 CH_3—**CHO** acetaldehyde (ethanal)
	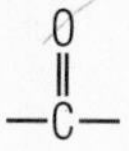 —C(=O)—	ketones	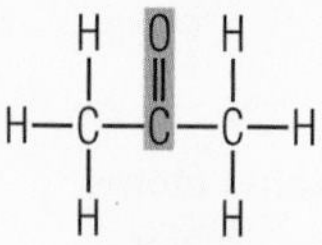 CH_3—**C(=O)**—CH_3 acetone
carboxyl	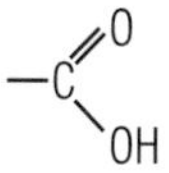 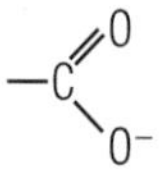 —C(=O)OH (non-ionized) —C(=O)O^- (ionized)	carboxylic acids	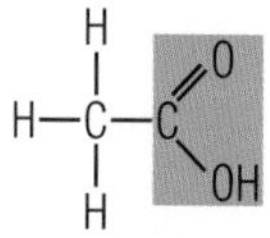 CH_3—**COOH** acetic acid
amino	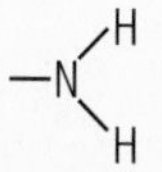 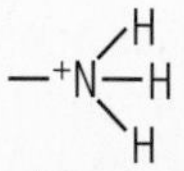 —NH_2 (non-ionized) —$^+NH_3$ (ionized)	amines	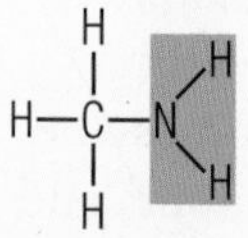 CH_3—**NH_2** methylamine
sulfhydryl	—SH	thiols	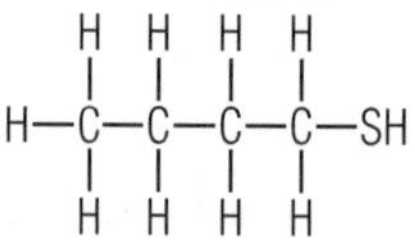 CH_3—CH_2—CH_2—CH_2—SH butanethiol
phosphate	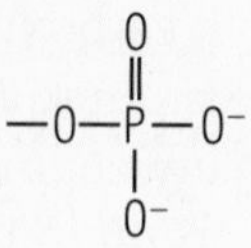 —O—P(=O)(O^-)—O^-	organic phosphates	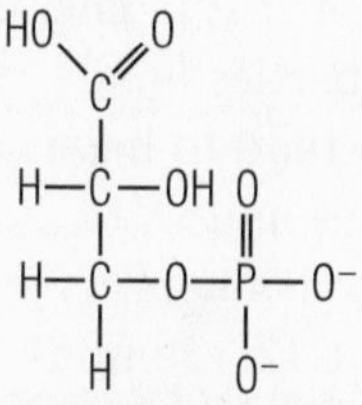 HOOC—CH(OH)—CH_2—O—P(=O)(O^-)—O^- 3-phosphoglyceric acid

Figure 1.13 Functional groups of organic compounds

Monomers and Macromolecules

As you know, atoms can join together — bond — to form small compounds called molecules. Similarly, molecules can join together to form large structures called **macromolecules**. The small, molecular subunits that make up macromolecules are called **monomers**. The macromolecules themselves are built up of long chains of monomers. These chains are called **polymers**.

Table 1.2 lists the main types of macromolecules and their monomer subunits. Figure 1.14 depicts the subunits that comprise carbohydrates, selected lipids, proteins, and nucleic acids. Chemical reactions in cells synthesize macromolecules from these subunits, and break the molecules apart to release their subunits. Refer to Figure 1.14 often as you examine these chemical reactions in the final section of this chapter.

WEB LINK

www.mcgrawhill.ca/links/biology12

To learn how researchers use molecular models to help them investigate the structure of complex molecules, go to the web site above, and click on **Web Links**. Identify a scientific discovery in which molecular modelling played a significant role. What techniques or technologies were used in the construction of the model? Present your findings in a class workshop on methods of molecular modelling.

Table 1.2
Macromolecules and their subunits and functions

Macromolecule	Sub-unit	Function
Carbohydrates		
glucose (monosaccharide)	–	energy storage
starch, glycogen (polysaccharides)	glucose	energy storage
cellulose (polysaccharide)	glucose	component of plant cell walls
chitin (polysaccharide)	modified glucose	cell walls of fungi; outer skeleton of insects and related groups
Proteins		
globular	amino acids	catalysis
structural	amino acids	support and structure
Lipids		
fats	glycerol + three fatty acids	energy storage
phospholipids	glycerol + two fatty acids + phosphate	component of cell membranes
steroids	four carbon rings	message transmission (hormones)
terpenes	long carbon chains	pigments in photosynthesis
Nucleic acids		
DNA	nucleotides	encoding of heredity information
RNA	nucleotides	blueprint of heredity information

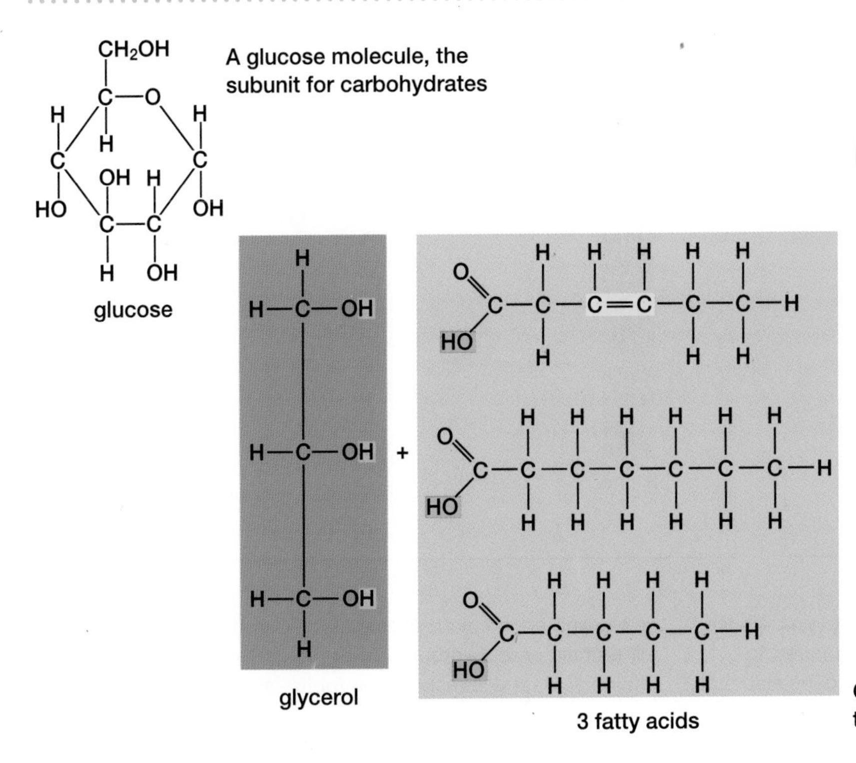

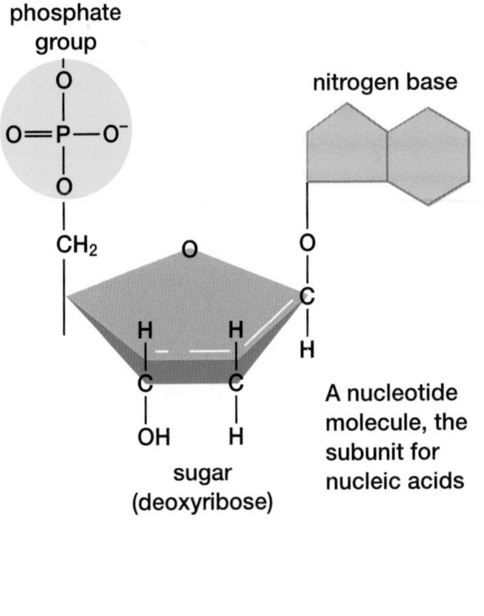

Glycerol and three fatty acid molecules, the subunits for fats (triglycerides)

Figure 1.14A Notice the prevalence of carbon, hydrogen, and oxygen atoms in the monomers that comprise the macromolecules of life.

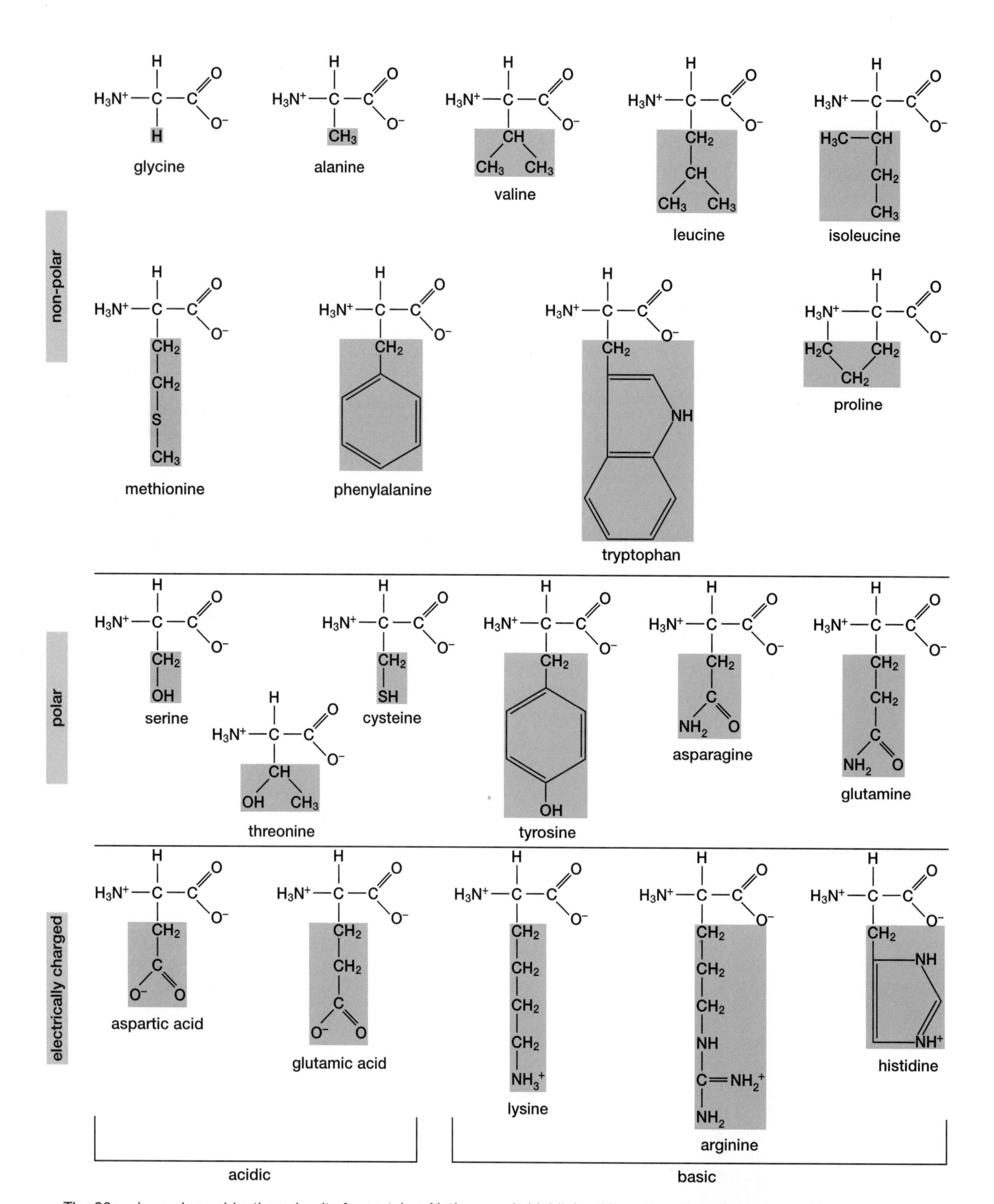

The 20 major amino acids, the subunits for proteins. Notice that each amino acid has one carbon that is bonded to both an amino group and a carboxyl group. Notice also that bonded to this same carbon atom is a molecular group that is highlighted in colour. Biologists refer to this group as a side-chain or an *R*-group. The *R*-group is different for each amino acid, and is responsible for its chemical properties.

Figure 1.14B Notice that each amino acid has an amine at one end and a carboxyl group at the other.

The Thalidomide Dilemma

In the late 1950s, a new drug called thalidomide was widely sold as a sleeping pill and cure for morning sickness during pregnancy. Research had shown that thalidomide is remarkably non-toxic, even in very high doses. Tragically, tests did not show that the drug affects the normal development and growth of a fetus. As a result, thousands of women who had taken thalidomide when pregnant gave birth to babies with missing or malformed limbs, facial deformities, and defective internal organs. The drug was banned for general use in 1964.

In the years since the catastrophe caused by thalidomide, researchers have discovered more effects that the drug has on the body. Not only does thalidomide induce sleep and reduce nausea, but also it is a powerful anti-inflammatory agent. It can also moderate extreme and damaging reactions of the immune system. These effects make it a valuable tool for treating leprosy, rheumatoid arthritis, lupus, certain conditions associated with AIDS, and other diseases. The dilemma is how to control and market the use of a drug that can cause great damage, yet has great benefits as well.

One Drug, Many Effects

It is quite common to find that pharmaceuticals developed for one purpose have other applications. For example, Aspirin™ was originally prescribed as a painkiller, but much later was found to help prevent the formation of blood clots. Prozac™ was marketed as an appetite suppressant before it was recognized as an effective antidepressant. Minoxidil™, used to control hypertension, is now used to treat baldness. The interactions between a particular drug and the body's cells and organ systems are often complex and poorly understood, even after a drug has been in use for many years.

Thalidomide and Disease

How does thalidomide produce its effects? Since the mid-1990s, scientists have learned that:

- Thalidomide inhibits the movement of cells needed to form new blood vessels. This is the property of the drug that affects fetal development and results in malformed limbs and organs. However, inhibition of blood vessel growth also has important clinical value. For example, cancerous tumours can only grow by developing new blood supplies to provide them with oxygen and nutrients and to carry away wastes. By preventing the growth of new blood vessels, thalidomide starves tumours and stunts their growth.
- Thalidomide suppresses the production of a chemical messenger called tumour necrosis factor, or TNF. This chemical is made by blood cells as part of the body's immune response. However, large quantities of TNF result in harmful inflammation — a common symptom of autoimmune disorders such as rheumatoid arthritis, AIDS, and lupus. Thalidomide is the most effective drug known to relieve this symptom.

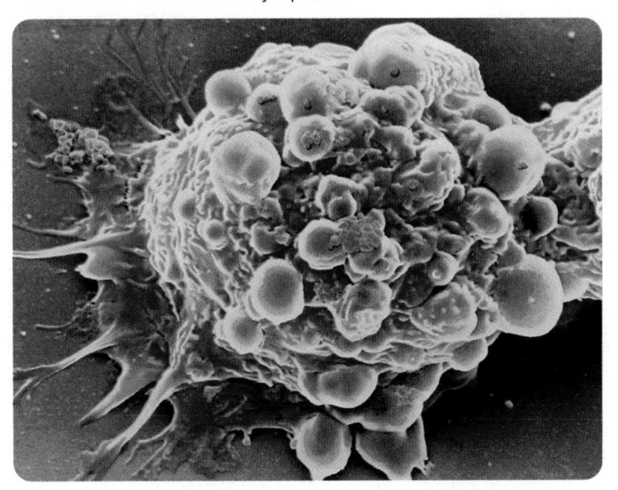

Thalidomide can be used to prevent the growth of cancerous tumours.

The Bottom Line

With little or no toxicity and a wide range of applications, thalidomide is a valuable drug for medical use. However, the drug's effect on babies born a generation ago still makes its use controversial. The Thalidomide Victims Association of Canada is a group of individuals born with physical defects caused by this drug. They lobby drug companies and governments to help ensure that another generation does not experience the same effects as they did.

Follow-up

1. Debate in class the arguments for and against prescribing a therapeutic drug that has been proven to pose serious health risks. Who should decide on its use? What might be the view of a person who has physical disabilities that were caused by thalidomide? What might be the view of a person who suffers from life-threatening symptoms of a disease (such as AIDS) that thalidomide can relieve?
2. Thalidomide is dangerous to fetal development for only a very short time in early pregnancy, three to five weeks after conception. Other approved drugs on the market, including an acne drug, also have side effects that include producing fetal malformations. What other examples can you think of where lack of complete information might distort the evaluation of a drug?

MINI LAB

Reviewing Macromolecules

You have studied carbohydrates, lipids, proteins, and nucleic acids in previous science courses. Use the molecules shown in Figure 1.14 as a starting point for designing a reference resource on the structure and function of these four main types of macromolecules. Your reference resource can incorporate one or several media, and should include the following information, as well as any additional details you deem appropriate:

- types of polymers formed from the monomer subunits shown in Figure 1.14
- principal functions of these polymers in living systems
- sources in nature from which cells and other living systems may obtain these essential compounds

Analyze

1. Cells metabolize macromolecules in a fluid environment. Why is water necessary for metabolic processes? Would these processes be possible in a non-fluid environment? Defend your answer.

SECTION REVIEW

1. C Biologists usually consider inorganic compounds as part of the non-living world and organic compounds as part of the living world.
   (a) What property of water makes it an inorganic compound?
   (b) Why is water, nevertheless, an essential component of all living systems?
2. C In this section you read, "The diversity of life relies greatly upon the versatility of carbon." Explain your understanding of the significance of this statement. Use examples to support and enhance your explanation.
3. K/U Explain why glucose, fructose, and galactose are isomers.
4. K/U Describe the differences between a stereoisomer and a structural isomer. Give an example of each.
5. K/U What is a functional group? State two characteristics of a functional group that make it important to biological systems.
6. C Name three macromolecules and sketch the monomer subunit for each.
7. MC Like carbon, atoms of the element silicon can form bonds with as many as four other atoms. Ammonia is a more polar substance than water. In your opinion, why is life on Earth based on carbon chains in water, instead of silicon chains in ammonia?
8. I Examine the carbon "skeletons" shown below. How many additional atoms can bond with the highlighted carbon atom in each case?

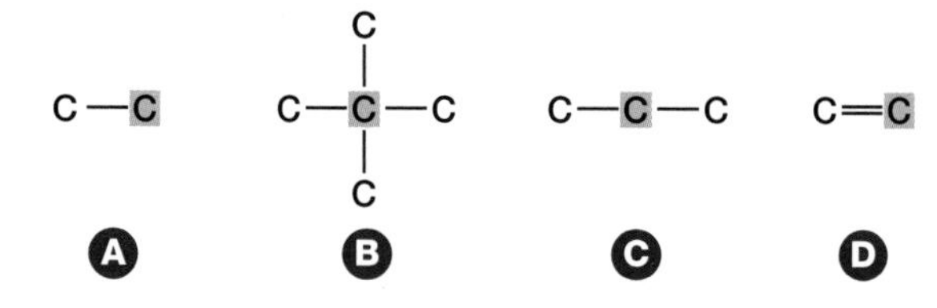

Making and Breaking Macromolecules

EXPECTATIONS

- **Identify and describe the four main kinds of biochemical reactions.**
- **Use molecular models to infer the polarity of molecules.**
- **Investigate the structures and properties of macromolecules and functional groups using models.**

Large molecules can be broken down to release energy. Alternatively, they can be formed to build cellular structures or store information. In biological systems there are four major types of chemical reactions involved in breaking apart and building molecules:

- acid-base or neutralization reactions, which transfer hydrogen ions between molecules,
- redox, or oxidation-reduction reactions, which transfer electrons between molecules,
- hydrolysis reactions, in which molecules react with H_2O to form other molecules, and
- condensation reactions, in which molecules react to form H_2O and other molecules.

These types of chemical reactions are described below.

Acids, Bases, and Neutralization Reactions

Acids and bases are compounds that may be inorganic or organic. Hydrochloric acid, found in the mammalian stomach, is an inorganic acid. Acetic acid and amino acids are examples of **organic acids**. Sodium hydroxide, a key component of oven cleaners, is an inorganic base. Purines and pyrimidines, the molecules that form part of the subunits of nucleic acids, are examples of **organic bases**; they are often referred to as nitrogenous bases, because they include the nitrogen-containing amine group.

What is it, however, that makes one substance an acid and another a base? In biology, acids and bases are understood in relation to their behaviour in water. Under normal conditions, pure water exists in the form of H_2O molecules. A small number of these molecules dissociate, which means that they break up into ions. When a water molecule dissociates, it forms a positively charged hydrogen ion, H^+, and a negatively charged hydroxide ion, OH^-. Since very few water molecules dissociate, the concentration of these ions is low. In pure water at 25°C, the concentration of each of these ions is the same: 1×10^{-7} mol/L. Because hydrogen and hydroxide ions are very reactive, changes in their concentrations can drastically affect cells and the macromolecules within them. Acids and bases, and more specifically the concentrations of hydrogen and hydroxide ions within cells, determine how effectively cellular processes are carried out.

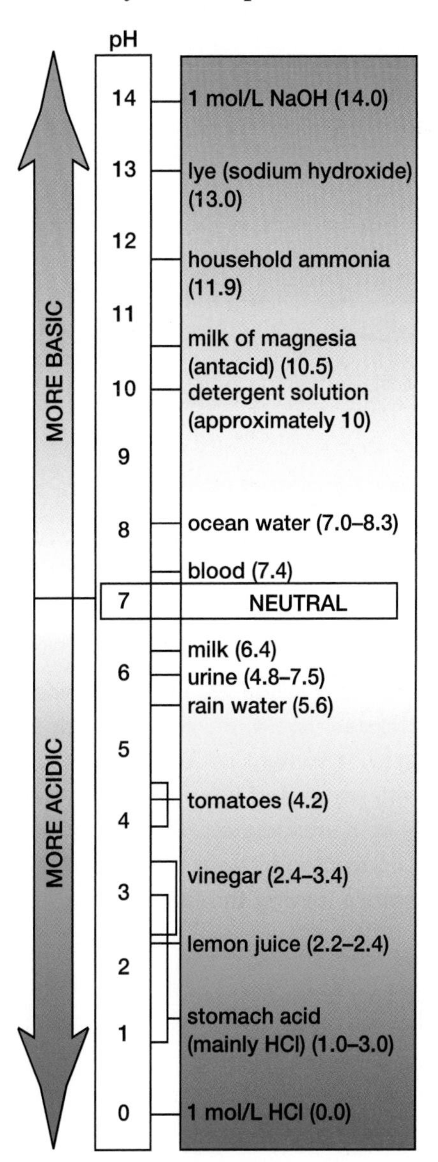

Figure 1.15 Substances that are acidic have a pH less than 7, while substances that are basic have a pH greater than 7.

An acid is any substance that donates H^+ ions when it dissolves or dissociates in water. Therefore, acids increase the concentration of H^+ ions in water solutions. Bases, on the other hand, decrease the concentration of H^+ ions in solution. Usually this occurs because bases attract H^+ ions, thus reducing their concentration. As a result, the concentration of OH^- ions increases when bases dissolve or dissociate in water. The pH scale, shown in Figure 1.15, is a means for ranking substances according to the relative concentrations of their hydrogen and hydroxide ions. Water, with equal concentrations of these ions, is considered neutral and has a pH of 7. Substances with a pH that is lower than 7 have higher concentrations of H^+ ions (and lower concentrations of OH^- ions), so they are acids. Substances with a pH that is higher than 7 have lower concentrations of H^+ ions (and higher concentrations of OH^- ions), so they are bases.

When acids and bases react, they produce two products: water and a salt (an ionic compound). This

Investigation 1 • A

SKILL FOCUS

- Predicting
- Performing and recording
- Analyzing and interpreting
- Communicating results

Acids, Bases, and Cellular pH

The pH value indicates whether a substance is acidic or basic. Acids have a pH value less than 7 and bases have a pH value greater than 7. Most cells function at around pH 7, which is considered to be a neutral environment. To maintain this neutral environment, cells must control, or buffer, the pH level so it does not become too acidic or too basic. In this investigation you will determine the effects of adding an acid and a base to several solutions. You will use a commercially prepared buffer solution and a solution made from living cells.

Pre-lab Questions

- What factors might influence the pH of cells?
- Why might cells need to maintain a constant pH environment?
- How do cells regulate pH?

Problem

How do acids and bases affect living cells?

Prediction

Make a prediction about how cells maintain a constant pH.

CAUTION: **Acids and bases are corrosive and caustic substances. Avoid any contact with skin, eyes, or clothes. If contact does occur, rinse thoroughly with water and inform your teacher. Clean up any spill immediately. Dispose of any materials as instructed by your teacher. Wash your hands before leaving the laboratory.**

Materials

universal indicator paper or pH meter
forceps
pH scale
0.1 mol/L HCl
0.1 mol/L NaOH
commercial buffer solution, pH 7
10% homogenized potato solution
50 mL beaker
distilled water in squirt bottle
medicine dropper
tap water
graphing paper

Procedure

1. Work in a small group. Read steps 2 to 7, then design a data table to record your results.
2. Add 20 mL of tap water to a clean beaker, and measure the pH of the water. If you are using universal indicator paper, use only a small piece. Immerse the paper into the water using the forceps. Compare the colour of the paper against the pH chart and determine the pH.
3. To the 20 mL of tap water, add one drop of the HCl solution. Gently swirl the contents, and then measure the pH. Continue adding drops of the HCl solution to the beaker, recording the pH after each drop, until a total of five drops has been added. Gently swirl the contents of the beaker after each drop is added.

chemical process in which acids and bases react to product a salt and water is called a **neutralization** reaction. In such a reaction, the acid no longer acts as an acid and the base no longer acts as a base; their properties have been neutralized.

Buffers

Many biological processes require specific pH levels in order to function properly. For example, pH and the control of pH play an integral role in both photosynthesis and cellular respiration. Many proteins require a certain pH in order to take on their characteristic shapes. Therefore, it is important for pH in organisms to be maintained at specific levels. Certain chemicals or combinations of chemicals known as **buffers** minimize changes in pH. Buffers maintain pH levels by taking up or releasing hydrogen ions or hydroxyl ions in solution. You will investigate the effect of a buffer in living cells in the next investigation. In Chapter 4, you will see how buffers play an important role in maintaining blood pH.

4. Dispose of the contents of the beaker. Rinse the beaker, medicine dropper, forceps, and pH meter (if used) with distilled water.
5. Repeat step 3 using the NaOH solution instead of HCl. Record the results in your data table.
6. Add 20 mL of commercial buffer to a clean beaker. Repeat steps 3 to 5 using the commercial buffer solution. Measure the pH of the buffer solution before you start to add drops of HCl or NaOH. Record the results in your data table.
7. Add 20 mL of potato solution to a clean beaker, and measure the pH of the solution.
8. Repeat steps 3 to 5 using the potato solution. Record the results in your table.

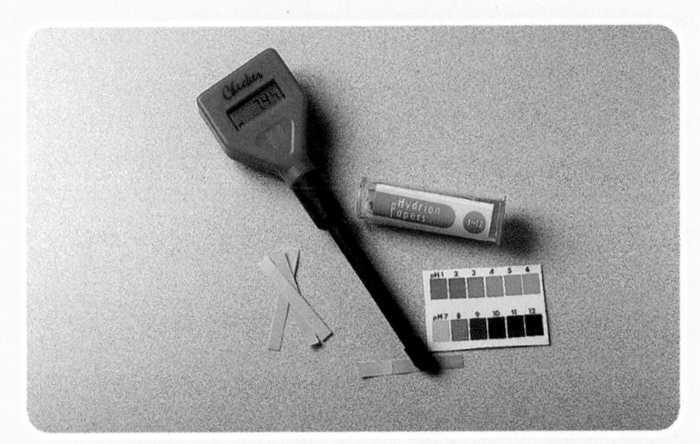

Post-lab Questions

1. Which do you think gave more accurate measures of pH, the universal indicator paper or the pH meter?
2. What was the total change in pH for the buffer and potato solutions?
3. How does this differ from the results you obtained using tap water? How do you account for these differences?
4. How do you account for the differences you observed between the buffer and potato solutions?

Conclude and Apply

5. Make line graphs showing the relationship between pH and added drops of HCl or NaOH. What will be your dependent variable? What will be your independent variable? For ease of comparison, use one set of axes to draw three graphs showing the effect on pH of adding HCl in each solution. Use another set of axes for the addition of NaOH. Label each line and your axes.
6. Describe and explain the trends you observe for each line.
7. Which solutions show the greatest similarity in change of pH? Explain briefly.
8. In terms of the pH changes you observed, how do living cells compare with the commercial buffer solution?
9. How might cells regulate pH?
10. What is the advantage of a buffering capacity in cells?

Exploring Further

11. Make a prediction about how the trend in each graph would continue if more drops of HCl or NaOH were added to each solution. Design an experiment to test your prediction. Based on your knowledge of buffers and acid-base reactions, account for the trends you observe.
12. Test the effects of adding drops of HCl and NaOH to solutions made from other plant material or from animal tissues such as liver or muscle tissue. Graph your results and compare these trends to what you obtained using the potato solution. Account for any similarities or differences you observe in trends of pH among the solutions tested.

Redox Reactions

Almost every element on Earth can react with oxygen. For instance, if oxygen combines with calcium, the oxygen receives electrons and forms negatively charged ions.

O	⟶	O^{2-}
oxygen atom no charge	gains two electrons	charge of atom decreased by two

The addition of two electrons has decreased the charge of the oxygen atom by two. The gain of electrons is referred to as **reduction**. The calcium loses electrons and forms positively charged ions, as shown here:

Ca	⟶	Ca^{2+}
calcium atom no charge	loses two electrons	charge of atom increased by two

The loss of electrons is called **oxidation**.

Investigation 1 • B

SKILL FOCUS

- Initiating and planning
- Predicting
- Analyzing and interpreting
- Communicating results

Finding the Products of Hydrolysis

Indicators are often used to test for the presence of certain compounds. For example, Benedict's solution is a reagent that can be used to test for the presence of monosaccharides, such as glucose. The free aldehyde group, which is present only in the ring form of glucose, reduces the copper (II) ion (Cu^{2+}) ion in Benedict's solution. Thus, monosaccharides are referred to as "reducing sugars." As more copper ions are reduced, the colour of the Benedict's solution changes, thus confirming the presence of glucose in a test solution. The ring structure configuration of glucose forms only in solution (when glucose molecules are dissolved in water). In a similar way, Lugol solution can be used to test for the presence of starches. In this investigation, you will conduct an experiment to identify the products formed by the hydrolysis of disaccharide and polysaccharide molecules.

Pre-lab Questions

- How does the starch molecule change as a result of hydrolysis?
- What products are formed by the complete hydrolysis of polysaccharides?
- What test can be used to identify these products?

Problem

How can the products of hydrolysis of carbohydrates be identified?

Predictions

Predict which factors will affect the rate of carbohydrate hydrolysis. Also, predict what types of products are formed by the hydrolysis of disaccharide and polysaccharide molecules.

CAUTION: Hydrochloric acid is very corrosive. Other chemicals are toxic. Avoid any contact with skin, eyes, or clothes. Flush spills immediately with copious amounts of cool water and inform your teacher. Exercise care when heating liquids and using a hot plate. Wash your hands before leaving the laboratory.

Materials

- test tubes
- test tube markers
- test tube brush
- test tube rack
- test tube holder
- spot plates
- beaker for hot water bath
- hot plate
- beaker tongs
- 30 mL graduated cylinder
- medicine droppers
- distilled water
- 1 mol/L HCl solution
- Benedict's solution
- Lugol solution
- potato starch
- 1% solutions of glucose, fructose, sucrose, and lactose
- labels
- marker

Procedure

Part A: Testing for the Presence of Monosaccharides

1. Add 3 mL of each of the following 1% solutions to separate test tubes: glucose, fructose, sucrose, and lactose. Add 3 mL of distilled water to a fifth test tube.
2. Add five drops of Benedict's solution to each test tube.

The terms "oxidation" and "reduction" are applied to many reactions involving ions whether or not oxygen is involved. For instance, in the reaction Na + Cl → NaCl, chlorine is reduced (gains an electron to form Cl^-) and sodium is oxidized (Na loses an electron to form Na^+). Because reduction and oxidation are both involved in the process, the entire reaction is called a **redox reaction**. Figure 1.16 is a generalized schematic representation of a redox reaction.

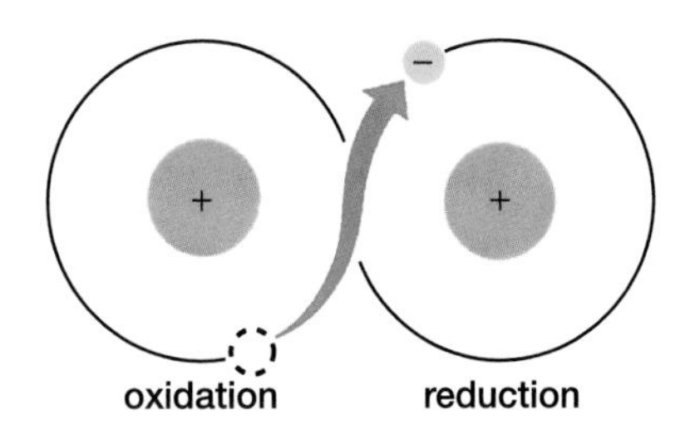

Figure 1.16 Oxidation is the loss of one or more electrons. Reduction is the gain of one or more electrons.

3. Shake each test tube gently to mix the solutions, and place them in a hot water bath for 5–10 min. Note any change in the colour of the solutions.

Part B: Testing for the Presence of Polysaccharides

1. To one test tube, add 5 mL of starch and 5 mL of cold, distilled water.
2. To another test tube, add 10 mL of cold, distilled water.
3. Add three drops of Lugol solution to each test tube. Note any changes in the colour of the solutions.

Part C: Hydrolysis of Disaccharides

1. Add 3 mL 1% solution of sucrose to each of two test tubes, labelled 1 and 2.
2. Add 3 mL of distilled water to a third test tube, labelled 3.
3. Add five drops of Benedict's solution to each test tube. Note any changes in the colour of the solutions in your data chart.
4. Repeat steps 1 and 2 using clean test tubes.
5. Add three drops of hydrochloric acid (HCl) to test tubes 1 and 3.
6. Heat the test tubes in hot water for about 3 min.
7. Add five drops of Benedict's solution to each test tube. Note any changes in the colour of the solutions.

Part D: Hydrolysis of Polysaccharides

1. To one test tube, add 5 mL of potato starch and 5 mL of water. Shake gently.
2. Now add 3 mL of concentrated hydrochloric acid to this test tube, and stir the solution.
3. To a second test tube, add 5 mL of potato starch to 8 mL of water. Shake gently.
4. To a third test tube, add 13 mL of distilled water.
5. Place the three test tubes in a hot water bath.
6. Using a clean dropper each time (or one rinsed with distilled water), remove five drops of solution from each test tube. Place each sample in a separate well in the spot plate.
7. Immediately add one drop of Lugol solution to each sample. Note any changes in colour.
8. Repeat steps 6 to 9 every minute, until no further colour changes are observed for three consecutive tests.

Post-lab Questions

1. What do these laboratory procedures indicate about the composition of polysaccharide molecules such as starch?
2. How does the presence of HCl change the structure of polysaccharide molecules?
3. What do the colour changes observed in the Lugol and Benedict's solution indicate about the concentrations of the various substances tested?

Conclude and Apply

4. What environmental factors cause hydrolysis? What evidence did you use to determine this?
5. Find the structural formula for starch. Use diagrams to show how hydrolysis changed this molecule.
6. Identify other molecules that could react in a similar way with these environmental conditions.

Cellular respiration is an important example of a redox reaction that takes place in biological systems. The overall reaction is:

$$\underset{\text{glucose}}{C_6H_{12}O_6} + 6O_2 \rightarrow 6CO_2 + 6H_2O + \text{energy}$$

In cellular respiration, high-energy electrons are removed from food molecules, which oxidizes them. These high-energy electrons are transferred to increasingly electronegative atoms, and help the cell manufacture energy-rich molecules used by cells to do work.

Hydrolysis and Condensation Reactions

Macromolecules in living systems are built and broken down by hydrolysis and condensation reactions (see Figure 1.17). In **condensation** (or dehydration synthesis), the components of a water molecule are removed to bond two molecules together. Because the organic molecule formed is bigger than the two organic molecules that reacted, condensation is an anabolic process. In the process of **hydrolysis**, the components of a water molecule are added to a molecule to break it into two molecules. Because the organic molecules produced are smaller than the organic molecule that reacted, hydrolysis is a catabolic process. Read on to see how hydrolysis and condensation work to break down and build carbohydrates, nucleic acids, proteins, and lipids.

Making and Breaking Carbohydrates

Carbohydrates are important macromolecules because they store energy in all organisms. Carbohydrates are groupings of C, H, and O atoms, usually in a 1 : 2 : 1 ratio. Often, carbohydrates are represented by the chemical formula $(CH_2O)_n$, where n is the number of carbon atoms in the carbohydrate.

Carbohydrates can be simple, such as the monomer glucose. Glucose is a hexose (six-carbon) sugar with seven energy-storing C-H bonds. If the number of carbon atoms in a carbohydrate molecule is low (from three to seven), then it is a **monosaccharide**. Greek prefixes for the numbers three through seven are used to name these sugars. For example a five-carbon sugar is a pentose, and a six-carbon sugar is a hexose. The glucose, fructose, and galactose isomers you studied in the previous section are all hexoses. Glucose is the primary source of energy used by cells.

Two monosaccharides can bond to form a **disaccharide**. For example, two glucose molecules can join to form the disaccharide maltose, as shown in Figure 1.18.

Organisms store energy in molecules known as **polysaccharides**. Polysaccharide molecules, such as starch and glycogen, are polymers made up of chains of linked monosaccharides. The long chains of glucose molecules, which make up starch, glycogen,

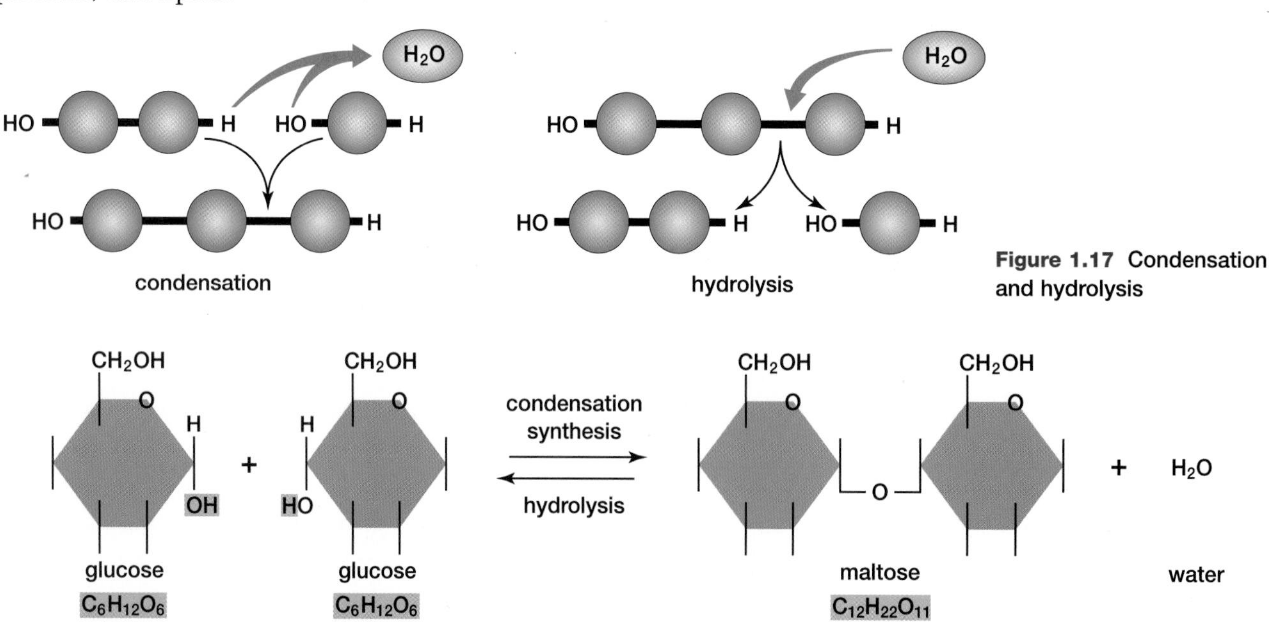

Figure 1.17 Condensation and hydrolysis

Figure 1.18 Maltose is a disaccharide. During condensation synthesis of maltose, a bond forms between the two glucose molecules and the components of water are removed. During hydrolysis, the components of water are added, and this bond is broken.

and some other polysaccharides, are formed by a condensation reaction, which removes water from 2 –OH functional groups or neighbouring monosaccharides. Because of its chemical composition, cellulose (a polysaccharide found in all plants) is indigestible for animals. The bonds in cellulose are difficult to break by normal metabolic means. In contrast, other polysaccharides, such as the amylopectin found in potatoes, rice, and wheat, serve as convenient and accessible forms of stored energy. The bonds that bind their high-energy glucose molecules together are easily broken and easily formed.

In living cells and tissues, polysaccharides and disaccharides can be broken into smaller units by the process of hydrolysis. The complete hydrolysis of most forms of starch produces a form of glucose, which is a simple sugar that cannot be decomposed by hydrolysis. In the investigation on page 24, you can determine the products of hydrolysis reactions.

Nucleotides and Nucleic Acids

Nucleic acids such as DNA and RNA are huge polymers of nucleotides. These are molecules composed of one, two, or three phosphate groups, a five-carbon sugar (deoxyribose or ribose), and a nitrogen-base (see Figure 1.19). DNA contains genetic information about its own replication and the order in which amino acids are to be joined to form a protein. RNA is the intermediary in the process of protein synthesis, conveying information from DNA regarding the amino acid sequence in a protein. There are four different bases in DNA — adenine, thymine, guanine, and cytosine. In RNA, uracil replaces thymine as a base. Adenine not only helps code genetic material and build proteins, but it also has important metabolic functions. You will investigate the structure and functions of nucleic acids further in Unit 3.

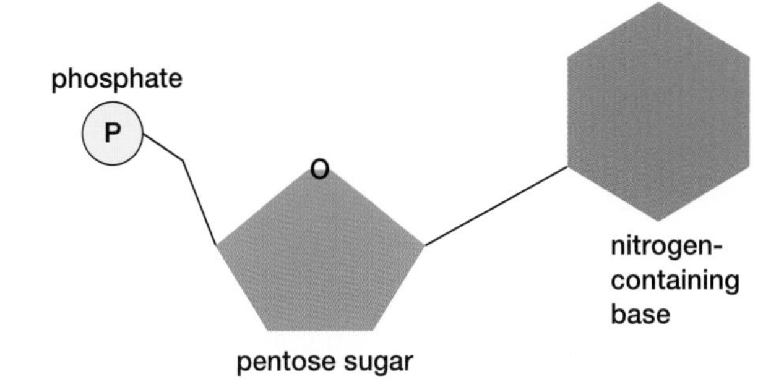

Figure 1.19 Structure of a nucleotide

ATP, adenosine triphosphate, is composed of adenosine (adenine joined to ribose, as in RNA) and three phoshate groups (see Figure 1.20). The hydrolysis of ATP results in the formation of ADP and a phosphate (P_i), and in the release of a large quantity of energy for cellular work. After ATP breaks down, it can be rebuilt by the addition of the phosphate to ADP by condensation.

Condensation Synthesis and Hydrolysis of Proteins

Proteins are important as structural components, sources of nutrition, and for their role in speeding up metabolic processes in the cell. **Peptide bonds** formed in condensation reactions link amino acids in proteins (see Figure 1.21). Each amino acid is composed of a carbon atom bound to a hydrogen atom and three additional groups — an amino group, $-NH_2$, a carboxyl group, –COOH, and an *R*-group that is different in each amino acid. When two amino acids join, they become a dipeptide. A chain of amino acids is called a **polypeptide**. Try the Thinking Lab to model a polypeptide. Polypeptides may join to form proteins. The sequence of these polypeptides, their particular orientations in space, and their three-dimensional shapes determine the type of protein they form. Enzymes, essential to metabolism (as you will see in Chapter 2), are proteins that are shaped in different ways depending

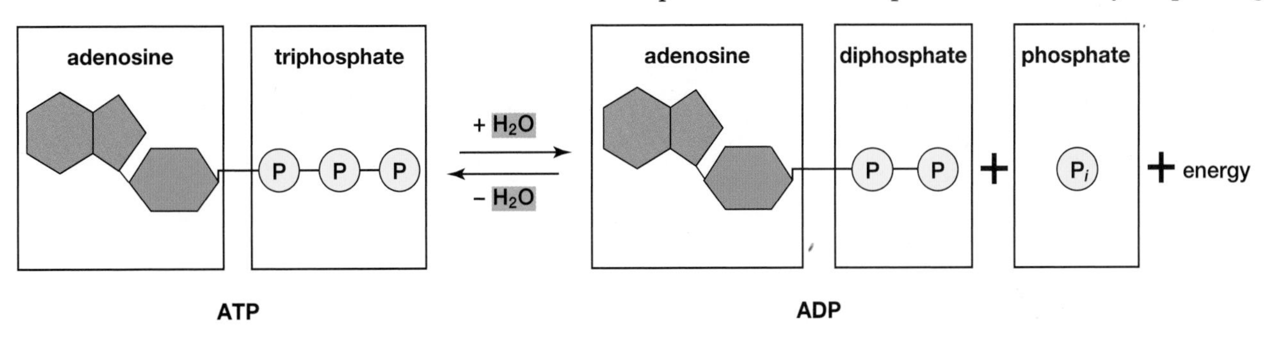

Figure 1.20 When cells require energy, ATP undergoes hydrolysis, thereby producing ADP, a phosphate, and energy.

on their function. Some proteins are composed of many polypeptides. These polypeptides can be broken during metabolism by hydrolysis.

Breaking Lipids

Lipids include fats and phospholipids (such as those in the cellular membrane), steroids, and terpenes (lipid pigments that operate during photosynthesis). Fats are composed of glycerol and three fatty acids; steroids and terpenes are composed of carbon rings and carbon chains respectively.

Fat is usually of animal origin and is solid at room temperature. Within animal bodies it is used for long-term energy storage. Fat also insulates against external heat and cold and protects major organs. Oil, the plant equivalent to fat, is liquid at room temperature. Fats and oils are often called triglycerides because of their structure. Fats and oils are insoluble in water because they are non-polar.

Both fats and oils are composed of two types of molecules: glycerol and fatty acids. Glycerol is a three-carbon alcohol in which each carbon is attached to a hydroxyl group (–OH), as shown in Figure 1.22. This three-carbon molecule is the core of the fat or oil molecule. In a condensation reaction, three fatty acids are attached to this core to form a fat. A fatty acid is a hydrocarbon chain that ends with the carboxyl group (–COOH). Most of the fatty acids in cells contain 16 or 18 carbon atoms per molecule. Saturated fatty acids have no double bonds between their carbon atoms; the carbon chain is "saturated" with as many hydrogen

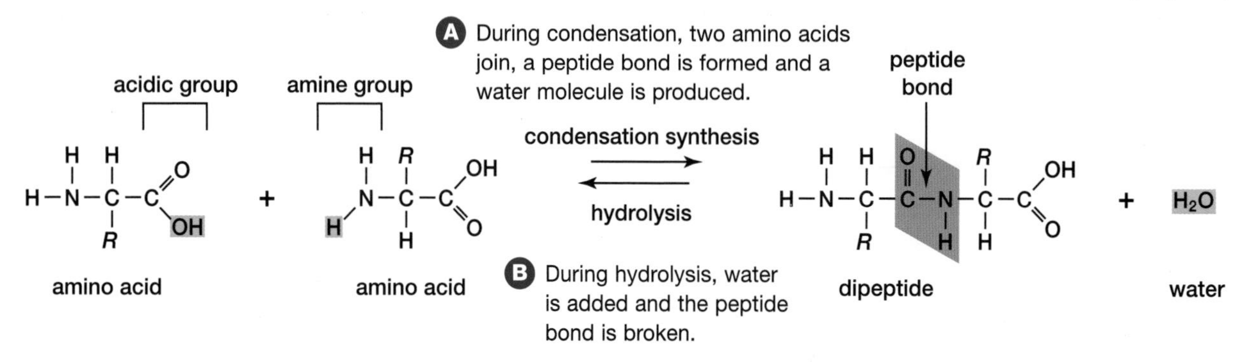

Figure 1.21 Condensation synthesis and hydrolysis of a dipeptide

THINKING LAB

Polypeptides and Polarity

Background

A protein is composed of a chain of amino acids connected together by peptide bonds. This forms a structure known as a polypeptide. In total, there are 20 different amino acids that make up proteins. The diagram shows a polypeptide made of three amino acids: alanine, serine, and glutamic acid. The *R*-groups are highlighted.

The molecule is held together with covalent bonds, which involve shared electrons. You have learned that some covalent bonds can be polar, based on how electrons are distributed between atoms. These polar bonds may result in polarity of the molecule or portions of the molecule.

You Try It

1. Working in a small group, use a model kit to construct the polypeptide shown in the diagram.
2. Where would you expect to find polarity in the molecule? Explain briefly.
3. Where might you expect hydrogen bonds to form with another molecule? Explain briefly.
4. Share your molecule with another group. Arrange the two molecules to show attraction through hydrogen bonding.
5. Notice how the *R*-groups are arranged in the molecule. How might this arrangement affect how the *R*-groups react with adjacent molecules?
6. Which *R*-group(s) would be most reactive? Explain briefly.
7. How would substituting valine for glutamic acid affect the properties of the polypeptide?

Alanine Serine Glutamic Acid

atoms as it can hold. Saturated fatty acids are generally solid at room temperature. In contrast, unsaturated fatty acids have one or more double bonds between carbon atoms. Therefore, the fatty acid is not saturated with hydrogen atoms. Fat molecules are split by hydrolysis for use in cells.

Figure 1.23(a) shows another lipid macromolecule with a different function in cells.

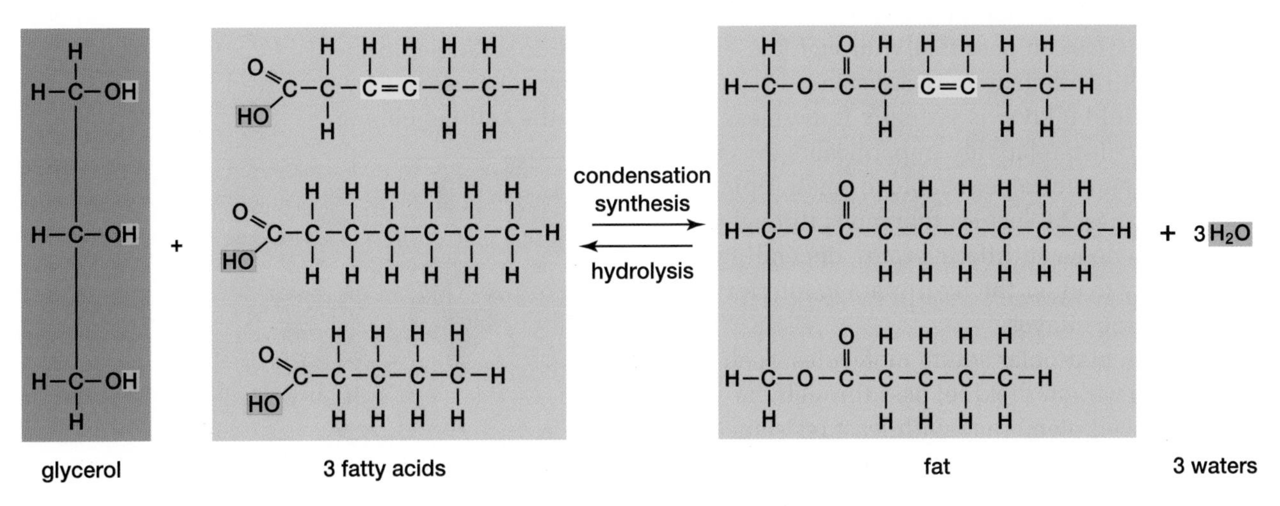

Figure 1.22 Condensation and hydrolysis of a fat molecule. When a fat molecule forms, three fatty acids combine with glycerol and three water molecules are produced. Unsaturated fats have double bonds (shown in yellow) between carbon atoms.

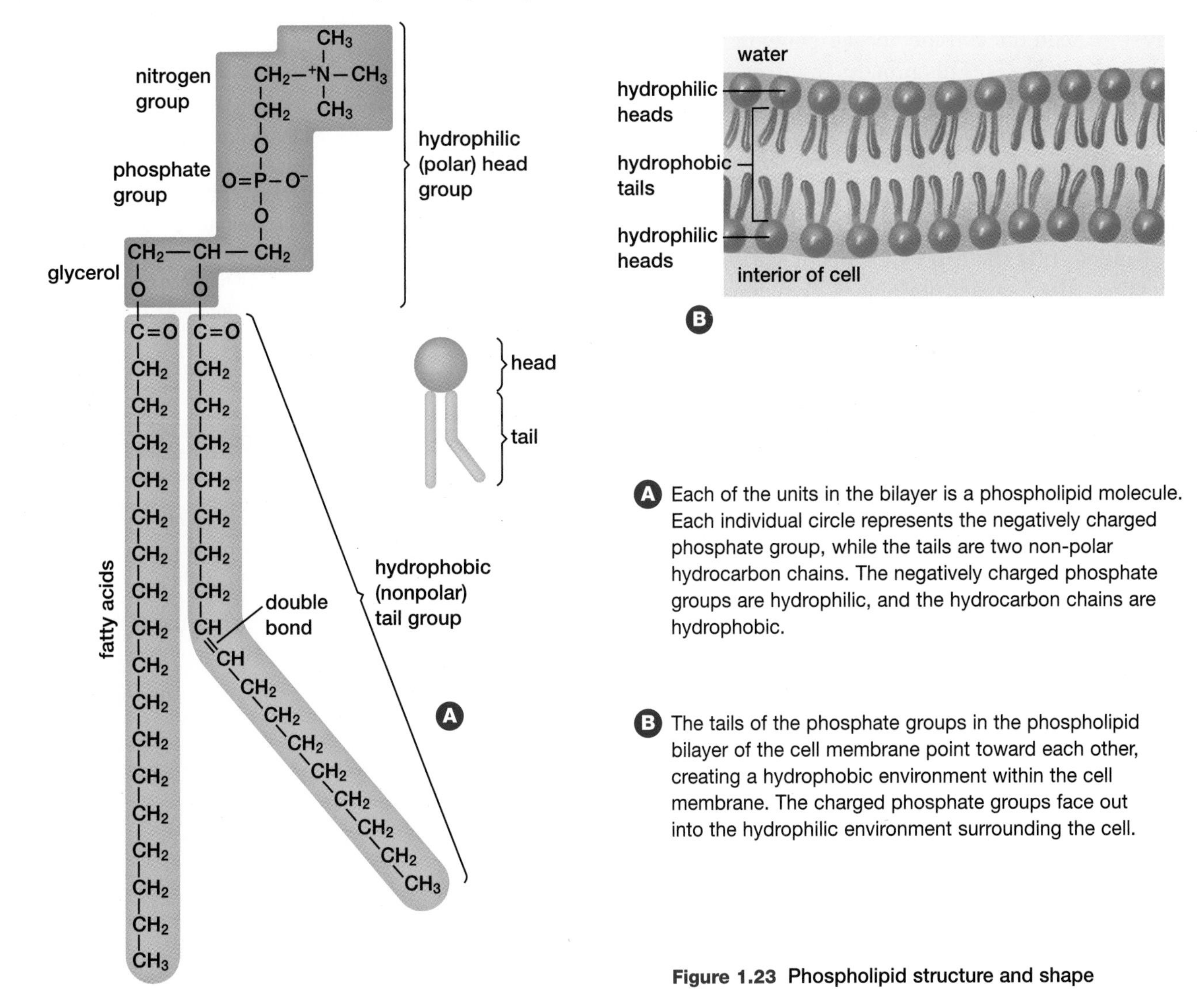

A Each of the units in the bilayer is a phospholipid molecule. Each individual circle represents the negatively charged phosphate group, while the tails are two non-polar hydrocarbon chains. The negatively charged phosphate groups are hydrophilic, and the hydrocarbon chains are hydrophobic.

B The tails of the phosphate groups in the phospholipid bilayer of the cell membrane point toward each other, creating a hydrophobic environment within the cell membrane. The charged phosphate groups face out into the hydrophilic environment surrounding the cell.

Figure 1.23 Phospholipid structure and shape

Called a phospholipid, this molecule interacts with water in a way that spontaneously results in the structure shown in Figure 1.23(b). This phospholipid bilayer is the foundation for the semi-permeable membrane that surrounds cells.

Some molecules can pass freely through the membrane, while others require assistance to enter. The phospholipid bilayer is virtually impermeable to macromolecules, relatively impermeable to charged ions, and quite permeable to small, lipid-soluble molecules. Molecules that move through the membrane do so at differing rates, depending on their ability to enter the hydrophobic interior of the membrane bilayer.

Many small, non-polar solute molecules, such as oxygen and carbon dioxide, pass through the bilayer of the cell membrane with least resistance. They enter by means of diffusion, a form of passive transport. As you learned in previous studies, in this method of cellular transport, molecules move from regions of high concentration to those of low concentration. Water, a small polar molecule, can travel through the cell membrane freely in the process of osmosis. This process involves the movement of the solvent water from an area of higher concentration of water to an area of lower concentration of water.

Some molecules are too large to diffuse unassisted across the cell membrane. These molecules enter the cell by means of specialized proteins called carrier proteins — they move and change shape to create an opening into the cell. Large uncharged hydrophilic molecules such as glucose make use of these proteins in order to enter cells (see Figure 1.24). No cellular energy is required for this facilitated diffusion process, so it is a form of passive transport. Appendix 5 shows several other examples of passive transport through the cell membrane. In the next chapter, you will see how cells use energy to move larger molecules across the cell membrane.

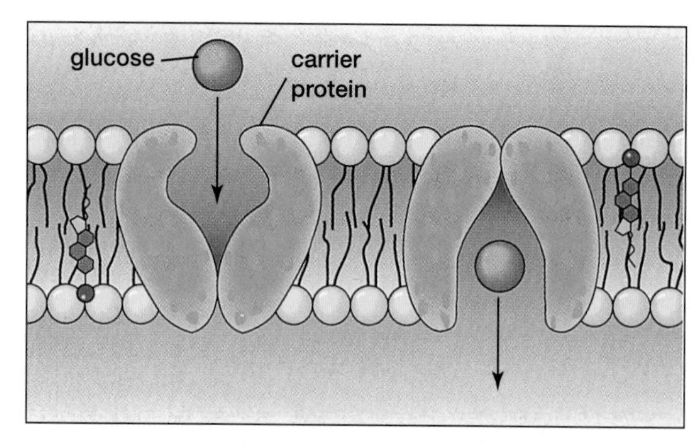

Figure 1.24 Molecules such as glucose move into the cell when carrier proteins in the cell membrane change shape.

BIO FACT

There may be a new way to move peptides and nucleic acids in synthetic DNA through the hydrophobic inner region of the cell membrane. Researchers in Pennsylvania have created hydrocarbon "umbrellas" that close around molecules when they are within the hydrophobic area of the membrane, but open in the hydrophilic areas, helping to move the molecules into the cell where they can do their work. This development holds promise for medical procedures such as gene therapy.

SECTION REVIEW

1. C Describe a redox reaction. Draw a simple diagram of such a reaction between a sodium atom and a chlorine atom. Provide an example of a redox reaction that takes place in all cells.

2. MC Explain how molecular models have helped us to better understand atoms, bonds, and molecules. Prepare a list of possible applications of molecular modelling in scientific research.

3. K/U Identify the functional group involved in the condensation reaction that forms the long chains of glucose molecules that make up starch.

4. K/U What is the monosaccharide product of the hydrolysis of starch?

5. C Explain what happens in a neutralization reaction.

6. K/U What are buffers, and why are they important for cellular processes?

Summary of Expectations

Briefly explain each of the following points.

- Chemical bonds join atoms to form molecules. (1.1)
- The core of every organic molecule contains carbon. (1.2)
- Functional groups attach to carbon cores of biological molecules and contribute to their function. (1.2)
- Acid-base reaction (neutralization) is one of the four major biochemical reactions in cells. (1.3)
- Redox reactions are important to cellular metabolism. A redox reaction involves: oxidation (the loss of electrons) and reduction (the gain of electrons). (1.3)
- Metabolic processes require a fairly constant pH. (1.3)
- Hydrolysis is a reaction that splits macromolecules through the addition of a water molecule. (1.3)
- Condensation is a reaction that removes water from monomers, synthesizing a polymer. (1.3)

Language of Biology

Write a sentence including each of the following words or terms. Use any six terms in a concept map to show your understanding of how they are related.

- metabolism
- atom
- nucleus
- protons
- neutrons
- electrons
- element
- isotopes
- valence shell
- valence electrons
- chemical bonds
- covalent bond
- ionic bond
- ion
- electronegativity
- non-polar covalent
- polar covalent
- polar molecule
- hydrogen bonds
- hydrophilic
- hydrophobic
- organic
- isomers
- structural isomers
- stereoisomers
- functional groups
- macromolecules
- monomer
- polymer
- organic acids
- organic bases
- neutralization
- buffers
- reduction
- oxidation
- redox reaction
- condensation
- hydrolysis
- monosaccharide
- disaccharide
- polysaccharides
- nucleic acid
- peptide bonds
- polypeptides
- lipid

UNDERSTANDING CONCEPTS

1. Proteins perform many functions in biological systems. Identify one protein and its function.

2. A protein is a polymer or chain of sub-units.
  (a) What is the subunit of a polymer?
  (b) What kind of bond links these subunits together?

3. Which is smaller, a hydrogen atom or a hydrogen molecule? Explain briefly.

4. How is a molecule of carbon dioxide different from a molecule of oxygen gas?

5. Define the word "isotope."

6. Which part of the atom is most involved in chemical reactions?

7. What is a valence electron?

8. Describe the structure and function of a chemical bond.

9. Use examples to explain how an ionic bond is different from a covalent bond.

10. How does the polarity of water molecules account for the ability of water to dissolve many substances?

11. **(a)** What is a hydrogen bond?
  (b) Identify one case where hydrogen bonds are important to a biological system.

12. Define the terms "hydrophobic" and "hydrophilic," and provide an example of each.

13. What role do buffers play in a cell?

14. Identify two isomers of glucose. How are these molecules similar? How are these molecules different?

15. How do functional groups change the nature of hydrocarbon chains?

16. What is meant by oxidation?

17. Define the terms "polymer" and "monomer." Use these terms in a description of carbohydrates.

18. Identify two nucleic acids.

19. Copy and label this diagram of condensation synthesis and hydrolysis.

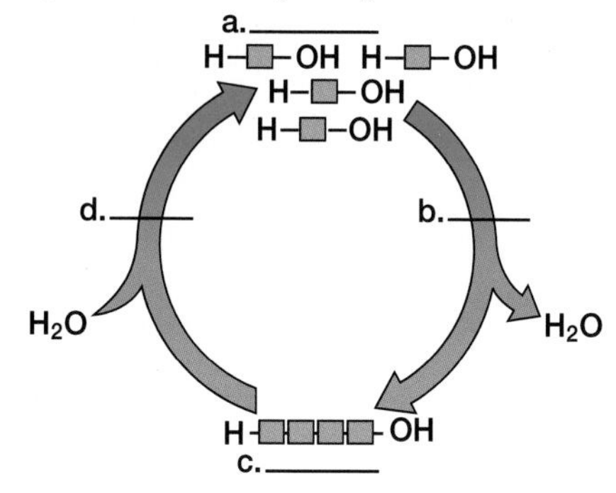

INQUIRY

20. Scientists must pay close attention to the concentration of solutions when reproducing conditions for living systems in an experiment. To make 1 L of a 1 mol/L solution of glucose, 1 mol of glucose (180.18 g) is added to a 1 L volumetric flask. Some distilled water is added and the contents swirled to dissolve the glucose. More distilled water is added up to the etch line of the volumetric flask.

    (a) What mass of glucose (in grams) should be used to make a 0.5 mol/L solution using the same volumetric flask?

    (b) What mass of glucose (in grams) should be used to make a 0.5 mol/L solution using a 500 mL flask?

    (c) What mass of glucose (in grams) should be used to make a 0.1 mol/L solution using a 500 mL flask?

21. Cellular respiration uses O_2 and releases CO_2. Released into water, CO_2 forms carbonic acid (H_2CO_3), which dissociates to produce a hydrogen (hydronium) ion. To determine the quantity of CO_2 produced by germinating seeds, a titration can be performed that neutralizes the hydronium ions produced. A 0.01 mol/L solution of NaOH is used to neutralize the acid. In one trial, 9.7 mL of NaOH was added to neutralize the water containing germinating seeds. Outline how to set up a control for this experiment and the results you would expect.

22. Dialysis tubing is a cellulose material used to make model cells for experiments in osmosis and diffusion. These model cells do not use living cell membranes. A student made a model cell by pouring distilled water into some dialysis tubing. The student placed the model cell in a salt solution and waited 10 minutes. The cell was then removed from the salt solution and the contents poured into a test tube. The student added silver nitrate to the test tube and a white precipitate formed. This indicated that chloride ions had entered the cell. Would this be true for living cell membranes? Design an experiment to test this.

COMMUNICATING

23. Use Internet resources to find the chemical make-up of a heme, a key component of hemoglobin.

24. Create a chart to summarize the bonds between atoms studied in biological systems.

25. Make a series of diagrams (similar to a cartoon strip) showing how buffers help maintain a near-constant environment in the cell.

26. Draw diagrams of two amino acids. Circle the *R*-group for each. Make a generalized diagram for all amino acids.

27. Some molecules can exist in different configurations that are important in a variety of metabolic processes. For example, glucose, a common monosaccharide, occurs in open chain and ring-shaped forms. The flexible nature of the chain form allows glucose to twist into a ring shape. The ring structure forms when the carbonyl functional group reacts with one of the alcohol groups within a single glucose molecule. In this reaction, one hydrogen atom is transferred from an alcohol group to the carbonyl group. This process

results in the formation of a chemical bond between the oxygen of the alcohol and the carbon of the carbonyl group. This new bond creates the ring structure of the glucose molecule. Glucose molecules constantly change back and forth from the ring to the chain form. To keep track of which carbons are involved, each carbon in the chain is assigned a number. With a partner, use a molecular model kit to construct models of the two isomers of the glucose molecule — chain and ring-shaped. Refer to the diagrams to help you plan and construct your models. Begin by constructing two identical chain versions of the glucose molecule. Make the appropriate changes to one of the chain models to transform it into the ring version.

Note that the carbon ring that forms the backbone of the glucose molecule is not a flat structure. Two of the carbons in the molecule are either above or below the other four. This will be apparent when you have completed your models.

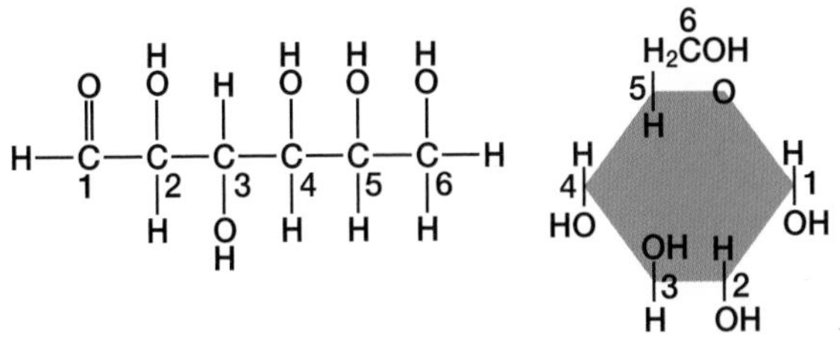

MAKING CONNECTIONS

28. Develop a series of questions that should be asked and answered before a new drug is approved for public use.

29. Lysosomal diseases, such as Tay-Sachs disease, are linked to large macromolecules called muccopolysaccharides that are produced in growing children. Defective enzymes in the lysosomes do not break down these macromolecules, which then accumulate in cells. The result is a deterioration of cell function. Use Internet resources to identify three other lysosomal diseases. Prepare a brief outline of each.

30. Many people who avoid eating beef or pork because they contain fats will eat cold-water fish (such as salmon or Arctic char), which have high levels of lipids. Why are these fish considered to be part of a healthy diet?

CHAPTER 2

Enzymes and Energy

Reflecting Questions

- How do biological processes follow the laws of thermodynamics?
- How do enzymes catalyze reactions and reduce the amount of energy required to fuel them?
- How are enzymes used in industry and pharmaceutical research?

Prerequisite Concepts and Skills

Before you begin this chapter, review the following concepts and skills:

- explaining how a redox reaction works (Chapter 1, section 1.3), and
- describing the structure and function of ATP (Chapter 1, section 1.3).

In the summer of 2001, a forest fire that had been started by a lightning strike raged through Kootenay National Park in British Columbia. Park officials allowed the fire to progress because the area was scheduled for a prescribed burn. Fires are a natural part of forest ecology and are important in forest regeneration. For example, some species of pine, such as the jack pine, drop cones that need the heat from a fire to open them and release their seeds. The Kootenay fire quickly consumed the dry grasses and trees; it soon spread beyond the area park officials could manage, threatening nearby communities.

Many firefighters risked their lives to control the spread of the flames. By the time the fire was contained and eventually extinguished, thousands of hectares of forest had burned.

The chemical reaction that occurred in the fire involved oxygen and the wood that formed the trees. While the forest fire was an example of a reaction that occurred with oxygen outside cells, reactions with oxygen also occur inside cells.

Energy is necessary to perform all cellular reactions, including redox, hydrolysis, and condensation reactions. Enzymes aid reactions within cells. Enzymes are necessary because they speed up the synthesis of energy-rich molecules needed for cellular processes.

In this chapter, you will learn how chemical reactions within cells are used to make energy-rich molecules. Energy from these molecules is used for various cellular processes. The bonds that hold atoms together store energy in molecules. This energy can be used by a cell to do work. You will explore various factors that influence how molecular bonds are formed and broken. You will also discover which molecules are involved in cellular processes and how the energy from one reaction can be used to drive another reaction.

How do these heart muscle cells obtain energy to keep the heart beating?

Chapter Contents

2.1 Thermodynamics and Biology

EXPECTATIONS

- **Describe the flow of energy in biological systems.**
- **Apply the laws of thermodynamics to the transfer of energy in the cell.**
- **Interpret quantitative data to learn how energy can be used in living systems.**

Many reactions occur inside every cell. These reactions, collectively known as **metabolism**, have been at the centre of much scientific investigation. For example, manufacturers of dietary supplements for athletes seek to isolate chemicals that increase metabolic activity. Creatine phosphate is one such chemical — it is a nitrogenous molecule that is stored in muscle cells. Enhanced stores of creatine phosphate in muscles have been shown to increase muscle mass and efficiency. The compound was synthesized and used in the former Soviet Union by elite athletes in the 1960s to increase their metabolic activity and performance. What are metabolic reactions, and why are they important? To understand this, you must first understand how energy flows through systems.

Figure 2.1 Metabolic reactions make possible all functions of cells. Some of these reactions fuel the mechanical work that a cell undertakes.

Energy and the Laws of Thermodynamics

To survive, all living things require **energy**, which is the capacity for doing work. Energy comes in different forms. For instance, energy comes from the Sun as light, and thermal energy from a furnace can be used to heat a home. All moving objects, such as falling water and pistons in an internal combustion engine, have **kinetic energy**. Energy can also be stored as **potential energy**. A molecule of glucose has potential energy. The potential energy stored in the bonds of a molecule is called **chemical energy**. If a molecule of glucose is broken down into carbon dioxide and water, the energy released can be used to do work. If a phosphate group is removed from a molecule of ATP, the chemical energy can be used to fuel various cellular processes.

Energy continually flows through living and non-living systems. The study of this flow of energy is called **thermodynamics**. Physicists and chemists have studied thermodynamics since the days of Sir Isaac Newton. Biologists also apply thermodynamics when they study metabolic processes and the energy transformations that take place within living systems. Scientists use the term **system** to identify a process under study, and they refer to it in relation to the rest of the universe. For instance, a hot drink in a sealed vacuum bottle is considered a **closed system** because the liquid is isolated from its surroundings — thermal energy cannot move from the liquid to outside the bottle. Removing the lid from the bottle results in an **open system**, because energy (thermal, in this case) can now move between the liquid and its surroundings — it moves from the liquid to outside the bottle.

All living organisms are open systems; energy moves two ways, both in and out of cells. For example, a green plant absorbs energy from the Sun and uses this energy for building structures, transporting materials, growth, and reproduction. The plant also releases energy into the environment in the form of thermal energy when the plant is forming metabolic products, such as water and carbon dioxide.

How energy flows between organisms and the environment is governed by the **laws of thermodynamics**. You have already encountered these laws in previous studies. The first law, or law of conservation of energy, states that *energy can neither be created nor destroyed, but can be transformed from one form to another.* For example, during photosynthesis, a green plant absorbs light energy from the Sun. This energy is transformed into chemical energy, which is stored in bonds that hold together atoms in a molecule of sugar. An internal combustion engine converts the chemical energy stored in gasoline molecules into kinetic energy — the motion of the car.

Some chemical reactions, such as burning a fuel, release energy. Some of this energy is useful because it is available to do work. The energy available to do work is known as free energy. Free energy can be used to do the work of building molecules in a cell. However, whenever energy is transformed from one form to another, some of it is lost. This lost energy is the portion that is not free energy and therefore is not available for useful work. The amount of free energy that can be harnessed by a green plant or car is much less than the total amount of light or chemical energy present in the sunlight or gasoline. This fact is the basis of the second law of thermodynamics, which states that *energy cannot be transformed from one form to another without a loss of useful energy.* The energy that is lost eventually escapes into the atmosphere largely as waste thermal energy. There are many transformations of energy that occur inside a cell. During each transformation, some energy is lost as thermal energy. Eventually, all forms of useful energy are transformed into thermal energy. After thermal energy dissipates, it can never be transformed back into a useful form, such as chemical energy, that can be used to do work. Therefore, biological systems require a constant supply of energy from the Sun to function.

A measure of the tendency of a system to become unorganized is called **entropy**. Every transformation of energy creates more disorder in the universe. Therefore, we can restate the second law of thermodynamics as follows: *every energy transformation increases the entropy of the universe.*

The conversion of chemical energy into thermal energy does not violate the first law of thermodynamics. If thermal energy is produced during a chemical reaction, it is still a form of energy. Although some of this energy is not available to do work, energy is still conserved.

Consider the following example as a case study of thermodynamic principles. Stacked beside the fire pit at your campsite are a stack of newspapers and a bundle of kindling that you intend to ignite to start a fire. The stack of paper and the wood are composed of cellulose, which is made up of complex carbon-based molecules. These molecules contain potential chemical energy. When you light the paper, the chemical bonds in the molecules are broken in a reaction with oxygen. During the reaction, thermal energy and light are released. Recall from your study of Chapter 1 that this is a redox reaction. Once the reaction begins, the paper quickly burns, forming the products of the oxidation of cellulose: carbon dioxide and water. If energy is released from the reaction of paper with oxygen, the paper and oxygen must contain more chemical energy than the products (see Figure 2.2).

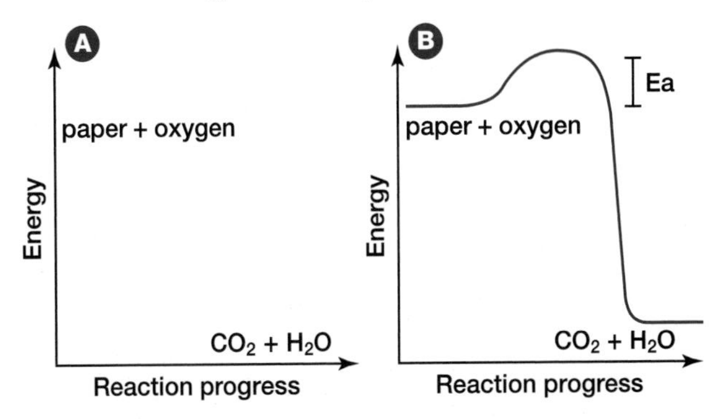

Figure 2.2 (A) Products (carbon dioxide and water) contain less energy than the reactants in a reaction between paper and oxygen. (B) An activation energy, Ea, is needed to initiate the reaction.

During the reaction, the chemical energy stored in the paper and in the oxygen molecules is transformed into thermal energy and light energy. You can feel the thermal energy that is released if you reach toward the fire to warm your hands.

Why does the paper require an initial input of energy to start the fire? Chemical bonds hold atoms and molecules together. These bonds maintain the chemical energy in the molecules. In order to destabilize the bonds, and thereby release the energy they hold, an initial input of extra energy is needed. This extra energy is known as **activation energy**. Figure 2.2 shows the activation energy required to ignite paper. Different substances require different amounts of activation energy to start a reaction. The activation energy needed to start a reaction within cells is governed by special proteins. Without these proteins, metabolic processes could not occur. Next, you will examine two types of metabolic reactions that occur within cells.

Exothermic and Endothermic Metabolic Reactions

Recall that metabolic reactions encompass all the reactions that occur within cells, including anabolic reactions (such as condensation) and catabolic reactions (such as hydrolysis), and redox reactions. Complex carbohydrates, fats, and proteins can be broken down in catabolic reactions, thereby forming molecules such as simple sugars and amino acids. Anabolic processes then join up these products and their functional groups to form various macromolecules needed by cells for maintenance and growth.

Investigation 2 • A

SKILL FOCUS

- Predicting
- Performing and recording
- Analyzing and interpreting
- Communicating results

Exothermic and Endothermic Changes

The living processes of cells involve numerous chemical reactions involving a wide variety of reactants and products that involve the making and breaking of bonds. Such chemical reactions can be exothermic and release energy or they can be endothermic and absorb energy. Within the cell, there appears to be no net change. In isolation, you can observe how each reaction exchanges energy with the external environment. In this investigation, you will observe energy changes that are characteristic of these reactions.

Pre-lab Questions

- Before a chemical change occurs, where is energy stored?
- How is a cell an open system with respect to energy?
- How is a reaction in an Erlenmeyer flask an open system?

Problem

What energy changes can you observe during different chemical reactions?

CAUTION: **The reactants and products formed in this experiment may be toxic. Avoid any contact with skin, eyes, or clothes. Flush spills on your skin immediately with copious amounts of water and inform your teacher. Exercise care when handling very hot or very cold objects. Handle thermometers with care. Wash your hands before leaving the laboratory.**

Materials

test tubes
test tube markers
test tube brush
test tube rack
test tube holder
250 mL Erlenmeyer flask
50 mL beaker
250 mL beaker
stirring rods
balance
beaker tongs
alcohol thermometer
 or temperature probe
squirt bottle
distilled water
ammonium thiocyanate
barium hydroxide
 (octahydrate)
ammonium chloride
ammonium nitrate
sodium chloride
calcium chloride

Procedure

Part A

1. Measure 9 g of barium hydroxide (octahydrate) and add it to a 250 mL Erlenmeyer flask.
2. Insert an alcohol thermometer into the solid barium hydroxide and record the temperature.
3. Measure 3 g of ammonium chloride and add it to a 50 mL beaker.
4. Using the thermometer or temperature probe, record the temperature of the ammonium chloride.
5. Use a squirt bottle to squirt a few drops of water onto the external surface of the flask. Observe any change in appearance of the water as you continue with the procedure.
6. Add the ammonium chloride to the Erlenmeyer flask containing the barium hydroxide and stir gently to thoroughly mix the ingredients.
7. Immediately insert the thermometer or probe into the mixture and record the temperature. Note any changes in the appearance of the materials in the flask or any other evidence of a chemical reaction occurring in the mixture.
8. Record the temperature every 30 s until there is no change for at least three minutes.

A reaction can be classified based on whether it releases or uses energy. A reaction that is accompanied by a release of energy is called an **exothermic reaction**, as shown in Figure 2.3 on the next page. For example, recall the overall reaction for cellular respiration:

$$C_6H_{12}O_6 + 6O_2 \rightarrow 6CO_2 + 6H_2O + \text{energy (useful and waste)}$$

glucose oxygen carbon dioxide water

For each molecule of glucose oxidized in cellular respiration, energy is released. Some of this energy is useful and available to do work and

9. Record the temperature and observational data in a table.
10. After all measurements and other observations have been completed, dispose of any chemicals according to your teacher's instructions.
11. Clean flasks, beakers, and thermometers thoroughly with distilled water to remove all traces of chemicals used in this procedure.

Part B

1. Using the balance, measure 3 g of sodium chloride.
2. Pour a small amount of water into a test tube. Record the temperature of the water.
3. Pour the sodium chloride into the test tube and gently swirl the contents until the sodium chloride dissolves in the water.
4. Immediately insert the thermometer into the solution and record the temperature. Note any changes in the appearance of the substances in the test tube or any evidence of a chemical reaction occurring in the mixture.
5. Record the temperature every 30 s until there is no change for at least three minutes.
6. Repeat steps 1 to 5, first using an equal mass of ammonium chloride, and then using an equal mass of calcium chloride. These steps may be performed by other class members or groups.
7. Record your data in a table. Include data of other class members who performed the investigation using the alternate chemical compounds.
8. After all measurements and observations have been completed, dispose of all chemicals according to your teacher's instructions.
9. Clean flasks, beakers, and thermometers thoroughly with water to remove all traces of the chemicals used in this procedure.

Post-lab Questions

1. Describe the energy change of each reaction.
2. Which reaction is endothermic? Which reaction is exothermic?
3. In a cell, energy from one reaction is often used to do work within the cell. Which of these reactions can be used to do work? Justify your choice.
4. The temperature of the surrounding environment may have had an effect on the rate of each reaction. If the temperature in the surrounding environment were very cold, as in a refrigerator, which reaction would be most affected? Justify your response.

Conclude and Apply

5. As mentioned before, cells can maintain a near-constant temperature despite a wide variety of reactions that take place. Some reactions release thermal energy to the surrounding environment, and others absorb thermal energy from the surrounding environment. Design a model cell that incorporates both reactions safely to demonstrate this.

Extend Your Knowledge

6. Examine a "heat pack" and a "cold pack." Identify the chemicals contained in each package. Follow the instructions on the package label to initiate the reaction. Measure the temperature change and time required to attain the maximum or minimum temperature. Observe and record any other change in the properties of the substances in the package. If you know the chemicals involved, write out the endothermic or exothermic chemical reactions that occurred in each package.

some is waste thermal energy. This means that the products (carbon dioxide and water) contain less energy than the reactants (glucose and oxygen).

In contrast, an **endothermic reaction** involves an input of energy. For example, the synthesis of glucose by plants during photosynthesis is as follows:

$$6CO_2 + 6H_2O + \text{energy (sunlight)} \rightarrow C_6H_{12}O_6 + 6O_2$$

carbon dioxide, water, (sunlight), glucose, oxygen

Synthesis involves building molecules — it uses more energy than it gives off (see Figure 2.3). For example, green plants require energy in the form of sunlight to produce glucose. Because this endothermic reaction stores chemical energy in molecules, there is a gain in energy.

As you can see in the two equations above, oxidation and synthesis of glucose are two reactions that are the reverse of each other. If two reactions are the reverse of each other, one reaction is endothermic and the other is exothermic.

Exothermic and endothermic reactions both involve energy transformations. How do cells control the flow of energy so that they do not overheat and destroy themselves? In the next section, you will learn how cells are able to lower the amount of activation energy necessary to carry out a variety of metabolic reactions.

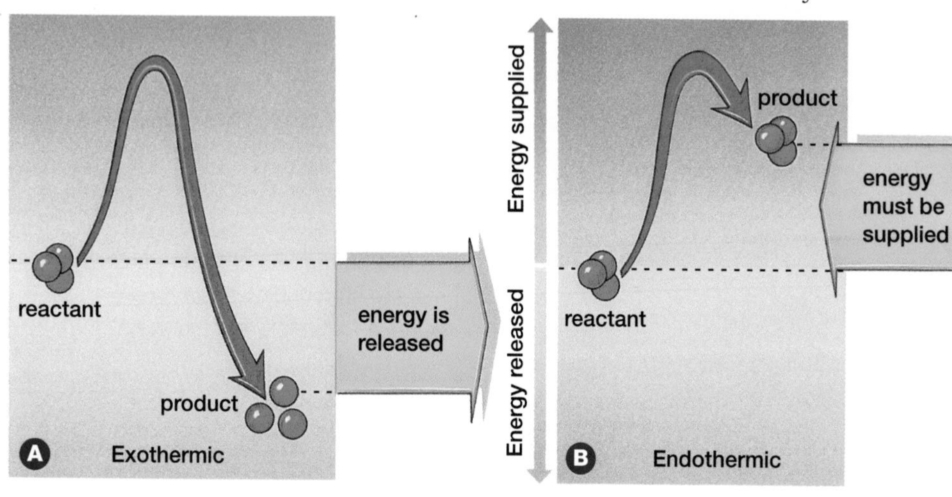

Figure 2.3 In an exothermic reaction (A), the reactants contain more chemical energy than do the products, so there is a release of energy. In an endothermic reaction (B), the reactants contain less energy than do the products, so energy must be added for the reaction to proceed.

SECTION REVIEW

1. K/U How is energy important to a living organism?
2. K/U A molecule of glucose has potential energy.
   (a) What does this mean to a cell?
   (b) Where is the energy in the glucose molecule?
3. K/U Why is a living cell considered to be part of an open system?
4. C Draw diagrams to show how energy flows in an open system and in a closed system.
5. K/U Define the term "free energy."
6. MC You turn on a CD player and the disc begins to rotate. How is this an example of energy transformation?
7. K/U Define and provide examples of exothermic and endothermic reactions.
8. C An endothermic reaction is often referred to as an "uphill" process. Use an analogy from everyday life to explain why this is a good description.
9. C Make a chart to contrast exothermic and endothermic reactions in terms of energy.
10. K/U Suppose you start a fire by using a magnifying glass to focus the Sun's image on a piece of paper. What is the source of activation energy for this reaction? What is the source of activation energy after the fire has started and the magnifier is taken away?
11. I A research scientist wishes to study the breakdown of glucose in a solution to form carbon dioxide and water. Design an investigation to determine if this is an exothermic or an endothermic reaction.
12. C Create a flowchart that shows the series of steps that take place starting with reactants and ending with products formed in a chemical reaction. Indicate where energy is involved in the reaction.

UNIT INVESTIGATION PREP

In the investigation at the end of the unit, you will design your own experiment. Be familiar with correct experimental procedure. How will you form your hypothesis? How will you plan your procedure? How will you determine your dependent and independent variables?

2.2 Enzymes as Catalysts

EXPECTATIONS

- **Describe the structure and function of enzymes in cellular respiration.**
- **Design and carry out an experiment involving enzyme activity.**
- **Describe the use of enzymes in the food and pharmaceutical industries.**

Metabolic reactions need activation energy to either build or break down molecules. Cells also use special proteins that aid metabolic reactions. These proteins, called **enzymes**, work by speeding up a chemical reaction. This chemical activity increases the *reaction rate*, or rate at which a reaction occurs, measured in terms of reactant used or product formed per unit time (while existing conditions remain unchanged). Some of the earliest studies on enzymes were performed in 1835 by Swedish chemist Jon Jakob Berzelius, who termed their chemical action "catalytic."

Enzymes and the Catalytic Cycle

The acceleration of a chemical reaction by some substance, which itself undergoes no permanent chemical change, is called **catalysis**. The catalysts of metabolic reactions are enzymes, which are involved in almost all chemical reactions in living organisms. Without enzymes, metabolic reactions would proceed much too slowly to maintain normal cellular functions. Consider the hydrolysis of sucrose, an exothermic reaction. A solution of sucrose dissolved in water could sit for years without showing signs of hydrolysis. If the enzyme sucrase is added to the solution, the enzyme speeds up the reaction millions of times, so that all of the sucrose will be hydrolyzed in several seconds. Enzymes speed up reactions by lowering the amount of activation energy needed. Thus, less energy is required for the reaction to begin. The action of an enzyme on an exothermic reaction is illustrated in Figure 2.4.

Cells carry out a large number of different biochemical reactions; many of these reactions require a specific enzyme in order to take place. Different sets of enzymes are responsible for catalyzing different chemical reactions. **Oxidative enzymes** (oxidoreductases) work to catalyze oxidation-reduction reactions. **Hydrolytic enzymes**

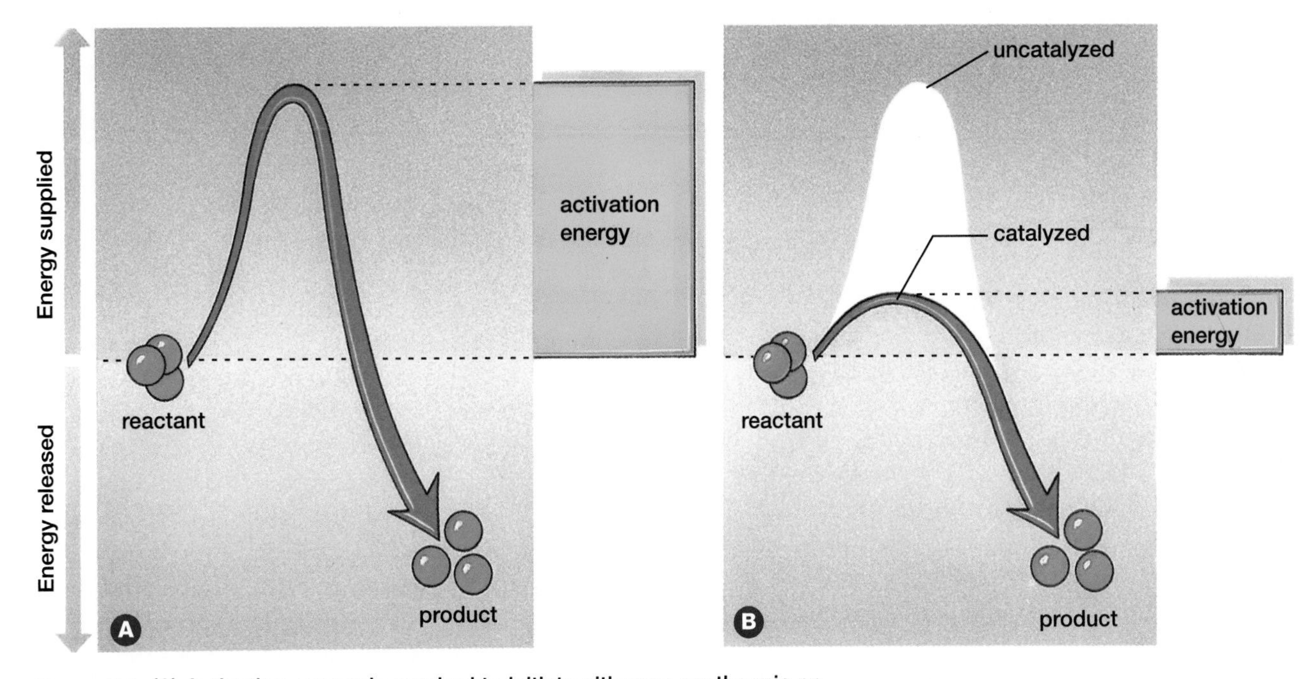

Figure 2.4 (A) Activation energy is required to initiate either an exothermic or an endothermic reaction. (B) Enzymes catalyze certain reactions by lowering the activation energy needed to start the reaction.

(hydrolases) catalyze the addition of water in reactions and split molecules into simpler forms. These simpler molecules may be used to build other molecules or may be excreted from the cell. For example, the lysosomes of cells contain many hydrolytic enzymes. Tasks such as breaking down nucleotides, proteins, lipids, and phospholipids are each carried out by a specific hydrolytic enzyme. Other enzymes remove carbohydrate, sulfate, or phosphate groups from molecules.

Synthesis reactions that build structures such as proteins, nucleic acids, hormones, glycogen, and phospholipids all require the use of enzymes. The enzyme DNA polymerase, for example, is needed for DNA replication, which precedes mitosis. Each chemical reaction in cellular respiration requires a specific enzyme. Deaminases remove the amino groups from amino acids so the remainder of the molecule can be used as an energy source. Enzymes also help to split long-chain fatty acids into smaller compounds, which are used as an energy source and broken down by the process of cellular respiration. Blood clotting, the formation of angiotensin II to increase blood pressure, and the transport of carbon dioxide in the blood all require specific enzymes. Tables 2.1 and 2.2 show categories of enzyme specificity and modes of action.

Table 2.1
Categories of enzyme specificity

Enzyme specificity	Action
absolute	catalyzes only one reaction
group	acts only on molecules that have specific functional groups
linkage	acts only on a particular chemical bond, regardless of the rest of the molecular structure
stereochemical	acts only on a particular optical isomer

A reactant in any given enzymatic reaction is called a **substrate** for that specific enzyme. Some enzymes catalyze one individual reaction; this is the case with peroxidase, an enzyme that decomposes hydrogen peroxide into water and oxygen. Reactions within cells, however, are often part of a **metabolic pathway** (series of linked reactions), beginning with one substrate and ending with a product. Such metabolic pathways can involve many reactions, which often include other pathways. Each step of a metabolic pathway, or each constituent reaction of the pathway, needs its own specific enzyme.

Table 2.2
Enzymes are classified (by an International Classification of Enzymes) according to the kind of chemical reactions they catalyze. The six classes of enzymes and their actions are given.

Action	Type of enzyme	Examples
add or remove water (hydrolysis or condensation)	hydrolases and hydrases	hydrolases: esterases, carbohydrases, nucleases, deaminases, amidases, proteases hydrases: fumarase, enolase, aconitase, carbonic anhydrase
transfer electrons (redox reactions)	oxidoreductases	oxidases, dehydrogenases
split or form a C–C bond	transferases	desmolases
change geometry or structure of a molecule	isomerases	glucose phosphate
form carbon to carbon, carbon to sulfur, carbon to oxygen, or carbon to nitrogen bonds by condensation and hydrolysis reactions coupled to ATP	ligases	pyruvate carboxylase, DNA ligases
add groups to a C=C double bond or remove groups to form a C=C double bond	lyases	aldolase, decarboxylases

To understand how enzymes work, consider that the key to enzyme function is enzyme structure. Enzymes are globular proteins with depressions on their surfaces, as shown in Figure 2.5. These depressions are called **active sites**. Active sites are places where substrates fit and where catalysis occurs. Active sites are not static receptacles. Substrates fit closely into active sites because enzymes can adjust their shapes slightly to accommodate the substrate. This process involves a subtle change in conformation, or three-dimensional shape, of the enzyme when the substrate binds to it. Multiple weak bonds between the enzyme and the substrate are involved in this process. The change in shape of the active site to accommodate the substrate is called **induced fit**. This process may bring specific amino acid functional groups on the enzyme into the proper orientation with the substrate to catalyze the reaction (see Figure 2.5 on the next page).

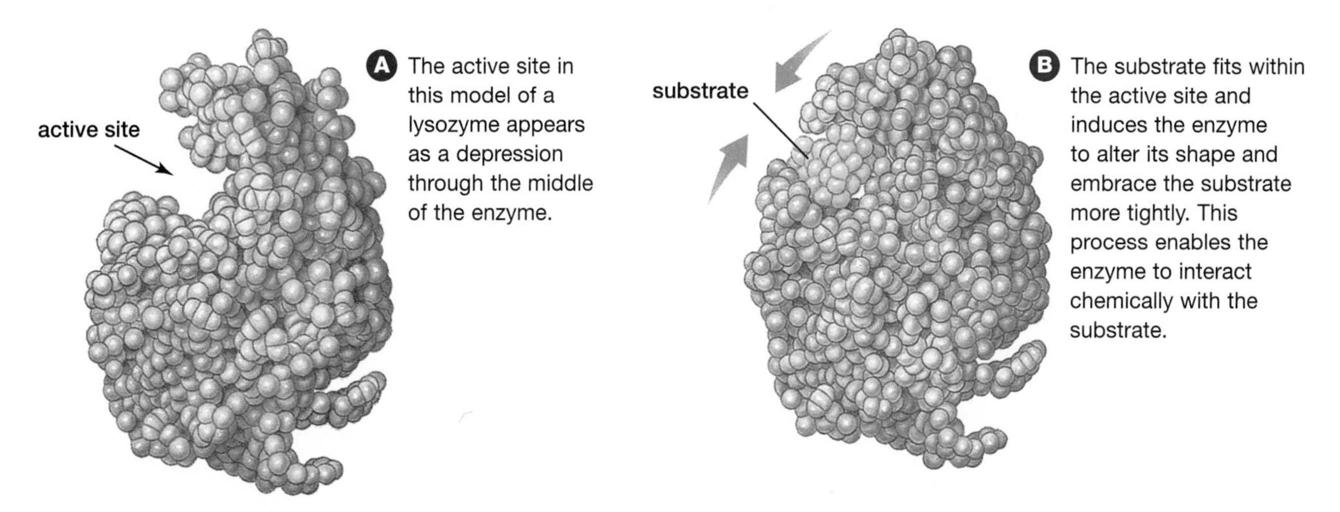

Figure 2.5 The induced-fit model of enzyme action

The combination of the substrate and the enzyme itself forms a compound called an **enzyme–substrate complex**. Swedish chemist Svanté Arrhenius first hypothesized about the enzyme–substrate complex in 1888, proposing that there must be a stage during catalysis when the enzyme and the substrate join together. Modern laboratory experiments have confirmed his hypothesis. In many cases, the enzyme–substrate complex is held together by such bonds as hydrogen bonds and weak ionic bonds. The polar and non-polar groups of the active site attract compatible groups on the substrate molecule. These attractions effectively lock the substrate molecule in the active site. Once in the active site, the substrate is subject to necessary collisions, bond breaks, and bond formations that must take place to form the product molecule. This reaction can be anabolic or catabolic, depending upon the enzyme. Once the product molecule has been formed, it is released from the enzyme–substrate complex. The enzyme is now able to accept another substrate, and begin the process anew. This cycle is known as the **catalytic cycle**. Figure 2.6 shows the catalytic cycle involving sucrose and the enzyme sucrase.

There are several methods by which enzymes reduce the activation energy needed to break the bonds in a substrate. In the enzyme–substrate complex, the substrate molecules experience

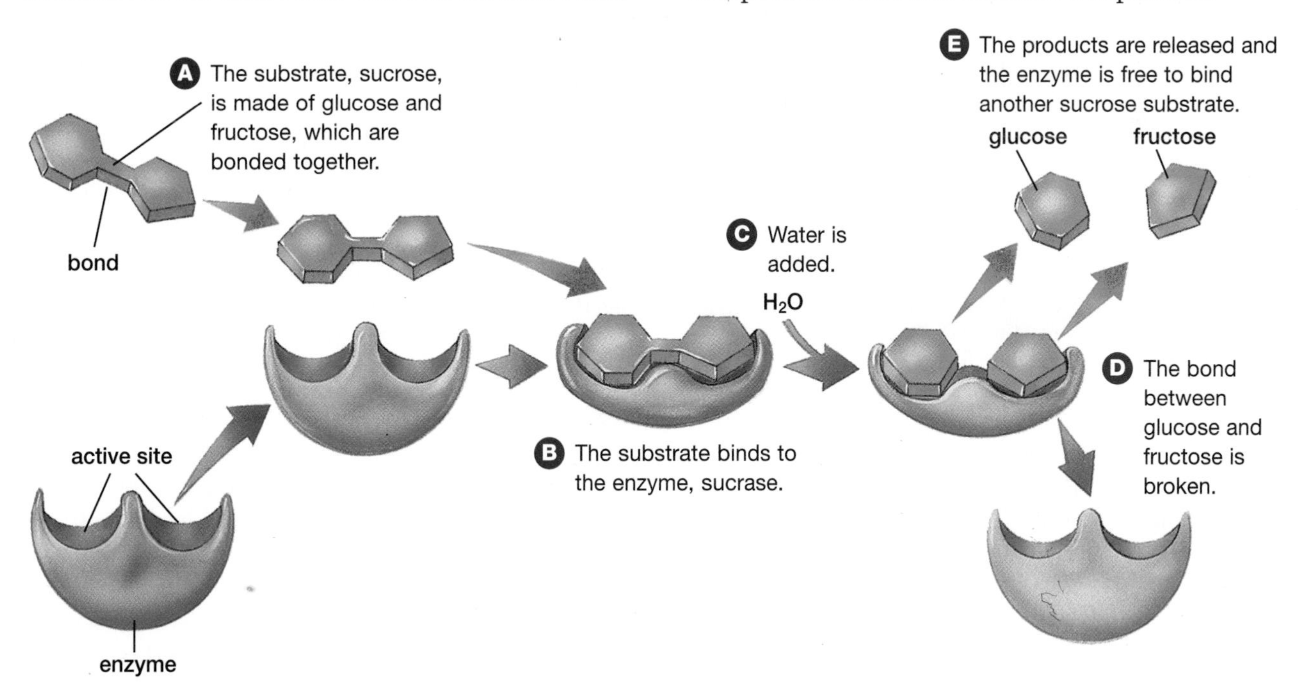

Figure 2.6 The catalytic cycle of this enzyme splits the sugar sucrose into two simpler sugars, glucose and fructose.

physical stress. The *R*-groups in the active site of an enzyme are able to stress the bonds of the substrate. There is bending and stretching of bonds that hold the molecule and the active site together. In this case, the activation energy is lowered because the bonds within the molecule have become weaker, reducing the amount of energy needed to break them.

Another way in which the active site of an enzyme may lower activation energy involves special amino acids that line the active site. These amino acids have reactive *R*-groups that can aid in the transfer of hydrogen ions to or from the substrate. For example, the active sites of hydrolytic enzymes, such as those within the lysosome, often provide acidic and/or basic amino acid groups at precisely the correct orientations required for catalysis. The yeast enzyme, intertase (also known as beta-fructofuranosidase), is a hydrolytic enzyme that speeds up the breakdown of sucrose into the products glucose and fructose.

Some other enzymes provide amino acid groups at their active sites that can accept electrons, while others are attracted to atomic nuclei of the substrate. This process can form a temporary attraction with the substrate. In this state, the substrate is less stable and can more easily react to form the product. Some enzymes may facilitate the correct reaction by bringing two different substrates together in the appropriate orientation to each other.

An oxidative enzyme (such as cytochrome P450s) catalyzes the transfer of electrons from substrates to oxygen molecules. Substrates for these enzymes are often referred to as hydrogen donors because hydrogen ions along with electrons are taken from the substrate. Cytochrome P450s is most common in the endoplasmic reticulum of liver cells. In these cells, the enzyme helps to metabolize toxins, as well as fat-soluble vitamins such as A, D, and E.

WEB LINK

www.mcgrawhill.ca/links/biology12

To find out more about lysosomal enzymes, go to the web site above, and click on **Web Links**. Prepare a three-column chart with the following headings: enzyme, substrate, product of reaction. Fill in your chart as you read about enzymes and their function. Predict what would happen if any of the lysosomal enzymes did not function properly.

Enzyme Activity

As you have learned, enzymes lower the activation energy required to start a chemical reaction. The activity of enzymes, however, can be influenced by environmental factors, such as pH and temperature.

The shape of an enzyme is determined by hydrogen bonds, which hold peptide chains in the enzyme in a specific orientation. (See Appendix 5, "Shapes of Selected Macromolecules," on page 559.) As well, all enzymes contain segments that are hydrophobic. The hydrogen bonds in an enzyme and any hydrophobic interactions that parts of the enzyme may experience are easily affected by changes in temperature. Enzyme activity increases as temperature increases, but only up to a maximum point. If the temperature increases beyond a critical point, enzyme activity declines rapidly (see Figure 2.7). When this occurs the enzyme has been denatured. When an enzyme is denatured by excessive heat, its shape changes and it can no longer bind to its substrate.

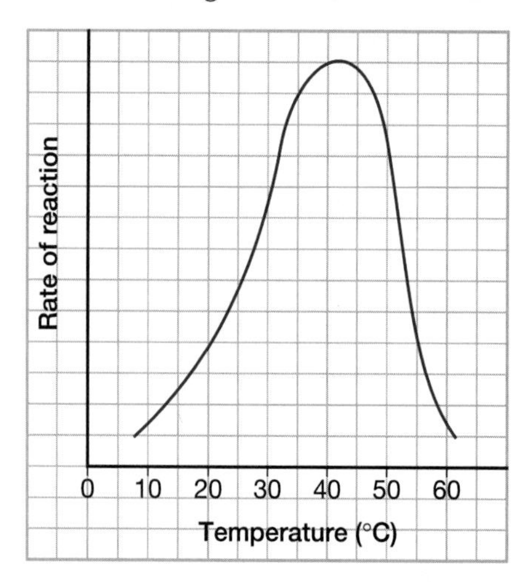

Figure 2.7 Rate of an enzymatic reaction as a function of temperature. The rate of the reaction shown here approximately doubles with every 10°C increase in temperature. For this reaction, the maximum rate occurs at 42°C; the rate then decreases and the reaction eventually stops.

Most human enzymes function best between 35°C and 40°C. Below this temperature range, enzymes are less flexible and therefore less able to provide an induced fit to substrates. Above this range, the bonds become weaker and less able to hold the peptide chains in the enzyme in the proper orientation. Some bacteria, however, can function at temperatures as high as 70°C. These bacteria live in and around **hydrothermal vents**, which are fissures in the Earth's crust on the ocean floor that release hot water and gases. The bacteria are able to survive in these environments because the bonds between peptide chains in their enzymes are relatively strong and able to withstand the extreme temperatures. These enzymes are therefore called thermostable enzymes.

Thermostable enzymes could operate above the growth temperature for pathogens that otherwise can contaminate foods. Potential applications of this knowledge might include development of food products that could be processed at higher temperatures, and are more resistant to microbial contamination (such as *E. coli*). Thermostable enzymes may also be useful in drug synthesis. Such enzymes may be able to catalyze reactions more effectively, affording higher productivity. They may also last longer and could possibly be re-used. In the next Thinking Lab, you will conduct research into the sources of enzymes in foods.

COURSE CHALLENGE

Perhaps the Thinking Lab on enzymes and diet can help you select a topic for the Biology Course Challenge. Look ahead to page 544 to see how you can start preparing for the Challenge.

THINKING LAB

Enzymes and Diet

Background

In 1963, Lithuanian-American Ann Wigmore spearheaded a dietary movement in the United States called Living Foods. She based her teachings about diet on her belief that cooked foods cause most mental and physical illnesses, saying that proteins and enzymes are destroyed during the cooking process. The diet she recommended is a vegetarian one, including sprouted grains and beans, vegetables, fruits, nuts, wheatgrass juice, and dehydrated snacks.

Wigmore's philosophy found an appreciative audience in North America. Bookstores now carry many raw foods diet cookbooks, restaurateurs have set up cafés serving raw foods, and alternative health-care practitioners often recommend raw food diets to help people lose weight and prevent disease. The raw foods diet has been extended to apply to pets — some veterinarians insist that dogs and cats are unhealthy because they do not eat enough raw foods. Vancouver veterinarian Julie-Anne Lee says, "Dogs and cats eat raw food in the wild and hunting cats naturally eat raw meat." She argues that only a raw foods diet produces the anti-bacterial enzymes in the gut and mouth that domestic animals such as dogs and cats need to be healthy.

While some, such as Dr. Lee, extol the virtues of raw foods for people and animals, others scoff at the idea, saying that people have been eating cooked foods for centuries. Skeptics point out that although people do need enzymes to digest their food, cells make enzymes for that very purpose. Some veterinarians also dismiss the need for raw foods for pets, saying that pets have become accustomed to a "domestic" diet, and would not function well on a "wild" one.

You Try It

1. Find out more about the Living Foods diet. Research web archives and dietary journals, and try to interview dietitians and health-care practitioners. Summarize the main concepts behind the raw foods diet.
2. Using your knowledge of the properties and functions of enzymes, review what you learned in your research. Prepare a report answering the following questions: Is a diet of raw food healthy for people? for dogs and cats? Why or why not?
3. Investigate a particular food and the enzymes it contains to ascertain whether it would be more beneficial to eat this food raw rather than cooked. Examples could be pineapples (containing the enzyme bromelain), tomatoes, soybeans, or broccoli. Consider also raw meats and fish.
4. Some enzymes break down as soon as they enter the digestive system. For example, lactose-intolerant people, who must ingest a lactase enzyme preparation before consuming dairy products (which contain lactose), should not do so too far in advance of eating a dairy product. This is because our own digestive enzymes begin to break down the lactase enzyme as soon as it enters the digestive system. Apply this knowledge, as well as a basic review of the digestive process, to the information you found for question 3. Do the enzymes in the raw food that you researched break down before they have a chance to work in the digestive process?

Each enzyme also works optimally (best) at a specific pH. Figure 2.8 shows the activity ranges for the enzymes pepsin and trypsin at different pH levels. At pH values where the enzymes work optimally, the enzymes have their normal configurations. The bonds that hold peptides in position in the enzyme are sensitive to hydrogen ion concentrations. A change in pH can alter the ionization of these peptides and disrupt normal interactions. Under extreme conditions of pH, the enzyme will eventually denature. Most enzymes function best in the pH range of 6 to 8. Pepsin, which digests proteins in the human stomach, works best under very acidic conditions (pH of 2).

In the next investigation, you will design an experiment to study the effects of temperature and pH on enzyme activity.

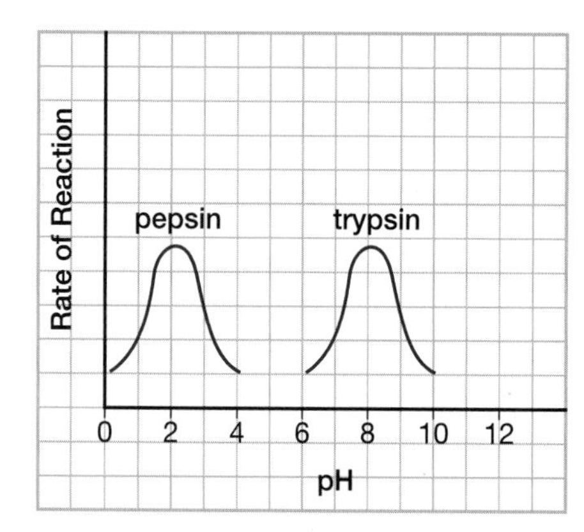

Figure 2.8 Rate of an enzymatic reaction as a function of pH. The maximum rate of reaction for the enzyme pepsin, which is found in the stomach, occurs at a pH of about 2. Trypsin, an enzyme found in the small intestine, performs best at a pH of about 8.

DESIGN YOUR OWN
Investigation 2 • B

SKILL FOCUS

- Initiating and planning
- Predicting
- Analyzing and interpreting
- Communicating results

Affecting Enzyme Activity

The compound hydrogen peroxide, H_2O_2, is a by-product of metabolic reactions in most living organisms. However, hydrogen peroxide is damaging to delicate molecules inside cells. As a result, nearly all organisms produce the enzyme peroxidase, which breaks down H_2O_2 as it is formed. Potatoes are one source of peroxidase. Peroxidase speeds up the breakdown of hydrogen peroxide into water and gaseous oxygen. This reaction can be detected by observing the oxygen bubbles generated.

Problem

How effectively does the enzyme peroxidase work at different temperatures and pH values?

Hypotheses

Make hypotheses about how you think temperature and pH will affect the rate at which the enzyme peroxidase breaks down hydrogen peroxide. Consider both low and high temperatures and pH values.

CAUTION: **Hydrochloric acid and sodium hydroxide are corrosive chemicals. Avoid any contact with skin, eyes, or clothes. Flush spills on your skin immediately with copious amounts of cool water and inform your teacher. Exercise care when heating liquids and using a hot plate. Take care when using sharp instruments. Do not taste any substances in the laboratory. Wash your hands before leaving the laboratory.**

Materials

- test tubes
- test tube markers
- test tube brush
- test tube rack
- beaker for hot water bath
- hot plate
- beaker tongs
- medicine droppers
- distilled water
- scalpel or sharp knife
- ruler
- forceps
- clock or timer
- ice
- thermometer or probe
- pH indicator paper or probe
- 0.1 mol/L hydrochloric acid (HCl)
- 0.1 mol/L sodium hydroxide (NaOH)
- 3% hydrogen peroxide (H_2O_2)
- raw peeled potato

Experimental Plan

1. Examine the materials provided by your teacher. As a group, list ways you might test your hypotheses.
2. Agree on a method(s) your group will use to test your hypotheses.

ELECTRONIC LEARNING PARTNER

To learn more about the function of enzymes, go to your Electronic Learning Partner now.

Enzyme Inhibitors and Allosteric Regulation

In addition to the environmental factors of pH and temperature, various substances can inhibit the actions of enzymes. **Inhibitors** are chemicals that bind to specific enzymes. This results in a change in the shape of the enzyme that causes the enzyme to shut down its activity. In cells, enzyme inhibition is usually reversible; that is, the inhibitor is not permanently bound to the enzyme. Inhibition of enzymes can also be irreversible. For example, hydrogen cyanide, a powerful toxin, is an inhibitor for the essential enzyme cytochrome *c* oxidase. Toxins, such as hydrogen cyanide, typically bind (either covalently or non-covalently) so strongly with an enzyme that the enzyme cannot bind with its substrate. Some poisons that result in irreversible enzyme inhibition do not combine with the enzyme; instead, they destroy enzyme activity by chemically modifying critical amino acid *R*-groups.

Other toxins, such as venom from the Malayan pit viper (*Calloselasma rhodostoma*) (shown in Figure 2.9 on the next page), are enzyme inhibitors that can help people overcome the effects of a stroke. Strokes are caused by blood clots in the brain, which can result in mental and physical debilitation. A substance called ANCROD, derived

3. Your experimental design should use a control and test one variable at a time. Will you be collecting quantitative or qualitative data?
4. Write a numbered procedure for your experiment that lists each step, and prepare a list of materials that includes the amounts you will require.

Checking the Plan

1. What will be your independent variable? What will be your dependent variable? How will you set up your control?
2. How will you determine peroxidase activity? How will you measure the amount of oxygen produced? Have you designed a table for collecting data?
3. Will you conduct more than one trial? How long will you allow each trial to run?
4. How will you analyze your data?
5. Before beginning the experiment, have your teacher check your plan.

Data and Observations

Conduct your experiments, record your data, and complete your data table. Design and complete a chart to present your results.

Analyze

1. What do these laboratory procedures indicate about the activity of peroxidase?
2. Make graphs showing the relationship between temperature and oxygen produced and between pH and oxygen produced.
3. How does the presence of HCl affect the activity of peroxidase?
4. How does the presence of NaOH affect the activity of peroxidase?

Conclude and Apply

5. Do your data support or reject your hypotheses?
6. At what temperature did peroxidase work best? at what pH?
7. What was the purpose of using control samples?
8. If you have ever used hydrogen peroxide as an antiseptic to treat a cut or scrape, you know that it foams as soon as it touches an open wound. How can you account for this observation?

Exploring Further

9. You may wish to use hydrogen peroxide to test for the presence of peroxidase in other foods, such as pieces of other vegetables or of meat. Which food shows the greatest peroxidase activity? How could you explain differences in enzyme activity among different foods?

Probeware 

If you have access to Probeware, do the activity "Enzymes and Rates of Reaction."

from this venom, contains enzyme inhibitors that prevent blood clots from forming. In 1999, pharmaceutical researchers found that more than 40% of stroke patients who received ANCROD recovered all of their mental faculties. Other venom, such as scorpion venom, is being used to treat autoimmune disorders.

Figure 2.9 Venom from this Malayan pit viper can be used to treat patients who have had a stroke.

There are two kinds of inhibition that can affect the activity of enzymes. In **non-competitive inhibition**, an inhibitor molecule binds to the enzyme at a site known as the **allosteric site**. As a result, the three-dimensional structure of the enzyme is altered, which prevents the substrate from binding to the active site (see Figure 2.10). Most metabolic pathways are regulated by **feedback inhibition**. This is a type of non-competitive inhibition in which the end product of the pathway binds at an allosteric site on the first enzyme of the pathway. In this way, non-competitive inhibitors can play a key role in the normal functioning and regulation of metabolic pathways. Study Figure 2.11 to learn how a metabolic pathway is regulated by feedback inhibition.

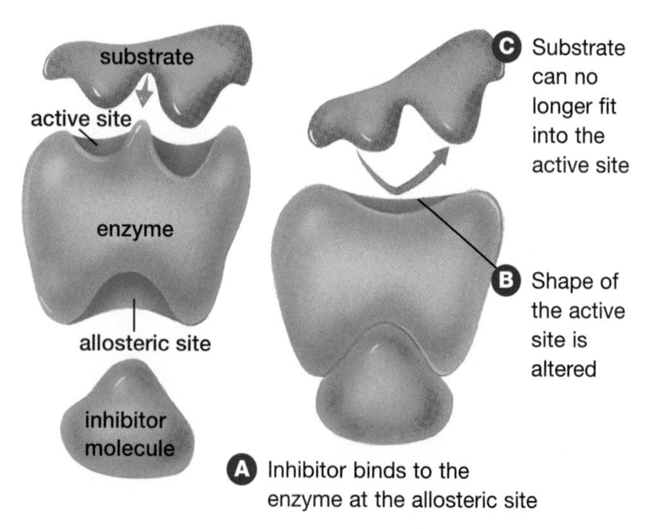

Figure 2.10 Non-competitive inhibition

A The substrate binds to the active site of the enzyme. The pathway is now active.

first reactant A

E_1

B **View of active pathway** The pathway remains active as long as there is insufficient end product.

E_1 E_2 E_3 E_4 E_5

A B C D E end product F

View of active pathway

E_1 end product F

C **View of inhibited pathway** Once there is sufficient end product, the product binds with the enzyme, changing the shape of the active site. This prevents the enzyme from binding with additional substrate and inhibits the pathway.

E_1

first reactant A stop

View of inhibited pathway

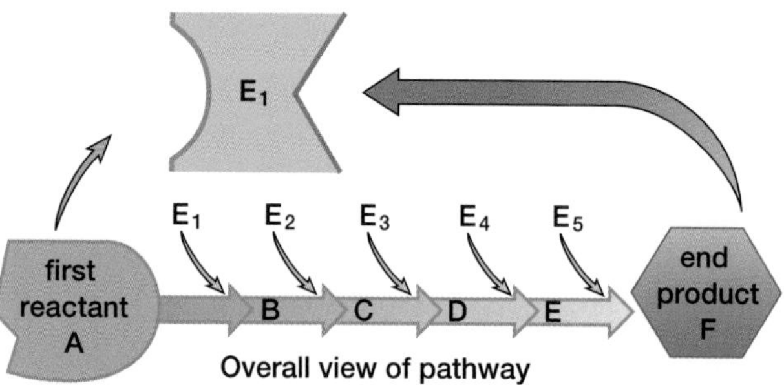

D **Overall view of pathway** Once enough end product has been produced, it binds to the allosteric site of the enzyme. The pathway is no longer active and the end product is no longer produced.

Figure 2.11 This hypothetical metabolic pathway is regulated by feedback inhibition.

Molecules that promote the action of enzymes can also bind to the allosteric site. These molecules are known as **activators**. The activity of any enzyme can change, depending on the number of activators and inhibitors in its environment. The regulation of enzyme activity by inhibitors and activators is known as **allosteric regulation**.

Competitive inhibition involves chemical compounds that bind to the active site of the enzyme and inhibit enzymatic reactions. The compounds compete with the true substrate for access to the active site. This competition is possible because competitive inhibitors are very similar in shape and structure to the enzyme's substrate. The metabolic pathway can only be restored if the substrate concentration is increased so that the substrate is more likely to enter the active sites than is the inhibitor. Penicillin is a commonly used competitive inhibitor. It works by bonding to the active site of transpeptidase, the enzyme involved in bacterial cell wall construction. When penicillin transpeptidase inhibits, a bacterial cell cannot divide successfully, and infection is prevented.

Protease inhibitors are a relatively new class of competitive inhibitors that interfere with the normal activity of protease enzymes. Molecular modelling played a major role in the research and design of effective protease inhibitor molecules. Figure 2.12 shows the general appearance and behaviour of protease and protease inhibitors. These inhibitors have been used to dramatically reduce the level of human immunodeficiency viruses (HIVs) in AIDS patients. HIVs infect host cells, such as the T-cells of the human immune system. The virus does this by injecting its genetic material into the host cell. The virus DNA then commandeers the cell's cellular processes to make polyproteins. The protease HIV enzyme then cuts these polyproteins into smaller structural proteins and enzymes that will be used to make new HIVs. The snipping or cleavage of polyproteins involves a hydrolysis reaction that uses a water molecule for every bond that is broken in the substrate molecule.

HIV protease inhibitors are similar in chemical composition and structure to the HIV polyprotein. The inhibitor molecules bind tightly to the active site of HIV protease enzymes. This process prevents the enzymes from cutting the actual HIV polyproteins to form new HIVs. The HIV protease enzyme is composed of two identical peptide halves. The enzyme's active site is located in the depression formed where the two halves join.

Cofactors and Coenzymes: Non-protein Helpers

The final manner in which enzymes are regulated comes in the form of **cofactors**. Cofactors are inorganic ions and organic, non-protein molecules that help some enzymes function as catalysts. The inorganic ions are metals such as copper, zinc, or iron. Located in the active sites of enzymes, these ions attract electrons from substrate molecules. For instance, carboxypeptidase breaks down proteins using a zinc cofactor. This cofactor draws electrons away from bonds, which causes them to break. If cofactors are organic, non-protein molecules, they are also called **coenzymes**. Many vitamins, small organic molecules that the human body requires in trace amounts to function, are parts of coenzymes.

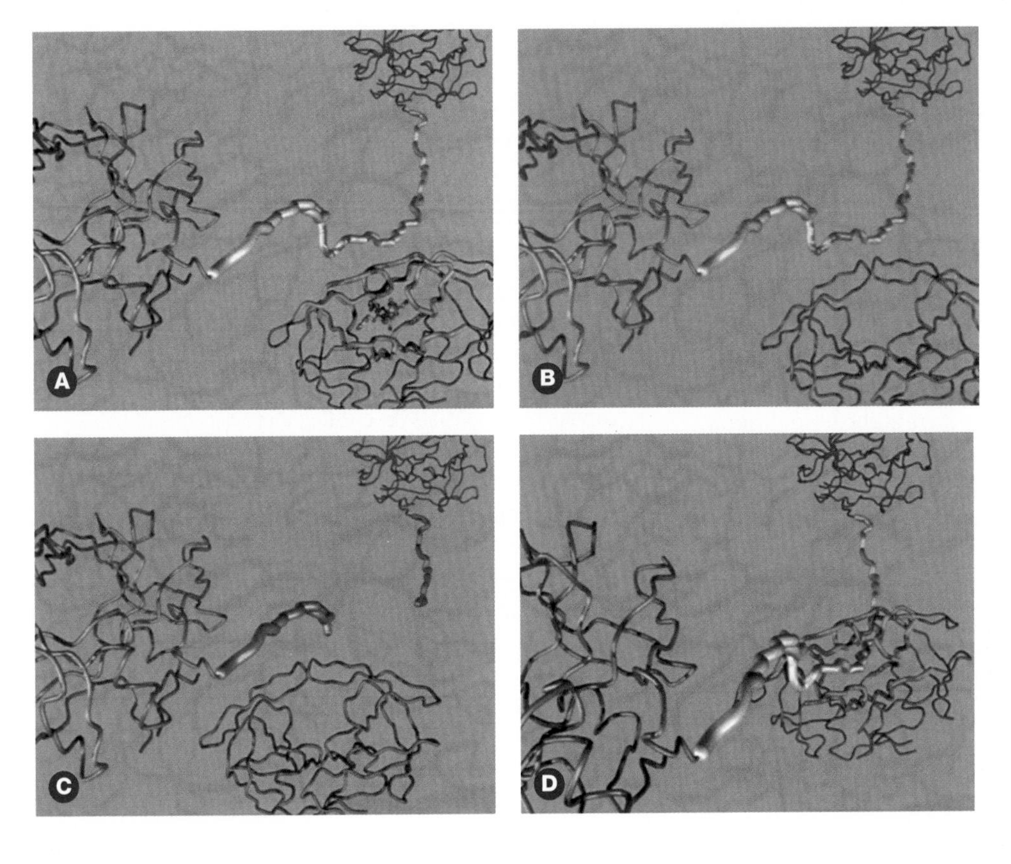

Figure 2.12 Protease and protease inhibitor molecules.
(A) Polyprotein blocked from the active site of a protease enzyme by a protease inhibitor molecule.
(B) Protease enzyme approaching an HIV polyprotein.
(C) Polyprotein that has been split by the protease enzyme.
(D) Polyprotein within the active site of a protease enzyme.

Table 2.3 shows vitamins necessary to the formation of specific coenzymes.

Deficiencies in any of these vitamins can affect the enzymatic reactions in cells. For example, lack of niacin may result in a lack of NAD^+ (nicotinamide adenine dinucleotide), which can affect enzymatic reactions in cellular respiration. Niacin deficiency can cause a skin disease called pellagra. At one time this disease was often mistaken for leprosy, but in the early 1900s American researcher Dr. Joseph Goldberger determined that pellagra is caused by a nutritional deficiency. To treat the disease, he recommended a diet that included meat, milk, fish, or a small portion of dried brewer's yeast.

Table 2.3
Coenzymes associated with some common vitamins

Vitamin	Coenzyme
niacin	NAD^+ (nicotinamide adenine dinucleotide)
B_2 (riboflavin)	FAD (flavin adenine dinucleotide)
B_1 (thiamine)	thiamine pyrophosphate
pantothenic acid	Coenzyme A (CoA)
B_{12} (cobalamin)	B_{12} coenzymes

Both coenzymes NAD^+ and FAD (flavin adenine dinucleotide) serve as electron acceptors in redox reactions. They carry electrons from one active site to another. Once the electrons have been released, the coenzymes return to the original enzyme for another complement of electrons.

The NAD^+ coenzyme takes the energy from the oxidation of nutritive molecules digested by animals to form NADH, a molecule with more chemical energy. NADH is then oxidized into NAD^+ again in order to collect more electrons. NAD^+ is the principal carrier of electrons in the oxidation of molecules that are used as an energy source in the cell. For example, NAD^+ accepts electrons from the products of the breakdown of glucose in one stage of cellular metabolism, and then transports them to a metabolic pathway that reduces oxygen to water. During such reactions, NAD^+ accepts two electrons, but only one hydrogen ion, as shown in the following equation:

$$NAD^+ + H^+ + 2e^- \rightarrow NADH$$

When NADH is oxidized back into NAD^+, energy is released. Similar in function to NAD^+, $NADP^+$ (NAD^+ plus an additional phosphate group) is a coenzyme in photosynthetic reactions.

Enzymes and Coenzymes for Human Health and Industry

Enzymes and coenzymes have proven useful in medical and industrial applications. Medical researchers have been conducting tests using NADH on patients with Alzheimer's disease or Chronic Fatigue Syndrome (CFS). In a study conducted in the 1990s at Georgetown University Medical Center, CFS patients who received injections of NADH experienced only one quarter of the symptoms experienced by patients who were given a placebo (a substance with no medical value). At the end of the twentieth century, six out of 10 individuals who were taking NADH used it to improve their energy level; two out of 10 used it to control Alzheimer's symptoms; and one out of 10 took it to relieve CFS.

Enzymes are also used in the process of DNA fingerprinting, which you will learn more about in Chapter 9. DNA fingerprinting has been used in a variety of circumstances, including paternity tests, murder trials, and identifying people. In one step of the DNA fingerprinting process, special enzymes called **restriction enzymes** are used to cut the DNA at specific places. DNA restriction enzymes recognize short, specific sequences of DNA bases and make breaks in the sugar–phosphate backbone of the DNA molecule in the region of the recognized sequence. Without these enzymes, the process of DNA fingerprinting would be much more involved. DNA fingerprinting also uses a process called PCR, polymerase chain reaction, which you will learn about in Chapter 9.

BIO FACT

Deposits of the protein fragment called beta-amyloid are always found in large amounts in the brains of Alzheimer's patients. Many scientists believe it is these protein deposits that damage brain cells, thereby producing memory loss and other symptoms of the disease. The beta-secretase enzyme is responsible for creating these deposits of beta-amyloid protein. This is accomplished by breaking a larger molecule called amyloid precursor protein (APP). Another enzyme, called gamma-secretase, also contributes to breaking APP, but beta-secretase is the first to act on APP. Scientists believe that drugs created to inhibit the action of beta-secretase will reduce the number of beta-amyloid deposits in the brains of Alzheimer's patients.

Biology At Work

Cheese Plant Manager

Elaine de Rooy

Elaine de Rooy did not intend to make a full-time career of her summer job at the cheese factory in Winchester, Ontario, 60 km south of Ottawa. She had been training as a medical lab technologist; however, after she graduated in 1979, she accepted a position in the factory's quality control lab. She rose through the ranks, eventually becoming the plant manager.

Today, the factory where deRooy works processes 250 000 kg of milk into 25 000 kg of cheese annually. The cheese is shipped to many countries, including the United States, Great Britain, Japan, Mexico, and countries in the Caribbean.

De Rooy enjoyed biology as a high-school student, and she feels that her scientific training is an asset in her work. She and the medical technologists she has hired for the lab bring a detailed scientific understanding to the process of cheese making.

An Ancient Process

Cheese making is a very old craft. At least 6000 years ago, people discovered that milk curdles in the stomach of a young mammal. This process occurs in order for the milk to be digested properly. Bacteria initially curdle the milk (separate it into solid curds and liquid whey). Enzymes in the animal's stomach accelerate this process. These enzymes, traditionally called rennet, are chymosin and pepsin.

Dried calf stomachs, called vells, were the traditional source of enzymes for cheese making. Today, enzymes from fungi and enzymes made through recombinant DNA are used to make cheese. Cheese can be made without using enzymes, but the results (fresh cheeses, such as cottage cheese) are "unripened" and should be consumed within a few days. Cheese made using enzymes can be stored for longer periods of time without spoiling. According to de Rooy, cheeses made using rennet obtained from calf stomachs are "a better quality of cheese for aging." Around the world, thousands of varieties of cheese have been developed, depending on the precise methods and bacteria employed.

Managing a cheese factory makes heavy demands on de Rooy's time. The factory operates 24 hours a day, seven days a week. "If everything is running really well, you've got a really nice mix between home life and work," she explains. "When there are problems, it's a huge demand on your time. During the past month, I have been here every day. I have been here in the middle of the night and gone home to catch a few hours sleep and come back to work again. That's my biggest challenge." However, de Rooy enjoys solving the variety of problems she encounters on the job. For instance, any day might involve dealing with equipment failure, bacteriological problems, or safety issues.

Career Tips

Managing a factory in the food industry is a multifaceted job. Managers organize and oversee the work of many employees. In addition, managers must be aware of what is involved in each step of producing a product, so they can ensure production goes according to schedule. This type of work involves leadership, organizational, social, and troubleshooting skills.

Select a food industry and research the managerial challenges specific to that industry, with respect to the biological processes involved in making the product.

DNA can play a role in determining whether or not an individual's enzymes are functioning normally. For example, Hurler syndrome is a genetic disorder caused by a defective gene. A child born with Hurler syndrome cannot manufacture the enzyme alpha-L-iduronidase. This enzyme is one of 10 lysosomal enzymes responsible for breaking down complex carbohydrates called mucopolysaccharides (MPS). Mucopolysaccharides are largely responsible for building connective tissues in the human body. If mucopolysaccharides cannot be broken down properly, they build up in body cells and form excess tissue. A child diagnosed with Hurler syndrome will become afflicted with various cardiac or respiratory ailments by the age of five and not survive long thereafter.

WEB LINK

www.mcgrawhill.ca/links/biology12
To find out more about the kinds of symptoms associated with MPS disorders, go to the web site above, and click on **Web Links**. Why might a buildup of mucopolysaccharides cause such symptoms? What kinds of treatments have been found to alleviate the symptoms of Hurler syndrome? Prepare a brief report of your findings.

People have also found ways to exploit enzymes and coenzymes for industry and profit. One of the most obvious ways that enzymes can be used in industry is in wine-making. Before 1897, scientists believed that enzymes required living material to function. The first to discover that a cell-free, or non-living, extract of yeast could cause alcohol fermentation was the German chemist Eduard Buchner (shown in Figure 2.13). His experiments led to the use of enzymes in industries as diverse as wine production, leather tanning, food production, textiles, pulp and paper, and pharmaceutical manufacturing.

Figure 2.13 Chemist Eduard Buchner (1860–1917)

Enzymes are essential to the pharmaceutical industry in making products — from chemotherapy treatments to common painkillers. Many of these products are composed of enzymes or make use of enzymatic reactions. As well, they often affect the activity of enzymes within the body. A new form of chemotherapy, Antibody-directed Enzyme Prodrug Therapy (ADEPT) uses enzymes to improve the efficiency of the drugs being used in the treatment of common solid tumours. This process involves using tumour-associated antibodies directed against tumour antigens. Doctors link the antibodies to enzymes and administer them to the patient. A prodrug is administered separately. A **prodrug** is an inactive drug that is only converted into its active form in the body by metabolic activity. At the site of the tumour, the enzyme converts the prodrug into an active compound that is toxic to the tumour.

Painkillers, for example, affect enzymes in order to relieve headaches, inflammation, or swollen tissues. Aspirin™ and similar painkillers reduce inflammatory pain by inhibiting enzymes called cyclo-oxygenase (Cox) 1 and 2. Cox-1 is located in the stomach, protecting it from hydrochloric acid in the digestive juices. Cox-1 is also found in blood platelets, where it aids in clotting reactions. Cox-2 is produced in the skin or joints following inflammation. Cox-2 is necessary in catalyzing the formation of prostaglandin E2 (PGE2), which increases the sensitivity of nerves to pain.

Until recently, biochemists believed that inhibition of PGE2 at the site of inflammation accounted for both the anti-inflammatory and painkilling actions of Aspirin™ and similar painkillers. Although Cox-2 is produced at the inflamed site of the body, recent studies have shown that nerve cells in the spinal cord and brain also begin to produce it. This results in the production of PGE2 throughout the central nervous system. Biochemists revised their knowledge of how and where Aspirin™ works. Aspirin™ reduces inflammatory pain not only at the inflamed site but also in the entire central nervous system.

Because PGE2 increases nerve sensitivity to pain, its manufacture throughout the central nervous system accounts for the tenderness surrounding inflamed tissues. Researchers suspect that the presence of Cox-2 and PGE2 may explain why people with inflamed tissues experience aches and pains and even appetite loss and depression.

Canadians in Biology

"Marvellous" Mutants: The Enzymatic Synthesis of Oligosaccharides

Dr. Stephen G. Withers

Until quite recently, scientists thought carbohydrates played only simple biological roles as sources of energy or as polymeric building blocks. Now many scientists are discovering that oligosaccharides, or carbohydrates in small clusters, can be used to recognize microbial infection, cancer metastasis, and cellular inflammation. Oligosaccharides also hold tremendous potential as therapeutics for bacterial infections, inflammation, and AIDS.

A major obstacle in glycobiology (the science of understanding the role of carbohydrates in biological events) has been to obtain large enough quantities of oligosaccharides, from natural sources or by synthetic methods, for research. The chemical synthesis of oligosaccharides has not been economical on a large scale, so biochemists have turned to enzymes as a cheaper means for large-scale synthesis.

Engineered Enzymes

Dr. Stephen Withers, a biochemist at the University of British Columbia, has developed an enzyme to synthesize oligosaccharides. Dr. Withers and his colleagues have developed and patented "glycosynthases," which are created through genetically engineered DNA. This DNA programs an organism to produce mutant glycosynthase enzymes that can synthesize high yields of oligosaccharides.

Glycosynthases are mutated from glycosidase enzymes, which normally function to hydrolyze oligosaccharides. Under certain conditions, glycosidases can be made to synthesize oligosaccharides, but only in low yields. The problem with glycosidases is that they hydrolyze the oligosaccharide product too quickly. Withers says, "We designed glycosynthases as mutant enzymes that will not hydrolyze the oligosaccharide product, but can still form it. So, in essence, we shut down the hydrolysis pathway."

Dr. Withers focusses his research on determining the fundamental mechanisms of enzymes. "How do these marvellous catalysts accelerate reactions by such huge amounts?" he asks. "We really want to understand how the enzymes do this." Withers and his colleagues are now trying to generate a library of glycosynthases to synthesize a vast array of oligosaccharides.

By his own account, Withers has "always been fascinated by science." He grew up on a dairy farm outside the village of Horton in Somerset, England. He traces his interest in biological chemistry to his experiences medicating sick animals on the family farm, and to the old chemistry set he found in the attic of his childhood home. He completed a Bachelor of Science degree at the University of Bristol in the United Kingdom. He went on to complete his doctoral degree in chemistry at both the University of Bristol and at the Université de Paris-Sud in Orsay, just south of Paris, France.

Withers came to Edmonton, Alberta, in 1977 for post-doctoral studies. He says, "I came here because of the high quality of science in Canada, and also because I wanted to see some of the fantastic scenery, and kayak some of the amazing white-water rivers!"

What are the long-term benefits of glycosynthase research? Withers says it will allow "access to cheaper oligosaccharides, which may then become viable as new therapeutics, functional foods, and maybe even as bulk commodities." For example, he says there is interest in manufacturing oligosaccharides as ingredients in infant milk formula. Scientists have discovered that breast milk contains certain oligosaccharides that are believed to play a role in protecting infants against bacterial infections. Specific oligosaccharides could also be used as a treatment for *E. coli* poisoning to bind to or absorb the toxin produced by these bacteria.

Recent studies on synthetic oligosaccharides (carbohydrates composed of a relatively small number of monosaccharides) indicate they have great potential as therapeutic agents. These compounds, which interfere with carbohydrate–protein reactions, are difficult to create in the laboratory. However, a new technology has been discovered using glycosidases, which are produced by genetically altering DNA. These altered enzymes catalyze the synthesis but not the hydrolysis of oligosaccharides, making them easier and less expensive to construct. The altered enzymes have been termed glycosynthases and can be used to make anti-ulcer agents, therapeutic drugs for middle-ear infections, and infant formula additives, to name a few. Dr. Stephen Withers, a scientist at the University of British Columbia, and his co-workers were the first to develop glycosynthases. They continue to be in the forefront of developing new ways to use these enzymes and their substrates in industry and medicine. The "Canadians in Biology" profile on the previous page provides a more complete account of the work accomplished by Dr. Withers and his team.

By lowering the activation energy needed by cells to start metabolic reactions, enzymes allow biological systems to undertake necessary processes at the temperatures that exist inside the cell. People have learned a great deal about enzymes and taken them from such diverse sources as yeast and organisms living in hydrothermal vents in order to manufacture foods and pharmaceuticals. The following section discusses another aspect of enzyme function and metabolic reactions within cells — coupled reactions and the production of ATP.

SECTION REVIEW

1. K/U Define the term "catalysis."
2. C (a) Draw and label a diagram that shows the nature of the catalytic cycle.
   (b) How is the cyclic nature of this process energy-efficient for the cell?
3. K/U What are the different ways that hydrolytic and oxidative enzymes change chemical reactions?
4. C Make a three- to five-panel cartoon to show an example of a metabolic pathway. Explain the terms "substrate" and "product" as part of this cartoon.
5. C Explain how the enzyme sucrase can participate in a reaction involving sucrose, but not in a reaction involving maltose. Use diagrams in your answer.
6. K/U (a) Label the diagram with the following terms: substrates, enzyme, active site, product, enzyme–substrate complex.
   (b) How is enzyme shape important to the enzyme–substrate complex?

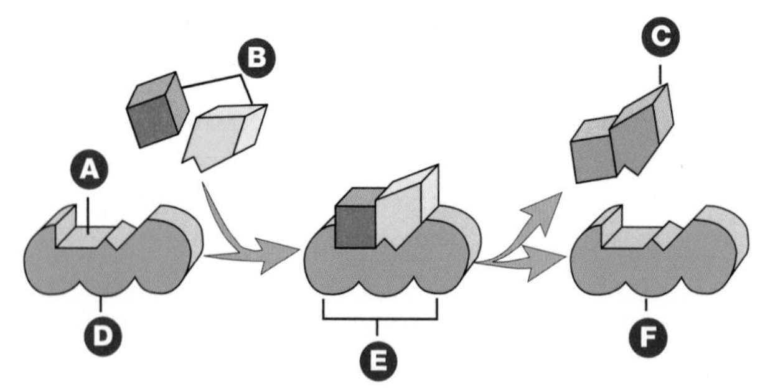

7. K/U Identify two ways in which enzymes reduce the activation energy necessary to break the bonds of the substrate.
8. C Make a chart that lists four factors that influence enzyme activity. For each factor, provide an example and a brief explanation of how the factor influences enzyme activity.
9. K/U Most enzymes function at a pH range of 6 to 8. What is one notable exception?
10. MC Describe how thermostable enzymes may be helpful to the pharmaceutical industry.
11. C Use diagrams to show how non-competitive inhibitors act differently from competitive inhibitors.
12. K/U Enzymes can be dissolved in the cytosol or embedded on a membrane. Identify one advantage and one disadvantage of each situation.
13. I Pepsin is an enzyme that breaks down protein.
    (a) A student has a test tube that contains pepsin, egg white, and water. What conditions would you recommend to ensure breakdown of the egg white?
    (b) How could you increase the rate of the reaction?

ATP and Coupled Reactions

EXPECTATIONS

- Identify how ATP molecules function within energy transformations in the cell.
- Describe the importance of ATP molecules in technological applications.

Coupled Reactions and ATP

Enzymes are crucial for both endothermic and exothermic metabolic reactions. Many reactions in cells are endothermic, such as protein synthesis. The energy released by an exothermic reaction can be used to drive an endothermic one. Such a combination of reactions is called a **coupled reaction**. The energy used by endothermic reactions comes from the breakdown of ATP molecules, which is an exothermic reaction (shown in Figure 2.14). **ATP**, adenosine triphosphate, is the molecule that all cells use as an energy source. The process of cellular respiration converts energy (for example, the energy stored in the chemical bonds of fats and starch) into chemical energy stored in ATP molecules. As you will soon discover, all organisms can produce ATP via a process called glycolysis, the first step in cellular respiration. Cellular respiration is dealt with extensively in Chapter 3.

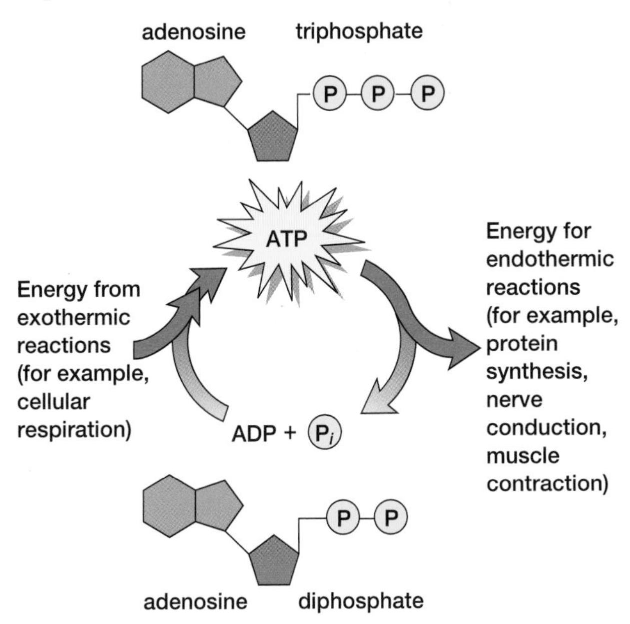

Figure 2.14 In the ATP cycle, hydrolysis removes a phosphate group from ATP, resulting in a release of energy used for various metabolic processes.

Figure 2.15 Like you, this Arctic fox uses ATP molecules to fuel cellular reactions.

ATP is a molecule composed of the sugar ribose, the base adenine, and three phosphate groups, as shown in Figure 2.16 on the next page. The sugar ribose and the base adenine together are called adenosine. Ribose is a five-carbon sugar that makes up the core of the molecule. Adenine is an organic molecule, composed of two carbon–nitrogen rings. Because of this, adenine is referred to as a nitrogenous base. As you may recall, adenine is also one of the four nitrogenous bases that make up RNA. The ribose and adenine molecules, together with one phosphate group, form a molecule called AMP (adenosine monophosphate). AMP is a component in the coenzyme, NAD^+, which is important in cellular respiration.

The final component that completes ATP is two more phosphate groups, linked by covalent bonds in a chain formation. The covalent bonds that join these phosphate groups to the rest of the molecule are often called high-energy bonds, because 31 kJ/mol (7.3 kcal/mol) of energy is released when the exothermic reaction of ATP hydrolysis occurs. This energy value was determined in a laboratory. The high-energy bonds of ATP do not require a large quantity of activation energy to break, so little energy is required to initiate hydrolysis. In the hydrolysis of ATP, a typical metabolic reaction, only the outer high-energy bond is broken, releasing the end phosphate group. Figure 2.17 shows that the products of this reaction are ADP (adenosine diphosphate), P_i (an inorganic phosphate molecule), and the release of 31 kJ/mol of energy.

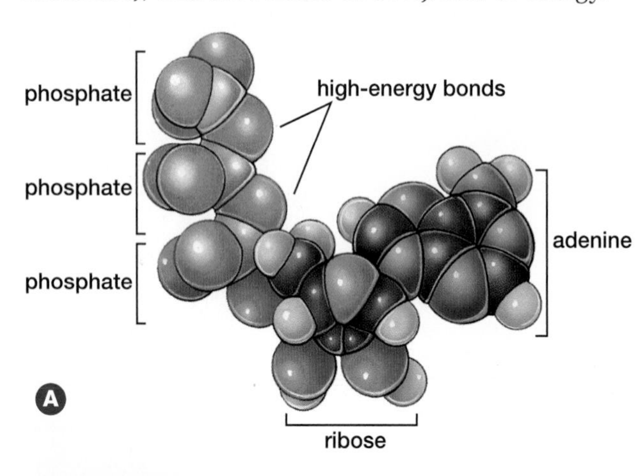

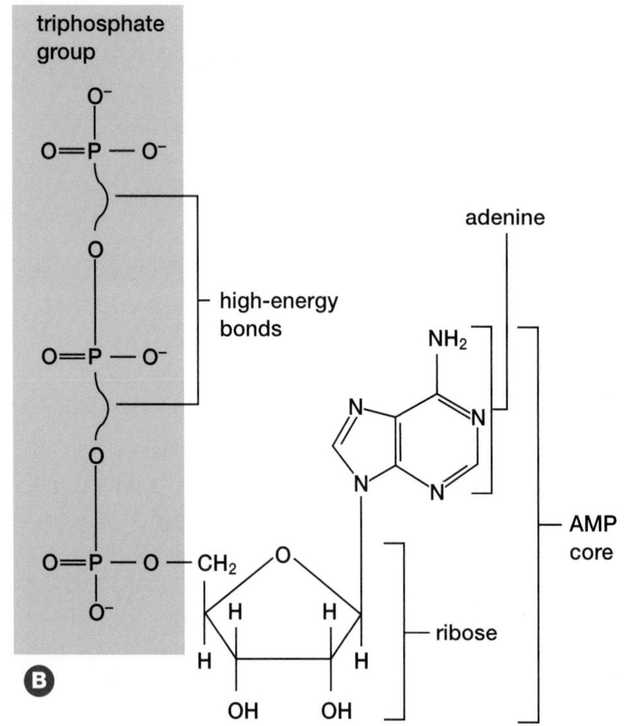

Figure 2.16 Diagram of the ATP molecule (A) and its chemical structure (B). ATP is composed of a ribose sugar, adenine, and a triphosphate group. High-energy bonds link the three phosphate groups. Energy is released from the molecule when these bonds are broken. AMP is adenosine monophosphate.

When cells require energy, they use ATP. Much ATP is needed as fuel for cellular work. A person who consumes 10.46 MJ (2500 Calories or 2500 kcal) per day manufactures and uses about 180 kg of ATP per day. ATP is constantly being generated from ADP and P_i. Consequently, cells do not keep a stockpile of ATP; it is more efficient for them to make ATP as the need arises.

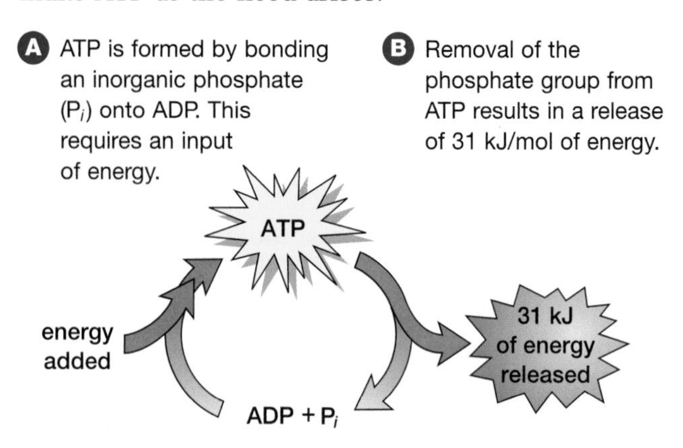

Figure 2.17 The ATP/ADP + P_i cycle

BIO FACT

Recent research indicates the molecule adenosine can induce sleep. Adenosine binds to specific receptors on the surface of nerve cells in the brain that promote sleep. The action of adenosine can be interrupted by caffeine, a stimulant found in foods, which blocks receptors on the nerve cells. As a result, a drink containing caffeine can make you feel more alert. Some researchers are investigating if sleep is influenced by ATP molecules, which are composed of an adenosine molecule and phosphate groups.

ATP Working Inside and Outside Cells

There are at least three distinct uses for ATP in cells:

1. **Chemical work** ATP supplies the energy needed to synthesize the macromolecules that comprise the cell.
2. **Mechanical work** ATP supplies the energy needed to permit muscles to contract, cilia and flagella to beat, chromosomes to move, and other functions.
3. **Transport work** ATP supplies the energy cells need to pump substances across the cell membrane.

Although some compounds (such as water) can move easily across the cell membrane, other substances require energy to move into the cell. This process is called **active transport**, and it requires the help of special proteins (often called pumps) and ATP. While facilitated diffusion and diffusion move particles along a concentration gradient, active transport moves particles *against* a concentration gradient — a process that requires energy. A common example of such a process is the **sodium-potassium pump**. This pump, which involves special carrier proteins, maintains an imbalance of sodium and potassium in cells, particularly nerve and muscle cells. The pump moves potassium ions to the inside of the cell and sodium ions to the outside of the cell. Study Figure 2.18 to learn how ATP and carrier proteins work in the sodium-potassium pump.

Outside

Inside

A The carrier protein has a shape that allows it to take up three sodium ions (Na^+).

3 Na^+

carrier

B ATP is split, and a phosphate group is transferred to the carrier protein.

ATP

3 Na^+

~P

ADP

C A change in shape of the carrier protein causes the release of three sodium ions (Na^+) outside the cell. The altered shape permits the uptake of two potassium ions (K^+).

3 Na^+

2 K^+

~P

D The phosphate group is released from the carrier protein.

P_i

2 K^+

E A change in shape of the carrier protein causes the protein to release the potassium ions (K^+) in the cell. The carrier protein is once again able to take up three sodium ions (Na^+).

2 K^+

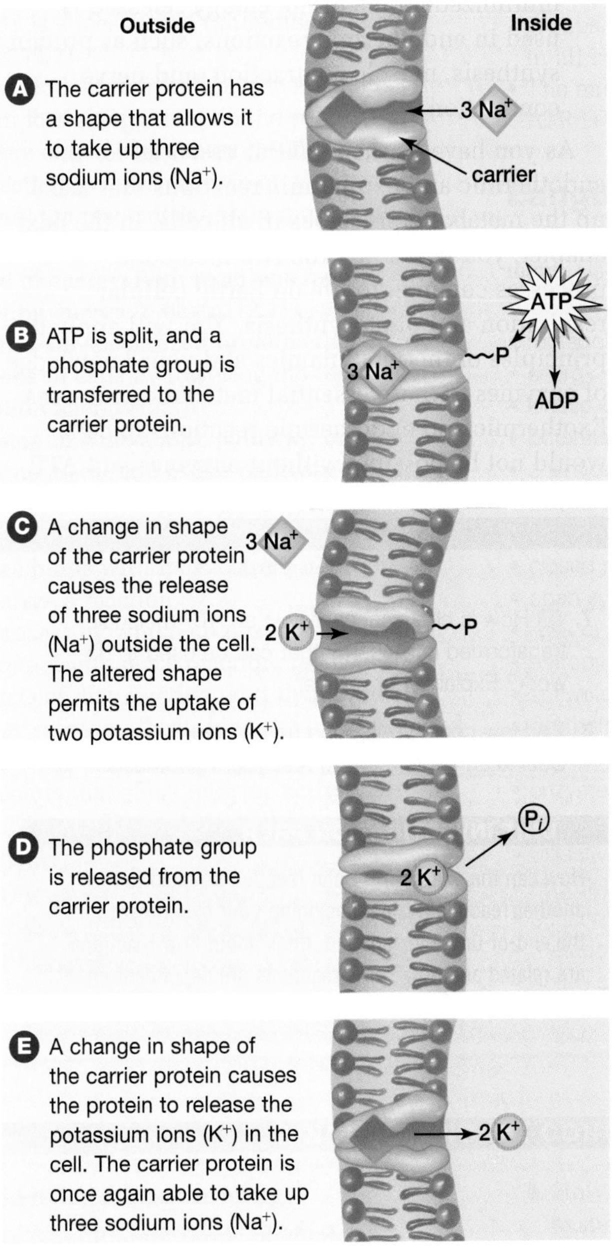

Figure 2.18 The sodium-potassium pump uses a carrier protein to move three sodium ions (Na^+) outside of the cell for every two potassium ions (K^+) moved inside the cell. Energy from ATP is required to accomplish this task.

ATP in Medicine and Industry

While ATP is important in cellular processes, it has also found a place in medical and industrial applications. Injections of ATP have helped some lung cancer patients by slowing down weight loss due to radiation therapy and helping to arrest the growth of tumours. Anesthetists use low doses of ATP in anesthesia to reduce pain, in much the same way as they administer morphine following surgery. Medical researchers also use ATP to treat patients with pulmonary hypertension, which is characterized by abnormally high blood pressure in the arteries of the lungs. Injections of ATP dilate the arteries, thereby lowering the blood pressure.

Food industries can monitor ATP to ensure food quality. ATP is only made by living organisms and can be found in low concentrations in nature (having leaked from living cells). The food industry scans food for abnormally high concentrations of ATP, which could indicate the presence of micro-organisms such as bacteria. Quality control technicians can make sure that foods with increased levels of ATP are not shipped to stores to be sold.

ELECTRONIC LEARNING PARTNER

To learn more about the transport of substances across cell membranes, go to your Electronic Learning Partner now.

WEB LINK

www.mcgrawhill.ca/links/biology12

In the mid-1980s, researchers discovered the enzyme kinesin, a motor protein that converts the energy of ATP into mechanical work. Kinesin and other motor proteins that have been identified are found in most eukaryotes. Motor proteins are responsible for such work as mechanical transport and chromosome movement during meiosis and mitosis. To learn more about kinesin and related motor proteins, go to the web site above, and click on **Web Links**. Select a motor protein and prepare an abstract outlining its function.

7. How do enzymes affect the rate of biological reactions?
8. How is an endothermic reaction different from an exothermic reaction?
9. Energy flows through living things, but matter is cycled. What does this mean? Explain briefly.
10. Distinguish between hydrolytic and oxidative enzymes.
11. What is a thermostable enzyme?
12. How do amino acid *R*-groups in enzymes affect chemical reactions?
13. Identify two environmental factors that affect the ability of a protein to catalyze a reaction.
14. How can an enzyme be regulated with an inhibitor?
15. Metabolic inhibition can reduce the waste of energy in a cell. Explain how energy is transformed in a metabolic pathway and how stopping the process conserves energy.
16. What is a competitive inhibitor?
17. Describe how allosteric regulation controls enzyme activity.
18. Identify one cofactor and one coenzyme.
19. Describe how snake venom can affect cellular processes in the human body.
20. How do cofactors and coenzymes contribute to cell processes?
21. How does a coupled reaction use energy?
22. Why is ATP an effective energy transfer molecule in the cell?
23. List three uses for ATP in cells.
24. Why does active transport require energy?

INQUIRY

25. Within the cell, some thermal energy released from chemical reactions causes convection of the cytosol. Design a model of a cell and cell contents to investigate how thermal energy within the cell affects the rate of enzyme reactions in the cell.
26. The enzyme lactase breaks down milk sugar (lactose) into glucose and galactose. Is this enzyme hydrolytic or oxidative? Use molecular model kits to demonstrate your choice.
27. Photosynthesis is an endothermic reaction that is affected by the temperature in the environment. Develop a testable hypothesis that explores how temperature affects photosynthesis.
28. Which process produces more thermal energy, photosynthesis or respiration? Design an experiment to explore this question.
29. Students studied the effectiveness of an enzyme at different temperatures by determining the concentration of product after equal elapsed times. The data table shows their results.
    (a) Make a graph using these values.
    (b) Interpret the graph to predict the concentration of the product at 47°C and 80°C.
    (c) At what temperature does the enzyme function best? Explain your reasoning.

Temperature (°C)	Concentration of product (μg/L)
24	5
27	6
30	8
32	10
35	15
36	18
40	11
45	9
50	4
60	0

COMMUNICATING

30. Earth is an open system for energy but a closed system for matter. Explain what this means in terms of cell processes.

31. In the ATP cycle, the breakdown of glucose is coupled to the buildup of ATP. Use words and diagrams to describe this process in terms of stored energy, free energy, and thermal energy.

32. Energy flows through living things, but matter is cycled. Explain how enzymes demonstrate efficient use of matter in the cell.

33. Make a flip-book animation to show the induced fit of a substrate on an enzyme's active site.

34. Enzymes are generally substrate-specific. However, some enzymes can catalyze the reactions of a family of substrates. Use diagrams to show how an enzyme could be specific to two substrates that are similar yet different in structure.

35. Make a diagram of a polypeptide chain of five amino acids to explain how *R*-groups:
    (a) interact with a substrate, and
    (b) are affected by a decrease in pH.

36. This diagram shows the metabolic pathway used by eukaryotic cells to break down proteins. Ubiquitin is a polypeptide chain that joins to certain proteins. Existing enzymes can break down the protein when ubiquitin is connected.
    (a) How is this pathway energy-efficient?
    (b) Ubiquitin prepares the protein for the enzyme. How does adding ubiquitin change the protein?

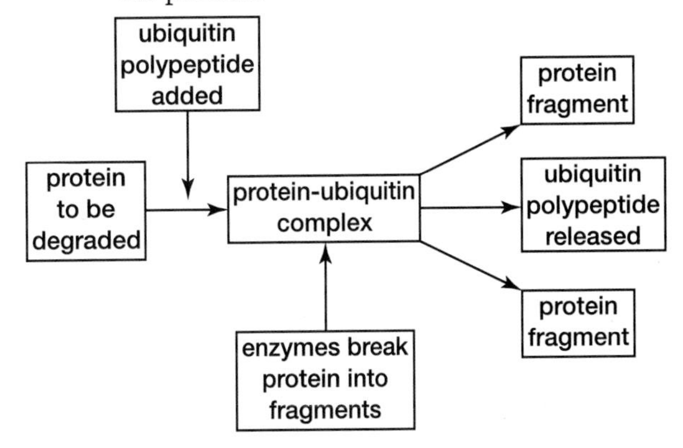

MAKING CONNECTIONS

37. Is a cell a closed system or an open system? Justify your response.

38. About 95 percent of the electric energy supplied to an incandescent light bulb is lost as waste thermal energy. If this were a biological reaction, how would an enzyme affect the use of energy?

39. Glucose is one of the products of photosynthesis. Glucose can also be produced artificially, but in this case both optical isomers are produced (that is, left-handed and right-handed glucose). Explain how biological enzymes produce only one isomer.

40. Milk is heated when it is pasteurized. The pasteurization process involves heating milk to kill bacteria without denaturing the milk itself. In Canada, milk is heated to 72.8°C for 16 s and then cooled rapidly to 4°C. Ultra High Temperature (UHT) milk products can be heated to 135°C for a shorter period of time, usually 2–5 s. How is each process beneficial to the producer and the consumer?

CHAPTER 3

Cellular Energy

Reflecting Questions

- How do cells effectively transfer energy between reactions?
- Under what conditions are aerobic and anaerobic cellular respiration efficient?
- How do electron transport chains contribute to metabolic reactions?
- What are the similarities and differences between work done in mitochondria and work done in chloroplasts?

Prerequisite Concepts and Skills

Before you begin this chapter, review the following concepts and skills:

- reviewing carbohydrate structure (Chapter 1, section 1.2),
- explaining how a redox reaction works in terms of electronegativity (Chapter 1, section 1.3),
- applying the laws of thermodynamics to cellular respiration and photosynthesis (Chapter 2, section 2.1),
- describing how enzymes affect the activation energy of a reaction (Chapter 2, section 2.2).

Imagine yourself running in a distance marathon. With each step you take, you breathe deeply. Although you are running steadily you do not get tired because you have been training throughout the year. As a result of this training, your lung capacity has increased and your blood can absorb a greater amount of oxygen. Your cells, too, are ready to use the oxygen you take in. In the presence of oxygen, cellular organelles called mitochondria produce energy-rich molecules of ATP. To accomplish this, the mitochondria use glucose molecules. Glucose is obtained from the food you eat and can be stored in cells in the form of glycogen. How is glucose used to produce molecules of ATP? How is enough ATP produced to carry out all necessary cellular activity?

Photosynthetic organisms, such as green plants, also produce ATP. Plants, however, make their own food — carbohydrate molecules — using energy from the Sun, water, and carbon dioxide. To accomplish this, cellular organelles called chloroplasts, shown in the inset photograph, transform light energy into chemical energy. This energy is used to make carbohydrate molecules. These molecules are then used to fuel metabolic reactions that take place inside mitochondria.

In this chapter you will learn how animal and plant cells produce ATP, the energy molecule that cells use to do work. This process, known as cellular respiration, involves a series of reactions with enzymes that occur in several systems. You will study these systems and discover how they function together. You will learn how many molecules of ATP are produced during cellular respiration. You will also learn which organisms use cellular respiration to produce ATP. Finally, you will explore how diet and exercise can influence metabolic processes in people.

This chloroplast captures solar energy, which is then used to make food in green plants.

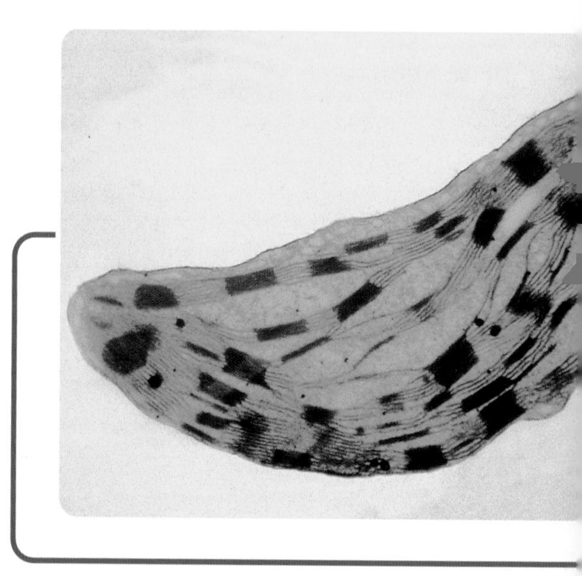

Chapter Contents

3.1 Cellular Respiration

EXPECTATIONS

- **Identify and describe the four stages of aerobic cellular respiration.**
- **Outline the key steps of glycolysis.**
- **Identify the products of glycolysis.**

You are running late for school. The bus stop is five minutes' walking distance away, and the bus is due to arrive in only three minutes. Taking a shortcut, you sprint the full distance, arriving just as the bus comes to the stop. You board the bus and collapse into a seat, breathing heavily. During your sprint, molecules of glucose were broken down in your cells. This process provided your muscles with the energy they needed to carry you to the bus. As you recall from Chapter 2, the breakdown of glucose is an exothermic reaction that releases energy. In this reaction, hydrogen atoms and electrons are removed from glucose and added to oxygen. Glucose is oxidized, and oxygen is reduced. Energy from this reaction is used for the production of ATP molecules, which are the energy source for cells. Metabolic pathways that contribute to the production of ATP molecules in cells are collectively referred to as **aerobic cellular respiration**. The term **aerobic** means that the process requires oxygen. The overall equation of aerobic cellular respiration is shown in Figure 3.1.

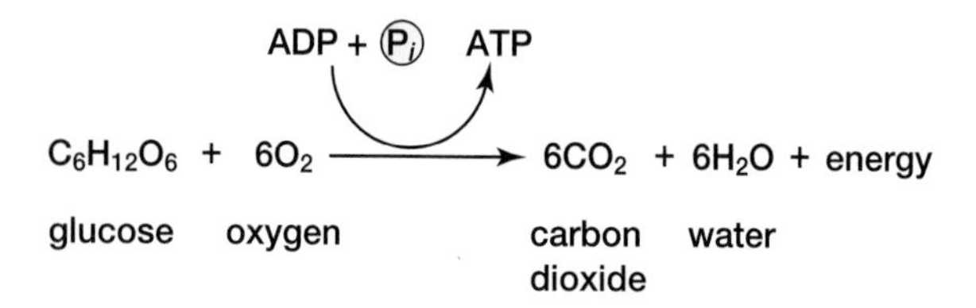

Figure 3.1 Aerobic cellular respiration most often involves the breakdown of glucose, coupled with the manufacture of ATP.

When a molecule of glucose undergoes aerobic cellular respiration, 36 molecules of ATP are produced. Glucose is an energy-rich molecule. The breakdown of glucose results in the formation of low-energy molecules and energy. ATP synthesis requires energy; it involves a series of endothermic reactions. The exothermic breakdown of glucose is coupled (linked) to the endothermic reactions involved in the synthesis of ATP. This coupling of reactions results in about 40 percent of the chemical energy in the glucose molecule being transformed into energy in ATP molecules. The rest of the energy is waste thermal energy.

Releasing energy from a stable molecule, such as glucose, within the cell requires controlled oxidation. Controlled oxidation is made possible by a series of reactions involving various enzymes and metabolic pathways, as well as coupled reactions. In this way, each reaction releases only a small portion of energy, while other reactions conserve this energy in molecules of ATP.

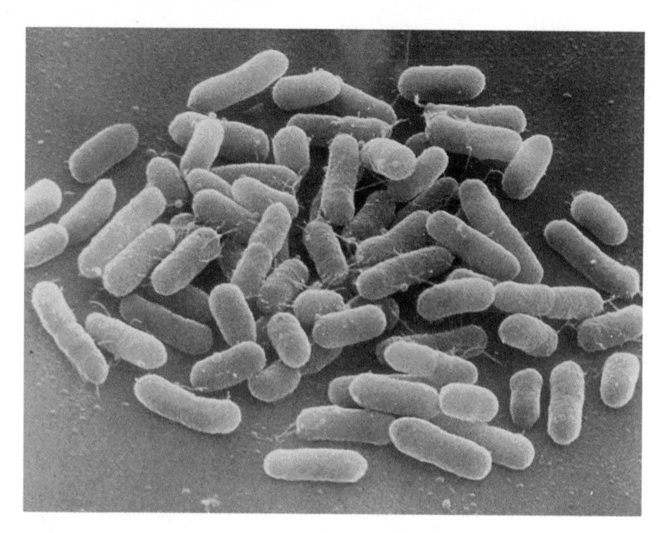

Figure 3.2 All organisms, including these *E. coli* bacteria, manufacture ATP to fuel cellular functions.

Stages of Aerobic Cellular Respiration

The process of aerobic cellular repiration can be divided into four distinct stages. These stages are summarized in Figure 3.3. Refer to the figure as you read the following overview. Section 3.1 describes glycolysis in detail. Section 3.2 examines the three remaining steps in aerobic cellular respiration.

Overview of Cellular Respiration

The first step in this process is a chain of reactions called **glycolysis**. "Glycolysis" means breaking sugar. This process is **anaerobic** (without oxygen) and occurs in the cytosol of cells, outside the organelles. The process of glycolysis produces

ATP molecules. During this stage, the six-carbon glucose is broken down into molecules of three-carbon **pyruvate**. Two pyruvate are produced from each molecule of glucose. The pyruvate can be used without oxygen in the process of **fermentation**, but no further ATP is produced during this process. You will learn more about fermentation in section 3.2. If oxygen is present, the pyruvate molecules enter the mitochondria and the process of aerobic cellular respiration can occur. Aerobic cellular respiration is a series of redox reactions that produce water, carbon dioxide, and additional ATP molecules.

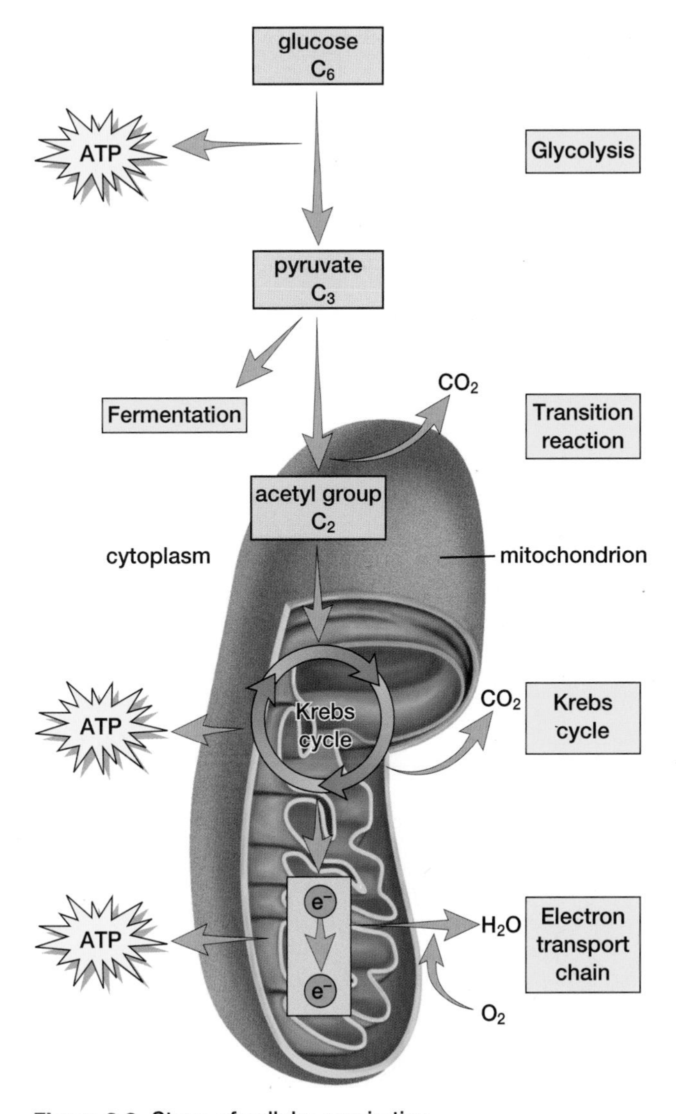

Figure 3.3 Steps of cellular respiration

The next stage in the process is the **transition reaction**, also called *oxidative decarboxylation*. In this reaction each pyruvate loses a carbon atom, or is decarboxylated, by the oxidative activity of NAD^+. This reaction changes a three-carbon pyruvate to a two-carbon acetyl group. This smaller molecule combines with coenzyme A to form **acetyl-CoA**, thus connecting glycolysis to the next stage, the Krebs cycle.

The **Krebs cycle**, also known as the citric acid cycle, is a cyclical metabolic pathway located in the matrix of a mitochondrion. The Krebs cycle occurs twice (once for each acetyl-CoA molecule) to oxidize the products of the transition reaction to carbon dioxide. Only one ATP molecule results from one cycle of this metabolic pathway.

The final stage, **oxidative phosphorylation**, requires oxygen to produce ATP by **chemiosmosis**, the movement of concentrated H^+ ions through a special protein complex. Oxidative phosphorylation relies on the **electron transport chain**. This is a series of molecules that are embedded on the inner membrane of the mitochondrion. The molecules in the electron transport chain are sequentially reduced and oxidized to move electrons to a final step where water is produced.

You will now learn about each stage of cellular respiration. Each stage is complex. The diagrams in sections 3.1 and 3.2 offer detailed descriptions of each stage. Study the diagrams as you read through the descriptions.

Glycolysis: Reactions in the Cytoplasm

Glucose is the primary reactant for glycolysis. The source of glucose may be from either carbohydrates or from glycogen (a molecule made of many glucose molecules) stored in muscle and liver cells. Glycolysis occurs in the cytoplasm of all cells, and it produces two pyruvate molecules and two ATP molecules. To accomplish this process, 11 different enzymes are used. Both prokaryotes (cells without nuclei) and eukaryotes (cells with nuclei) use glycolysis in some stage of ATP production. A few eukaryotes (yeast and mature human red blood cells) and many prokaryotes (some bacteria) can survive on the energy produced by glycolysis alone. However, this amount of energy is not sufficient for most eukaryotes, which use aerobic respiration in the mitochondria to increase ATP production. (Aerobic respiration will be covered in the next section.) Cellular respiration starts with glycolysis, which has two main phases:

- Glycolysis I: the endothermic activation phase, which uses ATP
- Glycolysis II: the exothermic phase, which produces ATP molecules and pyruvate

Glycolysis I

Glycolysis I involves a series of endothermic reactions. In order for glycolysis to begin, activation energy, from an ATP molecule, must be provided. This is accomplished in the first reaction of glycolysis by **substrate-level phosphorylation**, the transfer of an inorganic phosphate group (P_i) from one substrate to another by way of an enzyme, as shown in Figure 3.4. This process can remove (dephosphorylate) a phosphate from ATP or it can add (phosphorylate) a phosphate to ADP. As shown in Figure 3.5, one ATP is used to phosphorylate glucose to form glucose-6-phosphate. This molecule is then rearranged to form fructose-6-phosphate. At this point, another ATP molecule must phosphorylate the fructose-6-phosphate, producing fructose-1,6-diphosphate. In turn, this molecule is split into two **PGAL** (glyceraldehyde-3-phosphate). These PGALs act as the reactants for glycolysis II.

Glycolysis II

Glycolysis II is a sequence of exothermic reactions that provides energy for the cell. Following glycolysis I, each PGAL is oxidized. When PGAL is oxidized, energetic electrons move to NAD^+, which is reduced. A hydrogen ion attaches to the reduced NAD^- to form NADH. The oxidized form of PGAL is now able to attract a free phosphate ion in the cytosol, forming **PGAP** (1,3-biphosphoglycerate), as shown in Figure 3.6. Two PGAP are produced for each glucose molecule that enters glycolysis.

Following the formation of PGAP, two ADP molecules (with the help of enzymes) each remove one phosphate group from each PGAP to form **PGA** (3-phosphoglycerate), as shown in Figure 3.6. Here, substrate-level phosphorylation produces two ATP, one for each PGAP. At this point, because two molecules of ATP have been made, and two molecules of ATP were used to start glycolysis, the net change in the number of ATP molecules is zero.

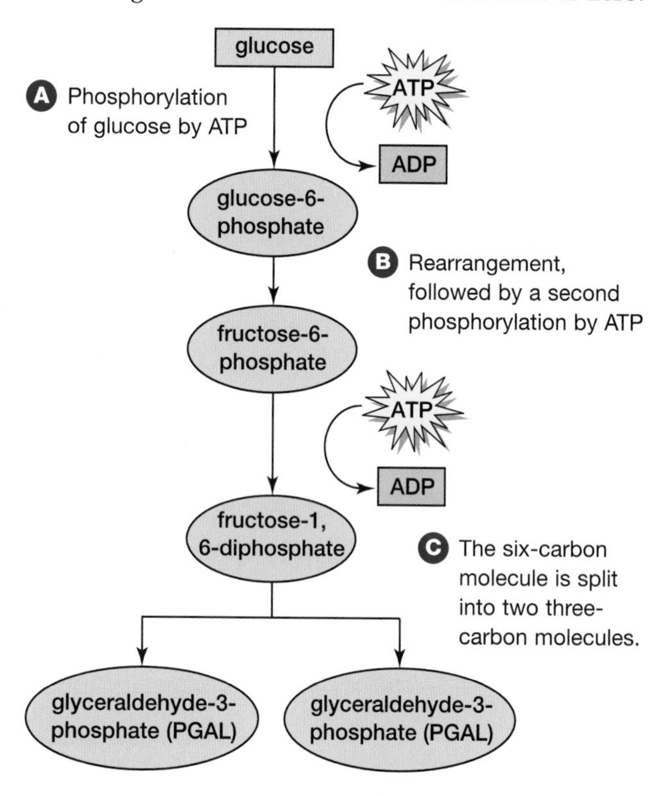

Figure 3.5 Glycolysis involves the formation of two PGAL molecules.

Next, the two PGA molecules are each oxidized, forming two water molecules and two **PEP** (phosphoenolpyruvate) molecules. Finally, another substrate-level phosphorylation occurs — two ADP molecules each remove the remaining phosphate group from each PEP molecule. The result is the production of two ATP molecules and two pyruvate molecules.

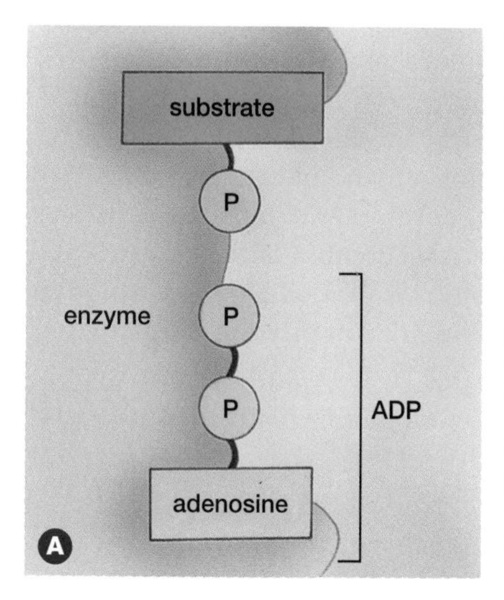

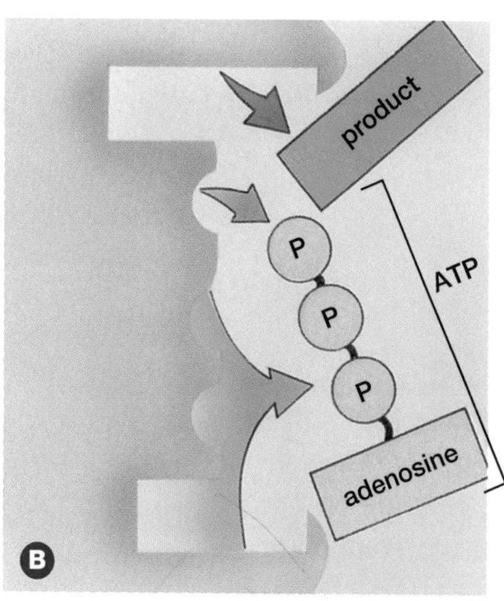

Figure 3.4 Substrate-level phosphorylation. Here, substrate 1 donates one phosphate group to substrate 2 (ADP), making ATP. The reverse reaction dephosphorylates substrate 1, producing ADP.

The entire glycolysis process, including glycolysis I and glycolysis II, produces a net gain of two ATP molecules. The energy stored in these ATP can now be used for cellular respiration in mitochondria.

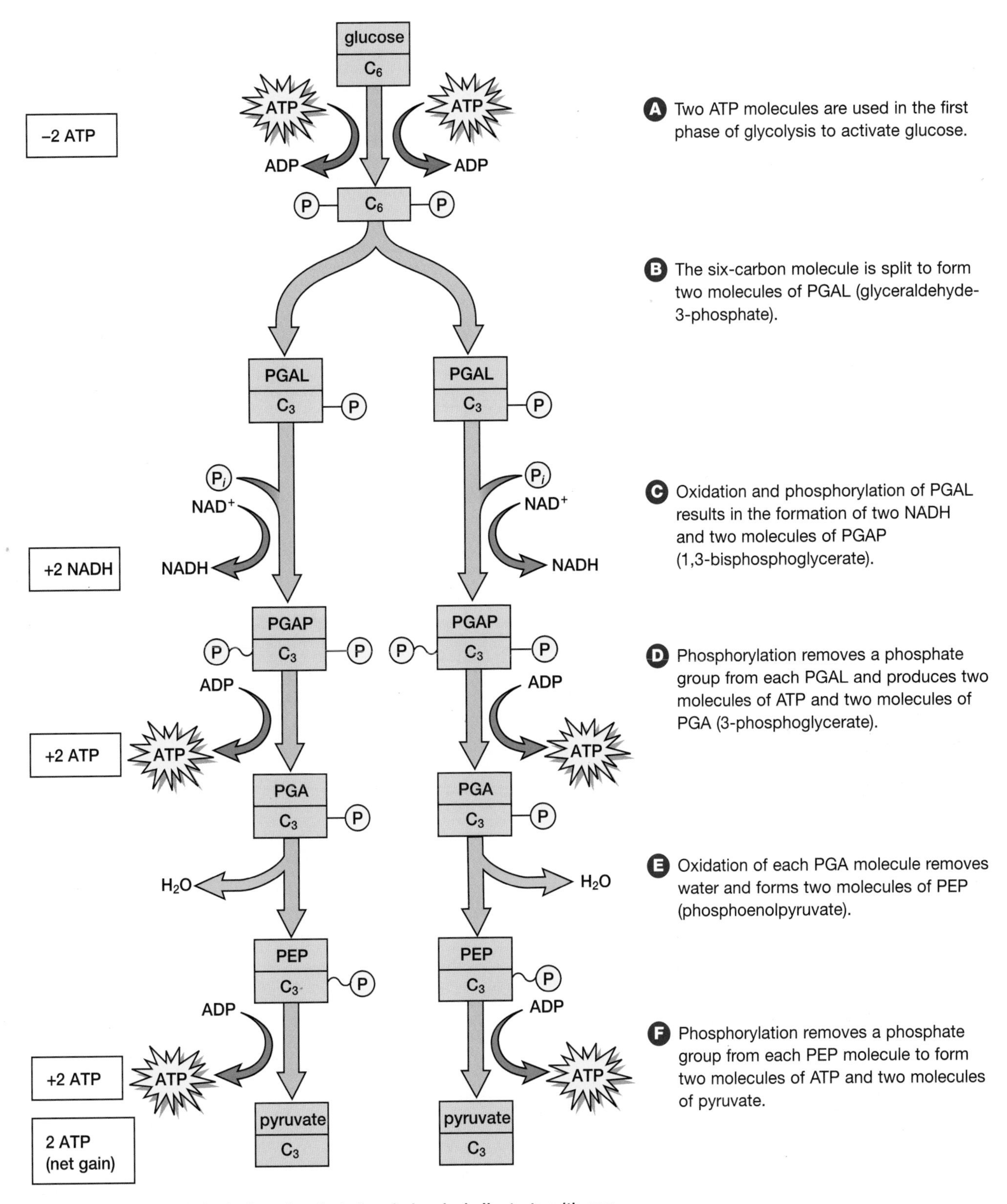

Figure 3.6 Glycolysis including glycolysis I and glycolysis II, starts with one molecule of glucose and produces two pyruvate molecules. There is a net gain of two NADH and two ATP molecules from glycolysis.

SECTION REVIEW

1. K/U List in order the four stages of aerobic cellular respiration. For each molecule of glucose that enters the respiration process, identify the number of ATP molecules produced in each stage.
2. K/U What is the purpose of glycolysis?
3. K/U List the products of glycolysis.
4. K/U How is NAD^+ important to glycolysis?
5. K/U How is ATP important to start glycolysis?
6. C Use Figures 3.5 and 3.6 as a guide to make a flowchart for glycolysis. Include the names of the intermediate molecules.
7. K/U Give an example of substrate-level phosphorylation.
8. C Using diagrams, explain the process of substrate-level phosphorylation.
9. K/U List the coupled reactions that transfer energy to or from glycolysis.
10. C The process of glycolysis is thought to have emerged very early in the origins of life. What evidence to support this theory can you find in the material presented in this section? Discuss your ideas with a partner, and then write a brief summary of your findings.
11. I The total amount of energy released from the chemical bonds of glucose is 2870 kJ/mol, after the initial investment of two ATP. The total energy that can be harvested to form ATP molecules is about 1200 kJ/mol glucose.
    (a) After the initial investment of two ATP, 38 ATP are produced. How much energy was used to phosphorylate one ADP?
    (b) What percentage of the total energy contained in glucose is captured during glycolysis?
12. C Glycolysis is often described as being an inefficient process. List points in support of this statement and points against it. Devise a new statement using a term other than "inefficient" to describe the process of glycolysis.
13. I

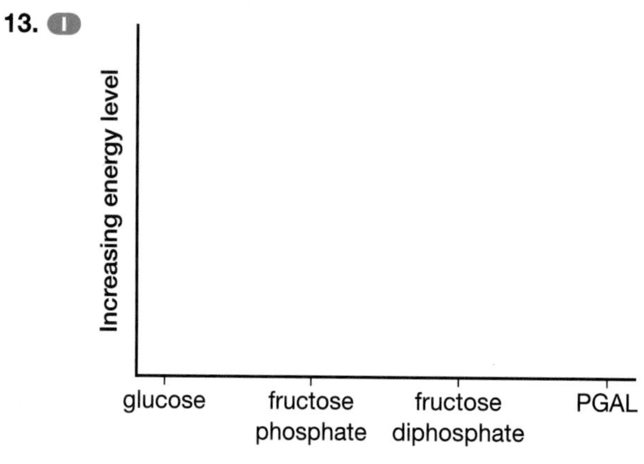

    (a) Copy and complete the graph by plotting the relative energy level of each molecule.
    (b) Which molecule has the highest amount of stored energy?
    (c) Which molecule has the least amount of stored energy?

Moving to the Mitochondrion

EXPECTATIONS

- **Describe the energy transformations that occur in aerobic cellular respiration.**
- **Explain the role of enzymes in metabolic reactions in the mitochondrion.**
- **Interpret qualitative data in a laboratory investigation of enzyme activity in the Krebs cycle.**
- **Explain the process of anerobic cellular metabolism.**
- **Research and describe how cellular processes are used in the food industry.**

Whereas glycolysis occurs in the cell cytosol, the remaining reactions of aerobic cellular respiration take place inside the mitochondria. As you have learned in previous studies, mitochondria are small organelles found in eukaryotic cells. A mitochondrion is composed of an outer and inner membrane, separated by an intermembrane space, as shown in Figure 3.7. The inner membrane folds as shelf-like cristae and contains the matrix, an enzyme-rich fluid. The cristae and the matrix are the sites where ATP synthesis occurs. More mitochondria are found in cells that require more energy, such as muscle and liver cells.

The two mitochondrial membranes have important differences in their biochemical composition. The outer membrane contains a transport protein called **porin** that makes it permeable to all molecules of 10 000 u (atomic mass units) or less. It also contains enzymes that help convert fatty acids to molecules that can pass into the interior of the mitochondrion to be broken down further. The inner membrane contains a high proportion of **cardiolipin**, a phospholipid that makes this membrane especially impermeable to ions. This relative impermeability, as you will discover when you learn about chemiosmosis, is very important in the production of ATP.

The intermembrane space is a fluid-filled area containing enzymes that use ATP. During the synthesis of ATP, the intermembrane space serves as a hydrogen ion (H^+) reservoir, storing the hydrogen ions that will be used for ATP synthesis. Finally, the inner mitochondrial compartment, comprised of the matrix and the cristae, remains relatively isolated from the outer mitochondrial membranes. The inner membrane contains the multienzyme complexes and the electron carriers of the electron transport chain. During aerobic cellular respiration, the transfer of electrons through this chain creates a high hydrogen ion (H^+)

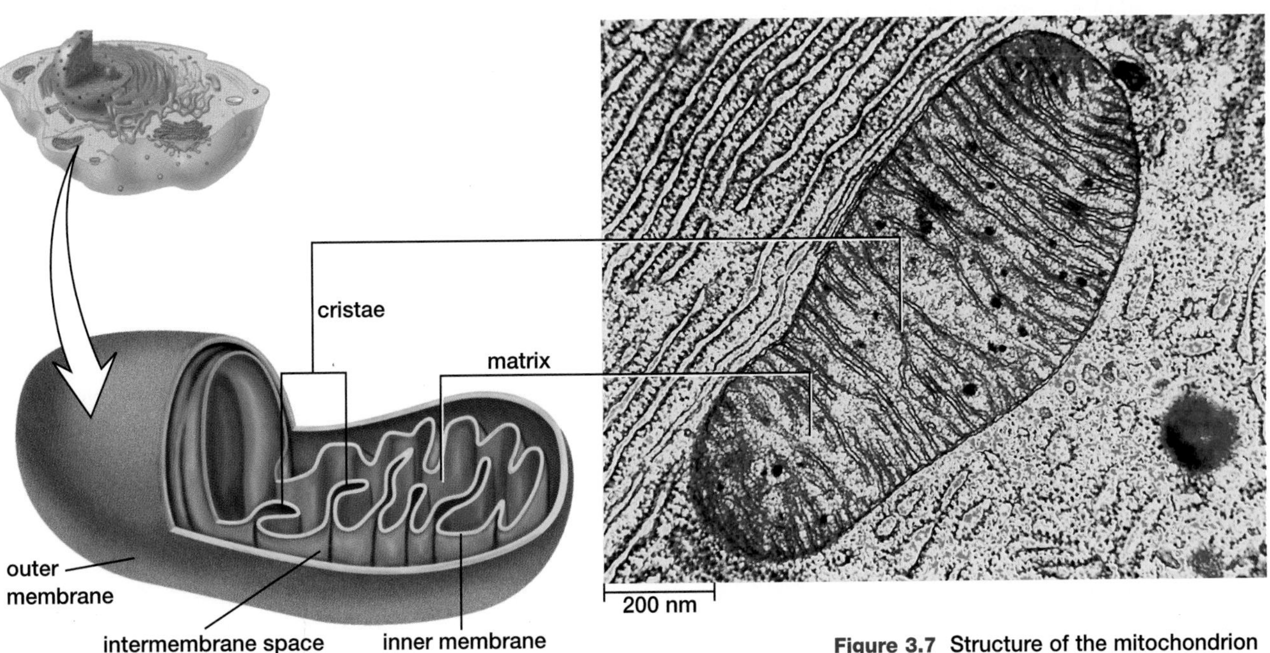

Figure 3.7 Structure of the mitochondrion

concentration in the intermembrane space. This process creates a positively charged gradient in which there is a greater concentration of hydrogen ions on one side of the membrane than the other. The passage of these protons to the inner mitochondrial compartment drives the synthesis of ATP.

BIO FACT

Eukaryotic cells are believed to have arisen from a relationship, called endosymbiosis, between two prokaryotic cells, or bacteria. In this relationship, one cell, the endosymbiont, lived inside the other cell, called the host cell. The endosymbiont provided the host cell with a surplus of ATP molecules, while the host cell provided some metabolic functions for the endosymbiont. The endosymbiont eventually took the role of the mitochondrion found in eukaryotic cells of today.

Mitochondria are the main source of ATP molecules in cells. When the mitochondria are not functioning properly, a depletion of ATP molecules can occur in tissues, such as muscle, brain, and heart. These tissues require large amounts of energy to function properly. Insufficient amounts of ATP can result in cell damage and even cell death. Eventually, entire body systems can fail, and the health of the organism may be severely compromised.

The intact membranes of the mitochondria are important to the overall health of the cell. Mitochondria are believed to be key activators for *apoptosis*, or programmed cell death. When the mitochondrion produces ATP, the membranes are polarized by the high concentration of H^+ ions in the intermembrane space. One of the early steps towards cell death is *depolarization*, the loss of this concentration gradient. Pores in both the outer membrane and inner membrane cause ions to escape. This compromises the electron transport chain, which is indirectly involved in ATP production. Aerobic respiration stops. Cytochrome c, a key component of the electron transport chain, is released into the cytosol. Here, cytochrome c may combine with ATP and a protein, forming an *apoptosome*, or apoptosis activator. Endonuclease and protease enzymes in the cytosol are activated and break down the nucleus and the rest of the cell.

ELECTRONIC LEARNING PARTNER

To learn more about how oxygen and nutrient levels can affect ATP production in mitochondria, go to your Electronic Learning Partner now.

The Transition Reaction

The transition reaction, which occurs in the matrix of the mitochondrion, is the first step in the process of aerobic cellular respiration. This process continues as long as sufficient levels of oxygen are available in the mitochondrion. If little or no oxygen is available, pyruvate in the cytosol can be oxidized through one of two fermentation processes. These processes will be described later in this section.

Pyruvate crosses the mitochondrion's outer membrane and then enters the matrix by way of a transport protein in the inner membrane (see Figure 3.8). Once in the matrix, the **pyruvate dehydrogenase complex** (a complex of enzymes) aids the process of oxidative decarboxylation. NAD^+ removes two electrons, oxidizing pyruvate. Carbon dioxide is removed, leaving a two-carbon acetyl group that combines with coenzyme A to form acetyl-CoA. Coenzyme A is a compound that

A One carbon atom and two oxygen atoms are removed from pyruvate as a CO_2 molecule.

B The remaining two-carbon fragment is oxidized to form an acetate ion. Electrons from this reaction are picked up by NAD^+, which is reduced to form NADH.

C The acetyl group of the acetate ion is transferred to coenzyme A, forming acetyl CoA.

Figure 3.8 In the transition reaction, pyruvate is converted to acetyl-CoA. This reaction marks the junction between glycolysis and the Krebs cycle.

contains a sulfur-based functional group. This electronegative sulfur-based functional group binds to a carbon in the acetyl group to make the reactive acetyl-CoA.

This product, acetyl-CoA, is a key reactant in the next step in aerobic respiration, the Krebs cycle. At this point, you have seen that carbohydrates provide the energy for ATP production. Lipids and proteins can be broken down to acetyl-CoA in the cell, and produce ATP in the mitochondrion. If ATP levels are high, acetyl-CoA can be directed into other metabolic pathways, such as the production of fatty

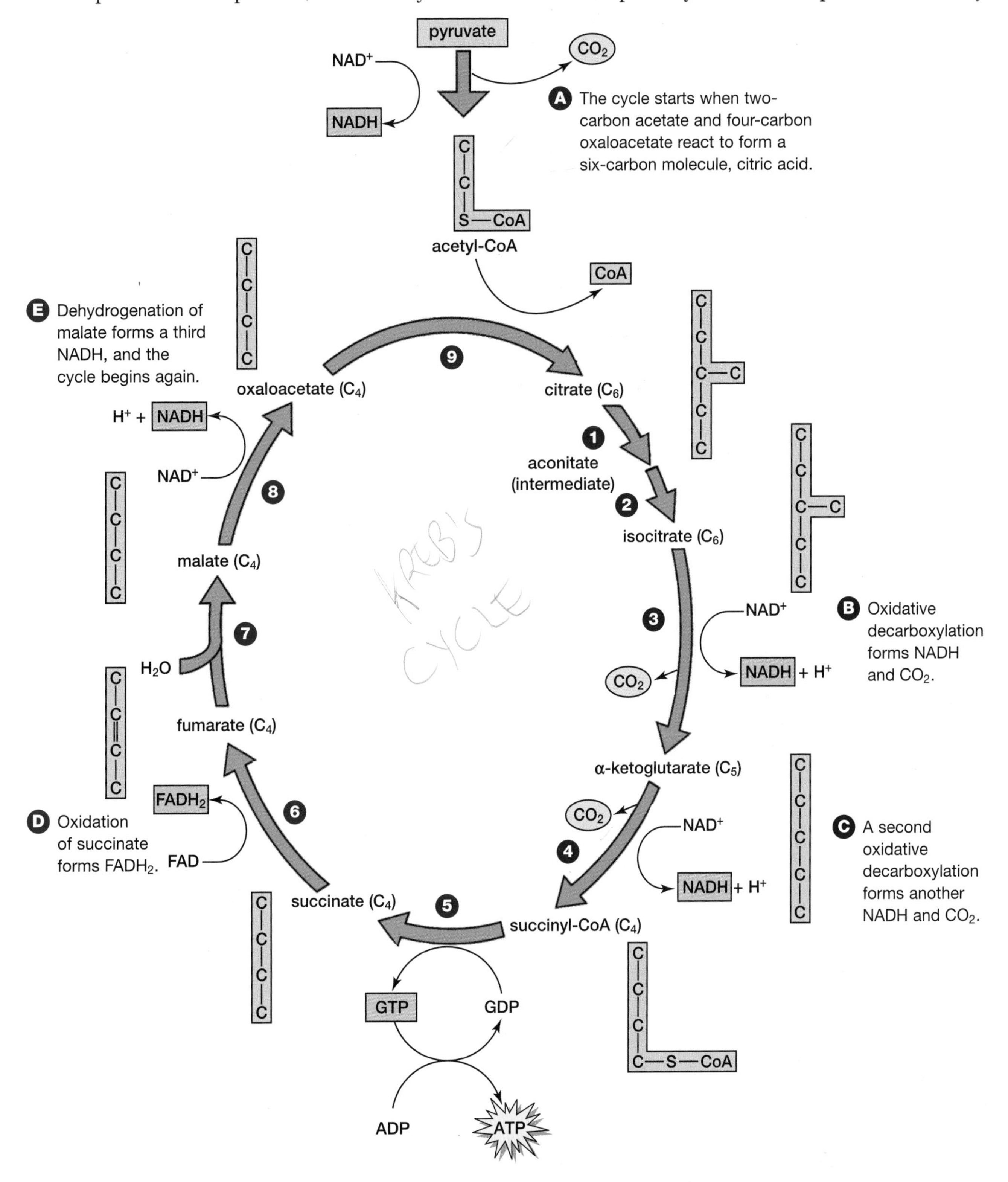

Figure 3.9 The Krebs cycle, which includes nine (numbered) reactions, oxidizes acetyl-CoA. Can you identify where substrate-level phosphorylation takes place?

acids that are required to produce lipids. The energy stored in these molecules can be released later, as needed. You can find a more detailed description of these other metabolic pathways at the end of this section.

As the product of the transition reaction from glycolysis, acetyl-CoA enters the Krebs cycle to produce ATP.

The Krebs Cycle

The Krebs cycle, named after scientist Hans Krebs, is a cyclical metabolic pathway that oxidizes acetyl-CoA to carbon dioxide and water, forming a molecule of ATP (see Figure 3.9 on page 71). In addition to producing one ATP molecule, the series of redox reactions that form the Krebs cycle involve the transfer of electrons. This electron transfer leads to the formation of three NADH molecules and one $FADH_2$ molecule.

The Krebs cycle involves a total of nine reactions. First, an enzyme removes the acetyl group from acetyl-CoA and combines it with a four-carbon oxaloacetate molecule to produce a six-carbon citrate molecule. With the acetyl group removed, coenzyme A is released to participate in another reaction in the matrix.

At this point, a series of oxidation-reduction reactions begins that will result in the formation of

Investigation 3 • A

SKILL FOCUS

- Predicting
- Performing and recording
- Analyzing and interpreting
- Communicating results

Enzyme Activity in the Krebs Cycle

The enzyme succinic dehydrogenase catalyzes the oxidation of the four-carbon molecule succinic acid to fumaric acid. This is an essential step in the Krebs cycle. In this investigation, you will use a solution of methylene blue to observe the action of succinic dehydrogenase. Fresh beef heart tissue will serve as a source of succinic dehydrogenase.

Pre-lab Questions

- What is the purpose of oxidizing succinic acid in the Krebs cycle?
- How does methylene blue indicate succinic dehydrogenase activity?

Problem

Where is succinic acid oxidized in a cell? In other words, where does succinic dehydrogenase activity occur?

Prediction

Predict how you could observe succinic dehydrogenase activity.

CAUTION: **The indicator methylene blue is a dye. Avoid any contact with skin, eyes, or clothes. Flush spills on your skin immediately with copious amounts of water and inform your teacher. Exercise care when handling hot objects. Handle thermometers with care. Follow your teacher's instructions on the safe use of scalpels. Dispose of all chemicals according to your teacher's instructions and wash your hands before leaving the laboratory.**

Materials

500 mL beaker
3 test tubes
scalpel or sharp knife
3 medicine droppers
hot plate
thermometer
grease pencil
colour chart
blue coloured pencils (several shades)
0.5 mol/L succinic acid
methylene blue (0.01%)
distilled water
mineral oil
two pieces of beef heart (each about 2–3 cm^3)

Procedure

1. Set up a hot water bath in the 500 mL beaker, and maintain it at 37°C.
2. Using the grease pencil, number each test tube.
3. Using the scalpel, cut away any fat tissue from the beef muscle. Cut away two small pieces of tissue and set aside.
4. In test tube 1, add a piece of beef heart, 4 drops of succinic acid, 8 drops of methylene blue, and 8 drops of distilled water.
5. In test tube 2, add 4 drops of succinic acid, 8 drops of methylene blue, and sufficient distilled water to equal the volume of the mixture in test tube 1.

two carbon dioxide molecules and one ATP molecule (produced by substrate-level phosphorylation).

During the cycle, energetic electrons reduce NAD^+ and FAD, which combine with H^+ ions to form NADH and $FADH_2$. Remember that an additional NADH was produced in the transition reaction. As you learned in Chapter 2, FAD is a coenzyme involved in redox reactions. Figure 3.9 shows the various reactions involved in the Krebs cycle.

Four NADH molecules and one $FADH_2$ molecule are produced for each molecule of pyruvate that enters the mitochondrion for aerobic respiration. Because two molecules of pyruvate enter the matrix for each molecule of glucose oxidized, eight NADH and two $FADH_2$ are produced for each molecule of glucose. Of these, six NADH and two $FADH_2$, along with two ATP molecules, result from the reactions of the Krebs cycle. At this point, oxygen has not been used in the reactions described. Oxygen plays a crucial role in the electron transport chain, during oxidative phosphorylation. NADH and $FADH_2$ molecules that have been formed by redox reactions in the Krebs cycle will donate electrons to the electron transport chain. Energy from these electrons fuels ATP synthesis by aerobic respiration.

At the end of the cycle, the glucose molecule that entered glycolysis has been completely

6. In test tube 3, add a piece of beef heart, 8 drops of methylene blue, and sufficient distilled water to equal the volume of the mixture in test tube 1.

7. Gently swirl the contents of each test tube. Then, quickly but carefully, use a medicine dropper to add about a 2 cm layer of mineral oil to each test tube. To do this, tip the test tube slightly to one side and allow the mineral oil to flow down the inside of the test tube.

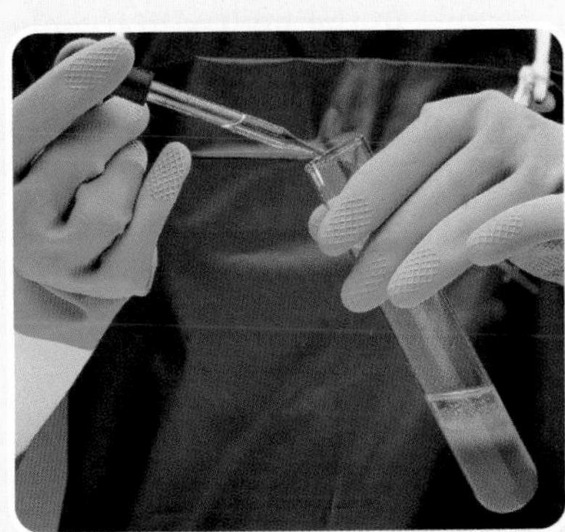

8. Place the test tubes in the water bath. Record the initial colour of the mixture in each test tube. Use a colour chart or make a series of categories, such as dark blue, medium blue, light blue, colourless. You can also make a series of sketches using coloured pencils. If possible, take a series of photographs of the samples to augment your written observations.

9. Observe the colour of the mixture in each test tube every 5 min. for about 30 min. Use a data chart to record your observations.

Post-lab Questions

1. Which sample(s) showed the most pronounced change in colour? Why?

2. Methylene blue is a dye (indicator) that changes from blue to colourless when it is reduced by other substances in a chemical reaction. Describe the chemical reaction that was responsible for changing the colour of the methylene blue in the test samples.

3. What other controls could be incorporated into this procedure?

4. Why was it necessary to add a layer of oil to the surface of each mixture?

Conclude and Apply

5. What cell structures (organelles) contain succinic dehydrogenase? Why would you expect to find high concentrations of these cell organelles in mammalian heart tissue?

6. Would this procedure work with other types of animal tissue? Explain briefly.

7. How could you modify this procedure to obtain more accurate evidence of succinic dehydrogenase activity?

Exploring Further

8. Repeat the investigation using the additional controls you outlined in answer to question 3. How would these controls help you to interpret the results?

9. Does plant tissue produce similar succinic dehydrogenase activity? Repeat the procedure using germinating white beans in place of beef heart tissue. What other plant parts might be suitable for this investigation? Explain briefly.

catabolized. As Table 3.1 shows, by the end of the Kreb's cycle, energy contained in the original molecule of glucose has been used to form four ATP molecules and 12 electron carriers.

Figure 3.9 shows that succinate, the ionic form of succinic acid, is oxidized to produce one molecule of $FADH_2$. The enzyme that catalyzes this reaction is succinic dehydrogenase. You will study the action of this enzyme in Investigation 3-A.

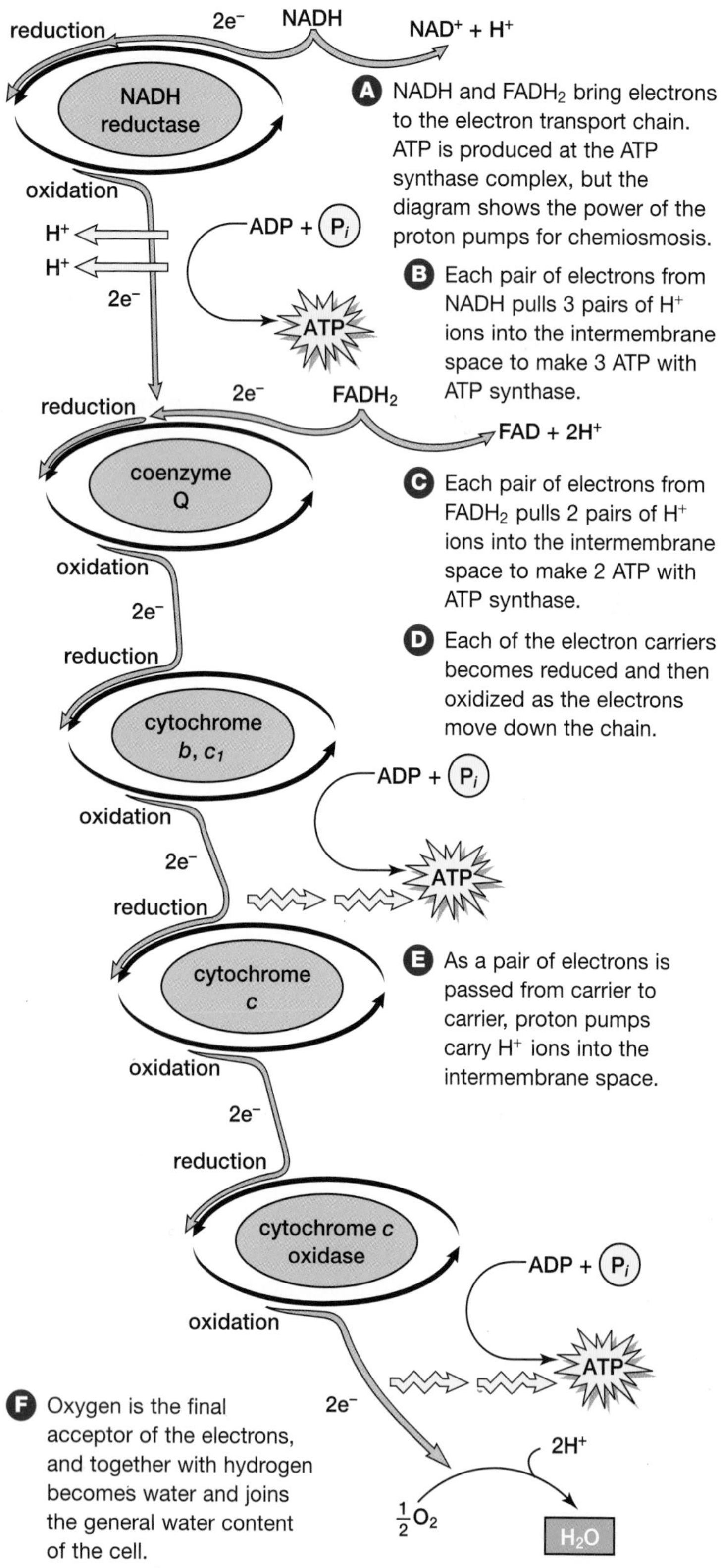

Table 3.1
The output of energy molecules up to the end of the Krebs Cycle

Metabolic process	ATP produced	Energy molecules
glycolysis	2 ATP	2 NADH (in cytosol)
oxidation (decarboxylation) of pyruvate (× 2)		2 NADH
Krebs cycle (× 2)	2 ATP	6 NADH 2 $FADH_2$
Total	4 ATP	10 NADH 2 $FADH_2$

A NADH and $FADH_2$ bring electrons to the electron transport chain. ATP is produced at the ATP synthase complex, but the diagram shows the power of the proton pumps for chemiosmosis.

B Each pair of electrons from NADH pulls 3 pairs of H^+ ions into the intermembrane space to make 3 ATP with ATP synthase.

C Each pair of electrons from $FADH_2$ pulls 2 pairs of H^+ ions into the intermembrane space to make 2 ATP with ATP synthase.

D Each of the electron carriers becomes reduced and then oxidized as the electrons move down the chain.

E As a pair of electrons is passed from carrier to carrier, proton pumps carry H^+ ions into the intermembrane space.

F Oxygen is the final acceptor of the electrons, and together with hydrogen becomes water and joins the general water content of the cell.

Figure 3.10 Overview of the electron transport chain

Chemiosmosis and ATP Production

The final stage of energy transformation in aerobic cellular respiration includes the electron transport chain and oxidative phosphorylation of ADP by chemiosmosis. The reduced coenzymes NADH and $FADH_2$ shuttle electrons and H^+ ions from the Krebs cycle in the matrix to the electron transport chain embedded on the cristae, the folds of the mitochondrion's inner membrane. The electron transport chain involves a series of electron carriers and multienzyme complexes. These carriers and complexes oxidize NADH and $FADH_2$ molecules. With their extra electrons removed, the additional H^+ ions also leave. As a result, the oxidized molecules NAD^+ and FAD can participate in a redox reaction in the matrix, such as the Krebs cycle.

The oxidation and reduction of the electron carriers in the electron transport chain releases small amounts of energy. This energy is then used to power proton pumps that pull H^+ ions across the inner membrane into the intermembrane space. These H^+ ions are now trapped between two membranes, building a concentration gradient between the intermembrane space and the matrix. The movement of these H^+ ions through special channels drives the production of ATP. (As you learned in the previous section, this movement of H^+ ions through a special protein complex is called chemiosmosis.) Figure 3.10 shows how electron transfer moves H^+ ions.

Recall that during glycolysis and the Krebs cycle, ATP molecules are produced through substrate-level phosphorylation. In this process, the ADP molecule is phosphorylated. A phosphate group is moved from another substrate (like PEP) to ADP to make ATP. In the electron transport chain, the carriers are reduced (accept electrons) and then oxidized (lose electrons to the next carrier) in a sequence. At each step, some energy is liberated and used to pump H^+ (against a concentration gradient) across the inner membrane to the intermembrane space. In addition, some of the liberated energy is lost to the environment as thermal energy. At the end of this process, the spent (low energy) electrons must be removed. Oxygen must be present in the matrix to oxidize the last component of the electron transport chain. When oxygen is combined with available H^+ ions in the matrix, water is formed. This allows additional electrons to enter the electron transport chain and release the energy needed to pump more H^+ ions into the intermembrane space. ATP is produced when the high concentration of H^+ ions diffuses through the channel of the **ATP synthase complex** that is embedded in the inner membrane, as shown in Figure 3.11. Because of the crucial role played by oxygen, this process of chemiosmosis is also called *oxidative phosphorylation*. In the next section, you will see that a similar process is used to make ATP in photosynthesis.

When NADH is oxidized, electrons enter the electron transport chain. The electrons first transfer to **NADH dehydrogenase complex**, a multienzyme system that oxidizes NADH. The liberated hydrogen ions are released into the matrix, leaving NAD^+. The electrons are passed along from one carrier to another, as shown in Figure 3.10. Some of these carriers are **cytochromes**, proteins with a heme group containing an iron atom that can be oxidized or reduced reversibly.

The two electrons from the $FADH_2$ molecules, which have less energy than the electrons carried by NADH, enter the electron transport chain at a different point. As a result, the less energetic electrons from $FADH_2$ are responsible for the production of fewer ATP molecules.

Energy from the electrons powers the multienzyme complexes, called **proton pumps**, to move hydrogen ions (H^+) into the intermembrane space, as shown in Figure 3.11. As a result of this high concentration of H^+ ions, the inner membrane of the mitochondrion becomes positively charged. At the same time, negative ions are attracted to the exterior of the outer membrane, helping the movement of protons through the proton pumps.

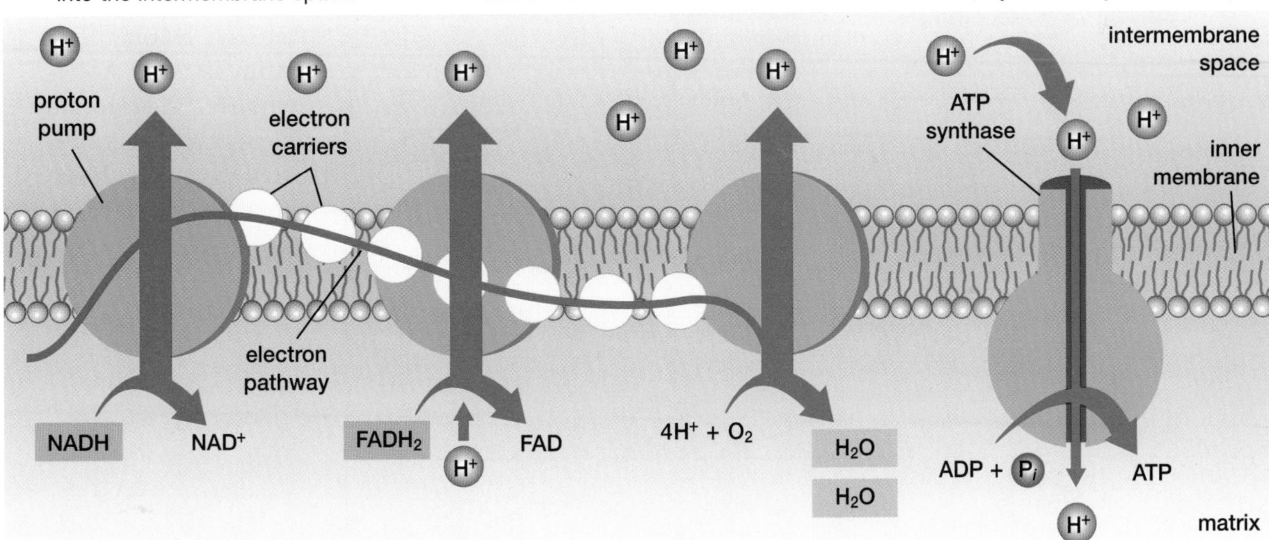

Figure 3.11 Mitochondria synthesize ATP by chemiosmosis. ATP production is based on a gradient of hydrogen ion (H^+) concentration. The gradient is established by pumping hydrogen ions into the intermembrane space of the mitochondrion.

The charge difference creates an electrical gradient, while the concentration difference creates a chemical gradient. The H^+ ions can leave the intermembrane space through a special channel in the ATP synthase complex, as shown in Figure 3.11. Chemiosmosis, the movement of these H^+ ions, creates an electric current of positively charged particles. This current provides the energy needed to phosphorylate ADP with inorganic phosphate ions in the matrix, forming ATP.

The multienzyme ATP synthase complex is not part of the electron transport chain. The electrons moving along the electron transport chain are not directly involved in the production of ATP. These electrons activate the proton pumps that move the H^+ ions into the intermembrane space. ATP synthase complex is both a collection of enzymes that phosphorylate ADP, and a tunnel that allows the H^+ ions to move from the intermembrane space back into the matrix. This flow of charged particles provides energy that is used by the enzymes to phosphorylate ADP, and release H^+ ions back into the matrix for other reactions. These reactions include the formation of water at the end of the electron transport chain. This process produces about 90 percent of the ATP molecules in a cell.

Most ATP production takes place during the reactions of the electron transport chain and chemiosmosis, because of the input of NADH and $FADH_2$ molecules (see Figure 3.12). NADH that is produced by glycolysis in the cytosol can donate electrons to the electron transport chain by way of the outer membrane of the mitochondrion. This additional step has a cost, however, and the electrons from the cytosol NADH can only pump enough protons for two ATP. In contrast, the NADH molecules that are produced in the matrix can produce three ATP.

WEB LINK

www.mcgrawhill.ca/links/biology12

Since Boyer and Walker's work on ATP synthase, biochemists worldwide have been studying the workings of ATP synthase complex. To find out more about the intricacies of ATP synthase, go to the web site above, and click on **Web Links**. Prepare an abstract about research into one aspect of the ATP synthase complex.

Substrate-level phosphorylation during glycolysis and the Krebs cycle produces four ATP for each molecule of glucose oxidized. Oxidative phosphorylation from the electron transport chain and chemiosmosis accounts for the remaining 32 ATP molecules. The total number of ATP molecules produced from one molecule of glucose is 36, as shown in Figure 3.12. The concept organizer, shown in Figure 3.13 on the next page, summarizes the important concepts you have learned that are involved in ATP synthesis. Refer to the chapters and sections listed to review these concepts.

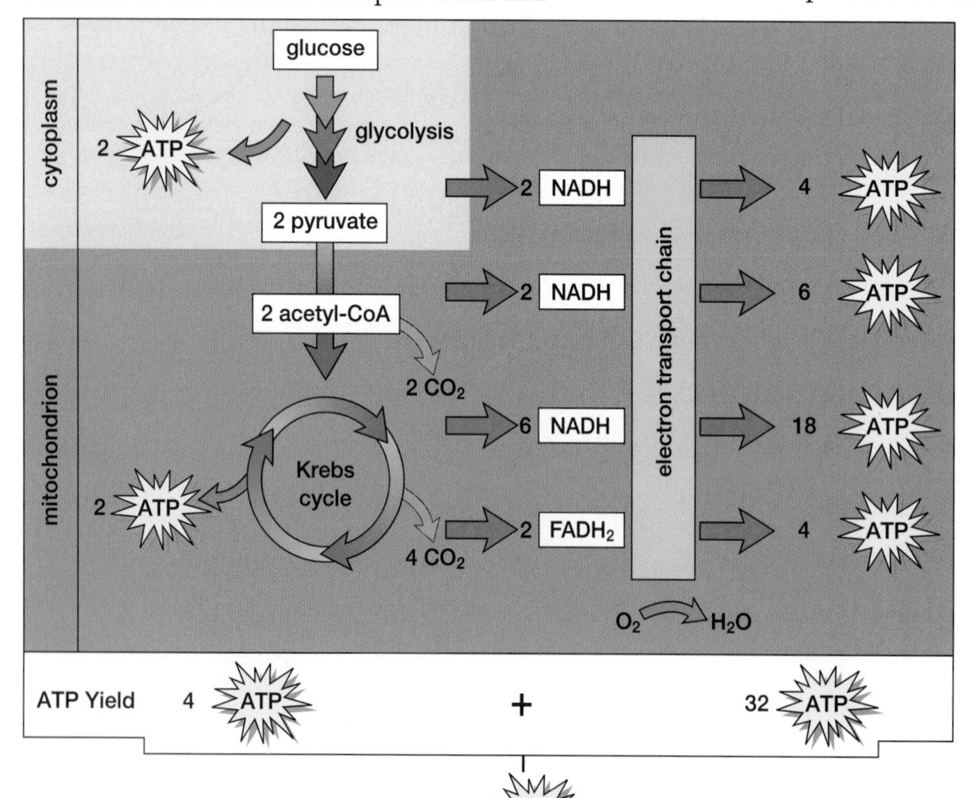

Figure 3.12 Number of ATP molecules produced for each molecule of glucose used in aerobic cellular respiration

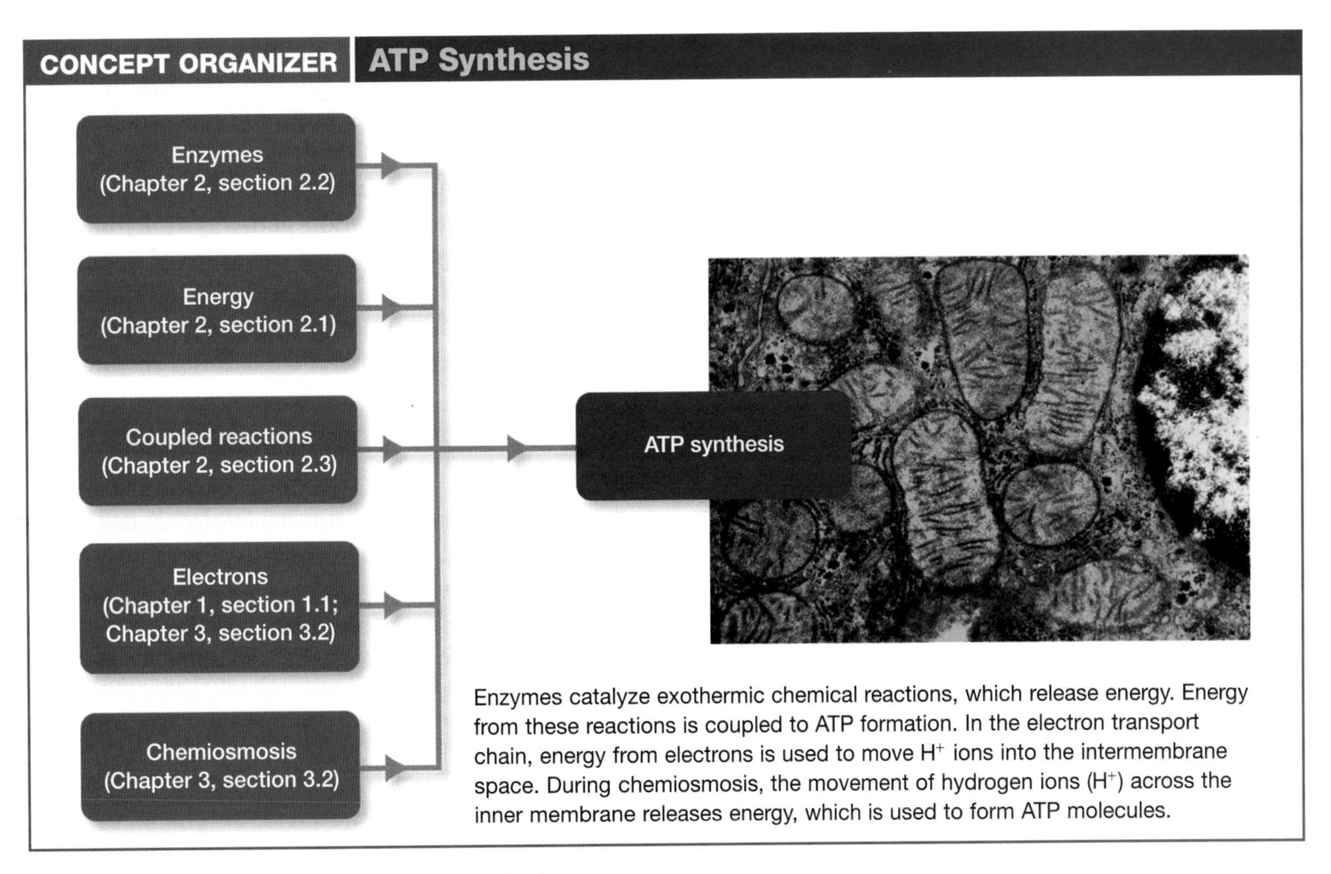

Figure 3.13 ATP formation during cellular respiration

THINKING LAB

Metabolic Rate and Exercise

Background

Researchers working on the impact of exercise and diet on human metabolic processes have come up with six actions to improve resting or basal metabolic rate (BMR). These six actions increase the efficiency with which cells metabolize food molecules into ATP and anabolize macromolecules for cellular work.

1. Exercise frequently, for a long period of time (between 30 and 60 minutes), and with sufficient intensity.
2. Increase metabolically active tissue (or muscles) through total-body exercise.
3. Eat well between five and six times per day.
4. Eat within 30 to 60 minutes after exercising.
5. Split your exercise routine into two sessions per day, and drink a carbohydrate drink 30 to 60 minutes after each session.
6. Concentrate on exercising large muscle groups.

You Try It

1. Using the library and/or Internet resources, conduct further research on human basal metabolic rate focussing on the actions listed above.
2. How would any of these actions increase the rate at which you metabolize food? Record your findings in a notebook.
3. How could you calculate your basal metabolic rate?
4. If you are able, experiment with any two or more of the above actions to improve your basal metabolic rate. First, determine your basal metabolic rate before you begin your study. Then practise the actions for at least three weeks, recording your exercise and eating regimen daily. Calculate your BMR again and compare the result to your original BMR. Did a three-week period of prescribed exercise and diet increase or decrease your BMR?

pyruvate — to make ATP. The organisms that use aerobic cellular respiration also rely, to a certain extent, on the anaerobic process of lactic acid fermentation. The fermentation of lactic acid can continue to provide muscles with energy in ATP molecules when oxygen is not available.

In the next section, you will learn that green plants also use aerobic cellular respiration to convert pyruvate to ATP molecules. However, unlike animals, plants must first manufacture glucose through photosynthesis. Although often considered the reverse of aerobic cellular respiration, photosynthesis involves many different enzymes and metabolic pathways to produce glucose. You will study photosynthesis in some detail in the next section.

SECTION REVIEW

1. **K/U** Explain how enzymes within the mitochondrion catalyze metabolic reactions that involve the products of glycolysis.
2. **K/U** How many turns of the citric acid cycle are required to catabolize one molecule of glucose?
3. **C** Draw a diagram that shows three possible reaction pathways for pyruvate following glycolysis. Identify the main products of each reaction.
4. **K/U** Under what conditions does fermentation take place in an animal muscle cell? Explain.
5. **K/U** The conversion of pyruvate to lactic acid does not produce any ATP. How, then, does this reaction contribute to the production of energy by a cell?
6. **C** Working with a partner or in a small group, use an analogy to explain the role of the mitochondrion in cellular respiration. Outline how various details in the analogy relate to various components of respiration in the mitochondrion.
7. **MC** In your local grocery store, compare the prices of foods that are rich in fat with those rich in carbohydrates. What general pattern can you see? How could you explain this pattern in terms of cellular metabolism?
8. **MC** A baker wishes to make a loaf of bread. According to the recipe, she should first prepare a yeast culture by mixing some dried yeast with warm water and a little sugar. The other ingredients are added to this mixture later.
   (a) Draw a diagram that illustrates the process of respiration taking place in the yeast cells. Why is the yeast necessary for the bread to rise?
   (b) If a strain of yeast existed that employed lactic acid fermentation, could this yeast be used in place of ordinary baker's yeast? Explain.
9. **I** A strain of cells undergoes a mutation that increases the permeability of the inner mitochondrial membrane to hydrogen ions.
   (a) What effect would you expect this mutation to have on the process of cellular respiration?
   (b) Assuming the mutant cells can survive, how might the metabolic requirements of these cells differ from those of a non-mutant strain of the same variety?
10. **I** The "four-minute mile" is often cited as an example of the limit of physical performance. That is, no matter how much athletes train, there will always be a limit to their endurance.
    (a) Why does this limit exist?
    (b) Design an experiment that you could conduct to test your hypothesis.
11. **MC** Crash diets that focus on highly regimented eating routines often produce yo-yo syndrome, in which weight lost is quickly regained. In many cases, dieters gain back more weight than they lost. How does an understanding of basal metabolic rate explain this? What kind of advice would help someone in this situation?

UNIT INVESTIGATION PREP

As you have learned, different types of cells can use different types of cellular respiration. What experiment could you perform to distinguish between yeast and muscle cells in an oxygen-free environment?

3.3 Photosynthesis

EXPECTATIONS

- **Describe the energy transformations that occur in photosynthesis.**
- **Describe the role of enzymes in metabolic reactions in chloroplasts.**
- **Investigate and explain how the structure of molecules can influence metabolic rate.**
- **Compare the structure and function of chloroplasts and mitochondria.**

Photosynthesis is one of the most important chemical processes on Earth. **Photosynthesis** involves the use of energy from light to form carbohydrates. Organisms that manufacture their own food (autotrophs), such as plants, algae, cyanobacteria, and photosynthetic bacteria, do so through photosynthesis. Autotrophs form the base of the food chain for virtually all communities of heterotrophs, which must eat to obtain nutrients. The process of photosynthesis also produces the oxygen found in the atmosphere. Oxygen is used by organisms for many processes, such as aerobic cellular respiration. The overall equation for photosynthesis is as follows:

$$\underset{\text{carbon dioxide}}{6CO_2} + \underset{\text{water}}{6H_2O} + \text{energy} \rightarrow \underset{\text{glucose}}{C_6H_{12}O_6} + \underset{\text{oxygen}}{6O_2}$$

The process of photosynthesis is believed to have originated in bacteria. Some of these bacteria were able to produce oxygen. Other bacteria were able to carry out a different form of photosynthesis but did not produce oxygen. In 2000, biochemists led by Dr. Carl Bauer at the University of Indiana found that non-oxygen-producing species (purple and green bacteria) are the most ancient photosynthetic bacteria. The oxygen-producing cyanobacteria that exist today (see Figure 3.19) evolved from a non-oxygen-producing bacteria called heliobacteria.

As you may have noticed, the overall equation for photosynthesis is exactly the opposite of the equation for aerobic cellular respiration. This does not mean, however, that the reactions follow the same course in reverse. Photosynthesis requires structures and metabolic processes similar to those used in mitochondria: electron transport chains, dissolved enzymes, and a membrane-enclosed space for chemiosmosis.

Figure 3.18 The process of photosynthesis is one of the most important of all life processes. Green plants use carbon dioxide and water to produce oxygen and food.

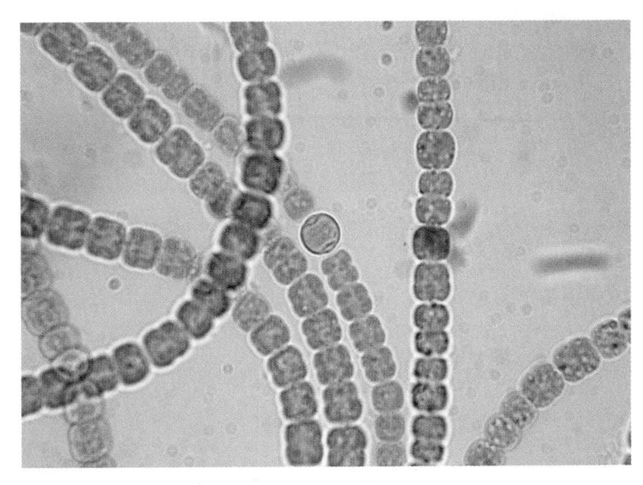

Figure 3.19 Cyanobacteria eventually gave rise to the structures that carry out photosynthesis in the algae and green plants of today.

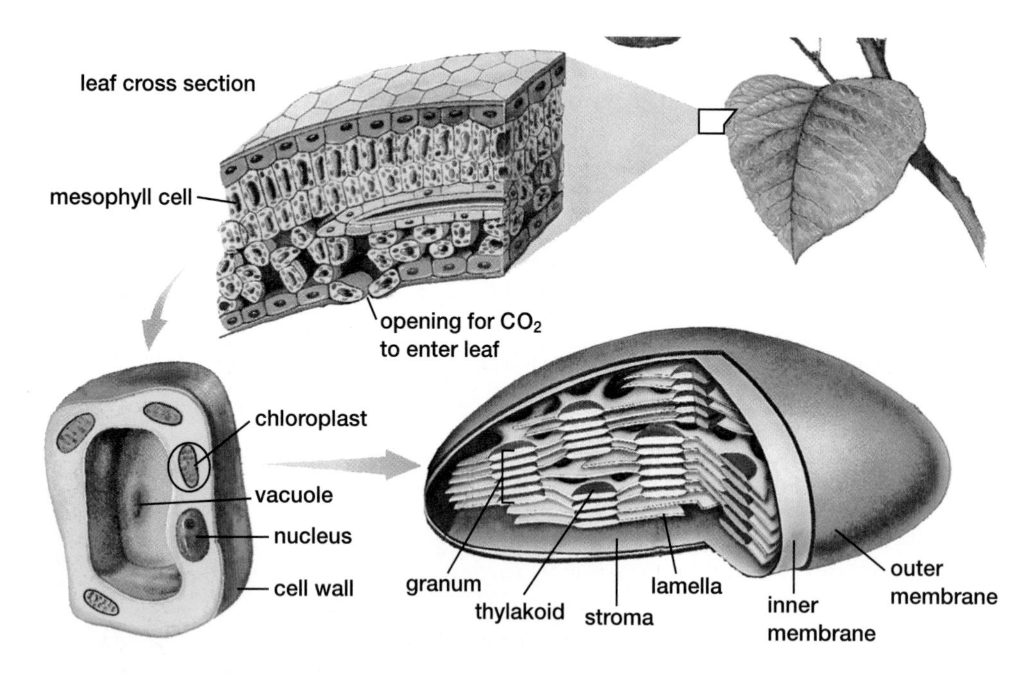

Figure 3.20 Structure of a chloroplast

Structure of Chloroplasts

In plant cells, photosynthesis occurs within **chloroplasts**. Chloroplasts have a double membrane and contain membrane pockets called **thylakoids** (see Figure 3.20). Thylakoids occur in stacked, parcel-like structures called grana (singular granum), which are held together by support structures called lamellae. The stroma, a thick, enzyme-rich liquid, fills the interior of each chloroplast.

Mesophyll cells in the leaves of plants are specialized for photosynthesis and contain numerous chloroplasts. These cells provide the chloroplasts with the two important ingredients necessary for photosynthesis — carbon dioxide and water. Gas exchange (oxygen and carbon dioxide) occurs through pores on the underside of leaves, and water is delivered via veins that extend to the roots of the plant.

Within the grana, solar light energy is captured by the thylakoids. This energy is used to form ATP molecules, which fuel the production of carbohydrates. These carbohydrate molecules are then used to synthesize glucose — the molecules used in cellular respiration. The thylakoid membrane in the chloroplast is the site of ATP production, using chemiosmosis and complex structures functionally similar to those found in mitochondria.

Stages of Photosynthesis

As the previous section suggests, there are two main stages of photosynthesis: the *photo* and *synthesis* stages. The first stage of photosynthesis converts solar energy into chemical energy. The second stage uses this energy to produce PGAL, which is then used to form glucose (see Figure 3.21). The *photo* reactions require light and are called light-dependent reactions. The *synthesis* reactions do not require light directly, and are called light-independent reactions. However, light seems to be important in activating enzymes in both the *photo* and *synthesis* reactions.

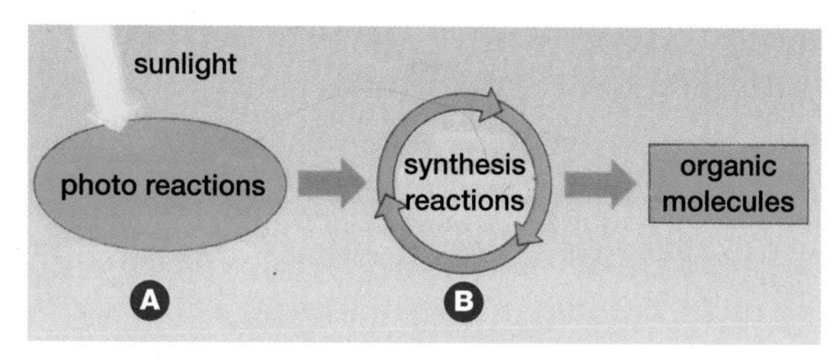

Figure 3.21 Two stages of photosynthesis. The first stage (A) consists of reactions that require light energy. The second stage (B) involves the synthesis of glucose molecules.

As light strikes the leaf of a plant, the energy is captured by pigments in the chloroplasts. These pigments, known as **chlorophylls**, absorb various wavelengths of visible light (see Figure 3.22). The two most important types of chlorophyll are chlorophyll *a* and chlorophyll *b*. Photosynthesis is most active at light wavelengths of about 400 nm to 450 nm and 650 nm to 700 nm. The colour of chlorophyll, green, is a result of the absorption of mainly blue and red parts of the visible light spectrum. In the following MiniLab, you will extract chlorophyll from leaves and examine the colour and properties of both types of chlorophyll.

B Chlorophylls *a* and *b* absorb certain wavelengths of visible light.

A Visible light represents only a small segment of the electromagnetic spectrum.

yellow and green transmitted

visible spectrum

prism

white light

solution of chlorophyll

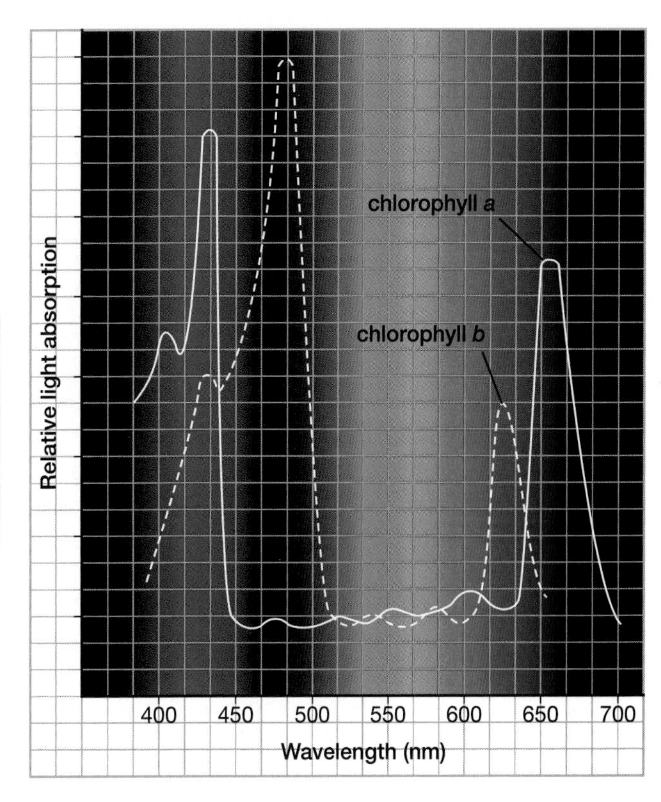

Figure 3.22 The relative absorption of light by chlorophylls *a* and *b*.

MINI LAB

Photosynthetic Pigments

In this lab you will investigate the colours of chlorophylls, the photosynthetic pigments found in many plants. You will need to produce a concentrated extract of chlorophylls.

The materials you will require are: 15 g fresh or frozen spinach, 50 mL isopropyl alcohol, food blender, 100 mL beaker, funnel, filter paper, strong light source (for example, slide projector).

Using the food blender, grind the spinach with 50 mL of isopropyl alcohol. Filter the extract through several layers of filter paper in the funnel. Then, in a darkened room, shine a strong beam of light at a sample of the filtered extract. Observe the colour of the chlorophylls by viewing the sample at a slight angle and then at a right angle to the beam of light. Describe the appearance of the chlorophylls as viewed from both angles.

Analyze

1. What colours of the visible light spectrum are absorbed by chlorophyll as part of the *photo* reaction?
2. What wavelengths of visible light are absorbed by the chloroplasts?
3. When you viewed the solution, the effect you saw is called fluorescence. What colours and wavelengths were produced?

Photosynthetic pigments

Photosystems

Light energy is absorbed by a network of chlorophyll molecules known as a **photosystem** (see Figure 3.23). These chlorophyll molecules are known as **antenna pigments** because they collect and channel energy. This energy causes electrons in the chlorophyll molecules to become energized. Energy from these electrons is passed from one chlorophyll molecule to another in the photosystem. Eventually the energy reaches the **reaction centre**, a specific chlorophyll *a* molecule. Only one in 250 chlorophyll molecules forms a reaction centre. A unit of several hundred antenna pigment molecules together with a reaction centre is called a **photosynthetic unit**. The large number of antenna pigment molecules in each photosynthetic unit allows the reaction centre to be supplied with the greatest possible amount of energy. Once the energy has reached the reaction centre, an **electron acceptor** receives the energized electron. Energy from these electrons is used to move H^+ ions into the thylakoid interior for ATP production.

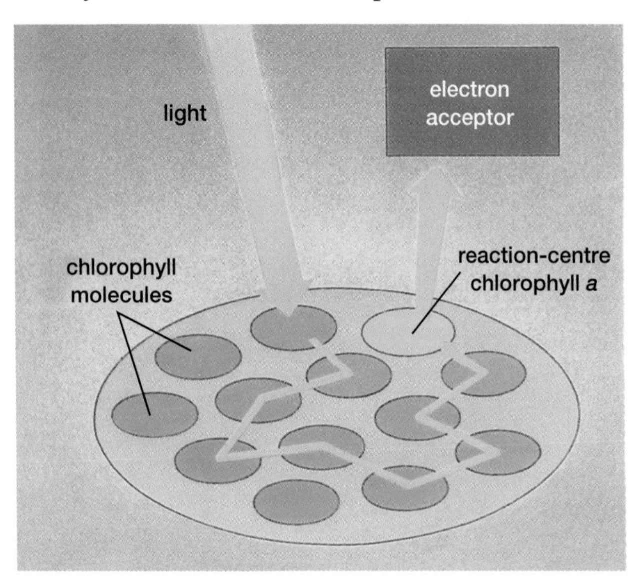

Figure 3.23 A photosystem works by passing light energy from one molecule of chlorophyll to another.

Cyclic Electron Pathway

There are two types of photosystems. Photosystem 700, which absorbs light 700 nm in wavelength, is used by some photosynthetic bacteria. This photosystem contains molecules of chlorophyll *a*, which is found in cyanobacteria and all photosynthetic eukaryotes (such as green plants). Figure 3.24 shows how electrons pass through photosystem 700. After the electron acceptors receive the energized electrons from the reaction centre, the electrons flow through an **electron transport system**. Here the electrons are passed from one electron carrier to another. Some of these carriers are cytochrome molecules. As the electrons pass through the system, they release energy that is used to phosphorylate ADP molecules to produce molecules of ATP. This process is called **cyclic photophosphorylation**, because after the ATP molecules are produced the electrons are cycled back into the photosystem. Only ATP molecules are produced by photosystem 700.

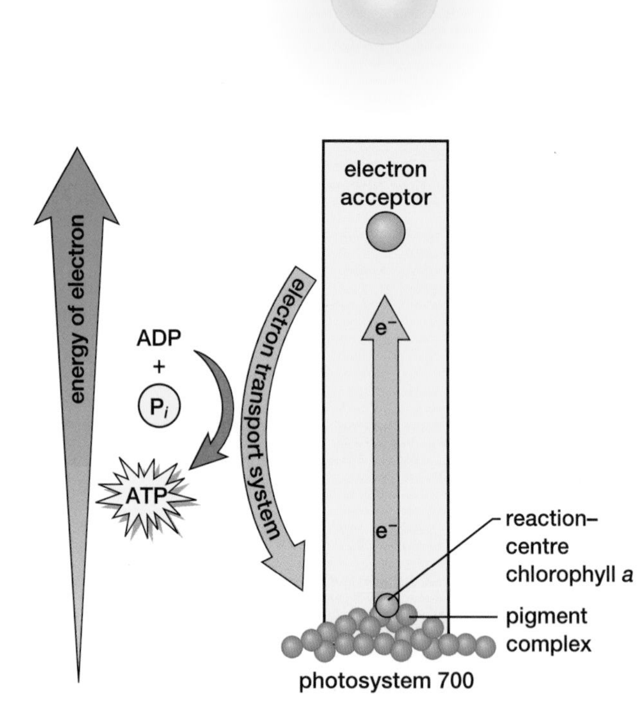

Figure 3.24 The cyclic electron pathway. In photosystem 700, electrons are recycled after their energy is used to form ATP molecules.

WEB LINK

www.mcgrawhill.ca/links/biology12
To find out more about how photosynthetic bacteria perform photosynthesis, go to the web site above, and click on **Web Links**. Note the types of bacteria that use photosynthesis to produce ATP. Prepare a chart that compares and contrasts the methods by which bacteria and plants perform photosynthesis.

Non-cyclic Electron Pathway

Photosystem 680 absorbs light 680 nm in wavelength. The shorter the wavelength of light, the higher its energy. Therefore, photosystem 680 is more powerful than photosystem 700 because photosystem 680 can capture higher-energy light. In addition to chlorophyll *a*, photosystem 680 contains molecules of chlorophyll *b*. It also contains molecules of chlorophyll *c*, chlorophyll *d*, and accessory pigments such as carotenes, xanthophylls, and anthocyanins. The pigments other than chlorophyll *a* aid in absorbing wavelengths of light not absorbed by chlorophyll *a*. Green plants, algae, and cyanobacteria (unlike other bacteria species) use both photosystems 680 and 700 to carry out photosynthesis. In this case, electrons from photosystem 680 are shunted to photosystem 700, as shown in Figure 3.25. The energy from electrons in photosystem 680 is used to produce ATP molecules. These electrons then move to photosystem 700 where, after becoming energized, they are taken up by $NADP^+$ (nicotinamide adenine dinucleotide). After $NADP^+$ accepts two electrons and a hydrogen ion (H^+), it becomes the coenzyme NADPH. The production of NADPH and ATP are endothermic reactions, which require an input of energy. The ATP and NADPH molecules are then used in the synthetic steps to produce glucose.

After ATP molecules are produced by photosystem 680, electrons that have passed through the electron transport system are not cycled back into photosystem 680. This type of ATP production is called **non-cyclic photophosphorylation**. However, photosystem 680 requires electrons to keep the photosystem operating. After photosystem 680 transfers an electron to the electron acceptor, photosystem 680 captures an electron from a Z enzyme. This enzyme is responsible for splitting water molecules into hydrogen ions and oxygen molecules and

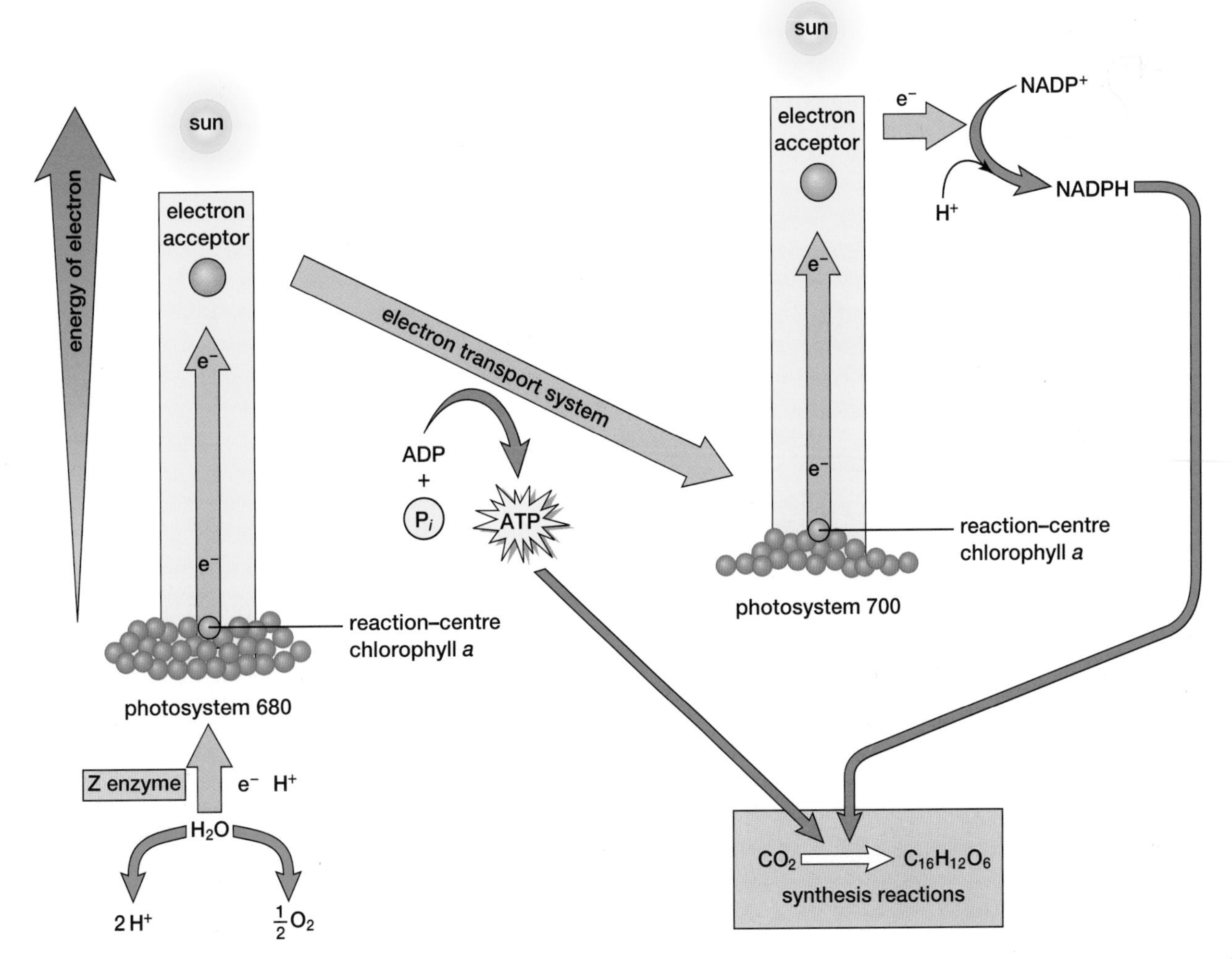

Figure 3.25 The non-cyclic electron pathway. Electrons from water move from photosystem 680 to photosystem 700 and then to $NADP^+$. The ATP and NADPH molecules that are produced by these reactions fuel the synthesis reactions that form glucose.

channelling electrons to the electron acceptor (see Figure 3.25). This process is called **photolysis** because light energy is required to split bonds within the water molecule. All of the oxygen that we breathe, and all the oxygen in Earth's atmosphere, has been generated through the photolysis stage of photosynthesis.

In addition to passing electrons from water to chlorophyll molecules, the Z enzyme that performs photolysis also donates a hydrogen ion from the same water molecule to the reaction-centre of photosystem 680. This hydrogen ion joins the electron in its journey along the electron transport chain. The electron–hydrogen ion combination supplies energy to an electron transport chain comprised of cytochrome enzymes. This chain of enzymes in turn drives a proton pump, similar to the one you learned about in chemiosmosis in the mitochondrion. The photosynthetic proton pump, like proton pumps in the electron transport chain of the mitochondrion, moves H^+ ions out of the stroma, into a membrane-enclosed space, as illustrated by Figure 3.26. Just as the inner membrane of the mitochondrion contains an ATP synthase complex that opens to the matrix, the thylakoid membrane of the chloroplast contains an ATP synthase complex where H^+ ions flow through to the stroma and energize the phosphorylation of ADP. This process is called *photophosphorylation*.

ELECTRONIC LEARNING PARTNER

To learn how the intensity and wavelength of light can affect ATP/NADPH production in chloroplasts, go to your Electronic Learning Partner now.

The thylakoid space serves as a reservoir for hydrogen ions. Every time the Z enzyme splits water to form two hydrogen ions, the thylakoid space receives them. Whenever photosystem 680 donates an electron to the electron transport system, giving up energy along the way to drive the proton pump, hydrogen ions move in from the stroma.

A hydrogen ion gradient is formed when the thylakoid space contains more hydrogen ions than the stroma. The movement of hydrogen ions across the thylakoid membrane releases energy that is used in ATP synthesis. This gradient forces the hydrogen ions through the ATP synthase complex that resides on the membrane of the thylakoid body. This movement of hydrogen ions provides the energy required to join ADP and P_i in the chemiosmotic synthesis of ATP.

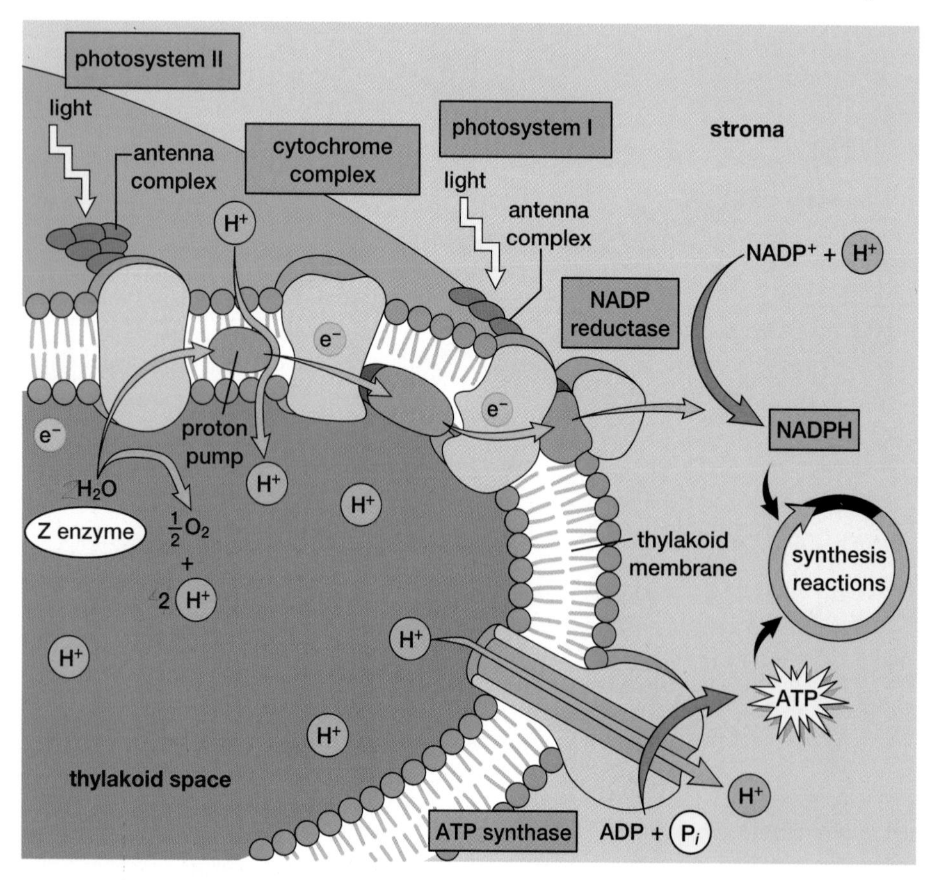

Figure 3.26 Within the thylakoid membrane, enzyme complexes pump hydrogen ions from the stroma into the thylakoid space. This process forms a hydrogen ion gradient. As hydrogen ions flow down the gradient and back into the stroma through the ATP synthase complex, ATP molecules are formed. As you can see in this diagram, photosystem 680 is also called photosystem II. Photosystem 700 is also called photosystem I.

The electrons from photosystem 680 energize electrons that travel through the electron transport chain in the thylakoid to pump protons. The electrons lose energy after they move through the electron carriers. At photosystem 700, the electrons are re-energized by light energy. These two electrons now move along the final carriers of the electron transport chain to the NADP reductase complex. Here, two electrons are transferred to $NADP^+$, which also combines with a hydrogen ion to form the reduced NADPH, as shown in Figure 3.26.

Both chloroplasts and mitochondria use chemiosmosis to produce ATP. These organelles also rely on an electron transport chain to power proton pumps and move electrons to an electron acceptor that removes them (such as water or NADPH). The proton pumps create the hydrogen ion gradient that both organelles use to make ATP. Chloroplasts and mitochondria even share the basic construct of an ATP synthase complex, which is remarkably similar in both structures. However, there are some differences in the way that phosphorylation occurs in the two organelles, as summarized in Table 3.3.

The Calvin Cycle

In photosynthesis, both the NADPH and the ATP produced by the *photo* reactions in the thylakoid membrane are used during the *synthesis* reactions to produce organic molecules from carbon dioxide. ATP and NADPH molecules are formed on the thylakoid membrane by means of the ATP synthase complex and the NADP reductase complex, respectively (see Figure 3.26). The ATP and NADPH

Table 3.3
Differences in phosphorylation between mitochondria and chloroplasts

Oxidative phosphorylation in mitochondria	Photophosphorylation in chloroplasts
■ Electrons in the electron transport chain are supplied by the oxidation of food molecules. ■ food energy → ATP molecules	■ Electrons in the electron transport chain are extracted from water during photolysis and passed on to pigment molecules (driven by captured solar energy) to donate them to the chain. ■ light energy → ATP molecules
■ Inner membrane pumps hydrogen ions from the matrix to the intermembrane space. Intermembrane space serves as proton reservoir.	■ Thylakoid membrane pumps hydrogen ions from the stroma to the thylakoid interior. Thylakoid interior pools ions.
■ ATP synthase resides between the membrane and the matrix, producing ATP molecules as hydrogen ions move back into the matrix from the intermembrane space.	■ ATP synthase bridges the thylakoid membrane and its interior, producing ATP molecules as hydrogen ions move into the stroma.

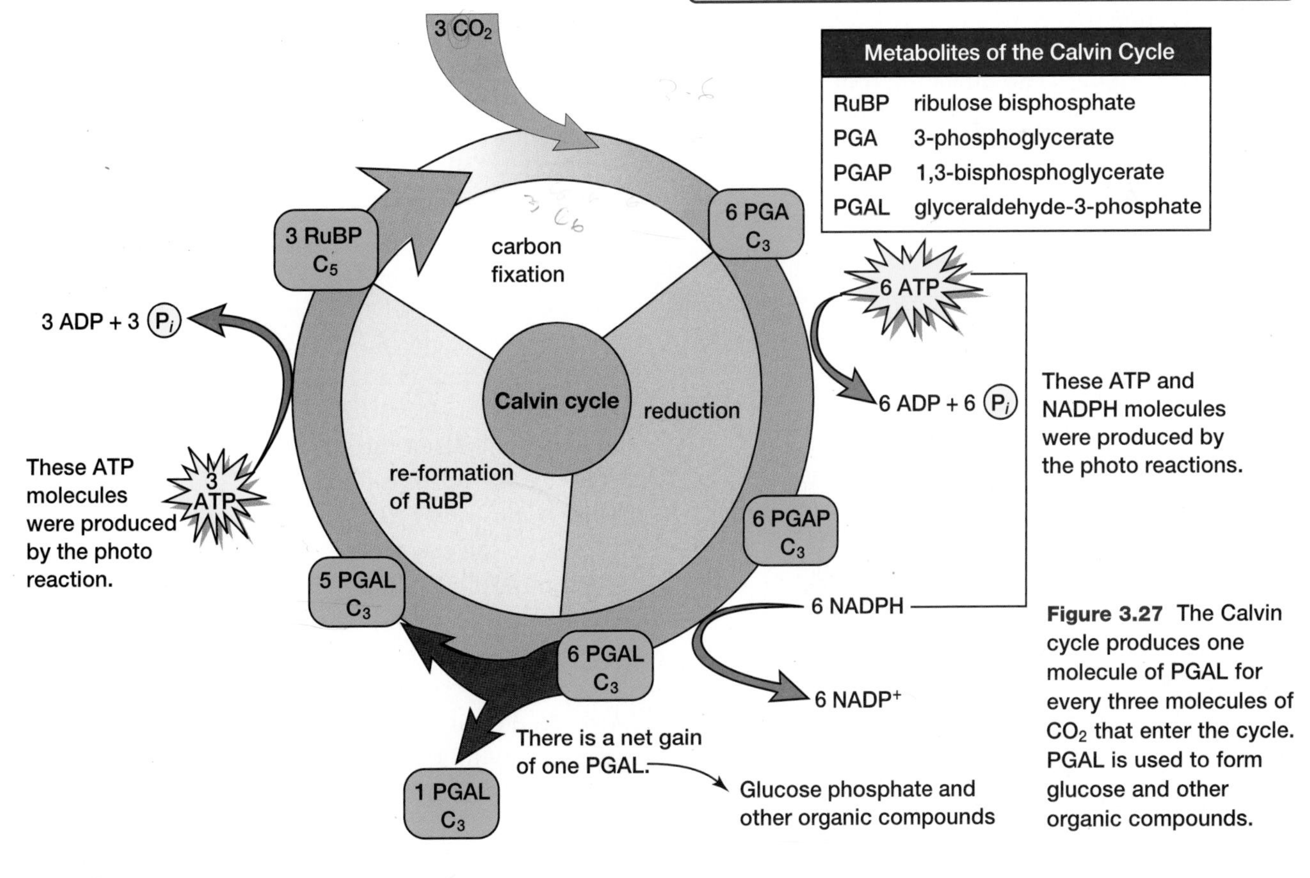

Figure 3.27 The Calvin cycle produces one molecule of PGAL for every three molecules of CO_2 that enter the cycle. PGAL is used to form glucose and other organic compounds.

molecules formed then leave the thylakoid membrane and enter the stroma, where a series of enzymes perform synthesis reactions in the **Calvin cycle**. The Calvin cycle is named after biochemist Melvin Calvin. In the late 1940s, Calvin led a team of researchers to determine the steps of this synthesis reaction.

Every photosynthetic plant uses the Calvin cycle to form PGAL. PGAL is then used to synthesize many different molecules. Using PGAL as the building block, plants can synthesize amino acids and fatty acids. Other molecules that can be formed from PGAL include fructose phosphate, glucose, sucrose, starch, and cellulose. Although plants synthesize these molecules, not every plant uses the same metabolic pathway.

The Calvin cycle has three distinct stages, as shown in Figure 3.27, on the previous page:

1. Stage 1: carbon fixation
2. Stage 2: reduction
3. Stage 3: re-formation of RuBP (ribulose 1,5 bisphosphate)

These three stages will now be discussed.

Stage 1: Carbon Fixation

Carbon fixation is the initial incorporation of carbon into organic molecules. To eventually build complex molecules, such as glucose, plants must first attach carbon to smaller carbon-containing molecules. They do this by taking carbon dioxide from the atmosphere and attaching it to RuBP, ribulose bisphosphate, as shown in Figure 3.28. A six-carbon molecule is the product of this reaction, but this molecule is extremely unstable and immediately splits into two molecules of three-carbon PGA (phosphoglycerate). The enzyme **RuBP carboxylase** catalyzes this reaction, as shown in Figure 3.28. This reaction is called C_3 fixation because it produces two three-carbon molecules of PGA. This molecule then passes into the next stage of the Calvin cycle. C_3 fixation is used by plants, such as rice, wheat, and oats, which occur mainly in temperate regions.

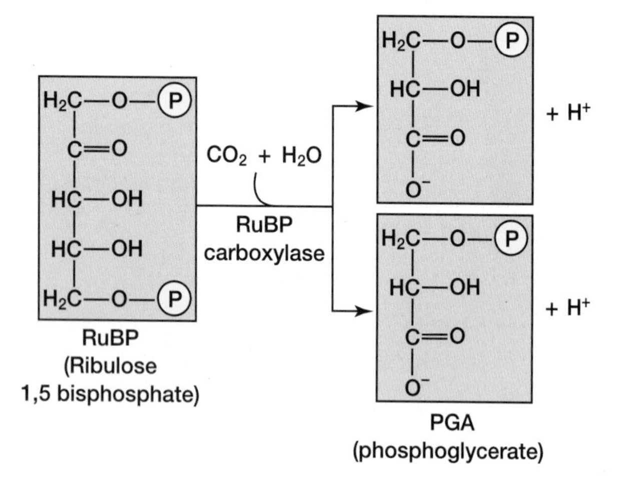

Figure 3.28 In the Calvin cycle, C_3 fixation produces two three-carbon PGA. The reaction is catalyzed by the enzyme RuBP carboxylase.

To form a molecule of glucose ($C_6H_{12}O_6$), six carbon atoms must be fixed. Figure 3.27 shows that nine molecules of ATP are required to fix the three carbon atoms in the PGAL that is available to be used for glucose production. Therefore, 18 molecules of ATP are needed to fix the six carbon atoms required to form a glucose molecule.

In addition to carbon fixation, RuBP carboxylase oxidizes RuBP with O_2 to form CO_2 by a process called **photorespiration**. Photorespiration creates an inefficiency in the carbon fixation process, since both the oxidation of RuBP and carbon fixation are catalyzed by the same enzyme — RuBP carboxylase. Both oxygen and carbon dioxide compete to bind with RuBP. The Calvin cycle is an ancient process that developed in an atmosphere with little free oxygen.

The rate of reactions in the Calvin cycle increases with temperature to about 25°C. Reaction rate levels out and declines when temperatures approach or exceed 37°C. At warmer temperatures, RuBP carboxylase is mainly involved in oxidizing RuBP, and very little carbon fixation occurs. Thus, plants that live in warmer climates have developed a different approach to fixing carbon. For example, C_4 fixation is used by plants, such as sugarcane and corn. In these plants, the Calvin cycle takes place in bundle-sheath cells, as shown in Figure 3.29. Plants that use C_4 fixation form the four-carbon oxaloacetate and malate in parenchyma cells. The malate moves into the bundle-sheath cells and a carbon is removed as CO_2. Inside the bundle-sheath cells, there is a greater concentration of CO_2 and a lower concentration of oxygen than in parenchyma cells at the surface of the leaf. This difference in concentration allows CO_2 to have a greater opportunity to bind with RuBP carboxylase. As a result, the plant can fix sufficient amounts of carbon to produce glucose in the Calvin cycle.

In tropical climates, where the temperature often exceeds 28°C, food crops such as corn and sugarcane are commonly grown. Crops that use C_3 fixation, however, do not survive well in tropical climates because they fix relatively less carbon and form fewer glucose molecules. Thus, the types of crops that can be grown in warmer climates are limited mainly to plants that use C_4 carbon fixation.

WEB LINK

www.mcgrawhill.ca/links/biology12

Deserts plants, such as cacti and aloe vera, use a third method of carbon fixation called CAM (crassulacean-acid metabolism) fixation. Why do desert plants require a different method of carbon fixation? What processes are involved in CAM fixation? To find out more about CAM fixation, go to the web site above, and click on **Web Links**. Prepare an illustrated information handout that explains the process of carbon fixation in desert-dwelling species of plants.

Stage 2: Reduction

In the second stage of the Calvin cycle, the stroma performs the necessary enzymatic reactions that reduce PGA to form PGAL. This happens in two stages. First, ATP molecules donate phosphate groups to the PGA molecules, converting them to bisphosphoglycerate, or **PGAP** molecules (see Figure 3.31). Secondly, an NADPH molecule, which was produced during the *photo* reactions, donates a hydrogen ion and two electrons to PGAP. This reduces PGAP to glyceraldehyde phosphate, or PGAL — the building block for anabolic processes including the synthesis of glucose. The oxidized $NADP^+$ can return to the thylakoid membrane to be reduced again.

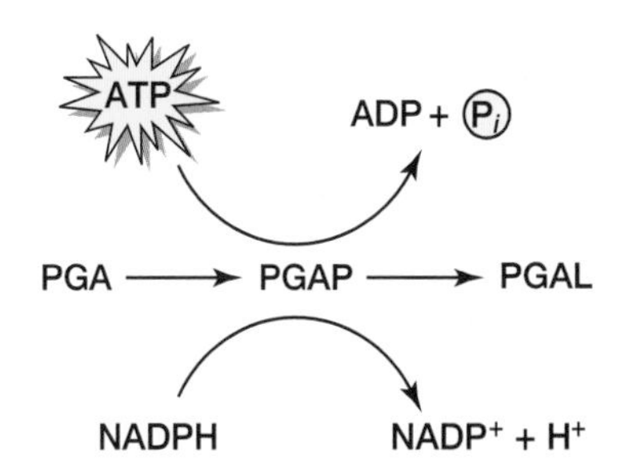

Figure 3.31 During the reduction stage, PGAP is reduced to become PGAL.

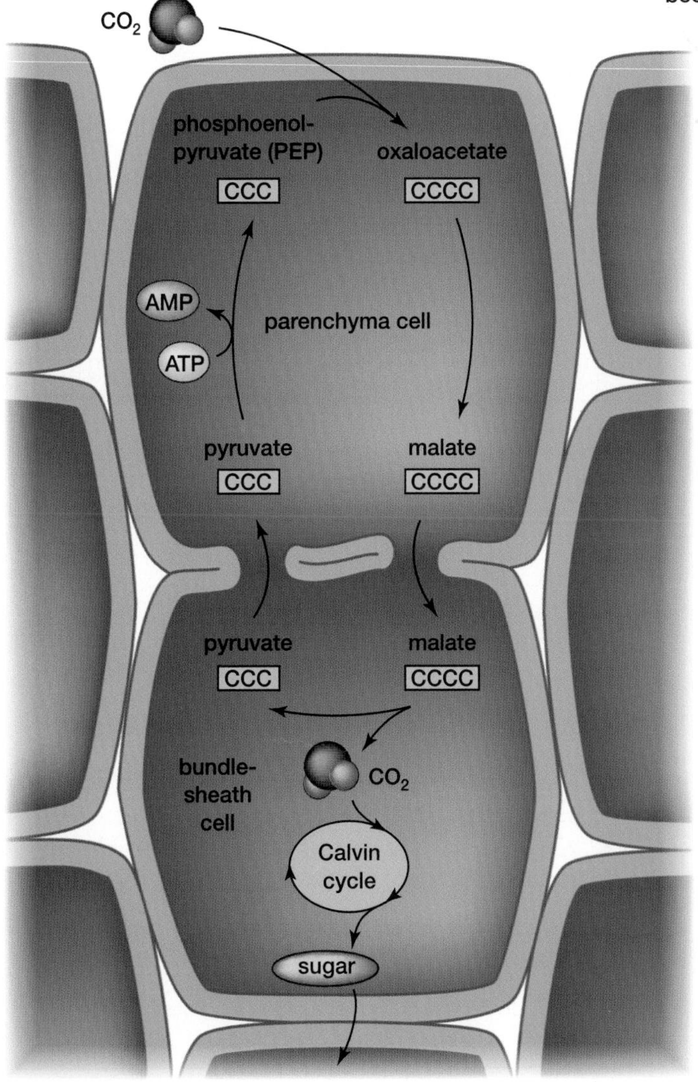

Figure 3.29 Carbon fixation in C_4 plants

Figure 3.30 Wheat (A), and corn plants (B) have adapted in different ways to their climates to circumvent the problem posed by active-site competition in the fixation of carbon dioxide.

Stage 3: Re-formation of RuBP

Recall from Figure 3.27 and Figure 3.28 that RuBP, ribulose bisphosphate, is required in the carbon fixation stage of the Calvin cycle. RuBP is used to produce PGA, needed for the reduction stage of the cycle. Because PGAL is needed to reform RuBP, the majority of PGAL molecules, do not contribute to glucose production. The Calvin cycle reactions must occur twice to create one molecule of glucose. This is because for every three times that the Calvin cycle reactions occur, five PGAL are used to re-form three RuBP, ribulose bisphosphate, as shown in Figure 3.32. Notice from Figure 3.27 that 5 three-carbon PGALs contain the same number of carbon atoms as 3 five-carbon RuBPs.

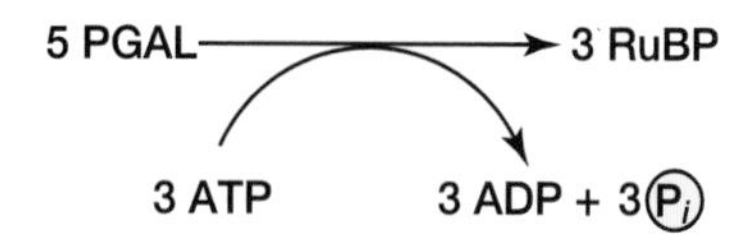

Figure 3.32 Re-formation of RuBP. As five molecules of PGAL become three molecules of RuBP, three molecules of ATP become three molecules of ADP + P_i.

To summarize the synthesis reactions of the Calvin cycle:

- Stage 1: Carbon fixation, which takes carbon atoms from atmospheric carbon dioxide molecules and incorporates these atoms into organic molecules.
- Stage 2: Reduction, which involves the formation of PGAP and its reduction to PGAL.
- Stage 3: Re-formation of RuBP, which uses most of the PGAL molecules formed in the reduction stage to produce RuBP. This is then used to form more PGA in the Calvin cycle.

Glucose: The Ultimate Food Source

After glucose is produced in the *synthesis* reactions, plant cells can use glucose for glycolysis, followed by aerobic respiration in the mitochondria. The products and intermediary molecules of aerobic respiration provide the carbon-based molecules necessary to build amino acids, as well as the precursors to nucleic acids and lipids. However, there are many other ways that plants use glucose, for example,

- the conversion of glucose to starch,
- the formation of cellulose from glucose, and
- the conversion of glucose to sucrose.

THINKING LAB

Metabolic Rate and the Structure of Molecules

Background

The molecule used by plants, such as corn and potatoes, to store energy is called starch. Starch is a large polymer composed of about 1000 glucose molecules. The starch is formed through condensation reactions, which link together individual glucose molecules. The starch molecule may contain side-branches, as shown in the illustration. Before a plant can use starch in aerobic cellular respiration, the starch must be broken down into individual glucose molecules. Recall that glucose is the molecule that first enters glycolysis. The rate at which glucose is available to be used in respiration can affect how quickly the cell will carry out metabolic processes. In other words, the availability of glucose can determine a plant's metabolic rate.

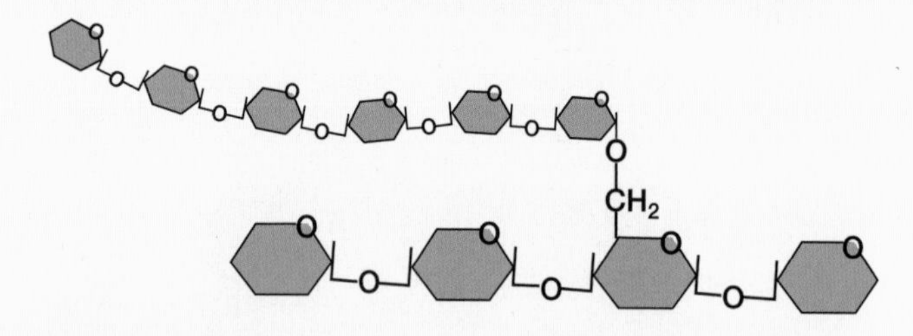

A starch molecule

You Try It

1. What reaction is needed to break down a starch molecule into individual glucose molecules?
2. If sweet corn contains mainly glucose molecules and starchy corn contains mainly starch molecules, in which type of corn would you expect cells to have a slower metabolic rate? Explain briefly.
3. Discuss how the structure of molecules might affect metabolic rate.
4. How could you determine if corn is sweet or starchy without tasting it?
5. What is the energy storage molecule in animal cells?

These three ways will now be described.

Autotrophs, such as green plants, produce a molecule used for energy storage, called starch. The starch is a large, branched polysaccharide composed of hundreds of glucose molecules linked by condensation reactions. Plants convert glucose to starch in the stroma. During peak hours of bright daylight, plants may produce more starch than they can use. This starch is stored in cells and is ready to be broken down into glucose for use in cellular processes. In the Thinking Lab on page 92, you will consider how the structure of starch can influence metabolic processes.

In another series of reactions, plants may form another kind of polysaccharide that is the building block of cell walls — cellulose. PGAL is first exported from the chloroplast into the cytoplasm where condensation reactions take place to link glucose molecules.

The formation of sucrose (the transport sugar in plants) also occurs in the cytoplasm. In order for glycolysis and cellular respiration to take place in the cytosol and mitochondria of plants, glucose is required. Because plants cannot move glucose molecules through the phloem (vascular tissue that transports organic material), they convert PGAL to glucose in the cytoplasm of leaf mesophyll cells. Glucose and fructose are then converted to sucrose. Sucrose is a molecule of fructose covalently bonded to a molecule of glucose. After sucrose is formed it is actively transported to the phloem, and then moved to locations in the plant that metabolize glucose.

Photosynthesis Versus Aerobic Cellular Respiration

Both plant and animal cells have mitochondria and carry out aerobic cellular respiration. However, only plants use photosynthesis. The cellular organelle for photosynthesis is the chloroplast, while the cellular organelle for aerobic cellular respiration is the mitochondrion. Figure 3.33 compares the processes of photosynthesis and respiration. Both processes have an electron transport chain located on membranes in the chloroplast and mitochondrion. ATP is produced on these membranes through the process of chemiosmosis. In photosynthesis, water is oxidized and oxygen is produced. In aerobic cellular respiration, oxygen is reduced to form water. Reactions in the chloroplast and mitochondrion are catalyzed by enzymes. These enzymes help to reduce CO_2 to glucose in the chloroplast and oxidize glucose to CO_2 in the mitochondrion.

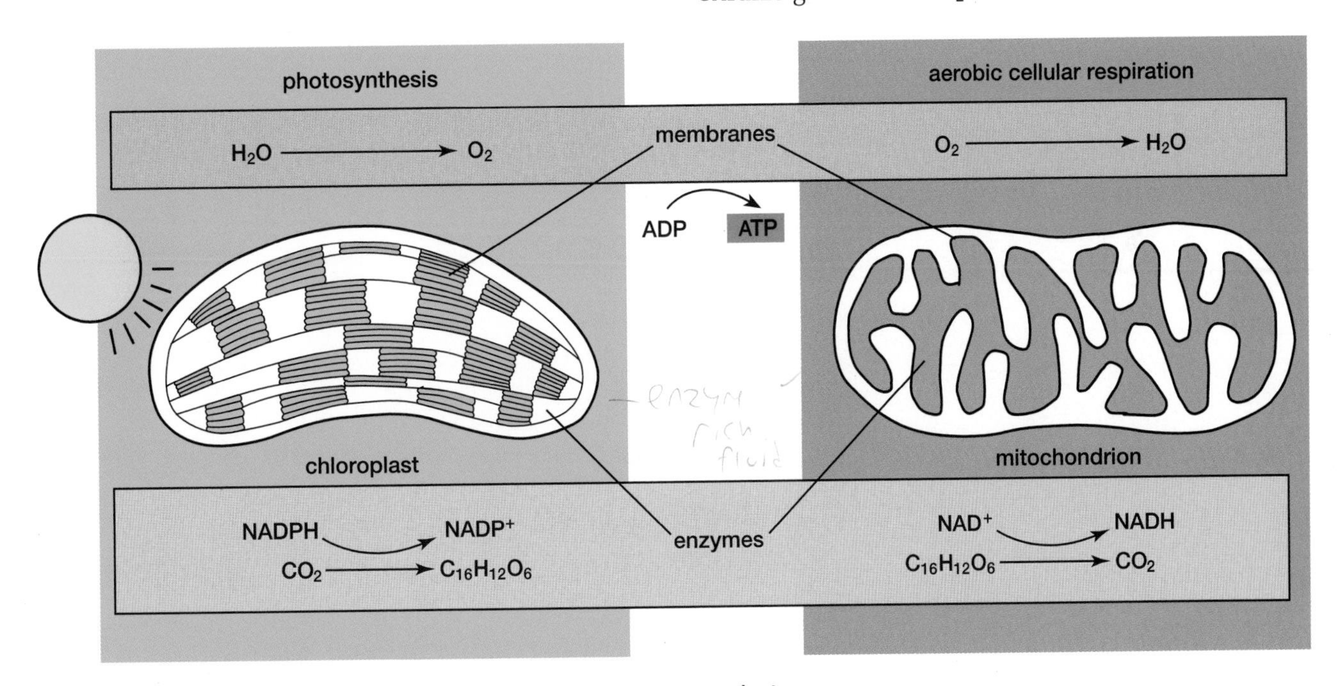

Figure 3.33 Photosynthesis versus cellular respiration. How are products produced by one organelle used by the other organelle? The organelles shown are not drawn to scale, a chloroplast is about four times larger than a mitochondrion.

SECTION REVIEW

1. K/U Draw a chloroplast and label the key structures.
2. K/U Explain how plants capture solar energy. Why does photosystem 680 provide some adaptive advantages over photosystem 700?
3. K/U Write an equation to summarize the photo reaction of photosynthesis.
4. K/U What are the components of a photosynthetic unit, and what roles do they play? In which part of the chloroplast are photosynthetic units located?
5. K/U What role does water play in the *photo* reactions of photosynthesis?
6. K/U List the three stages of the Calvin cycle. For each stage, identify the energy input and the product(s).
7. K/U What becomes of the atmospheric carbon fixed by a plant cell?
8. C Explain, with diagrams, how high levels of oxygen reduce the effectiveness of the Calvin cycle. Explain how alternative forms of carbon fixation avoid this problem.
9. K/U How many CO_2 molecules need to be fixed to produce 2 PGAL molecules that can leave the Calvin cycle to form glucose?
10. I In an experiment, a team of researchers uses a heavy isotope of oxygen-18 to track the passage of oxygen through the process of photosynthesis. What results would you expect to find if the researchers initiated the reaction in an environment in which
    (a) the carbon dioxide contained the heavy oxygen?
    (b) the water contained the heavy oxygen?
11. MC You are hired to advise strawberry producers on ways to increase their harvest. Assume the plants are grown under fully controlled conditions, so you can alter the temperature, lighting, atmosphere, water supply, and nutrients.
    (a) What conditions would you recommend to maximize the plants' productivity?
    (b) Can you think of any mutation that would help increase the rate or efficiency of photosynthesis? Discuss your reasoning.

Summary of Expectations

Briefly explain each of the following points.

- Stages of aerobic cellular respiration include glycolysis, transition reaction, the Krebs Cycle, and the electron transport chain. (3.1)
- Glycolysis occurs in the cytoplasm of all cells and produces two pyruvate molecules and two ATP molecules. (3.1)
- Substrate-level phosphorylation involves the transfer of a phosphate group to ADP to form ATP. (3.1)
- The transition reaction links the processes of glycolysis and the Krebs cycle. (3.2)
- The Krebs cycle is a cyclical metabolic pathway that oxidizes acetyl-CoA to carbon dioxide and water, forming two molecules of ATP. (3.2)
- The electron transport occurs in the cristae of the mitochondrion and involves a series of electron carriers and multienzyme complexes. (3.2)
- Oxidative phosphorylation requires oxygen in order to form ATP molecules. (3.2)
- Most ATP molecule production takes place during the reactions of the electron transport chain. (3.2)
- Chemiosmosis involves the production of ATP molecules when a hydrogen ion gradient is formed across a membrane by an electron transport chain. (3.2)
- When oxygen is not present, cells use anaerobic cellular respiration to produce ATP molecules. (3.2)
- Photosystems are made up of a network of chlorophyll pigments. (3.3)
- In cyclic photophosphorylation, electrons are cycled back into the photosystem. (3.3)
- ATP synthesis occurs within the thylakoids. (3.3)
- The Calvin cycle has three distinct stages — fixation of CO_2 , reduction of CO_2, and re-formation of the molecule RuBP. (3.3)
- The Calvin cycle produces one molecule of PGAL for every three molecules of CO_2 that enter the cycle. (3.3)
- There are three methods of carbon dioxide fixation in plants — C_3, C_4, and CAM fixation. (3.3)

Language of Biology

Write a sentence including each of the following words or terms. Use any six terms in a concept map to show your understanding of how they are related.

- aerobic cellular respiration
- aerobic
- glycolysis
- anaerobic
- pyruvate
- fermentation
- transition reaction
- acetyl-CoA
- Krebs cycle
- electron transport chain
- oxidative phosphorylation
- chemiosmosis
- substrate-level phosphorylation
- PGAL
- PGAP
- PGA
- PEP
- porin
- cardiolipin
- pyruvate dehydrogenase complex
- ATP synthase complex
- NADH dehydrogenase complex
- cytochromes
- proton pumps
- anaerobic cellular respiration
- deamination
- β-oxidation
- photosynthesis
- chloroplast
- thylakoids
- chlorophylls
- photosystem
- antenna pigment
- reaction centre
- photosynthetic unit
- electron acceptor
- electron transport system
- cyclic photophosphorylation
- non-cyclic photophosphorylation
- photolysis
- Calvin cycle
- carbon fixation
- RuBP carboxylase
- photorespiration
- PGAP

UNDERSTANDING CONCEPTS

1. Why do cells use energy?
2. What are the four steps of aerobic cellular respiration?
3. Which of the four steps of cellular respiration acts on glucose?
4. Which of the four steps of cellular respiration take place in the mitochondrion?
5. Where is pyruvate produced?
6. Glycolysis I involves substrate-level phosphorylation.
   (a) What is substrate-level phosphorylation?

(b) Which molecules gain phosphate groups in glycolysis I?

7. How is NADH formed during glycolysis?
8. List the key products of the transition reaction.
9. Where in the mitochondrion does the Krebs Cycle take place?
10. Acetyl Co-A is produced by the transition reaction. What happens to each part (acetyl and Co-enzyme A)?
11. What is meant by the term decarboxylation?
12. List the key products of the Krebs cycle.
13. The Krebs cycle uses matter efficiently. How is this useful to the cell?
14. Is the Krebs cycle an aerobic or anaerobic process?
15. Which molecules dissolved in the matrix are essential in the reactions in the Krebs cycle?
16. What are the roles of NADH and $FADH_2$ in the mitochondrion?
17. What is chemiosmosis? Where does it take place?
18. (a) Where is the high concentration of hydrogen ions (H^+) located in the mitochondrion?
    (b) How did the hydrogen ions get there?
19. Which product of the Krebs cycle has the most energy? Explain.
20. What is the role of oxygen in aerobic respiration?
21. If oxygen is not available for the electron transport chain, how is the Krebs cycle affected?
22. Pyruvate moves into a mitochondrion if the concentration of pyruvate inside the mitochondrion is lower than the concentration outside. What conditions will prevent the net movement of pyruvate into a mitochondrion?
23. Compare the net gain of ATP for glycolysis to the net gain of ATP for fermentation. How is fermentation more efficient for cells?
24. Muscle cells can switch to lactic acid fermentation if oxygen levels are very low. How is this useful to the body?
25. Where does the light-dependent reaction take place?
26. What is photolysis and why is it important for photosynthesis?
27. What is the product of the Calvin cycle?
28. The Calvin cycle is not very efficient. Explain this statement.

INQUIRY

29. Mature human red blood cells do not have any mitochondria, yet they live for weeks. Predict which respiration processes red blood cells most likely use. What metabolic products would you expect to find in red blood cells that would support your prediction?
30. Which type of molecule can act as a limiting factor for reactions that take place within the mitochondrion? Assume you have access to a specific toxin that can bind with this molecule and make it inactive. Predict how the toxin would affect metabolic processes. What key product (or absence of product) would support your prediction?
31. Special centrifuges can be used to separate various components found in mitochondria and chloroplasts. The components separate in a centrifuge tube. For each organelle, which component is likely to be at the bottom of the centrifuge tube (because it is the heaviest component)? Assume that the process of separating these components has not disrupted the enzyme structures involved in the different metabolic reactions. Identify ways to verify your answer.
32. During carbon fixation in the Calvin cycle, CO_2 combines with RuBP. Oxygen can also combine with RuBP, and can prevent CO_2 from reacting. If the ambient temperature is increased, will this increase or reduce glucose production? Make a data chart to indicate factors important to consider when investigating this question.
33. Photosynthesis is an endothermic reaction. Consider that the temperature in the environment affects this process. Develop a testable hypothesis that explores how temperature affects photosynthesis.
34. Which process produces more thermal energy: photosynthesis or respiration? Design an experiment to explore this question.

COMMUNICATING

35. Glycolysis is a series of reactions. Some of these reactions are endothermic and some of these reactions are exothermic. Endothermic reactions form products have more energy than the reactants. Exothermic reactions form products that have less energy than the reactants. Show the relative energy of each of the following molecules on the energy graph provided: glucose, fructose-diphosphate, PGAL, PGA, pyruvate.

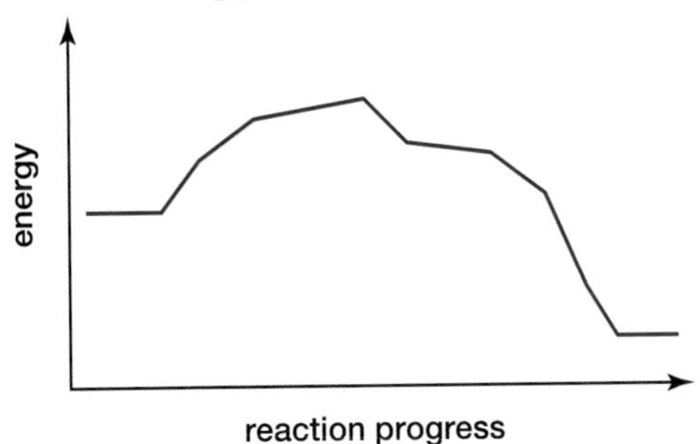

36. Make a diagram of the cell showing the cell membrane, nucleus, cytosol and at least one mitochondrion. Use one colour to show the path of one molecule of glucose into the cell, through glycolysis and into the mitochondrion for respiration. Use a second colour to show where energy is used to drive cellular respiration. Use a third colour to show where energy is released by cellular respiration.

37. Glycolysis describes the process of breaking glucose into two pyruvate molecules. Use words and diagrams to describe this process in terms of stored energy, free energy and thermal energy.

38. Acetyl Co-A is a crucial component of cellular respiration. Use diagrams and captions to explain how the cell uses acetyl Co-A efficiently.

39. Make a diagram to show the movement of hydrogen ions (H^+) into the intermembrane space of a mitochondrion. Also show how the energy from the hydrogen ions is used to make ATP. Indicate the movement of electrons along the electron transport chain of a crista. Show how water is made as a final product of aerobic cellular respiration.

40. Make a diagram of a mitochondrion and a chloroplast. Indicate which reactions take place on membranes and which reactions take place dissolved in solution. Indicate how the following molecules move between the two organelles: glucose, CO_2, O_2, H_2O.

41. Make a rough diagram of a cell. Label locations of cell processes that use ATP. Label locations where glycolysis produces ATP. Label locations where aerobic cellular respiration produces ATP. Explain how a cell can increase the efficient transfer of ATP within the cell.

42. Use diagrams to show how non-cyclic photophosphorylation is different from cyclic photophosphorylation. Explain how one system uses matter and energy more efficiently.

MAKING CONNECTIONS

43. Cells require energy in the form of ATP molecules. As you have learned, cellular respiration describes a metabolic pathway for the carbohydrate glucose. How does decreasing carbohydrate (glucose) intake affect cellular energy?

44. In a multicellular organism, different organ systems, tissues, and specialized cells contribute to maintaining the health of the body.
    (a) List three molecules that must enter cells to support cellular metabolism.
    (b) List three molecules that must be produced inside cells to support cellular metabolism.
    (c) Add the following to the list you made for part (b), and give reasons for their inclusion: PGAL, fructose-phosphate, oxaloacetate, and coenzyme A.
    (d) Which substance given in (c) would you consider adding to your breakfast? Explain.

45. Pyruvate is available as a dietary supplement. Explain the possible results of adding pyruvate to your diet.

Enzymes and Reaction Rate

Background

In Unit 1, you have seen that living cells rely on enzyme activity for metabolic processes. The different enzymes allow a reaction to move forward more quickly by lowering the activation energy required for the reaction to proceed. Enzymes are proteins that catalyze the rate of the reaction by providing a lower-energy pathway.

The study of catalysts and reaction rates provides insight into how cells use enzymes effectively. The rate of a reaction is measured in one of two ways: by finding the amount of reactant (substrate) consumed, or by the amount of product formed. You used this second method in your experimental work with peroxidase and hydrogen peroxide.

Peroxidase is an enzyme that is found in peroxisomes, specialized organelles in the cell. The enzymes responsible for glycolysis are dissolved in the cytosol. Other enzymes are found embedded in membranes. Enzymes on the inner membrane of the mitochondrion or the thylakoid membrane of the chloroplast are crucial to the energy-transforming reactions that take place here.

One strategy that maximizes the efficiency of enzyme activity is extensive folding of the membrane structures. The simple outer membranes do not contribute to the energy reactions. The active membranes are extensively folded, however, allowing many copies of the enzymes to participate in reactions at the same time. This strategy effectively speeds up the rate of the reaction. Embedded in the membrane, these enzymes are in a different phase from the substrate. The effect is similar to increasing the surface area of a reactant, like chopping wood before burning it to make it burn faster.

As proteins, enzymes each have a distinctive shape. Factors affect enzyme shape by disrupting bonds and promoting different folding patterns. A differently folded enzyme may become insoluble, precipitating out of the cytosol and out of the cell's metabolic processes.

Environmental factors that can affect enzyme shape include temperature, radiation, pH, and ion concentration. While a raw potato has working peroxidase enzymes, cooked potatoes do not. The concentration of ions, including pH, can affect the bonds within an enzyme. Metal ions can act as cofactors, but adding more metal ions, such as Cu^{2+} or Ag^{2+} can disrupt the natural folding pattern of the enzyme.

ASSESSMENT

After you complete this investigation,
- **assess your procedure by having a classmate try to duplicate your results;**
- **assess your results by comparing your data with those found by other students.**

Pre-lab Focus

- How can you measure the rate of an enzyme reaction?
- What is a substrate and what conditions in the cell control the amount of substrate available?
- How is the shape of an enzyme important and what are factors that could affect it?

Problem

Cell metabolism relies on enzymes that catalyze biochemical reactions in the cell. Many factors contribute to the enzyme's effective activity within the cell, including temperature and pH. What is another factor?

Hypothesis

Based on your understanding of metabolic processes, formulate a hypothesis that predicts the impact of a specific factor on the rate of an enzyme reaction.

Materials

- $CuSO_4$ (copper(II) sulfate)
- $AgNO_3$ (silver nitrate)
- H_2SO_4 (sulfuric acid)
- HCl (hydrochloric acid)
- NaCl (sodium chloride)
- safety goggles
- aprons
- gloves
- hot water bath
- ice bath
- thermometers or temperature probes
- graduated cylinders
- droppers or pipettes
- pH paper or probes
- test tubes
- test tube rack
- test tube holder
- test tube brush
- beakers
- beaker tongs
- substrate
- enzyme

Safety Precautions

- Check WHMIS charts to identify appropriate safety equipment when using chemicals.
- Avoid allowing a hot water bath to boil vigorously, which can cause test tubes to break. Use a moderate heat setting.
- Use test tube holders and beaker tongs to move hot glassware.

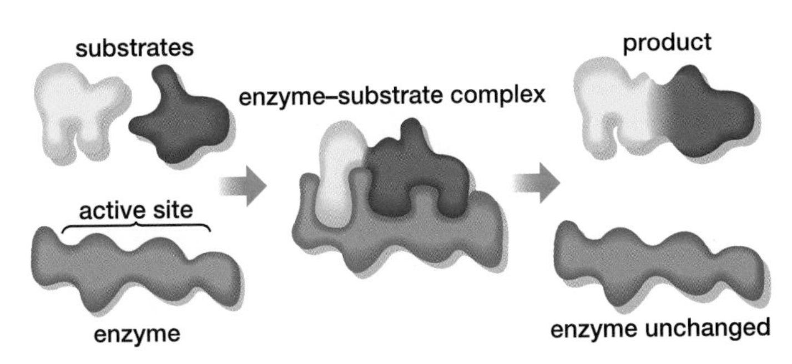

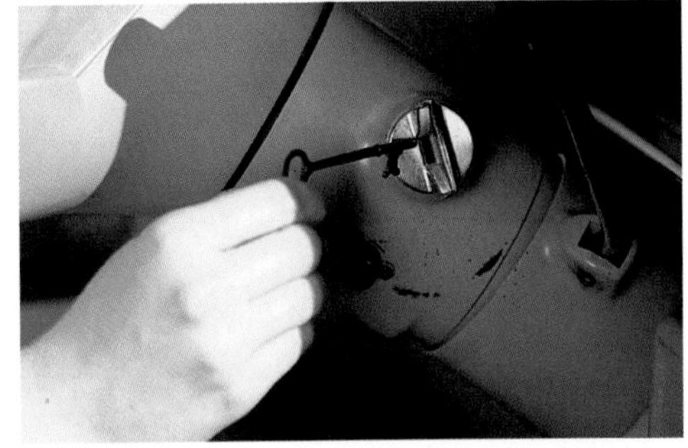

Substrates are brought close together in the active sites of an enzyme, which lowers the activation energy of the reaction by facilitating the bonding of the substrates to form a product. After the substrates have reacted, the product is released. The enzyme is then able to bind more substrate molecules and continue catalyzing the reaction. Substrates with shapes that do not match the shape of the active site will not undergo the catalyzed reaction, just as a house key cannot be used to start a car.

- Avoid contact between $AgNO_3$ and your skin or clothing because $AgNO_3$ will cause stains. If you accidentally spill chemical solutions on your skin, immediately wash the affected area with copious amounts of cool water. Inform your teacher.
- Wash your hands thoroughly with soap after the lab.

Experimental Plan

1. Design a checklist or rubric that assesses the completeness of the lab design and the lab report. Compare your checklist or rubric with those of two other groups. Submit a copy of your checklist or rubric to your teacher.
2. Identify an enzyme reaction that you can study. What are the names of the enzyme, the substrate(s), and the product(s)?
3. List some factors that affect enzyme reactions. Identify those factors that can be varied to study changes in the rate of the reaction. Support your thinking by calling upon your understanding of the way enzymes work.
4. Design an experimental procedure to test your hypothesis. Identify the control, and the dependent and independent variables. The steps must clearly explain how the experiment will be carried out. Where appropriate, use diagrams for clarity. Explain how you will measure the rate of the reaction. Identify the chemical you are testing to measure the reaction rate.

Checking the Plan

1. Develop a flowchart relating the procedure to the hypothesis. Include lists of materials. Explain the purpose of listed chemicals. Identify safety precautions required. Prepare a data chart.
2. Share your plan with one or two other groups for feedback. Revise your plan and submit it to your teacher for approval.
3. Collect the materials you need.
4. Determine a benchmark reaction rate as a control. Compare other reactions to this rate.

Data and Observations

1. Set up and perform the experiment. If necessary, modify the experimental design. Check the new design with your teacher.
2. Record data and observations. Present your data in an effective format that is easy to interpret and understand.

Analysis and Conclusions

1. Prepare a poster report of your work. Include all of the report elements you identified in your checklist.
2. Prepare an abstract, a brief summary of the experiment and findings.

Assess Your Experimental Design

3. Use the checklist or rubric you developed to assess your experimental design.
4. Compare your data with those of other groups. How might you account for any discrepancies? List ways you might improve your experimental design and explain any modifications you think would enhance it.

UNIT 1 REVIEW

UNDERSTANDING CONCEPTS

Multiple Choice

In your notebook, write the letter of the best answer for each of the following questions.

1. Most materials dissolve in water because
   (a) water is a small molecule
   (b) hydrophobic bonds form
   (c) water can break into hydronium and hydroxyl ions
   (d) hydrophilic bonds form
   (e) ionic bonds form

2. If a molecule is oxidized, it always
   (a) loses an electron
   (b) gains an electron
   (c) loses a proton
   (d) gains a proton
   (e) combines with oxygen to form water

3. The pH scale for acids and bases is a measure of
   (a) the strength of hydrophilic bonds
   (b) the strength hydrophobic bonds
   (c) the concentration of hydronium ions
   (d) the concentration of hydroxyl ions
   (e) the ability to form ions

4. Which process is an example of a hydrolysis reaction?
   (a) Peptide bonds are formed.
   (b) Lactase breaks lactose into glucose and galactose.
   (c) Catalase breaks hydrogen peroxide into less reactive molecules.
   (d) Chemiosmosis combines ADP with a phosphate group to make ATP.
   (e) Carbon dioxide molecules pass through the cell membrane.

5. Proteins are macromolecules made from amino acids. Read these statements carefully. Which combination of statements best describes an enzyme?
   I The primary structure is linear.
   II *R*-groups may form ionic bonds or hydrogen bonds.
   III Polypeptides can fold to form pleated sheets.
   IV Hydrophilic amino acids are attracted to water.
   V Polypeptides can link to form a larger structure.

   (a) I, II, and III
   (b) I, II, III and IV
   (c) II, IV and V
   (d) I, III and V
   (e) I, II, III, IV and V

6. Enzymes are important because they
   (a) reduce the activation energy for a reaction
   (b) increase the activation energy for a reaction
   (c) are flexible, and can combine with different substrates.
   (d) use most of the ATP in the cell.
   (e) act as neurotransmitters.

7. The active site on an enzyme is responsible for
   (a) binding with an allosteric effector
   (b) feedback control of the metabolic pathway
   (c) binding with the substrate for a reaction
   (d) the structure of the protein
   (e) binding with toxins to eliminate them from the cell

8. Which of the following is *not* a factor that can affect enzyme shape?
   (a) temperature
   (b) pH
   (c) concentration of the enzyme
   (d) ions
   (e) amino acid substitution

9. Which molecule has the lowest potential energy stored in its chemical bonds?
   (a) glucose
   (b) glucose phosphate
   (c) fructose di-phosphate
   (d) phosphoglycerate
   (e) pyruvate

10. ATP is produced during which reaction?
    (a) glycolysis
    (b) the production of citric acid from oxaloacetate
    (c) photosystem 700
    (d) osmoregulation in the nephron
    (e) the Calvin cycle

11. The Calvin cycle is responsible for producing
    (a) carbon dioxide
    (b) PGAL
    (c) oxygen
    (d) pyruvate
    (e) glucose

12. Lactic acid fermentation starts when
    (a) there is too much lactose in the cell
    (b) oxidative phosphorylation stops and NAD^+ is trapped as NADH

(c) oxygen is present
(d) glyceraldehyde-3-phosphate (PGAL) is phosphorylated
(e) all the co-enzyme A is used up

13. The *synthesis* reaction of photosynthesis takes place in
(a) the thylakoid membrane
(b) the stroma
(c) the matrix
(d) the intrathylakoid space
(e) the lamellae

14. Which of the following statements best explains why photosynthesis is consistent with the laws of thermodynamics?
(a) Most of the Sun's energy is reflected into space.
(b) More energy is stored in glucose than is captured by chlorophyll.
(c) Heat is stored from the Sun.
(d) NADPH uses energy in the Calvin cycle.
(e) Energy from the Sun is transformed into chemical energy.

15. If oxygen levels are increased, which best describes the resulting activity of a green plant?
(a) Photosynthesis rates increase and then drop off.
(b) Carbon dioxide levels increase.
(c) Glucose production decreases.
(d) More ATP will be produced by the *photo* reaction.
(e) Photosynthesis rates increase and level off.

16. Pyruvate can be broken down to form
(a) RuBP
(b) carbon dioxide and ethanol
(c) fructose-di-phosphate
(d) malate
(e) $FADH_2$

17. C_4 plants can continue photosynthesis in a warm climate because
(a) the heat speeds up the Calvin cycle
(b) carbon dioxide is captured by carrier molecules
(c) high oxygen concentrations in cells help glucose production
(d) these cells make ATP directly from light
(e) oxidation-reduction reactions have fewer steps

18. An electron transport chain is best described as a
(a) carrier molecule that is changed by enzymes
(b) cycle of substrate level phosphorylation reactions
(c) chemiosmotic gradient
(d) biochemical pathway
(e) sequence of oxidation-reduction reactions

Short Answers

In your notebook, write a response to answer each of the following questions.

19. How are valence electrons important for chemical reactions?
20. What is a hydrogen bond?
21. Use isomers of glucose to distinguish between structural isomers and stereoisomers.
22. An acid is often called a proton donor. What does this mean?
23. Give examples for each of these reactions: hydrolysis, condensation (dehydration synthesis), and neutralization.
24. Give an example of an oxidation-reduction reaction. Identify which reactant is more electronegative than the other.
25. How is ATP different from the nucleotides that make up DNA?
26. Identify three polar amino acids.
27. How can an acidic environment affect polymers?
28. How is potential energy stored in a biological system?
29. Identify two laws of thermodynamics. Give one biological example for each.
30. How is activation energy important to understanding how reactions work?
31. Using some part of cellular respiration, identify one example of an endothermic reaction and one example of an exothermic reaction.
32. Explain the difference between an oxidative enzyme and a hydrolytic enzyme.
33. How does the active site make an enzyme substrate specific?
34. Enzymes work best under optimal conditions. In terms of temperature, what does this mean?
35. How can an inhibitor affect enzyme activity in a cell?
36. Identify a coenzyme and outline how it participates in a reaction.
37. What is a restriction enzyme and how is it used in industry?
38. How are coupled reactions useful to biological processes?

39. How is the sodium-potassium pump important to cells?

40. Outline glycolysis in terms of phosphorylation and dephosphorylation.

41. List the chemical products and energy output of glycolysis.

42. How is oxidative phosphorylation different from substrate-level phosphorylation? Use an example of each process in your answer.

43. What is the purpose of anaerobic fermentation?

44. How do NADH and $FADH_2$ carry energy within the mitochondrion?

45. How can proteins be used for energy in a cell?

46. How does light energy break the bonds of water molecules in photosynthesis? What is the purpose of this process?

47. List the products of the *photo* reaction in photosynthesis. Which products participate in the *synthesis* reactions?

48. What is the product of the Calvin cycle?

49. How do C_4 plants improve the efficiency of carbon fixation in the Calvin cycle?

INQUIRY

50. Identify the safety supplies you need as you prepare for a laboratory activity. What materials should be on hand in case of a spill?

51. The second law of thermodynamics states that when energy is transformed from one form to another, some energy is lost as waste heat. Bean seeds are large seeds that germinate easily. The germinating seed uses energy stored in the seed, to grow.
    (a) How do you predict this process is consistent with the second law of thermodynamics? Explain.
    (b) Write a procedure for an investigation to test your prediction.
    (c) Make a diagram of the set-up for your investigation.
    (d) List the materials that you need and include all safety precautions you should take.

52. Lactase breaks the milk sugar lactose into glucose and galactose. This enzyme can be considered a catalyst. Catalysts help chemical reactions go forward, but catalysts are not used up in the reaction. Design a laboratory experiment to demonstrate the catalytic ability of an enzyme. Include a hypothesis, basic materials, and a simple procedure.

53. Cells can be broken and then centrifuged to separate organelles, such as mitochondria. If isolated mitchondria were then broken and centrifuged, the membranes would separate from the matrix. Knowing this, develop a hypothesis about the properties of either the matrix or the membranes. Clearly identify what you intend to discover and briefly outline how you would test your hypothesis.

54. You are planning to build a greenhouse for the school. To ensure maximum growth of plants, the greenhouse needs extra carbon dioxide and heat at a reasonable cost. This graph shows the effect of temperature and carbon dioxide on plant growth (as measured by glucose production.) Write a summary of these data, indicating the best temperature and carbon dioxide concentration for the greenhouse. Be clear and precise.

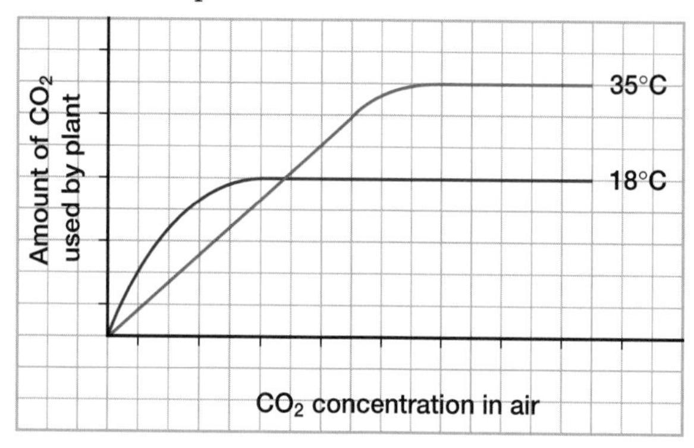

COMMUNICATING

55. Use structural formulas to draw one polar molecule and one non-polar molecule. Explain how the bonding electrons contribute to the polar and non-polar nature of the molecules.
56. Make a chart listing the major macromolecules, an example for each, and a list of the functional groups that are characteristic of each macromolecule.
57. Use diagrams to show how a phospholipid can be partially dissolved in water.
58. Explain how functional groups on glucose are important for the condensation reactions that build a polysaccharide.
59. A polypeptide consists of a long chain of amino acids. Some amino acids have polar or ionizable *R*-groups while other amino acids have non-polar *R*-groups. Use one or more diagrams to show how the orientation of *R*-groups can make a polypeptide insoluble.
60. Make a series of diagrams showing how ATP stores and releases energy in biological reactions.
61. The transition reaction is also called oxidative decarboxylation. Use diagrams to explain what this means.
62. Make diagrams to represent aerobic cellular respiration and anaerobic cellular respiration. Indicate how matter is cycled.
63. Make a diagram of the chloroplast and indicate where each of the following processes takes place: hydrolysis, ATP is produced, NADPH is produced, PGAL is produced.
64. Make a diagram of a mitochondrion and a chloroplast. Show where specific products of metabolism are formed. Draw arrows to link the products of one organelle to the reactants of the other organelle.

MAKING CONNECTIONS

65. Enzymes increase the rates of reactions that take place in the cell. Cells live in near constant conditions. In particular, the temperature of a cell is moderate and steady. The commercial production of biologically important molecules takes place in conditions that are often very different from cell conditions. For example, industrial production of biological molecules often takes place at relatively high temperatures.
    (a) How do enzymes help prevent dramatic changes in the cell's condition?
    (b) How could enzymes help companies that produce biological molecules save energy?
    (c) Identify a disadvantage to commercial molecule production that is *not* similar to cellular molecule production.
66. Glucose is the final product of photosynthesis. Consider that glucose is like a compressed spring. Explain how releasing the spring is like cellular respiration. What factors contribute to the loss of energy in a spring? What factors contribute to the loss of energy in cellular respiration?
67. Identify three snack options: one that is high in fat, one that is high in protein, and one that is high in carbohydrates. Explain how each snack contributes to specific metabolic conditions in the body.

UNIT

2

Unit Preview

In this Unit, you will discover

- the physiological and biochemical mechanisms involved in the maintenance of homeostasis,
- the feedback mechanisms that maintain chemical and physical homeostasis in animal systems, and
- how environmental factors and technological applications affect and contribute to the maintenance of homeostasis, and how they relate to certain social issues.

Unit Contents

UNIT PROJECT PREP

Look ahead to pages 208–209.

- You will be examining the topic of treatments for diabetes at the end of this chapter.
- Start gathering newspaper clippings, magazine articles, videos of television specials, or web site addresses that pertain to this topic.

Homeostasis

In 1775, Dr. Charles Blagden of the Royal Society of London tested the human body's ability to withstand heat. He had a special room heated to 126°C, well above the boiling point of water. He then entered the room with a pet dog and a piece of raw meat. After 45 minutes, he left the room with no ill effects other than a pulse rate that had increased to 144 beats per minute (about twice its normal rate). The dog, too, was fine, but the meat had been cooked.

As the results of Dr. Blagden's experiment indicate, the body has developed physiological and biochemical mechanisms that allow it to maintain its internal physiological environment in a relatively stable state. It is able to do so in the face of external stresses such as high or low temperatures and internal stresses such as fever and infection. There are limits to the extent of stress with which these mechanisms can cope, however. For example, long workouts in unusually hot, humid conditions can result in heat exhaustion, which, in one case, led to the death of a football player.

Individual differences also play a significant role in determining the effectiveness of the body's mechanisms for regulating its internal environment. Young children, the elderly, and heavy individuals, for instance, are more likely to suffer from heat exhaustion than others. In this unit, you will see how these mechanisms on which we all depend, in spite of individual differences, work together to maintain the stable state within us.

The very young and the very old are more susceptible to internal and external stress.

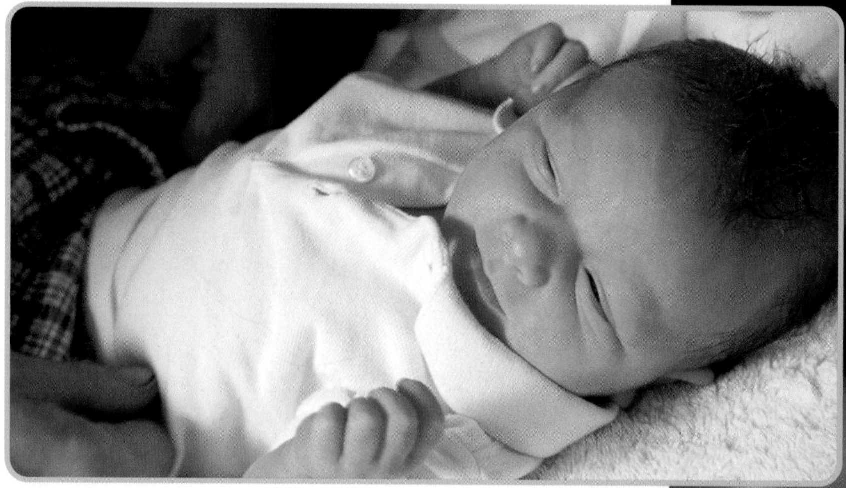

CHAPTER 4

Homeostatic Mechanisms

Reflecting Questions

- How do our bodies maintain a stable internal physiological state despite constant changes in our external environment?
- How do kidneys help keep our blood in a healthy state?
- How does the body recover from illnesses?
- Why do some people develop serious allergic reactions to environmental substances?
- Why do people with diabetes have to control the amount of sugar they consume each day?

Prerequisite Concepts and Skills

Before you begin this chapter, review the following concepts and skills:

- describing the unique composition and properties of macromolecules found in cells and tissue fluids (Chapter 1, section 1.2),
- identifying and explaining how some substances move across cell membranes by passive or active transport (Chapter 1, section 1.5; Chapter 2, section 2.3),
- describing how buffers maintain stable pH in solutions (Chapter 1, section 1.3).

As shown in the photograph on the opposite page, some people live in extreme conditions: the dry heat of the Sahara desert, the humid heat of the Amazon jungle, or the freezing cold of the Canadian Arctic in winter. Humans manage to survive freezing cold and blistering hot temperatures — not a bad claim for a thin-skinned, warm-blooded, relatively hairless species that is composed of about 70 percent water!

The ability of humans to survive almost anywhere on the planet is one of the things that make us successful as a species. In spite of the external environment, the human body strives to keep its internal environment relatively constant — a tendency known as **homeostasis**. This tendency, coupled with the aid of appropriate food, clothing, and shelter, helps us to survive even under extreme conditions.

Healthy people around the world share certain bodily constants:

- a blood glucose concentration that remains at about 100 mg/mL,
- a blood pH near 7.4,
- a blood pressure of about 160/106 KPa (120/80 mm Hg), and
- a body temperature of approximately 37°C.

Of course, there are slight differences. A normal body temperature for one person in one situation is not necessarily normal for all people in all situations. Metabolism varies from person to person, and therefore so does temperature. Body temperature can also vary with time, place, and activity level, as shown below, and still be considered normal.

"Normal" blood pressure also differs. Children tend to have lower blood pressure than adults, due to the flexibility of their arteries. As people age, their arteries lose some of this flexibility, and their blood pressure increases.

In this chapter, you will see how the human body adapts to changing internal and external environments, and examine some of the organs and systems that help us maintain homeostasis. In Chapters 5 and 6, you will explore the role played by the nervous and endocrine systems.

Normal body temperature fluctuates slightly according to a variety of factors.

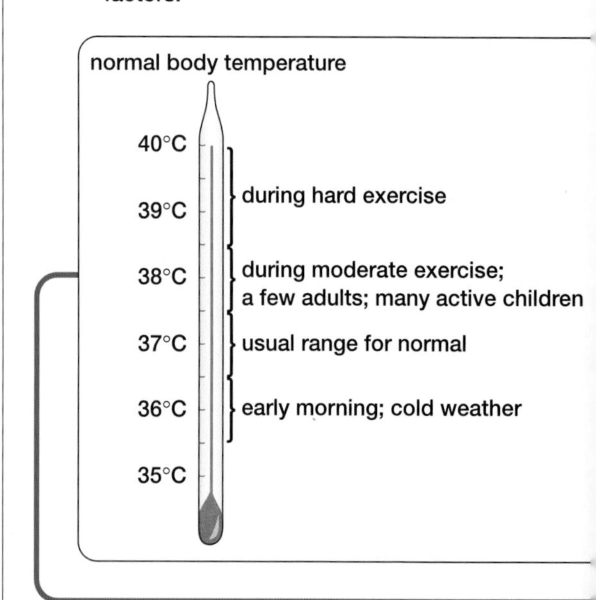

Chapter Contents

4.1 Feedback Systems

EXPECTATIONS

- **Describe and explain homeostatic processes at work in the body and the systems that govern them.**
- **Begin to develop a model that illustrates the essential components of the homeostatic process.**

The cells of the human body are surrounded by a fluid having a salinity, or salt concentration, nearly equal to that of the Earth's oceans from which life began. The bearer of this evidence of our evolutionary past is called interstitial fluid. The salinity and pH of the **interstitial fluid**, or fluid in the pores between our cells, must be relatively constant in order for the cells of the body to remain healthy. This maintenance of constant extra-cellular conditions is a function of homeostasis, which is an example of dynamic equilibrium.

As used here, **dynamic equilibrium** is a state of balance achieved within an environment as the result of internal control mechanisms that continuously oppose outside forces that tend to change that environment. For example, consider a house equipped with a furnace, an air conditioner, and a thermostat that controls both (see Figure 4.2). The furnace comes on when the temperature drops. When the temperature increases, however, the air conditioner turns on instead. The thermostat measures the temperature in the house and turns either appliance on or off as needed. Although the temperature in the house remains relatively constant, this constancy is achieved through a series of small temperature changes. Equilibrium is thus maintained as long as the system is active (dynamic). While the system functions to maintain a desired temperature, large temperature changes inside and outside the house might cause minor fluctuations above and below that level.

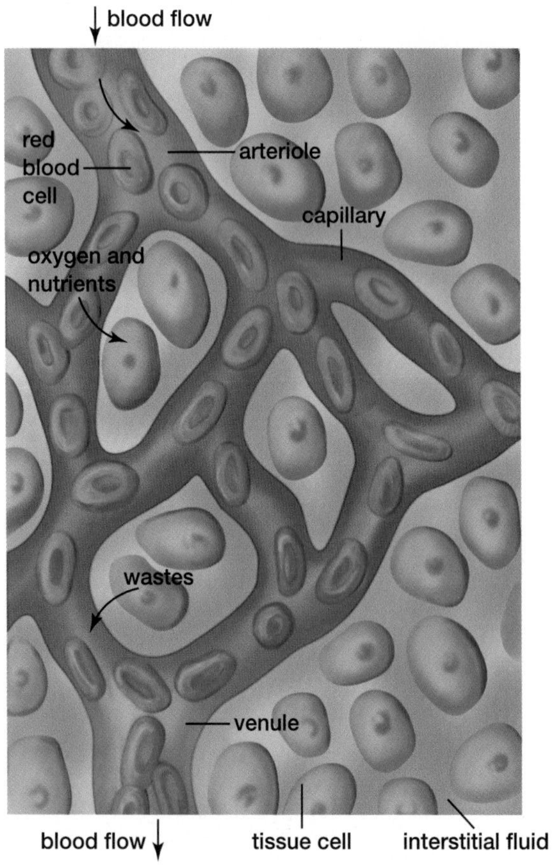

Figure 4.1 Cells are surrounded by interstitial fluid, which is continually refreshed. As shown, oxygen and nutrient molecules constantly exit the bloodstream, and waste molecules continually enter the bloodstream.

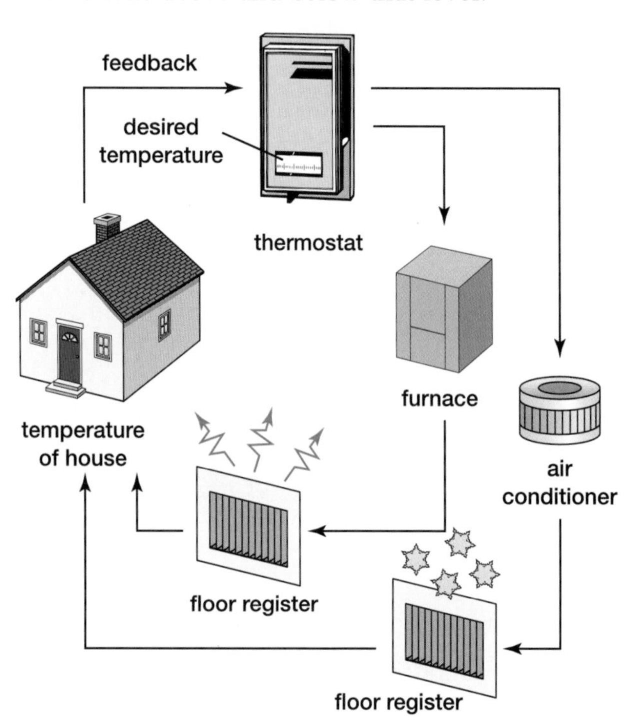

Figure 4.2 While a house may have a thermostat, furnace, and air conditioner to keep its internal temperature constant, the human body has a considerably more complex system to maintain homeostasis.

Human body temperature is also an example of dynamic equilibrium. However, homeostasis is much more complex than the temperature control mechanisms in a house. The human body features a series of control mechanisms to keep its temperature at approximately 37°C. For example, when your skin is cold, "goose bumps" appear. Goose bumps are an attempt by the body to fluff up non-existent fur. The fluffed-up fur of other mammals traps air that acts as an extra layer of insulation. This reaction is similar to birds fluffing up their feathers in winter. In humans the reaction is not very effective because we have minimal body hair.

A second control mechanism is shivering. When we shiver, much of the energy processed by the muscle action is given off as heat, which is transferred throughout the body. Shivering is not something we consciously control. Rather, it is the involuntary contraction of muscles to generate heat when the body detects a drop in temperature (see Figure 4.3).

A third control mechanism employed to maintain internal temperature is the transfer of heat energy to the body's most critical parts. When this mechanism, known as **vasoconstriction**, occurs, the diameter of blood vessels near the skin is reduced and blood circulation is concentrated in the core of the body to keep the major organs functioning. If the body's temperature continues to drop, **hypothermia** results. At this point, the body no longer has the energy to shiver; extremities such as fingers, toes, and ears begin to freeze; and blood flow to the brain decreases, resulting in impaired judgement, sleepiness, and eventual loss of consciousness. Ultimately, death may occur.

To maintain homeostasis, the body also frequently needs to rid itself of excess heat. It accomplishes this in two ways. First, as shown in Figure 4.4, sweat is released onto the surface of the skin. Energy in the form of heat is then given off as the sweat evaporates. Second, the body reverses the mechanism of vasoconstriction. In a process called **vasodilation**, blood vessels near the skin dilate to transport more blood (and thus heat) to the skin's surface, where the heat is released. If excess heat cannot be eliminated fast enough, heat exhaustion and **hyperthermia**, or an unusually high body temperature, may result. As with hypothermia, hyperthermia can cause death.

BIO FACT

Occasionally hypothermia can save lives. Some people who have fallen into very cold water have been revived up to one hour later. In most cases, drowning victims suffer brain damage after five minutes due to oxygen deprivation. However, hypothermia slows down bodily processes, reducing the demand for oxygen by the brain. Victims of near-drowning in cold water may thereby escape brain damage.

Figure 4.3 The muscle action of shivering generates heat, which is then distributed throughout the body.

Figure 4.4 To rid itself of excess heat, the body releases sweat onto the skin. Cooling occurs as the sweat evaporates.

Feedback Loops

The body's reactions to increased or decreased temperature are examples of negative feedback loops. A **negative feedback loop** is a process that detects and reverses deviations from normal body constants. This process involves three parts: a receptor, an integrator, and an effector.

Sensory receptors are found throughout every body organ and tissue. The function of the **sensory receptors** is to send nerve impulses (stimuli) to the brain in response to environmental information. They monitor the body's internal conditions, such as temperature, blood pH, blood sugar, and blood pressure, on a continual basis. For instance, when the body's temperature begins to drop or rise, specialized receptors in the skin detect the change and signal the hypothalamus (a part of the brain) accordingly.

The brain, similar to the thermostat in a house, is an **integrator** — it sends messages to effectors (analogous to the furnace or air conditioner in a house). **Effectors**, in turn, cause a change in internal conditions. In this case, the brain sends messages to various tissues and organs that cause the body to either generate or conserve heat (see Figure 4.5). These messages can be transmitted by the nervous system (the topic of Chapter 5) or by chemical messengers known as hormones, which you will study in greater depth in Chapter 6.

Negative feedback loops exist throughout the body to maintain homeostasis. These systems prevent blood sugar, blood pressure, temperature, and other body constants from becoming too high or too low (see Figure 4.6).

Positive feedback loops also exist, but they are usually associated with disease or change (for example, drug addiction) and are therefore rare in healthy bodies. Another example of a positive feedback loop is high blood pressure. Damage to arteries due to high blood pressure results in the formation of scar tissue. This scar tissue traps cholesterol, which impedes the flow of blood through the arteries and thereby increases blood pressure even more.

Positive feedback loops act to increase the strength of the stimuli, whereas negative feedback loops decrease or reverse effects of the stimuli. A negative feedback loop moves a system toward balance and stability, while a positive feedback loop pushes a system away from balance and

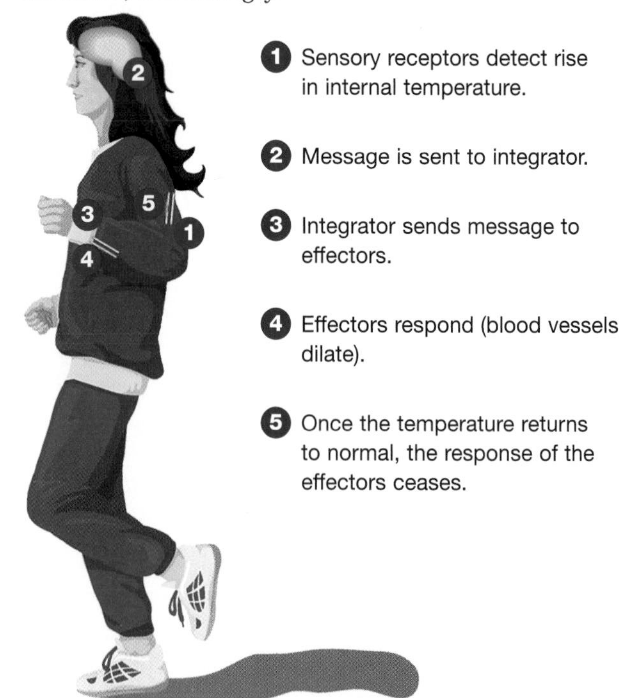

Figure 4.5 When you exercise, sensory receptors detect your body's rise in internal temperature and send a message to the brain (the integrator). The brain then sends a message to the blood vessels (the effectors) to dilate. The dilation causes more blood to course through the blood vessels. The double lines in the diagram indicate that once the temperature returns to normal, the effectors' response stops.

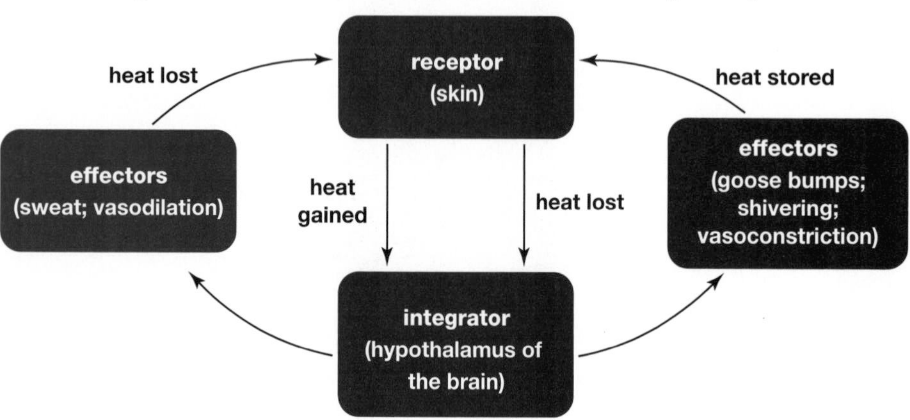

Figure 4.6 Negative feedback loops prevent internal body temperature from becoming too high or too low.

stability. For cells, tissues, and organisms that require constant conditions to function normally, positive feedback is usually disastrous, causing serious health issues. An exception occurs in the process of childbirth, as shown in Figure 4.7. Near the end of a pregnancy, the baby's head pushes against the opening of the uterus. The pressure on the uterine muscles triggers the contraction of these muscles and sends messages to the hypothalamus, which releases a hormone (oxytocin) that increases the strength of the next contraction. Each subsequent contraction pushes the baby's head farther into the uterine opening, which triggers successively stronger contractions until the baby is born. (You will learn more about the role of the hypothalamus and hormones in Chapter 6.)

In this section, you have looked at how the body's negative feedback systems affect its ability to maintain a stable state. In the next section, you will learn about the kidney's role in homeostasis.

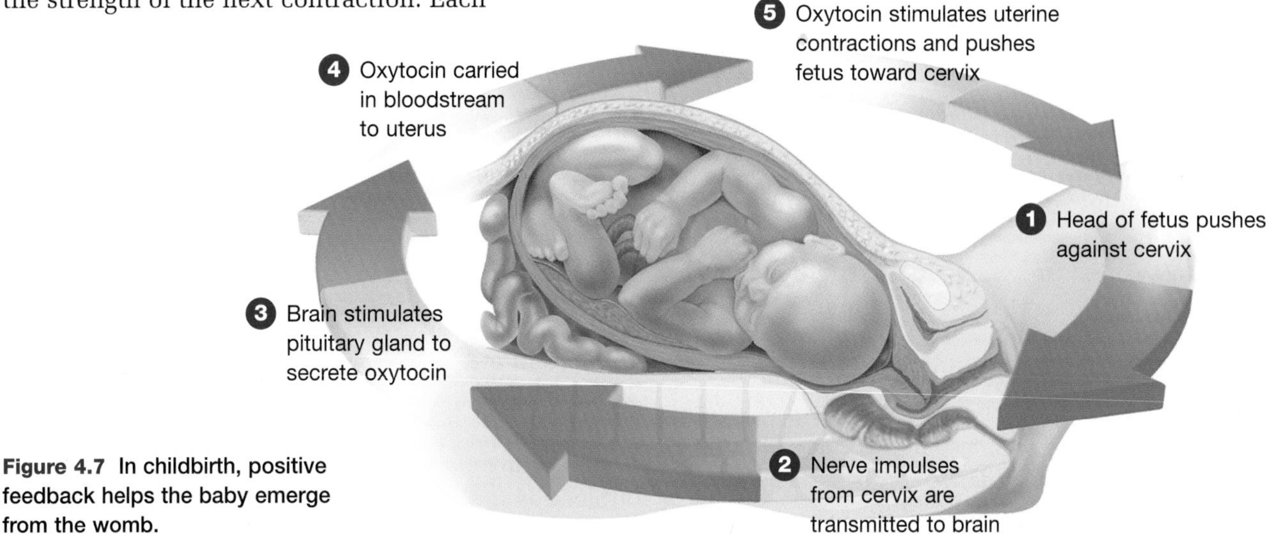

Figure 4.7 In childbirth, positive feedback helps the baby emerge from the womb.

SECTION REVIEW

1. K/U Explain how the brain helps to regulate body temperature.
2. K/U Provide one example of dynamic equilibrium in the body.
3. K/U Describe the role of the skin in maintaining homeostasis in the body.
4. K/U Explain the difference between hypothermia and hyperthermia.
5. I Examine the graph on the right showing the relationship between body temperature and ambient (environmental) temperature in a cat and a lizard. How is the cat able to maintain a more constant body temperature than the lizard?
6. K/U Explain the difference between a negative feedback loop and a positive feedback loop.
7. C Explain how illness can upset the dynamic equilibrium of the body. Use a flowchart to show how feedback mechanisms help or hinder the body to recover from illness.
8. MC Some people experience symptoms of mountain sickness when they travel to high altitude locations. Symptoms can range from mild headaches and decreased appetite to more serious symptoms, including confusion, hallucinations, shortness of breath, and (in extreme cases) coma. In most cases, people eventually adjust to the change in altitude without lingering side effects. Do some research to determine the homeostatic mechanisms involved in adjusting to high altitude conditions. If you were planning a mountain climbing expedition, what precautions would you recommend to help you and the other travellers avoid developing mountain sickness?

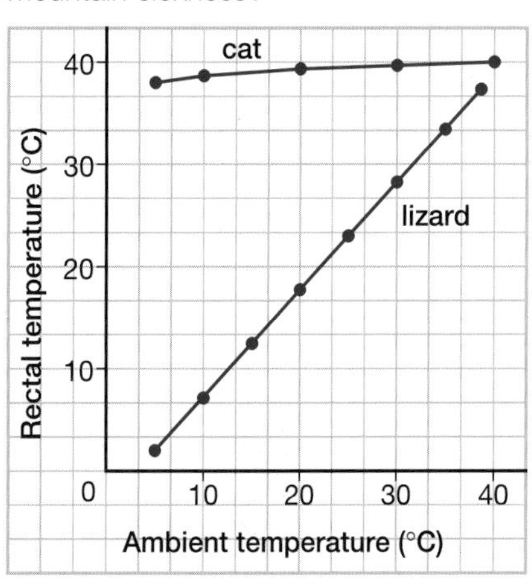

4.2 The Kidney

EXPECTATIONS

- Explain the role of the kidney in maintaining water and ion balance.
- Design and carry out an experiment to investigate the physiological effects experienced by people who consume coffee.
- Describe issues and present informed opinions about problems related to kidney functions and kidney transplants.
- Describe the contribution made to knowledge and technology in the field of homeostasis by Dr. Gordon Murray's development of the kidney dialysis machine.

Humans have two fist-sized kidneys, which are found in the lower back on either side of the spine. The kidneys release their waste product (urine) into tubes called **ureters**, which carry the fluid to the urinary bladder where it is temporarily stored (see Figure 4.8). The bladder can hold a maximum of about 600 mL of fluid. When there is about 250 mL of urine in the bladder, we become aware of it, and at 500 mL we become very uncomfortable. Drainage from the bladder is controlled by two rings of muscles called **sphincters**. One sphincter is involuntarily controlled by the brain. During childhood we learn to voluntarily control the other. Urine exits the bladder through a tube called the **urethra**. In males, the urethra is approximately 20 cm long and merges with the vas deferens from the testes to form a single urogenital tract. In females, the urethra is about 4 cm long and the reproductive and urinary tracts have separate openings.

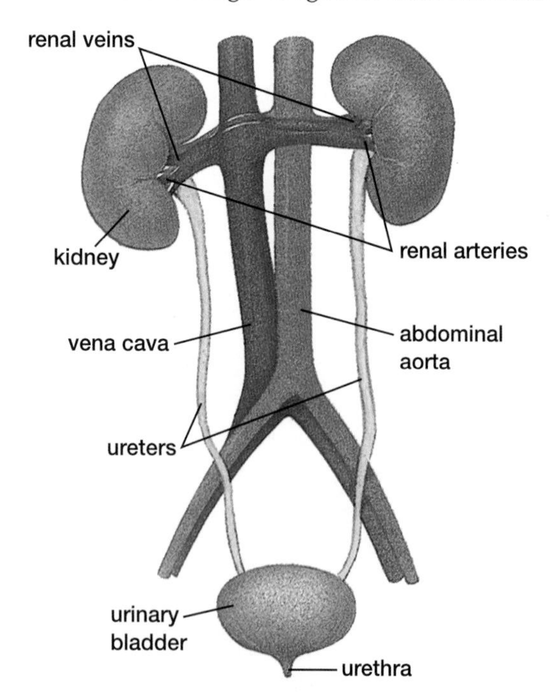

Figure 4.8 The human urinary system

Kidney Functions and Structure

The kidney's principal function is to filter the blood in order to remove cellular waste products from the body. The essential connection between the kidney and the blood is illustrated by the fact that at any given time, 20 percent of the body's blood is in the kidneys. Although most people have two kidneys, the human body is capable of functioning with only one. If one kidney ceases to work or if a single kidney is transplanted into a patient, the functioning kidney increases in size to handle the increased workload.

Although the kidney has other important functions, this organ is usually associated with the excretion of cellular waste. The main metabolic wastes are urea, uric acid, and creatinine, all of which have nitrogen as a major component. Figure 4.9 shows the formula by which urea is produced in the liver from the breakdown of excess amino acids that are the building blocks of proteins. The amine group (NH_2) is removed to release the rest of the amino acid molecule, which can then be converted to carbohydrates or fats. Although the amine group can combine with a hydrogen ion to form toxic ammonia, the ammonia is transformed in the liver into the less toxic urea before being released into the bloodstream.

$$\underset{\text{ammonia}}{2\,NH_3} + \underset{\text{carbon dioxide}}{CO_2} \longrightarrow \underset{\text{urea}}{H_2N-\overset{\overset{\displaystyle O}{\|}}{C}-NH_2} + H_2O$$

Figure 4.9 The liver combines ammonia with carbon dioxide to form urea and water.

Canadians in Biology 

Cellophane and Imagination — Dr. Gordon Murray

On December 6, 1946, Dr. Gordon Murray was called into Toronto General Hospital. A female patient in a coma was wheeled into the room on a stretcher. She was "uremic," which means her kidneys were not functioning and the toxins in her blood were poisoning her body. Staff from all over the hospital watched as Murray set up an innovative apparatus. He inserted one end of 46 m of tubing into a vein in one of the patient's legs and the other end into an artery in her other leg. He then turned on a pump. The patient's blood immediately began to circulate through the tubing, which was strapped to a cylinder, immersed in a bath, coiled around and around, and then diverted back to the patient. Regular readings were taken of both the patient's blood and the bath solution to see if the toxins were leaving the blood. As time passed, the readings proved the device was working. The patient regained consciousness in six hours.

Dr. Gordon Murray

The Steps to Success

Dr. Gordon Murray was born in 1894 in Stratford, Ontario. He was awarded the Order of Canada in 1967 as the first North American to develop and use kidney dialysis (also known as "hemodialysis"). Dr. Murray became renowned as a hard-working and imaginative scientist.

An impediment to successful kidney dialysis was how to get impure blood outside the body and keep it from clotting. In 1935, Dr. Murray pioneered the use of the drug heparin, an anticoagulant that keeps blood from clotting and keeps sutures pliable after surgery. Heparin brought Dr. Murray one step closer to successful dialysis.

The second problem to overcome was how to filter impurities from the blood. Dr. Murray considered using the principle of osmosis. He looked for a material that was porous enough to let the smaller, toxic particles in the blood pass through into another solution while keeping in the larger particles (plasma).

In a series of experiments, Murray tried a variety of materials, including leather and nylon, to form the tubing for his machine. Finally, he tried cellophane designed for use as sausage casing. It was the ideal filter; it retained the important contents of blood while allowing the toxins to escape into a solution in which the tubing was immersed.

In Dr. Murray's first experiment on an animal, he attached one end of his tubing to an artery and the other to a vein, letting the heart pump the blood from the artery through the tubes and back into the vein. The experiment was successful — the toxins left the animal's blood. However, recalling that blood returning *to* the heart *from* the body's extremities (venous blood) is the most impure blood, Murray developed a pump that would simulate the heart's action and pump the blood the opposite way through the tube.

A Partial Victory

Today, kidney dialysis (or the "artificial kidney") enables people with kidney disease to live relatively normal lives. However, these patients must undergo regular dialysis sessions — usually three six-hour sessions per week. Researchers are striving to invent a machine to make this labour-intensive process obsolete.

Another waste product found in the blood is uric acid, which is usually produced by the breakdown of nucleic acids such as DNA and RNA. Creatinine is a waste product of muscle action. All of these waste products are potentially harmful to the body and therefore must be removed.

The kidneys are more than excretory organs; they are one of the major homeostatic organs of the body. In addition to filtering the blood to remove wastes they also control the water balance, pH, and levels of sodium, potassium, bicarbonate, and calcium ions in the blood. They also secrete a hormone (erythropoietin) that stimulates red blood cell production, and they activate vitamin D production in the skin. Since the kidneys are involved with so many of the body's functions, the analysis of a urine sample can tell a physician a great deal about a patient. For example, diabetes and pregnancy can be determined using a urine test.

Each kidney is composed of three sections — the outer **cortex**, the **medulla**, and the hollow inner **pelvis** where urine accumulates before it travels down the ureters. These sections are shown to the left of Figure 4.10. Within the cortex and medulla of each kidney are about one million tiny filters called **nephrons**. As Figure 4.10 illustrates, each nephron consists of five parts — the **Bowman's capsule**, the **proximal tubule**, the **loop of Henle**, the **distal tubule**, and the **collecting duct**. The upper portions of the nephron are found in the renal cortex, while the loop of Henle is located in the renal medulla. The tubes of the nephron are surrounded by cells, and a network of blood vessels spreads throughout the tissue. Any material that leaves the nephron enters the surrounding cells and eventually returns to the bloodstream through the network of blood vessels. By controlling what leaves and what remains in the nephrons, the kidneys keep the levels of water, ions, and other materials nearly constant and within the limits necessary to maintain homeostasis.

Blood enters the cavity of the ball-shaped Bowman's capsule through a tiny artery that branches to form a network of porous, thin-walled capillaries called the **glomerulus**. Under the influence of blood pressure, some blood plasma and small particles are forced out of the capillaries and into the surrounding capsule. Larger blood components, such as blood cells and proteins, remain in the capillaries. The fluid in the Bowman's capsule is called **nephric filtrate**, and it is pushed out of the capsule into the proximal tubule. About 20 percent of the blood plasma that enters the kidney becomes nephric filtrate.

BIO FACT

People who are trying to increase muscle mass sometimes use a diet high in proteins or amino acids. The problem with a diet like this is that it creates an excess of amino acids, which are broken down in the liver to form the carbohydrates necessary for metabolism. The excess amine groups in turn produce high levels of urea, which is released into the bloodstream to be removed by the kidneys. However, high urea levels can damage the kidneys, so it is necessary to find ways to bring these levels down. The simplest way to accomplish this is to drink plenty of fluids.

When the nephric filtrate enters the proximal tubule, re-absorption begins. **Re-absorption** is the

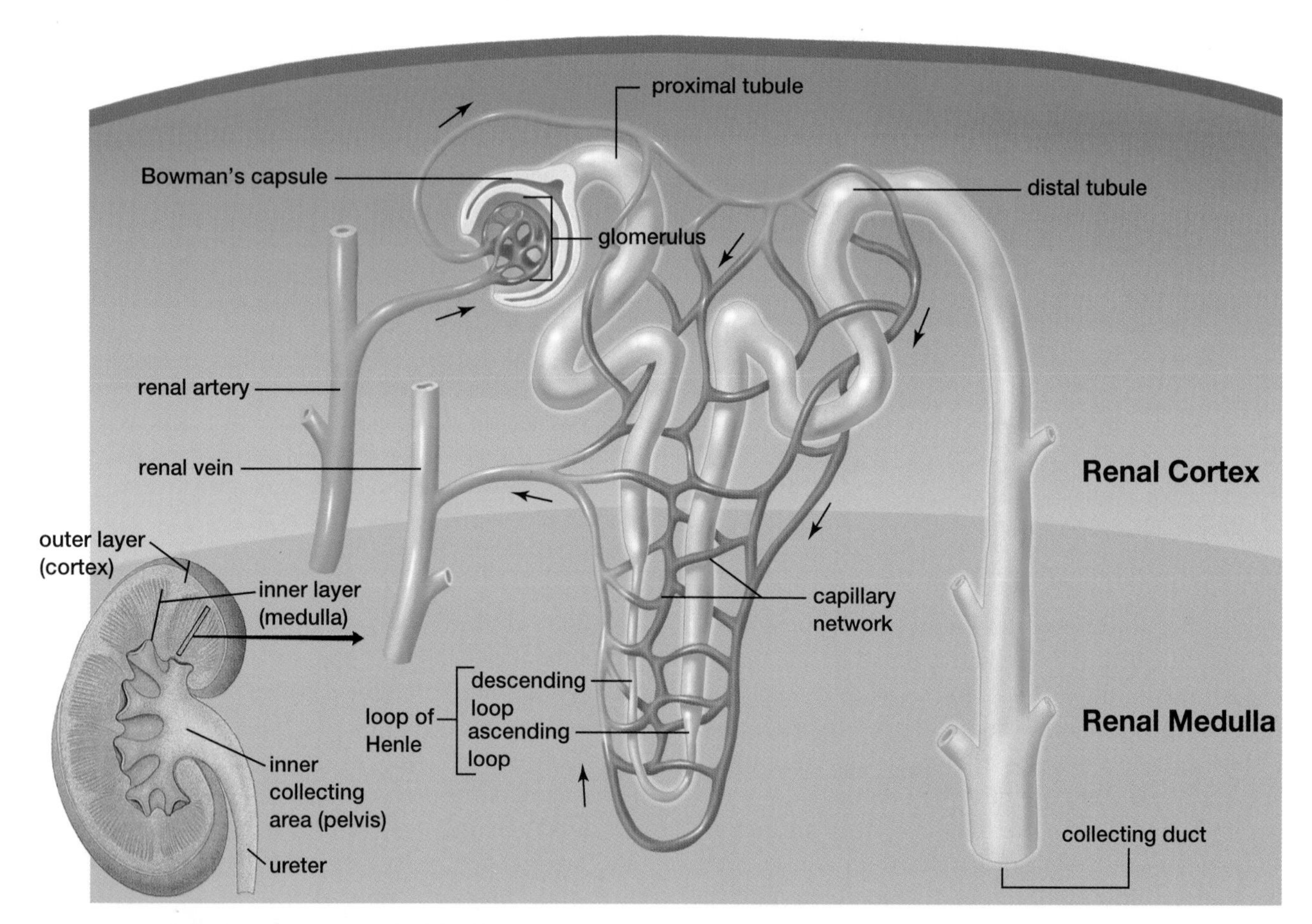

Figure 4.10 A nephron is composed of the Bowman's capsule, the proximal tubule, the loop of the nephron (called the loop of Henle), the distal tubule, and the collecting duct.

process by which materials required by the body are removed from the filtrate and returned to the bloodstream. Osmosis, diffusion, and active transport draw water, glucose, amino acids, and ions from the filtrate into the surrounding cells. From here the materials return to the bloodstream. This process is aided by active transport of glucose and amino acids out of the filtrate. The lining of the proximal tubule is covered with tiny projections (like the villi of the small intestine) to increase the surface area and speed up the process of re-absorption. When the filtrate reaches the end of the proximal tubule, the fluid is isotonic with the surrounding cells, and the glucose and amino acids have been removed from the filtrate. We say a fluid is **isotonic** when it has the same concentration of water and solutes as that in the cells surrounding it.

From the proximal tubule, the filtrate moves to the loop of Henle. The primary function of the loop of Henle, which first descends into the inner renal medulla and then turns to ascend back towards the cortex, is to remove water from the filtrate by the process of osmosis (see Figure 4.11). The cells of the medulla have an increased concentration of sodium ions (Na^+). These ions increase in a gradient starting from the area closest to the cortex and moving toward the inner pelvis of the kidney. This increasing gradient acts to draw water from the filtrate in the loop of Henle. This process continues down the length of the descending loop due to the increasing level of Na^+ in the surrounding tissue. You will observe a similar process in the MiniLab below.

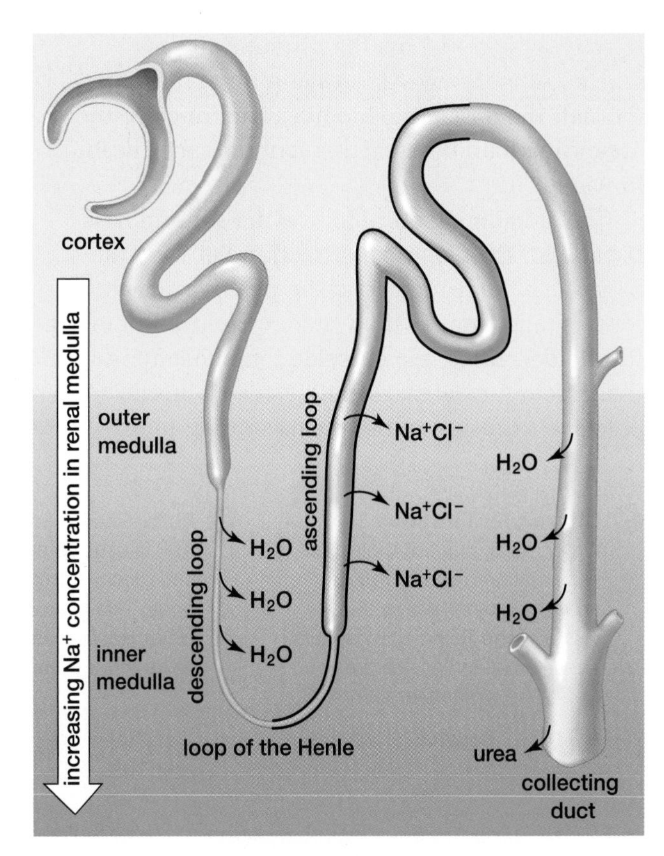

Figure 4.11 As the filtrate travels down the descending loop of Henle, water moves out by osmosis. What prevents the water from being re-absorbed into the ascending loop?

The high levels of Na^+ in the surrounding medulla tissue are the result of active transport of Na^+ out of the ascending loop of Henle. The amount of water removed from the filtrate by the time it reaches the bottom of the loop of Henle results in an increased concentration of all of the materials dissolved in the remaining filtrate, including Na^+. Thus, as the filtrate moves up the ascending loop of Henle, Na^+

MINI LAB

The Effect of Salt Concentration

Common table salt (NaCl) is an important component of cells and the fluids that surround them in the body. Life evolved in the salt water environment of the ocean and salt plays an important role in cellular function. The concentration of salt affects osmosis or the movement of water in the body's cells. This movement can be demonstrated in other cells, for example, in those of an onion. In this MiniLab, you will use a piece of coloured onion to observe the effect of increased salt concentration. Make a wet mount slide of a thin layer of red onion skin. Draw a diagram of two of the cells that you can see at 100x or 400x. Label the cell wall, cell membrane, cytoplasm, and nucleus (if visible). This usually works better if you use the microscope's diaphragm to decrease the amount of light. Lift the cover slip and put two drops of saturated salt (NaCl) solution on the onion cells. After two minutes, examine the onion cells again.

CAUTION: Handle the microscope slides and cover slips carefully. Wash your hands after completing the MiniLab.

Analyze

1. Draw two of the cells that have changed. Label the cell wall, cell membrane, cytoplasm, and nucleus.
2. Explain why the cytoplasm changed in the presence of the salt solution.

is actively pulled from the filtrate into the surrounding tissue. At the same time, the water that left the descending loop cannot re-enter the ascending loop because this loop is impermeable to water.

Chloride ions tend to follow the sodium ions because of the electrical attraction between the negative chloride ions and the positive sodium ions. In addition, as the water concentration in the filtrate decreases, the chloride ion concentration in the filtrate increases, resulting in still more chloride diffusion out of the ascending loop.

As shown in Figure 4.12, the distal tubule is responsible for a process called tubular secretion. **Tubular secretion** involves active transport to pull substances such as hydrogen ions, creatinine, and drugs such as penicillin out of the blood and into the filtrate. The fluid from a number of nephrons moves from the distal tubules into a common collecting duct, which carries what can now be called urine into the renal pelvis. At that point, 99 percent of the water that entered the proximal tubule as nephric filtrate has been returned to the body. In addition, nutrients such as glucose and amino acids have been reclaimed.

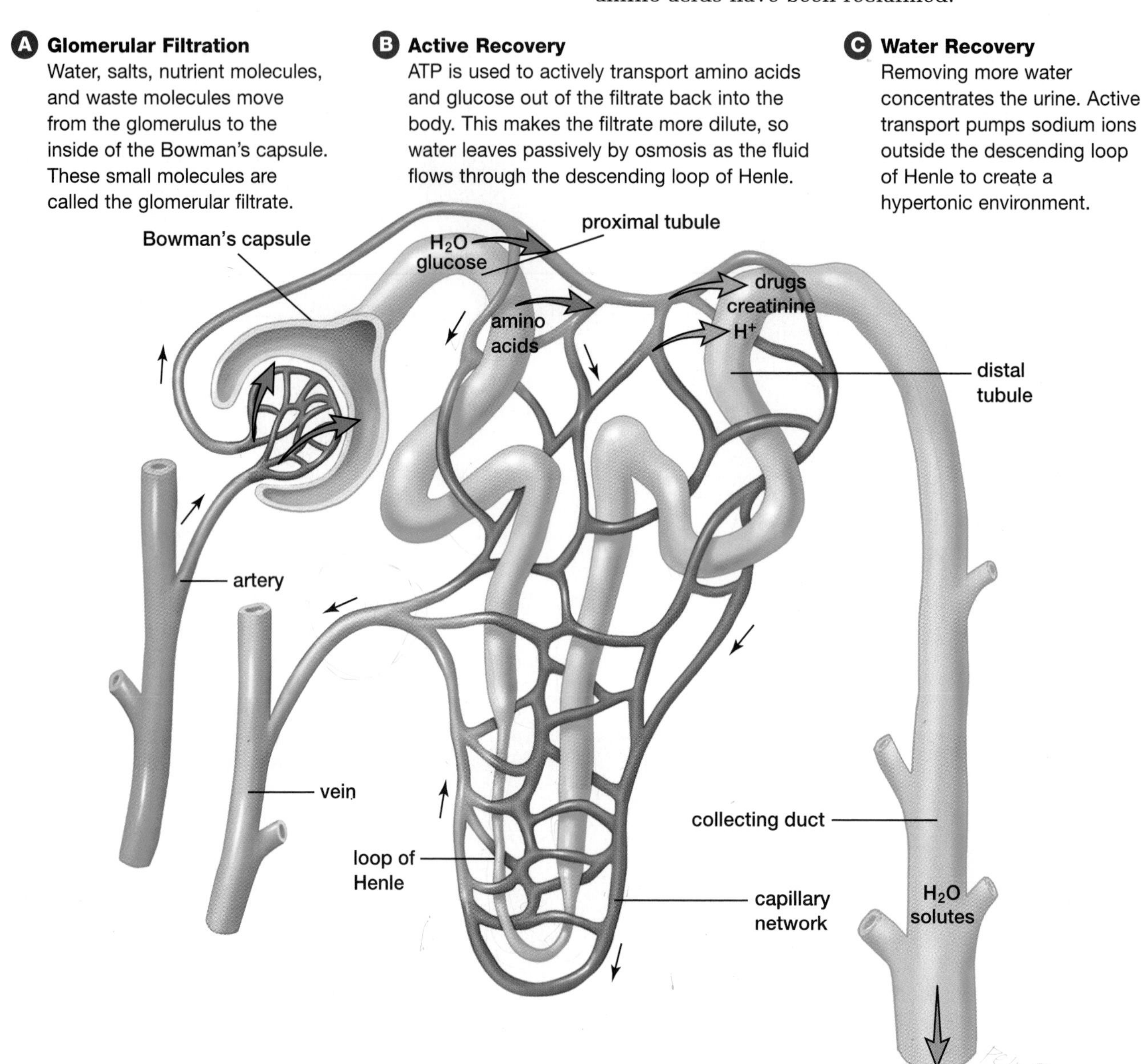

Figure 4.12 Active and passive transport are both used to maintain a balance of solutes and water. At A, the pressure of the blood flowing into the glomerulus pushes solutes and water into the Bowman's capsule. At B, active transport is used to recover amino acids and glucose. This makes the filtrate relatively dilute, so water also moves out of the nephron. At the bottom of the loop of Henle, the filtrate is almost isotonic. Diffusion of urea and the active transport of sodium ions out of the ascending loop of Henle creates a relatively hypertonic environment. At C, water can leave the distal tubule, resulting in more concentrated urine.

Urine Output

The permeability of the distal tubule and collecting duct is controlled by a hormone called anti-diuretic hormone (ADH). ADH is secreted by a gland attached to the hypothalamus called the pituitary gland. ADH increases the permeability of the distal tubule and collecting duct, thus allowing more water to be removed from the nephric filtrate when the body has a need to conserve water.

The pituitary, as you will see in Chapter 6, is controlled by the hypothalamus. As shown in Figure 4.13, the hypothalamus acts to regulate the body's feedback systems. When the body needs to eliminate excess water, ADH is inhibited and more water is excreted in the urine. Drugs such as alcohol and caffeine block the release of ADH and increase the volume of urine. Increased urine output can also be a symptom of conditions such as diabetes. In people who have diabetes, the increased level of blood sugar can overload the active transport system of the proximal tubule, which causes glucose to remain in the nephric filtrate as it moves through the loop of Henle, distal tubule, and collecting duct. The glucose retains water in the filtrate, offsetting the system that is designed to remove it. The result is that large volumes of sugary urine are produced, which is one of the major symptoms of diabetes.

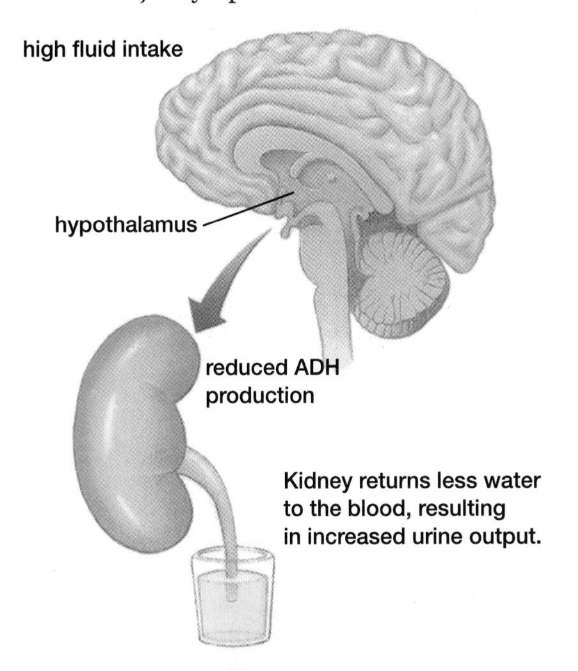

Figure 4.13 When you drink a large amount of water, the fluid level in your blood vessels increases. This increase triggers the hypothalamus to slow down production of ADH. As a result, you eliminate more water. When the water level in the blood drops too low, the hypothalamus produces more ADH.

BIO FACT

Penicillin is an acid that the body actively secretes into urine. About four hours after penicillin is ingested, 50 percent of the penicillin in the blood is secreted and removed from circulation. In the early days of penicillin use the drug was difficult to obtain, so hospitals recycled penicillin by collecting patients' urine and separating the penicillin for re-use.

ELECTRONIC LEARNING PARTNER

For more information about how the kidney functions, refer to your Electronic Learning Partner.

WEB LINK

www.mcgrawhill.ca/links/biology12

The kidney is a vital organ. Unfortunately, various diseases of the kidney and other medical conditions (such as diabetes) can seriously impair normal kidney function. Patients who experience loss of kidney function must maintain a continual program of regular hemodialysis.

You learned about hemodialysis in Canadians in Biology on page 113. To access articles describing the process of hemodialysis, go to the web site above, and click on **Web Links**. Describe the major causes of kidney failure. How does hemodialysis compensate for normal kidney function? Explain how an artificial kidney removes waste products from a patient's blood. Make a sketch to illustrate how substances are filtered out of the blood in an artificial kidney. Describe the health risks associated with hemodialysis treatment.

Blood pH and the Kidney

The kidneys regulate the acid-base balance of the blood. To remain healthy, our blood pH should stay around 7.4, which is slightly basic. One way in which blood pH is controlled at this level is by regulation of the active transport of hydrogen ions (H^+) into the nephric filtrate. If blood pH fluctuates, the secretion of H^+ either slows or increases until the pH returns to normal. As a result of this fluctuation, urine can have a pH as low as 4.5 or as high as 8.0. Normally, urine has a pH of about 6.0.

While the kidneys are ultimately responsible for the removal of excess hydrogen ions from the blood, the respiratory system works with the kidneys to help maintain the pH of the blood at 7.4. The two systems depend on chemicals called **buffers** to

Continued on page 120 ➡

Biology Magazine — TECHNOLOGY • SOCIETY • ENVIRONMENT

Kidney Transplants

When the first kidney transplant occurred 40 years ago, it was a major medical breakthrough. In the year 2000 there were 1112 kidney transplants performed in Canada. Of these, 724 used cadaveric (deceased) donors and 388 used living donors.

The major problem with transplant surgery is not in the operating room — it is in finding suitable donors. In Canada, most donors are victims of stroke or head trauma (often related to motor vehicle accidents) who are being maintained on a ventilator. Only two to three percent of all deaths in Canada are the result of brain death, and that pool of potential donors is further limited by the fact that hospitals require permission from the relatives of the donor to perform a transplant. Ninety-six percent of relatives agree to organ donation if they know the wishes of the potential donor, while only 58 percent agree if the issue has not been discussed in advance.

Canada has one of the lowest organ donation rates among industrialized countries. There are fewer than 14 donors per million in Canada, compared with more than 30 per million in Spain. Why? This disparity is partly due to Canada's "opting in" policy — donors and/or their relatives must give permission for transplantation of organs. In contrast, many European countries have an "opting out" policy, in which permission is assumed unless the potential donor has specifically requested not to be an organ donor.

What Are the Ethical Considerations?

The use of living donors is a growing trend. People can survive with only one kidney, but the donation of a kidney by a living donor creates a number of ethical problems for the medical community. There are risks associated with any surgery — should a doctor risk the life of a healthy person in an attempt to aid someone who is ill? In addition, there is the problem of "informed consent." Can a doctor be sure that a living donor is a willing participant and is not the victim of pressure from relatives? Can a doctor proceed with the operation if he or she suspects that the donor is receiving some form of benefit in exchange for donating a kidney?

An additional ethical problem is looming on the biotechnological horizon. Pig tissue (not organs) has been used in clinical trials to replace damaged human tissue. Pig organs are similar in size and shape to their human counterparts, which makes pigs good candidates for

DESIGN YOUR OWN **Investigation** 4 • A

SKILL FOCUS

- Hypothesizing
- Initiating and planning
- Performing and recording
- Analyzing and interpreting
- Communicating results

The Physiological Effects of Coffee

This investigation gives you an opportunity to explore one of the body's feedback systems. You will discover how coffee, which is consumed by millions of Canadians each day, affects the homeostatic processes of the human body. Coffee contains caffeine, a stimulant and diuretic that affects the body in a variety of ways. Begin your investigation by using the Internet or your library to research the positive and negative effects of coffee.

Problem

How does coffee affect the physiology of the body?

Hypothesis

Create a hypothesis related to one physiological effect of consuming coffee.

CAUTION: **Due to health concerns, it may not be appropriate for some students to participate as subjects in this investigation. Be sure that students do not exceed their normal coffee intake.**

Materials

Select your own materials.

Experimental Plan

1. After deciding which physiological reaction you want to measure, design an experiment that allows you to measure the effect of coffee.
2. Be sure to establish proper controls so you can compare your results before and after the ingestion of coffee.
3. Establish the amount of time required, and ask your teacher to approve your experimental design and to arrange for any equipment you may need.

organ donation. One of the major concerns with the use of animal organs has been the possibility of transmitting animal viruses into humans. These fears were somewhat diminished by a report that showed that none of the 160 people who had received heart valves or other tissue from pigs had become infected. Concern persists, however, about potential transfer of viruses.

Pigs are good potential candidates for organ donation.

As you will learn in Chapter 9, the next step in using pigs as organ donors will be to genetically modify the pig genome to decrease the risk of rejection after organ donation. The genetic manipulation of animals is ongoing, and the creation of transgenic animals (animals that have genes from more than one species) is producing animals with new characteristics. Some experts predict that clinical trials using pig organs as donors for humans could begin in less than two years.

Follow-up

1. Debate with classmates whether Canada should adopt an "opting out" policy to increase the number of cadaveric donors. What problems might this create?
2. Consider the following cost comparison of kidney transplant versus dialysis. The operation costs $20 000 and requires $6000 per year in follow-up treatments. Compare this with the $50 000 per year required to maintain a patient on dialysis (an artificial kidney). Over a five-year period, the costs are $50 000 for the transplant and $250 000 for dialysis. Kidney transplant operations have a 98 percent success rate using living donors and a 95 percent success rate with cadaveric donors. Should a destitute person be allowed to sell one of his or her kidneys to avoid starvation? Is this different from a family member donating a kidney?

Checking the Plan

1. What are your dependent and independent variables? What are your controlled variables?
2. What will be your control?
3. What will you measure and how?
4. How will you record and graph your data?

Data and Observations

Conduct your investigation and make your measurements. Graph your results and then enter the data in a summary table.

Analyze

1. **(a)** Was the variable you investigated affected by coffee?

   (b) If so, how was it affected?
2. Were your results consistent?
3. What factors of your population may have affected your results?

Conclude and Apply

4. What conclusions can you make about one physiological response to the intake of coffee?
5. Based on your results, predict what other measurable effects coffee would have on the human body.

Exploring Further

6. Using the Internet, find the results of various research studies that have explored the positive and negative effects of coffee. How might you account for the conflicting conclusions?

levels, which vary slightly from person to person. Variation is an important aspect of all species, and it is likely that some people are more prone to diabetes than others.

You have just learned about a disease, diabetes, that occurs when the body's immune system attacks itself. In the next section, you will learn how the immune system protects the body from foreign invaders such as bacteria and viruses.

COURSE CHALLENGE

There are many topics that are associated with homeostasis that would be a good choice for your Course Challenge. For example, this chapter has dealt with the impact of drugs, organ transplantation, and cancer on homeostasis. Once you choose your topic, begin gathering material as you also look for connections between new course material and your choice of topic.

SECTION REVIEW

1. K/U Explain how the pancreas is an organ that seems to have two different functions.
2. C Draw a negative feedback loop that illustrates how the beta cells regulate glucose blood sugar levels.
3. MC Investigate how people with diabetes monitor their blood sugar levels daily to manage their schedule of insulin injections.
4. K/U Describe the major difference between Type 1 and Type 2 diabetes. List some of the risks involved with both types.
5. MC What advice would you give athletic students with diabetes to help them avoid episodes of hypoglycemia (low blood sugar) when engaged in strenuous athletic activity?
6. C Diabetes is especially difficult for children. Devise a one-week diet plan for young children with diabetes that will allow them to occasionally enjoy some of their favourite foods while still maintaining a properly balanced blood sugar level.
7. I Banting and Best's original experiments did not produce usable insulin. Interpreting what you already know about the functions of the pancreas from previous studies or independent research, determine what might have complicated their first efforts. Explain the procedure you would follow and the materials you would use to overcome this complication.
8. MC The early source for insulin was the pancreases of farm animals such as pigs and cows. More recently, people with diabetes have the choice of using actual "human insulin" instead. Investigate how human insulin is produced and describe the advantages of this type of insulin for treating people with diabetes. Find out how genetic engineering may further improve the types of insulin used to treat people with diabetes.
9. I Having blood glucose levels tested for diabetes starts by fasting for a certain number of hours. Next, you drink a sweet drink. After waiting a specified time, blood is removed and tested.
   (a) What is the purpose of fasting before you start?
   (b) Does it matter how much of the drink you consume? Explain.
   (c) Does it matter how long you wait before blood is removed? Explain.
   (d) What results would you expect for a diabetic response?
   (e) What results would you expect for a normal response?

UNIT PROJECT PREP

Diabetes is becoming an epidemic, and researchers predict that the future incidence of this disease will put a burden on health systems around the world.

Start preparing for the project at the end of this unit by writing down the projected number of diabetics worldwide for the year 2025. Gather information on treatments, both accepted and experimental, for diabetes.

4.4 Homeostasis and the Immune System

EXPECTATIONS

- **Describe the mammalian immunological response to a viral or bacterial infection.**
- **Describe the role of the immune system in maintaining homeostasis in the body.**
- **Predict the impact of environmental factors such as allergens on homeostasis within an organism.**
- **Compile and display, based on data you collect on food allergies from an investigation, information on the impact of allergies on the homeostasis of those affected.**

Another homeostatic mechanism that we depend on for existence is the immune system. The immune system comprises a variety of white blood cells and proteins that attack foreign invaders such as bacteria and viruses. This system recognizes and destroys damaged cells and irregular growths such as tumours. The human immune system must distinguish harmless from harmful organisms and must deal with both in efficient ways that pose no risk to human health.

The air we breathe, the water we drink, and the food we eat are all inhabited by millions of unseen organisms. For many of these organisms, the human body is an ideal habitat. The cells of the body are kept moist, warm, and well fed by our homeostatic systems. Here many bacteria, viruses, and parasites are able to reproduce at incredible rates and spread throughout the body or beyond in search of new, favourable habitats. Over millions of years, many of these organisms have become adapted to allow them to penetrate the body's defences or live in particular places in or on the body. Many of these organisms cause little or no damage to us. In fact, it is estimated that the human body is host to more than 10 000 times more foreign cells (usually bacteria) than human cells. In some cases these close associations are mutualistic — both the host human and the micro-organism benefit. For example, the inner walls of the large intestine are coated with a beneficial strain, or form, of the bacteria *Escherichia coli* (*E. coli*) that helps its human host to absorb nutrients (see Figure 4.16).

Other invaders, including strains of the same species, can prove dangerous. Before the 1950's the most common cause of death in North America was communicable disease. The use of antibiotics and vaccines has diminished this risk, but the reprieve may prove temporary, as many bacteria and viruses develop resistance.

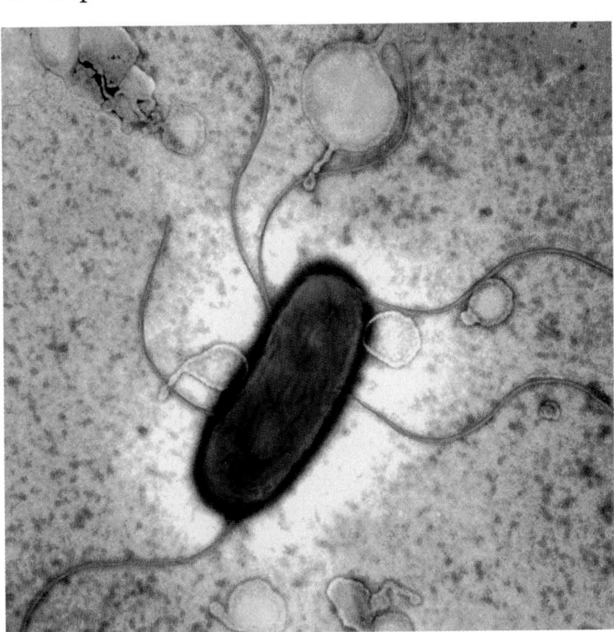

Figure 4.16 Some strains of the *Escherichia coli* (*E. coli*) bacteria are beneficial to humans; others can be harmful.

Non-specific Defences and Specific Immunity

The human body survives by means of various defences. These defences can be divided into two groups — those that provide non-specific defence and those that provide specific immunity. **Immunity** is the ability to resist a disease after being exposed to it in the past. **Non-specific defences** guard against a wide variety of **pathogens** (disease-causing agents). Their effectiveness does not depend on prior exposure to the pathogen. The first layer of defence is the skin. Although thousands of bacteria, fungi, and other pathogens can be found on the skin, it is not a hospitable environment for these organisms. The outer layer of the skin is dry and contains large amounts of tough, relatively indigestible keratin. The skin's oil contains bactericides, and sweat forms an acidic layer that is inhospitable for microbial growth. The linings of the respiratory and digestive pathways also contribute to non-specific defence. They are covered with thick mucus that continually traps and sweeps away potentially dangerous micro-organisms. In addition, the stomach's acid kills most of the micro-organisms in the food we eat.

Non-specific defence also includes the macrophages, neutrophils, and monocytes that kill bacteria by the process of **phagocytosis**, in which the cell ingests the bacteria to destroy it. **Macrophages** are phagocytic cells found in the liver, spleen, brain, and lungs; they also circulate in the bloodstream and interstitial fluid. **Neutrophils** and **monocytes** are white blood cells (leucocytes) that attack bacteria using phagocytosis. Non-specific defence also includes **natural killer cells**, which are other white blood cells that carry out phagocytosis. In this case, their targets are body cells that have become cancerous or infected by viruses.

The **specific immune system** includes a wide variety of cells that recognize foreign substances and act to neutralize or destroy them. Over time, and as a result of the variation in our genetic make-up, each of us develops an immune system unique in its capability to deal with a wide variety of possible infections. We are not all exposed to the same diseases, and some diseases require a stronger response than others. The specific immune system is primarily a function of the lymphocytes of the circulatory system. **Lymphocytes** are specialized white blood cells formed in the bone marrow. They are divided into two specialized groups, depending on where they mature. B lymphocytes (B cells) mature in the bone marrow, while T lymphocytes (T cells) mature in the thymus gland, which is near the heart. (Chapter 6 will examine the role of glands in homeostasis in greater detail.)

Cellular Immunity and Antibody Immunity

B cells and T cells work together to attack invaders. **Cellular immunity** is primarily a function of T cells (see Figure 4.17), while **antibody immunity** is performed by B cells. Both systems are controlled by T cells and are initiated by the action of macrophages.

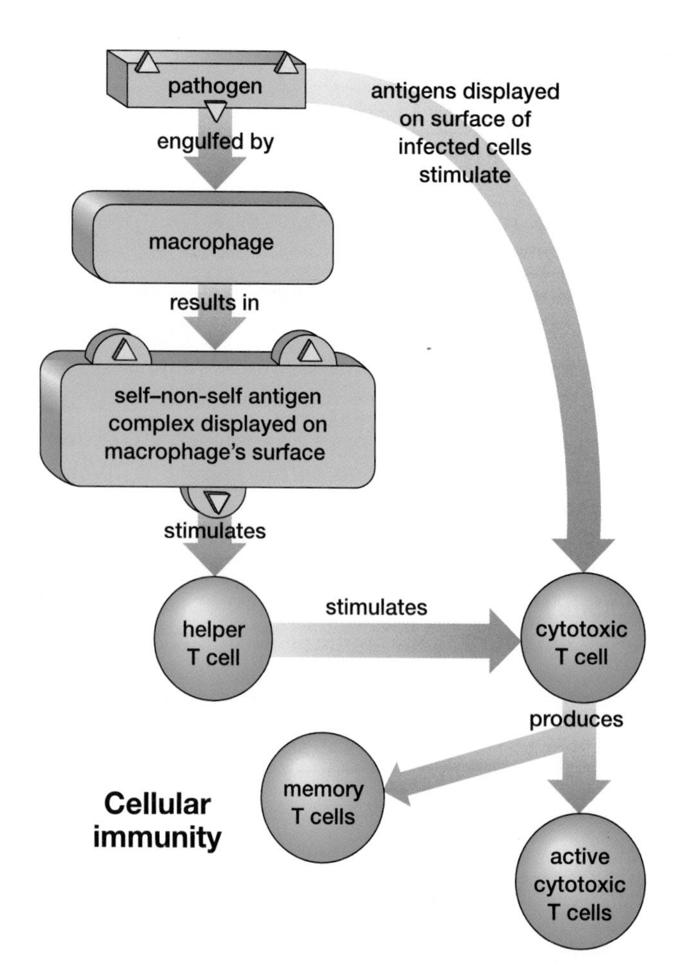

Figure 4.17 In cellular immunity, a macrophage engulfs an antigen, breaks it down, and places part of the foreign substance on its own cell surface. The macrophage then binds to the antigen receptor on helper T cells, activating the cytotoxic T cells to differentiate and produce identical clones. Some T cells remain behind in the lymph nodes as memory T cells. These T cells are able to respond rapidly to a second attack. Other T cells travel out to the infected tissue to destroy the pathogen. At the infection site, cytotoxic T cells may either destroy the pathogen directly or release chemicals that attract other macrophages to the site.

As mentioned earlier, the immune system depends on its ability to tell the difference between "self" (the normal cells of the body) and "non-self" (abnormal cells or invading organisms). "Non-self" cells are distinguished by having large molecules (such as proteins) on their surface. These molecules are called **antigens**. Macrophages are able to recognize these cells as "non-self" and engulf them. During this process, antigens from the foreign cell are moved to the surface of the macrophage. T cells and B cells have specialized receptors on their surface that match specific antigens.

In cellular immunity, a T cell that has a receptor for the particular antigen attaches to the macrophage

as "The Capital City of Allergies in Canada." Dr. Lyanga explains, "Allergy is a science of inflammation. As in organ transplants, human tissue becomes inflamed when it reacts against a foreign substance. All allergic reactions, from eczema to asthma, involve inflammation." In Windsor from spring until fall, thousands of people suffer symptoms such as itchy eyes, asthma, or "terrific paroxysmal sneezes." At his clinic, Dr. Lyanga obtains a detailed history of each patient. Since prick or scratch tests cause a visible reaction, Dr. Lyanga can do these tests on his patients and interpret the severity or type of allergy. He then tries to "fool" his patients' immune systems into accepting one foreign substance to protect them from another, by inoculating them with pre-season vaccines.

Self and Non-self

In searching for better treatments, Dr. Lyanga takes a keen interest in the Major Histocompatibility Complex (MHC), a set of genes located at chromosome 6. Our immune systems either accept or kill a foreign substance, depending on the make-up of an individual's MHC. These molecules decide whether the substance is "self" or "non-self." Therefore, being able to read a patient's MHC becomes essential in not only understanding allergic

3. When you have collected the completed questionnaires, create a chart or scatter plot to summarize the data.

Post-lab Questions

1. According to your data, which is the most common food allergy?
2. Analyze results according to the gender of the respondents.
3. What percentage of your class has food allergies?

Conclude and Apply

4. Based on your results (and keeping in mind your small sample size), would you conclude that when a person has one food allergy he or she is likely to have multiple food allergies?

Exploring Further

5. Write up a summary of your study, including your conclusions and any sources of error that you suspect may limit the strength of your conclusion.

and then goes through a process of rapid cell division (clonal expansion). This process produces a number of different types of T cells. **Helper T cells** give off chemicals that stimulate other macrophages, B cells, and other T cells. **Suppresser T cells** slow and stop the process of cellular immunity, while **memory T cells** remain in the bloodstream to promote a faster response if the same foreign antigen appears again. **Cytotoxic T cells** bind to other cells that have been infected and destroy them by puncturing a hole in their cell membrane. These destroyed cells may be invading bacteria, human cells that have been infected by a virus, or cancer cells (see Figure 4.18). The cytotoxic T cells can also be triggered directly by foreign antigens bound to a cell that has been attacked by a virus.

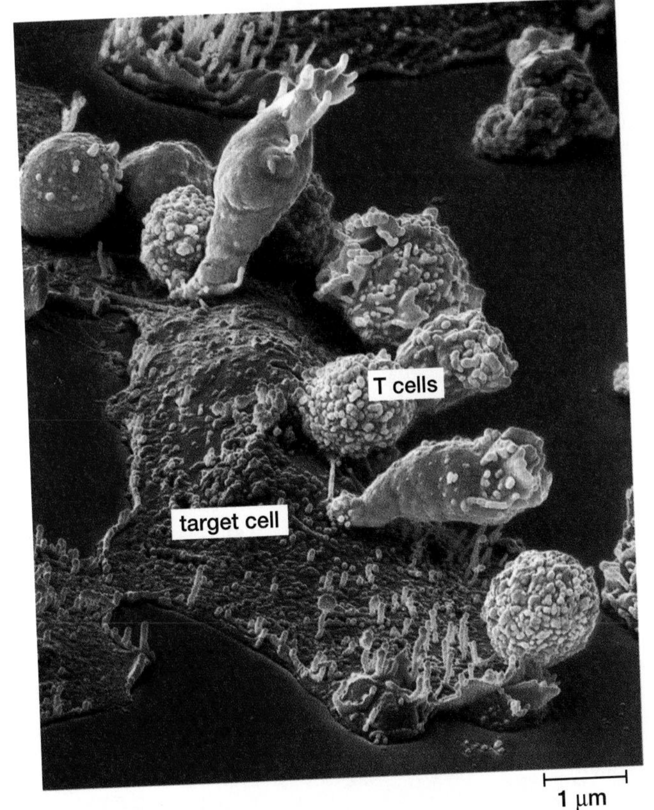

Figure 4.18 A scanning electron microscope shows cytotoxic T cells attacking and destroying a cancer cell.

As shown in Figure 4.19 on the following pages, antibody immunity begins the same way as cellular immunity does — with a macrophage ingesting a foreign cell and presenting antigens from that cell on the surface of the macrophage's cell membrane. In this case, helper T cells trigger specialized B cells (that recognize the antigen) to undergo rapid cell division. The dividing B cells form plasma cells and memory B cells. The plasma cells produce large quantities of antibodies that are matched to the foreign antigen. These antibodies attach to the foreign cells, which are ingested by macrophages. Antibodies can also attach to toxins, trigger the release of other defensive chemicals from surrounding cells, and prevent foreign cells from attaching to the digestive or respiratory systems. When the danger has passed, the plasma cells undergo **apoptosis** or programmed cell death.

As in the case of memory T cells, memory B cells stay in the body to speed up the response if the same antigen reappears. Vaccines contain antigens that trigger the immune system to respond, thus protecting us from future infection of a particular disease.

Allergies — A Good System Gone Bad

An allergy is an exaggerated response by the immune system to a harmless material such as pollen, mould, or cat dander. There are two major types of allergic reactions: immediate and delayed. An immediate or acute reaction is the most common type of allergic reaction. It occurs within seconds of exposure to the allergen and usually disappears within 30 minutes. In these reactions, specialized antibodies trigger certain cells to release **histamines**, which are chemicals that increase the permeability of blood vessels, making the area red and swollen. The specialized antibodies can also trigger the release of cellular fluids, which can result in watery eyes and a runny nose.

Food allergies can trigger vomiting, cramps, and diarrhea, although most of these symptoms can be treated with antihistamines.

Some forms of asthma, the most common chronic disease among North American children, is an acute reaction to allergens that are inhaled. The inhaled allergens trigger a massive release of histamines, which sets off spasms of the bronchioles, the tiny air passageways in the lungs. In people who have asthma, these passageways are particularly sensitive. Spasms can also be triggered by stimuli such as cold air and fatigue. The result can be coughing, wheezing, and sometimes fatal suffocation.

Asthma can be treated with anti-inflammatory drugs such as various steroids, and with bronchodilators. The latter contain medicines that can open the airways of the bronchia and thereby ease the symptoms of an acute attack. Current research is exploring new drugs that may provide long-term relief from the inflammation that

Biology At Work

Immunologist

Dr. John Jacob Lyanga

Back to the Roots

"Early in life," Dr. John Jacob Lyanga remembers, "I started on the road of one day becoming a healer." Dr. Lyanga grew up in Tanzania. "In our culture, dogs were very important because they were used for hunting and were responsible for the surveillance of cattle and premises. If

Investigation 4 • B

Food Allergy Survey

Most people who have food allergies are aware of the foods or f should avoid. In this investigation you will gather data on how w allergies are in your class and determine which are most commo

Pre-lab Questions

- How widespread are food allergies in your class?
- Which are the most common food allergies among students in your class?

Problem

What will the data you collect and analyze indicate about the importance of food allergies?

Prediction

Predict how many people in your class have food allergies.

5. Is temperature regulation an example of dynamic equilibrium? Explain your answer.
6. Explain how the process of active transport helps control the volume of urine produced during the day.
7. How does ADH affect urine concentration?
8. Why are proteins and blood cells normally not found in urine?
9. What parts of the nephron are most involved in regulating the pH of blood?
10. Identify four differences between the blood entering the kidney and the blood leaving the kidney.
11. Identify four types of dissolved substances found in the filtrate that forms in the Bowman's capsule of the kidney.
12. How does the kidney control water levels in the body?
13. An older pet cat starts to drink large quantities of water and urinates frequently. You think the cat may be diabetic. How could you confirm your suspicion?
14. Explain how islet cells in the pancreas respond to an increase in blood sugar levels.
15. Explain how the hormone glucagon counters the action of insulin.
16. Identify the differences between how islet cells in the pancreas regulate insulin and how the hypothalamus in the brain regulates body temperature.
17. Describe the role of T cells in the body's response to infectious agents.
18. The immune system helps resist pathogens. Identify three ways that pathogens can enter the system.
19. Identify the types of cells that are destroyed by the phagocytotic activity of macrophages.
20. How do the following processes help maintain body temperature: changes in blood flow, shivering, and sweating?
21. You have just consumed a big bag of salty snack food. Describe how your kidneys will restore normal water balance and concentration of salt in the blood and body fluids.
22. Explain how diuretic drugs decrease blood pressure.
23. Asthma may be genetic or it may be an autoimmune disorder. Select one option. What evidence would you expect to find to support your choice?

INQUIRY

24. Examine the following graph of insulin and glucose levels in the blood. Explain the apparent connection between the fluctuations in the levels of glucose and insulin.

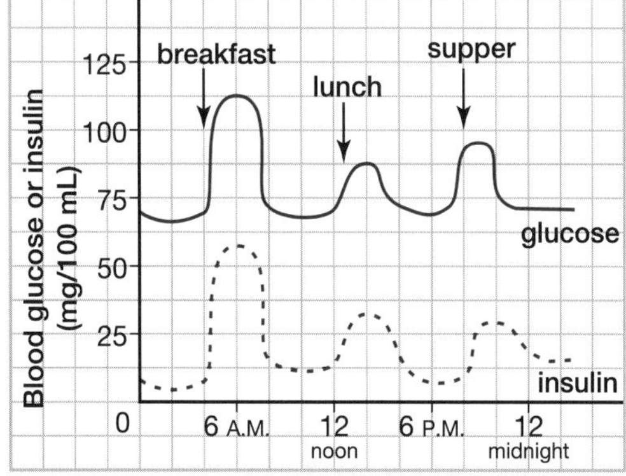

25. Examine the following graphs, which illustrate fluctuations in glucose levels in the blood of people with Type 1 diabetes and people without diabetes. Explain why there seem to be greater variations in blood sugar levels among people with diabetes.

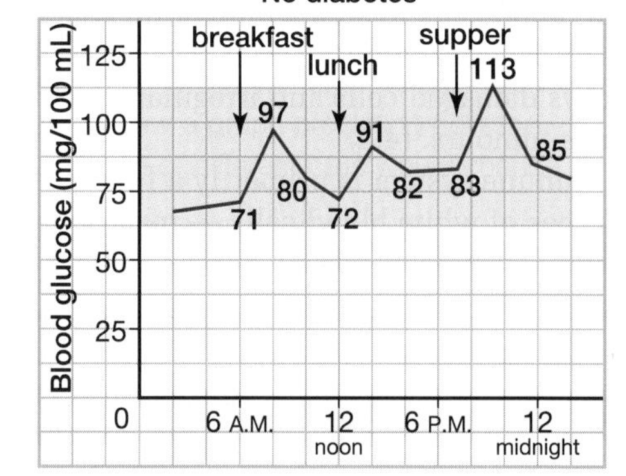

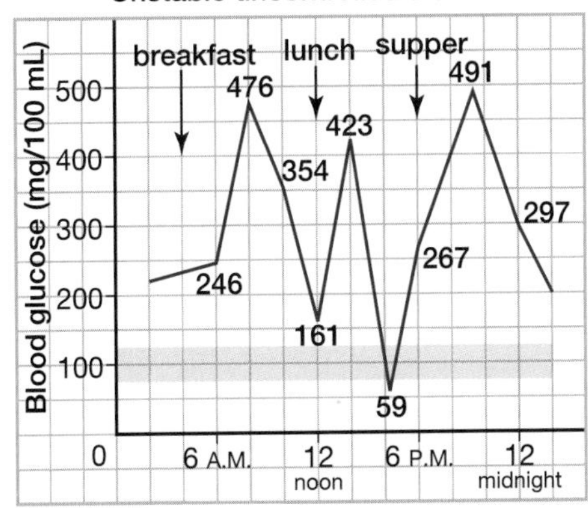

26. A particular chemotherapy prevents the formation of microtubules that form spindle fibres for mitosis. Predict which immune cells will be adversely affected by this therapy. How can you determine if your prediction is correct?

27. Design a test that could be used to confirm a diagnosis of diabetes in a patient who displays some of the common symptoms of the disease.

28. How could transplanting healthy islet cells into people with Type 1 diabetes help them manage their blood sugar levels more effectively? Develop a hypothetical clinical trial that would test the effectiveness of a new transplant procedure. Focus on creating appropriate qualifying criteria for prospective participants, and on establishing criteria for evaluating the procedure's effectiveness.

COMMUNICATING

29. Describe and illustrate two negative feedback loops that have opposite physiological effects.

30. Describe and illustrate a negative feedback loop in which bicarbonate ions help regulate the pH of blood.

31. Create a flowchart that illustrates the relationship between different immune system cells.

32. Create a flowchart that illustrates how the filtrate is modified by the processes of active and passive transport as it flows through a single nephron in the human kidney.

33. Create a flowchart that lists the steps that can lead to heat stroke in an athlete.

MAKING CONNECTIONS

34. Meal A consists of mostly simple carbohydrates with some proteins. Meal B consists of mostly complex carbohydrates with some proteins.
 (a) Compare the glucose levels you would expect to find soon after eating each meal.
 (b) Describe how each meal could affect the feedback loops created by the antagonistic hormones insulin and glucagon.

35. Someone with Type 2 diabetes is concerned about kidney failure. This person then decides to switch to a high protein, low carbohydrate diet. Is this a wise decision? Explain the impact of this diet on the kidney.

36. **(a)** Use a feedback model to show how diet and exercise can moderate the effect of Type 2 diabetes.
 (b) Explain why this model is less effective for Type 1 diabetes.

37. Explain how you might develop different types of allergies than the other members of your family do.

38. Explain why babies and young children might contract infectious diseases more frequently than adolescents do.

39. Explain how hypoglycemia (low blood sugar) episodes can turn into life-threatening situations for people with Type 1 diabetes.

40. How do cells that provide cellular immunity sometimes trigger allergic reactions when they come into contact with environmental allergens?

41. What are some daily habits you could adopt to help reduce your chances of contracting a contagious disease (such as a cold or flu) from people you come into contact with on a regular basis?

CHAPTER 5

The Nervous System

Reflecting Questions

- How does the nervous system help us cope with changes inside and outside the body?
- How does the structure and function of the neuron influence the activities of the nervous system?
- How is current research on the nervous system increasing our understanding of how the nervous system works?

Prerequisite Concepts and Skills

Before you begin this chapter, review the following concepts and skills:

- describing properties of negative and positive ions in solution (Chapter 1, section 1.3),
- identifying and explaining how ions move across cell membranes (Chapter 2, section 2.3), and
- describing the structure and properties of enzyme molecules (Chapter 2, section 2.2).

Humans have the most complex nervous system of all organisms on Earth (although some scientists give an honourable mention to certain members of the whale and dolphin family). The complex structure of the human nervous system is the result of millions of years of evolution, from the development of simple nervous systems in animals such as flatworms to the system we will examine in this chapter.

A simple brain is a collection of nerve cells that co-ordinate reactions to a limited number of stimuli. The evolution of the more complex vertebrate brain exhibits a number of trends. First, the ratio of the brain to body mass increases. Second, there is a progressive increase in the relative size of the area of the brain that is involved in higher mental abilities. The higher intellectual abilities, such as the capacity to learn and solve problems, are functions of an area of the brain called the cerebrum. In vertebrates such as fish and reptiles, the brain is small and the cerebrum is only a fraction of the total size of the brain. In more complex species such as the cat and the chimpanzee, the cerebrum is the dominant part of the brain. In humans, the cerebrum (as shown on the right) is so large that it almost covers the rest of the brain.

Over the past two million years of human evolution, the human brain has doubled in size. Newborn humans have a very large head (in relation to body size) when compared to other primates. The head is so large that birth is only possible because of the flexible connections of the bones that make up the skull. The head is compressed as it comes through the birth canal, and it regains its normal shape days after delivery. Some researchers speculate that humans may have reached their maximum brain size, which has been limited by the birth canal.

As you read through this chapter, learn more about the human brain and the nervous system's role in homeostasis.

The progressive enlargement and increasing level of complexity of the cerebrum has contributed most to the overall development of the brain.

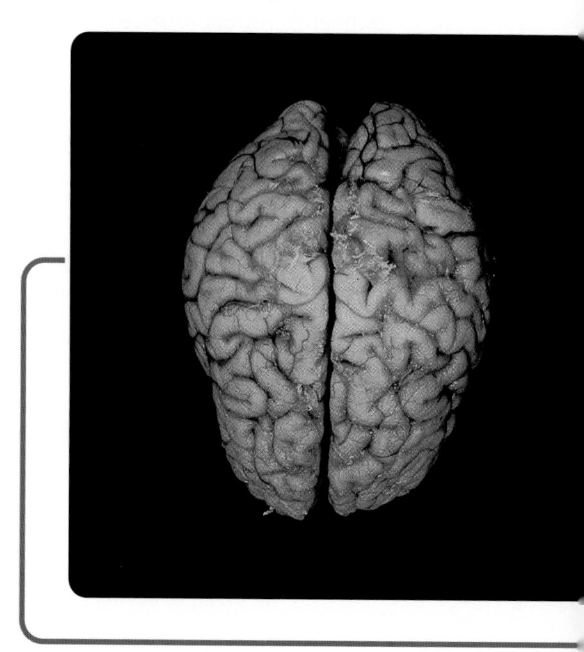

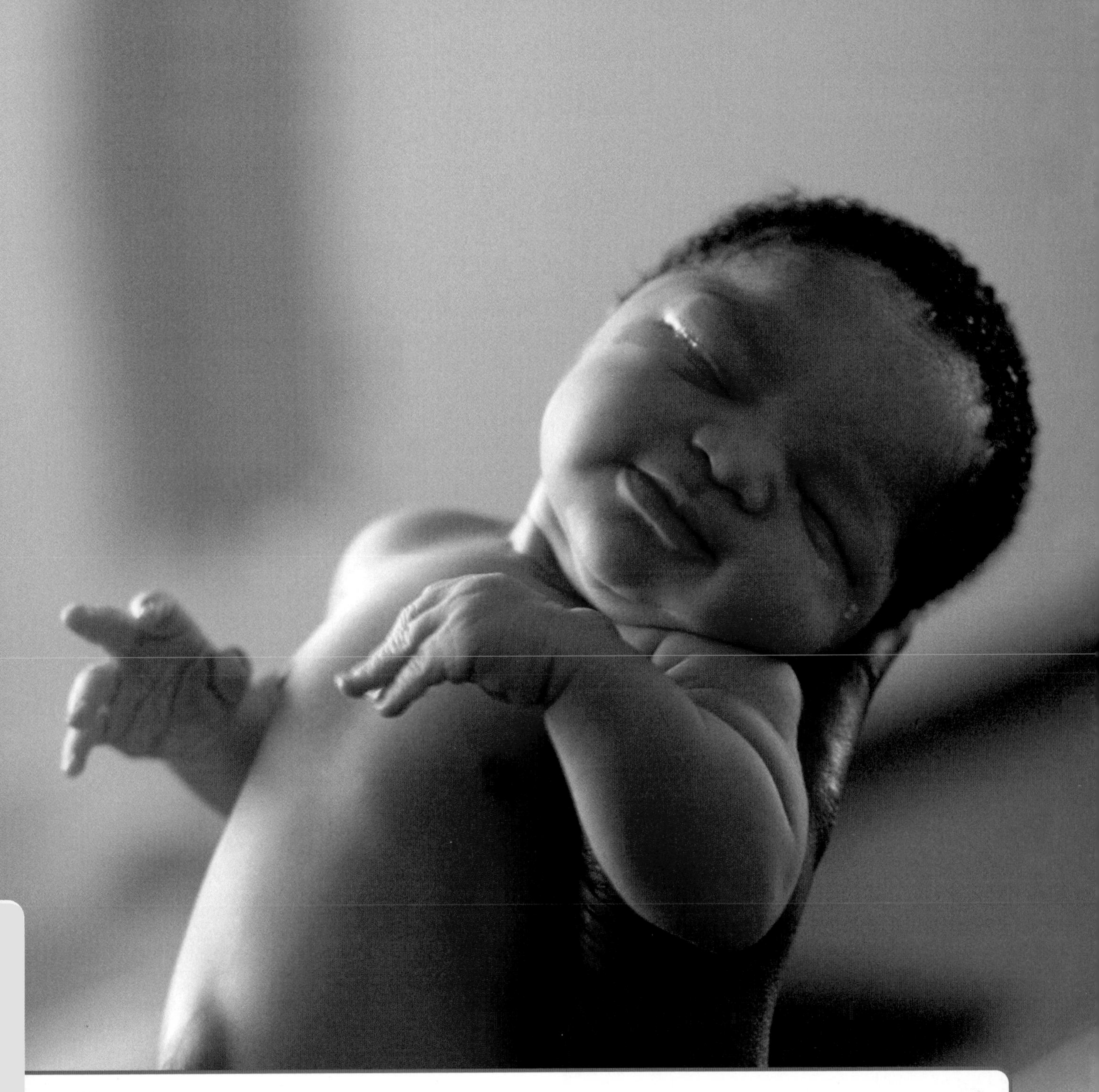

Chapter Contents

5.1 The Structure of the Nervous System

EXPECTATIONS

- **Describe the anatomy and physiology of the nervous system, and explain this system's role in homeostasis.**
- **Describe some Canadian contributions to knowledge of and technology in the field of homeostasis.**

How does the human nervous system help maintain homeostasis? In other words, how does the nervous system cope with changes both inside and outside the body? The nervous system provides a high-speed communication system to and from almost every part of the body. A series of sensory receptors provides information about changes in both the internal and external environment that may affect the body's responses. For example, you have already studied the sensory receptors in the skin that provide information about whether the body is gaining or losing heat. You know that this information is sent to the brain to stimulate the appropriate body response. In Figure 5.1, the basketball players could not focus on the game if they were conscious of all the information being processed by the various parts of their brains. Information about carbon dioxide, water, and glucose levels, as well as blood pressure, are all monitored by the hypothalamus. Other parts of the brain monitor the spatial orientation of the body. The eyes, ears, and nose provide information about the external environment.

Figure 5.1 While this player is concentrating on shooting the ball into the net, his hypothalamus is monitoring his blood pressure as well as carbon dioxide, water, and glucose levels.

The human nervous system, as shown in Figure 5.2, is actually a complex of interconnected systems, with larger systems comprised of subsystems that each have specific structures and functions.

Two major parts comprise the human nervous system. The **central nervous system** (CNS) is made up of the brain and spinal cord. The **peripheral nervous system** (PNS) includes the nerves that lead into and out of the central nervous system. The peripheral nervous system consists of the **autonomic nervous system** and the **somatic nervous system**.

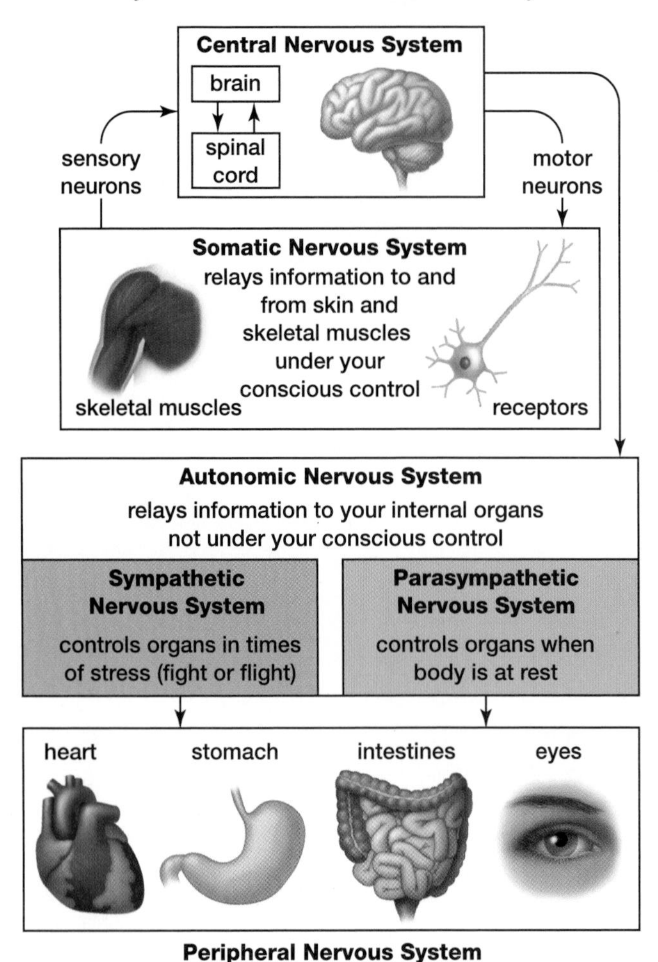

Figure 5.2 The organization of the human nervous system

The autonomic nervous system is not consciously controlled. As Figure 5.3 shows, the autonomic nervous system is made up of the sympathetic and parasympathetic nervous systems, which control a number of organs within the body. The **sympathetic nervous system** sets off the "fight-or-flight" reaction that prepares the body to deal with an immediate threat. When this system is stimulated, heart rate and breathing rate increase. Also, blood sugar is released from the liver to provide the energy required to deal with the perceived threat.

Many perceived threats today do not require the fight-or-flight response. In fact, this response may worsen a situation. For example, some students suffer from test anxiety. In these individuals, the stress of a test produces rapid heart and breathing rates that may interfere with higher levels of brain activity, such as concentration and memory. Learning relaxation techniques can often help people deal with stressful situations. (You will learn more about the "fight-or-flight" response in Chapter 6.)

The **parasympathetic nervous system** has an effect opposite to that of the sympathetic nervous system. When a threat has passed, the nerves of this system slow heart rate and breathing rate and reverse the effects of the sympathetic nervous system response.

The somatic nervous system is made up of **sensory nerves** that carry impulses from the body's sense organs to the central nervous system. This system also consists of **motor nerves** that transmit

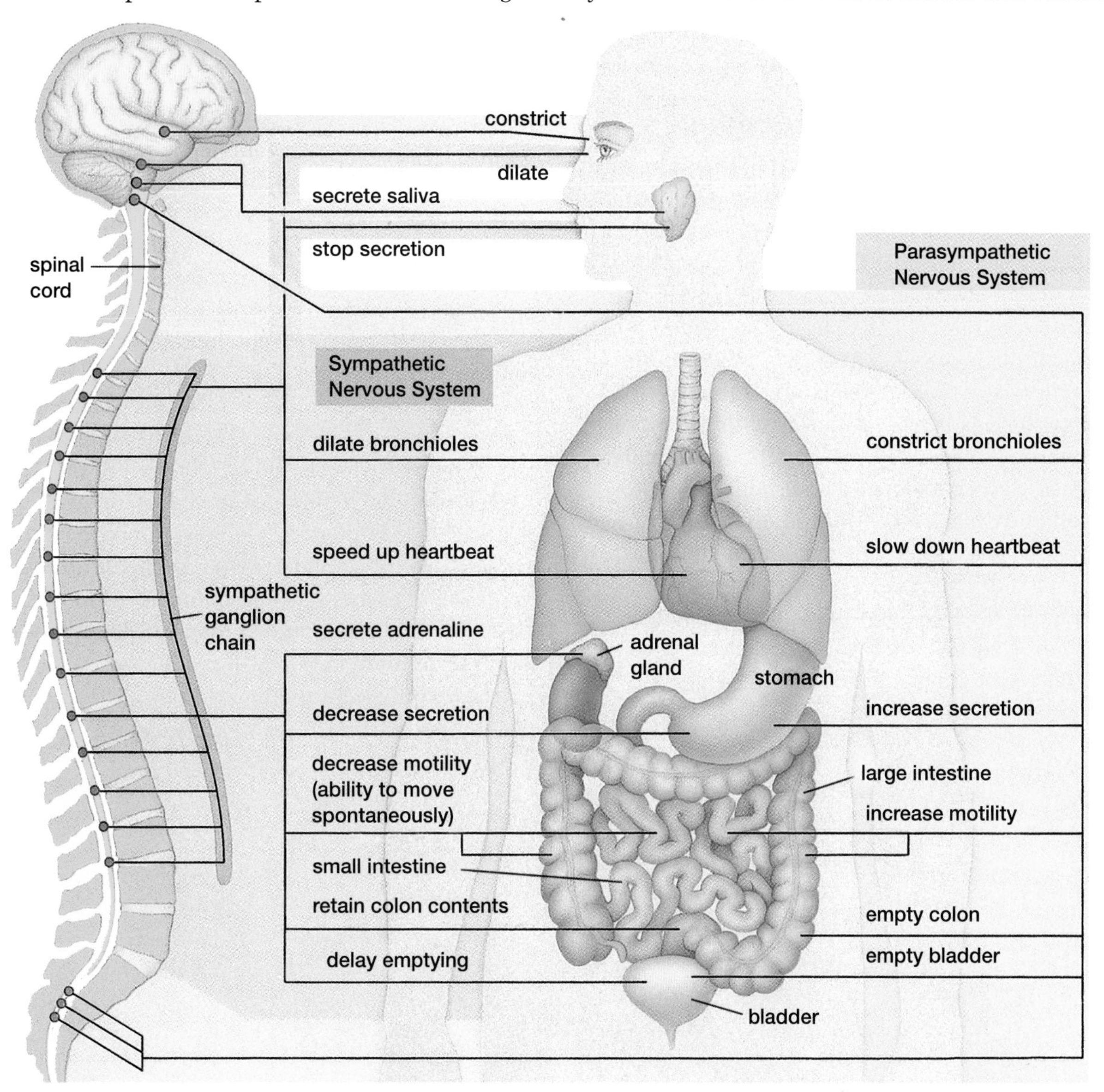

Figure 5.3 The ganglia of sympathetic nerves are located near the spine. The ganglia of parasympathetic nerves are located far from the spine, near the organs they affect. As you can see, a nerve pathway runs from both of the systems to every organ indicated, except the adrenal gland.

commands from the central nervous system to the muscles. This combination of sensory and motor nerves deals primarily with the external world and the changes in it. The somatic nervous system, to some degree, is under your conscious control. The information from your eyes and ears is processed by the brain and correlated with information you already have. You then make a decision that may or may not involve the movement of muscles.

Other functions of this system are more subtle. Decision making can be time-consuming, so in some cases the system is designed in such a way that a certain action sets off a specific reaction. This reaction is known as a **reflex**. An example is an eye blink when something moves close to your eye. Such reflexes do not require a conscious decision. Indeed, sometimes we are not even aware of their occurrence.

Neurons and Reflex Responses

The structural and functional unit of the nervous system is the **neuron**. The central and peripheral nervous systems are both composed of a series of interconnected neurons. The PNS consists of **nerves**, which are numerous neurons held together by connective tissue. The CNS is also made up of neurons; in fact, 90 percent of the body's neurons are found in the CNS. A **reflex arc** is the nerve pathway that leads from stimulus to reflex action.

A neuron consists of three parts: the cell body, dendrites, and an axon, as shown in Figure 5.4. The **cell body** has a large, centrally located nucleus with a large nucleolus. The cytoplasm contains numerous mitochondria and lysosomes, along with a Golgi complex and rough endoplasmic reticulum. Neurons are capable of surviving for over 100 years, since most do not undergo cell division after adolescence.

Dendrites are the primary site for receiving signals from other neurons. The number of dendrites can range from one to thousands, depending on the neuron's function. The **axon** is a long, cylindrical extension of the cell body that can range from 1 mm to 1 m in length. When the neuron receives a sufficiently strong stimulus, the axon transmits impulses or **waves of depolarization** along its length. At the end of the axon are specialized structures that release chemicals. These chemicals stimulate neighbouring neurons or muscle cells. The various functions of the neuron will be discussed in more detail shortly.

The three major types of neurons can be illustrated in a simple reflex. For example, if you touch something hot, what happens first? Do you pull your hand back or are you aware of the pain first? The hand usually moves first. Pulling your hand back is a reflex that involves your spinal cord rather than your brain. In the reflex arc shown in Figure 5.5, the heat triggers nerve endings in the skin of the hand. These are dendrites of a *sensory neuron*, which requires a strong stimulus to activate it. The impulse travels along this neuron, up the arm, and into the spinal cord. In the spinal cord the signal is passed on to interneurons. An *interneuron* is a nerve cell that acts as a link between a sensory neuron and a motor neuron. Found only in the CNS, a motor neuron is stimulated and transmits an impulse along its axon. When stimulated, the *motor neuron* triggers the contraction of the muscles in your arm, pulling your hand away. While this is happening, other interneurons in the spinal cord transmit a message to the brain, making you aware of what has just happened. This type of reflex is involuntary and can be triggered without input from the brain. (Section 5.3 features an activity exploring a reflex in the eye.)

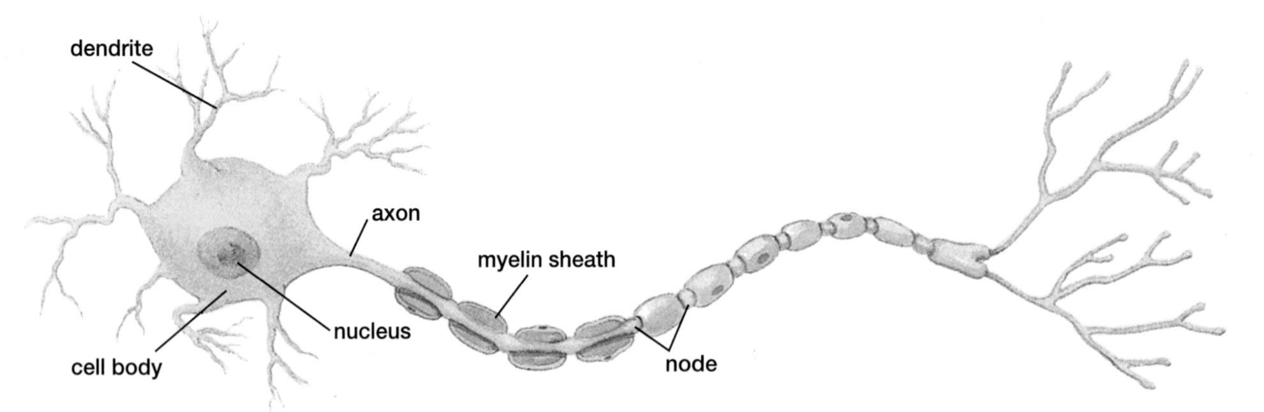

Figure 5.4 A typical neuron is composed of dendrites, a cell body, and an axon. Impulses are transmitted from the dendrites to the cell body. The axon carries impulses away from the cell body to the dendrites of the next neuron.

ELECTRONIC LEARNING PARTNER

To learn more about the reflex arc, go to your Electronic Learning Partner.

The Brain and Homeostasis

A neurologist once compared brain research to the following scenario. Aliens from outer space land on the roof of a hockey arena. They drill holes in the roof of the building and send down microphones. Some microphones pick up sounds in the crowd, some pick up sounds in dressing rooms, some pick up sounds near the popcorn stand, and some pick up sounds of the game itself. From this information the aliens try to figure out the rules of hockey. Scientists today have many pieces of information about what happens in the various parts of the brain, but they are striving for a complete and thorough understanding of precisely how the brain functions.

The brain co-ordinates homeostasis within the body. The brain processes the information that is transmitted through the senses so the body can deal with changes in the external and internal environment. The human brain makes up only 2 percent of the body's weight. However, at any given time it contains 15 percent of the body's blood supply and consumes 20 percent of the body's oxygen and glucose. Obviously, the brain is a high-energy organ. It contains about 100 billion neurons, which is roughly equal to the number of stars in our galaxy.

However, the complexity of the brain is not just a function of the large number of cells. The brain's complexity is also due to the variety of cells involved; to the brain's unique internal hormone system; and to complex interconnections between the various parts of the brain.

There is still much to learn about how complex operations of the brain, such as memory and decision making, are carried out. Early knowledge of brain function came from studying the brains of people with brain diseases or injury. Brain damage causes symptoms such as the loss of particular body functions or changes in behaviour. Researchers assumed that any abnormalities in the structure of the patient's brain must have been the source of the symptom. They believed that the area of the brain that was abnormal must control whatever body function was changed by the disorder or damage.

In general, early researchers were reluctant to probe healthy human brains for ethical reasons. This limited many areas of research to work with rats or monkeys. However, technological innovation has resulted in many new, benign ways of probing the structure and function of the human brain. For example, the electroencephalograph (EEG) was invented in 1924 by Austrian psychiatrist Dr. Hans Borger. This device measures the electrical activity of the functioning brain and produces a printout that allows doctors to diagnose disorders, such as epilepsy and to locate brain tumours. Figure 5.6

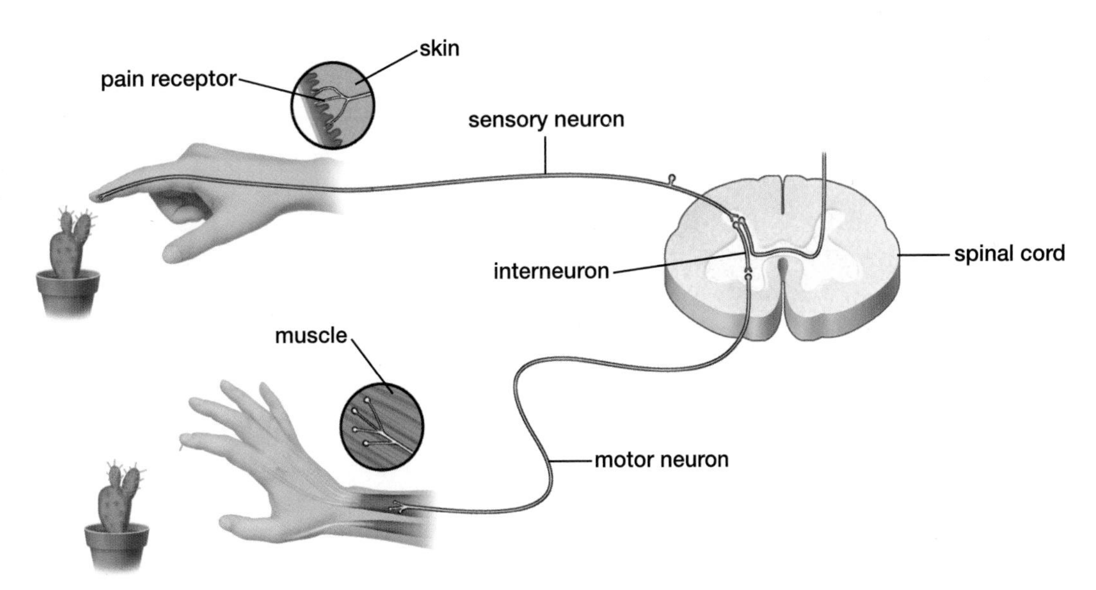

Figure 5.5 Reflex responses are carried out at the level of the spinal cord without assistance from the brain. The impulse travels directly to the spinal cord from the affected body part, crosses to a small interneuron, and then moves to a motor neuron that transmits the impulse to a muscle. The muscle contracts. The brain becomes aware of the reflex only after it occurs.

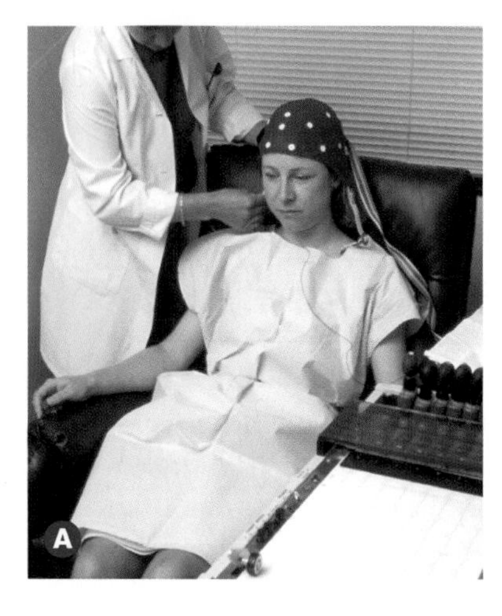

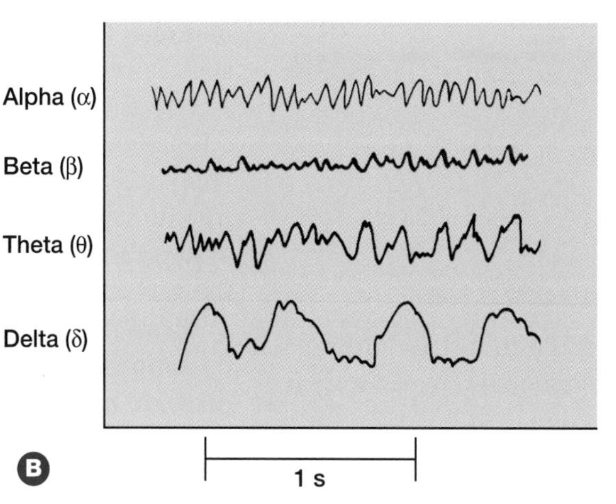

Figure 5.6 The electroencephalogram (EEG). (A) An EEG is recorded from an array of electrodes attached to the forehead and scalp. (B) The printout of brain waves help doctors diagnose certain diseases.

illustrates an EEG reading. Such readings have been used to study brain activity during sleep to help doctors diagnose and understand sleep disorders.

Another method of research is direct electrical stimulation of parts of the brain during surgery. This type of stimulation has been used to map the functions of the various areas of the brain. The brain has no pain receptors, so brain surgery can be carried out without anesthesia while the patient is fully awake. In the 1950s, Canadian neurosurgeon Dr. Wilder Penfield, shown in Figure 5.7, pioneered this method of mapping the functions of the brain. He won a Nobel Prize for his work. Dr. Penfield stimulated the temporal lobe of one patient, who then "heard" the voices of her mother and brothers. With similar brain stimulation, another patient "heard" a concert she had attended in the past.

Figure 5.7 Canadian neurosurgeon Dr. Wilder Penfield pioneered the mapping of the functions of the brain.

Modern advances in scanning technology now allow researchers to observe changes in activity in specific areas of the brain. Scanning techniques can be used to show which parts of the brain are involved in activities such as speaking or listening. Computerized tomography (CAT) scans and positron emission tomography (PET) scans continually enhance our knowledge of both healthy and diseased brains. CAT scans take a series of cross-sectional X-rays to create a computer-generated, three-dimensional image of a part of the body. Figure 5.8 shows how PET scans identify which areas of the brain are most active when the subject performs certain tasks.

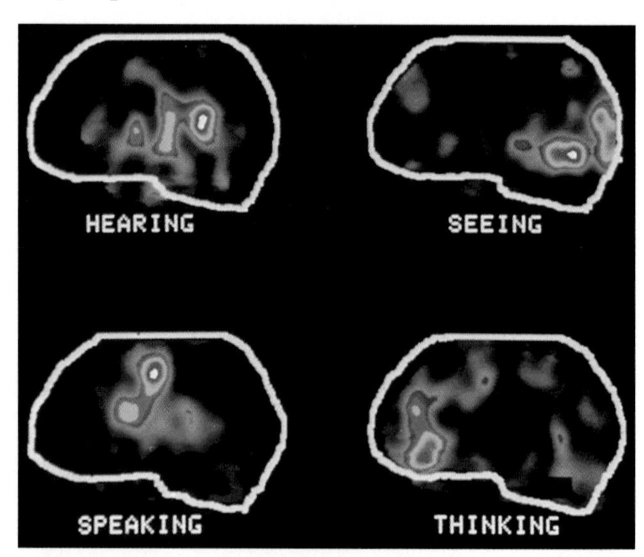

Figure 5.8 As these PET scans show, different areas of the brain are active when we hear a spoken word, see and read that same word silently, speak the word aloud, and think of and say a word related to the first.

Research based on scanning technologies has produced information about the functions of each part of the brain. The **medulla oblongata**, attached to the spinal cord at the base of the brain, has a number of major functions, each of which is related to a particular structure. The **cardiac centre** controls heart rate and the force of the heart's contractions. The **vasomotor centre** adjusts blood pressure by controlling the diameter of blood vessels, and the **respiratory centre** controls the rate and depth of breathing. The medulla oblongata also contains reflex centres for vomiting, coughing, hiccupping, and swallowing. Any damage to this part of the brain is usually fatal.

The **cerebellum**, shown in Figure 5.9, controls muscle co-ordination. Although the cerebellum makes up only 10 percent of the volume of the brain, it contains 50 percent of the brain's neurons. If you stand in one place, specific muscles are contracted while others are relaxed. As groups of muscle fibres become fatigued, others are contracted to compensate. The position of the head, the limbs, and other parts of the body all affect decisions as to which muscles should be involved. This series of decisions is complex and, like other forms of muscle co-ordination, develops over time. Have you ever watched a young child just learning to stand or walk? The child works hard to keep her balance. However, older children and adults do not even think about maintaining balance because the cerebellum takes over. Although we consciously decide to stand, walk, or run, we do not have to consciously control all the separate muscle actions involved.

Indeed, most of the physical skills that we learn are slowly taken over by the cerebellum. The difference between a beginner and a more accomplished player of any game may be related to the degree to which the basics have been taken over by the cerebellum and are no longer consciously controlled. For instance, do you think an experienced hockey player concentrates on the plays of the game or on his or her skating?

The **thalamus** is a sensory relay centre. It receives sensations of touch, pain, heat, and cold, as well as information from the muscles. If the sensations are mild, the thalamus relays the information to the appropriate part of the cerebrum

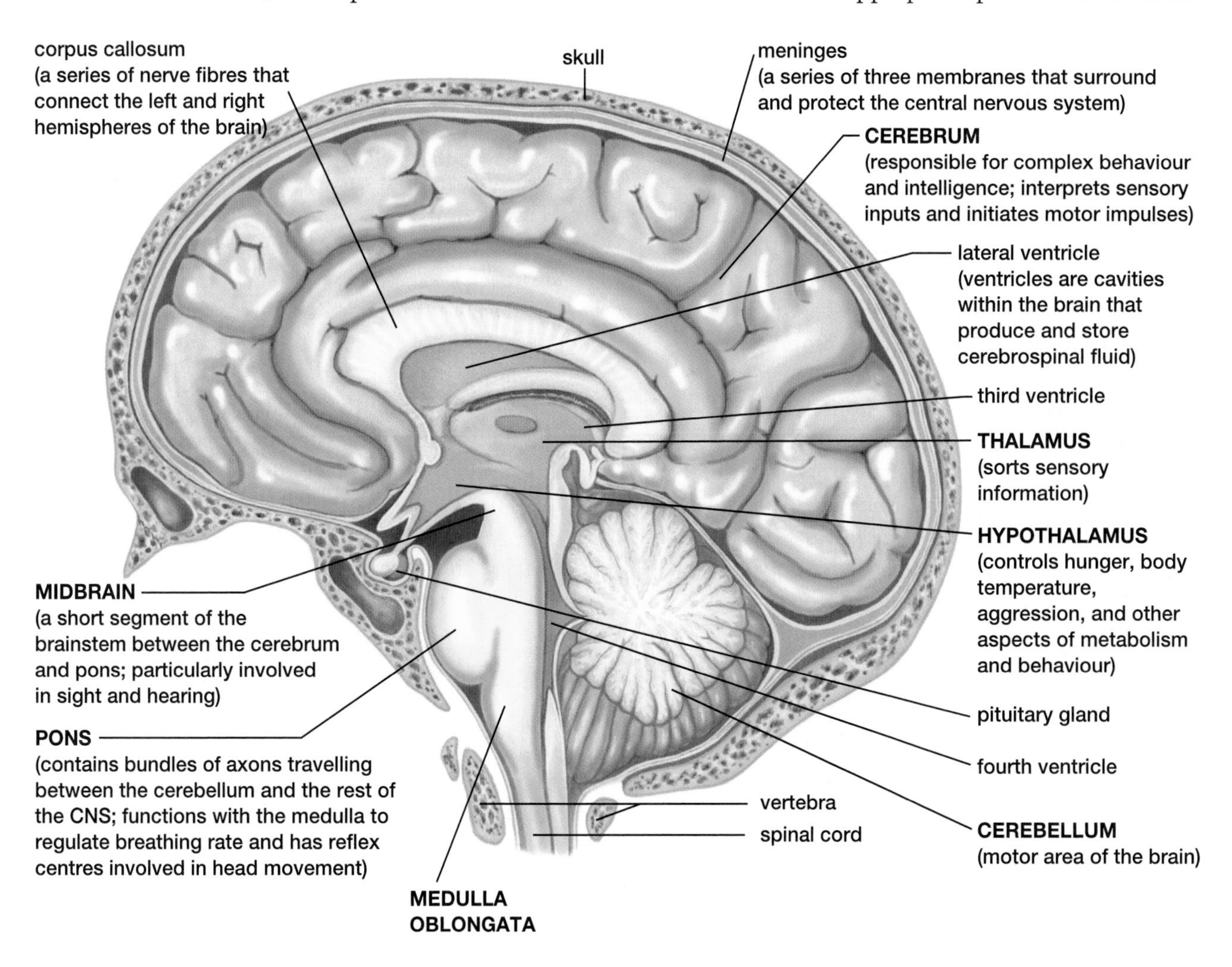

Figure 5.9 An overview of the structure and major functions of the brain

(the conscious part of the brain that will be discussed shortly). However, if the sensations are strong, the thalamus triggers a more immediate reaction while, at the same time, transferring the sensations to the homeostatic control centre — the hypothalamus.

The **hypothalamus** — to be studied in detail in Chapter 6 — is an incredibly complex and important bundle of tissues that acts as the main control centre for the autonomic nervous system. The hypothalamus enables the body to respond to external threats by sending impulses to various internal organs via the sympathetic nervous system. It re-establishes homeostasis after the threat has passed by stimulating the parasympathetic nerves.

The hypothalamus also controls the endocrine hormone system, the focus of Chapter 6. Attached to the hypothalamus is the pituitary gland, which is sometimes called the master gland. The pituitary gland produces hormones that control many of the other endocrine glands. For example, the pituitary gland produces the thyroid stimulating hormone (TSH) that stimulates the thyroid gland to make and release its hormone, thyroxine. (Thyroxine and other hormones will be discussed in greater detail in Chapter 6.)

The pituitary is actually two glands, both of which are controlled by the hypothalamus. The back, or posterior lobe of the pituitary is formed from the cells of the hypothalamus during fetal development. The hormones released from this lobe of the pituitary — ADH and oxytocin — are produced in the hypothalamus. The front, or anterior, lobe of the pituitary is controlled by stimulating factors — chemicals produced by the hypothalamus.

As you saw in Chapter 4, the hypothalamus controls water levels in the body through the release of ADH. It also monitors glucose levels to determine satiety or a feeling of fullness when we eat. The information on water and glucose levels is relayed from the hypothalamus to the brain, where it is combined with impulses from other organs (such as the stomach) to create our awareness of thirst and hunger.

As mentioned in Chapter 4, the hypothalamus also monitors body temperature and can trigger responses that increase or decrease that temperature. The hypothalamus maintains a biological clock as well. Generally, this clock is set to our 24-hour day and establishes when we feel tired and when we are alert. Have you ever experienced jet lag? If you have, then you know what it feels like when your biological clock is upset.

By monitoring and making us aware of all of these biological functions, the hypothalamus functions as the centre of homeostasis in the human body. However, this biological marvel cannot compare to the complexity of the human cerebrum.

Why is the cerebrum so complex? The **cerebrum** is the part of the brain in which all the information from our senses is sorted and interpreted. Voluntary muscles that control movement and speech are stimulated from this part of the brain. Memories are stored and decisions are made in this region as well. The cerebrum is what makes humans different from any other animal on the planet. It is the centre of human consciousness.

As shown in the photograph on page 136, the cerebrum is divided into two halves, called the left and right hemispheres. The surface of each half is covered with convolutions that increase the cerebrum's surface area.

The **cerebral cortex**, the thin layer that covers each hemisphere of the brain, contains over one billion cells. It is this layer that enables us to experience sensation, voluntary movement, and all the thought processes we associate with consciousness. The surface of the cerebral cortex is made up of grey matter, composed primarily of cell bodies and dendrites packed closely together for maximum interaction. The two hemispheres are joined by the **corpus callosum**, a layer of white

MINI LAB

Make a Model of a Brain

Some medical schools require students to build a model of the brain. You can use jelly, other types of food, modelling clay, or any other material that you think will be suitable to build a 3-D model of the brain. Refer to the diagram on page 145 and make each part of the brain (identified in bold capital letters) a different colour. Include a card that lists each part of the brain and its colour in your model.

Analyze

1. List the functions of each part of the brain that is shown in your model.
2. In what ways is your model similar to the human brain (in terms of size, texture, and so on)?
3. In what ways is your model different from the human brain?

matter made up of axons. The corpus callosum transfers impulses from one hemisphere to the other.

Figure 5.10 shows that the cerebrum is also divided into four lobes, each associated with different functions. The **frontal lobe** is involved in the control of muscles (motor areas) and the integration of information from other parts of the brain to help us reason. This area allows us to think critically and plan our actions. The **parietal lobe** receives sensory information from the skin and skeletal muscles, and is associated with our sense of taste. The **occipital lobe** receives information from our eyes, and the **temporal lobe** receives information from our ears.

In the next section, you will examine more closely the structure and function of the neurons in your body, as well as the high-speed communication network of which they are a part. You will see how chemical substances stimulate neurons to trigger the impulses that enable you to think, move, and respond to stimuli. In the final section of this chapter, you will study sensory receptors associated with sight, hearing, and touch.

WEB LINK

www.mcgrawhill.ca/links/biology12

Memory is an important cognitive function. What do you do to help remember important information for your next class test? To access procedures for memory tests, go to the web site above, and click on **Web Links**.

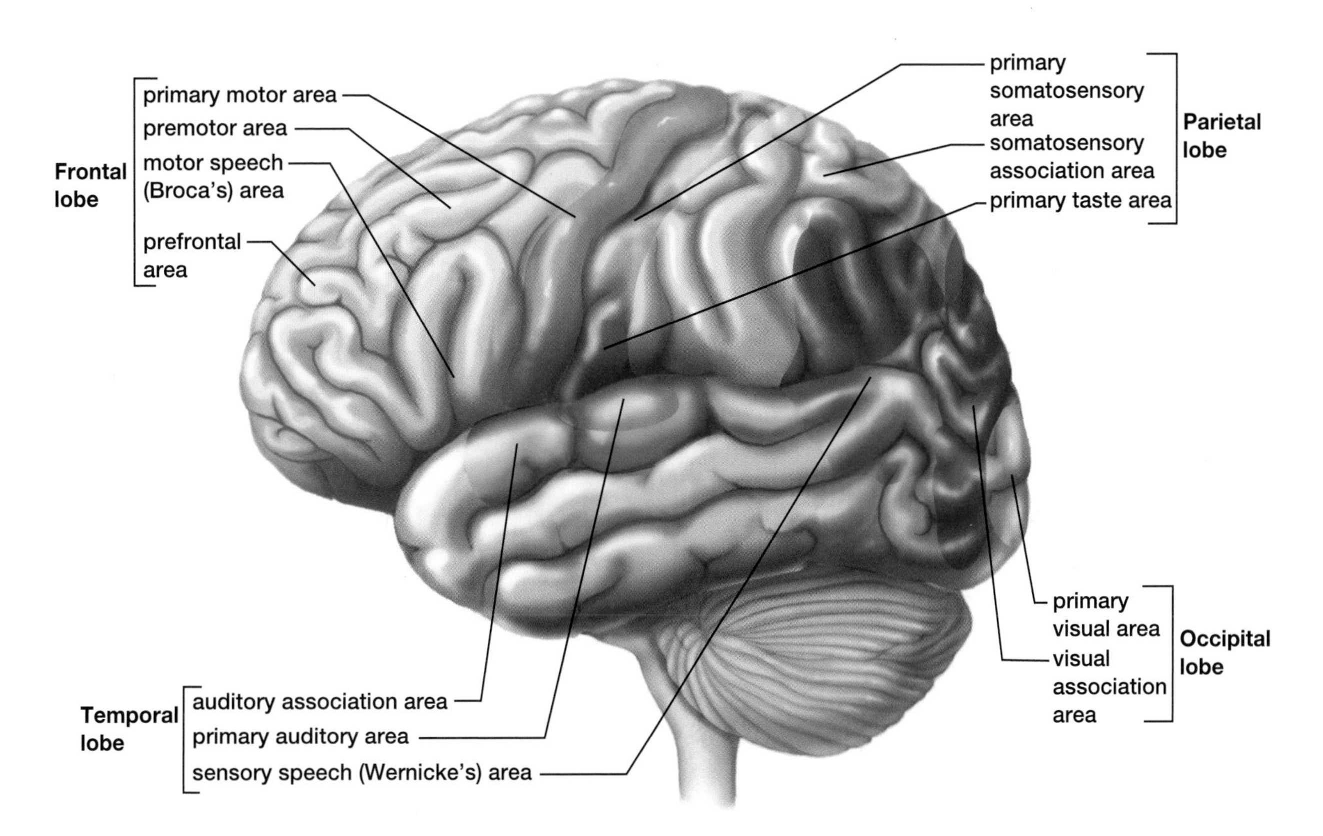

Figure 5.10 The cortex of the cerebrum is divided into four lobes: frontal, parietal, temporal, and occipital. The frontal lobe has motor areas and an association area called the prefrontal area. The other lobes have sensory areas and also association areas.

SECTION REVIEW

1. K/U Identify the two major parts of the human nervous system.

2. K/U How is the autonomic system different from the somatic system?

3. C Which parts of your nervous system are you using to complete this homework assignment? Make a diagram of yourself and label the parts of your nervous system that are working.

4. C In Chapter 4 you learned that insulin and glucagon are antagonistic hormones. Make a diagram similar to Figure 4.15 on page 122 to show that the sympathetic and parasympathetic nervous systems are antagonistic systems.

5. K/U What is a reflex arc? How is it useful to an organism like you?

6. C A sudden interruption in the blood supply to the brain (known as a stroke) can cause serious brain damage. Investigate how a stroke can affect a person's ability to speak naturally and clearly and to comprehend what others are saying (in other words, to communicate effectively). Present your findings in a brief report to the class.

7. K/U What factors contribute to the complexity of the brain?

8. K/U Explain how the EEG and PET scans help us to understand brain function.

9. MC How have studies of people with brain disorders helped us understand brain function? Explain your answer.

10. I Doctors sometimes conduct the "knee jerk" reflex test during a physical examination. This test involves quickly tapping a point just below the kneecap with a rubber mallet. In a normal response, the lower part of the leg jerks up and outward. What is the diagnostic value of this type of test? Try this test in class under the supervision of your teacher (use a ruler if mallets are not available and be careful to tap gently). How do individual responses to this test vary among your classmates? Make a sketch of the reflex arc involved in this type of test. In the sketch, include the sensory receptor, sensory nerve, motor nerve, effector, spinal cord, interneurons, and the brain (in outline). Use arrows to indicate the direction of the nerve impulses.

11. I Prions are proteins that can cause other proteins to change shape and lose normal function. Bovine spongiform encephalitis (BSE), also known as Mad Cow Disease, is caused by prion infection. BSE prions attack proteins in the brain, which eventually leads to deterioration of brain tissue. You are interested in conducting research about prions.
    (a) Identify some safety issues that need to be addressed before you start.
    (b) How could PET scans be helpful to monitor the progress of this disease in an infected person?
    (c) How could animal tests be useful?

12. MC Describe how damage to the CNS will have a different impact than damage to the PNS. As medical researchers study the effects of damage to both types of nervous systems, consider the use of implanting new stem cells or transplanting whole neurons to reverse damage to nerve tissue. Which strategy might be most effective for each system? Explain your thinking.

13. C Copy and complete the following chart.

Brain Structure	Location	Functions
Pons		
Midbrain		
Corpus callosum		
Cerebellum		
Cerebrum		

14. I If possible, find out the age at which you and your siblings took your first unassisted steps. Survey your classmates about this question, and plot the results on a graph. Use blue dots for boys and red dots for girls. Does there seem to be a gender difference with regard to the age at which infants learn to walk? Should parents or other caregivers encourage or prompt infant children to walk on their own?

15. C Discuss a situation in which you may have experienced the symptoms of a fight-or-flight reflex. How long did it take for your heart rate and breathing rate to return to normal? Compare your fight-or-flight experience with that of others in your class. Discuss the apparent differences, if any, in the way males and females experience the fight-or-flight reflex.

UNIT PROJECT PREP

Wilder Penfield was able to do research on the human brain because he had the "informed consent" of his patients and proof that his experiments did no harm. Gather some information on the ethics of both animal and human experimentation.

5.2 How the Neuron Works

EXPECTATIONS

- Describe the function of neurons.
- Explain the role of neurotransmitters in the central nervous system.
- Describe how neurons respond to a stimulus.

A neuron is formed by the same process of cell division as occurs in other cells. A neuron has a single nucleus, numerous mitochondria, ribosomes, lysosomes, and other organelles. However, neurons also have very specialized structures that make them different from other cells and that enable them to perform their unique functions.

As mentioned previously, a neuron is composed of dendrites, a cell body, and an axon. The axon sends a wave of depolarization along its length, which is part of the high-speed network that sends impulses from one part of the body to another. The wave of depolarization is primarily the movement of two positive ions (Na^+ and K^+) from one side of the axon's cell membrane to the other.

The Neuron at Rest

When a neuron is at rest, the outside of the membrane of the neuron is positively charged compared to the inside. This is the result of the uneven distribution of positively charged ions (cations) and negatively charged ions (anions). Outside the cell there are high concentrations of sodium ions (Na^+) and lower concentrations of potassium ions (K^+). Chloride (Cl^-) is the dominant anion exterior to the cell. Inside the cell there is a high concentration of K^+, a lower concentration of Na^+, and the dominant anions are negatively charged proteins, amino acids, phosphates, and sulfates. The membrane has specialized channels or gates for the movement of Na^+, K^+, and Cl^-, but the larger anions (such as proteins and amino acids) are trapped within the cell. As shown in Figure 5.11 on the following page, the movement of Na^+ and K^+ is critical to the wave of depolarization. At rest, the membrane is 50 times more permeable to K^+ than to Na^+. That is, while Na^+ is moving into the cell, there is more K^+ diffusing out of the cell. As this happens, the inside of the cell becomes increasingly negatively charged because the larger anions are trapped inside. Although the increasing negative charge within the cell attracts both the Na^+ and K^+, this force is offset by the Na^+/K^+ pump, which is found in the cell membrane. The Na^+/K^+ pump uses active transport to pull three Na^+ cations from the inside of the cell to the outside. In exchange, two K^+ cations are pulled from outside to inside the cell, thereby increasing the difference in charge. The final result is a relatively negative charge inside the cell compared to the outside. This charge (due to the unequal distribution of cations and anions) can be measured using tiny micro-electrodes placed inside and outside the membrane. At rest, the difference in charge is approximately –70 mV. This difference is referred to as the **resting potential**.

The All-or-none Principle

Sensory neurons can be stimulated by chemicals, light, heat, or the mechanical distortion of their membrane. Motor neurons and the neurons of the central nervous system are usually stimulated by **neurotransmitters**, which are chemicals secreted by other neurons. Neurons can also be stimulated experimentally using an electrical current. If the neuron is given a mild electrical stimulus, there is a brief and small change in the charge of the cell membrane near the point of stimulation. The axon itself does not send a wave of depolarization along its length. However, if the electrical stimulus is strong enough (that is, if it reaches the threshold of stimulus), a wave of depolarization will sweep along the surface of the axon.

An axon is governed by the **all-or-none principle**. If an axon is stimulated sufficiently (above the threshold), the axon will trigger an impulse down the length of the axon. The strength of the response is uniform along the entire length of the axon. Also the strength of response in a single neuron is independent of the strength of the stimulus. An axon cannot send a mild or strong response; it can only respond or not respond. The threshold of

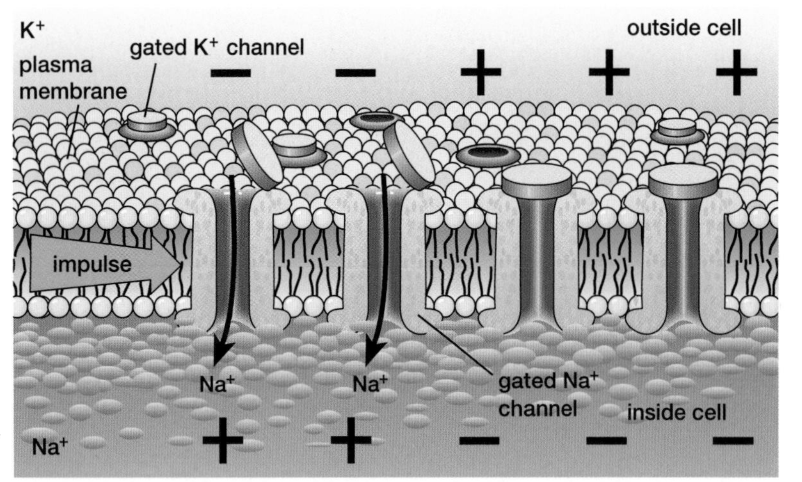

A Gated sodium ion channels open, allowing sodium ions to enter and make the inside of the cell positively charged and the outside negatively charged.

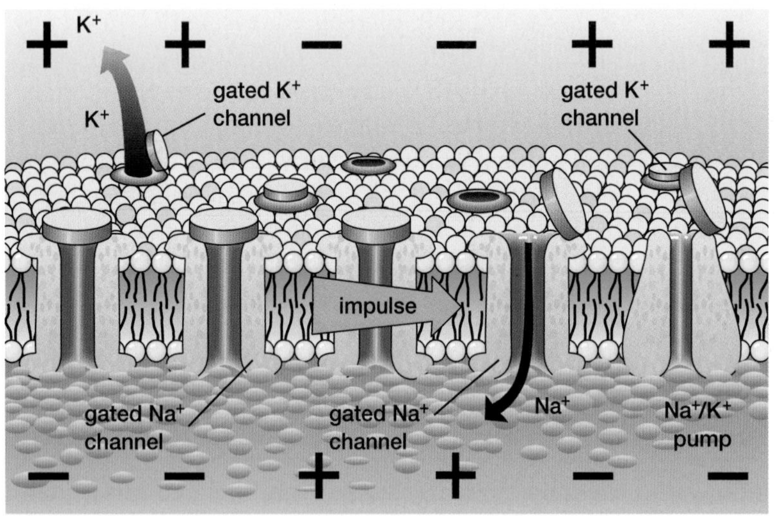

B As the impulse passes, gated sodium ion channels close, stopping the influx of sodium ions. Gated potassium ion channels open, letting potassium ions out of the cell. This action repolarizes the cell.

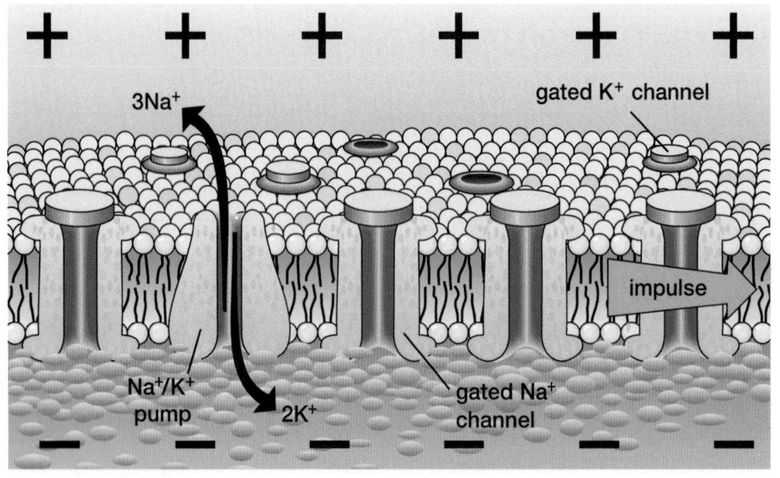

C As gated potassium ion channels close, the Na^+/K^+ pump restores the ion distribution.

Figure 5.11 The membrane of a neuron contains open as well as gated channels that allow movement of sodium ions (Na^+) and potassium ions (K^+) into and out of the cell. The gated channels open and close as a wave of depolarization moves down the axon of a neuron.

stimulus is like the trigger of a gun. Once the pressure on the trigger is strong enough, the bullet is on its way — pulling harder on the trigger has no effect on the speed of the bullet.

Depolarization

As mentioned above, when a neuron is sufficiently stimulated, a wave of depolarization is triggered. When this occurs, the gates of the K^+ channels close and the gates of the Na^+ channels open. Sodium ions move into the axon. This input of positively charged ions neutralizes the negative charge in the axon. This change in charge is called the **action potential**. The depolarization of one part of the axon causes the gates of the neighbouring Na^+ channels to open, and this depolarization continues along the length of the axon. Note that action potentials can occur in the dendrites and in the cell body as well.

Repolarization

Any specific region of the axon is only depolarized for a split second. Almost immediately after the sodium channels have opened to cause depolarization, the gates of the K^+ channels re-open and potassium ions move out. The Na^+ channels close at the same time. This process, combined with rapid active transport of Na^+ out of the axon by the Na^+/K^+ pump, re-establishes the polarity of that region of the axon. The speed with which this process occurs allows an axon to send many impulses along its length every second, if sufficiently stimulated. The brief time between the triggering of an impulse along an axon and when it is available for the next impulse is called the **refractory period**. For many neurons, the refractory period is approximately 0.001 s.

One advantageous effect of a wave of depolarization is that an impulse can move along the entire length of a neuron and the strength of the signal does not dissipate. The signal moves at about 2 m/s. However, in some cases it is important that a wave of depolarization travel faster than this. For a wave of depolarization travelling along a neuron to the small intestine, speed is not a priority. But if an object is moving rapidly toward your eye, hundredths or even thousandths of a second may count. The speed of a wave of depolarization is increased by the addition of a fatty layer called the **myelin sheath**. As shown in Figure 5.12, this layer is formed by **Schwann cells** lined up along the length of the axon.

ELECTRONIC LEARNING PARTNER

To learn more about channel behaviour in a neuron, go to your Electronic Learning Partner.

BIO FACT

In the PNS, the myelin sheath is made up of Schwann cells. In the CNS, the myelin sheath is made up of cells called oligodendrocytes.

Between each Schwann cell is a gap called the **node of Ranvier**, where the membrane of the axon is exposed. A nerve impulse that travels along a myelinated neuron is able to jump from one node of Ranvier to the next. This ability speeds up the wave of depolarization to 120 m/s. Myelinated nerve fibres are found in the central nervous system and in the peripheral nervous system, wherever speed is an important part of the function of a neuron.

The Schwann cells perform another important function. As mature cells, most neurons are incapable of reproducing themselves. This means that damage to the nervous system either by accident or disease cannot follow the same healing process that occurs in other parts of the body. However, neurons that have a neurolemma (the outer layer of the Schwann cells) are capable of regenerating themselves if the damage is not too severe. If a neuron is cut, the severed end of the axon grows a number of extensions or sprouts, and the original axon grows a regeneration tube from its neurolemma. If one of the sprouts from the severed section connects with the regeneration tube, the axon can re-form itself. Even if the muscle tissue that the axon was attached to has atrophied, the muscle will regrow when stimulated by the repaired axon.

Damaged neurons in the CNS cannot regenerate, but if an area of the brain itself is damaged, its functions can often be taken over by other parts of the brain. With extensive rehabilitation, the patient may be able to recover. Damage to the spinal cord is usually permanent, however, and can lead to paralysis, as was the case for actor Christopher Reeve, shown in Figure 5.13.

The repair of brain and spinal injuries is a major area of medical research. One recent study has identified a gene that inhibits spinal regeneration. This gene, designated Nogo, produces a protein that prevents neurons of the CNS from regenerating. It is believed that this protein is produced to prevent wild, uncontrollable growth of tissue. Researchers hope this discovery will lead to drug therapies that will enable damaged CNS tissue to regenerate. Research on mice at the University of Toronto is also showing promise in repairing spinal

axon
myelin sheath
Schwann cell cytoplasm
A
node of Ranvier
myelin sheath
Schwann cell nucleus
400 nm
B
axon

Figure 5.12 (A) A myelin sheath forms when Schwann cells wrap themselves around a nerve fibre. (B) Electron micrograph of a cross section of an axon surrounded by a myelin sheath.

Figure 5.13 Because of a spinal cord injury incurred in an equestrian accident in 1995, actor Christopher Reeve is confined to a wheelchair. He actively campaigns for funding on behalf of spinal cord research.

cord injuries. In addition, Peter Erikson from Göteborg University in Sweden and his co-worker Fred Gage found as many as 200 new neurons per cubic millimetre of tissue in the brains of some patients who had recently died of cancer. This discovery involved using a specialized drug used to trace the formation of new cells. The scientists estimate that up to 1000 new neurons may be created each day, even in the brains of people in their 50s and 70s. These new neurons apparently arise not by mitosis of mature neurons but from a reserve of embryonic stem cells. These cells are found in some parts of the brain and do not form into specialized cells during the development of the brain. Similar stem cells are found in bone marrow and are responsible for the formation of the wide variety of blood cells found in the body.

The Synapse

Neurons do not touch one another; there are tiny gaps between them. These gaps are called **synapses**. The neuron that carries the wave of depolarization toward the synapse is called the **presynaptic neuron**. The neuron that receives the stimulus is called the **postsynaptic neuron**. When a wave of depolarization reaches the end of a presynaptic axon, it triggers the opening of special calcium ion gates. The calcium triggers the release by exocytosis of neurotransmitter molecules. The neurotransmitter is then released from specialized vacuoles called **synaptic vesicles**, which are produced in the bulb-like ends of the axon. The neurotransmitter diffuses into the gap between the axon and dendrites of neighbouring postsynaptic neurons, as shown in Figure 5.14. The dendrites have specialized receptor sites and the neurotransmitter attaches to these receptors and excites or inhibits the neuron. The **excitatory response** involves the opening of sodium gates, which triggers a wave of depolarization. The **inhibitory response** makes the post-synaptic neuron more negative on the inside in order to raise the threshold of stimulus. This process is usually accomplished by opening chloride channels to increase the concentration of these negative ions in the neuron.

ELECTRONIC LEARNING PARTNER

To learn more about how the synapse works, go to your Electronic Learning Partner.

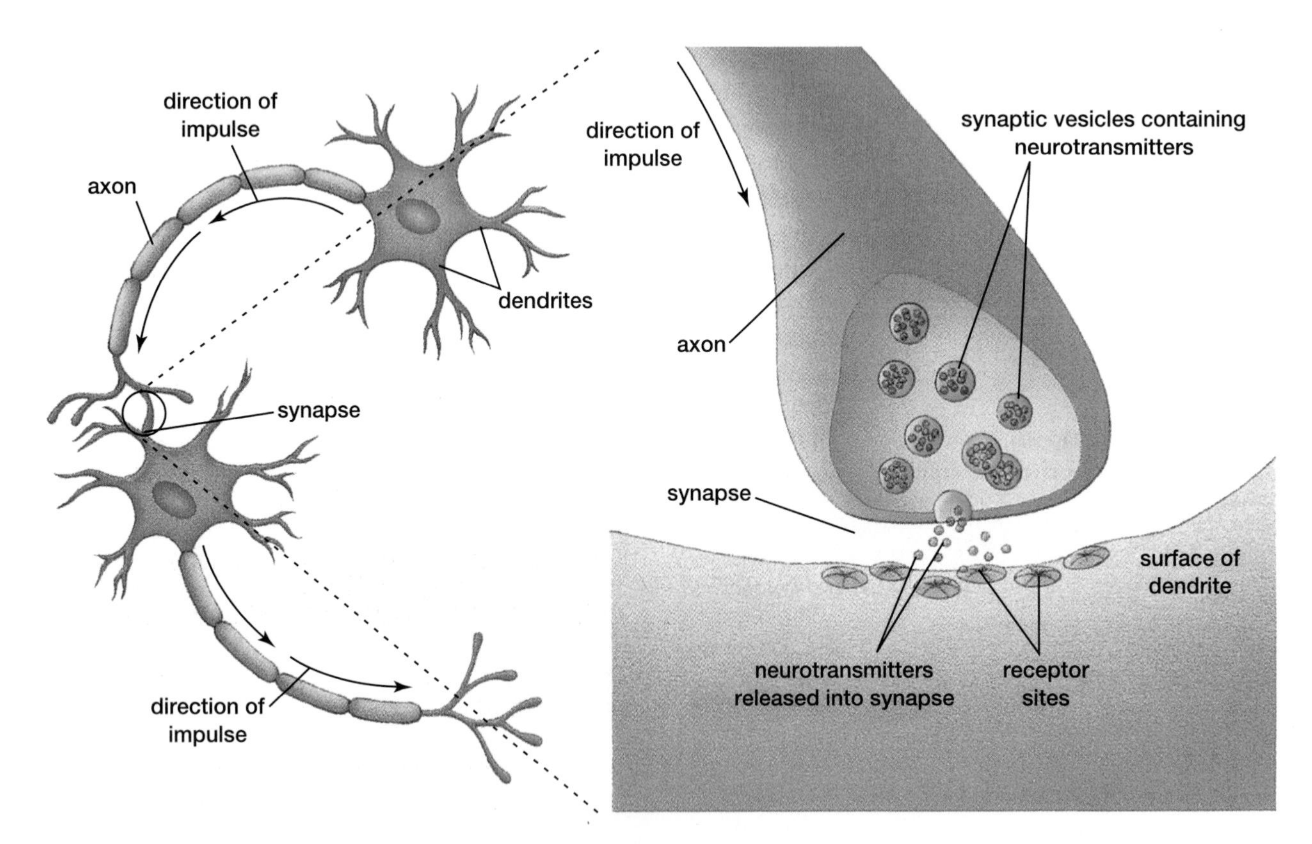

Figure 5.14 At a synapse, neurotransmitters pour into specific sites on nearby dendrites.

Neurotransmitters can also stimulate or inhibit cells that are not neurons. This applies particularly to muscles, where the neurotransmitter from the neuron triggers the contraction of the muscle. Also, the adrenal gland is composed of modified neurons of the sympathetic nervous system that release the neurotransmitters adrenaline and noradrenaline into the blood as hormones. They have a variety of effects that will be discussed in Chapter 6.

The neurotransmitter that enters the synapse and attaches to the postsynaptic receptors is broken down almost immediately by an enzyme released from the presynaptic neuron. For example, the enzyme **cholinesterase** breaks down the neurotransmitter **acetylcholine**. Acetylcholine is the primary neurotransmitter of both the somatic nervous system and the parasympathetic nervous system. Acetylcholine can have excitatory or inhibitory effects. This neurotransmitter stimulates skeletal muscles but inhibits cardiac muscles. **Noradrenaline**, also called norepinephrine, is the primary neurotransmitter of the sympathetic nervous system.

As is evident from the above discussion, the neurons of the brain involve a wide variety of neurotransmitters that have numerous functions. **Glutamate** is a neurotransmitter of the cerebral cortex that accounts for 75 percent of all excitatory transmissions in the brain. Gamma aminobutyric acid (**GABA**) is the most common inhibitory neurotransmitter of the brain. Many of the brain's neurotransmitters have multiple functions. **Dopamine** elevates mood and controls skeletal muscles, while **seratonin** is involved in alertness, sleepiness, thermoregulation, and mood.

The study of neurotransmitters forms a specialized branch of biology. New discoveries are being made almost daily, and today there are more than 100 neurotransmitters that have been confirmed or are inferred. These discoveries confirm that the brain is incredibly complex. The complexity of the brain has always formed a barrier to the treatment of disorders of the mind. An understanding of the brain's neurotransmitters allows scientists an opportunity to treat a number of complex neurological problems. Drugs have been developed to stimulate or inhibit specific neurotransmitters. For example, Valium™ increases the level of the neurotransmitter GABA to alleviate anxiety. Prozac™, an antidepressant, enhances the action of seratonin. Better understanding of the chemistry of the brain will pave the way to increasing our knowledge of how the brain works. Read section 5.3 to learn more about the physiology of the sensory receptors that channel information to the brain.

SECTION REVIEW

1. **C** Make a diagram of a neuron and label it.
2. **C** Referring to your diagram from Question 1, show where each of these ions is concentrated for a neuron at rest: Na^+, K^+, Cl^-, and other ions.
3. **K/U** The phospholipid bilayer is not very permeable to ions. How can ions diffuse into or out of a nerve cell?
4. **K/U** Explain what is meant by resting potential. What is the approximate measure of resting potential?
5. **K/U** An axon is governed by the all-or-none principle. What does this mean?
6. **K/U** What is an action potential? How is it started?
7. **C** Use diagrams to show the relative concentration of ions during depolarization and repolarization.
8. **K/U** How can some neurons grow new axons after injury?
9. **C** Make a diagram with labels to explain how the presynaptic axon responds to the action potential. Show that the dendrite of the next neuron responds if there are *not* enough neurotransmitters to meet the threshold.
10. **K/U** What is the role of an enzyme like cholinesterase?
11. **I** Much of what is known about the activity of motor neurons is based on research of the axon of the giant squid. This large cell extends from the head to the tail of the squid. The table shows how the external environment of the neuron can be changed, without changing the internal environment of the cell.
    (a) Copy and complete the table.

Solutes in the water outside the cell	Nerve impulse conducted?	Proposed explanation
Na^+ only		
K^+ only		
Cl^-		

(b) Suggest three other possible solutions you could use to alter the external environment of the neuron.

cones, but they are unable to distinguish colours. Cones require more light to stimulate them, but they are able to detect red, green, and blue.

The eye has two chambers divided by the lens. The anterior chamber between the cornea and the lens is filled with the fluid **aqueous humour**. The cornea and anterior chamber act as a pre-lens to initiate the process of focussing an image on the retina before it encounters the lens. The lens completes the process. The posterior chamber, behind the lens, is filled with a clear gel (**vitreous humour**) that helps maintain the shape of the eyeball.

How the Eye Functions

As light enters the eye, the pupil dilates if there is insufficient light or constricts if there is too much. The pupil also constricts when you focus on something close, to reduce the distortion that occurs around the outside field of view.

The shape of the lens changes in response to your distance from the object being viewed. Figure 5.16 illustrates that if you are looking at something far away, the ciliary muscles relax. The suspensory ligaments, which are attached to the ciliary muscles, become taut and the lens flattens. When you focus on something close, the ciliary muscles contract, pulling forward and releasing the tension on the suspensory ligaments. The lens then becomes more rounded. This adjustment is called **accommodation**.

BIO FACT

Some evidence shows that the pupil of the eye dilates when the person is interested in what he or she is viewing.

The image is focussed on the retina, which is composed of three layers — the ganglion cell layer, the bipolar cell layer, and the rod and cone cell layer (see Figure 5.17 on page 156). Bipolar cells synapse with rods or cones and transmit impulses to the ganglion cells. The ganglion cells join together and form the optic nerve as they exit the eye. The retina is composed of approximately 150 million rod cells and 6 billion cone cells. Both function using a purple pigment called **rhodopsin**. When light strikes this pigment, rhodopsin breaks down into two proteins — retinal (which is formed from vitamin A) and opsin (which releases the energy

Investigation 5 • A

SKILL FOCUS

- Predicting
- Performing and recording
- Analyzing and interpreting
- Conducting research

The Effect of Light on Pupil Size

The size of the pupil is controlled by ciliary muscles of the iris. The size of the pupil changes with the amount of light. This reaction is called the **pupillary reflex**.

Pre-lab Question

- What is the advantage of the pupillary reflex?

Problem

Does the colour of light affect the speed of the pupillary reflex?

Prediction

Predict whether the colour of light affects the speed of the pupillary reflex and, if it does, which colour will have the greatest effect.

Materials

stop watch
flashlight
pieces of coloured cellophane (red, yellow, green, blue) large enough to cover the flashlight lens

required to stimulate a bipolar cell). The two proteins are then rejoined in a process that requires energy from ATP.

Rods are very sensitive to light and are therefore ideal for night vision. However, rods cannot distinguish colours, resulting in images that are primarily made up of shades of grey. Cones require more light to be stimulated. They exist in three forms, which are characterized by slight changes in the structure of opsin. The three kinds of cones are sensitive to red, green, or blue wavelengths of light. Each cone is connected to a separate bipolar cell, which enables the brain to form a very detailed image from the information it receives. Many rods (sometimes as many as 100) can be attached to a single bipolar cell. This is why night vision is blurry and indistinct.

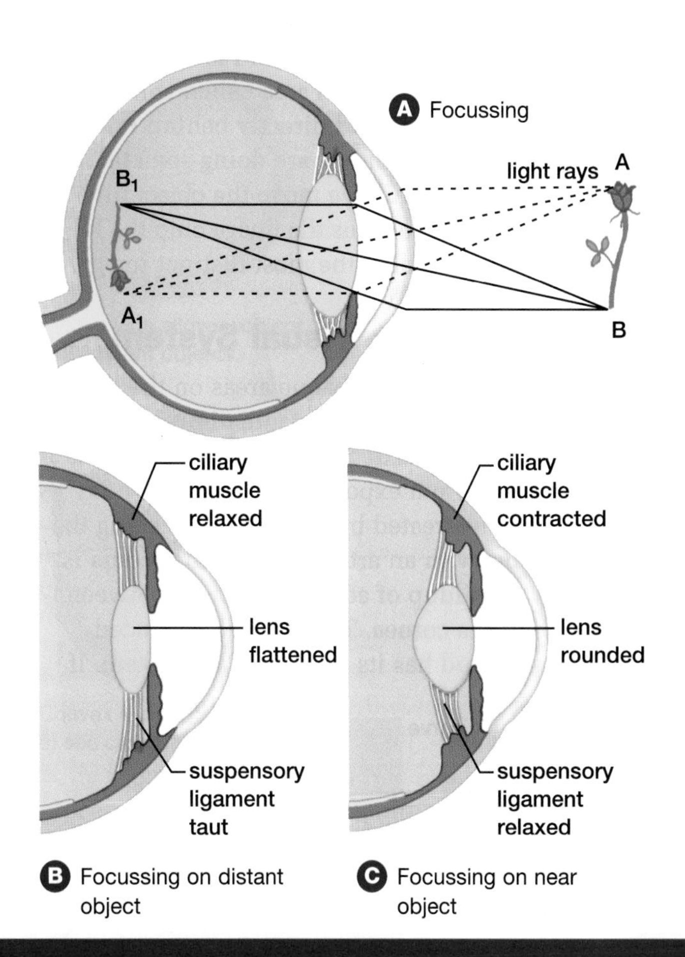

Figure 5.16 (A) Light rays from each point on an object are bent by the cornea and the lens in such a way that an inverted and reversed image of the object forms on the retina. (B) When focussing on a distant object, the lens is flat because the ciliary muscle is relaxed and the suspensory ligament is taut. (C) When focussing on a near object, the lens accommodates. It becomes rounded because the ciliary muscle contracts, thus causing the suspensory ligament to relax.

Procedure

1. This is a controlled experiment. The more variables you control, the more reliable your data will be.
2. To be consistent, select one member of each lab group to be the subject of the experiment, and allow at least 2 min between trials. Another member of the lab group should hold the flashlight, and a third should be in charge of observing the subject's eye and timing the pupillary reflex.
3. Prepare a data table.
4. The subject covers one eye with his or her hand for 1 min and then removes the hand. Time how long it takes for the pupil to constrict until it stops constricting. Repeat this step three times.
5. Repeat step 4 using the flashlight covered with one colour of cellophane to shine at the subject's eye after the eye is uncovered. Record your results. **CAUTION:** Do not shine the uncovered flashlight into the subject's eye.
6. Alternate colours of cellophane until you have completed three trials with each colour.
7. Graph your results. Be sure to use the format that will best illustrate your data.

Post-lab Questions

1. Why did you alternate colours during the experiment?
2. Would the experiment be more valid if you used more that one subject? Explain.
3. What are the advantages and disadvantages of doing this experiment in one session?
4. What is the purpose of taking the first set of readings in normal room lighting?

Conclude and Apply

5. Based on your data, does the colour of light affect the pupillary reflex?

Exploring Further

6. What other factors might affect the speed of the pupillary reflex?
7. What controls the pupillary reflex?

Merkel discs, have flattened endings that respond to light touch. There are also nerve endings wrapped around hair follicles — they detect movements of the hairs.

Other receptors, as shown in Figure 5.19, are dendrites that are associated with specialized structures. Meissner's corpuscles respond to light touch or texture. Usually found in sensitive areas such as the fingertips, they allow you to tell the difference between, for example, cotton and leather. Krause end bulbs are similar to Meissner's corpuscles but are found in mucous membranes. Pacinian corpuscles have a series of layers and respond to deep pressure, stretching, tickling, and vibration. Ruffini endings respond to heavy touch, pressure, and stretching of the skin as well.

Meissner's and Pacinian corpuscles are receptors that initiate a rapid series of nerve impulses when first stimulated. Other receptors may continue to send impulses, but all are subject to adaptation. **Adaptation** is the process by which a receptor responds to a change in the environment, but if the stimulus continues over a long period of time, we lose awareness of it. This response explains why we are aware of our clothes when we first put them on, but after a few minutes we do not notice them. In some cases the adaptation is a function of the receptor, and in other cases the brain filters out the sensations from our consciousness.

The fact that the skin is exposed to the outside world presents a challenge in terms of maintaining homeostasis. The skin is exposed to many hazards that result in cuts, burns, and bruises, but the body has a mechanism to repair them, as shown in Figure 5.20 on page 160.

Investigation 5 • B

SKILL FOCUS

- Performing and recording
- Analyzing and interpreting
- Communicating results

How Do We Hear?

Pre-lab Question

- How does the ear interpret sound waves?

Problem

How does distance affect hearing?

Prediction

Predict at what distance a person can no longer reliably hear a series of numbers.

Materials

paper
container without a lid

Procedure

1. Select a quiet area that has a large amount of space (such as an empty hallway).
2. On separate pieces of paper, write a series of five random numbers. Be sure there is an equal distribution of one-, two-, three-, and four-syllable numbers in each group.
3. Place the pieces of paper in a container.
4. The person who chooses the numbers will read out one group of five numbers to a partner who is 3 m away. It is important that the reader try to maintain the same tone and volume of voice throughout the experiment.
5. The partner will write down the numbers he or she hears.
6. Keep repeating the procedure at 3 m intervals until the partner can no longer hear anything intelligible.
7. Graph the number of mistakes versus the distance in a scatter plot. (Your teacher will distribute a handout related to scatter plots.)

Post-lab Questions

1. At what distance could your partner no longer hear the numbers reliably?
2. Compare your results with those of other lab groups in the class. What are some of the reasons for variation in the data?

Conclude and Apply

3. How does distance affect a person's ability to hear sounds?
4. What do you think would happen if you tried this experiment outside? Explain your answer.

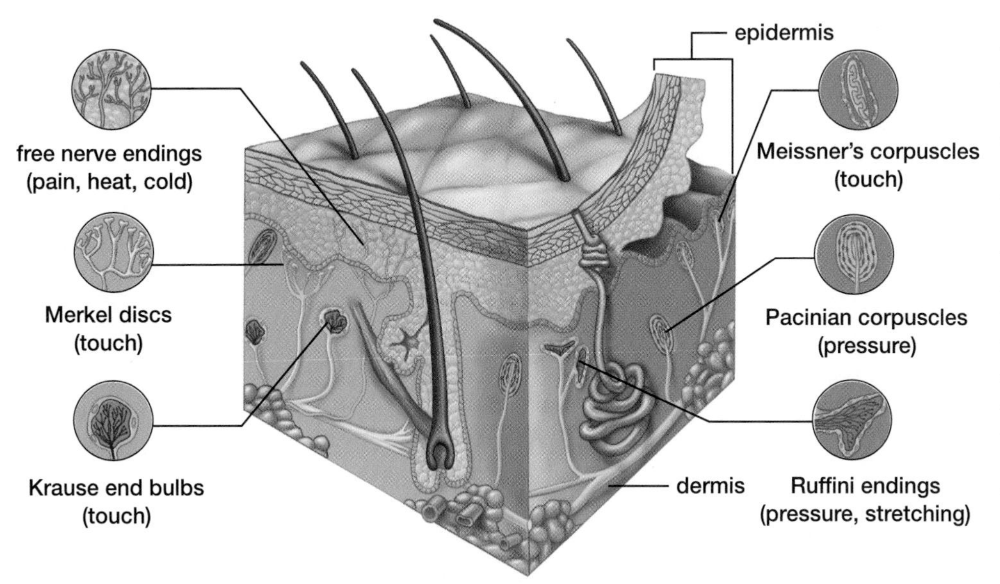

Figure 5.19 This classic view shows each receptor as having the main function indicated. However, scientific investigations indicate that matters are not so clear-cut. For example, microscopic examination of the skin of the ear shows only free nerve endings (pain receptors), and yet the skin of the ear is sensitive to all sensations. Therefore, it appears that the receptors of the skin are somewhat, but not completely, specialized.

Exploring Further

5. Use sources such as other texts, reference books, or the Internet to match the labels below to the numbers in the diagram. Record the function of each part.

auditory canal	pinna
auditory tube	round window
cochlea	semicircular canals
cochlear nerve	stapes
incus (anvil)	tympanic membrane
malleus (hammer)	vestibule
oval window	vestibular nerve

6. The ear contains structures that provide information to the cerebellum to maintain equilibrium. Use resources other than this textbook to answer the following questions.
   (a) What parts of the ear are used to maintain equilibrium?
   (b) What is the difference between dynamic and static equilibrium?
   (c) Explain which parts of the ear are responsible for dynamic equilibrium and which are responsible for static equilibrium. Describe how these structures work.

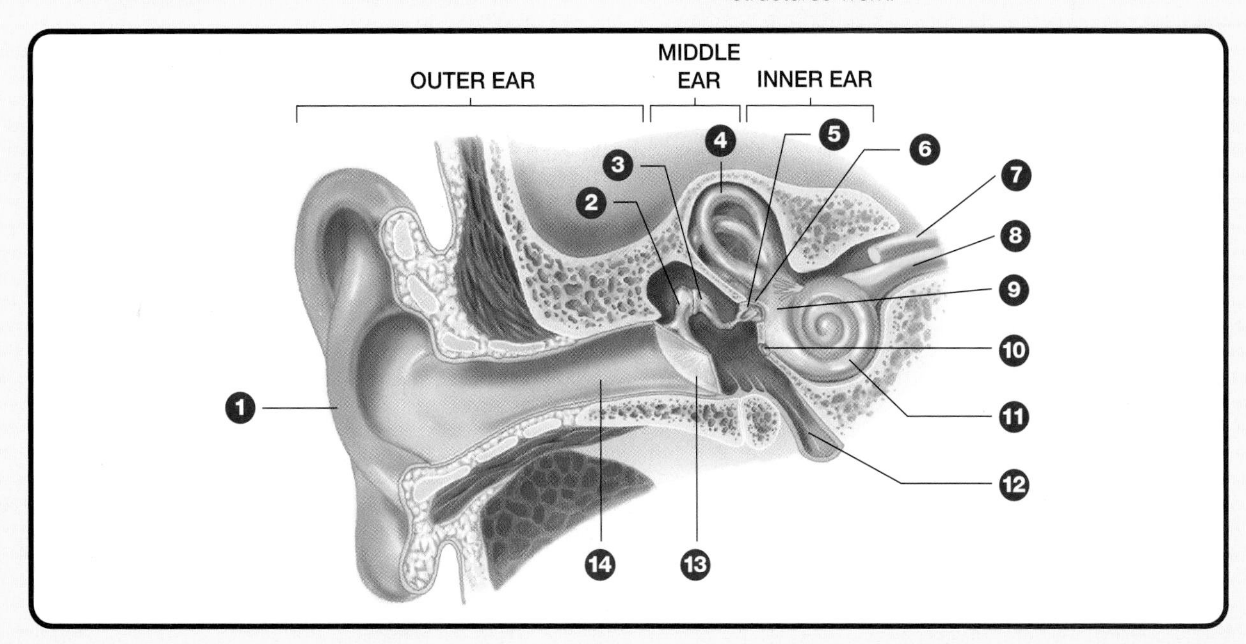

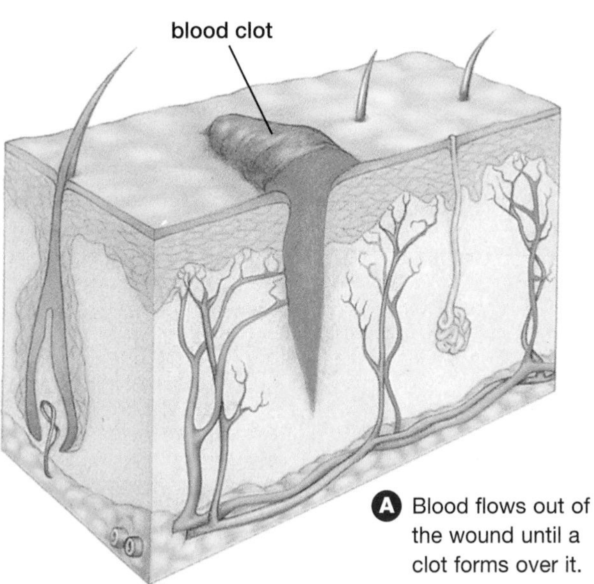

A Blood flows out of the wound until a clot forms over it.

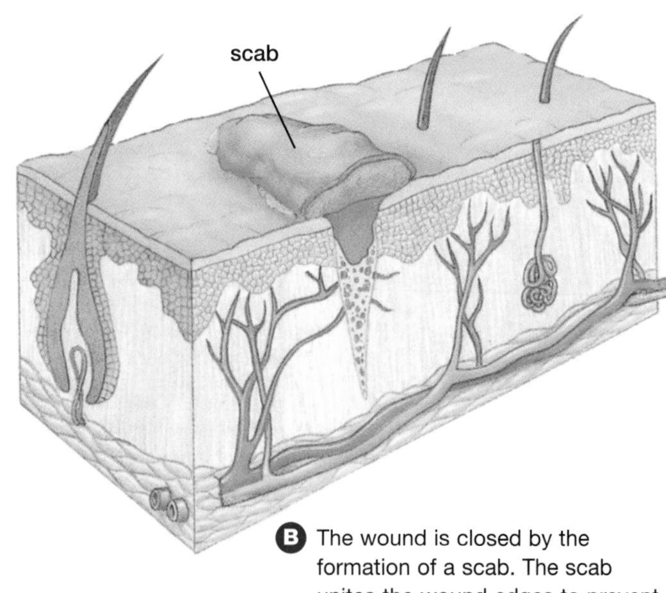

B The wound is closed by the formation of a scab. The scab unites the wound edges to prevent bacteria from entering. Blood vessels dilate, and infection-fighting cells speed to the wound site.

Figure 5.20 When a skin injury extends into the dermis, the first reaction of the body is to restore the continuity of the skin. This prevents the invasion of harmful bacteria that live on the skin.

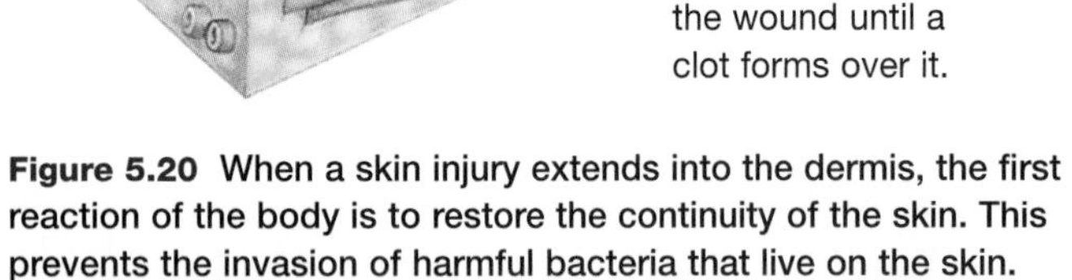

SKILL FOCUS

- Initiating and planning
- Predicting
- Identifying variables
- Performing and recording
- Analyzing and interpreting

Invertebrate Response to External Stimuli

Different species have developed diverse homeostatic systems in order to survive. These systems are adaptations that are the result of natural selection — they developed because they were successful in the organism's environment. Therefore, a particular stimulus may result in different responses in different species. To be able to conduct an experiment on the behaviour of an organism, you should know about its natural environment and physiology. For example, some invertebrate species are attracted to light, while others avoid it. The same applies to water, heat, and a variety of other stimuli. This activity can be carried out using any one of a number of invertebrate species, but one possibility is the sowbug (*Porcellio scaber*), shown below.

Sowbug

Problem

How can you demonstrate a particular invertebrate's responses to a chosen stimulus?

Hypothesis

Based on your research, formulate a hypothesis stating how the invertebrate will respond to the stimulus. The hypothesis will form the basis of your experimental design.

CAUTION: **Treat your invertebrate organism with respect. Wash your hands after completing the experiment.**

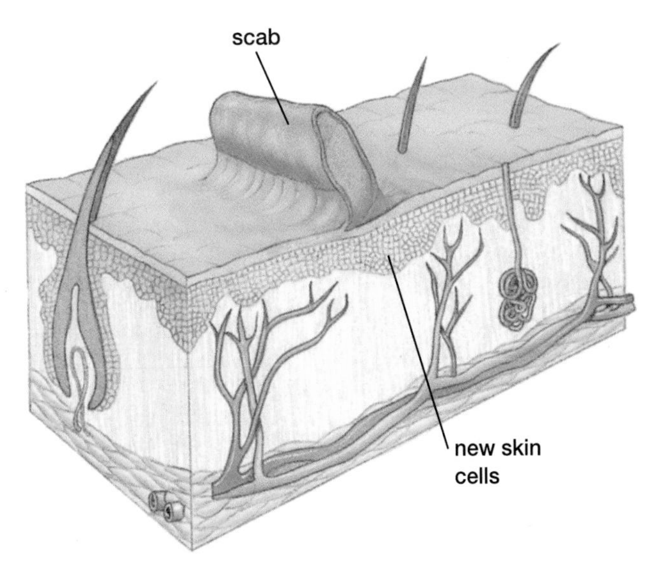

C Skin cells beneath the scab begin to multiply and fill in the gap. Eventually, the scab falls off to expose new skin. If a wound is large, a scar may result from the formation of large amounts of dense connective tissue fibres.

BIO FACT

In 1998, scientists at the Salk Institute in La Jolla, California, appeared to disprove a long-standing "fact" that the human brain cannot grow new neurons once it reaches adulthood. Since then, there is increasing evidence that stem cells may be able to produce neurons. As you learned in previous studies, stem cells are immature cells that can divide repeatedly, producing many different kinds of tissues. A team at McGill University in Montreal led by Freda Miller has forced stem cells to send out neuronlike arms and to express some of the same proteins that neurons do. They have culled these stem cells from the skin of rats and the scalps of adult humans. Perhaps one day researchers will grow petri dishes full of neurons from a patient's own skin! The goal is to use transplanted neurons culled from skin and bone marrow cells to repair damage caused by Parkinson's disease, Alzheimer's disease, multiple sclerosis, stroke, or head trauma.

Materials

These will vary according to your experimental design.

Experimental Plan

1. Select an invertebrate species for your study. Sample organisms may be found outdoors or bought at a pet store or bait shop. After you have made your choice, gather information on the species to provide the background you will need to design your experiment.
2. Choose a particular stimulus (such as heat, light, water, etc.) that you predict will produce a consistent response by the organisms.
3. Design an experiment that will show a positive or negative response to the chosen stimulus by the organisms. Be sure that your design has only one experimental variable and that you eliminate as many other variables as you can. Consider building an apparatus to provide the necessary controls.

Checking the Plan

1. What will your independent variable be? What will your dependent variable be?
2. How will you control other variables?
3. What measurements will you make?
4. How many trials will you carry out?

Data and Observations

Record your data in a suitable data table. Be sure you have enough repetitions to make your results reliable.

Analyze

1. Did the animals in your experiment always behave in the same way? Explain.

Conclude and Apply

2. Did the animals in your experiment show a consistent response to the stimulus?
3. Very few experiments are free from possible sources of error. What are some other factors that may have affected the behaviour of the animals?

Exploring Further

4. A learned behaviour is different from a stimulus–response reaction. How are they different?

ELECTRONIC LEARNING PARTNER

To learn more about how certain stimuli affect invertebrates, go to your Electronic Learning Partner.

SECTION REVIEW

1. K/U How is sensory information important to you?

2. K/U List three parts of your body that relay sensory information to the brain.

3. C Draw a diagram of the eye. Use a coloured marker to highlight those parts of the eye that are part of the nervous system. Use a differently coloured marker to highlight the part of the eye that is mostly muscle. Use a third coloured marker to highlight the parts of the eye that are vascular (tissue supplied by blood).

4. K/U Describe the path of light through each part of the eye, from the outermost structure to the retina.

5. K/U Explain why there is a blind spot in one portion of the retina.

6. K/U Identify those parts of the eye that are responsible for allowing light in.

7. K/U Identify those parts of the eye that are responsible for keeping light out.

8. C Draw a feedback loop showing how the eye responds to low levels of light.

9. K/U Distinguish between rod cells and cone cells and identify the advantage of each.

10. K/U List three disorders of the visual system.

11. C How would you explain to a young child the nature of the damage that can be done to the eye by staring directly at the Sun in the sky?

12. C Describe how your eye focusses on details of close images, such as the words on this page. Make a sketch that shows how the lens of the eye changes shape as you focus on near and distant objects.

13. K/U Identify how the skin collects information about the outside world.

14. I Try the following experiment to find out which parts of your hand are most sensitive to touch. Push two pins through a small card, about 2 mm apart. Draw an outline of your hand. Have a lab partner *lightly* touch the tip of the pins to different parts of your hands and fingers. Without looking, say whether you feel one or two pins for each test. Mark each test on the drawing, indicating whether you felt one or two pins. When finished, analyze the drawing to determine which parts of your hands and fingers are most sensitive to touch.

15. C If you wear eyeglasses or contact lenses, investigate the type and strength of the corrective lenses you are using. Find out if the prescription for each eye is the same or different. If possible (and if your teacher judges it appropriate to do so), compare the corrective lenses you use to those used by your classmates. What type of corrective lenses are most common in your class? Discuss and debate the pros and cons of wearing eyeglasses versus contact lenses.

16. I When you are outside on a sunny day, you feel heat, not light. How can you decide if a sowbug responds to light stimulus, but not heat stimulus?

17. MC A severe impact to the head can cause blindness. Understanding the mechanisms of sight involve study of brain activity as well as study of the eye. How can research into the neurology of vision be used to develop safe helmets?

18. MC Some activities put people at risk for injury. For one activity (such as hockey) identify how risk of injury can be reduced.

19. MC What can you do to prevent damage to your hearing from extremely loud or persistent noises in your environment? Is the school presently doing enough to protect student hearing? What types of policies could be adopted to help protect students from hearing loss while attending noisy school functions, such as school dances?

Summary of Expectations

Briefly explain each of the following points.

- An essential function of the nervous system is maintaining homeostasis in the body. (5.1)
- The human nervous system is composed of two parts, the central nervous system and the peripheral nervous system. (5.1)
- The fight-or-flight response to dangerous situations illustrates how the sympathetic nervous system regulates many essential physiological processes in the body. (5.1)
- The nervous system is composed largely of neurons, which are specialized body cells. (5.1)
- Reflex arcs are constructed to produce quick responses to certain stimuli without input from the brain. (5.1)
- The medulla oblongata is the region of the brain that is the primary control centre of the autonomic nervous system. (5.1)
- The cerebrum is the centre of human consciousness. (5.1)
- A nerve impulse is actually composed of a wave of electrical depolarization that travels down the cell membrane of a neuron. (5.2)
- The polarity of cell membranes of neurons is a result of unequal concentrations of positive ions on either side of the membrane. (5.2)
- The action of neurotransmitters determines the direction of nerve impulses travelling in the nervous system. (5.2)
- The all-or-none principle determines whether a particular stimulus will initiate a new nerve impulse in a stimulated neuron. (5.2)
- Even though all nerve impulses are essentially the same, the brain can still distinguish between weak and strong sensory stimuli, and between stimuli originating from different types of sensory structures and locations in the body. (5.1, 5.2)
- Schwann cells can repair some types of damage to nerve tissue. (5.2)
- A healthy eye is capable of focussing on and forming sharp images of both near and distant objects. (5.3)
- In bright light, we see clearly in colour. In low light, our colour vision is diminished. (5.3)
- The human ear is a multifaceted sensory structure that is sensitive to sound waves of various frequencies and that helps us maintain equilibrium. (5.3)
- The skin is the largest sensory structure of the human body. (5.3)

Language of Biology

Write a sentence using each of the following words or terms. Use any six terms in a concept map to show your understanding of how they are related.

- central nervous system
- peripheral nervous system
- autonomic nervous system
- somatic nervous system
- sympathetic nervous system
- parasympathetic nervous system
- sensory nerve
- motor nerve
- reflex
- neuron
- nerve
- reflex arc
- cell body
- dendrite
- axon
- wave of depolarization
- medulla oblongata
- cardiac centre
- vasomotor centre
- respiratory centre
- cerebellum
- thalamus
- hypothalamus
- cerebrum
- cerebral cortex
- corpus callosum
- frontal lobe
- parietal lobe
- occipital lobe
- temporal lobe
- resting potential
- neurotransmitters
- all-or-none principle
- action potential
- refractory period
- myelin sheath
- Schwann cell
- node of Ranvier
- synapse
- presynaptic neuron
- postsynaptic neuron
- synaptic vesicle
- excitatory response
- inhibitory response
- cholinesterase
- acetylcholine
- noradrenaline
- glutamate
- GABA
- dopamine
- seratonin
- sclera
- cornea
- conjunctiva
- choroid layer
- iris
- pupil
- ciliary body
- lens
- suspensory ligament
- retina
- rod
- cone
- aqueous humour
- vitreous humour
- accommodation
- rhodopsin
- pupillary reflex
- fovea centralis
- cataract
- glaucoma
- myopia
- hyperopia
- astigmatism
- mechanoreceptor
- adaptation

UNDERSTANDING CONCEPTS

1. Compare the structure and function of sensory neurons and motor neurons.
2. Describe the role of the Na^+/K^+ pump in restoring resting potential of a nerve cell membrane following a nerve impulse.
3. Explain how nerve impulses travel faster along axons covered by a myelin sheath than they do along non-myelinated neurons.
4. Compare how the parasympathetic and sympathetic nervous systems affect heart function.
5. Explain why the brain consumes more glucose and oxygen than any other body organ.
6. Is the medulla oblongata part of the somatic nervous system, part of the autonomic nervous system, or does it function as part of both systems? Explain your answer.
7. Some injuries to the body involve a severing of the spinal cord. Explain why all sensations and control of motor functions are lost in the parts of the body below the location of the injury.
8. The diagram on the right illustrates the distribution of sodium and potassium ions on either side of a nerve cell membrane at rest. Reproduce this diagram in your notebook and label the following: Na^+, K^+, gated Na^+ channel, gated K^+ channel, nerve cell membrane, outside neuron, inside neuron.
9. Describe how the nervous system functions when you perform the following activities.
   (a) Your hand grasps an apple and raises it to your mouth.
   (b) You take a bite out of the apple and swallow it.
10. Explain how light energy is transformed into nerve impulses in the retina.
11. Describe how your eyes adjust when you move from a brightly lit environment to a dimly lit area.
12. There are many forms of epilepsy, a disease in which spontaneous electrical discharges occur in the brain. Different types of epileptic seizures produce different symptoms, including unusual colours, sounds, or other types of sensations. What do these variations in symptoms tell you about the parts of the brain affected by each type of epilepsy?

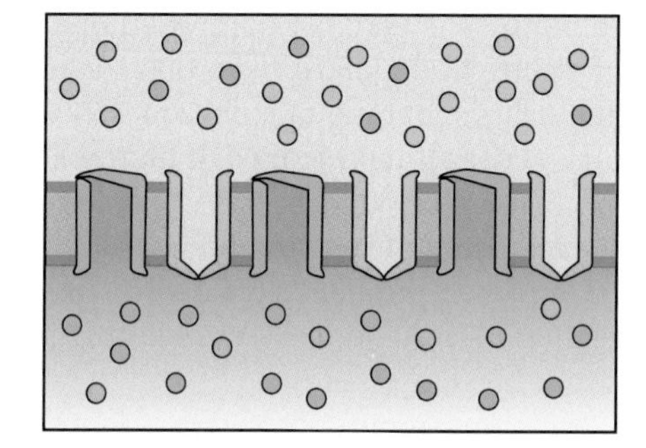

INQUIRY

13. Observe the optical illusion in the following figure. Draw what you see. Compare your observations to those of others in your class. What do these kinds of illusions tell us about how our brains interpret visual stimuli? Do people have identical or different perceptions of the world around them? Discuss possible factors that influence a person's perception of a particular experience.

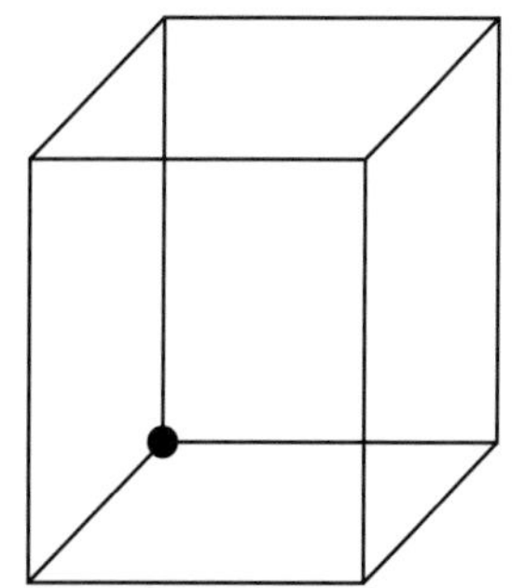

14. Design an experiment to test if caffeine has an effect on a person's ability to perceive different colours of light. Make a chart to record your data. Propose a control for your experiment. What variables will you include in your control? What kind of data would you expect to find? Justify your response.
15. Acupuncture is increasingly being used as a drug-free therapy for pain.
    (a) How could you determine the effect of acupuncture on alertness?
    (b) What kind of drugs would be inappropriate to take while under acupuncture therapy? Explain.
16. A breathalizer test can be used to determine how much alcohol is in a person's body. This test, however, gives no information about a person's neurological function. Design two

tests to determine if alcohol affects the somatic system or the autonomic nervous system.

17. The motor neuron produces an action potential or impulse as an all-or-none response to stimulus. Muscles respond to this impulse, and contract as an all-or-nothing action. The presynaptic axon is highly branched at the neuro-muscular synapse. If some of the presynaptic branches are damaged, will the muscle contract? Describe a procedure that would help you to answer this question.

COMMUNICATING

18. (a) Use wire to make a model of a neuron. Complete the circuit using dry cells, a switch, and a light bulb.
    (b) Record your circuit in your notebook. If the wire is a neuron, what do the other parts of the circuit represent?
    (c) Identify one neuronal disorder. Model this disorder using your circuit.

19. Find pictures of the brain of a fish, a turtle, and a bird. Make rough diagrams showing the major parts of the brain for each organism. What differences in proportion of the brain parts do you see among the organisms. Suggest an explanation for the differences you see.

20. Medication can be used to treat many disorders, including mood disorders. What is the impact of mood-disorder medication on systems in the body other than the nervous system? Use Internet resources to research a specific mood disorder. Make a chart to list the advantages and disadvantages of mood-disorder therapy.

21. A rarely performed treatment for epilepsy involves severing of the corpus callosum. This surgical procedure stops the flow of information from one side of the brain to the other. A more common therapy for epilepsy involves the use of drugs to block certain nerve impulses. Use diagrams to show how each method can limit the transmission of information in the brain.

MAKING CONNECTIONS

22. Both alcohol and caffeine affect the neurological system. Although alcohol is a controlled substance, caffeine is not. Develop an argument to make caffeine (coffee and other caffeinated beverages) a controlled substance.

23. Acupuncture and acupressure can be effective treatments to reduce pain. Traditional acupuncture involves the use of needles. Acupressure involves the use of pressure at specific points on the body. Both acupuncture and acupressure are not as well accepted as prescription therapy. Identify points to support more research into acupuncture and acupressure therapy. Identify cautions to further research into each therapy.

24. Some people increase their risk for stroke by eating a poor diet, smoking, and having little or no exercise. Develop a plan to educate people about maintaining a healthy lifestyle. How might a healthy lifestyle contribute to a healthy nervous system?

25. Your class has raised $200 to donate to charity and has narrowed the choice down to supporting Alzheimer research, Multiple Sclerosis research, or research into alcoholism. You and two classmates will debate the merits of donating the money to each group. After the debate the class will select which charity to support. Select one group and prepare for the class debate.

CHAPTER 6

The Endocrine System

Reflecting Questions

- How does the nervous system interact with and regulate endocrine activity?
- How do endocrine hormones maintain homeostasis in the body?
- What are some of the major treatable diseases of the human endocrine system?

Prerequisite Concepts and Skills

Before you begin this chapter, review the following concepts and skills:

- distinguishing between the structure and properties of protein and lipid molecules (Chapter 1, section 1.2),
- explaining the concepts of homeostasis and dynamic equilibrium (Chapter 4, section 4.1),
- explaining the difference between negative and positive feedback loops (Chapter 4, section 4.1).
- explaining how the kidneys control water balance in the body (Chapter 4, section 4.2),
- describing the function of the parasympathetic and sympathetic nervous systems (Chapter 5, section 5.1), and
- describing the nature of nerve impulses (Chapter 5, section 5.2).

When a gymnastic performer entertains us with a display of prolonged, intense physical activity, the performer's body is being pushed to the limits of its capabilities. During the performance, homeostatic mechanisms work to maintain the body's internal environment within tolerance limits — the narrow range of conditions within which cellular processes are able to function at a normal level.

The endocrine system works in parallel with the nervous system to maintain homeostasis. It does so by releasing chemical substances — hormones – which in turn trigger actions in specific target cells. For example, vigorous exercise triggers endocrine glands in the brain and elsewhere to release several different hormones. These hormones regulate oxygen consumption, basal metabolic rate, and the metabolizing of carbohydrates and fat for energy. As a result, the rate and depth of breathing increase, as do heart rate and muscle contraction. In addition, energy stores are quickly mobilized. These changes ensure that fuel is readily available for an increase in skeletal muscle, heart, and brain activity, while at the same time maintaining normal physiological processes.

In this chapter, you will learn about the components of the endocrine system. You will examine how the nervous system and the endocrine system interact to regulate physiology, and how these systems affect each other's functions. You will also explore how, in addition to maintaining homeostasis, the endocrine system regulates a wide range of other biological processes. These processes include the control of blood sugar, metabolism, growth, reproductive development and function, and other physiological activities. You will discover how the body responds to stress, and how abnormal endocrine function can result in disease. Finally, you will see how some major endocrine disorders can be treated.

Endocrine glands within the brain secrete hormones to influence metabolism in other target glands. This positron emission tomography (PET) scan relies on a radioactive tracer, injected into the bloodstream, to reveal the brain's metabolic activity.

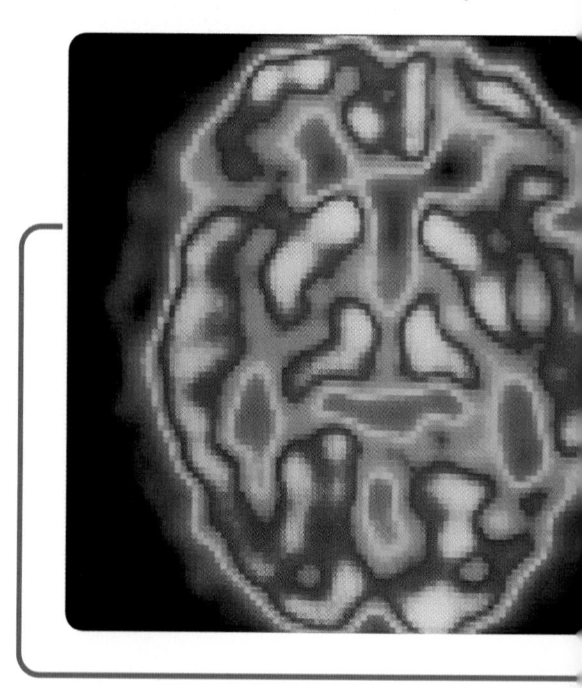

Chapter Contents

6.1 The Endocrine System and Homeostasis

EXPECTATIONS

- **Describe the anatomy and physiology of the endocrine system and explain its role in homeostasis.**
- **Describe and explain homeostatic processes involved in maintaining equilibrium in response to both a changing environment and medical treatments (for example, describe the effect of disorders of the endocrine system).**

In Chapter 5, you studied the role of the human nervous system in maintaining homeostasis. This chapter focuses on the **endocrine system**, which comprises the hormone-producing glands and tissues of the body. As mentioned previously, **hormones** are chemical substances that circulate through the blood and exert some measure of control over virtually every organ and tissue in the body.

In conjunction with the nervous system, the endocrine system acts as a complex internal communication network that continuously monitors and responds to the body's ever-changing internal environment. This system regulates critical physiological processes and plays a key role in homeostasis. Some of the regulatory functions of the endocrine system include the control of heart rate, blood pressure, immune response to infection, reproduction, emotional state, and the overall growth and development of the body. Hormone production and secretion fluctuate in response to nervous system activity, stimulation by other hormones, and changing concentrations of salt, glucose, and other essential constituents in the blood.

Abnormal endocrine function can seriously disrupt the body's normal metabolic functions. However, as you will see in this chapter, endrocrinology — the scientific study of the endocrine system — is a very active field of medical research. This field continually yields new and exciting discoveries about the unique functions of endocrine glands and the hormones they produce. New and effective medications and procedures are constantly being developed for many endocrine disorders.

Components of the Endocrine System

As Figure 6.1 illustrates, the endocrine system is composed of a number of glands and tissues. This system consists of the pituitary, thyroid, parathyroid, adrenal, thymus, and pineal glands, as well as the pancreas and reproductive tissues (ovaries and testes). Many other organs, such as the liver, skin, kidney, and parts of the digestive and circulatory systems, produce hormones in addition to their other physiological functions.

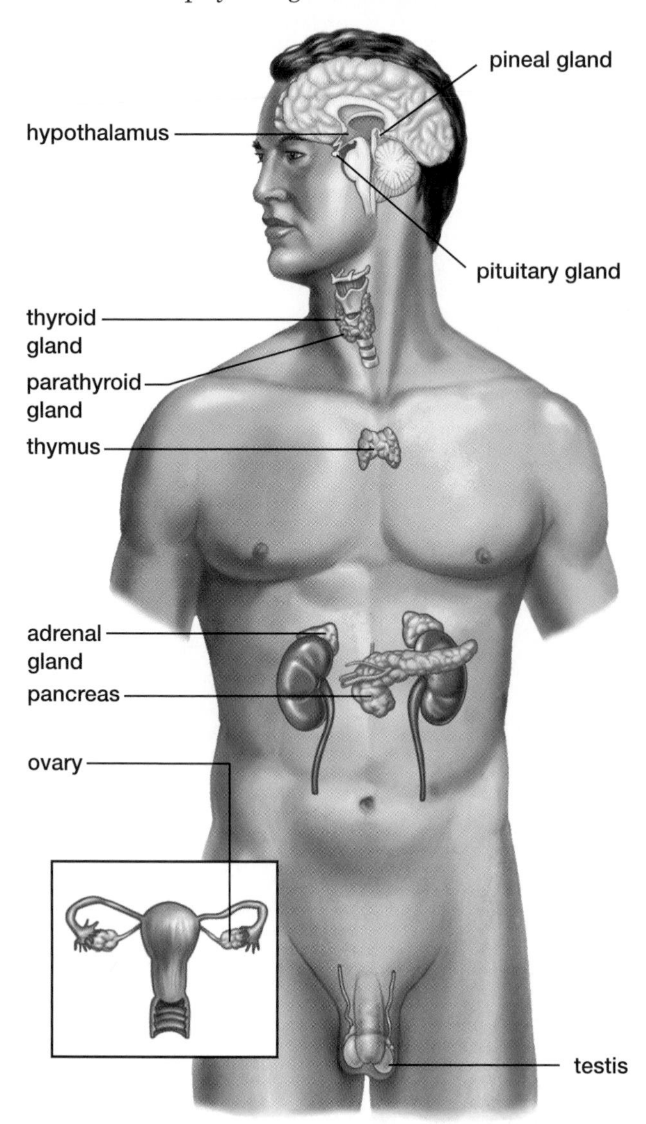

Figure 6.1 Anatomical location of major endocrine glands in the body

Endocrine glands are ductless glands that secrete hormones directly into the bloodstream, whereas **exocrine glands** release their secretions through ducts or tubes (as shown in Figure 6.2). Examples of exocrine glands are sweat glands, salivary glands, and tear (lacrimal) glands.

Hormones produced by the endocrine glands influence the activity of every organ and tissue in the body. The term "hormone" was introduced in 1908 by British physiologist Ernest Henry Starling. He identified hormones as "chemical messengers," substances that carry instructions or signals to one or more distant organs or tissues in the body. These signals instigate some type of specialized biochemical process in the target organ. For example, the pancreas produces hormones that can stimulate the liver to convert sugar to glycogen, or glycogen to sugar, depending on the body's immediate needs.

Very small quantities of each type of hormone are produced and secreted into the blood. The concentration of a hormone substance in blood may be no more than 10^{-12} mol/L (which can be compared to a drop of oil in a swimming pool full of water). However, the potency (or impact) of hormones is magnified many times by their ability to affect key metabolic processes in the target cells.

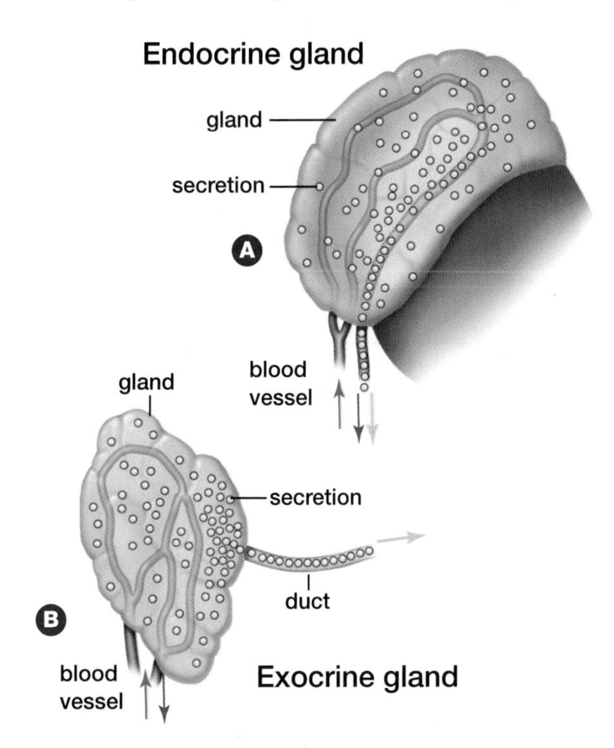

Figure 6.2 (A) Endocrine glands do not have ducts; they produce and secrete hormones directly into the bloodstream. (B) Exocrine glands have ducts and secrete sweat, milk, digestive enzymes, and other materials.

The adrenal gland has been the focus of many experimental breakthroughs in the field of endocrinology. One of the founders of endocrinology, a British country doctor named George Oliver, was among the first to demonstrate the physiological action of an extract of endocrine tissue. In 1894, he discovered that a preparation of tissue from the adrenal gland could raise blood pressure in test subjects.

In 1897, American scientist John Jacob Abel discovered adrenaline, the first hormone molecule to be isolated from an endocrine gland extract. Following Abel's discovery, Jokichi Takamine, a Japanese-born chemist working in the United States, independently isolated the same hormone molecule, which he named "adrenaline." With Takamine's assistance, a U.S. pharmaceutical company soon began to mass-produce and sell adrenaline as a treatment for a number of ailments. At the time, the company was unaware that their product was actually a mixture of two hormones, adrenaline and noradrenaline. Both hormones are produced by the adrenal gland.

Hormones such as adrenaline secreted by the adrenal gland come into contact with virtually all cells and tissues as they circulate throughout the body. However, they trigger a response only in cells (such as liver cells) that have specific receptor sites for the hormone. The combining of a hormone, such as adrenaline, with a specific receptor on the outer membrane of a liver cell sets off a cascade of chemical reactions, such as the conversion of glycogen to sugar.

Factors in Hormone Production and Function

In general, hormone production increases or decreases in response to changing metabolic needs of the body, such as fluid balance, and other factors such as infection, physical injury, and emotional stress. Hormone levels are also regulated by the activity of the nervous system and other endocrine glands. Hormone-secreting cells contain receptor molecules that are sensitive to regulatory hormones from other sources in the body. For example, thyroid-stimulating hormone molecules produced by the pituitary gland bind to receptors on cells of the thyroid gland. This action stimulates synthesis of a hormone called thyroxine, which you will study in greater detail later in this chapter.

The impact of a specific hormone on the activity of target tissues is a function of the rate of hormone production and secretion, hormone concentration in the blood, the rate of blood flow to a target organ or tissue, and the half-life of the hormone. The half-life refers to the length of time a hormone remains viable in the blood before it is degraded by the liver or other tissues. Half-life may range from several hours to several days.

Normal endocrine function can be disrupted by various medical problems such as tumours, infection, autoimmune diseases, and physical injury. Genetic disorders, industrial pollutants, and certain food additives have also been linked to abnormal endocrine function. Symptoms can range from mild discomfort to chronic, but manageable conditions, to more severe, potentially life-threatening complications.

The medical treatments for endocrine disorders include hormone replacement therapy, other medications that moderate endocrine activity, and changes in diet and other forms of behavioral modification. If required, treatment might involve more aggressive procedures such as surgery to remove the affected endocrine tissues or organs. Throughout this chapter, discussions of each type of endocrine gland will be followed by an overview of related hormonal disorders.

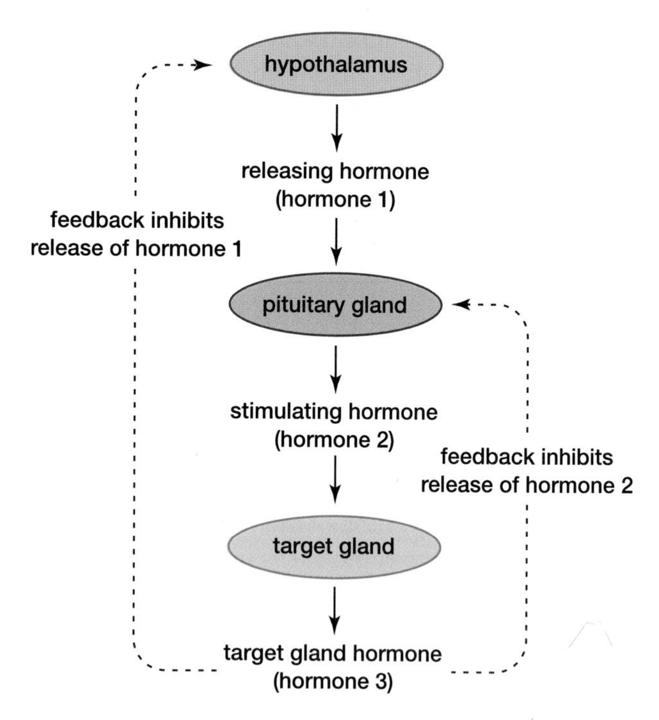

Figure 6.3 Hormones regulate endocrine activity by means of negative feedback loops.

Types of Hormones

Hormones produced by the endocrine system also interact with each other. In addition to the regulation of endocrine activity through the action of negative feedback loops (as shown in Figure 6.3), hormone levels can also be controlled by the interaction of hormones that have opposing physiological properties. Such contrary hormonal substances are referred to as **antagonistic hormones**. Recall, for example, the discussion of the opposite effects of insulin and glucagon on the levels of blood glucose in Chapter 4.

The endocrine system produces two main types of hormone product: steroid and non-steroid hormones. These hormone types can be differentiated by their chemical composition and their mode of action in target cells and tissues.

Steroid hormones, such as cortisol, are manufactured from cholesterol. Each type of steroid hormone is composed of a central structure of four carbon rings attached to distinctive side chains that determine the hormone's specific and unique properties (as shown in Figure 6.4).

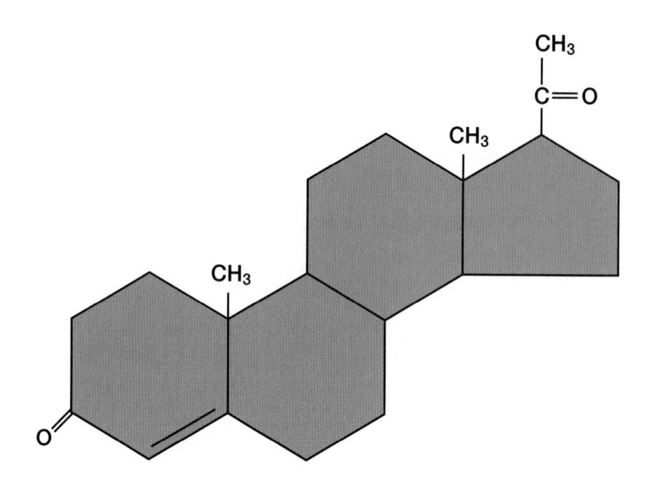

Figure 6.4 The structural formula of the steroid hormone progesterone.

Within the endocrine cells, steroid hormones are synthesized in the smooth ER. Most steroid hormones are secreted quickly into the blood by the endocrine organs that produce them. Since steroid hormones are hydrophobic, they combine with a protein carrier that transports them through the bloodstream.

Fat-soluble steroid hormones can pass through the membrane of a target cell. Once inside the target cell, steroid hormones attach to a protein receptor molecule in the cytoplasm. This hormone-receptor complex then enters the nucleus, where it binds with and activates a specific gene on the cell's DNA molecule. The activated gene then produces an enzyme that initiates the desired chemical reaction within the cell. This process is illustrated in Figure 6.5.

Non-steroid hormones, such as adrenaline, are composed of either proteins, peptides, or amino acids. These hormone molecules are not fat-soluble, so they usually do not enter cells to exert their effect. Instead, they bind to receptors on the surface of target cells. This combination substance then triggers a specific chain of chemical reactions within the cell. The structure of a non-steroid hormone is illustrated in Figure 6.6.

In 1971, Edward W. Sutherland, Jr., received the Nobel Prize for his discovery of the biochemical mechanism by which adrenaline, and other hormones, influence target cell activity. Normally, adrenaline stimulates the conversion of stored glycogen to glucose in the liver. The liver then releases the glucose into the bloodstream.

Sutherland and his team of researchers investigated the mechanism by which adrenaline regulated glucose synthesis in liver tissue. Their procedure involved breaking open liver cells and separating the cell membranes from the rest of the cellular material. They observed that adrenaline had no effect on glucose production in liver cells when cell membranes were removed from the inner cell contents. However, when adrenaline was added to isolated cell membranes, they found that adrenaline molecules bound to receptor molecules located on the surface of the membranes (as shown in Figure 6.5). This hormone-receptor combination triggered the synthesis of yet another molecule, called a "second chemical messenger." Sutherland's team identified this substance as cyclic AMP (cAMP). In this process, the hormone was referred to as the "first messenger."

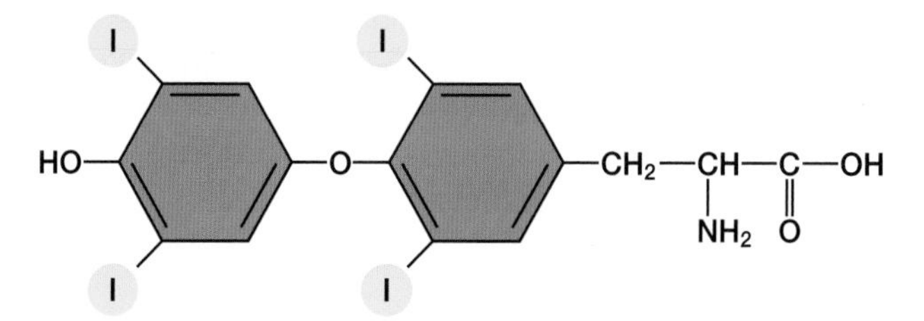

Figure 6.6 The structural formula of the non-steroid hormone thyroxine.

Researchers discovered that the binding of the adrenaline to the receptor activated the enzyme portion of the receptor molecule. The activated enzyme then catalyzed the production of cyclic AMP from ATP inside the cell. This second messenger triggered a series of reactions that influenced the synthesis and reactivity of intracellular enzymatic proteins involved in the conversion of glycogen to glucose. This chain reaction acts as a kind of "biological amplifying

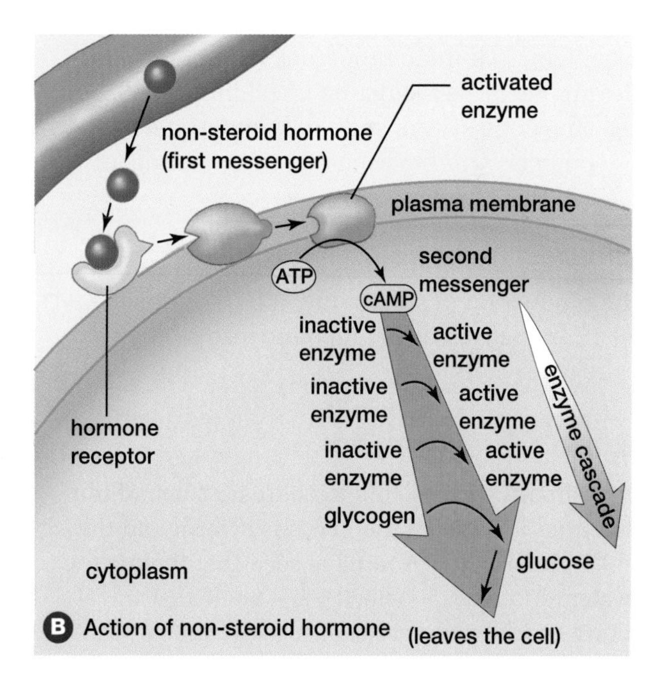

Figure 6.5 (A) After passing through the plasma membrane and nuclear envelope, a steroid hormone binds to a receptor protein inside the nucleus. The hormone-receptor complex then binds to DNA, and this leads to activation of certain genes and protein synthesis. (B) Non-steroid hormones, called first messengers, bind to a specific receptor protein in the plasma membrane. A protein relay ends when an enzyme converts ATP to cAMP (the second messenger), which activates an enzyme cascade.

system," in which relatively small amounts of hormone can substantially affect the biochemistry of target cells in the liver, heart, and other organs of the body.

Other hormones that exert their effect on cells through cyclic AMP-activating mechanisms include adrenocorticotropic hormone (ACTH), glucagon, luteinizing hormone (LH), follicle-stimulating hormone (FSH), and anti-diuretic hormone (ADH).

Sutherland and other researchers later discovered that similar biochemical pathways involving the formation of cyclic AMP were at work in many different types of cells in the body. Researchers also found that other cells used calcium, or an enzyme within the cell, as the second messenger.

The stimulating properties of caffeine, as discussed in Chapter 4, are the result of the way in which caffeine inhibits the breakdown of cyclic AMP in cells. This causes cyclic AMP to accumulate in the cell cytoplasm, which extends its amplifying effect on cell processes, such as the contraction of heart muscles. In this way, caffeine mimics the stimulating properties of hormones such as adrenaline. Due to its properties as a stimulant, caffeine is listed as a banned substance in the Olympic Movement Anti-Doping Code.

It is interesting to note that caffeine also acts as a diuretic, increasing urine production. This can result in increased calcium excretion, a contributing factor for osteoporosis. The physiological basis of caffeine's diuretic properties has yet to be determined. The regulation of calcium levels by the endocrine system and the symptoms of osteoporosis will be discussed in greater detail later in this chapter.

Another stimulant, nicotine, has a substantial impact on endocrine function. Nicotine stimulates the production of adrenaline, ACTH, cortisol, and ADH. In male smokers, nicotine also increases estrogen secretion.

Endocrine Glands

You should now be able to compare some of the essential features of the nervous system and the endocrine system. You have seen that the nervous system produces bioelectrical signals that travel along specialized nerve cells, while the endocrine system releases hormones into the bloodstream that circulates throughout the body. The nervous system elicits a rapid but short-lived response, illustrated by the body's reflex actions. Endocrine hormones produce a slower, but more sustained and enduring response in their target tissues.

The hypothalamus, a part of the brain connected to the pituitary gland, continuously monitors the state of the body's internal environment and regulates pituitary gland activity. Together, the hypothalamus and the pituitary gland control many critical physiological processes. These processes include metabolic rate, kidney function, appetite, mental alertness, reproduction, and growth and development. The hypothalamus and the pituitary gland secrete hormones that influence the activity of other hormone-producing glands. The constant interaction between the hypothalamus and the pituitary gland is a key factor in maintaining homeostasis.

An In-depth Look at the Pituitary Gland

Figure 6.7 shows how the pituitary gland is connected to the hypothalamus of the brain. A short but complex network of blood vessels, called a portal system, extends from the hypothalamus to the pituitary gland. This is the critical link by which the nervous system exerts its control over hormone production in the pituitary gland and other endocrine glands. The portal system carries small peptide molecules called "releasing hormones" secreted by neurosecretory cells in the hypothalamus directly to tissues in the pituitary gland.

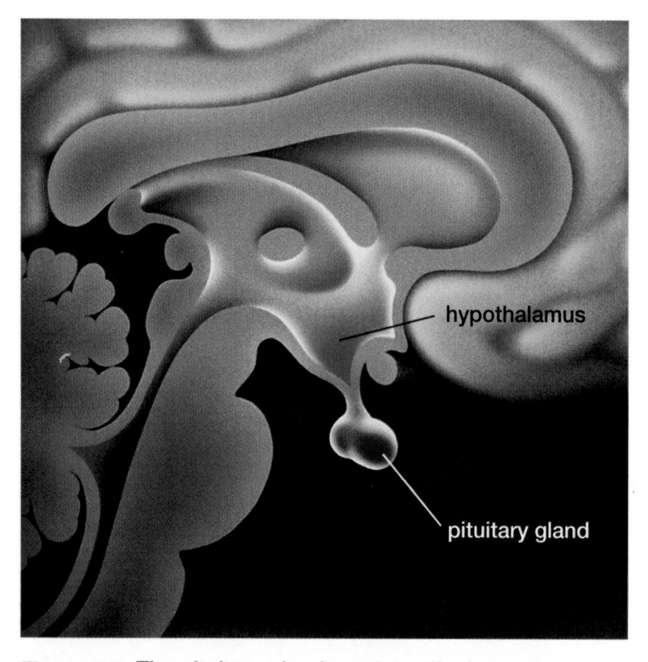

Figure 6.7 The pituitary gland regulates the hormone production of many of the body's endocrine glands.

As noted above, the pituitary gland produces hormones that regulate the hormone production of many other endocrine glands in the body. Such substances are referred to as tropic hormones. For example, the thyroid-stimulating hormone (TSH) is a tropic hormone produced by the pituitary gland that stimulates — as its name implies — the thyroid gland to produce and secrete thyroid hormone. TSH regulates thyroid secretion through a negative feedback system.

The pituitary gland is typically referred to as the "master gland" of the endocrine system. This designation reflects the role of the pituitary gland in regulating the activity of the other hormone-producing glands of the endocrine system. The pituitary gland is actually composed of two glands – the anterior and posterior pituitary, as shown in Figure 6.8. As the human embryo develops, the anterior pituitary is formed from cells from the roof of the mouth that migrate toward the brain. The posterior pituitary is composed of neural tissue.

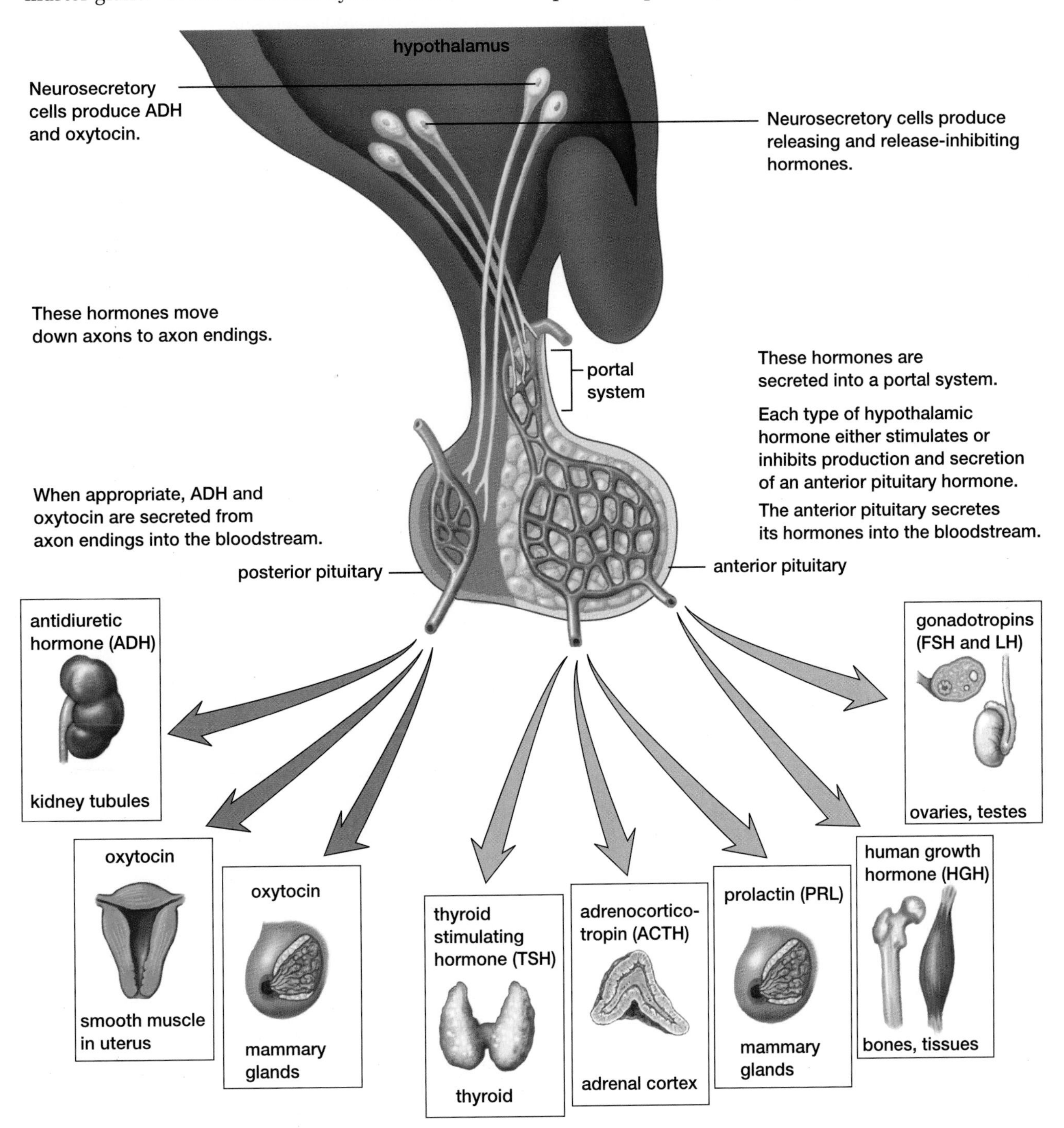

Figure 6.8 The hypothalamus produces two hormones, ADH and oxytocin, which are stored and secreted by the posterior pituitary. The hypothalamus controls the secretions of the anterior pituitary, and the anterior pituitary controls the secretions of the thyroid, adrenal cortex, and gonads (which are also endocrine glands).

The differing embryonic origins of the anterior and posterior lobes are reflected in their dissimilar functions.

The Anterior Pituitary

The anterior lobe of the pituitary gland produces six types of endocrine hormones, human growth hormone, and four tropic hormones. The four tropic hormones made by the anterior pituitary are the thyroid-stimulating hormone, the follicle stimulating hormone (FSH), the luteinizing hormone (LH), and the adrenocorticotropic hormone (ACTH), all of which will be discussed later in this chapter.

Human growth hormone The anterior pituitary regulates growth and development of the body through the production and secretion of a non-steroid hormone called human growth hormone (HGH). Human growth hormone (sometimes referred to as somatotropin) is a small protein molecule.

HGH spurs body growth by increasing intestinal absorption of calcium, increasing cell division and development (especially in bone and cartilage), and stimulating protein synthesis and lipid metabolism. HGH triggers the release of fatty acids from fat cells, and prompts the conversion of fatty acids into fragments that can then form acetyl CoA for use as an energy source for the body. HGH also suppresses glycolysis and increases glycogen production in the liver. In summary, HGH spares proteins and carbohydrates by enhancing the use of lipids as an energy source for cell functions.

HGH has a half-life of about 20 hours after secretion, after which it is no longer chemically active. HGH, acting as a tropic hormone, triggers the production of **growth factors** in the liver and other tissues. These growth factors (composed of protein molecules) prolong the effects of HGH on bone and cartilage tissues.

Levels of HGH tend to decrease with age. The resulting decline in protein synthesis may be responsible for some of the characteristic signs of aging, such as diminished muscle mass and wrinkles.

Insufficient HGH production during childhood results in a condition called **pituitary dwarfism**. This disorder results in abnormally short stature. But, unlike genetic dwarfism, body proportions (the length of arms and legs and the size of the head) are normal. Puberty may be delayed or not occur at all. Pituitary dwarfism may be the result of a pituitary tumour or the total absence of a pituitary gland. Measurement of growth hormone levels in the blood is used to confirm the diagnosis of pituitary dwarfism (see Figure 6.9).

In the past, treatment of pituitary dwarfism required the extraction of growth hormone from the pituitary glands of human cadavers. However, this source yielded insufficient quantities of the hormone. In addition, growth hormones from animal sources were not suitable for human use. However, current biotechnology techniques now provide a much more reliable supply of this hormone. These procedures involve inserting sections of DNA that code for HGH into certain strains of bacteria. The altered, rapidly reproducing bacteria are thus transformed into biochemical factories that produce HGH as a "waste product."

Figure 6.9 The amount of human growth hormone produced during childhood affects the height of an individual. The symptoms of both inadequate and excessive growth hormone are readily treated today.

An excess of HGH production prior to puberty causes a disorder known as **gigantism** (as illustrated in Figure 6.9). The symptoms of gigantism are primarily the result of abnormal growth of long bones in the skeleton. The disorder is easily treated by the microsurgical removal of a tumour from the pituitary gland, irradiation of gland tissue, or both.

Age 16

Age 33

Age 52

Figure 6.10 Acromegaly is caused by overproduction of HGH in the adult. It is characterized by an enlargement of the bones in the face, fingers, and toes of an adult. Today, various therapies are used to treat this disorder.

Excess HGH production during adult years produces **acromegaly**, symptoms of which include excessive thickening of bone tissue. This thickening causes abnormal growth of the head, hands, and feet as shown in Figure 6.10, as well as spinal deformities. Treatment of patients diagnosed with acromegaly involves surgical removal of the tumour, radiation therapy, injection of a growth hormone blocking drug, or a combination of these treatments. The development of a tumour within the pituitary gland is the most common cause of both gigantism and acromegaly.

HGH shares many structural and functional similarities with prolactin, the next pituitary hormone to be described in this section.

Prolactin This substance is a non-steroid hormone produced by the anterior pituitary and, in smaller quantities, by the immune system, the brain, and the pregnant uterus. Prolactin stimulates the development of mammary gland tissue and milk production (lactogenesis).

The hypothalamic regulation of prolactin production is somewhat unusual. The hypothalamus secretes the neurotransmitter dopamine, which inhibits rather than stimulates the production and secretion of prolactin by the pituitary. Severing the connection between the hypothalamus and the pituitary gland results in an increase in prolactin production. After birth, however, the stimulation of nerve endings in the nipples during infant feeding will trigger the release of prolactin-secreting hormones by the hypothalamus. This spinal reflex (known as a neuroendocrine reflex) stimulates the production of prolactin. Increasing estrogen levels also stimulate prolactin production in late pregnancy to prepare the mammary glands for lactation after the birth of a baby. Increased prolactin levels in pregnancy also inhibit ovulation by suppressing the production of LH. Figure 6.11 illustrates one common application of increasing prolactin levels, in milk-producing cows.

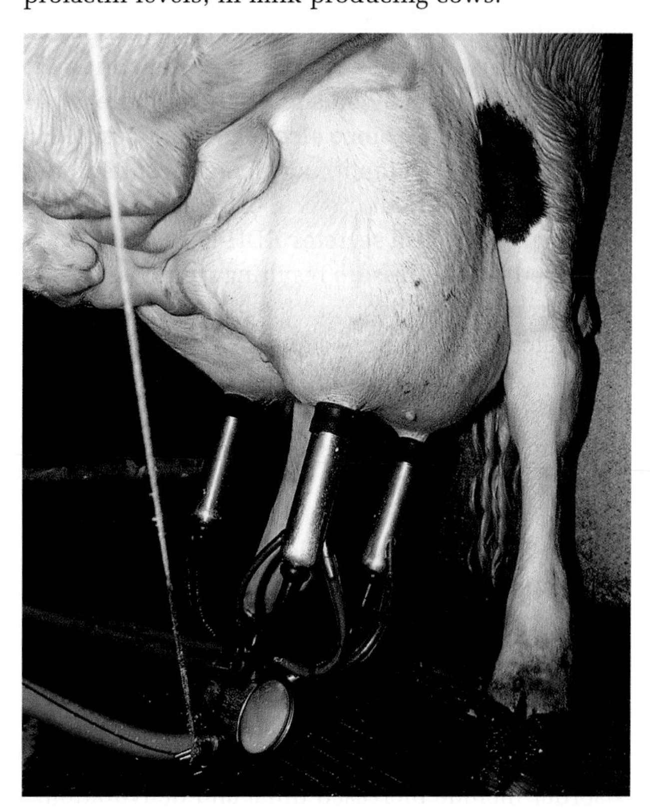

Figure 6.11 This milking machine stimulates prolactin release so milk production continues after the calf has been weaned.

Hyperthyroidism and Hypothyroidism

Various medical complications can arise from the production of abnormally high (hyper) or low (hypo) levels of thyroxine.

Hyperthyroidism An excess of thyroxine production is referred to as hyperthyroidism, also known as Grave's disease. **Grave's disease** is an autoimmune disorder in which antibodies attach to TSH receptors on thyroid cells. This attachment puts receptors in a "perpetually on" mode that stimulates cell division and production of thyroid hormone. The excessive hormone production causes enlargement of the thyroid, muscle weakness, increased metabolic rate, excessive heat production, and sweating and warm skin due to dilation of blood vessels in the skin (vasodilation). Patients also experience increased appetite despite continued weight loss. Grave's disease also causes the eyes to bulge out or protrude, due to edema (the buildup of fluid) and the entry of lymphocytes into orbital tissues.

Treatment involves surgical removal of the thyroid gland, thyroid-blocking drugs, treatment with radioactive iodine that destroys overactive thyroid tissue, and injections of thyroid hormone. All these therapies are effective in eliminating the symptoms of this disease.

Hypothyroidism A deficiency in thyroxine production is referred to as hypothyroidism, or myxedema. A decrease in thyroxine output can be caused by an iodine deficiency. Decreased thyroxine levels disrupt the negative feedback loop to the pituitary, resulting in continued production of TSH. TSH continues to stimulate cell division in thyroid tissue. The symptoms of hypothyroidism are like a mirror-image of hyperthyroidism. Typically, a hypothyroid condition results in reduced basal metabolic rate (which decreases heat production), reduced tolerance of cold temperatures, decreased heart rate and output, and weight gain despite decreased appetite. Hypothyroidism is also characterized by decreased mental capacity, general weakness and fatigue, and poor physical development.

Failure of normal thyroid development in infants results in a related disorder referred to as congenital hypothyroidism. Since this disorder appears in about 1 out of 4,000 infants, screening for hypothyroidism is recommended for all newborns within one week after birth. Screening involves testing for the presence of thyroid hormone (using radioactive isotopes) in a drop of a baby's blood, as shown in Figure 6.14. Children with congenital hypothyroidism typically have a short, stocky stature and are developmentally delayed. Failure to diagnosis and treat this disorder before two months of age can result in life-long mental impairment.

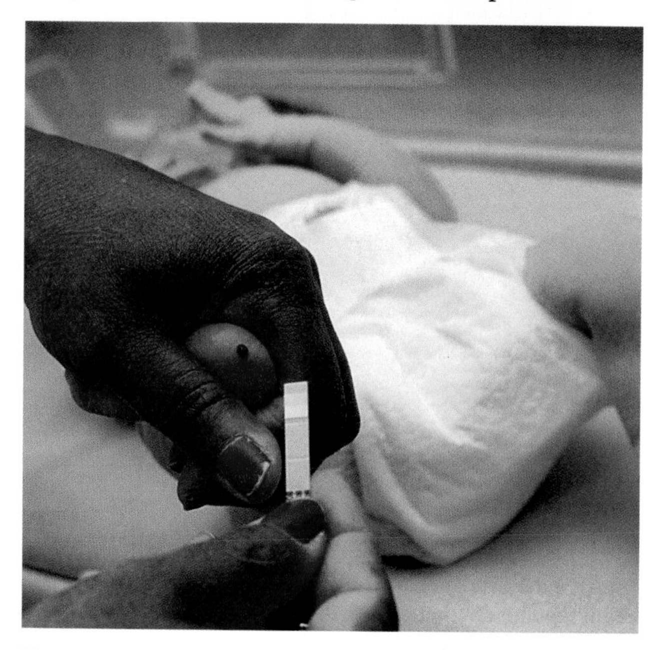

Figure 6.14 Testing for the presence of thyroid hormone

Goiter

Figure 6.15 illustrates a condition referred to as a "goiter." A goiter is a swelling of the thyroid gland caused by insufficient levels of dietary iodine. While this disorder had been well documented for many years, its cause remained a mystery. Ultimately, the puzzle was solved by studying geographical disparities in the incidence of goiter around the world. These studies suggested that goiter was more prevalent in regions where the soil was lacking in iodine. Locally produced food crops in these regions typically had low levels of iodine. A diet low in iodine increased the risk of developing an enlarged thyroid gland.

A lack of dietary iodine prevents the thyroid gland from producing sufficient thyroxine to meet the metabolic demands of the body. Reduced thyroxine levels lower the basal metabolic rate and stimulate the pituitary gland to increase TSH secretion. TSH stimulates cell division in the under-producing thyroid gland, causing the gland to expand. This swelling produces the characteristic bulge in the neck associated with a goiter. In more advanced cases, a goiter can become a disfiguring growth. A goiter can weigh as much as 200 g (a normal thyroid weighs approximately 20 g).

Early treatments for goiter involved the simple addition of small amounts of iodine to the diet. Later attempts to add iodine to drinking water were dropped in response to public opposition. The problem was finally dealt with by the addition of trace amounts of iodine to common table salt. In much of the world, the consumption of "iodized salt" has generally eliminated goiter as a public health problem.

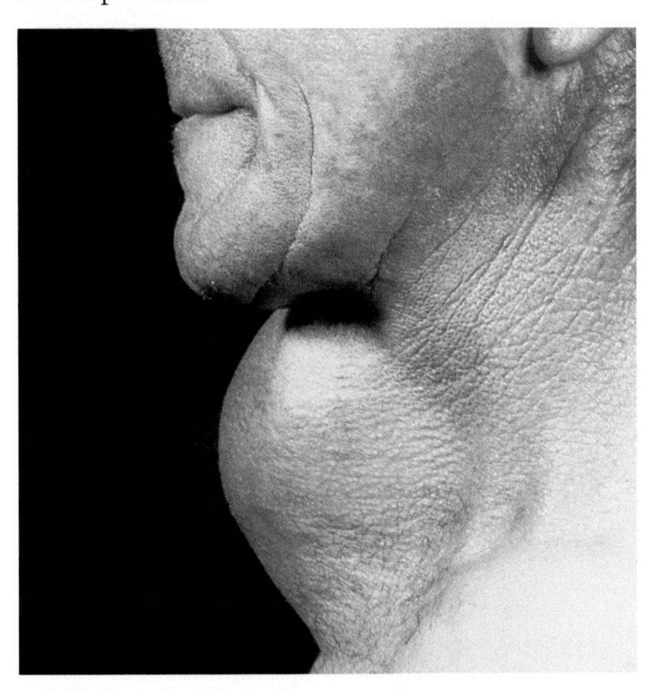

Figure 6.15 An enlarged thyroid gland can result from a lack of iodine in the diet. Without iodine, the thyroid is unable to produce thyroid hormones, and continued anterior pituitary stimulation causes the gland to enlarge.

Calcitonin and Parathyroid Hormone

Calcium is essential for healthy teeth and normal skeletal development. This mineral also plays a significant role in blood clotting, the formation of nerve impulses, and in muscle contraction. Calcium levels in the blood are regulated by **calcitonin**, a hormone which is produced by the thyroid gland, and **parathyroid hormone** (PTH), which is made by the parathyroid glands. Calcitonin and parathyroid hormone are antagonistic hormones. They have opposite effects on blood calcium levels.

High levels of calcium, obtained from dietary sources, stimulate an increase in calcitonin secretion, which then increases the rate at which calcium in blood is deposited into bone tissue of the skeletal system. This results in a lowering of blood calcium levels. Calcitonin acts by increasing the rate of calcium excretion in the kidney.

As illustrated in Figure 6.16 on the following page, a decrease in blood calcium prompts the parathyroids to produce more parathyroid hormone. When blood calcium falls below a critical threshold level, PTH secretion is stimulated by a negative feedback loop. Increased PTH stimulates bone tissue to release calcium into the blood and increases the rate of re-absorption of calcium from the kidneys and the duodenum of the digestive system.

BIO FACT

Osteoporosis is the loss of bone mass that results in the bones becoming brittle and subject to fractures. At least 20 different hormones, growth factors, and vitamins affect bone formation, along with diet and activity level. Postmenopausal women are at greatest risk of osteoporosis because they have less bone mass than men and begin losing it earlier (starting around age 35). By age 70, the average woman has lost 30 percent of her bone mass, and some have lost as much as 50 percent. In men, bone loss begins around 60 years of age and seldom exceeds 25 percent. Bone mass is acquired primarily during puberty and adolescence, when high levels of growth hormone, estrogen, and testosterone stimulate bone formation. This makes proper diet and exercise critical during adolesence.

In past generations, before they were recognized as distinct glands, one or more of the parathyroid glands might have been removed along with the surgical removal of the thyroid gland. In the absence of parathyroid glands, the amount of calcium in a patient's blood would fall to dangerously low levels. These patients were then at high risk of developing tetany, a potentially fatal condition characterized by uncontrollable, continuous muscular contraction sustained by the activity of hypersensitive nerve cells. Due to our current level of understanding of how these glands function, surgeons no longer use this procedure.

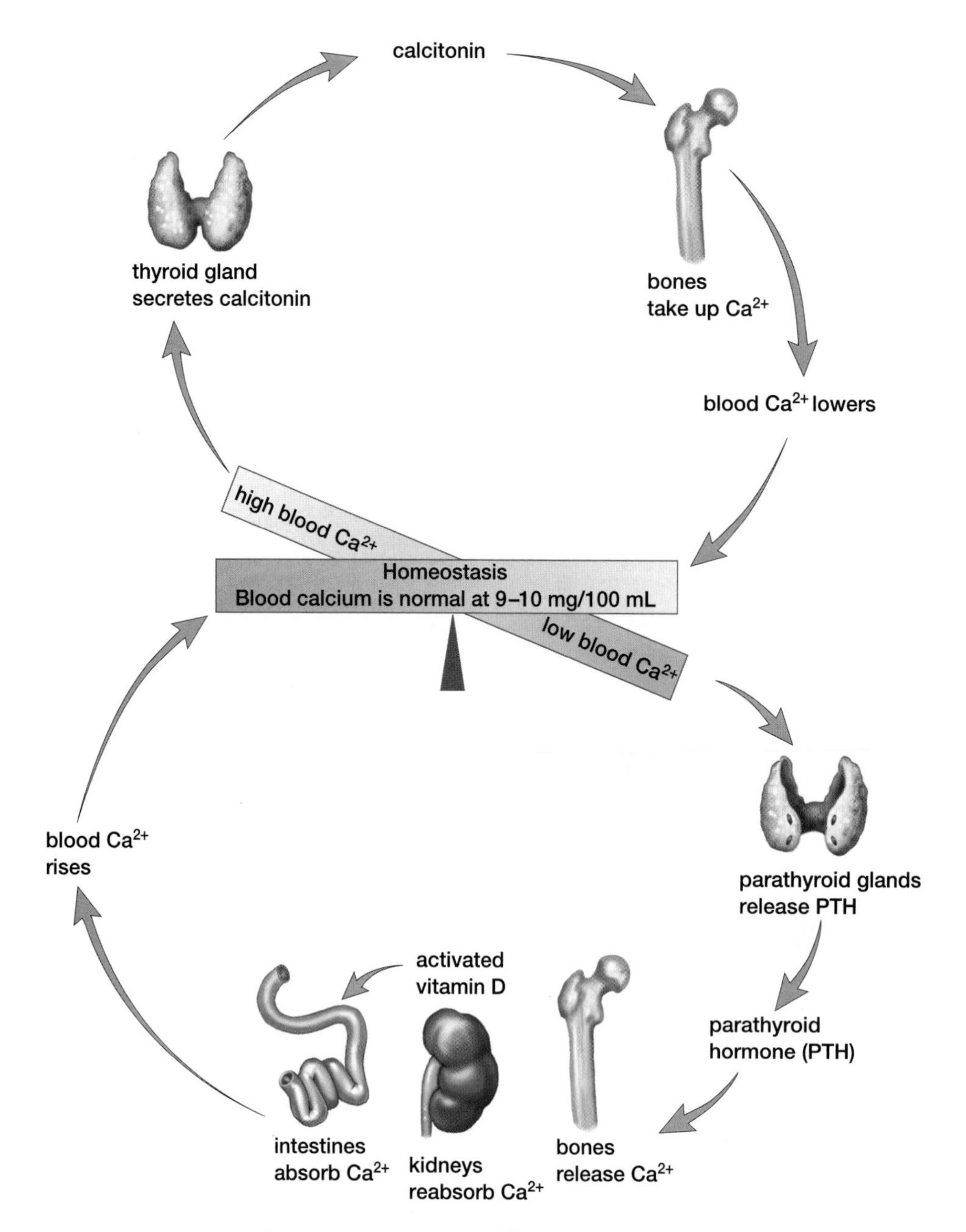

Figure 6.16 When the blood calcium (Ca^{2+}) level is high, the thyroid gland secretes calcitonin. Calcitonin promotes the uptake of Ca^{2+} by the bones, and therefore the blood Ca^{2+} level returns to normal. When the blood Ca^{2+} level is low, the parathyroid glands release parathyroid hormone (PTH). PTH causes the bones to release Ca^{2+}, the kidneys to re-absorb Ca^{2+}, and the intestines to absorb Ca^{2+}. As a result, the blood Ca^{2+} level returns to normal.

Vitamin D

Vitamin D is a steroid hormone involved in the regulation of blood calcium (as well as phosphate). Vitamin D is synthesized in a multi-step process starting in the skin and culminating in the kidney. An inactive precursor substance, vitamin D3, is first synthesized in the skin by the action of ultraviolet radiation on molecules of cholesterol. The liver converts this substance into an intermediate product, which the kidney then uses to produce the active form of vitamin D. This process is regulated by the parathyroid hormone.

PTH also promotes vitamin D synthesis in the proximal tubule of the kidney. The primary role of vitamin D is to maintain blood calcium levels. Vitamin D increases the release of calcium into the blood from bone tissue. In the kidney, it increases the retension of calcium.

In the small intestine, vitamin D increases the rate of calcium absorption by stimulating the growth of cells lining the intestine and by stimulating the synthesis of cellular proteins involved in calcium transport. Low levels of blood calcium due to a lack of vitamin D can impede mineralization of bone tissue. This problem can cause osteomalacia ("softness of bone") (in adults) or rickets (in infants). Symptoms include interruption of normal growth and development, skeletal deformities, and susceptibility to bone fractures. In adults, symptoms of osteomalacia also include skeletal pain and muscular weakness.

Lifestyle factors or geographical location may prevent sufficient production of vitamin D. In the past, rickets was sometimes described as a common disease of "smoky cities and cloudy skies." In developed countries, the addition of vitamin D to common foods such as dairy products has sharply reduced the incidence of this disease.

During pregnancy, adequate levels of vitamin D and dietary calcium are critical for normal bone development in the fetus. In the last few days of pregnancy, the fetus requires about 2 g of calcium per day. A vitamin D deficiency can result in the development of fetal rickets.

The Pancreas

The **pancreas** is a small gland located near the small intestine. It is made up of two kinds of tissues that independently function as exocrine and endocrine glands. As an exocrine organ, the pancreas secretes digestive enzymes into the duodenum. (Refer to Chapter 4 for a more detailed discussion of the exocrine function of the pancreas.)

The primary products of the endocrine portion of the pancreas are glucagon and insulin, two non-steroid protein hormones. These hormones regulate the body's metabolism of sugar and other carbohydrate molecules. They are produced by the islets of Langerhans, small groups of cells scattered throughout the pancreatic tissue.

Insulin is sometimes referred to as the "hormone of abundance" because it forces the body to store nutrients surplus to our immediate needs as glycogen in the liver, fat in adipose tissue, and protein in muscle tissue. Insulin and glucagon are antagonistic hormones. The secretion of glucagon, a catabolic hormone, triggers the cellular release of glucose, fatty acids, and amino acids into the bloodstream.

Low levels of blood glucose stimulate the secretion of insulin, an anabolic hormone, by the beta cells. Insulin increases the intake of glucose, fatty acid, and amino acids by adipose (fat) and muscle cells and activates enzyme systems that convert glucose to glycogen in liver and muscle cells. In addition, insulin stimulates protein synthesis and tissue growth throughout the body, and suppresses the metabolism of glucose in liver and muscle cells.

Insulin receptors are found on the surface of most cells in the body. When insulin attaches to a receptor, the insulin-receptor combination migrates into the cell. Part of the receptor molecule has enzymatic properties. Insulin activates these enzymes by attaching to the enzyme molecule. These receptor enzymes then activate protein molecules (carriers) that transport glucose into the cell by facilitated diffusion. These activated carriers significantly increase the rate of glucose intake.

Once in the bloodstream, insulin is broken down in a few minutes by the liver and kidneys. The number of insulin receptors on any cell varies according to the current physiological state of the body. Starvation tends to induce the production of more receptors, while obesity decreases the number of cell receptors on cell surfaces.

Recall the discussion of diabetes in Chapter 4. Type 1 diabetes, in which the pancreas can no longer manufacture insulin, may be the result of an autoimmune reaction that specifically targets the beta cells. Antibodies to specific components of beta cells have been found in the pancreases of Type 1 diabetes patients. Research suggests that hereditary factors may also play a role in the onset of this disease. Type 2 diabetes can involve insufficient insulin production or the production of insulin molecules that have reduced functionality because of some defect in chemical structure. Treatment for Type 2 diabetes includes dietary management and medication to boost insulin production.

The Pineal Gland

As you have seen, human physiological processes are generally in a constant state of flux, as the body adjusts to changes in the external and internal environment. For example, a stressful situation may trigger the release of hormones that cause a sudden increase in blood pressure and breathing rate. Similar effects may result from excessive exercise.

However, some hormone levels and physiological processes in the body seem to rise and fall in a regular pattern. Some of these biological cycles correlate with the seasons. Other processes follow a regular 24-hour cycle. Such daily cycles are referred to as **circadian rhythms**. Cortisol is one hormone that fluctuates in a circadian rhythm.

Cortisol levels tend to increase at night, peaking just before a person awakes. These levels then decrease sharply during the daytime. Another hormone that is subject to circadian rhythm is melatonin, a non-steroid hormone composed of a modified amino acid. Melatonin is produced by the **pineal gland**, a small, pine cone-shaped structure located deep in the centre of the brain. Melatonin production is highest during nighttime hours and diminishes considerably during the day.

As daylight fades, melatonin levels rise, producing the familiar feeling of sleepiness. Some studies indicate a connection between abnormal melatonin production and sleep disorders. In addition, medical research suggests that abnormal melatonin production could be a factor in the development of mood disorders and depression.

One example is seasonal affective disorder (SAD), a condition that typically produces symptoms of depression and an overwhelming desire for sleep. These symptoms generally appear at the onset of winter and are believed to affect about 20 percent of residents of northern countries. This disorder is much less prevalent in southern regions. For example, SAD appears in only about 2.5 percent of the population of Florida.

Above-normal levels of melatonin enhance SAD symptoms. Some researchers suggest that exposure to bright lights for two to three hours each day can diminish the symptoms of this disorder.

The secretion of thyroxine is another process that follows a seasonal pattern. Thyroxine levels tend to increase during winter months, boosting metabolic rate. This may explain why a 15°C day may feel cool in the fall but warm in the spring.

Research efforts continue to refine our understanding of the role of these cycles in maintaining homeostasis in the body. The many rhythmic patterns discussed in this section are a daily reminder that despite our technologically advanced way of life, our close association with the natural environment continues to be strong and pervasive.

The Thymus Gland

The **thymus gland** is located between the lobes of the lungs in the upper chest cavity. This part of the immune system is especially active in young

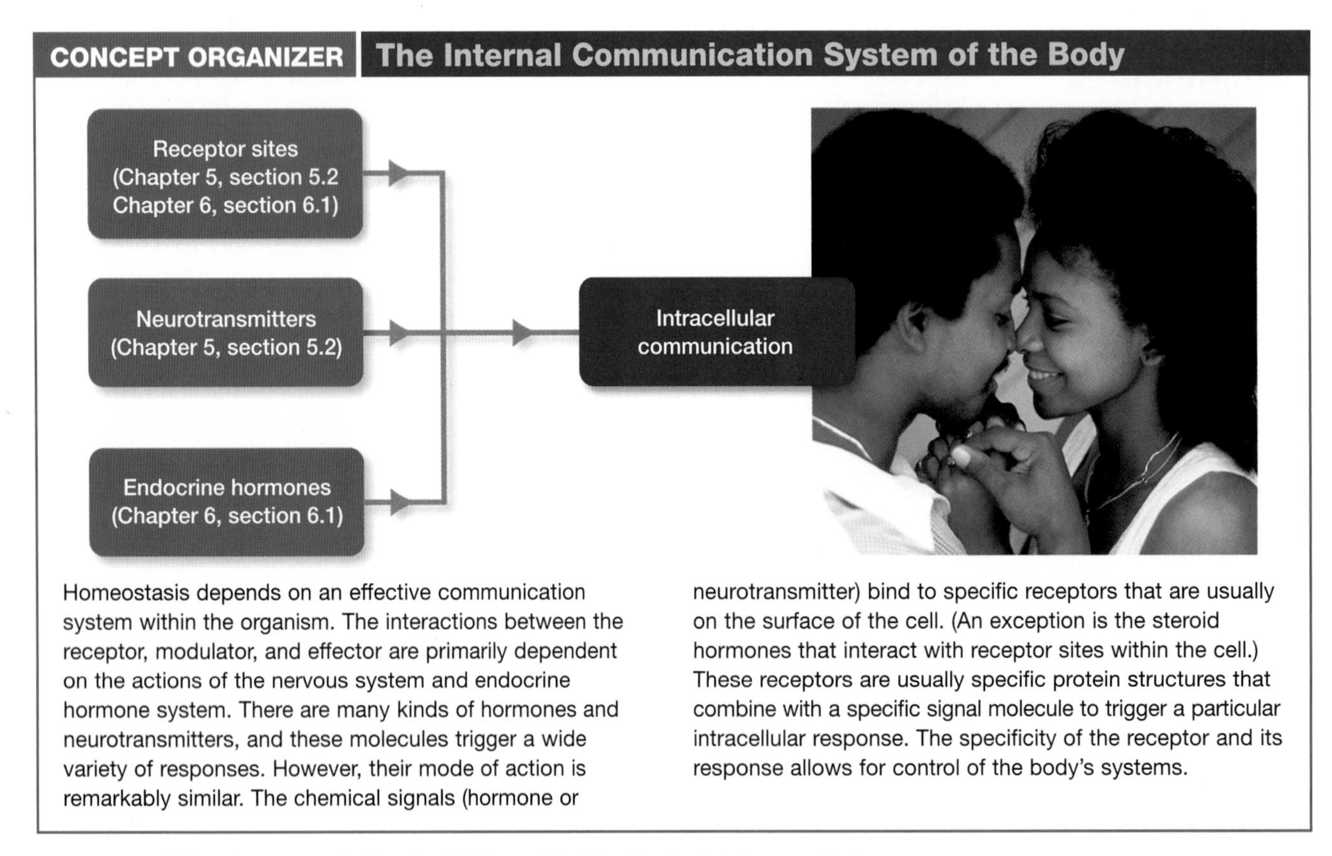

Homeostasis depends on an effective communication system within the organism. The interactions between the receptor, modulator, and effector are primarily dependent on the actions of the nervous system and endocrine hormone system. There are many kinds of hormones and neurotransmitters, and these molecules trigger a wide variety of responses. However, their mode of action is remarkably similar. The chemical signals (hormone or neurotransmitter) bind to specific receptors that are usually on the surface of the cell. (An exception is the steroid hormones that interact with receptor sites within the cell.) These receptors are usually specific protein structures that combine with a specific signal molecule to trigger a particular intracellular response. The specificity of the receptor and its response allows for control of the body's systems.

Figure 6.17 Internal communication is vital to maintaining the body's homeostasis.

children. The thymus gland produces **thymosin**, a hormone that stimulates the production and maturation of lymphocytes into T cells. The thymus gland normally disappears after puberty. (In adults, lymphocytes continue to be produced by the spleen and lymph glands.) The pituitary gland regulates hormone production in the thymus gland. If the pituitary gland is surgically removed, the thymus gland will atrophy (shrink).

In this section, you learned that the pituitary gland is aptly named the master gland. You saw that the pituitary gland is the essential link between the nervous system and the endocrine system. This gland regulates the activities of other endocrine organs in the body. The next section examines the adrenal glands and their role in managing the body's physiological response to stressful situations.

WEB LINK

www.mcgrawhill.ca/links/biology12

Nicotine, alcohol, and caffeine are considered to be drugs that can alter the production and effectiveness of hormones secreted by endocrine organs. To access information about how these substances act on the endocrine system, go to the web site above, and click on **Web Links**. Compare the effects of nicotine, alcohol, and caffeine on the production and secretion of the following hormones: thyroxine; cortisol; ACTH; and insulin. How do these changes in hormone function affect metabolic activity (for example, heart rate, absorption of nutrients, and basal metabolic rate)?

SECTION REVIEW

1. K/U Describe how hormones secreted by glands of the endocrine system regulate metabolic rate.
2. K/U **(a)** How is an exocrine gland different from an endocrine gland?
   (b) Identify two organs in the body that act as both endocrine and exocrine glands.
3. C In a chart, list the hormones produced in the human body. Divide the list into two sections: steroid and non-steroid hormones. Describe the substances used by the body to manufacture non-steroid hormones.
4. K/U Compare the way steroid and non-steroid hormones affect cellular activity.
5. K/U Explain why adequate lipid intake is essential for the normal function of some endocrine glands and hormones.
6. K/U Identify the endocrine glands and hormones responsible for regulating blood pressure in the circulatory system.
7. C Make a chart to list the similarities and differences between the roles of the endocrine system and the nervous system.
8. K/U Describe the difference between hyperthyroidism and hypothyroidism.
9. I Design a study to investigate how medications that inhibit the normal metabolism of iodine in the body can be used to treat hyperthyroidism.
10. C Make an outline diagram of the human body. In the diagram, draw the approximate shape of each endocrine gland (to scale) in its correct location within the body. Label the hormones produced by each gland and use arrows to show the target organs and tissues of each hormone.
11. MC Some studies indicate that normal endocrine function in people and other animals can be disrupted by exposure to industrial toxins, such as dioxins and PCBs, at the prenatal or infant stage of development. These toxins interfere with growth or lead to neurological disorders, including learning difficulties. Identify hazardous substances in your area. List strategies to minimize exposure to these substances. What long-term changes are needed to radically reduce the degradation of the environment?

6.2 The Adrenal Glands and Stress

EXPECTATIONS

- **Describe the anatomy and physiology of the endocrine system and explain its role in homeostasis.**
- **Compile and display, either by hand or computer, data and information about homeostatic phenomena in a variety of formats.**
- **Describe some Canadian contributions to knowledge and technology in the field of homeostasis.**

The body has two adrenal glands, one on top of each kidney. As shown in Figure 6.18, the **adrenal gland** is composed of two layers: an outer cortex and an inner medulla. Each layer produces different hormones and functions as an independent organ. While the cortex and the medulla do not interact, they are both regulated by the hypothalamus. In addition to many other functions, the adrenal gland produces adrenaline, noradrenaline, and cortisol as part of the body's response to stress.

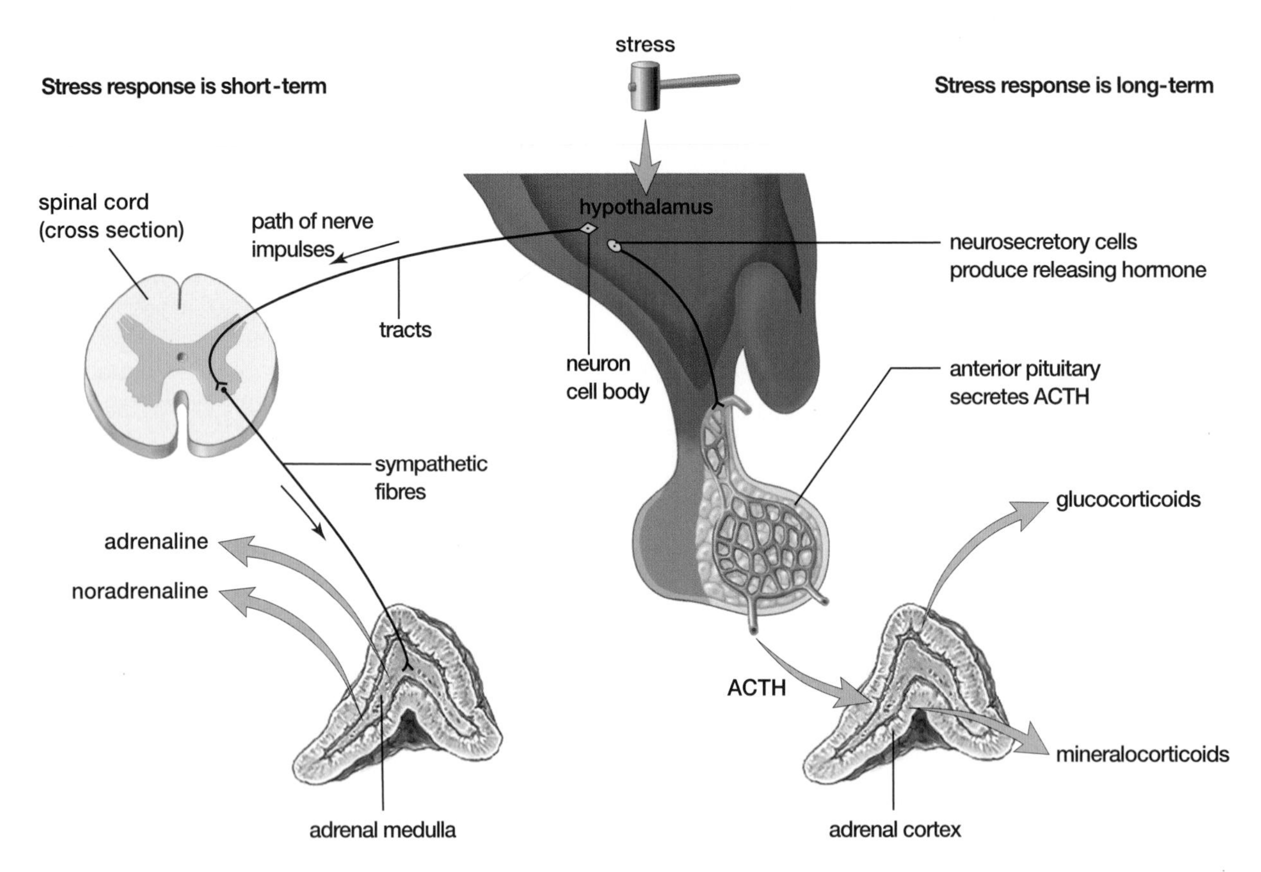

Stress response:
- heartbeat and blood pressure increase
- blood glucose level rises
- breathing rate increases
- muscles become energized
- digestive system shuts down

Stress response:
- protein and fat metabolism occur instead of glucose breakdown
- reduction of inflammation; immune cells are suppressed
- sodium ions and water are re-absorbed by kidney
- blood volume and pressure increase

Figure 6.18 Both the adrenal medulla and the adrenal cortex are under the control of the hypothalamus when they respond to stress. The adrenal medulla provides a rapid but short-lived emergency response, while the adrenal cortex provides a sustained stress response.

The Adrenal Cortex

The **adrenal cortex** produces two types of steroid hormones — the glucocorticoids (cortisol) and the mineralcorticoids (aldosterone). Cortisol stimulates carbohydrate synthesis and related metabolic functions. Aldosterone regulates salt and water balance, which in turn affects blood pressure. Both types of hormones contribute to the long-term stimulation of the immune system when the body is under stress. The adrenal cortex also produces male sex hormones (androgens) and female sex hormones (estrogens).

The production of cortisol and aldosterone is regulated by the **adrenocorticotropic hormone (ACTH)**, a polypeptide molecule synthesized by the anterior pituitary gland. ACTH production, in turn, is stimulated by a peptide substance, corticotropin-releasing factor (CRF), which is produced by the hypothalamus. Cortisol is secreted in "spurts" by the adrenal cortex.

Increased aldosterone and cortisol levels exert a negative feedback effect on the hypothalamus and anterior pituitary, which suppresses ACTH production. However, the synthesis of aldosterone is primarily controlled by changes in blood pressure and the production of angiotensin in the kidney (discussed below).

In healthy individuals, the secretion of CRF in the hypothalamus exhibits a diurnal pattern, reaching its lowest levels late at night (around midnight) and rising to a peak in early morning hours before awakening. This pattern is also reflected in the production of ACTH, aldosterone, and cortisol. Changing sleep patterns, caused by shift work, for example, will cause a corresponding change in this pattern of hormone production.

Cortisol

Cortisol secretion causes a dramatic rise (6 to 10 times normal levels) in the process of gluconeogenesis, the synthesis of carbohydrates from amino acids and other substances in the liver. Cortisol triggers the conversion of protein to amino acids in muscle tissues, and the release of amino acids into the blood. In the liver, cortisol triggers the uptake of amino acids and stimulates the production of enzymes active in glucogenesis. This increase in glucose synthesis leads to increased glycogen stores in the liver. Subsequently, under the influence of other hormones such as glucagon and adrenaline, this stored carbohydrate can then be converted back to glucose when needed (such as between meals). In addition, cortisol prompts the breakdown of lipids in fat tissues (for use as an alternative energy source in other tissues), inhibits metabolism, and suppresses protein synthesis in most organs in the body (with the exception of the brain and muscles).

Cortisol also has strong anti-inflammatory properties. In general, cortisol decreases the buildup of fluids in the region of inflammation by decreasing the permeability of capillaries in affected tissues. This hormone also suppresses production of T cells and antibodies, as well as other immune system responses that might cause further inflammation. It is therefore often used to treat and reduce inflammation caused by skin injuries, autoimmune disorders such as rheumatoid arthritis (which causes inflammation of the joints of the skeletal system), and asthma. As an asthma medication, corticosteroids are most effective when inhaled. They help alleviate inflammation of bronchial tissues during an asthmatic attack.

Physiological Response to Stress

The hypothalamus plays a key role in the body's physiological response to stress. Any form of physical or emotional stress stimulates a very rapid response in the hypothalamus. For example, during times of mental stress, increased signals from the brain stimulate the hypothalamus to produce more CRF. This increased production, in turn, prompts ACTH secretion in the anterior pituitary (as illustrated in Figure 6.19 on the following page). ACTH then triggers higher levels of cortisol production by the adrenal cortex. The extra cortisol may help to relieve some of the possible negative physiological effects of stress. As described above, higher levels of cortisol speed up gluconeogenesis and other metabolic functions that may provide additional energy sources for cell functions. Note that the precise role of cortisol in mediating the body's response to stress is still the subject of much speculation and research.

WEB LINK

www.mcgrawhill.ca/links/biology12

Think about how you react to situations that generate stress, such as preparing for a driver's test or watching a suspenseful movie. To access a stress test and supporting articles that will help you gauge your own reactions to stress, go to the web site above, and click on **Web Links**. Please note that this is not a medical diagnostic test. Consult your family health professional to discuss personal concerns about stress.

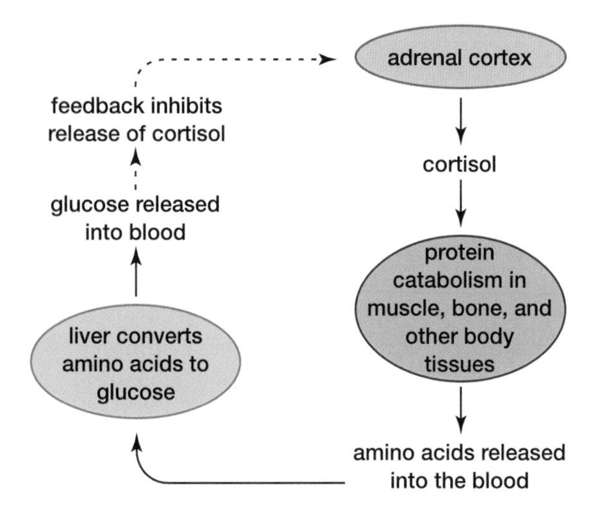

Figure 6.19 Cortisol appears to play a major role in the body's physiological response to stress.

During extended periods of stress, cortisol can interact with insulin to increase food intake and redistribute stored energy from muscle to fat tissues, primarily in the abdominal region. Abdominal obesity is a strong risk factor for Type 2 diabetes, coronary heart disease (such as arteriosclerosis), and stroke. Excessive cortisol production in times of stress may also depress immune function by reducing the availability of proteins needed for synthesis of antibodies and other substances produced by the immune system. Over time, depressed immune system function may increase the body's susceptibility to infection and the onset of some forms of cancer. The connection between cortisol and other adrenal secretions in times of stress will be discussed later in this chapter, in "Fight or Flight Syndrome."

Aldosterone

Two primary (and related) functions of aldosterone are osmoregulation (the process of regulating the amounts of water and mineral salts in the blood) and regulation of blood pressure. In the kidneys, aldosterone acts to increase sodium ion absorption and secretion of potassium ions, primarily in the collecting ducts of nephrons in the kidneys. Aldosterone also stimulates sodium re-absorption in the colon. This process raises sodium concentration in the blood. Recall from Chapter 4 that this action triggers the hypothalamus to release ADH, which in turn increases the absorption of water, leading to an increase in blood pressure.

Aldosterone production is primarily controlled by changes in blood pressure. A decrease in blood pressure will stimulate the kidneys to secrete the enzyme rennin. This enzyme secretion, in turn, triggers the activation of the protein angiotensin (which the kidney produces from blood proteins). Angiotensin raises blood pressure by triggering the constriction of arterioles and by stimulating the release of aldosterone from the adrenal cortex.

Abnormal secretion of mineralcorticoids or glucocorticoids can cause ailments such as Cushing's syndrome and Addison's disease. Addison's

THINKING LAB

Symptoms of Cushing's Syndrome and Addison's Disease

Background

Cushing's syndrome is caused by an excess of glucocorticoids (hypersecretion) due to either elevated levels of ACTH or a tumour on the adrenal gland. The disease is characterized by high blood pressure, high blood sugar, muscle weakness, and edema (the accumulation of fluid in the tissues). These symptoms can also appear if someone is treated with cortisone over a long period of time.

Addison's disease is the result of a hyposecretion (deficient secretion) of glucocorticoids and mineralocorticoids. The symptoms are the reverse of those for Cushing's syndrome. Addison's disease sufferers have low blood pressure and low blood sugar. They also tend to suffer from weight loss, weakness, and a loss of resistance to stress.

You Try It

1. Based on what you have learned in this chapter, explain each of the symptoms of these diseases.
2. What treatment(s) do you think should be provided?

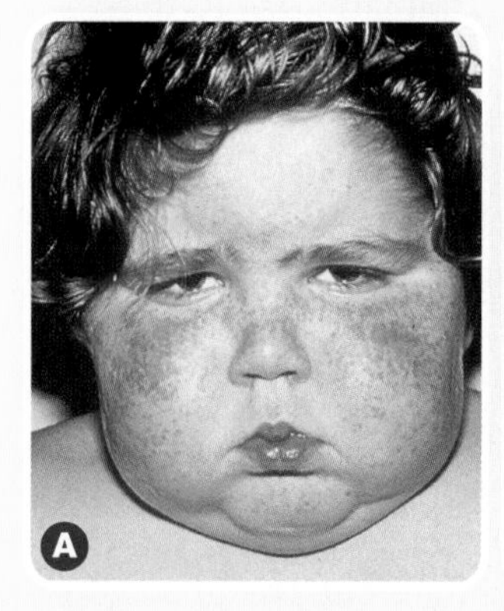

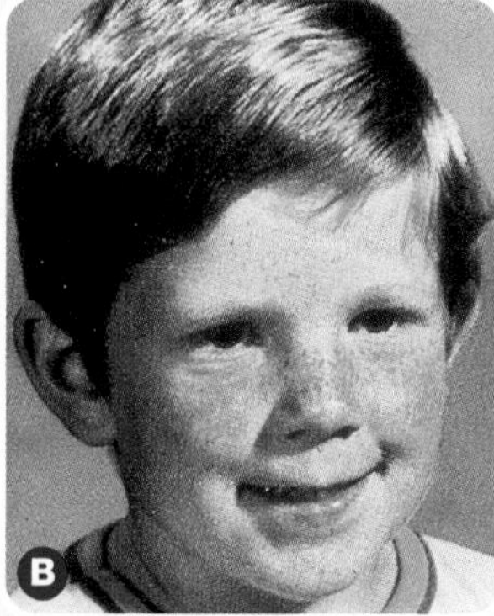

Cushing's syndrome can result from hormonal hypersecretion due to an adrenal cortex tumour. (A) First diagnosed with Cushing's syndrome. (B) Four months later, after therapy.

disease is caused by the autoimmune destruction of the adrenal cortex tissue, resulting in decreased cortisol and aldosterone production. Loss of cortisol typically leads to hypoglycemia, weight loss, and feelings of nausea. Depressed aldosterone levels result in lowered blood pressure due to decreased volume of body fluids. The loss of adrenal androgens also contributes to a reduced sex drive. In addition, Addison's disease sufferers may develop a "perpetual tan" appearance (hyperpigmentation) as a result of increased ACTH secretion.

BIO FACT

U.S. President John F. Kennedy, assassinated in 1963, was likely the most famous Addison's patient. Doctors controlled his disease with hormone replacement therapy, involving injections of glucocorticoids and mineralcorticoids.

Cushing's syndrome is the result of a spontaneous and chronic production of glucocorticoids by the adrenal cortex. Treatment for this disorder involves the administration of drugs that block the synthesis of glucocorticoids. Cushing's disease, a related disorder, is a condition in which a pituitary tumour results in the over-secretion of ACTH and the subsequent increase in cortisol production. This disease is generally treated by surgically removing the pituitary tumour. Both diseases cause increased protein breakdown and muscle wasting, as well as excessive fat buildup in the abdomen and elsewhere in the body.

In the Thinking Lab on page 186, you can consider these conditions in greater detail.

Sex Hormones

The adrenal cortex also produces small amounts of male sex hormones (androgens) and female sex hormones (estrogens). These hormones are found in both sexes, but males produce higher levels of androgens while females synthesize more estrogens. Because the testes in males produces high levels of androgens, the amount of this hormone secreted by adrenal glands have only a slight effect on body functions. The androgen hormones produced by the adrenal glands account for 50 percent of the total androgen output in females.

As you will see in section 6.3, androgens promote muscle and skeletal development in both males and females. The estrogen production by the adrenal glands remains insignificant until after menopause, when the ovaries cease production of these hormones.

The Adrenal Medulla

The adrenal gland secretes adrenaline (also called epinephrine) and noradrenaline (also called norepinephrine), two non-steroid hormones. Adrenaline, the first hormone discovered (in 1894) is often called the "stress hormone" because it is the major hormone secreted in response to stress.

The adrenal medulla is another example of the overlapping functionality of the nervous and endocrine systems. The **adrenal medulla** is composed of modified neurons of the sympathetic nervous system. The production of adrenaline and noradrenaline is under the control of the hypothalamus via this direct connection with the sympathetic nervous system. The hormones adrenaline and noradrenaline also serve as excitatory neurotransmitters in the sympathetic nervous system. When adrenaline and noradrenaline act as neurotransmitters, their effect on the body is limited and of short duration.

The adrenal medulla secretes a mixture of 85 percent adrenaline and 15 percent noradrenaline. Despite minor differences in molecular structure, these hormones produce very similar effects on target tissues.

Adrenaline and noradrenaline act to increase heart rate and blood pressure, and cause vasodilation (widening) of blood vessels in the heart and respiratory system. These hormones also stimulate the liver to break down stored glycogen and release glucose into the blood. When the body is "at rest," these two hormones sufficiently stimulate cardiovascular function to maintain adequate blood pressure without additional input by the sympathetic nervous system.

Adrenaline and Anaphylactic Shock

Some people suffer severe allergic reactions, called anaphylactic shock, to antigens from bee stings, peanuts or other foods, latex gloves, or intravenous medications such as penicillin. If these antigens enter the bloodstream, they can trigger a life-threatening chain reaction called "anaphylactic shock." Once in the circulatory system, the antigen stimulates the widespread release of histamine, an anti-inflammatory substance produced by the immune system. Histamine stimulates vasodilation of arterioles throughout the cardiovascular system, and leakage of fluid and proteins out of the capillaries. This precipitates a rapid decline in blood pressure and reduced flow of blood (and oxygen) to the organs and tissues of the body.

Without immediate treatment, this situation can result in death within minutes. Emergency treatment involves injections of adrenaline hormone. Adrenaline mimics the action of the sympathetic nervous system, increasing heart rate and the strength of heart contractions. This action helps to restore blood pressure.

People at risk for anaphylactic shock usually carry a syringe-like device called an EpiPen™ (as shown in Figure 6.20). This device can be used to quickly self-administer (if necessary) a life-saving dosage of adrenaline. Doctors recommend that at-risk individuals also wear emergency identification bracelets.

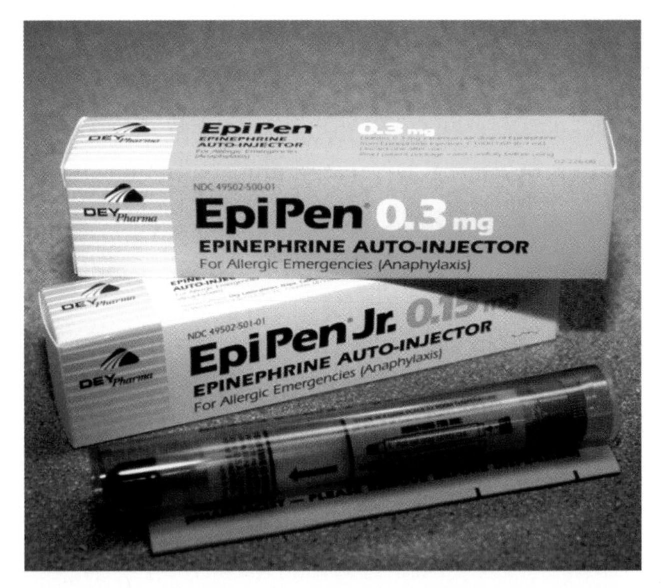

Figure 6.20 An EpiPen™ is used to counteract anaphylactic shock.

DESIGN YOUR OWN **Investigation** 6 • A

SKILL FOCUS

- Initiating and planning
- Hypothesizing
- Identifying the variables
- Communicating results

Crowding Stress on a Guppy Population

Animal populations react to the stress of overcrowding in various ways. Studies show that animals under stress exhibit changes in hormone levels, changes in behaviour, or a combination of both. In the 1970s, researchers at the University of Wisconsin performed a series of experiments on mouse populations.

In the first experiment, a set amount of food was given to a group of mice who were roaming freely in the basement of an old building. The researchers noted that when the food supply could no longer support the population, some mice moved to other parts of the building.

In the second experiment, the mice were confined to a large enclosure, so they could not leave. When the food ran out, the researchers noted that the birth rate dropped to zero. In this instance, the crowding resulted in changes to the reproductive system of the female mice.

The third experiment involved a population of mice in a closed enclosure supplied with unlimited food. This experiment was performed in a variety of locations and resulted in different observations. In Wisconsin, the crowded population of mice became more aggressive, and chasing and fighting behaviours increased. As well, females stopped caring for their young, and most of the young mice died. When the same experiment was performed in England, the result was a decline in birth rate.

Problem

How do guppy populations respond to crowded conditions?

Hypothesis

Make and record a hypothesis on the effect of crowding on guppy populations.

CAUTION: **Treat living organisms with care and respect.**

Materials

a number of male and female guppies
fish tanks
fish food

Experimental Plan

1. Working in a group, design an experiment to show the effects of crowding on guppies.
2. Choose the size of fish tank(s) you want to use, the amount of food, a method of keeping the amount of food constant, the number and sexes of fish, and so on.

Fight or Flight Syndrome

In Chapter 5, the "fight or flight" syndrome was discussed in the context of nervous system function. The fight or flight reaction is the way in which the body responds to a sudden, unexpected stressful stimulus, such as a physical threat or some other perceived emergency situation. This reaction is produced by the rapid release of adrenaline and noradrenaline by the adrenal gland (which reinforces the action of the sympathetic nervous system).

For example, if you were to notice that you were in the path of a rapidly oncoming car or if you came face to face with a bear in the woods, your breathing rate and heart rate would quickly increase. You would also experience a rapid elevation in blood sugar level.

In our human ancestors, this reaction would have been induced primarily by a perception of physical danger. However, stress can also be triggered by feelings of excitement, anxiety, psychological conflicts, environmental extremes (such as severe heat or cold), or a lack of sleep.

The phrase "fight or flight response" was coined by Walter B. Cannon and Hans Selye, a Canadian doctor, in the 1930s. The phrase defines the pattern of physiological responses that prepares the body for emergency situations. Cannon studied the effect of stress on dogs, observing that stress depressed digestive function. His work led him to develop the concept of homeostasis. Much of Selye's research focussed on how the adrenal gland affects the body's immune and inflammatory responses during times of stress.

Conduct Investigation 6-A to see how animals are affected by the stress caused by situations of overcrowding.

3. Design an experiment that can be carried out in the classroom.
4. Outline a procedure for your experiment, listing each step. Assemble the materials you will require.
5. Decide what observations you will make and a method to quantify them.

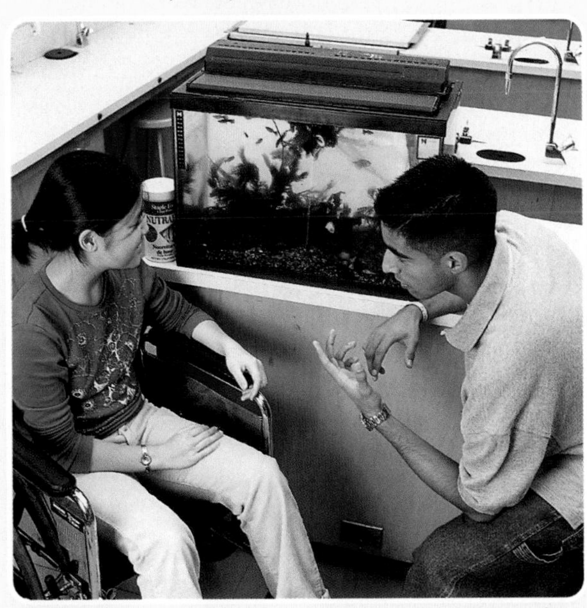

Checking the Plan

1. What will be your independent variable? What will be your dependent variable(s)? What will be your controlled variable(s)?
2. Does your method of collecting data allow for a number of different responses to crowding?
3. Has your teacher approved your plan?

Data and Observations

Conduct your experiment and record your results. Present your data in a graph or chart that will help you communicate your findings to other groups in the class.

Analyze

1. What was the size of the population(s) at the beginning of the experiment?
2. Did the population(s) increase? If so, at what rate?
3. Did the ratio of males to females change?
4. Were the fish randomly distributed throughout the tank?
5. Was there any observable aggressive behaviour?

Conclude and Apply

6. Were your results consistent with those obtained by other groups in your class? Explain.
7. Is there any evidence that crowding affected the endocrine system of the fish?

Exploring Further

8. Using the results from your experiment, form a new hypothesis and an experiment to test it.

Selye's Legacy

Selye described the anti-inflammatory properties of adrenal steroid hormones, such as cortisol. He discovered that steroid hormones regulated the activity of many lymphatic structures, such as the thymus, spleen, and lymph nodes.

In 1936, Selye, whose photograph appears in Figure 6.21, produced a paper that first described what he termed the General Adaptation Syndrome (GAS). He stated that the GAS was the body's "non-specific" response to any stressor that placed a demand on the sympathetic nervous system, adrenal gland, ADH production, and production of rennin and angiotensin. Selye suggested that the human body produces a similar response to most types of stressors.

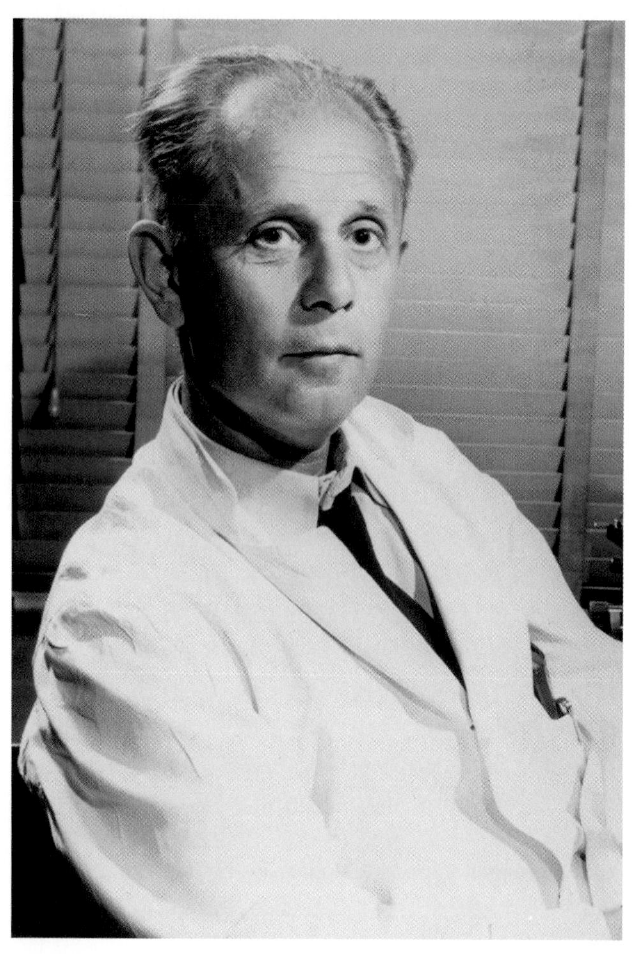

Figure 6.21 Dr. Hans Selye

Briefly, the GAS was a three-stage response by the body to stress. The first stage was the "Alarm Reaction," in which some form of physical or mental trauma would trigger an immediate stress response in the body. This reaction depressed immune system activity, leaving the body more susceptible to infection and disease. This was followed by the "Resistance" stage, in which the body adapted to a state of stress. In this stage, the body draws on fat and protein energy reserves to strengthen the body's defences. The final stage was "Exhaustion," in which there is a sharp decline in immune system function and other homeostatic mechanisms.

Chronic stress can lead to diarrhea, migraine headaches, insomnia, bruxism (grinding of the teeth), emotional instability, heart palpitations, trembling or nervous ticks, inability to concentrate, and increased frequency of urination. Ignoring such warning signs of stress can lead to the onset of serious illness.

Recently, the GAS concept has met with considerable criticism in the medical community. For example, it is now known that the body tends to produce specific, rather than general, responses to a particular type of stressor, such as infection, extreme exercise, or pain.

Selye was also unable to develop a physiological definition of "stress." One dilemma he faced was that stress is a highly subjective experience. He knew that the same situation could elicit a wide range of emotional responses, from pleasure to irritation, anger or fear, in different individuals. For some people, rock climbing is an exhilarating activity, while others may regard it as terrifying.

In addition, Seyle's theories were often based on relatively crude experimental observations, such as the visual study of stomach ulcers. Also, much of his research focussed on the physiological action of cortisol. Current research methods and technology are considerably more advanced than those on which Selye based many of his theories.

However, Selye's innovative ideas did prompt mainstream medicine to focus more intensely on the link between psychological stress and general health, including immune system function. Thus, despite its shortcomings, Seyle's body of work did make a major contribution to the scientific understanding of the impact of stress on the physiology of the endocrine and immune systems in the body.

In this section, you learned how hormones of the adrenal gland regulate metabolism and control the body's response to stressful situations. You also examined the negative effects of chronic stress on the immune system and other body functions. In the next section, you will see how the endocrine system controls development and function of the male and female reproductive systems.

SECTION REVIEW

1. K/U Describe how two different information pathways control hormone secretion in the adrenal glands.

2. K/U Compare the functions of the hormones produced by the adrenal medulla and the adrenal cortex.

3. C Make a concept map to describe the typical physiological symptoms of stress, the systems and tissues involved, and possible factors that may trigger a stress response in an individual.

4. MC Refer back to the stress test you performed in the Web Link on page 185. What are some coping strategies that might help you deal with stressful situations in the future?

5. K/U Compare the functions of adrenaline and noradrenaline in the nervous and endocrine systems.

6. C The body's response to stress involves different hormones. Is this a negative feedback loop or a positive feedback loop? Explain your answer.

7. C Describe how adrenal gland activity would be affected by the consumption of food with a high salt content.

8. C Explain why ACTH is sometimes used to treat problems associated with the adrenal gland.

9. C Make a chart listing the physical impact of too much or too little corticoid secretion.

10. C Create a flowchart showing the sequence of steps that lead to anaphylactic shock.

11. K/U Describe the "fight-or-flight response."

12. K/U How did Hans Selye contribute to our understanding of stress?

13. I Does caffeine contribute to stress or decrease stress? How could you determine this? Develop a testable hypothesis for this problem.

14. MC Studies with rats suggest that overcrowding causes behaviour changes.
    (a) Describe the impact of overcrowding in terms of stress.
    (b) Can you find evidence that people living in cities are more prone to stress than people living in less crowded environments?
    (c) If city dwellers are found to be more prone to stress, can you conclusively link this to population density, or are there other factors that may be more relevant?

15. I Current news stories from around the world claim that more people now have an unhealthy body size. Many factors contribute to this problem, making it difficult to correct.
    (a) Outline how a high level of stress may be a factor.
    (b) Suggest preventative strategies.
    (c) Briefly describe a research plan to test the relationship between stress using mice as the subjects.

6.3 Reproductive Hormones

EXPECTATIONS

- **Explain the action of hormones in the female and male reproductive systems, including the feedback mechanisms involved.**
- **Compile and display, either by hand or computer, data and information about homeostatic phenomena in a variety of formats.**
- **Analyze how environmental factors (physical, chemical, emotional, and microbial) and technological applications affect the maintenance of homeostasis, and examine related societal issues.**
- **Synthesize case study information about the effects of taking chemical substances to enhance performance or improve health (for example, explain the effect of steroids on health).**
- **Present informed opinions about problems related to the health industry, health legislation, and personal health.**

Previously you learned that sexual reproduction is the result of the combination of gametes, one from a male and one from a female. On a biological level, the process is necessary for the survival of the species. Sexual reproduction produces the variations necessary to maintain homeostasis as a species. As the environment changes, natural selection determines which individuals will survive to breed and pass on their variations to the next generation. A species with few variations faces possible extinction. The human species, with its reasonably high degree of variation, has proven to be very adaptable as it has populated the globe. In fact, the incredible reproductive success of the billions of *Homo sapiens* will be at the heart of many of the challenges of the twenty-first century.

Figure 6.22 As a species, *Homo sapiens* is characterized by a relatively high degree of variation.

The biology of human reproduction is one of the factors that has allowed us to inhabit almost every part of Earth. In most mammals, reproductive cycles are seasonal and mating occurs so that the offspring are born when environmental conditions are most favourable. Humans, along with some non-human primates, elephants, and giraffes, are examples of continuous breeders. That is, human reproduction is not influenced by the seasons or by location. However, as you will see, human reproduction does have a cyclical component, which is controlled by reproductive hormones.

The Male Reproductive System

The male reproductive system begins to form when the fetus is eight weeks old and becomes functional at the end of puberty, when a boy is about 13 years old. **Puberty** in males is the stage of life during which reproductive hormones are formed and reproductive development begins, until the first viable sperm are formed. From the end of puberty, the male reproductive system is usually capable of producing sperm 24 hours a day, seven days a week until death. Figure 6.23 shows the major parts of the male reproductive system.

The testes and the penis are outside the body cavity. The testes produce the sperm and reproductive hormones, and the penis transfers the sperm into the female reproductive system during sexual intercourse. The testes hang outside the body cavity within the scrotum so that they have the cooler temperatures (by two or three degrees) required for the formation of healthy sperm. The testes actually develop inside the body but descend into the scrotum during the last two months of fetal development. If the testes fail to descend, the result is usually sterility. This occurs in about three percent of all male births, but this condition can be corrected surgically.

The testes are composed of long, coiled tubes called the **seminiferous tubules**, which are surrounded by the **interstitial cells**. As shown in Figure 6.24 on the following page, the formation of sperm (**spermatogenesis**) occurs within the seminiferous tubules, which can be up to 250 m long. The lining of each of these tubules is composed of cells undergoing cell division (**meiosis**), with sperm being continuously produced and released from the inner lining. **Sertoli cells**, which support, regulate, and nourish the developing sperm, are also found within the tubules. As sperm are formed, they move to the **epididymis** for maturation, where they become motile.

During sexual arousal, blood flows into the penis and is prevented from leaving; this forms an erection. The sperm move out of the epididymis and the vas deferens and are mixed with fluid from a series of glands. The **seminal vesicles** produce a mucus-like fluid containing the sugar fructose, which provides energy for the sperm. The **prostate gland** and the **Cowper's gland** secrete an alkaline fluid to neutralize the acids in the female reproductive tract. This combination of sperm and fluid enters the urethra from the urinary bladder and exits the body. The movement of the **semen**, which includes the sperm and the fluid from the glands, is the result of a series of interactions between the sympathetic, parasympathetic, and somatic nervous systems. Sensory stimulation, arousal, and the co-ordinated muscular contractions combine to trigger the ejaculation of sperm.

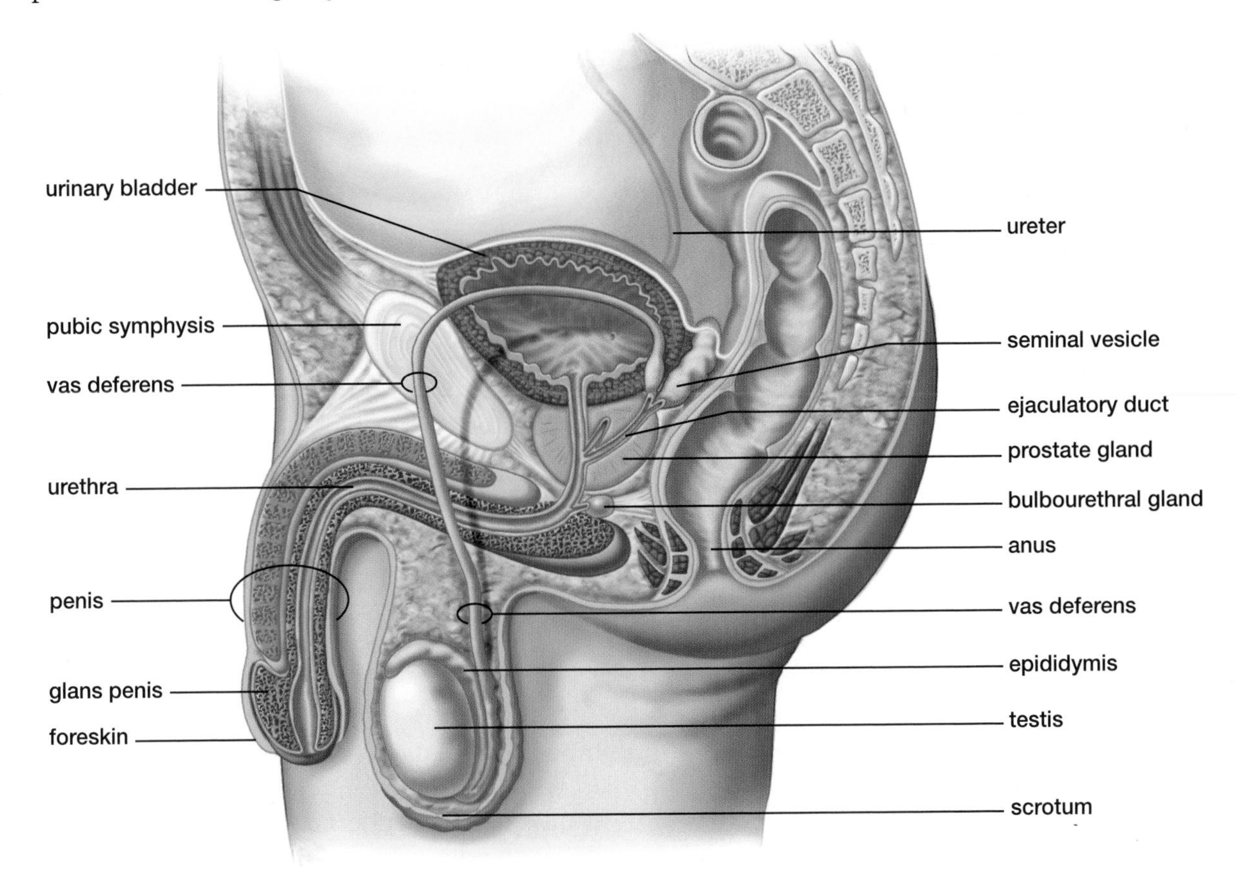

Figure 6.23 The testes produce sperm. The seminal vesicles, the prostate gland, and the bulbourethral gland provide a fluid medium. Circumcision is the removal of the foreskin. Notice that the penis in this drawing is not circumcised because the foreskin is present.

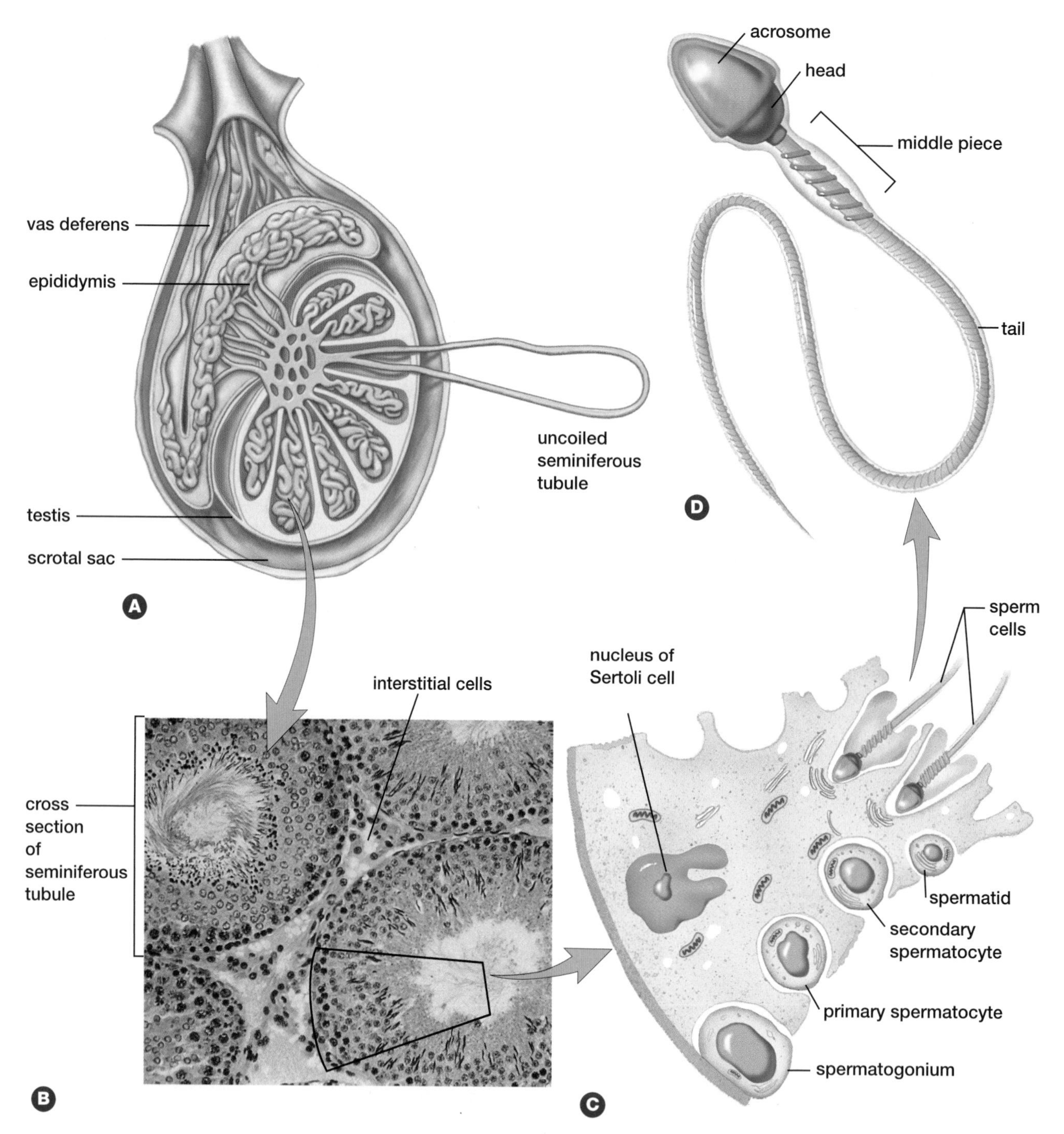

Figure 6.24 (A) The lobules of a testis contain seminiferous tubules. (B) Light micrograph of a cross section of seminiferous tubules, where spermatogenesis occurs. (C) Diagram of spermatogenesis, which occurs in the wall of the tubules. (D) A sperm has a head, a middle piece, and a tail. The nucleus is in the head, capped by the enzyme-containing acrosome.

Male Reproductive Hormones

The process of spermatogenesis is stimulated by the FSH from the anterior pituitary gland. The seminiferous tubules, in addition to producing sperm, release the hormone **inhibin**, which forms a negative feedback loop with FSH (shown in Figure 6.25). Inhibin acts on the hypothalamus to slow the production of the releasing factors that control the release of FSH. The interaction of FSH and inhibin controls the rate of formation of sperm.

Another hormone from the anterior pituitary, LH (luteinizing hormone), stimulates the interstitial cells of the testes that surround the seminiferous tubules. These cells produce the male sex hormones. The steroid hormone **testosterone** is the major androgen and is responsible for the development of the male secondary sexual characteristics. The secondary sexual characteristics include the enlargement of the primary sexual characteristics (the penis and testes) and the enlargement of the larynx (the Adam's apple). Testosterone also inhibits fat while promoting the development of muscle tissue, and stimulates the formation of hair on the face, chest, under the arms, and around the genitals. These characteristics begin to appear during puberty, with an increase in testosterone levels around ages 10 to 12. As Figure 6.25 shows, the level of testosterone in the blood inhibits the production of LH, forming the second negative feedback loop.

The following Thinking Lab looks at how testosterone levels vary in dominant and subordinate baboons.

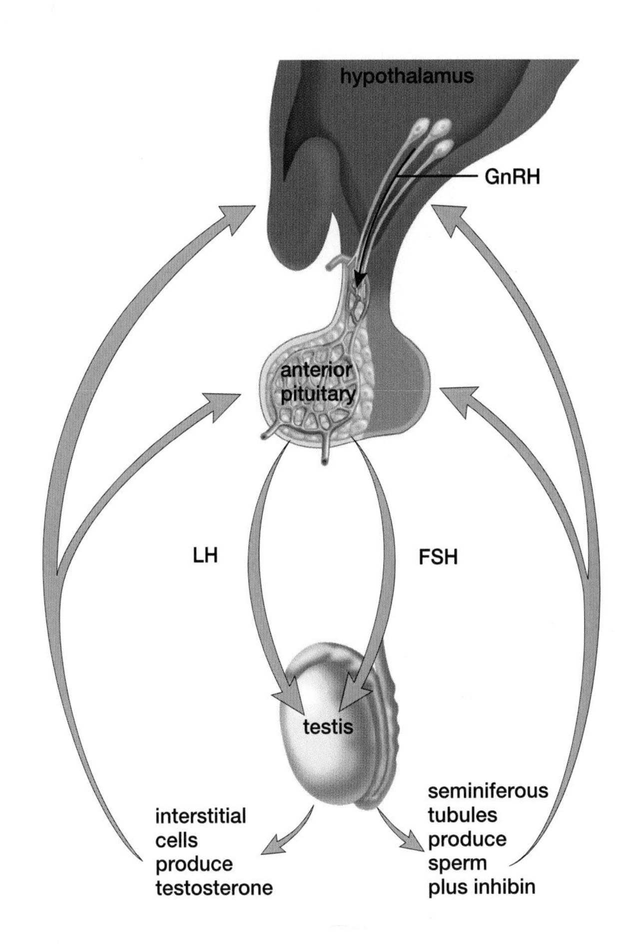

Figure 6.25 GnRH (gonadotropin-releasing hormone) stimulates the anterior pituitary to secrete the gonadotropic hormones FSH and LH. FSH stimulates the testes to produce sperm, and LH stimulates the testes to produce testosterone. Testosterone and inhibin exert negative feedback control over the hypothalamus and the anterior pituitary; this regulates the level of testosterone in the blood.

THINKING LAB

Testosterone and Baboon Behaviour

Background

In baboon tribes, a social structure of dominant and subordinate males exists. The dominant males have better access to food, the best resting spots, and the female baboons. In contrast, the subordinate male baboons must laboriously search for food, often only to have it stolen by a dominant male.

In males, the hormone testosterone regulates sexual behaviour and aggression and increases the rate at which glucose reaches the muscles. This graph shows testosterone levels of dominant and subordinate male baboons. When the male baboons are at rest, the testosterone levels are essentially equal. After being exposed to the same stress, however, the reactions of the dominant and subordinate males differ sharply for the first few hours.

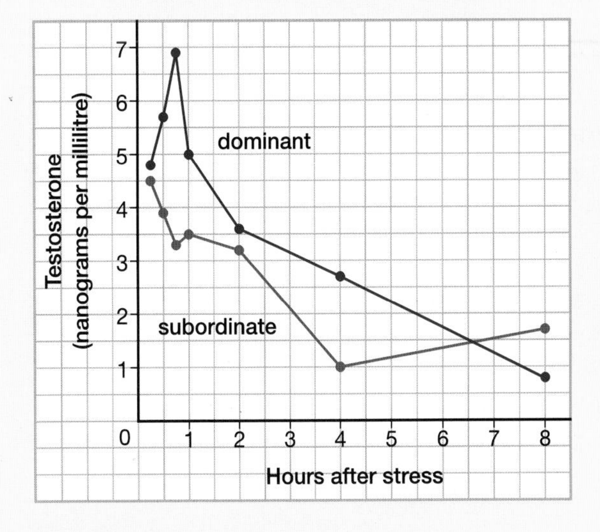

You Try It

Explain the adaptive advantage of higher levels of testosterone in the dominant male during times of stress.

The Female Reproductive System

The female reproductive system (illustrated in Figure 6.26 on page 198) performs many functions — it produces the female reproductive cell (the ovum), maintains a fertilized egg through its development as an embryo and fetus, and allows for the birth of the baby.

The vagina acts as an entrance for the erect penis to deposit sperm during sexual reproduction and an exit for the fetus during childbirth. The cervix forms the opening/exit to the uterus, a thick-walled muscular organ about the size and shape of an inverted pear. Although the uterus is usually the size of a fist, it must be capable of expanding to six times its usual size to accommodate a fetus. The lining of the uterus, called the **endometrium**, is richly supplied with blood vessels to provide nutrients to a developing fetus. The endometrium is affected by the changing hormone levels of the menstrual cycle.

The two **ovaries** produce the ova and are suspended in the abdominal cavity. They are each held in place by two ligaments; one extends from the ovary to the abdominal wall while the other

Biology Magazine TECHNOLOGY • SOCIETY • ENVIRONMENT

Steroid Use

Would you want to take a drug that builds your muscle mass, increases your strength, reduces your body fat, improves your endurance, and makes you feel powerful? All these effects are produced by the male sex hormone testosterone, and by synthetic versions of the hormone called anabolic steroids ("anabolic" means growing or building). Many people, especially athletes of both sexes, find the positive effects of steroids appealing. But adding more hormones to the body's normal level unbalances hormonal regulating cycles and may lead to health problems. Typically, abusers of steroids use levels that are 10 to 10 000 times higher than those prescribed by doctors for therapeutic purposes. Would you still want to take steroids if you knew they can also result in shrunken testes, reduced sperm count, breast enlargement, and impotence in males, and menstrual irregularities, breast reduction, masculine features, and sterility in females? In addition, steroids can lead to acne, mood disturbances, and increased risk of diseases of the heart, liver, and kidneys in both sexes. If taken by teenagers and young adults, steroids can also stunt growth.

What Are Steroids?

Steroids form a class of about 30 hormones normally secreted by the adrenal cortex, testes, ovaries, placenta, and corpus luteum (a group of cells that produces progesterone). Testosterone is a steroid hormone produced by the testes from puberty throughout adult life, in quantities of 4 mg to 10 mg per day. Testosterone has three main functions:

1. It stimulates the development of male secondary sex characteristics such as beard growth, deep voice, and the maturation of sperm. These are called androgenic functions.
2. It helps the body retain dietary protein, which accelerates muscle growth, increases the formation of red blood cells, and speeds up regeneration and recovery time after injuries or illness. These are called anabolic functions.

3. It provides feedback for the sex hormone regulating cycle. Excessive levels of testosterone in the blood signal the hypothalamus to release less LH-releasing hormone (also known as GnRH or gonadotropin-releasing hormone). GnRH stimulates secretion of LH (leutenizing hormone) from the anterior pituitary, which in turn stimulates the testes to produce testosterone. In other words, a negative feedback loop ensures that if you have high levels of testosterone, your body will reduce or even stop production of this hormone until levels subside to normal. This regulating cycle produces pulses in the levels of testosterone circulating in the bloodstream throughout the day.

The feedback function of testosterone explains many of the negative side effects experienced by steroid users.

reaches from the ovary to the outside of the uterus. The two **fallopian tubes** transfer an ovum from the ovary to the uterus. The lining of each tube is ciliated to create a current that moves the ovum toward the uterus. The sperm fertilizes the ovum as it travels down the fallopian tube. Ova are released from different parts of the ovaries, so the openings of the fallopian tubes consist of finger-like projections called **fimbriae**, which sweep over the ovaries. The fimbriae are also ciliated to sweep the ovum into the fallopian tube for its trip to the uterus.

BIO FACT

Cervical cancer occurs most often among women 30 to 50 years of age. A Pap test involves removal and examination of cells from the cervix and vagina to determine the presence of cancerous and pre-cancerous cells. Most cases of cervical cancer are related to a sexually transmitted disease called human papillomavirus (HPV).

Extra testosterone added to the body means less of the following:

- LH. This results in lower testosterone production in men and disruption of the menstrual cycle in women.
- testosterone. This leads to impotence and shrunken testes.
- follicle-stimulating hormone (FSH). This hormone is responsible for the maturation of sperm and eggs. Lower levels of FSH lead to sterility.

Synthetic Steroids

Synthetic anabolic steroids were first developed in the 1930s. They were used medically to help the body:

- rebuild tissues weakened by injury or disease,
- regain weight after illness, and
- recover more quickly from breast cancer and osteoporosis.

By the 1940s, synthetic steroids were being used in sports that relied on bulk and strength, such as weight lifting, football, and shot-put. Steroid use soon spread to other sports, and by the 1980s there was a growing illegal market for steroids for non-medical purposes. The International Olympic Committee lists over 17 anabolic steroids and related compounds that are banned for use by athletes in competition.

The synthetic steroids used by bodybuilders and athletes are designed to mimic the anabolic traits of testosterone while minimizing the androgenic or masculinizing effects. To try to combat the unwanted effects produced by the feedback loop, steroid users are increasingly sophisticated with respect to the quantities of steroids they take and the timing of use. As well, they often take several drugs in combination, including other types of hormones, vitamins, enzymes, and protein supplements.

For example, DHEA (dehydrepiandosterone) is a natural, weakly androgenic steroid hormone secreted by the adrenal glands. It is converted in the body to androstenedione (a testosterone-like hormone), testosterone, or estrogen, depending on the user's sex, age, and physical condition. But attempts to influence testosterone production in this way can also backfire. DHEA can displace testosterone at androgen binding sites, while enzymes convert excess DHEA to the female sex hormone estrogen.

Do They Work?

There are few reliable clinical studies on the long-term effects of anabolic steroids and related drugs. With so much still unknown, the regulations governing their use and legality vary widely. What is freely sold as a harmless food supplement in the United States may be illegal or available only by prescription in Canada.

There are uncertainties and conflicting claims about the benefits and dangers of steroids. However, a knowledge of the body's regulatory mechanisms suggests that long-term use of hormones and other potent biological molecules by a healthy individual is likely to have cumulative harmful side effects. This is especially true of young people, whose bodies are still developing. With a balanced diet, rest, and exercise, most people have sufficient levels of all the hormones they need.

Follow-up

1. Make a sketch of the feedback loops governing the production of testosterone by the testes.
2. One survey of professional athletes reported that about half would be willing to take a drug that ensured their success, even if it shortened their lives. In groups, discuss the arguments for and against the use of synthetic anabolic steroids by athletes in competition.

Female Reproductive Hormones

Typically, puberty begins at age nine or 10 in North American girls but significantly later in many countries. A 1997 study showed some girls begin puberty as early as three years old. Although this is unusual, there is a trend toward earlier onset of puberty than has historically been the case. To a certain extent, this lowered age can be attributed to better diet, but many scientists are concerned. They are looking for other social or environmental factors that may be influencing this trend.

Puberty is triggered by the hypothalamus, which secretes releasing factors to begin the production of the reproductive hormones. Hormone levels typically rise gradually from ages eight to 12, and then rise sharply in the early teens. These hormones stimulate the development of the female secondary sexual characteristics, which include the development of breasts, the growth of hair around the genitals and under the arms, the widening of the hips, and an increase in body fat.

BIO FACT

The first menstruation does not occur until a girl has at least 17 percent body fat. This accounts for the fact that some young athletes and dancers have delayed menstruation. In older women, menstruation stops if body fat is less than 22 percent. This prevents pregnancy if a woman's body is unable to carry a baby.

While the female reproductive system is not affected by season or location, the reproductive hormones follow a cyclical pattern. The release of an ovum is timed to coincide with the changes in the uterus that make it receptive to the fertilized ovum. This **menstrual cycle** is usually about 28 days long, although it commonly varies from 20 to 45 days and can differ from month to month.

The menstrual cycle is controlled by the hypothalamus, which excretes releasing factors for the FSH and LH that are released from the anterior pituitary gland. The ovary is composed of many groups of cells called **follicles**, each of which contains a single ovum. Although a woman is born with over two million follicles, only approximately

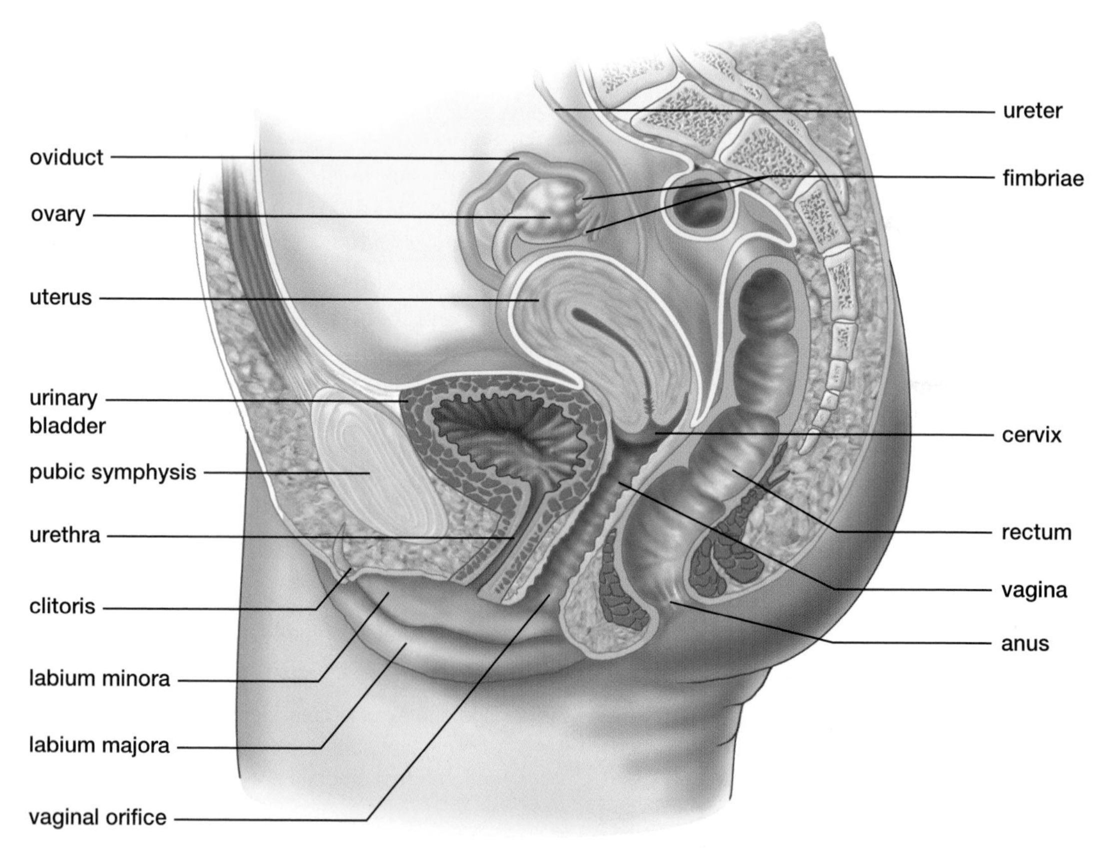

Figure 6.26 The ovaries release one egg a month; fertilization occurs in the oviduct, and development occurs in the uterus. The vagina is the birth canal and the organ of sexual intercourse.

400 will mature to release its ovum during her reproductive life.

As shown in Figure 6.27, during the initial **follicular stage** of the menstrual cycle, increased levels of FSH stimulate the follicles to release increased quantities of estrogen into the bloodstream. Estrogen stimulates the endometrium of the uterus to thicken and increase the blood supply in preparation for a possible pregnancy. Estrogen also inhibits the levels of FSH, forming a negative feedback loop. At the same time, estrogen stimulates the hypothalamus to release large amounts of LH, which trigger the release of an ovum from one of the developing follicles.

Ovulation usually occurs at the midpoint (day 14) of a 28-day menstrual cycle. The release of the ovum triggers a rapid biochemical change in the follicle that released the ovum. The follicle changes to become the **corpus luteum**, a group of cells that produces the hormone **progesterone**.

Ovulation signals the beginning of the **luteal stage** of the menstrual cycle. LH stimulates the corpus luteum to produce progesterone, which inhibits the development of other follicles. Thus,

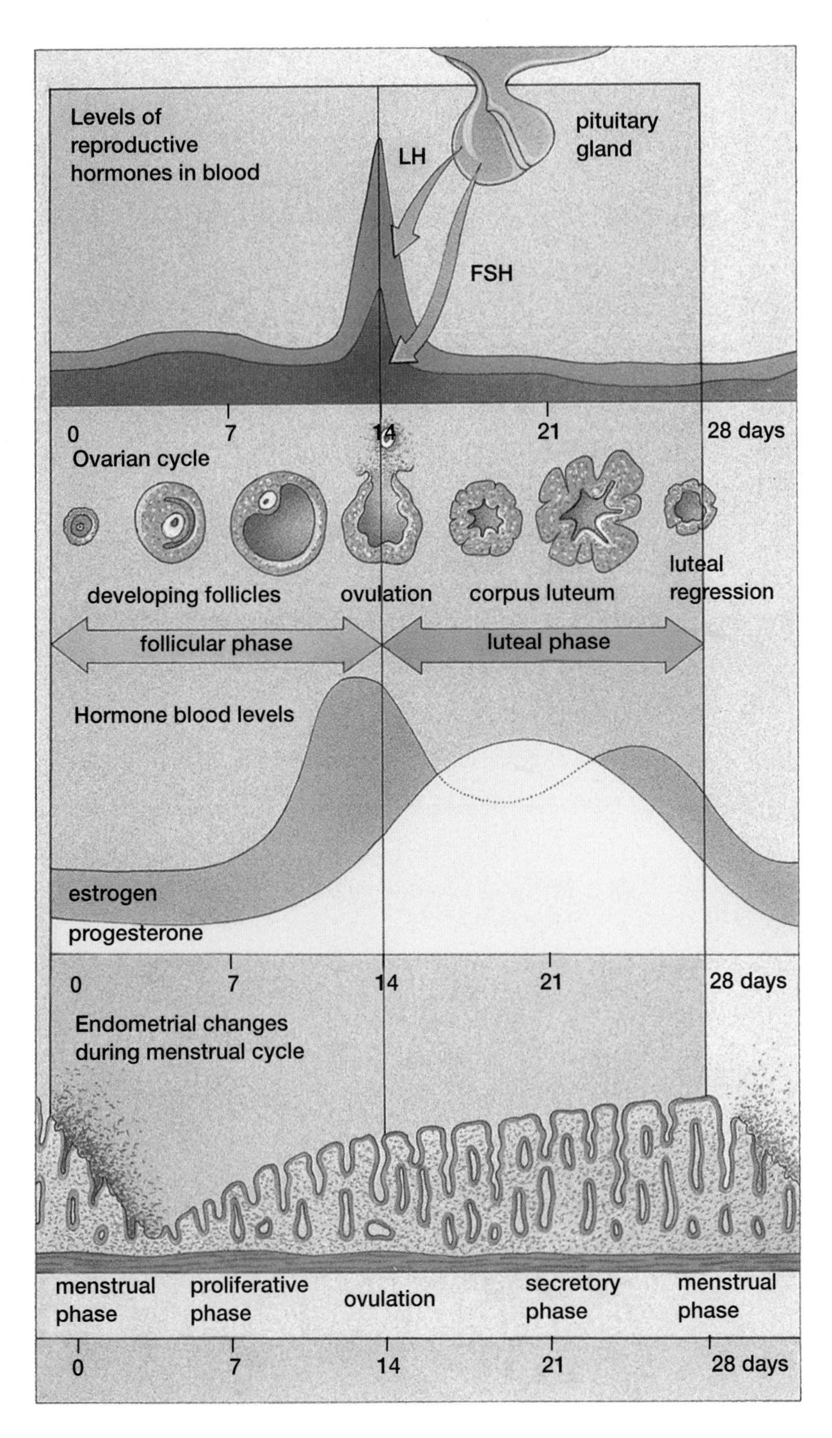

Figure 6.27 The growth and thickening of the uterine lining (the endometrium) is governed by rising levels of progesterone. Menstruation (the sloughing off of blood-rich tissue) is initiated by falling progesterone levels.

only one ovum is released during each cycle. In addition, progesterone inhibits the production of LH. As the level of LH decreases, the corpus luteum begins to degenerate and progesterone levels decrease, ending the luteal stage. This decrease in progesterone decreases the blood supply to the endometrium, which leads to menstruation. During **menstruation**, the endometrium disintegrates, its blood vessels rupture, and the tissues and blood flow out the vagina. The beginning of menstruation signals the first day of the follicular stage of the cycle.

In the Thinking Lab below, you will identify the changes in hormone levels during the menstrual cycle.

Menopause

With age, there is a reduction in the number of functioning follicles, causing a decrease in the amount of estrogen and progesterone in the blood. The decrease in these hormones signals the onset of **menopause**, which is characterized by cessation of menstruation. During and after menopause, cholesterol levels rise and bone mass declines. Also during menopause, blood vessels alternately constrict and dilate, resulting in uncomfortable sensations known as "hot flashes." As well, some women experience mood changes. While some doctors prescribe low levels of estrogen and/or progesterone (**hormone replacement therapy**) to alleviate the symptoms, this practice is the subject of considerable debate. The benefits of hormone replacement therapy include:

- relief of menopausal symptoms such as hot flashes, night sweats, and sleep disturbance;
- prevention of bone loss (osteoporosis);
- protection of cardiac health and a decreased risk of stroke;
- improved memory;
- decreased chance of urinary tract infections; and
- decreased rate of macular degeneration (a leading cause of blindness).

The risks associated with hormone replacement therapy vary with the hormone(s) prescribed. The potential side effects of estrogen replacement include:

- irregular vaginal bleeding;
- stomach upset;
- severe headaches;
- formation of blood clots;
- increased risk of breast cancer; and
- increased risk of uterine cancer.

THINKING LAB

The Menstrual Cycle

Background

Use the data table below to draw a line graph showing the changes in hormone levels for FSH, LH, estrogen, and progesterone during the menstrual cycle. Use different colours to show the changes in each hormone. Be sure to include a title and a labelled scale for each axis.

You Try It

1. Describe the pattern of change for each hormone.
2. Which hormone affects the development of the follicle at the beginning of the cycle?
3. What causes the increase in the level of LH on day 12? What is the effect of this increase?
4. What causes the increase in the level of progesterone on day 16?
5. In what ways does the menstrual cycle illustrate homeostasis?
6. Compare and contrast the menstrual cycle with the changes in the male reproductive hormones.

Day	2	4	6	8	10	12	14	16	18	20	22	24	26	28
LH	17	17	17	17	17	46	35	20	19	18	17	16	14	13
FSH	14	14	14	13	10	8	15	8	7	7	6	6	6	7
estrogen	4	4	5	6	10	13	13	10	9	10	11	11	11	8
progesterone	1	1	1	1	1	1	2	4	7	12	14	14	9	3

The potential side effects of progesterone replacement include:

- stomach upset;
- irregular vaginal bleeding; and
- edema (water retention).

The potential side effects of combined estrogen and progesterone replacement include:

- formation of blood clots;
- headaches;
- fluctuations of blood sugar level;
- edema; and
- premenstrual-like syndrome.

The wide variety of side effects points to the interconnection between hormones and the various homeostatic mechanisms of the body. As you have seen in this unit, the body's control systems play a vitally important role in maintaining the narrow range of conditions within which living cells can survive. All of the organ systems of the body work together to preserve homeostatic equilibrium.

MINI LAB

Sex Hormones in Your Salad

Did you realize that you may have estrogen-mimicking hormones in your salad veggies?

All plants make hormones that control growth and development of plant tissue. However, many plants (about 300 varieties) also make phytoestrogens, substances that mimic the structure and function of estrogen in animals. (The prefix "phyto" is derived from the Greek word for "plant.") Vegetables containing phytoestrogens include: beans (and other legumes); alfalfa and clover sprouts; soybeans, tofu, and other soy products; cereal bran; and flax seeds.

Preliminary studies indicate that these substances may help reduce the incidence of breast cancer. Some research studies show that phytoestrogens, acting as weak estrogen mimics, block human estrogen molecules from binding to estrogen receptors on breast cells. This action may reduce the rate of cell division in breast tissue, which, in turn, may also inhibit the development of tumours in the breast.

While no health risks have been associated with consumption of phytoestrogens so far, more research is needed to better understand the health benefits of these compounds.

Analyze

Find out if your diet includes foods with phytoestrogens.

1. Make a list of food types that contain phytoestrogens.
2. Write out a detailed list of the types of food you would consume over a typical three-day period (include ingredients of mixed foods such as burgers, soups, and salads).
3. Identify the foods that contain phytoestrogens. How many different sources of phytoestrogens do you consume?
4. Find out why it may be more beneficial to obtain phytoestrogens from a number of different food sources.
5. Collect class data on phytoestrogen consumption.
6. Speculate about how plants might benefit by producing estrogen mimics.

THINKING LAB

Hormone Replacement Therapy

Background

For many years, estrogen has been the preferred form of medication for hormone replacement therapy. However, estrogen is a potent medication that has been associated with increased risk of adverse reactions and various forms of cancer and other health problems.

You Try It

Compare the effectiveness of two types of drugs used in hormone replacement therapy, a treatment program designed to alleviate post-menopausal symptoms in women.

In the search for a safer alternative to estrogen, pharmaceutical companies recently developed a new class of prescription medications referred to as "Selective Estrogen Receptor Modulators," or SERMs. Two examples of these medications are "tamoxifen" and "raloxifene."

Find out more about how estrogen and SERMs are used in hormone replacement therapy by researching sources on the Internet (see the McGraw-Hill Biology 12 web site for specific web site references). Then answer the following questions about these two medications.

1. Describe the physiological action of SERMs on cells and tissues.
2. Why are SERMs referred to as "estrogen antagonists"?
3. Make a chart in your notebook similar to the one below and compare the health benefits and risks associated with estrogen and SERMs.

Comparing effectiveness and side effects of estrogen and SERM medications

Effectivness Side effects	Post-Menopausal Medication	
	Estrogen	SERMs
Symptoms of Menopause		
Bone Density		
Uterine Cancer Risk		
Breast Cancer Risk		
Heart Disease Risk		
Other Side Effects		

4. Based upon your analysis, is it possible to claim that one of these medications is safer than the other?
5. If you were a doctor who was developing a hormone replacement program for a post-menopausal patient, would you prescribe estrogen, a SERM, or, perhaps, some combination of both medications? What criteria would you use to decide which drugs to use for an individual patient?

SECTION REVIEW

1. K/U Describe the difference between the primary and secondary sex characteristics in males and females.
2. K/U Define the term "sex hormone."
3. K/U Identify the organs that produce sex hormones in males and females. List the hormones produced by each structure.
4. K/U Use Appendix 4 to compare spermatogenesis to oogenesis. Identify reasons for the difference in cell size between a mature sperm cell and a mature egg cell.
5. K/U Describe the role of the adrenal cortex in prompting the development of secondary sex characteristics in males.
6. K/U Describe the difference between hormones produced prior to the onset of puberty and after puberty in males and females.
7. C Some women suffer from the painful disorder endometriosis, which is a condition in which some of the endometrium is found outside the uterus, trapped during menstruation. One treatment for this condition is hormone therapy. Explain how testosterone can affect the natural cycle of hormones and help this condition.
8. K/U Describe how the hypothalamus and pituitary gland regulate the changes that occur in the female reproductive system during a typical menstrual cycle.
9. K/U Describe how female sex hormones prepare the reproductive system for a possible pregnancy during a normal menstrual cycle.
10. MC While osteoporosis is a bone disease that can afflict both males and females, it is generally associated with post-menopausal women. Research how hormones are sometimes used to help prevent the onset of osteoporosis. Identify the criteria used to develop treatment protocols for patients diagnosed with this problem. If possible, research the occurrence of osteoporosis in your family.

Summary of Expectations

Briefly explain each of the following points.

- The endocrine glands produce hormones that influence the activity of every organ and tissue in the body. (6.1)
- Endocrine hormones are secreted directly into the bloodstream, whereas exocrine hormones are secreted through ducts or tubes. (6.1)
- Steroid hormones are fat-soluble and can pass through the membranes of target cells; non-steroid hormones are not fat-soluble, so they bind to receptors on the surface of target cells. (6.1)
- The nervous system produces bioelectrical signals that travel along specialized nerve cells, while the endocrine system releases hormones into the bloodstream. (6.1, 6.2)
- Abnormal endocrine function can upset a number of metabolic processes in the body. (6.1, 6.2)
- Hormones of the adrenal gland control the body's response to stress in several ways. (6.2)
- Testosterone is one of the sex hormones produced by an adult male. Estrogen and progesterone are hormones produced by an adult female. These hormones have a variety of effects in the male and female reproductive systems. (6.3)
- In the male, sperm release the hormone inhibin, which forms a negative feedback loop with FSH. In the female, estrogen inhibits the levels of FSH, forming a negative feedback loop. (6.3)

Language of Biology

Write a sentence using each of the following words or terms. Use any six terms in a concept map to show your understanding of how they are related.

- endocrine system
- hormones
- endocrine glands
- exocrine glands
- antagonistic hormones
- steroid hormones
- non-steroid hormones
- human growth hormone
- growth factors
- pituitary dwarfism
- gigantism
- acromegaly
- prolactin
- anti-diuretic hormone
- oxytocin
- thyroid gland
- hyperthyroidism
- Grave's disease
- hypothyroidism
- calcitonin
- parathyroid hormone
- pancreas
- circadian rhythms
- pineal gland
- thymus gland
- thymosin
- adrenal gland
- adrenal cortex
- adrenocorticotropic hormone (ACTH)
- adrenal medulla
- puberty
- seminiferous tubules
- interstitial cells
- spermatogenesis
- meiosis
- Sertoli cells
- epididymis
- seminal vesicles
- prostate gland
- Cowper's gland
- semen
- inhibin
- testosterone
- endometrium
- ovaries
- fallopian tubes
- fimbriae
- menstrual cycle
- follicles
- follicular stage
- ovulation
- corpus luteum
- progesterone
- luteal stage
- menstruation
- menopause
- hormone replacement therapy

UNDERSTANDING CONCEPTS

1. Describe the relationship between the endocrine system and each of the following.
   (a) nervous system
   (b) immune system
   (c) reproductive system
2. Using the list below, make a chart with at least three pairs of antagonistic hormones. Describe the nature of their antagonistic functions.
   (a) insulin
   (b) thyroxine
   (c) calcitonin
   (d) glucagon
   (e) testosterone
   (f) aldosterone
   (g) PTH
   (h) cortisol
   (i) estrogen
3. Some disease conditions are due to abnormal endocrine function. Name the glands and hormones associated with each of the following conditions. Describe the symptoms associated with each problem.
   (a) acromegaly
   (b) SAD
   (c) diabetes
   (d) hypothyroidism
   (e) goiter
   (f) gigantism

4. Why is caffeine banned for Olympic athletes? What hormones would you expect would be banned? Explain your answer.
5. Explain how hormones regulate the level of calcium in the blood.
6. Explain why, in most regions of the world, goiter is less common today than in past generations.
7. How do levels of HGH change as we age? How do these changes in hormone level affect our bodies?
8. Give some examples of hormone levels changing in response to:
   (a) the nervous system
   (b) other hormones
   (c) changes in body chemistry
9. Explain and give an example of a negative feedback loop in the human body.
10. Compare the general roles of the nervous system and the endocrine system in maintaining homeostasis in the body.
11. Explain the difference between steroid and non-steroid hormones in terms of their chemical structure and how they alter the chemistry of a cell.
12. Compare how cortisol and insulin affect blood sugar levels.
13. How does the body maintain an internal clock?
14. What are the causes and symptoms of SAD?
15. Explain why the side effects associated with prolonged use of cortisone medications are similar to the symptoms of Cushing's syndrome.
16. Describe how the levels of various hormones in your body might change during a typical day — from the time you wake until you fall asleep.
17. Explain how impaired adrenal function can lead to hypotension (low blood pressure).
18. What hormones are involved in the "fight or flight" syndrome and what hormones adapt the body to long-term stress?
19. Describe anaphylactic shock and the conditions that can trigger it.
20. Describe the role of the pituitary gland in initiating the physiological and anatomical changes associated with puberty.
21. Outline the path of a sperm from where it is formed to where it fertilizes an egg. What glands contribute fluids to semen?
22. Describe the changes in the endometrium through the various stages of the menstrual cycle.
23. Compare the effects of testosterone and estrogen on the body.
24. Compare the levels of female sex hormone secretion in females prior to and after the onset of menopause. Discuss how these changes in hormone production affect the reproductive system.
25. What could be some effects of the increased levels of estrogen that can be found in male smokers?
26. Immediately after the Chernobyl nuclear disaster, people in Sweden began to buy and consume iodine. Explain why.
27. Many athletes "psych" themselves up before a competition by visualizing themselves in competition. Do you think this strategy might enhance their performance? Explain your answer.
28. Menopausal women have elevated levels of FSH and LH when compared to premenopausal women. Explain why these hormones would be elevated and what hormones could be used to treat this problem.

INQUIRY

29. Describe procedures or tests that could be used to evaluate stress levels in an individual. Think about and evaluate how you respond to stressful situations. Discuss with others in your class the types of situations that cause short-term stress and various methods of dealing with stress.
30. Human growth hormone, thyroid hormones, and reproductive hormones that are produced during puberty are all important in human growth at various ages. Together, these hormones stimulate the growth of bone and cartilage, protein synthesis, and the addition of muscle mass. Because the reproductive hormones are involved in human growth, perhaps there is a difference in the growth rate between males and females.

(a) Construct a graph that plots mass on the vertical axis and age on the horizontal axis.
(b) Plot the data in the table for the average female growth in mass from ages 8 to 18. Connect the points with a red line.
(c) On the same graph, plot the data for the average male growth in mass from ages 8 to 18. Connect the points with a blue line.
(d) Construct a separate graph that plots height on the vertical axis and age on the horizontal axis.
(e) Plot the data for the average female growth in height from ages 8 to 18. Connect the points with a red line. Plot the data for the average male growth in height from ages eight to 18. Connect the points with a blue line.
(f) During what ages do females and males increase the most in mass? in height?
(g) How can you explain the differences in growth between males and females?
(h) Interpret the data to find if the average growth rate is the same in males and females.

Averages for Growth in Humans

	Mass (kg)		Height (cm)	
Age	Female	Male	Female	Male
8	25	25	123	124
9	28	28	129	130
10	31	31	135	135
11	35	37	140	140
12	40	38	147	145
13	47	43	155	152
14	50	50	159	161
15	54	57	160	167
16	57	62	163	172
17	58	65	163	174
18	58	68	163	178

31. The following graph represents the average blood concentration of four circulating hormones collected from 50 healthy adult women who were not pregnant.

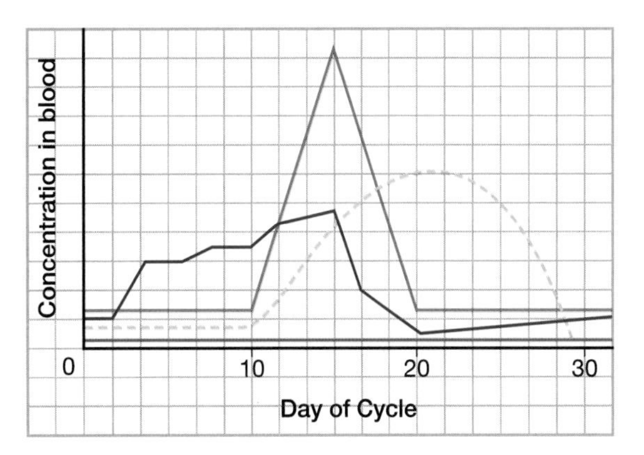

Use the graph to answer the following questions:
(a) Which line represents luteinizing hormone?
(i) red line
(ii) blue line
(iii) yellow line
(iv) green line
(b) Which hormone increases during the last half of the menstrual cycle?
(i) estrogen
(ii) progesterone
(iii) LH
(iv) FSH
(c) Which hormone is responsible for stimulating the egg development each month?
(i) estrogen
(ii) progesterone
(iii) LH
(iv) FSH

32. Many women do not begin to ovulate until after they finish nursing their newborn babies. What hormones would be involved in this phenomenon? Formulate a hypothesis related to how these hormones would interact.

33. Caffeine is not a controlled substance, and yet it can affect the body in many ways.
(a) Review the various ways caffeine can affect the body.
(b) Identify an endocrine system that may be affected by caffeine. Prepare a prediction.
(c) Propose a research program to investigate your prediction.

34. In addition to the reproductive hormones you have studied in this unit, there are a number of hormones and neurotransmitters that contribute to both physical and emotional changes. Chocolate contains some of these chemicals. Is chocolate the food of love? What information do you require before you can consider this a problem that can be tested through scientific inquiry?

35. Scientists in Western countries have been searching for a chemical that will curb an alcoholic's need for alcohol. Recently, some scientists have taken a look at a treatment used in China for over two thousand years. Chinese healers have given alcoholics an extract from the root of the kudzu vine, which they claim is about 80 percent effective in patients who have been treated with it for two to four weeks. Dr. Wing-Ming Keung of Harvard Medical School in Boston visited China to find out what modern researchers thought of the herbal remedy. He spoke to physicians who claimed to have treated 300 alcohol abusers with the extract. They were convinced that the chemicals in the extract effectively suppressed the appetite for alcohol.
After returning to Harvard, Dr. Keung and Dr. Bert L. Vallee decided to try the drug on Syrian golden hamsters in their laboratory. These hamsters were specifically selected because they are known to drink large amounts of alcohol when it is available to them. Imagine you are a member of the research team. Your job is to design an experiment that will prove the effectiveness of the extract from the root of the kudzu vine in suppressing the hamsters' craving for alcohol.
(a) Dr. Keung and Dr. Vallee discovered two active ingredients in the root extract. They found that each extract alone had the effect of lessening alcohol use in the hamsters by 50 percent. The two compounds appeared to affect enzymes involved in alcohol metabolism. Hypothesize how this discovery might help people overcome alcohol dependency.
(b) How would you follow up on your experiment?
(c) Predict the results of the experiment.
(d) What will be the variable in the experiment?
(e) What will you use as your control in the experiment?

COMMUNICATING

36. Acupuncture, an ancient and important part of Eastern medical practice, has only recently come under serious scrutiny by Western doctors and scientists. Investigate how this method of treatment is used in Chinese medicine. Research the current understanding of the possible physiological processes that might explain the apparent pain-relieving effects of acupuncture treatment.

37. Write a brief essay on the major biological factors that can lead to infertility in males and females.

38. Create a pamphlet that explains how cortisone medications can be used to treat some forms of skin inflammations and injuries.

39. Most people know about the impact of smoking on the respiratory system. However, smoking also negatively affects the male and female reproductive systems. Investigate how smoking can interfere with male and female reproductive function at various stages of life. Conduct a survey to determine the number of male and female smokers in your class or grade and the level of awareness about the various health risks associated with smoking.

40. Current research studies suggest a link between the levels of various hormones in the body, such as ACTH from the pituitary gland, and certain forms of depression. By consulting print resources, the Internet, or a medical expert, summarize the latest research findings on how hormone production may be connected to depression.

41. In a chart, list the endocrine glands found in the human body, describe their location, identify the hormones produced by each gland, and describe the hormones' function.

42. Draw a negative feedback loop to illustrate the control of water levels in the human body.

43. Make a sketch that illustrates the structural and functional relationships between the hypothalamus and the two lobes of the pituitary gland.

44. Draw a rough line graph of the changes in a person's blood sugar level over a 24-h day. Show the effects of meals, exercise, sleep, and so on. Label the time of each event on the horizontal axis.

MAKING CONNECTIONS

45. Research the three most common sexually transmitted diseases and their treatment.

46. Some students use anabolic steroids to enhance athletic performance and improve their appearance. What risks are associated with the use of these hormones?

47. Hormones are sometimes used to increase the productivity of cattle, chickens, and/or pigs. Do you think that the labels on food products should include information on the use of supplements such as hormones? What would be the advantages and disadvantages of this type of labelling?

48. Many types of animals (such as mice, rats, and monkeys) are used in some scientific and medical research studies that investigate diseases of the endocrine system, such as abnormal thyroid or pituitary function. How do such studies benefit human life? In your opinion, should animals be used for such research studies? Are there effective alternative research protocols that could replace research procedures involving laboratory animals?

49. Canada's pharmaceutical companies submit a steady stream of new prescription medications to Health Canada for approval each year. However, only 10 percent of these applications receive government approval. The other 90 percent are rejected, mostly because of safety concerns.

Suppose you were responsible for developing the regulations regarding control of prescription drugs in Canada. Some of the major issues concerning prescription drug use and safety involve providing timely information about new medications to Canadian doctors and the reporting of adverse reactions to new medications. What changes would you make to current legislation to address these issues?

50. Funding medical research often involves funding research in basic science, such as zoology. In this situation, the researcher usually has no medical training, but is a specialist in his or her own field of study. For one hormonal or neurological disorder, identify how research in basic science has helped develop understanding of the medical condition.

51. When conducting drug trials, a test group of people is divided into two groups. One group receives the new drug, and the other group receives a placebo, a treatment that has no active ingredients. While testing a new asthma medication, a test subject claims that she is using the new drug. She is pleased with the product and agrees to be interviewed on television to promote it. Has this person compromised the study? Justify your response.

52. People have easier access to diet supplements such as vitamins, minerals and various plant extracts such as *Echinacea* than to prescription drugs. *Echinacea* is promoted as an immune-system boost. Some proponents claim that it should be used daily, while others suggest that people add it to their diets only when they are most at risk, such as during flu season. Since *Echinacea* is an uncontrolled substance, might it be abused, thus creating problems for our health care system and society in general? Explain your response.

53. Competitive events such as the Olympics test athletes for banned substances such as anabolic steroids. Some professional athletes, like Major League Baseball players, are not restricted by these rules. How does the use of these substances change the nature of the sporting event? Should these substances be banned? Explain your responses.

Is There a Cure for Diabetes?

Background

In Canada, it is estimated that five to six percent of the population over the age of 12 has diabetes. About one third of these cases are undiagnosed. Approximately 779 000 Canadians, however, have actually been diagnosed with this disease.

Diabetes is the seventh leading cause of mortality in Canada. Typically, damage to small blood vessels in people who have diabetes can also lead to serious, and potentially life-threatening, complications involving the heart, eyes, and/or kidneys. Currently, people who are diagnosed with Type 1 diabetes may have their life expectancy shortened by 15 to 27 years. Type 2 diabetes reduces life expectancy by five to 10 years, depending on when the disease is diagnosed.

Despite these rather gloomy statistics, there is great optimism that science will one day triumph over diabetes. People who have diabetes are encouraged by current advances in medical research that are providing a more complete understanding of the physiological, genetic, and lifestyle factors that contribute to the onset of this disease. While the goal of developing a proven cure for diabetes has yet to be achieved, newly developed medications and technology are helping many patients with diabetes to manage their blood sugar levels more effectively. Managing blood sugar levels will, in turn, reduce the risk of developing some of the more serious complications typically associated with this disease.

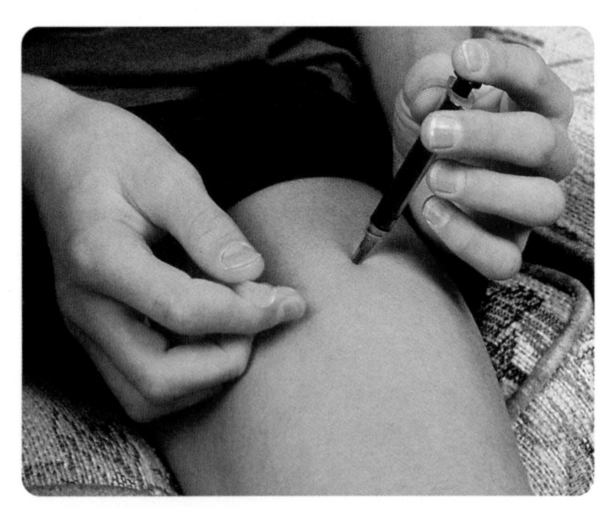

New medications are being developed that may replace insulin injections for people who have diabetes.

ASSESSMENT

After you complete this project,

- **Assess your report or presentation based on how clearly your information was conveyed;**
- **Assess your research skills during the development of your project. How have your skills improved?**
- **Use the rubric you developed to assess your project.**

Before you begin this project, reread section 4.3 in Chapter 4, which introduced the topic of diabetes.

Challenge

Many people believe that Canadian scientists Dr. Fredrick Banting and Dr. Charles Best "cured" diabetes when they isolated insulin. However, as you have seen, the number of people who have diabetes is increasing rather than decreasing. Nevertheless, many researchers are convinced that modern science and technology will one day find a cure for diabetes.

Working individually or in a small group, research, design, and present a report on the future of diabetes treatment. Your report should include a section on the problems the disease presents to individuals who have diabetes and to those involved in the treatment of it. You may wish to include interviews with patients, doctors, and/or researchers. Your teacher and team members can help you determine the format of your report, which may include any of the following:

- an oral, video, or computer presentation
- a written report
- a web site
- any combination of the above approved by your teacher

Materials

Print sources and/or the Internet; a tape recorder; appropriate materials to present your project (such as video equipment, computers, audio/visual equipment, computer presentation software, and so on).

Design Criteria

A. Choose a format for a report on the treatment of diabetes.

B. The introduction to your report should include a discussion of the differences between Type 1 and Type 2 diabetes and how each of these disorders affects the homeostatic systems of the body.

C. Gather and present information on current treatments for each type of diabetes and their effectiveness.

D. Gather and present information on current research into the cure and treatment of diabetes.

Action Plan

1. Develop a plan to find, collect, and organize the information you will need.
2. Arrange for interviews with people who have diabetes or who are doing research on diabetes.
3. Identify the materials you will need to prepare your report and presentation.
4. After you have collected enough information, prepare a series of questions to guide your interviews.
5. Interview your subjects and record their responses. Be sure to include their opinions on and perceptions of current and proposed methods of treatment. Note that people who are directly involved with diabetes treatment (either administering or receiving it) may have a more realistic view of its possible effectiveness.
6. Select pertinent sections of the interviews to include in your report. Depending on the format of your presentation, you may include taped segments (either video or audio) or transcribed sections of the interviews in your report.
7. As you gather information, think about how you will incorporate some of the following points.
   - Type 1 and Type 2 diabetes have very different causes. How does this difference affect attempts to cure or treat these diseases?
   - Cures and treatments do not always involve complex technology or cutting-edge science; sometimes they involve changes in lifestyle. How might people be persuaded to adopt healthier and more constructive lifestyle choices?
   - The many interactions between blood sugar levels and the nervous, endocrine, and circulatory systems create a wide range of symptoms. How will treatments affect these interactions?
   - Many proposed forms of treatment are known primarily through press releases that may exaggerate their success. How can you get accurate, balanced information on the reliability of new forms of treatment?
   - Some new forms of treatment raise ethical questions. What criteria could be used to determine if some forms of scientific research should be curtailed or regulated for ethical reasons?
8. With your group, design a rubric for assessing your project. Ask for your teacher's help in deciding on the assessment categories you will use.
9. Obtain your teacher's approval, then carry out your plan and make modifications as necessary.

Evaluate

1. Before you hand in your report, ensure that you have covered the points in the Design Criteria and Action Plan sections.
2. If you are doing a class presentation, practise it in front of an audience.
   (a) How long does your presentation take?
   (b) Do all the technological aids work the way you intend?
   (c) Does the audience understand all of your presentation?
3. Using the rubric you prepared, evaluate your project. How effective do you think it was?
4. How did working on this project help you think about what you learned in this unit?

UNIT 2 REVIEW

UNDERSTANDING CONCEPTS

Multiple Choice

In your notebook, write the letter of the best answer(s) for each of the following questions. (You may select more than one item per question.)

1. Which of the following substances are not normally found in urine?
   (a) sodium
   (b) water
   (c) glucose
   (d) urea
   (e) uric acid
2. Which of the following is/are normally not part of the fight-or-flight reflex?
   (a) increase in blood glucose level
   (b) increase in oxygen level in blood
   (c) increase in thyroid hormone level in blood
   (d) increase in testosterone level in blood
   (e) increase in pH of blood
3. Which of the following structures is/are not part of the peripheral nervous system?
   (a) sensory receptors in skin
   (b) spinal cord
   (c) sensory neuron
   (d) motor neuron
   (e) hypothalamus
4. Which of the following structures, if any, could be considered part of both the peripheral and central nervous systems?
   (a) sensory receptors in skin
   (b) spinal cord
   (c) sensory neuron
   (d) motor neuron
   (e) hypothalamus
5. Which of the following organ(s) act(s) as both an exocrine and endocrine gland?
   (a) pituitary
   (b) thyroid gland
   (c) pancreas
   (d) parathyroid gland
   (e) hypothalamus
6. Which of the following organs secrete(s) pairs of antagonistic hormones?
   (a) pituitary
   (b) thyroid gland
   (c) pancreas
   (d) parathyroid gland
   (e) hypothalamus
7. Which of the following substances is/are used by the body to make hormones?
   (a) proteins
   (b) glucose
   (c) cholesterol
   (d) polypeptides
   (e) vitamins
8. Which of the following illnesses are not initially caused by an endocrine disorder?
   (a) osteoporosis
   (b) cancer
   (c) high blood pressure
   (d) arthritis
   (e) Type 1 diabetes

Short Answer

In your notebook, write a sentence or a short paragraph to answer each of the following questions.

9. Which system plays the dominant role in managing the body's response to stress: the central nervous system or the endocrine system? Explain your answer.
10. Which is more common in the human body: steroid or non-steroid hormones? In general, what are the major differences in the way each type of hormone interacts with target cells?
11. How does the body restore and maintain a normal homeostatic state after exposure to extreme environmental temperatures?
12. Which organs in the body secrete tropic hormones? How does the action of tropic hormones differ from that of other types of hormones?
13. Explain why the secretion of oxytocin during childbirth is normally regulated by a positive feedback loop.
14. Can women become pregnant after menopause? Explain your answer.
15. What prevents a second pregnancy in a woman who is already pregnant?
16. What part of the eye is referred to in each of the following descriptions (the diagram on the facing page may help you identify the parts)?
    (a) regulates the amount of light entering the pupil of the eye
    (b) the pigmented inner layer made up of photoreceptor cells
    (c) a protective layer that extends over the lens
    (d) the photoreceptor cells that function in colour vision
    (e) the part of the retina that is the centre of the visual field and is made up of cones
    (f) adjusts the shape of the lens
    (g) the photoreceptor cells most active in low light conditions

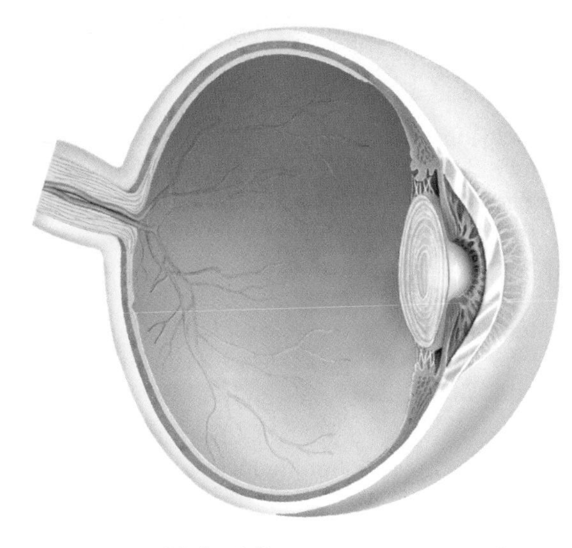

17. How would the following processes in the kidney respond to serious dehydration in the body due to excessive sweating on a hot day?
    (a) glomerular filtration
    (b) tubular re-absorption
    (c) tubular secretion
18. Describe the primary functions of each of the following sections of a human kidney.
    (a) cortex
    (b) medulla
    (c) collecting duct
19. Describe the type of stimulus that would trigger the following homeostatic reactions.
    (a) constriction of blood vessels in the skin
    (b) increase in shivering activity
    (c) increase in breathing rate
    (d) increase in sweating
20. Identify the endocrine glands, hormones, and symptoms associated with the following diseases or disorders. (Remember that more than one type of gland or hormone may be involved with a single type of ailment.)
    (a) gigantism
    (b) seasonal affective disorder
    (c) Cushing's syndrome
    (d) congenital hypothyroidism
    (e) diabetes
    (f) acromegaly
    (g) goiter
    (h) Addison's disease
    (i) Graves' disease
21. Construct a chart of the following components of the human immune system. Match each component with one or more of the following types of immune function: non-specific immunity, specific immunity, cellular immunity, antibody immunity.
    (a) the epidermis of the skin
    (b) memory B cells
    (c) neutrophils
    (d) lining of the respiratory tract
    (e) monocytes
    (f) T lymphocytes
    (g) lining of the digestive tract
    (h) memory T cells
    (i) macrophages

INQUIRY

22. Investigate the types of performance-enhancing substances that are banned in major athletic competitions, such as the Olympics. Distinguish between hormone and non-hormone products. Describe the physiological effects of each substance on the body. How do they disrupt normal homeostatic function? Which effects are temporary or transitory and which can result in permanent alteration of a body structure or function?
23. As you have learned, people with Type 1 diabetes must self-administer insulin injections daily. There are various types of insulin (such as short-acting, intermediate-acting, and long-acting) that patients may need to use to manage their blood sugar. Some examples of these medications are listed below. Investigate the properties of each type of insulin, and describe how each is used in the treatment and management of Type 1 diabetes.
    (a) Ultralente insulin
    (b) Lente insulin
    (c) NPH insulin
    (d) regular insulin
24. Anaphylaxis is a general term that refers to life-threatening allergic reactions to various types of environmental triggers, such as peanuts or other foods, insect stings, and latex. Investigate four examples of specific stimuli that trigger anaphylactic shock and the recommended treatments for each type of reaction.
25. Most types of living cells, from bacteria to human cells, secrete a substance called "heat shock protein" when they are under stress. The chemical structures of all heat shock proteins are quite similar, regardless of their origin. In humans, T cells may respond to the presence of this protein by destroying the cells that

produced it, whether these cells are infectious agents or actually part of the body. Medical researchers believe this action may be part of what causes auto-immune diseases such as diabetes and arthritis. Investigate the latest findings of current research into the potential use of heat shock proteins to treat or even prevent some types of auto-immune diseases in humans. Describe how vaccines containing these proteins may some day be able to inhibit destructive auto-immune activity associated with Type 1 diabetes and other diseases.

26. Investigate the type of lifestyle factors that can contribute to the onset of Type 2 diabetes in individuals who previously had no visible symptoms of this disease. Investigate how oral hypoglycemic drugs can manage some symptoms of Type 2 diabetes.

27. What is the difference between analgesic and anesthetic medications? How are they used to manage pain? Investigate the differences in the way analgesic and anesthetic medications affect the function of the central and peripheral nervous systems.

28. Narcolepsy, one of many sleep disorders, is a condition in which individuals experience an overwhelming need to sleep at any time during a 24-hour day. People with narcolepsy may also experience abnormal "dreaming sleep" patterns. Investigate the specific causes, symptoms, and management programs associated with this illness.

29. Cyclosporine is the medication most often used in kidney transplant procedures to help prevent rejection of the new organ by the host's immune system. Investigate the physiological action of cyclosporine. How does this medication interact with the immune system to protect a transplanted organ such as a kidney? Also, find out why (in transplant cases) organs from living donors are preferred to cadaver organs.

30. Investigate how artificial kidneys filter blood in patients suffering from reduced kidney function or kidney failure. Specifically, compare how the process is carried out by dialysis machines (artificial kidneys) and real human kidneys.

31. Explain how a vasectomy can make a man sterile and unable to impregnate a woman. How does this procedure affect the function of other parts of the male reproductive system? Is it reversible? What is the equivalent operation for a woman?

32. Colony stimulating factors (CSFs) are proteins that stimulate the immune system to produce more white blood cells. They are sometimes given to cancer patients to help their immune systems recover from the destructive effects of chemotherapy. Investigate how CSFs help boost immune system activity in cancer patients.

COMMUNICATING

33. Draw a flowchart that illustrates the sequence of changes in the internal concentrations of sodium and potassium in neurons during depolarization and repolarization of a neuron.

34. Starting at the Bowman's capsule, draw a flowchart that illustrates the sequence of changes in the composition of the nephric filtrate as it travels through a nephron.

35. Make an outline drawing of the human body and draw in the organs and tissues that are part of the immune system. Indicate which parts of the immune system create the symptoms associated with an allergic reaction to an environmental allergen, such as plant pollen.

36. Draw a flowchart that illustrates the sequence of events that make up an allergic response to an environmental allergen, such as pollen from flowering plants.

37. Construct a series of feedback loops that illustrate: (a) normal thyroid gland activity; (b) activity of the thyroid gland in individuals with hypothyroidism; and (c) activity of the thyroid gland in individuals with hyperthyroidism. Include the following components in each feedback loop: effector, receptor, integrator. Describe any differences and similarities among these feedback loops.

38. Draw a feedback loop that illustrates how the buffer carbonic acid regulates the pH level of blood (normally about 7.4) as it circulates through the kidney.

39. Some people categorize other individuals as left-brained or right-brained on the basis of their predominant interests and abilities. What parts of the brain are involved in this characterization of mental function? What is the perceived distinction between right- and left-brained abilities and aptitudes? Is there any scientific evidence to support such a classification scheme of cognitive function? If so, would you classify yourself as a left- or right-brained individual? Explain your answer. If possible, conduct a survey to determine the number of people who consider themselves to be left- or right-brained. Is such a survey scientifically valid? Might it be used to predict someone's future academic performance?

40. Draw a reflex arc that involves reflexively removing your foot after stepping on a sharp object, such as a tack. Label each part of the diagram and use arrows to show the direction of nerve impulses along each nerve in the reflex arc.

MAKING CONNECTIONS

41. Sleep is a normal part of human life. Investigate the neurological basis of sleep — what happens in your nervous system when you fall asleep and wake up? Find out why a good night's sleep is essential for good health. For one week, keep track of the amount of sleep you get each night. If possible, compare your data with those of other people in your class (including, perhaps, your teacher). What percentage of your class gets eight hours of sleep on an average night? Some studies show that most North Americans are sleep-deprived. What do you think are some of the social implications of this?

42. Research the history of the medical definition and treatment of schizophrenia. Is there a cure for schizophrenia? What are the current medical theories regarding the causes of this form of mental illness? How common is schizophrenia in Canada (as a percentage of the total population)? Explain why there may be a difference between the incidence of schizophrenia in male and female populations in Canada.

43. Compare the availability of effective treatment programs for infectious diseases, such as tuberculosis, in Canada and in developing countries. How does the average incidence and death rate for TB in Canada compare to those of developing nations? To what extent does the quality of medical care TB patients receive depend on where they live? Should Canadians be concerned about the incidence of various infectious diseases, such as TB, in developing countries? Why or why not?

44. Some common over-the-counter medications, such as various types of analgesics, come with warnings that their use by people with diabetes should be restricted and monitored by a health professional. Why should people with diabetes, or others at risk for kidney disease, be especially cautious about regular or repeated consumption of such products? Should these products be available by prescription only?

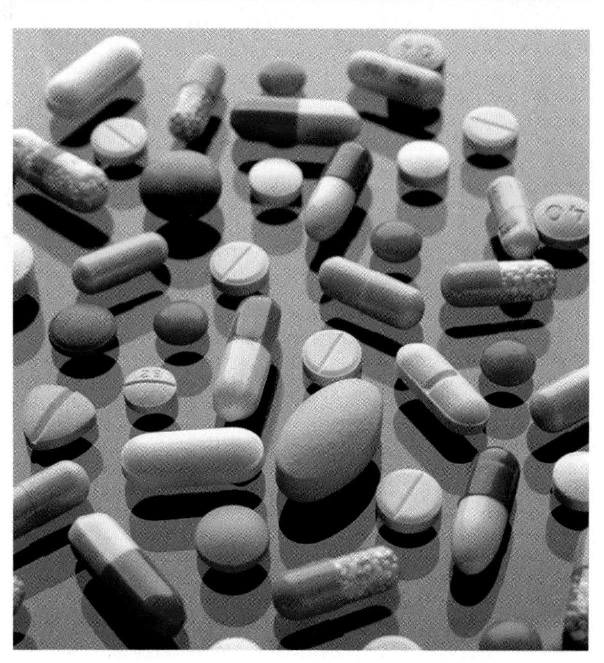

45. Many people obtain flu shots each year. What are flu vaccines composed of and what type of immunity do they provide to the recipient? Why can't a single flu vaccine injection provide lifetime immunity? What are the health risks associated with flu vaccinations? Currently, flu vaccines are free for everyone in some provinces, but in other provinces some people have to pay. Find out what the eligibility rules are in a province that does not provide free flu shots to everyone. Should the flu vaccine be made available free to all Canadians?

COURSE CHALLENGE

Consider the following as you continue to build your biology research portfolio.

- Review the information you have gathered in preparation for the Biology Course Challenge. Consider any new findings to see if you want to change the focus of your project.
- Add important concepts, interesting facts, and diagrams from this unit.
- Scan magazines, newspapers, and information on the Internet to enhance your project.

UNIT

3

Unit Preview

In this Unit, you will discover

- how living cells store, transmit, and express genetic information,
- how the outside environment affects the nature and expression of genetic material within cells, and
- how new genetic technologies are transforming science and society.

Unit Contents

UNIT PROJECT PREP

Read pages 322–323 before beginning this unit.

- Throughout the unit, think about the roles played by DNA and its associated proteins in the development and treatment of cancer.
- Start planning for your project now by setting up a system to organize the information you gather.

Molecular Genetics

The chicks of this white-crowned sparrow will learn the song of their species only if they hear it at least a few times within the first month of their lives. They will not learn the song of any other bird, however, no matter how often they hear it.

What genetic mechanisms are at work here? Clearly there is a song-learning skill programmed into the cells of the chicks. There is also an element of timing, since this skill is only activated during a specific stage of the chicks' lives. Finally, there is an environmental component, since the chicks' innate ability to learn the song can be fulfilled only if they hear the correct song.

All these components can be traced to the expression of stored hereditary information through proteins created by specialized cellular structures like the ones shown below. The genetic information inherited by each chick defines which proteins will be created in each of its cells throughout its life. In turn, these proteins set the stage for the chick's development. But how can the information leading to a behaviour like singing be stored in and transmitted by molecules? How do these processes interact with each other, and with the outside world?

The task of molecular genetics is to examine how genetic information stored on molecules is expressed in the world — whether that expression takes the form of an illness, a new strain of rice, or the song of a bird. In this unit, you will also consider some of the practical, legal, and ethical issues that arise from our increasing understanding of the processes of life.

Specialized structures use the information encoded in hereditary material to construct proteins.

CHAPTER 7

Nucleic Acids: The Molecular Basis of Life

Reflecting Questions

- How did the discovery of the role of DNA change our understanding of cellular processes and heredity?
- How does the structure of DNA contribute to its function as the material of heredity?
- How do the structure and function of genetic material in bacteria compare with those in plants and animals?

Prerequisite Concepts and Skills

Before you begin this chapter, review the following concepts and skills:

- distinguishing between the cell structure and metabolic processes in prokaryotes and eukaryotes (Chapter 3, section 3.2),
- understanding cell life cycle and cell division (Appendix 4).

The bear on the opposite page is neither a polar bear nor an albino; it is a Kermode bear (*Ursus americanus kermodei*). About 10 percent of Kermodes, like the individual shown here, sport the white or cream fur colour that has earned these animals their popular "Spirit bear" name. The Kermode is a subspecies of the black bear, found only in a small portion of coastal British Columbia and Alaska. In contrast, the salmon on which this bear is about to feed travelled thousands of kilometres through the open ocean before returning to spawn in the same stream in which it hatched four years earlier.

The Kermode's colour is due to genetic traits. Similarly, the salmon's ability to navigate out to sea and home again is also due in large part to genetic traits. The molecular structures and processes that govern the development of each animal are the same — as they are in all living organisms. But how can the same structures and processes be responsible for features as different as fur colour and a homing instinct? How can these traits be transmitted so accurately from one generation to the next?

The photomicrograph shows a molecule of DNA in the process of replicating. In 1943, the atomic physicist Erwin Schrödinger predicted that the secret to life would be found in a special crystal. A decade later, Watson and Crick discovered the structure of DNA, a crystalline molecule capable of infinite internal variation, yet reproducible within cells with astonishing accuracy. This combination of rich variety and stability allows DNA to guide an infinite number of hereditary traits while ensuring that each trait is passed on generation after generation.

Over the twentieth century, research on inheritance took scientists from the observation of physical characteristics into the inner workings of the cell nucleus, and from the first identification of nucleic acids to an understanding of the precise structure and arrangement of DNA within living cells. In this chapter you will follow the same path of exploration as you study the molecular structures and processes that work together to encode and transmit hereditary information.

The unique molecular properties of DNA allow it to both store and transmit hereditary information.

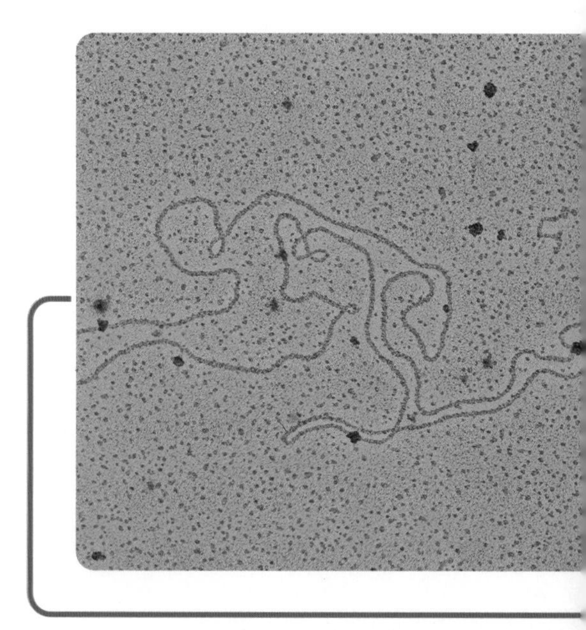

Chapter Contents

Isolating the Material of Heredity

EXPECTATIONS

- **Explain the roles of evidence, theories, and paradigms in the development of scientific knowledge about genetics.**
- **Demonstrate an understanding of the process of discovery that led to the identification of DNA as the material of heredity.**
- **Interpret the findings of key experiments that contributed to this process.**

In 1865, the Austrian monk Gregor Mendel presented the results of his research on patterns of inheritance in garden peas to the Natural Science Society in Brunn, Austria. He proposed a number of hypotheses that challenged much of the thinking of his day about heredity. He argued, for example, that the maternal and paternal gametes contributed equally to the development of the offspring. He also held that the information contributed by each parent was not blended but rather passed on to the offspring as discrete bits of information or "factors of inheritance." He went on to state that, while two factors will exist for any one visible trait, one of them (known as the recessive factor) might not be expressed.

Mendel's findings on the transmission of hereditary information were not widely recognized at the time. This was partly due to the strong divisions that existed among scientific disciplines then, which meant that the work of a botanist was not likely to be noticed by zoologists or by medical doctors. The apparently fixed nature of Mendelian factors of inheritance also seemed to be at odds with the newly emerging theory of evolution. Over the next few decades, however, scientists began to recognize the many similarities among cellular processes in bacterial, plant, animal, and human cells (including the processes you studied in Unit 1). They also found that Mendel's principles were consistent with the idea that species change and evolve over time. Today, Mendel's work is recognized as the foundation of modern genetics.

Only four years after Mendel's presentation in Brunn, and less than 300 km away, the young Swiss physician and scientist Friedrich Miescher isolated a substance he called "nuclein" from the nuclei of white blood cells. Miescher, shown in Figure 7.1, determined that nuclein was made up of an acidic portion (which he termed "nucleic acid") and an alkaline portion (which was later shown to be protein). Shortly thereafter Miescher turned to the study of chemical properties of other cellular structures. Almost a century passed before scientists established the connection between the nucleic acid isolated by Miescher and Mendel's factors of inheritance.

Figure 7.1 Friedrich Miescher was 25 years old when he isolated nucleic acids from the nuclei of white blood cells in 1869. He was working in a hospital treating wounded soldiers, and he was able to collect white blood cells from their bandages.

The Components of Nucleic Acids

Following the work of Miescher, Phoebus Levene studied nucleic acid in more detail. During a career that stretched from the early 1900s to the 1930s, Levene isolated two types of nucleic acids that could be distinguished by the different sugars involved in their composition. One acid contained the five-carbon sugar ribose, so Levene called it "**ribose nucleic acid**" (**ribonucleic acid** or **RNA**). The other acid contained a previously unknown five-carbon sugar molecule. Since this sugar was similar in structure to ribose but lacked one oxygen molecule, Levene called it deoxyribose. He went on to call the nucleic acid containing this sugar "**deoxyribose nucleic acid**" (**deoxyribonucleic acid** or **DNA**). Figure 7.2 shows the structures of ribose and deoxyribose sugars. Levene is pictured in Figure 7.3.

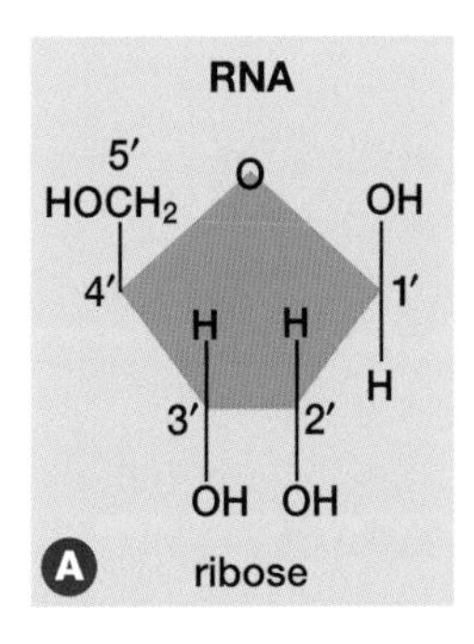

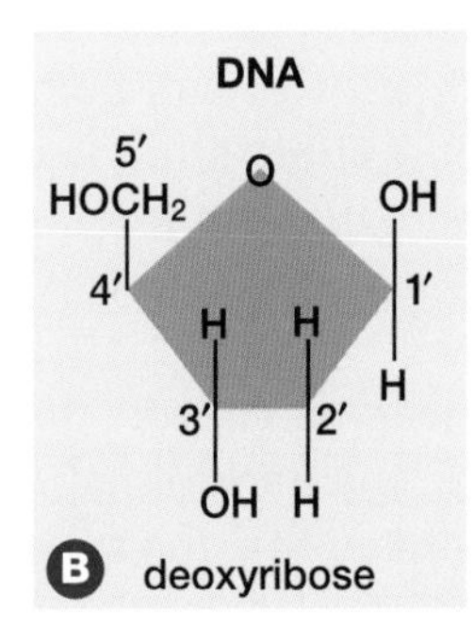

Figure 7.2 The structure of (A) ribose, found in RNA, and (B) deoxyribose, found in DNA. In ribose, the 2′ carbon is bonded to a hydroxyl group. In deoxyribose, this carbon is bonded to a single hydrogen molecule.

During the period when Levene was conducting his studies on nucleic acids, other experimenters demonstrated that Mendel's factors of inheritance were associated with the nuclein substance first isolated by Miescher. By that time, nuclein had been shown to be made up of individual structures known as **chromosomes**, strand-like complexes of nucleic acids and protein tightly bound together. Thus, the finding that the factors of inheritance were associated with nuclein drew increased attention to both the protein component and the properties of nucleic acids.

Figure 7.3 Phoebus Levene made some important discoveries about the properties of nucleic acids.

After distinguishing between DNA and RNA, Levene went on to show that nucleic acids are made up of long chains of individual units he termed **nucleotides**. Both DNA and RNA contain a combination of four different nucleotides. As shown in Figure 7.4, each nucleotide is composed of a five-carbon sugar, a phosphate group, and one of four nitrogen-containing (**nitrogenous**) bases. The bases found in DNA nucleotides are **adenine** (A), **guanine** (G), **cytosine** (C), and **thymine** (T). In RNA, the base **uracil** (U) is found instead of thymine. The only difference between the nucleotides in each nucleic acid is in their bases. As a result, scientists studying nucleic acids soon began to identify the nucleotides simply by their bases or, more commonly, by their initials: A, G, C, T, and U.

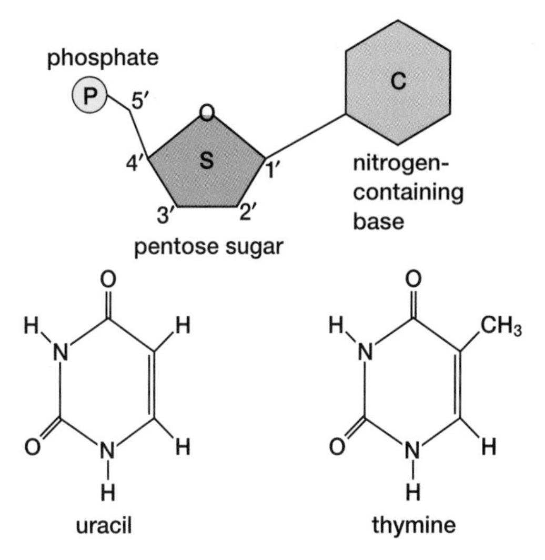

Figure 7.4 The general structure of a nucleotide. In DNA, the sugar is deoxyribose, and the nitrogenous base is one of the following: adenine (A), guanine (G), cytosine (C), or thymine (T). In RNA, the sugar is ribose and the nitrogenous base uracil (U) appears instead of thymine.

While A, G, C, T, and U are the major bases found in nucleic acids, there are also some minor ones. These are usually slightly altered forms of the major bases. In many cases the minor bases serve as specific signals involved in programming or protecting genetic information.

At this point, the results of Levene's work led him to conclude incorrectly that nucleic acids contained equal amounts of each of these nucleotides. Based on this finding, he suggested that DNA and RNA were made up of long chains in which the nucleotides appeared over and over again in the same order; for example, ACTGACTG ACTG and so on. This, in turn, caused most scientists to conclude that DNA could not be the material of heredity because it was not complex enough to account for the tremendous variation in inherited traits. It was generally accepted that DNA could be a structural component of hereditary material, but scientists thought the primary instructions for inherited traits must lie in the proteins that are also found in chromosomes.

Several decades passed before Levene's conclusion was finally corrected.

Mounting Evidence for the Role of DNA in Heredity

One important piece of evidence that DNA was, in fact, the material of heredity came in 1944, when the team of Oswald Avery, Colin MacLeod, and Maclyn McCarty published the results of their experiments with bacteria. These experiments, which built on the 1928 work of British researcher Fred Griffith, were conducted over a period of nearly 15 years. As illustrated in Figure 7.5, Griffith showed that when a heat-killed, pathogenic (disease-causing) strain of the bacterium *Streptococcus pneumoniae* was added to a suspension containing a non-pathogenic strain, the non-pathogenic strain was somehow transformed to become pathogenic.

Avery and his colleagues undertook several important steps to isolate the agent behind this transformation, which Griffith had called the transforming principle. When they treated a suspension of the pathogenic bacteria with a protein-destroying enzyme, they noticed that the transformation of non-pathogenic bacteria into a pathogenic strain still took place. When the pathogenic bacteria were treated with a DNA-destroying enzyme, however, the transformation did not take place. Finally, when the bacteria were treated with an enzyme that destroyed RNA but not DNA, the transformation occurred again. This demonstrated that the substance responsible for the transformation of the non-pathogenic bacteria into a pathogenic strain was DNA.

Although the work of Avery, MacLeod, and McCarty provided strong evidence for the role of DNA in determining cell function, the results were

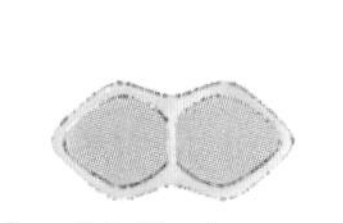

A When a heat-killed, pathogenic strain of *Streptococcus pneumoniae* is injected into mice, the mice live.

B When live, pathogenic *S. pneumoniae* bacteria are injected into mice, the mice die.

C When a live, non-pathogenic mutant strain of the same *S. pneumoniae* bacteria is injected into mice, the mice live.

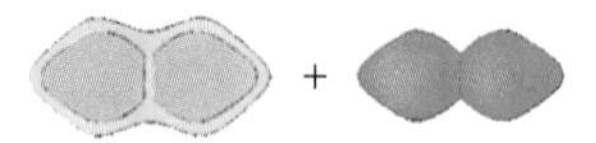

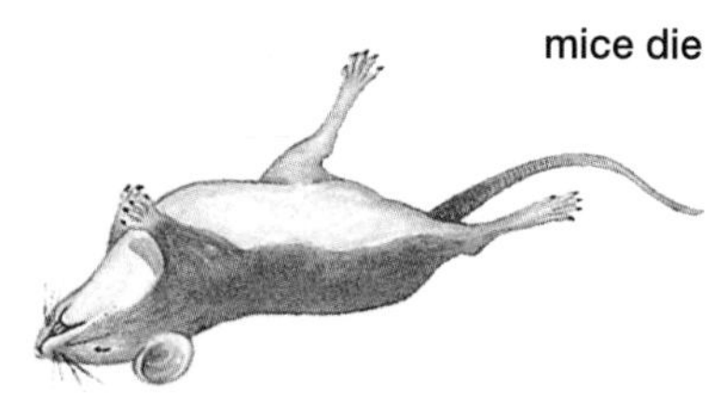

D When heat-killed, pathogenic bacteria are added to a suspension containing the live, non-pathogenic strain of bacteria, transformation occurs and the colony of non-pathogenic bacteria become pathogenic. When these bacteria are injected into mice, the mice die.

Figure 7.5 Griffith's discovery of the "transforming principle" in 1928 was accidental. He employed heat-killed, pathogenic bacteria as a control in an experiment on infection, but did not treat the cells at a high enough temperature to denature their DNA. In so doing, he discovered that the dead cells' pathogenic properties could be passed on to living bacterial cells. Griffith died of injuries suffered in an air raid during World War II before he could discover what caused this transformation. In 1944, Avery and his team were the first to demonstrate that the transforming principle was DNA.

not widely accepted. Many scientists who had accepted Levene's theory of the structure of nucleic acids simply refused to believe that the apparently simple, repetitive DNA molecule could play a key role in heredity. Others maintained that while DNA might be an agent of heredity in bacteria, prokaryotes were not a reliable model for genetic mechanisms in more complex organisms. It was not until many years later that scientists determined that the encoding of genetic information works in very similar ways in all living cells.

During the same years that Avery and his team were trying to pin down the identity of the transforming principle, other experimental evidence for the role of DNA in heredity began to accumulate. One key discovery was that in any given species, the quantity of DNA in somatic cells is both constant and double the quantity of DNA in gametes. Since at each mating two gametes come together to produce a zygote with a full complement of hereditary material, you would expect reproductive cells to have only half as much hereditary material as the cells of the body. However, it was found that the amount of protein varies widely from the cells of one tissue to another, and is not necessarily any lower in reproductive cells.

In the late 1940s, Erwin Chargaff, shown in Figure 7.6, revisited the results of Levene's experiments on the nucleotide composition of DNA. A more careful study, made possible in part by more advanced equipment, led Chargaff to overturn one of Levene's main conclusions. Chargaff argued that the four nucleotides were not present in equal quantities, but rather were found in varying but characteristic proportions. Chargaff demonstrated that although the nucleotide composition of DNA varies from one species to another, DNA specimens taken from different animals of the same species (or from different tissues collected from one animal) have the same nucleotide composition. He also found that this base composition remains consistent despite changes in the age of the specimen, its physical state (including nutrition and health), or its environment. Perhaps the most significant of Chargaff's findings was his discovery that, in any sample of DNA, the amount of adenine present is always equal to the amount of thymine, and the amount of cytosine is always equal to the amount of guanine. This constant relationship is known as **Chargaff's rule**.

Figure 7.6 Erwin Chargaff clearly refuted the theory that DNA was made up of a single sequence of nucleotides repeated over and over again. The possibility that DNA had a more complex structure helped scientists accept that DNA could play a role in heredity and development.

Further, and largely conclusive, evidence that DNA and not protein is the genetic material emerged in 1952. In an experiment, Alfred Hershey and Martha Chase used radioactive labelling techniques to follow the process of a virus known as T2 infecting a bacterial host. The T2 virus, which infects the bacteria *Escherichia coli*, is made up of a protein coat housing a strand of DNA. As shown in Figure 7.7, when the virus infects a bacterium, it first attaches to the wall of the bacterium and

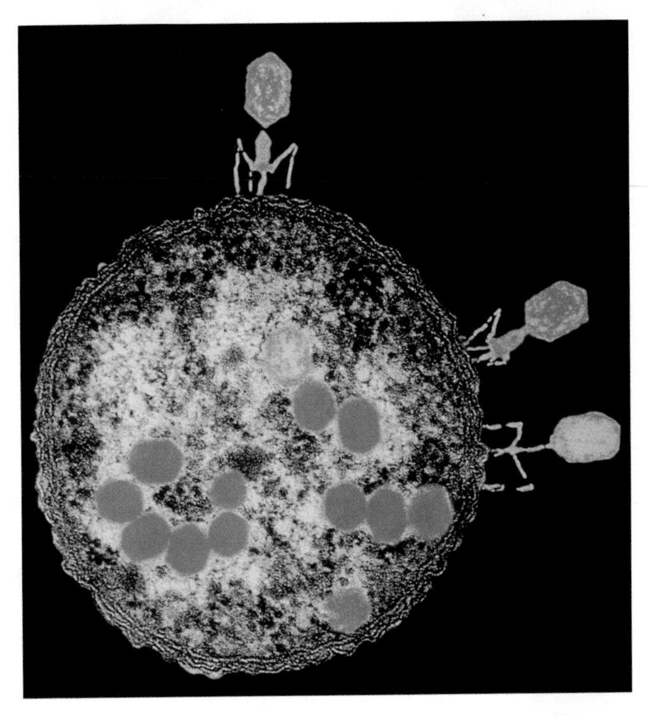

Figure 7.7 Looking somewhat like space capsules, these T2 phages use leg-like structures to bind to the cell wall of a bacterium.

7.2 The Structure of Nucleic Acids

EXPECTATIONS

- **Describe and compare the molecular structure and function of DNA and RNA.**
- **Compare and contrast the arrangement of genetic material in prokaryotes and eukaryotes.**
- **Explain how changes in the molecular arrangement of genetic material are linked to particular stages in the cell cycle.**

By the late 1940s, it was known that DNA was made up of a strand of nucleotides, and that each nucleotide was made up of a sugar, a phosphate group, and a particular nitrogenous base. Exactly how the strand was arranged, however, remained a mystery. The methodical work undertaken by British scientists Rosalind Franklin, pictured in Figure 7.9, and Maurice Wilkins to photograph and analyze X-ray diffraction images of DNA molecules added a number of new observations that helped other scientists to finally deduce the molecule's structure.

Figure 7.9 Rosalind Franklin's work was a major factor in the effort to determine the structure of DNA. Her contribution was not widely recognized at the time, in part because of prevailing attitudes toward women in science in the 1950s. Franklin died of cancer at age 38, shortly before the Nobel prize was awarded to Watson and Crick. Her many years of work with X rays may have contributed to her illness.

The pattern of shaded areas in the image shown in Figure 7.10, for example, indicated that DNA had a helical structure. From the nature of these X-ray "shadows," Franklin was able to identify two distinct but regularly repeating patterns in the structure — one pattern recurring at intervals of 0.34 nm, and another at intervals of 3.4 nm. As she prepared her samples for photographing, Franklin also observed how DNA reacted to water. From this evidence she deduced that the hydrophobic nitrogenous bases must be located on the inside of the helical structure, and that the hydrophilic sugar-phosphate backbone must be located on the outside, facing toward the watery nucleus of the cell. Her observations proved to be important keys to understanding the structure of DNA.

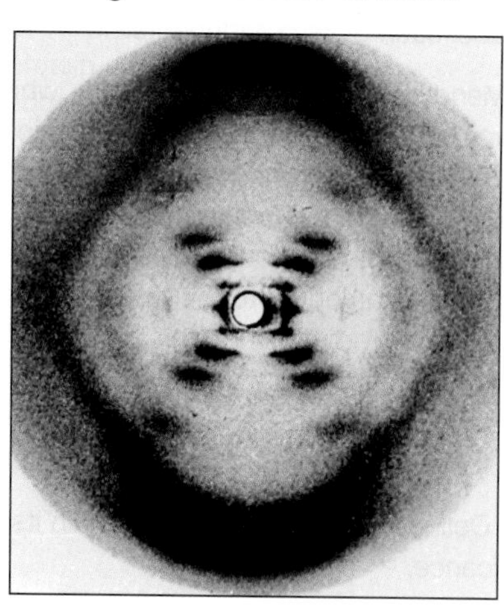

Figure 7.10 The shaded areas in this deceptively simple image indicate the pattern formed by X rays as they diffract through crystallized DNA. This photograph was made by Rosalind Franklin in 1953, and provided a number of important clues about DNA's molecular structure.

The partnership between the American geneticist James Watson and the British physicist Francis Crick was the first to produce a structural model of DNA that could account for all the experimental evidence at hand. Watson and Crick worked with physical models, as shown in Figure 7.11, trying different arrangements until they decided on the double-helix model that soon became established as the definitive structure for DNA. They published their results in a two-page paper in *Nature* magazine in 1953.

Figure 7.11 James Watson (left) and Francis Crick in 1953 with their model of a DNA molecule.

DNA — The Double Helix

As you probably know, DNA is a thread-like molecule made up of two long strands of nucleotides bound together in the shape of a double helix. If the helix were unwound, the molecule would look something like a ladder, as shown in Figure 7.12. The sugar-phosphate "handrails" form the two sides, while the paired nitrogenous bases form the rungs. The space between the rungs is 0.34 nm, while the strand as a whole makes one complete turn of the helix for every 10 base pairs, or 3.4 nm. Thus, the structure accounts for the two different recurring patterns indicated by Franklin's X-ray diffraction photograph. The following paragraphs explain how the nucleotides are joined within each

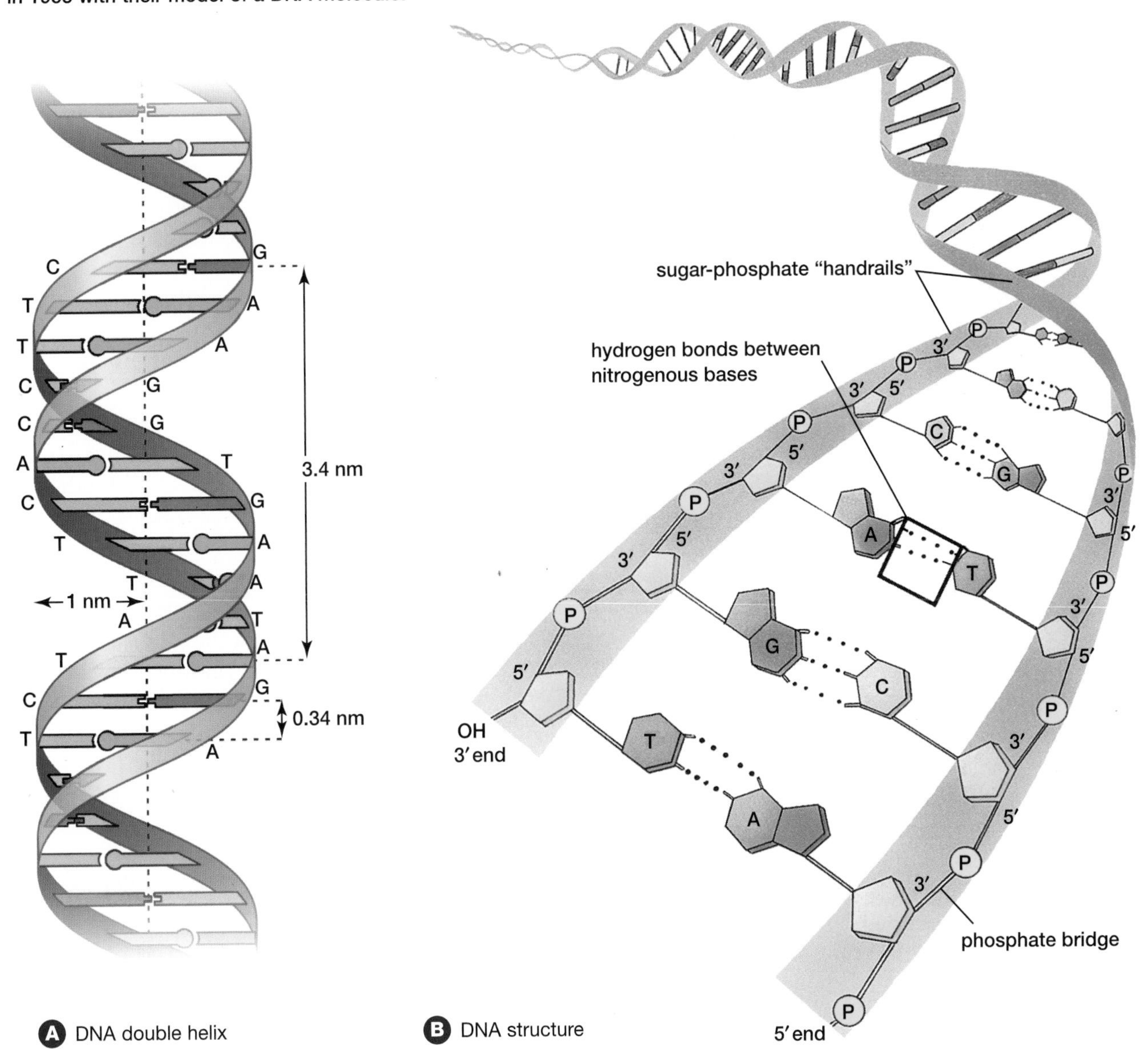

Figure 7.12 The DNA molecule is made up of two chains of nucleotides wound around each other. The "handrails" of the molecule are made up of alternating sugar and phosphate groups, with the phosphate groups serving as bridges between nucleotides. The nitrogenous bases protrude at regular intervals into the interior of the molecule.

individual strand, how the two strands are bound together in the double helix, and how the molecule as a whole remains stable.

Figure 7.12 shows both the primary and secondary structure of the DNA molecule. On each strand, the nucleotides are joined to form a long chain. The 5′ carbon of the pentose (five-carbon) sugar of one nucleotide is connected to the 3′ hydroxyl group on the next, with the phosphate group serving as a bridge between the two nucleotides. All the phosphate bridges have the same orientation, so each strand of DNA also has a specific orientation or **directionality**, which is opposite in direction to the other strand of the double helix. Thus each DNA strand (and any severed fragment of a DNA strand) has a 5′ and a 3′ end. By convention, the sequence of nucleotides along a strand of DNA is always read in the 5′ to 3′ direction.

As Figure 7.12 also shows, the nitrogenous bases protrude at regular intervals from the sugar-phosphate handrails into the interior of the DNA molecule. One of the challenges facing Watson and Crick was to determine how the bases could be arranged in such a way that the distance between the two handrails remained constant. They knew that the four bases fell into two different categories. Adenine and guanine are derived from the family of nitrogenous compounds known as **purines**, which have a double ring structure. Thymine and cytosine are derived from **pyrimidines**, which have a single ring structure. Watson and Crick hit upon the idea that if a purine always bonded with a pyrimidine, the base pairs would have a constant total width of three rings.

It was not until after they had been experimenting with models for some time that Watson and Crick examined the molecular structure of the bases themselves. When they did so, they discovered that the structure of the bases allows only certain **complementary** base pairings: Adenine (A) can only form a stable bond with thymine (T), and cytosine (C) with guanine (G). The complementary bases are linked by hydrogen bonds, as shown in Figure 7.13. These pairings provided for the constant width of the molecule and, equally importantly, also supported Chargaff's rule. Thus, wherever an A nucleotide appears on one DNA strand, a T must appear opposite it on the other, and wherever a C nucleotide appears on one strand, the other strand will have a G nucleotide.

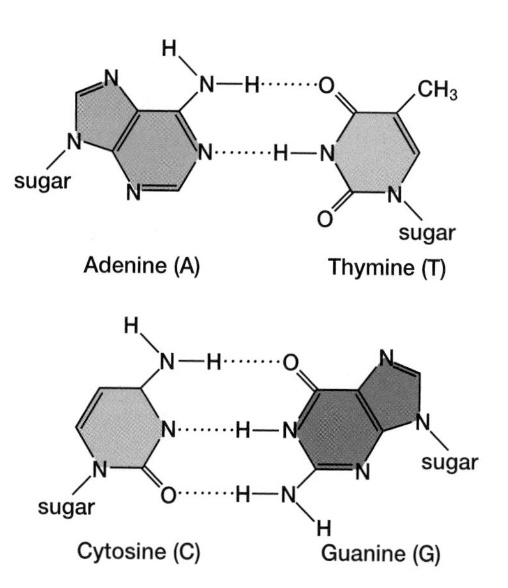

Figure 7.13 Complementary base pairing in DNA. C-G pairs are joined by three hydrogen bonds, while A-T pairs are joined by two hydrogen bonds.

In all, three kinds of forces contribute to the molecular stability of the DNA molecule. First, the phosphate bridges link the sugar-phosphate handrails. Second, hydrogen bonds between base pairs keep the two strands together in the helix. Finally, hydrophobic and hydrophilic reactions cause the bases to remain on the inside of the molecule and the handrails to face out into the watery nucleus of the cell.

As illustrated in Figure 7.14, the two strands of DNA that make up each double helix are not identical but rather complementary to each other. The strands are also **antiparallel** — that is, the phosphate bridges run in opposite directions in each strand. This means that the end of each double-stranded DNA molecule contains the 5′ end of one strand and the 3′ end of the other. These two properties have important implications for DNA replication and protein synthesis, as discussed later in this chapter and in Chapter 8.

You will have a chance to test your understanding of the composition of DNA and apply Chargaff's rule in the Thinking Lab that follows.

BIO FACT

If you could arrange all the DNA strands contained in your body end to end, their total length would stretch 2×10^{10} km. This is well over 100 times the distance between Earth and the Sun.

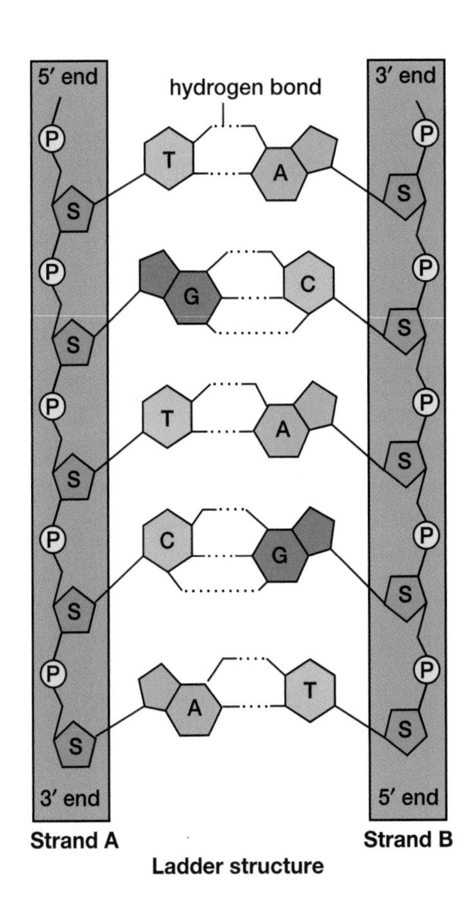

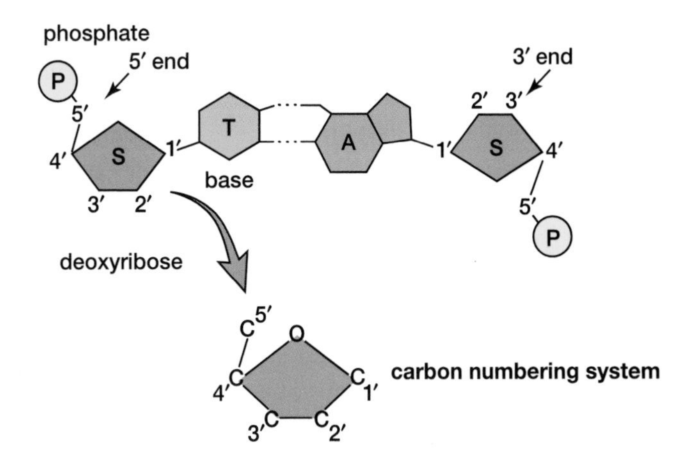

Figure 7.14 The two strands of DNA have complementary base sequences. Strand A has the nucleotide sequence TGTCA. How would the sequence of strand B be written, according to convention?

THINKING LAB

DNA Deductions

Background

Erwin Chargaff discovered that although the nucleotide composition of DNA varies from one species to another, that composition always follows certain rules. While the variation in nucleotides helps to explain the complexity of life, the physical structure of the DNA itself can also help certain organisms adapt to particular environments.

Imagine you are working with a research team sampling the ocean floor near a hot vent that releases a steady stream of hot water. The hot water has a temperature of about 45°C, while the surrounding ocean has a temperature of 6°C.

Nucleotide	Presence in DNA of bacterial sample 1 (percent)	Presence in DNA of bacterial sample 2 (percent)
adenine	31	18
cytosine		
guanine		
thymine		

Your team collects two samples of bacteria — one from the mouth of the hot vent, and one from the ocean floor about 20 m away. When you return to the lab, you isolate the DNA from these bacteria to determine their nucleotide composition. The table shows the results of your test for the adenine content of the DNA.

You Try It

Apply what you have learned about Chargaff's findings and DNA composition to solve the following problems.

1. Complete the table to determine the amounts of the other nucleotides found in each DNA sample.
2. For each DNA sample, draw a linear stretch of DNA about 15 nucleotides long, with a nucleotide composition that corresponds to its data set. With a dotted line, illustrate the hydrogen bonds between complementary base pairs.
3. Considering the bonds between base pairs, which of these DNA samples is most likely taken from the bacteria collected at the mouth of the hot vent? Explain your answer.

RNA

Along with DNA, RNA is the other main nucleic acid. Both DNA and RNA are found in most bacteria and in eukaryotic cell nuclei. The molecular structure of RNA is similar to that of DNA, with three key differences.

- As Levene observed, the sugar component of RNA is ribose rather than deoxyribose.
- As noted previously, the nucleotide thymine is not found in RNA; in its place is the nucleotide uracil.
- RNA remains single stranded, although at times this single strand can fold back on itself to produce regions of complementary base pairs.

The different structures that may be assumed by the RNA molecule result in several different types of RNA, each serving a particular function. The specific structures and roles of these different molecules are described in more detail in Chapter 8.

Organization of Genetic Material

So far, you have examined the primary structure of DNA — that is, the way in which nucleotides are joined together to form a chain. You have also looked at its secondary structure, in which the chain of nucleotides forms a stable double helix. How is this material organized in three-dimensional space within a cell?

Although viruses are typically described as containing only a short strand of either DNA or RNA, this strand is still many hundreds of times longer than the virus itself (see Figure 7.15). This material must be arranged so that it fits within the protein coat. Figure 7.16 shows the variety of forms that the genetic material of viruses can assume.

Living cells face additional challenges in arranging their DNA — they have far more DNA than a virus, and the organization of this hereditary material must also allow for two key considerations. First, the material must be arranged in a compact manner to keep the long strands of DNA from interfering with one another or with other cellular processes. Second, the hereditary material of the organism must be protected from enzymes within the cell that are designed to break down free DNA into its component nucleotides. Prokaryotes and eukaryotes each have distinct ways of arranging their DNA to meet both requirements.

Figure 7.15 The protein coat of this virus has been broken, enabling its single molecule of DNA to escape. In an intact virus, the entire length of the DNA molecule is packed within the head of the protein coat.

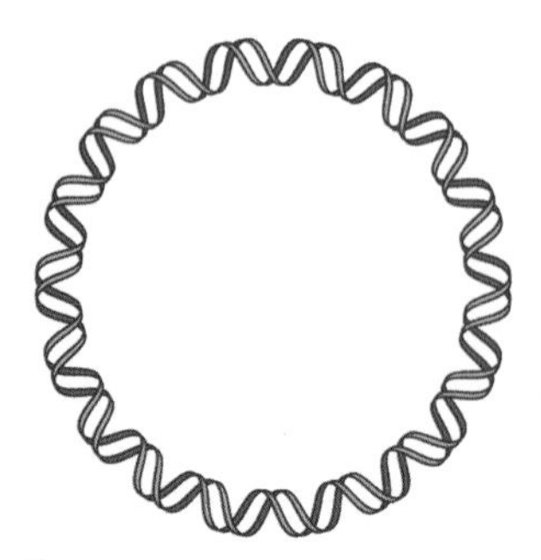

A The 5′ and 3′ ends of a DNA double helix can bind to each other, producing a closed loop.

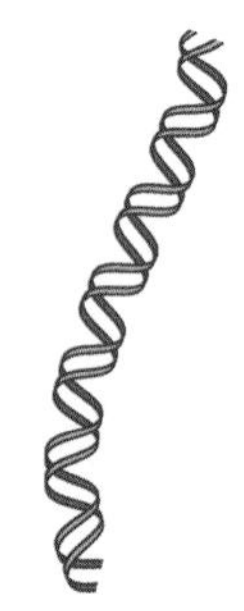

B The double helix may remain as a linear strand.

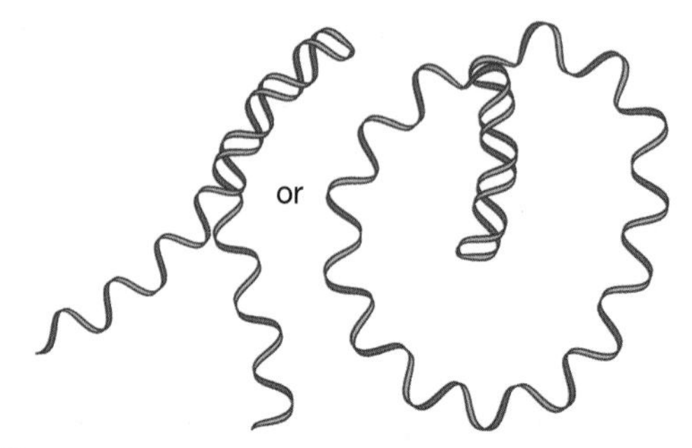

C In some viruses, the genetic material is a molecule of RNA or of single-stranded DNA. In either case, this strand may form a closed loop or remain as a linear strand, and may show regions of complementary base pairing.

Figure 7.16 The relatively short strand of either DNA or RNA found in a virus contains the information required to direct the infected cell to produce new viruses. As shown, this short strand can assume a variety of forms.

Genetic Material in Prokaryotes

Most prokaryotes have a single, double-stranded DNA molecule. One of the characteristic features of prokaryotes is that they have no nucleus. Therefore, there is no nuclear membrane to keep the DNA strand contained in a particular location within the cell. Instead, the prokaryote uses an arrangement that helps to pack genetic material tightly into a specific region known as the **nucleoid** or nuclear zone of the cell. To accomplish this, the ends of the DNA molecule bind together to form a closed belt. As shown in the photomicrograph and simplified illustration in Figure 7.17, this belt is then further twisted in on itself like a necklace that is coiled into a series of small loops. Such a structure is said to be **supercoiled**. The loops of the supercoiled DNA are held in place by proteins.

In addition to the relatively large, supercoiled DNA molecule found in the nucleoid, prokaryotes often have one or more small, circular double-stranded DNA molecules floating free in the cytoplasm of the cell. These additional DNA molecules are called **plasmids**. Although plasmids are not physically part of the nucleoid DNA, these small strands of genetic material can contribute to cell metabolism and to the hereditary mechanism. For example, the genes that confer resistance to antibiotics may be found on plasmids rather than within the nucleoid DNA. Plasmids can be copied and transmitted between cells, or they can be incorporated into the cell's nucleoid DNA and reproduced during cell division. As a result, the hereditary information contained on a plasmid can spread within bacterial colonies.

Plasmids have proven to be a valuable tool in genetic engineering techniques. You will learn more about the use of plasmids in DNA sequencing processes and in other applications later in this unit.

Genetic Material in Eukaryotes

Supercoiling is an effective way of arranging DNA within a prokaryote. However, even the simplest eukaryotic cell has over 10 times the DNA of a bacterial cell, and mammalian cells may have many thousands of times the quantity of genetic material found in a prokaryote. Each human cell nucleus contains about 2 m of DNA, or six billion base pairs. By way of comparison, this is roughly the equivalent of trying to pack 400 km of wet spaghetti into a bathtub — yet these DNA fibres never become entangled. A highly structured arrangement of proteins and DNA helps to pack and organize this material within the cell.

The nuclei of plant and animal cells contain double-stranded DNA. This DNA is organized into a number of separate chromosomes. Each chromosome contains one linear double-stranded DNA molecule together with different types of the protein **histone**. Overall, the composition of a chromosome is about 60 percent protein, 35 percent DNA, and five percent RNA. These components are organized into the long fibres that Miescher called "nuclein," and which are now known as **chromatin**.

Figure 7.18 on the following page shows the arrangement of genetic material in a eukaryotic cell. The DNA molecule wraps tightly around groups of histone molecules in a regular pattern

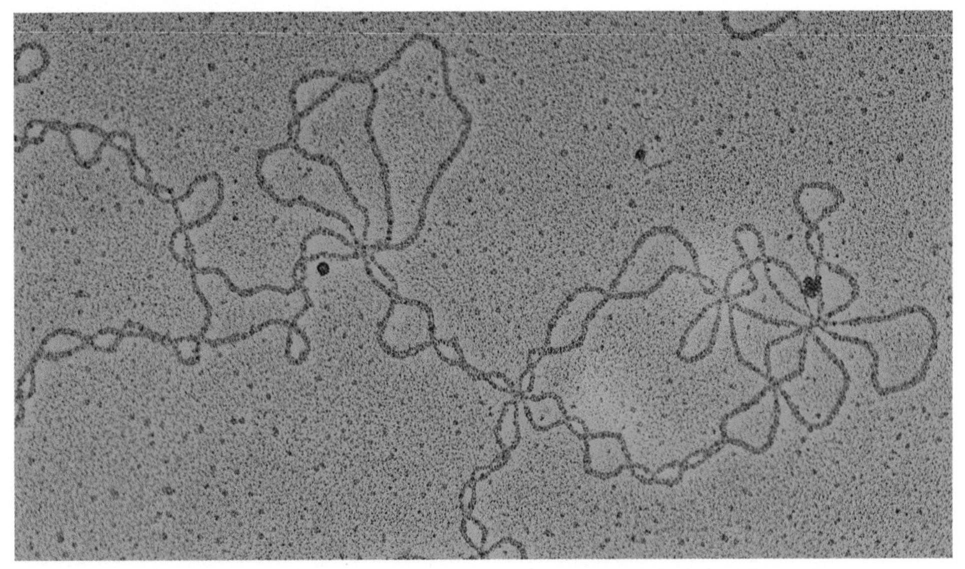

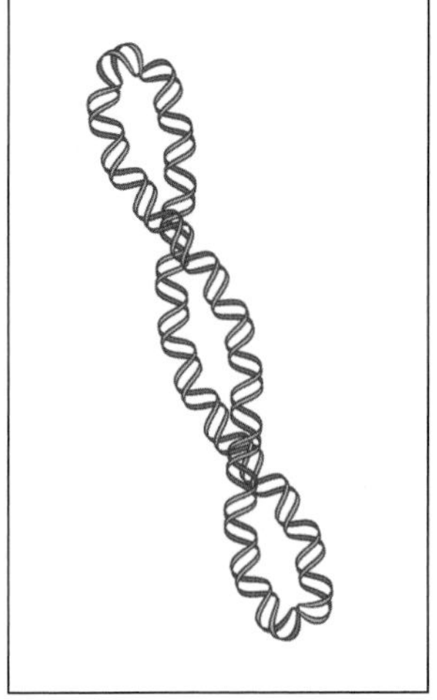

Figure 7.17 The supercoiled molecule of bacterial DNA shown at left is coiled into a series of small loops, as shown in the simplified illustration at right.

to produce bead-like structural units called **nucleosomes**. Each nucleosome bead is a short segment of DNA (about 200 base pairs) wrapped twice around a cluster of eight histone molecules. The attraction between the acidic DNA and the highly alkaline histone molecules helps to keep the arrangement in place.

A short stretch of DNA extends between each nucleosome, as shown in Figure 7.19. This short segment of DNA is bound to a single molecule of a type of histone known as H1. The interaction among H1 molecules helps to draw the arrangement into a tight, regular array.

In turn, this array (which has a total thickness of about 30 nm) undergoes another level of packing. It forms loops that attach to a supporting structure of non-histone proteins. As the cell prepares to reproduce, the protein structure folds back on itself to condense the chromatin even further. The result is the short, thick chromosomes you have seen in typical karyotypes. In a living cell, some regions of chromatin will be found in the 30 nm arrays during the bulk of the cell life cycle, while other regions are more condensed. You will learn more about the effect of this arrangement on cell functions in the next chapter.

In the next section, you will learn about the processes involved in DNA replication and how they ensure the accurate transmission of hereditary material.

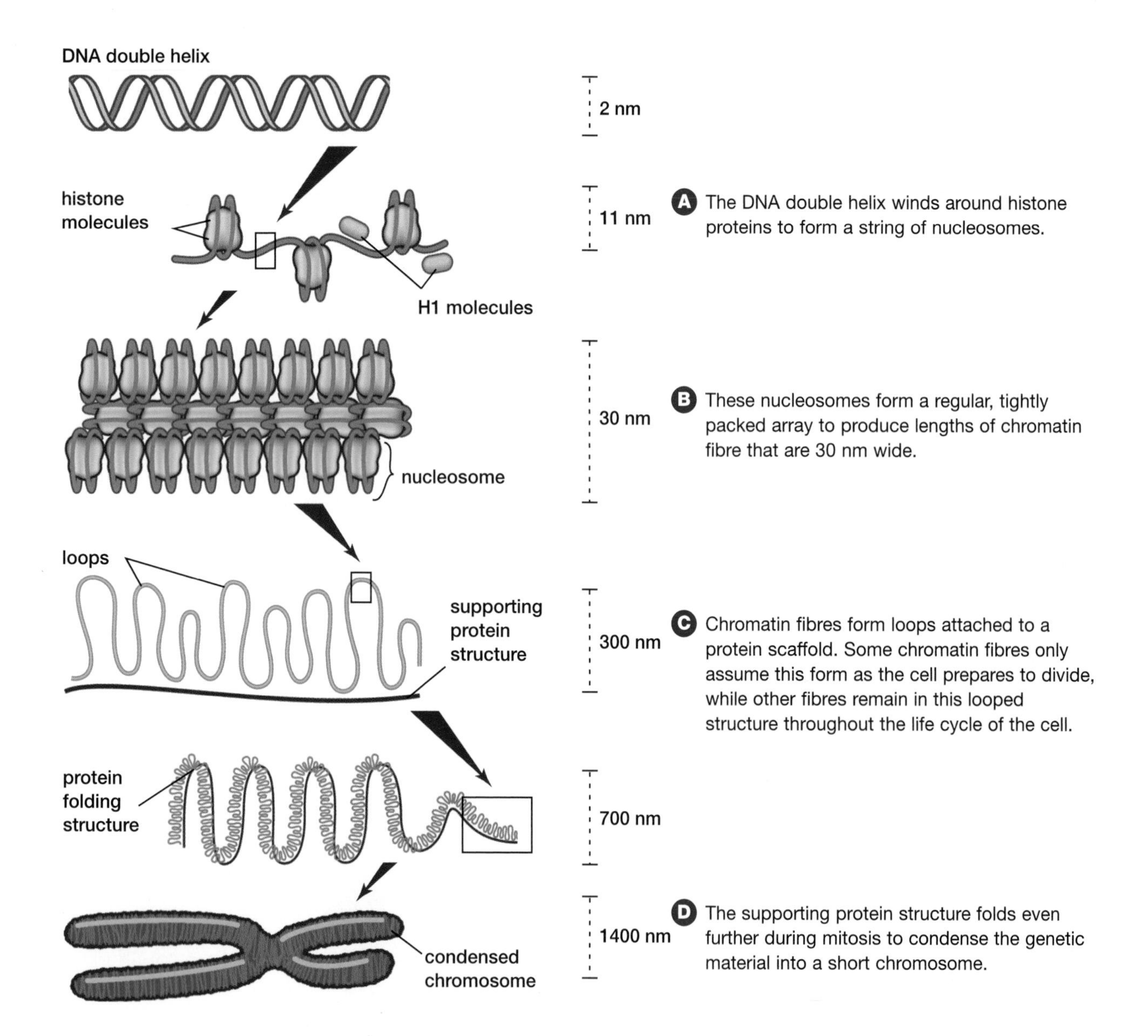

Figure 7.18 The successive ordering of genetic material within a eukaryotic cell.

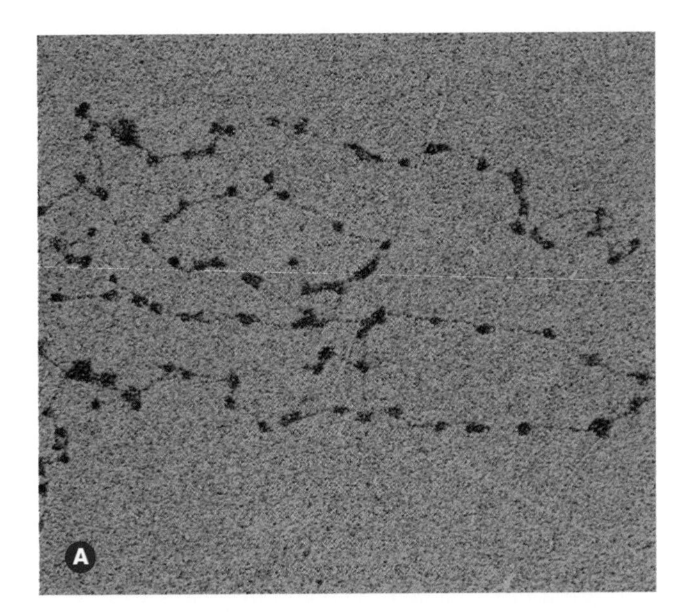

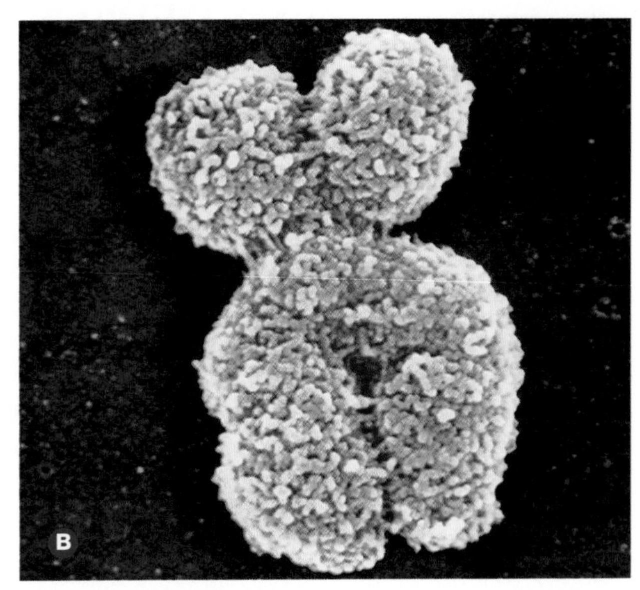

Figure 7.19 The total length of the DNA in a single human chromosome reaches approximately 10 cm. The successive ordering of genetic material begins with the formation of nucleosomes (A) and culminates with the condensed chromosome (B), shown here as a pair of sister chromatids during mitosis. This ordering compacts the DNA by a factor of tens of thousands — the equivalent of coiling 50 km of thread into a piece of rope that fits across the palm of your hand.

SECTION REVIEW

1. K/U Describe the observations made by Franklin that proved to be keys to understanding the molecular structure of DNA.
2. K/U Describe the evidence that led Watson and Crick to their particular model of DNA.
3. K/U DNA is a large double helix, resembling a twisted ladder. Describe the component molecules that make up the ladder uprights and the pattern of their arrangement.
4. C Create a chart that compares and contrasts the similarities and differences in the structure of DNA and RNA.
5. C The four nitrogenous bases are paired to make up the rungs of the DNA ladder. Differentiate between the purine and pyrimidine structures of each of these nitrogenous bases, and explain the significance of the pattern of their arrangement.
6. K/U Watson and Crick concluded that DNA consisted of two individual strands arranged in antiparallel directions. Explain the significance of this arrangement.
7. K/U Identify and describe the three different kinds of forces that contribute to the molecular stability of the DNA molecule.
8. K/U Describe the organization of genetic material in prokaryotic organisms. What purpose is served by having DNA incorporated into plasmids?
9. K/U Describe the organization of DNA in eukaryotic organisms. What is the reason for the compact nature of chromosomes?
10. I Scientists have extracted DNA from the different types of cells within one species and from the same kinds of cells among many different species. Decide whether the results from this research alone will or will not support the hypothesis that DNA is the genetic material. Justify your decision.
11. K/U DNA is a doubled-stranded helical molecule. RNA is a single-stranded molecule. Based on the structural characteristics of each molecule, determine which is more efficient at coding genetic information.
12. MC Should scientists be compelled to share information out of a spirit of fairness? Linus Pauling was one of the favoured competitors in the race to discover the molecular structure of DNA, due to his superior knowledge of organic chemistry. However, Watson and Crick, who had access to Franklin and Wilkins' X-ray pictures and knowledge of Pauling's theories from his son, claimed the victory. Based on your independent review of the history of this event, how do you think the race might have ended differently?

UNIT PROJECT PREP

Start gathering statistics on the type of cancer you have chosen to study. Who is most likely to be affected by this cancer?

7.3 DNA Replication

EXPECTATIONS

- **Describe the current model of DNA replication and methods of repair following an error.**
- **Demonstrate an understanding of the roles played by the key enzymes involved in the process of replication.**
- **Explain how differences between the molecular structure of DNA in prokaryote cells and eukaryote cells affect the process of replication.**
- **Relate the key mechanisms for regulating DNA replication to the accurate transmission of hereditary material during the process of cell division.**

The eight cells in the human blastula shown in Figure 7.20 arose from the single-celled zygote formed by the merging of sperm and egg. During the 240-day gestation period, the cells of the blastula will divide over and over again to produce about one hundred trillion more cells. These trillions of cells will differentiate into the hundreds of different structures and tissues that make up a human baby, yet each cell will have exactly the same genetic complement as the original zygote. The success of this process of development depends on two factors — the genome must be copied relatively quickly, and it must be copied accurately.

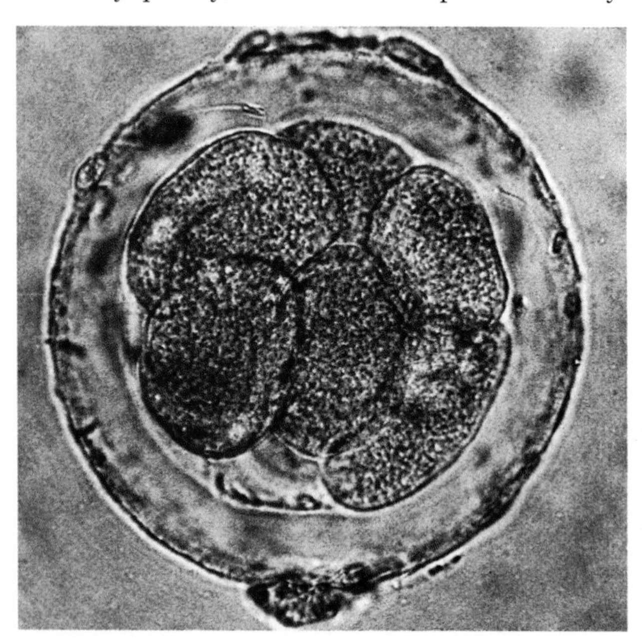

Figure 7.20 The process of DNA replication balances the need for speed and the need for accuracy.

Imagine you are asked to type out one-letter codes for each of the six billion base pairs in the genome of a single human cell. If you type at a rate equivalent to 60 words per minute and work without a break, it will take you over 30 years to complete the sequence. The cell, on the other hand, needs only a few hours to copy the same material. The error rate of the cell's replication process is about one per billion nucleotide pairs, which is the equivalent of you making a one-letter error once in every five years of steady typing. The remarkable speed and accuracy of the replication process relies on both the structural features of DNA and the action of a set of enzymes.

In their landmark paper on the structure of DNA, Watson and Crick made the passing observation that the complementarity of the DNA strands pointed to a means of accurate replication of DNA molecules. What they meant was that each strand of the double helix can serve as a template for the production of a new complementary strand. A replication process based on this principle was suggested by Watson and Crick in a second paper published not long after the first. In this paper, Watson and Crick proposed that the two strands of the DNA double helix molecule unwind and separate, after which each nucleotide chain serves as a template for the formation of a new companion chain. The result would be a pair of daughter DNA molecules, each identical to the parent molecule.

After the publication of Watson and Crick's papers, researchers began exploring the question of how DNA replicates. They identified three main possibilities, as illustrated in Figure 7.21. The **conservative theory** proposed that replication involves the formation of two new daughter strands from the parent templates, with the two new strands then joining to create a new double helix. The two original strands would then re-form into the parent molecule. The **semi-conservative theory**, the model suggested by Watson and Crick, proposed that the

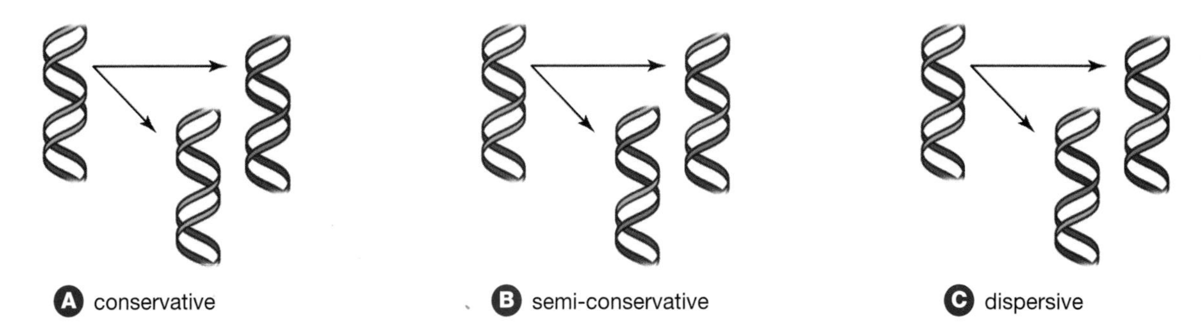

Figure 7.21 The three main theories advanced for DNA replication. According to the conservative theory (A), the parent molecule is re-established intact after replication. In the semi-conservative theory (B), the individual strands of the parent molecule remain intact but are separated, one forming half of each of the two daughter molecules. In the dispersive theory (C), both strands of the parent molecule are broken into fragments, copied, and then reassembled, with the fragments shared among parent and daughter molecules.

daughter DNA molecules were each made up of one parental strand and one new strand. A third hypothesis, the **dispersive theory**, proposed that the parental DNA molecules were broken into fragments and that both strands of DNA in each of the daughter molecules were made up of an assortment of parental and new DNA.

The issue was resolved in 1957 as the result of an experiment conducted by Matthew Meselson and Franklin Stahl. The steps in the experiment are illustrated in Figure 7.22. The two scientists grew a colony of bacteria in a medium containing ^{15}N, a form of nitrogen that has a higher molecular mass than regular nitrogen (^{14}N). As the bacteria developed, their DNA incorporated the heavier

heavy DNA

direction of sedimentation

regular DNA

results

hybrid DNA after one replication

DNA after two replications

regular DNA

hybrid DNA

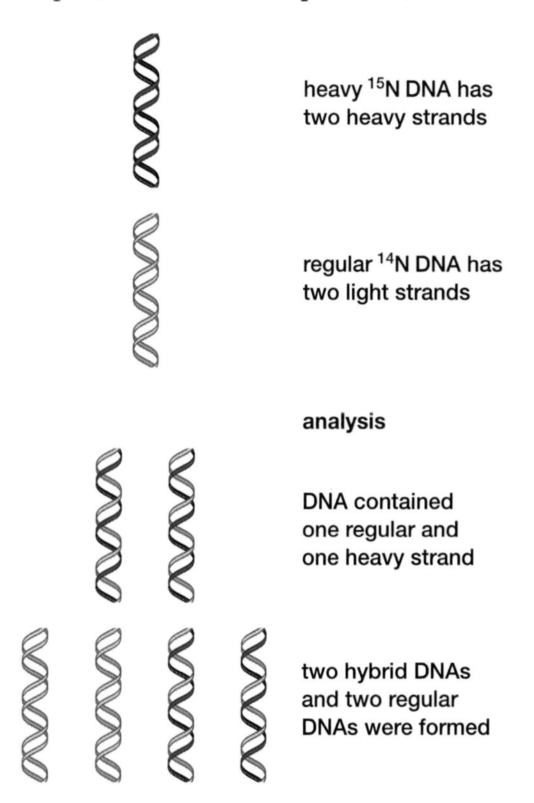

Figure 7.22 The experiments conducted by Meselson and Stahl tested all three hypotheses for DNA replication. First, the scientists cultured a colony of *E. coli* on a medium containing the heavy nitrogen isotope ^{15}N. As a result, these bacteria had uniformly heavy DNA. The bacteria were then transferred to a medium containing the more common and lighter ^{14}N, to allow them to incorporate this isotope during future DNA synthesis. After one replication on the ^{14}N medium, results showed a single band with a density that indicated hybrid DNA. After two replications on the ^{14}N medium, results showed two distinct bands, indicating the presence of equal quantities of both hybrid and regular DNA. These results supported the semi-conservative replication hypothesis.

nitrogen. The colony of bacteria was then transferred to a new medium containing regular nitrogen. After the colony had doubled in size (indicating approximately one complete round of cell division), Meselson and Stahl isolated the DNA from the bacterial cells and spun it in a centrifuge. Afterward, they observed a single band marking a density midway between the expected densities of DNA containing ^{15}N ("heavy" DNA) and DNA containing ^{14}N ("regular" DNA). This indicated that the DNA after one round of replication was a hybrid — that is, a mixture of heavy and regular DNA. This result ruled out the possibility of conservative replication, since that would have resulted in one band of heavy DNA and another of regular DNA. This left semi-conservative and dispersive replication as potential models. To determine which was correct, a second round of experimentation was required.

When Meselson and Stahl left the colony on the second medium for two generations before extracting the DNA and spinning it in a centrifuge, they found two distinct bands. One of these bands appeared at the same midway point, while the other appeared at the expected density for regular DNA. This result was consistent with the expected pattern for semi-conservative replication. It also ruled out the possibility of dispersive replication, since the random assortment of DNA fragments would result in the appearance of only one density band. With both conservative and dispersive replication ruled out, semi-conservative replication became the accepted hypothesis for DNA replication. This hypothesis has since been supported by further experiments and by microscopic images of replicating DNA.

The Process of Replication

The process of DNA replication is discussed in the following pages as a series of three basic phases. In the initiation phase, a portion of the DNA double helix is unwound to expose the bases for new base pairing. In the elongation phase, two new strands of DNA are assembled using the parental DNA as a template. Finally, in the termination phase, the replication process is completed and the new DNA molecules — each composed of one strand of parental DNA and one strand of daughter DNA — re-form into helices. In actual practice, all of these activities may take place simultaneously on the same molecule of DNA.

Initiation

In bacteria, the circular DNA strand includes a specific nucleotide sequence of about 100 to 200 base pairs known as the **replication origin**. This nucleotide sequence is recognized by a group of enzymes that bind to the DNA at the origin and separate the two strands to open a replication bubble. After a replication bubble has been opened, molecules of an enzyme called **DNA polymerase** insert themselves into the space between the two strands. Using the parent strands as a template, the polymerase molecules begin to add nucleotides one at a time to create a new strand that is complementary to the existing template strand.

For most of the life cycle of the cell, DNA is a tightly bound and stable structure. Because the bases face into the interior of the molecule, the helix must be unwound for the individual chains of nucleotides to serve as templates for the formation of new strands. The points at which the DNA helix is unwound and new strands develop are called **replication forks**. One replication fork is found at each end of a replication bubble, as shown in Figure 7.23.

A set of enzymes known as **helicases** cleave and unravel short segments of DNA just ahead of the replicating fork. As the helicases work their way along the DNA, the replication forks move around the circular DNA molecule until they meet at the other side. At this point the two daughter DNA molecules separate from each another.

In *E. coli*, replicating forks move at a rate of over 45 000 nucleotides per fork per minute, and the entire bacterial genome is replicated in less than an hour. Given that a eukaryotic cell contains hundreds or thousands of times more DNA than a bacterium, and that this DNA must be unpackaged from its complex array of nucleosomes before it can be replicated, you would expect the process of replication to take much longer in a eukaryotic cell. Nevertheless, the complete replication of the genome of a eukaryotic cell is accomplished within a few hours.

BIO FACT

In the bacteria *E. coli*, unwinding DNA spins at a rate of over 4500 r/min — almost twice as fast as the engine speed of an average car cruising on an expressway.

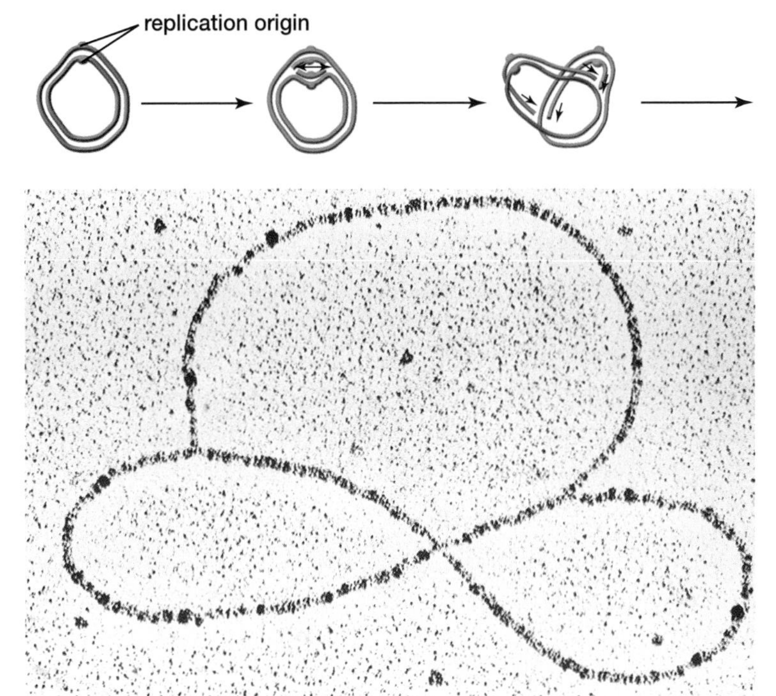

Figure 7.23 The movement of the replication forks around the circular chromosome in a prokaryote. Note that one replication bubble incorporates two replication forks, and that replication proceeds in both directions around the circular chromosome.

Figure 7.24 shows the pattern of replication along a linear strand of eukaryotic DNA. Replication is initiated at hundreds or even thousands of replication origins at any one time. Replication continues until all the replication bubbles have met and the two new DNA molecules separate from each other.

The packaging of chromatin means that individual replication forks proceed much more slowly in eukaryotic cells than in prokaryotes. In mammals, for example, the rate of movement of replication forks is less than one tenth the rate in *E. coli*. Nevertheless, the presence of multiple replication forks means that the whole process can be accomplished very quickly. All the chromosomes in a eukaryotic cell are replicated simultaneously during the S1 phase of the cell cycle.

Elongation

Recall that the formation of a new DNA strand relies on the action of DNA polymerase. This enzyme has a very specific role in catalyzing the elongation of DNA molecules — it attaches new

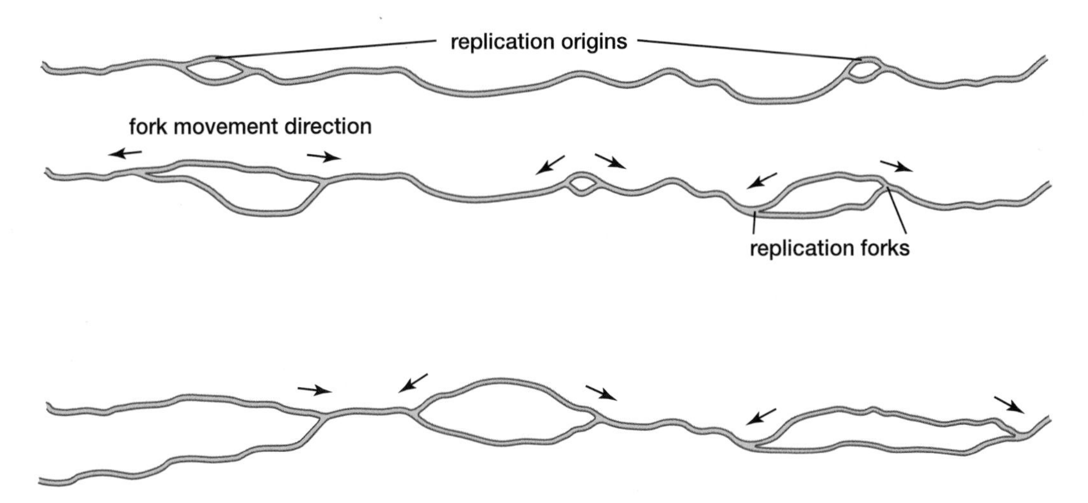

Figure 7.24 Replication bubbles open simultaneously at many sites along a linear DNA strand. As replication proceeds along the strand, the bubbles grow until they meet and the daughter strands separate from each other.

nucleotides only to the free 3′ hydroxyl end of a pre-existing chain of nucleotides. This imposes two conditions on the elongation process. First, replication can only take place in the 5′ to 3′ direction. Second, a short strand of ribonucleic acid known as a **primer** must be available to serve as a starting point for the attachment of new nucleotides. Each of these conditions helps shape the process of building new DNA strands.

The fact that polymerase can only catalyze elongation in the 5′ to 3′ direction appears to conflict with the observations that both DNA strands are replicated simultaneously and replication proceeds in both directions simultaneously along the template strand. This puzzle was solved with the discovery that, during replication, much of the newly formed DNA could be found in short fragments of one to two thousand nucleotides in prokaryotes (and a few hundred nucleotides in eukaryotes). These are known as **Okazaki fragments** after Japanese scientist Reiji Okazaki, who first observed the fragments and deduced their role in replication in the late 1960s. Okazaki fragments occur during the elongation of the daughter DNA strand that must be built in the 3′ to 5′ direction. These short segments of DNA are synthesized by DNA polymerase working in the 5′ to 3′ direction (that is, in the direction opposite to the movement of the replication fork) and then spliced together.

As illustrated in Figure 7.25, replication must thus take place in a slightly different way along each strand of the parent DNA. One strand is replicated continuously in the 5′ to 3′ direction, with the steady addition of nucleotides along the daughter strand. On this strand, elongation proceeds in the same direction as the movement of the replication fork. This strand is called the **leading strand**. In the other strand, which is first made in short pieces, nucleotides are still added by DNA polymerase to the 3′ hydroxyl group. However, elongation takes place here in the opposite direction to the movement of the replicating fork. DNA polymerase builds Okazaki fragments in the 5′ to 3′ direction. The fragments are then spliced together by an enzyme called **DNA ligase**, which catalyzes the formation of phosphate bonds between nucleotides. This strand is called the **lagging strand**, because it is manufactured more slowly than the leading strand.

Remember that DNA polymerase is unable to synthesize new DNA fragments — it can only attach nucleotides to an existing nucleotide chain. This means that a separate mechanism is required to establish an initial chain of nucleotides that can serve as a starting point for the elongation of a daughter DNA strand. In fact, a short strand of RNA that is made up of a few nucleotides with a base sequence complementary to the DNA template serves as a primer for DNA synthesis. The formation of this primer requires the action of an enzyme called **primase**. Once the primer has been constructed, DNA polymerase extends the fragment by adding DNA nucleotides. Then DNA polymerase chemically snips out the RNA molecules with surgical precision, starting at the 5′ end of the molecule and working in a 5′ to 3′ direction.

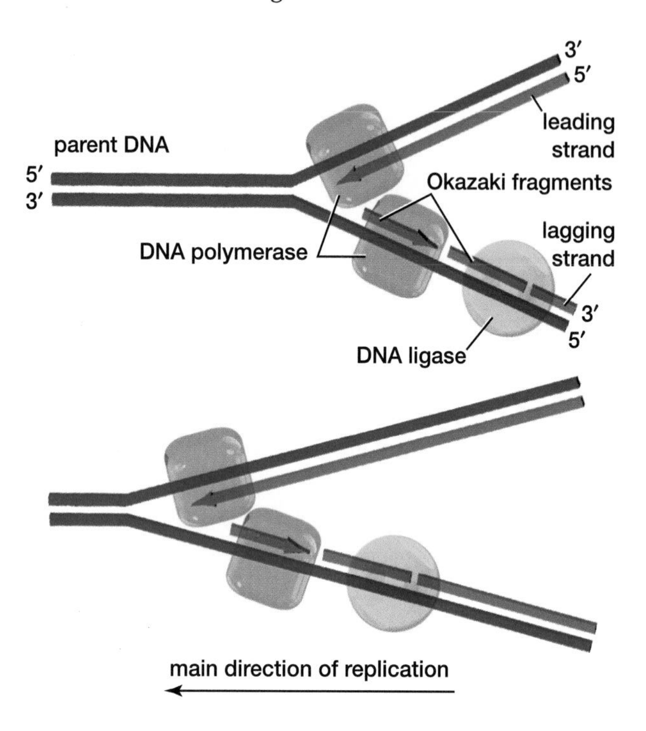

Figure 7.25 During DNA synthesis the overall direction of elongation is the same for both daughter strands, but different along each of the parent DNA strands. Along the lagging strand, DNA polymerase moves in a 5′ to 3′ direction, while DNA ligase moves in the 3′ to 5′ direction to connect the Okazaki fragments into one daughter strand.

On the leading strand, only one primer has to be constructed. On the lagging strand, however, a new primer has to be made for each Okazaki fragment. Once these primers have been constructed, DNA polymerase adds a stretch of DNA nucleotides to the 3′ end of the strand. Then another molecule of DNA polymerase attaches to the 5′ end of the fragment and removes each RNA nucleotide individually from the primer stretch. At the same time, this DNA polymerase extends the preceding Okazaki fragment, working in the 5′ to 3′ direction to replace the excised RNA nucleotides with DNA nucleotides. Finally, the two fragments are joined together by the action of DNA ligase. This process is illustrated in detail in Figure 7.26.

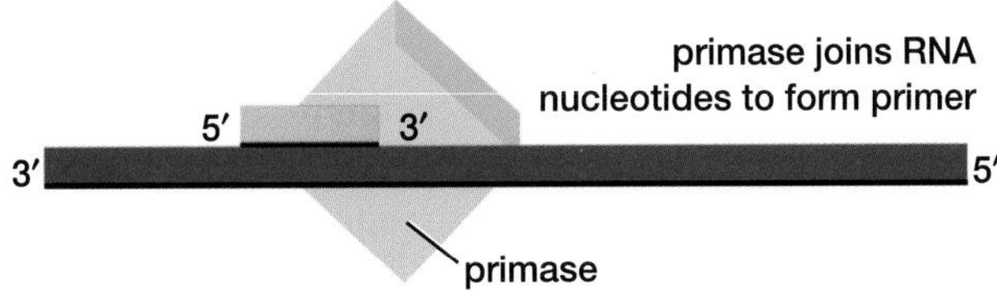

A Primase catalyzes the formation of an RNA primer.

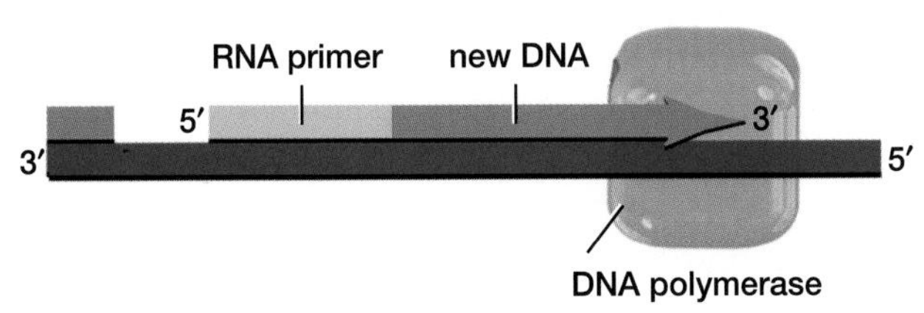

B Working in the 5′ to 3′ direction (that is, away from the movement of the replicating fork), DNA polymerase adds nucleotides to the primer.

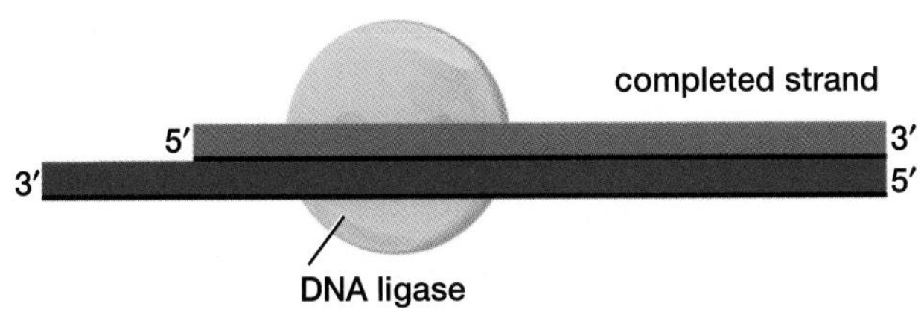

C A second molecule of DNA polymerase binds to the previous Okazaki fragment adjacent to the RNA primer. It then excises the RNA nucleotide, replacing it with a DNA nucleotide.

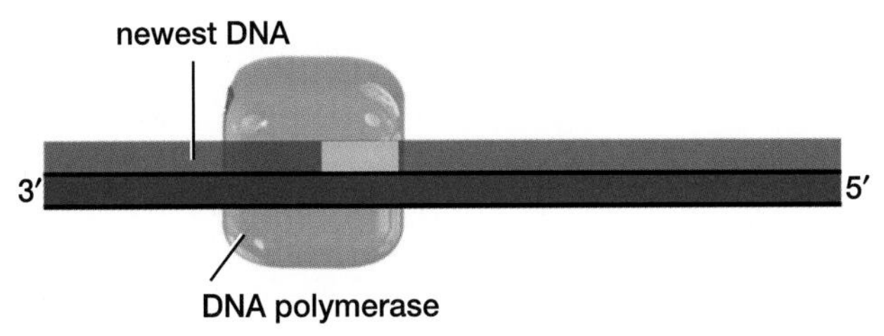

D When the last RNA nucleotide is replaced with a DNA nucleotide, DNA ligase binds the two Okazaki fragments.

Figure 7.26 Elongation of the lagging strand

ELECTRONIC LEARNING PARTNER

Your Electronic Learning Partner has an animation on DNA replication.

Termination

Once the newly formed strands are complete, the daughter DNA molecules rewind automatically in order to regain their chemically stable helical structure. This rewinding process does not require any enzyme activity. However, the synthesis of daughter DNA molecules creates a new problem at each end of a linear chromosome.

As you saw earlier, each time DNA polymerase excises an RNA primer from an Okazaki fragment, the resulting gap is normally filled by the addition of nucleotides to the 3′ end of the adjacent Okazaki fragment. But what happens once the RNA primer has been dismantled from the 5′ end of each daughter DNA molecule? There is no adjacent nucleotide chain with a 3′ end that can be extended to fill in the gap, and the cell has no enzyme that can work back in the 3′ to 5′ direction to complete the 5′ end of the DNA strand. Furthermore, the nucleotides on the complementary strand are left unpaired, and they eventually break off from the new strand. As shown in Figure 7.27, the result is that each daughter DNA molecule is slightly shorter than its parent template. With each replication, more DNA is lost. Human cells lose about 100 base pairs from the ends of each chromosome with each replication. Prokaryotes, which have circular DNA, do not have the same problem.

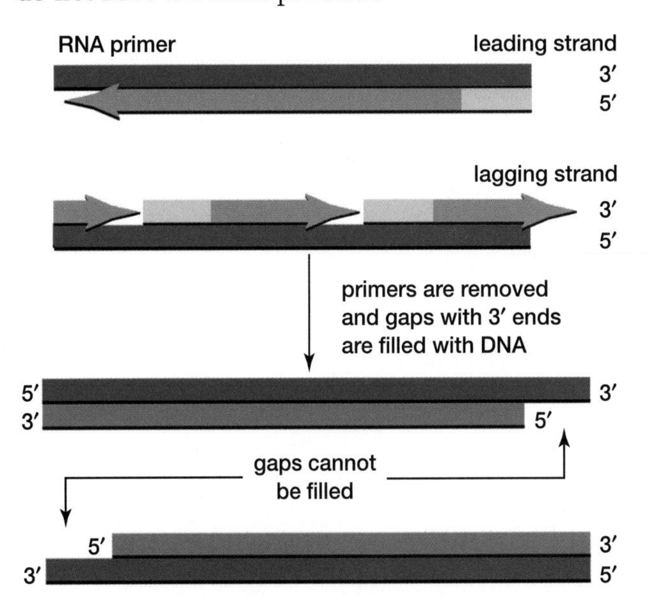

Figure 7.27 Each end of a linear chromosome presents a problem for the DNA replication process. Once the RNA primer has been removed from the 5′ end of each daughter strand, there is no adjacent fragment onto which new DNA nucleotides can be added to fill the gap. Therefore, each replication results in a slightly shorter daughter chromosome.

The loss of genetic material with each cell division could prove disastrous for a cell, since this lost material might code for activities that are important for cell functions. Special regions at the end of each chromosome in eukaryotes help to guard against this problem. These regions, called telomeres, serve as a form of buffer zone. **Telomeres** are stretches of highly repetitive nucleotide sequences that are typically rich in G nucleotides. In human cells, telomeres are composed of the sequence TTAGGG repeated several thousand times. These regions do not direct cell development. Instead, their erosion with each cell division helps to protect against the loss of other genetic material.

As you might expect, the erosion of the telomeres is related to the death of the cell. Conversely, the extension of telomeres is linked to a longer life span for the cell. Studies published in 2001 by a Canadian-led team of scientists working at the University of British Columbia found that the activity of a gene that codes for telomerase (an enzyme that extends telomeres) is directly linked to longevity in organisms such as worms and fruit flies. Cancer cells, which continue to divide well beyond the normal life span of a somatic cell, also contain telomerase. This finding has led scientists to explore the possibility of controlling cancer by pinpointing the trigger for the production of this enzyme.

Proofreading and Correction

The illustrations shown in this chapter present DNA replication as an orderly, step by step process. In reality, the setting at a molecular level is nothing short of chaotic. Imagine a sea of small and large molecules — nucleotides, free phosphate groups, dozens of different enzymes, Okazaki fragments, DNA helices, proteins, and more — all involved in a complex series of molecular collisions and chemical reactions. In this dynamic environment, it is hardly surprising that the wrong base is occasionally inserted into a lengthening strand of DNA. Studies suggest that if the replication process relied only on the accuracy of the base pairing function of DNA polymerase, errors would occur with a frequency of about one in every 10 000 to

Investigation 7 • A

SKILL FOCUS
- Performing and recording
- Analyzing and interpreting
- Conducting research

DNA Structure and Replication

James Watson and Francis Crick did not conduct any experiments in their efforts to discover the structure of DNA. Instead, they worked with physical models, trying to build a structure that could account for all the available evidence. In this investigation, you will design and build a DNA model and use this model to simulate the process of DNA replication.

Pre-lab Questions

- What happens during DNA replication?
- How does the structure of DNA contribute to the accurate transmission of hereditary material?

Problem

How can you use physical models to simulate molecular interactions?

Prediction

Predict how closely your model will resemble the one constructed by Watson and Crick.

Materials

DNA model-building supplies
sketching supplies
notebook

Procedure

1. Working with a partner, make a list of all the facts that were known about DNA when Watson and Crick began their work.

100 000 nucleotides. In fact, the accuracy of the process is up to 10 000 times better. An additional process must therefore be involved in ensuring the accuracy of replication. This function is also performed by DNA polymerase.

After each nucleotide is added to a new DNA strand, DNA polymerase can recognize whether or not hydrogen bonding is taking place between base pairs. The absence of hydrogen bonding indicates a mismatch between the bases. When this occurs, the polymerase excises the incorrect base from the new strand and then adds the correct nucleotide using the parent strand as a template. This double check brings the accuracy of the replication process to a factor of about one error per billion base pairs.

In total, the process of DNA replication involves the action of dozens of different enzymes and other proteins. These substances work closely together in a complex known as a **replication machine**. Figure 7.28 on the following page shows a simplified version of a replication machine, while Table 7.1 summarizes the roles of the key enzymes.

Table 7.1
Key enzymes in DNA replication

Enzyme group	Function
helicase	cleaves and unwinds short sections of DNA ahead of the replication fork
DNA polymerase	three different functions: – adds new nucleotides to 3' end of elongating strand – dismantles RNA primer – proofreads base pairing
DNA ligase	catalyzes the formation of phosphate bridges between nucleotides to join Okazaki fragments
primase	synthesizes an RNA primer to begin the elongation process

In this section, you saw how the molecular structure of DNA contributes to its role as the material of heredity. In the next section, you will see how DNA is organized into the functional units that make up the genetic material of an organism.

2. Use the materials available to construct a short strand of DNA. Make a note of how each fact on your list is supported by your model.
3. Write down the nucleotide sequences for each strand of DNA in your molecule, using the correct conventions.
4. Now use your model to simulate the process of DNA replication. Keeping in mind the specific action of the enzyme DNA polymerase, use your model to demonstrate:
   (a) replication along the leading strand;
   (b) replication along the lagging strand; and
   (c) the problem created at the ends of linear chromosomes.

Post-lab Questions

1. Which base pairs in a DNA molecule will be least resistant to heat? Why?
2. Are there any aspects of DNA replication that your model cannot illustrate? Explain.

Conclude and Apply

3. Make a list of the key replication enzymes in the order in which they are involved. For each enzyme, write a brief description of what would happen if that enzyme were not present in the replication medium. (For the purpose of this exercise, assume that the absence of any one enzyme does not affect the activity of others.) Compare your findings with those of another group.
4. Draw a flowchart or concept map relating events at the molecular level to the observed changes in chromosomes during cell division. You may wish to refer to Appendix 4 for a review of cell division.
5. In one of the early models tested by Watson and Crick, the sugar-phosphate handrails were located on the inside of the helix while the nitrogenous bases protruded outward. In what ways is this model inconsistent with experimental evidence about the structure of nucleic acids?

Exploring Further

6. In the late 1940s and early 1950s, before the publication of Watson and Crick's paper, other researchers proposed different structures for the DNA molecule. Conduct research on one of these early models. Prepare a short written report that compares this model with Watson and Crick's. How did Watson and Crick's model fit better with the scientific evidence?

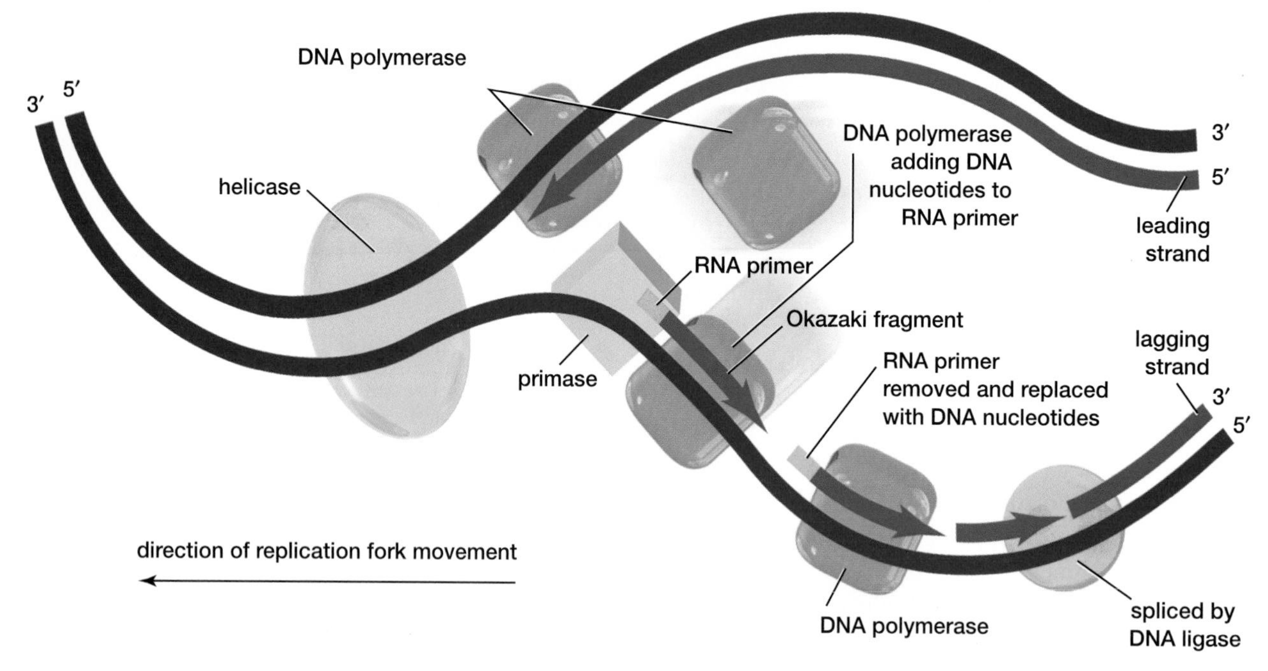

Figure 7.28 As this illustration of the replication machine indicates, only a very short region of either the parent or daughter DNA strand is ever left in a non-base-paired form as the replication fork progresses.

SECTION REVIEW

1. K/U Differentiate between conservative, semi-conservative, and dispersive theories of replication. Which theory was supported by experimental evidence?

2. C Summarize the experiment conducted by Meselson and Stahl to establish the nature of DNA replication.

3. C In a series of sketches, briefly outline the three phases of DNA replication.

4. K/U Replication of DNA strands can only take place in one direction. Find some analogies that could be used to explain the significance of this for living cells.

5. K/U Explain the role of the following enzymes in DNA replication.
   (a) helicase
   (b) DNA polymerase
   (c) DNA ligase
   (d) DNA primase

6. K/U What is the purpose of the Okazaki fragments? What happens to them during replication?

7. K/U Explain how replication errors are corrected.

8. MC Some scientists studying telomeres hope their research will eventually lead to a way of treating cancer. Give two examples of additional applications that could arise from a better understanding of these structures.

9. I Suppose mammalian cells are cultured in a medium containing radioactive thymine. They grow and divide many times, until eventually every chromosome contains radioactive thymine. The cells are then removed and allowed to replicate several more times in a culture medium containing normal thymine. Daughter chromosomes are tested with each successive generation to determine whether they carry the radioactive thymine.
   (a) Predict the radioactive status of the daughter chromosomes after one, two, and three rounds of division in the normal medium.
   (b) Explain how your predictions are consistent with the Watson-Crick explanation of semi-conservative DNA replication.

10. I Lacking knowledge of Franklin's X-ray analysis of the DNA molecule, Linus Pauling proposed a DNA structure in which the phosphate groups were tightly packed on the inside of the molecule, thus leaving the nitrogenous bases sticking outward. If DNA replication occurred in this structure, how do you think it would differ from what you know is the actual process?

11. MC Could you use what you have learned about the replication of DNA to develop a drug that kills bacteria but not eukaryotic cells? Explain your answer.

Genes and the Genome

EXPECTATIONS

- Relate the evolution of functional definitions of the term "gene" to advances in scientific understanding of the structure and function of DNA.
- Describe how the structure of genes varies among different organisms.
- Identify two features of the eukaryote genome that allow for greater developmental complexity than is allowed by the prokaryote genome.

How much do you have in common with a small, spiny fish like that shown in Figure 7.29? Studies of DNA from such diverse organisms as pufferfish, fruit flies, yeast, and humans demonstrate a number of shared patterns in the way hereditary information is organized at the molecular level. For instance, there are similarities in how individual **genes** — specific sequences of DNA that have the potential to be expressed to guide an organism's development — are organized. There are also similarities in the organization of that organism's **genome** — the sum of all the DNA carried in its cells.

Figure 7.29 Although pufferfish are separated from humans by millions of years of evolution, their DNA contain thousands of genes that are nearly identical to genes found in humans.

Genes

The gene is the major functional sub-unit of DNA. Each chromosome in any living cell carries a particular set of genes. The specific number, type, and arrangement of genes are unique to each species, but even organisms that are only distantly related may carry very similar genes. The pufferfish genome, for example, includes genes that are almost identical to the human genes associated with Huntington's disease and Alzheimer's disease.

Although Mendel was unaware of the existence of DNA or chromosomes, his factors of inheritance underlie the traditional definition of a gene. According to this definition, a gene is the portion of inherited information that defines one particular trait of an organism's physical characteristics. The more precise functional definition of a gene has evolved since then, keeping pace with researchers' growing understanding of the role of DNA in directing development.

In the 1940s, after DNA had been identified as the material of heredity but before its structure had been discovered, George Beadle and Edward Tatum studied patterns of development in a particular species of *Neurospora* bread mould. They identified a number of different strains of this mould that had different nutritional needs. The wild variety of *Neurospora* could be grown on an agar medium containing minimal nutrients. Various mutant strains would grow only on a medium that contained additional nutrients such as particular sugars or amino acids.

Beadle and Tatum hypothesized that while the wild variety of *Neurospora* can synthesize all the amino acids it requires from the minimal medium, different mutations disrupt the metabolic pathways by which the mould synthesizes the proteins it requires for development. With further research,

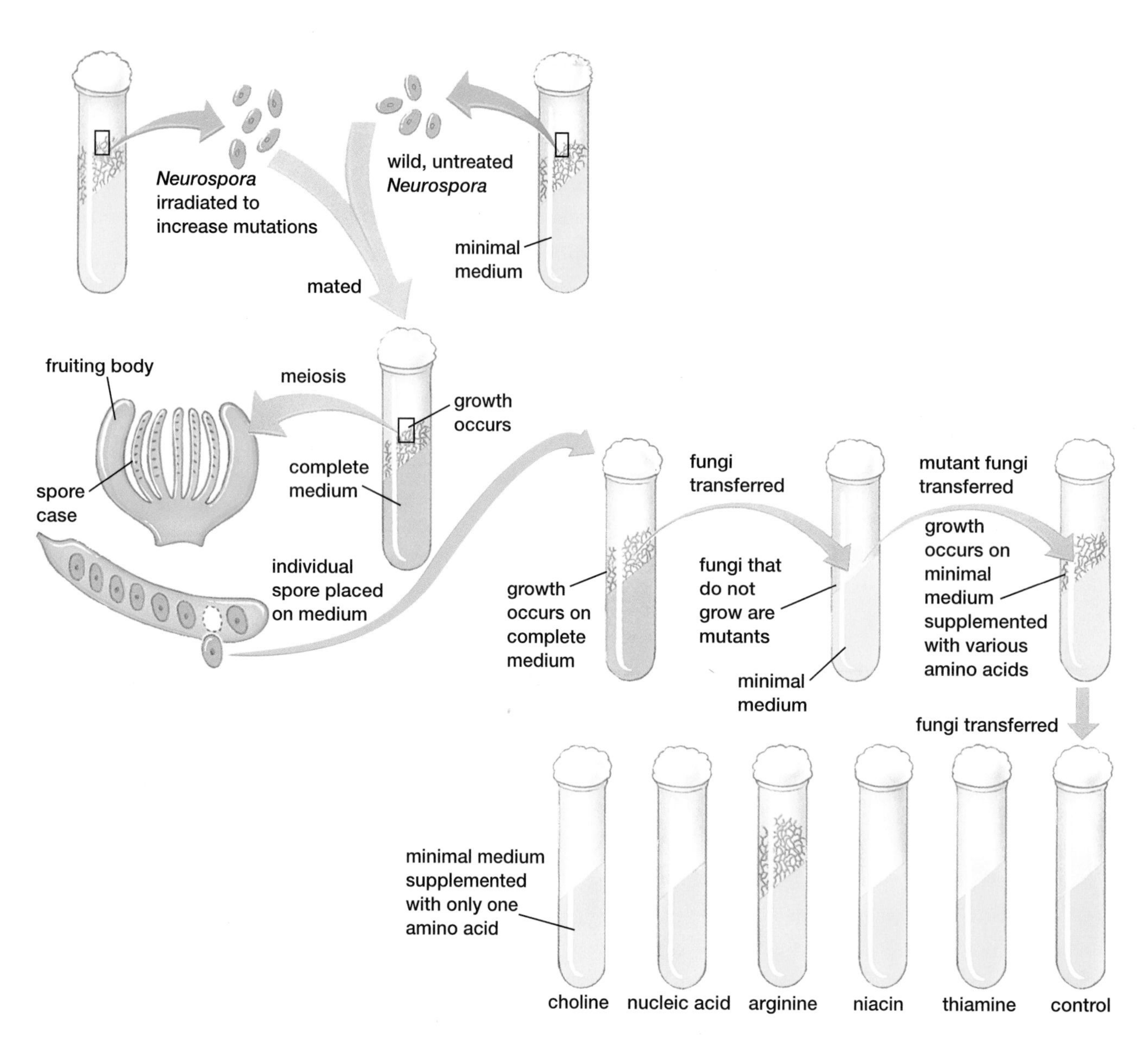

Figure 7.30 Beadle and Tatum isolated mutant strains of the mould that could grow on a complete medium but not on a minimal medium. By gradually adding one nutrient at a time to the minimal medium, they were able to isolate different mutant strains. Here, the mutant strain isolated lacks the ability to produce arginine, and so will grow only on a medium in which this nutrient has been added.

they were able (in many cases) to identify the specific stage in a metabolic pathway that was blocked by a particular mutation. This work, illustrated above in Figure 7.30, led them to conclude that each mutant variety of the mould had one defective gene that caused the mould to be deficient in one enzyme that catalyzed a particular step in a given metabolic pathway. Their hypothesis became known as the one gene-one enzyme theory of gene function.

In the years that followed, this functional definition was broadened to one gene-one protein, since some genes were found to code for proteins other than enzymes. (Examples include structural proteins such as collagen and the silk of spider webs.) This definition was later modified to one gene-one polypeptide when scientists found that the different polypeptides in a single protein complex may be coded for by entirely separate genes. An example of this is hemoglobin — different genes code for each of the two types of polypeptide sub-units that make up one hemoglobin molecule.

In short, the precise functional definition of a gene has become more complex as scientists have learned more about how genes work. The one gene-one polypeptide theory still does not account for all aspects of gene function. In eukaryotes, a single gene can code for several different polypeptide products. Other genes code for the synthesis of non-polypeptide products, such as the various types of RNA molecules that play a role in protein synthesis and other cellular processes. In addition to genes, chromosomes contain **regulatory sequences**, which are strands of DNA that help determine when various genetic processes are activated. You will examine some of these processes in more detail in Chapter 8.

BIO FACT

In the human genome, genes make up just over one percent of the total length of DNA.

Arrangement of the Genome

Genes are not spaced regularly along chromosomes. In any eukaryotic organism, the density of genes varies from one chromosome to another. In humans, for example, chromosome 4 is close to 1300 million bases long and has about 200 genes, while chromosome 19 is only 72 million bases long and has about 1450 genes. This makes chromosome 19 approximately three times richer in genes than chromosome 4. There is, in fact, no set relationship between the number of genes on a chromosome and the total length of the chromosome.

The same variation holds true in the relationship between the number of genes in any organism and the overall size of its genome. The single-celled protozoan *Amoeba dubia* has an enormous genome of over 650 billion base pairs, but fewer than 7000 genes. The human genome, in contrast, contains about three billion base pairs and an estimated 35 000 genes. A roundworm has 30 times less DNA than a human, but over half as many genes. This means that the genomes of different organisms contain varying quantities of DNA that do not serve as genes or regulatory sequences. These non-coding sequences may be interspersed with coding sequences in a variety of ways.

WEB LINK

www.mcgrawhill.ca/links/biology12
Comparative genomics is the study of the similarities and differences among the genomes of different organisms. Use the Internet to compare human DNA with that of another organism. Go to the web site above, and click on **Web Links**. What percentage of the genes in the human genome are found in the genome of your comparison organism, and vice versa? What do these numbers indicate? What are some of the practical applications of studying the DNA of the comparison organism?

Exons and Introns

In general, prokaryotes have only one copy of DNA, which in turn contains only one copy of any given gene. Structural genes and regulatory sequences make up most or all of the bacterial genome. This means that very little of the DNA is not actively used in metabolic processes over the course of the cell's life cycle. Figure 7.31 shows some of the genes found on the *E. coli* chromosome.

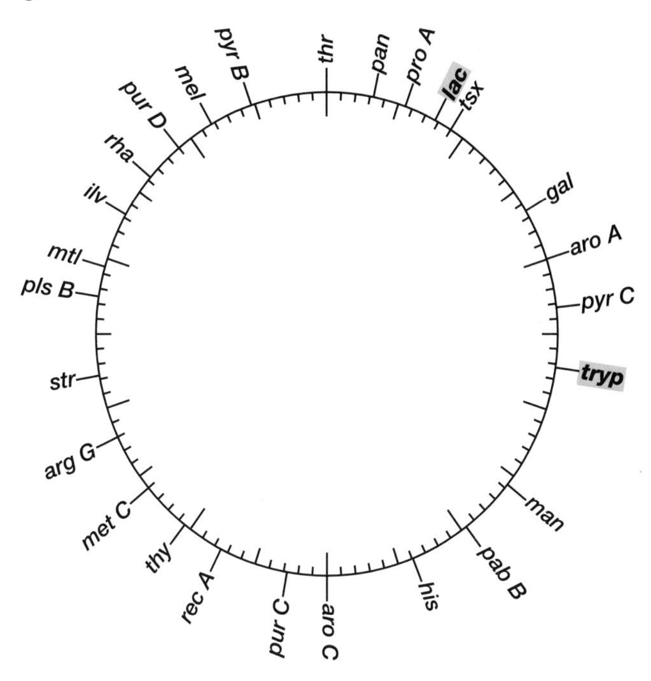

Figure 7.31 This reference map of the *E. coli* genome shows only a few of the several thousand genes on its circular DNA strand. In Chapter 8, you will learn more about how the *lac* and *tryp* genes govern the metabolism of lactose by an *E. coli* cell.

In contrast, eukaryote genes have a much more complex structure. Each gene is typically composed of one or more coding regions, known as expressed regions or **exons**. These coding regions are interspersed with a number of intervening non-coding sequences, or **introns**. As shown in Figure 7.32 on the following page, introns may make up well over half of the total length of any given gene.

Table 7.2 on the next page shows the average number of introns in relation to the average length of genes for several different types of organisms. In general, the frequency and length of introns is loosely related to the developmental complexity of the organism. Only five percent of yeast genes have introns, and it is rare for any one yeast gene to contain more than one intron. Vertebrates, on the other hand, have introns in about 95 percent of their genes.

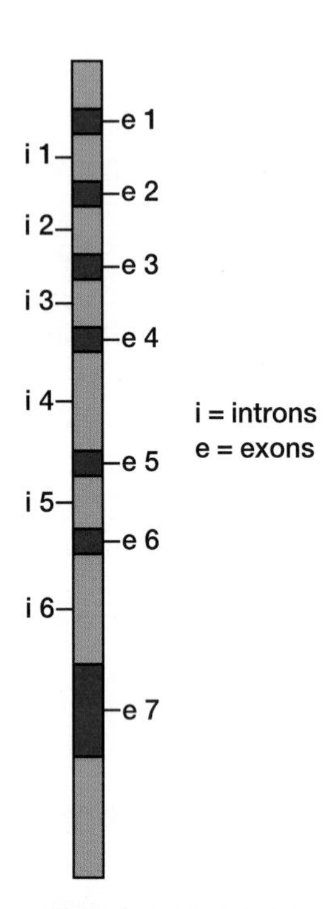

Figure 7.32 The gene for ovalbumin, the protein found in egg whites, is divided into seven exons. The total length of the exons (the expressed portions of the gene) is much smaller than the total length of the six introns. Some eukaryote genes have many more introns. The human gene for the muscle protein titin, for example, contains almost 180 introns.

Table 7.2
Number of introns compared to gene length

Organism	Average number of introns per gene	Average gene length (1000 bases)
bacteria	0	1.0
fungi	1	1.5
roundworm	3	4.0
fruit fly	3	11.3
chicken	8	14.0
mouse	7	16.5

Where similar genes are found in different species, the order of introns and exons tends to be the same from one species to another. The length of the introns, however, can be quite variable. Human introns typically contain about 100 base pairs, although some can be tens of thousands of

Canadians in Biology

Chromosomal Abnormalities in Sperm

We know that women's eggs deteriorate as women age, but what about men's sperm? Do they also deteriorate with age? If men who are diagnosed with testicular cancer receive chemotherapy, will their sperm be affected? Do sperm have a greater-than-normal proportion of chromosomal abnormalities (in either their structure or number) after being stored in sperm banks at very low temperatures? Do men who smoke have more defective sperm than men who do not? Do infertile men, who now have the option of using reproductive technologies, run a higher risk of having unhealthy offspring?

Dr. Renee Martin

Dr. Renee Martin, a professor with the department of Medical Genetics at the University of Calgary, has researched these questions. When Dr. Martin began her medical career 20 years ago, fertility studies focussed almost entirely on women even though half of the genetic material in human embryos is contributed by men. Consequently, Dr. Martin decided to study the causes of chromosomal abnormalities in humans with an emphasis on the abnormalities that could be traced to sperm. After all, if we are ever to prevent the profound problems people sometimes inherit as a result of chromosomal abnormalities, we need to determine what causes those abnormalities.

New Molecular Technologies at Work

Dr. Martin's lab, which uses new molecular technologies to study sperm chromosomes and the genes on them, was the first in the world to demonstrate that a large proportion of some men's sperm can have abnormal chromosomes. For instance, men exposed to radiotherapy and chemotherapy during cancer treatment have more chromosomal abnormalities in their sperm. Some of these chromosomal abnormalities result in children who are born with abnormal chromosomes. Elevated levels of sperm with chromosomal abnormalities are also associated with infertile men, who can now take advantage of a new reproductive technology technique called Intracytoplasmic Sperm Injection (ICSI). ICSI is used when natural fertilization cannot take place due to, for example, low sperm count or poor motility (locomotion). During ICSI, sperm is injected directly into an egg in a test tube.

Among the new molecular techniques at the disposal of Dr. Martin's lab are those involving Fluorescence in situ

base pairs long. Figure 7.33 shows the pattern of introns and exons in the same gene found in three different species.

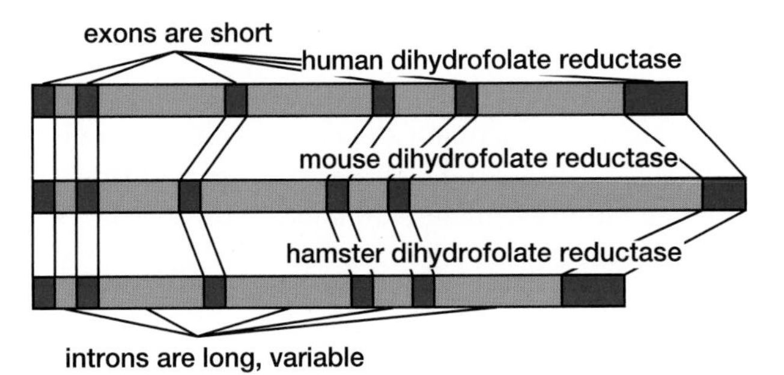

Figure 7.33 The same gene that codes for the protein dihydrofolate reductase (an enzyme involved in the metabolism of folic acid) shows a different structure in three different vertebrates. The exons are of the same length and appear in the same order, but the introns vary in length. In the human gene, the intron length varies from about 360 base pairs to nearly 12 000 base pairs.

When introns were first discovered in the 1970s, they were often referred to as junk DNA because they had no known function. Introns do, however, support a variety of developmental and regulatory functions. The existence of introns allows some genes to code for more than one polypeptide product by using different combinations of introns and exons. This means that even though humans have only about twice as many genes as a roundworm, human cells can produce far more than twice as many different proteins. Introns can also be the location of various regulatory sequences that initiate or silence gene activity. You will learn more about these processes in Chapter 8.

Multi-gene Families

A eukaryotic cell typically has many copies of certain genes. Genes may be found in **multi-gene families**, which contain from a few hundred up to

Hybridization (FISH) and Polymerization Chain Reactions (PCR). FISH uses DNA probes tagged with fluorescent molecules to detect both simple and complex chromosomal re-arrangements (such as aberrations and abnormalities) over small areas of a chromosome. The fluorescent molecules are activated by light, and they emit light of different wavelengths (and therefore different colours). Using FISH, the structure and behaviour of different chromosomes can be studied. PCR, a technique created by Kary B. Mullis in 1984, creates multiple copies of a specific sequence of nucleotides within a segment of DNA. You will learn more about this technique in Chapter 9.

A Career in Genetics

A career in genetics was not one of Dr. Martin's original goals. Although she enjoyed science courses in high school, her focus was on physical education and dance. At one point, however, she had to spend a week in hospital. When she returned to school, her Grade 11 science teacher informed her that she had missed the opportunity to learn about DNA, a very important topic in the course. After the teacher explained DNA's role as a blueprint for genetic inheritance and information exchange within the body, Dr. Martin found herself intrigued and fascinated by, to use her own words, the "beauty and simplicity of it." After one year pursuing a bachelor's degree in Physical Education at the University of Toronto, she switched fields and eventually received her bachelor's degree in Science with honours (Zoology) from the University of British Columbia. Continuing in her studies, she then earned her Ph.D. in Medical Genetics from UBC by studying chromosomal abnormalities that result in too few or too many chromosomes. Examples of conditions caused by such abnormalities are Down syndrome and Turner syndrome, which are known as trisomy 21 and monosomy, respectively.

During her career, Dr. Martin has published many research papers that investigate whether age, smoking, prolonged exposure to pesticides, or cryopreservation (low-temperature storage or cryogenics) produce chromosomal abnormalities in sperm. She has also studied the ethical issues involved in assisted reproductive technologies; and the chromosomal abnormalities in sperm before, during, and after cancer treatment involving chemotherapy and radiotherapy. She has been invited to address many international conferences and symposiums on fertility and genetics in Europe, the United States, Scandinavia, the Middle East, Australia, India, and Canada. She has also served as president of the Canadian College of Medical Geneticists and the Canadian Society of Andrology and Fertility.

What Dr. Martin enjoys most about her job is the flexibility it gives her to plan her day as she chooses, which occasionally allows her to schedule an exercise session or a volunteer activity at her children's school. She also loves the opportunity it gives her to travel and meet with other researchers from around the world. About the only unwelcome side of her work is the continuous need to apply for research grants. Even in this activity she finds a silver lining, as it encourages her to focus clearly on her research goals.

hundreds of thousands of copies of the same or very similar genes. These families may be clustered together on the same chromosome, or they may be distributed among a number of different chromosomes. In some cases, multi-gene families code for products that are in very high demand in the cell. (Examples include the genes that code for histone proteins and RNA products.) In other cases, however, the genome contains only a single copy of a gene even though it codes for a product in high demand. The existence of multi-gene families is partly the result of the action of **transposons** or jumping genes, which are sequences of DNA that are inserted and copied randomly throughout the cell genome. You will examine the effect of transposons more closely in Chapter 9.

Pseudogenes

In some multi-gene families, some copies of the gene may have mutated to the point that they no longer work as genes. These sequences are known as **pseudogenes**, since they are nearly identical to functional genes but are never expressed during the life cycle of the cell. It is not currently known whether such pseudogenes have a metabolic function.

Repetitive Sequences

Along with introns and multi-gene families, eukaryotic cells also contain long strands of **repetitive sequence DNA**. These regions contain short sequences of nucleotides repeated thousands or millions of times. About 30 percent of mouse DNA consists of repetitive sequences of about 10 nucleotides. While such structures do not have a coding function, they can play an important role in processes such as DNA replication. Telomeres are an example of such a repetitive sequence. A similar repetitive sequence occurs on chromosomes where the centromere forms, suggesting that this sequence helps in the attachment of spindles during cell division.

Genome Size Versus Organism Complexity

It was once thought that the human genome contained approximately 100 000 genes. This figure was based partly on the estimated number of proteins produced by human cells, as well as on estimates of the ratio of genes to the length of the overall genome. The Human Genome Project, however, found that the actual figure is much smaller — about 30 000 to 35 000. This finding indicates that the complexity of mammals is not only the result of information stored in the genes themselves. In fact, a wide range of different mechanisms affect when and how genes are translated into their protein products. The structures and processes involved in moving from the molecular sequence of DNA to functional proteins in an organism are explored in more detail in Chapter 8.

SECTION REVIEW

1. K/U Explain the one gene-one enzyme theory proposed by Beadle and Tatum. Why was this theory later modified first to one gene-one protein and then to one gene-one polypeptide?
2. K/U Describe the arrangement of genes in the genome of organisms. Account for the large amount of DNA that does not code for genetic or regulatory sequences.
3. K/U Explain the differences in the number of introns and exons in organisms of differing complexity. What is the significance of introns in protein production?
4. K/U What is a pseudogene?
5. C In small groups, debate whether eukaryotic cells enjoy selective advantages because they contain long stretches of repetitive DNA sequences. If so, what might these advantages be?
6. MC How might humans be affected if our genome really did contain 100 000 genes? What if it contained only 10 000 genes? Evaluate some of the possibilities, and give reasons for your conclusions.
7. I Do research to discover some of the nucleotide sequences that code for specific characteristics in humans or other eukaryotic organisms. Using everyday materials, build one or more standing models of the DNA double helix, each incorporating one of these sequences on one strand and its corresponding sequence on the other strand. Do the corresponding sequences also encode useful information? Explain your answer in a summary paper.

UNIT PROJECT PREP

Find out what, if any, gene is associated with the cancer you are studying. Try to find this gene on a map of the human genome.

Summary of Expectations

Briefly explain each of the following points.

- For many years, scientists believed that proteins rather than DNA were the material of heredity. (7.1)
- The bonding of nucleotide base pairs proposed by Watson and Crick accounts both for Chargaff's rule and for the observed structure of the DNA molecule. (7.1)
- The two strands in a DNA molecule are complementary and antiparallel. (7.2)
- Any fragment of a DNA strand has a constant orientation or directionality. (7.2)
- DNA is arranged in different ways in prokaryotes and eukaryotes, but these different arrangements serve some of the same needs. (7.2)
- The base pairing properties of DNA provide a way to replicate the molecule accurately. (7.2)
- The accuracy of DNA replication is much higher than that which could be attained through base pairing mechanisms alone. (7.3)
- DNA can only be synthesized in one direction, but replication of DNA strands proceeds in two directions at once. (7.3)
- Organisms that have many similar genes may physically appear to be very different. (7.4)
- The mammalian genome contains a great deal of DNA other than that contained in genes. (7.4)

Language of Biology

Write a sentence using each of the following words or terms. Use any six terms in a concept map to show your understanding of how they are related.

- ribose nucleic acid (ribonucleic acid or RNA)
- deoxyribose nucleic acid (deoxyribonucleic acid or DNA)
- chromosome
- nucleotide
- nitrogenous
- adenine
- guanine
- cytosine
- thymine
- uracil
- Chargaff's rule
- directionality
- purine
- pyrimidine
- complementary
- antiparallel
- nucleoid
- supercoiled
- plasmid
- histone
- chromatin
- nucleosome
- conservative theory
- semi-conservative theory
- dispersive theory
- replication origin
- DNA polymerase
- replication forks
- helicases
- primer
- Okazaki fragments
- leading strand
- DNA ligase
- lagging strand
- primase
- telomere
- replication machine
- gene
- genome
- regulatory sequence
- exon
- intron
- multi-gene family
- transposon
- pseudogene
- repetitive sequence DNA

UNDERSTANDING CONCEPTS

1. Identify and describe three experiments that helped pave the way to the discovery of DNA as the hereditary material.
2. What are the components of a single nucleotide?
3. Using symbols to represent the four nitrogenous bases, illustrate the molecular structure of both strands of a portion of a DNA molecule. Use each base at least twice.
4. What is the base sequence of the DNA strand that is complementary to a strand with the sequence ACGTTGCTA?
5. What contribution do hydrophilic and hydrophobic reactions make to the stability of a DNA molecule?
6. The following terms describe different arrangements of DNA. Organize them by distinguishing which are found in prokaryotic versus eukaryotic cells, and by placing them in order from least compact to most compact.
   - nucleosome
   - chromatin
   - supercoil
   - plasmid
   - 30 nm array
7. What role does histone play in arranging DNA within eukaryotic cells? What properties of this protein contribute to this role?
8. Do all your body tissues contain the same amount of DNA? Explain.
9. The replication of DNA is said to be semi-conservative. What does this mean?
10. What is the function of primase in DNA replication?
11. Briefly compare the roles played by the following:
    (a) leading strand and lagging strand

(b) pyrimidine and purine
(c) DNA polymerase and DNA ligase

12. After one round of replication in a bacterial colony, Meselson and Stahl's analysis showed one band of hybrid DNA. What is the most useful interpretation they could draw from this result?
   (a) that DNA replication can be neither conservative nor dispersive
   (b) that the results are consistent with semi-conservative replication
   (c) that the results indicate either conservative or semi-conservative replication
   (d) that the results are not consistent with conservative replication

13. If Meselson and Stahl continued their study into a fourth generation of replication, what result would they see?

14. Refer to Appendix 4 to review the cell life cycle. Does DNA replication occur during mitosis or meiosis, or both?

15. Arrange the following events in the order in which they occur during the replication of a single portion of a DNA molecule.
   (a) primase synthesizes new RNA strand
   (b) helicases cleave DNA
   (c) 30 nm array loosens
   (d) ligase catalyzes formation of a phosphate bond
   (e) recognition of replication origin
   (f) DNA polymerase adds nucleotides to a fragment of DNA

16. Fill in the following table to show the molecules and enzymes involved in DNA replication.

Molecule or enzyme	Function	Involved in leading strand or lagging strand synthesis, or both
primer		
DNA ligase		
DNA polymerase		
Okazaki fragments		
helicase		

17. A researcher combines in a solution all the molecules required for DNA replication except DNA ligase. What result would you expect to see after a complete round of replication?

18. How does the base-pairing property of DNA contribute to the proofreading function of DNA polymerase?

19. Compare the process of replication in prokaryotic and eukaryotic cells. How do differences in the structure of DNA contribute to differences in replication?

20. In what ways is the one gene-one enzyme definition of a gene inaccurate?

21. During the process of cell division in a human cell, the chromosomes are visible under a microscope as short, thick strands. Describe how a single molecule of DNA is organized within one of these chromosomes.

22. A gene coding for the same enzyme is found in both earthworms and rats. Which gene is likely to have a greater total length? Explain.

INQUIRY

23. Provide evidence to support the conclusion that DNA is, in fact, the genetic material.

24. As part of your research you isolate the DNA from a particular strain of virus. Your analysis indicates the following base composition:

base	A	C	G	T
concentration (%)	36	24	18	22

   (a) What can you conclude about the DNA of this virus?
   (b) What results would you expect to see if you conducted the same analysis immediately after the viral DNA had replicated?

25. Chargaff concluded that in any sample of DNA, the amount of adenine would be equal to the amount of thymine, and the amount of guanine would be equal to the amount of cytosine. Watson and Crick concluded that adenine paired with thymine, and cytosine paired with guanine. Explain how you think the structure and function of DNA would change if cytosine paired with adenine instead?

26. A particular species of bacteria can normally exist on a minimal medium. In your research, however, you find a mutant variety that will only grow if the amino acid valine is added to the medium. You know that the metabolic

pathway for valine in this bacteria includes the following steps: precursor A → precursor B → valine.

(a) What might you conclude about the mutant bacteria?

(b) How would you design an experiment to test your hypothesis?

(c) Could a similar experiment be used if you were studying a plant cell rather than a bacterium? Explain.

27. A biochemist develops a chemical that interferes with histone-histone bonding. What effect would you expect this chemical to have on the following:

(a) the organization of DNA within a prokaryotic cell?

(b) the organization of DNA within a eukaryotic cell?

(c) the process of replication of eukaryotic DNA?

28. Examine the photograph at right, taken by an electron microscope.

(a) What events and processes are taking place in this image?

(b) Make an illustration showing what you would expect to see if a second photograph were taken a short time after this one.

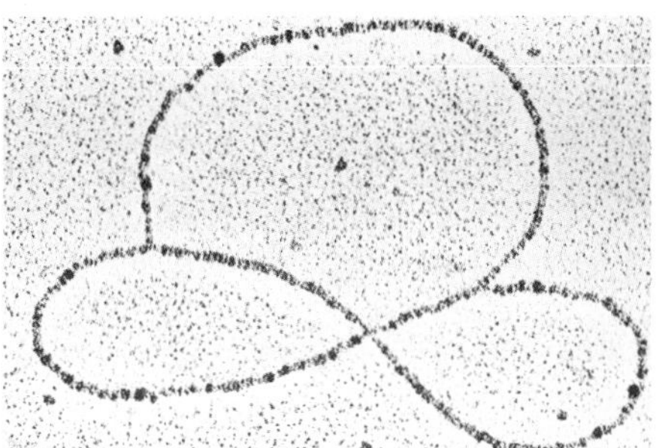

29. During cell division in a eukaryotic cell, replication can be simultaneously initiated in thousands of places along a single linear chromosome. Propose a mechanism that the cell could use to ensure that each portion of the chromosome is only replicated once during any given round of cell division.

COMMUNICATING

30. Draw a diagram that illustrates the hydrogen bonds that occur between two different base pairs of DNA. Why is base pairing so specific?

31. Create a table that contrasts the structure and function of DNA and RNA.

32. Develop a flowchart that illustrates how the discoveries of the following researchers contributed to an understanding of the role of DNA in heredity: Hershey and Chase; Chargaff; Franklin and Wilkins; Watson and Crick; Meselson and Stahl.

33. Imagine a conversation between Phoebus Levene and Erwin Chargaff about the role played by nucleic acids in heredity. What evidence might each researcher rely on to support his point of view?

34. DNA is sometimes said to be like a language. In a short essay, explain the ways in which this comparison is valid.

35. "DNA replication involves a careful balancing act." Prepare a poster demonstration that could be used to explain what this statement means.

36. What examples can you provide from everyday life that could be used to illustrate the concepts of conservative, semi-conservative, and dispersive replication?

MAKING CONNECTIONS

37. In his work on patterns of inheritance, Gregor Mendel coined the term "factors of inheritance." How do Mendel's factors relate to today's definition of a gene?

38. Only a small portion of the mammalian genome is made up of genes. The rest was originally referred to as "junk DNA." How do popular terms such as this help or hinder society's understanding of the need for scientific procedures and ongoing research? What are some of the potential advantages and disadvantages to an organism of having large quantities of non-gene DNA?

39. A certain planarian has a genome 6000 times larger than that of a particular yeast cell.

(a) What conclusions, if any, could you draw about the relative complexities of the two organisms?

(b) What practical applications could there be from a study that compared the genomes of the two organisms?

40. A researcher develops a form of DNA polymerase that adds nucleotides only to the 5′ hydroxyl group of an existing DNA strand. Is this a useful discovery? Explain.

CHAPTER 8

Protein Synthesis and Gene Expression

Reflecting Questions

- What are the main features of the genetic code?
- How does the information stored in an organism's DNA guide its development?
- How do differences in the protein synthesis pathway contribute to the differences in how prokaryotic and eukaryotic cells develop?
- How can changes in an organism's environment affect the expression of its genetic information?

Prerequisite Concepts and Skills

Before you begin this chapter, review the following concepts and skills:

- describing the molecular structure of nucleic acids (Chapter 7, section 7.2),
- identifying the main steps and enzymes involved in the replication of DNA (Chapter 7, section 7.3), and
- explaining the differences in arrangements of genetic material in prokaryotes and eukaryotes (Chapter 7, section 7.4).

The eastern massasauga rattlesnake (*Sistrurus catenatus catenatus*) is the only poisonous snake still found in Ontario. This timid reptile lives in marshy habitats bordering Georgian Bay and the Niagara Escarpment. Of the following features, which do you think are the result of this snake's genetic information: its unique markings, the chemical composition of its venom or its mating behaviour? The answer is all of the above. The snake's DNA contains information that determines the structure and function of all its different cells.

Occasionally snakes are born with partial limbs. This shows that the information that codes for legs is present in the snake genome, even though this information is not normally expressed. In other words, a snake has genes that carry the instructions to produce legs even though it has no legs on its body.

In the same way, the genetic information contained in the apple tree shown here determines which proteins are produced by each of the cells of its different tissues, from roots to petals. However, if the root cells contain the same genetic information as the petal cells, why are these cells so different? The answer lies in the process of gene expression. Every cell in any given plant or animal contains an identical set of genetic information, but in any one cell only a fraction of this information is ever used.

In this chapter, you will learn how the information stored in the molecular structure of DNA is expressed in the living processes of the cell. You will see how nucleic acids carry out the synthesis of the enzymes and other proteins that give a cell its particular properties. You will also examine the close partnership between DNA and proteins. This relationship makes it possible for a cell to respond to its environment by changing its patterns of gene expression.

One set of genetic instructions produces all the various cells found in the roots and petals of this apple tree. Another set produces the very different cells found in the body of the eastern massasauga rattlesnake.

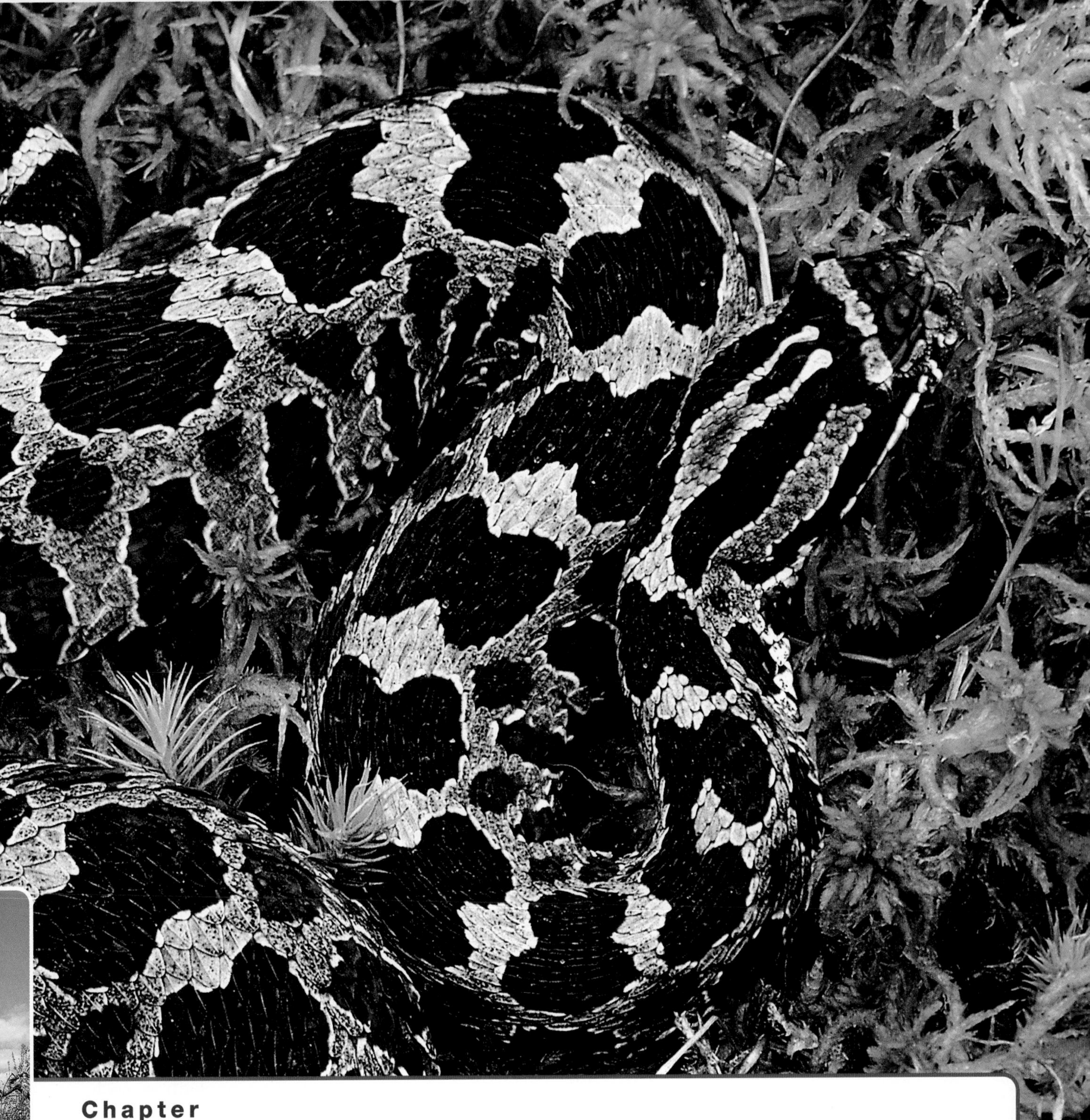

Chapter Contents

8.1 The Central Dogma and the Genetic Code

EXPECTATIONS

- **Explain the general roles of DNA and RNA in protein synthesis.**
- **Describe the process of discovery that helped researchers decipher the genetic code, and outline their contributions.**
- **Discuss the significance of the main features and universality of the genetic code.**

When you look at a set of building plans (such as those in Figure 8.1), it may seem obvious that they provide a set of instructions for building a house. Consider, however, the processes by which the two-dimensional information contained on those plans ultimately achieves expression in the form of a three-dimensional structure made of wood, bricks, and concrete. In the case of houses, we know that these processes include concrete work, carpentry, plumbing, electrical wiring, and roofing. Genetics researchers in the 1950s knew that genes contained the instructions for building proteins, and that these genes were made up of stretches of double-helical DNA. The challenge was to determine the processes by which the information contained in DNA (an essentially linear arrangement of nucleotides) could be transformed into three-dimensional proteins (structures made out of polypeptides).

Figure 8.1 In much the same way that building plans contain the instructions to build a house, genes contain the instructions needed to build proteins.

By a fortunate coincidence, British scientist Frederick Sanger established that proteins are made up of an identifiable sequence of amino acids in 1953 — the same year that Watson and Crick published their model for the structure of DNA. Sanger, who was working at the same university as Watson and Crick, showed that each molecule of the protein insulin contains precisely the same linear arrangement of amino acids. He went on to demonstrate that the linear arrangement of amino acids can lead automatically to the characteristic three-dimensional, folded structure of the functioning protein. This structure results from chemical interactions among the amino acids that cause the polypeptide chain to fold in specific ways. Once scientists understood that a given set of amino acids arranged in a specific order could produce a particular protein, they began to consider a new possibility. Perhaps there was a connection between the sequence of nucleotides along a DNA molecule and the sequence of amino acids in a protein.

The Triplet Hypothesis

Crick took up the challenge of cracking the genetic code. He knew that proteins are made of 20 different amino acids but that DNA contains only four different nucleotides. Crick thus hypothesized that the code must be made up of "words," or **codons**, consisting of a minimum of three nucleotides each. He reasoned that if one nucleotide corresponded to one amino acid, the code could account for only four amino acids. If a combination of any two nucleotides was needed to code for a single amino acid, the code could generate a maximum of $4^2 = 16$ combinations. However, if the code relied on a combination of three nucleotides to specify one amino acid, it could generate a maximum of $4^3 = 64$ different combinations. These combinations would be more than enough to account for the 20 amino acids. This theory, which suggests that genetic code codons are made up of nucleotide triplets, is known as the **triplet hypothesis**.

Crick tested his hypothesis by inserting different numbers of nucleotides into bacterial DNA and then observing the effects on the resulting bacterial colonies. If his triplet hypothesis was correct, the insertion of a new nucleotide triplet would produce only a minor disruption in the coding sequence. In contrast, the insertion of a single nucleotide or a pair of nucleotides would cause a major change in the triplet sequence following the insertion point, as shown in Figure 8.2. The results of Crick's experiments, which were reported in 1961, supported the triplet hypothesis.

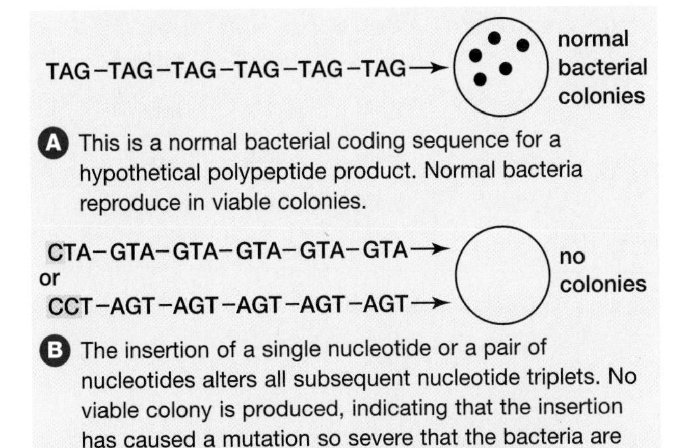

Figure 8.2 A schematic of Crick's experiments to test the triplet hypothesis

The Transfer of Genetic Information

Before they discovered the structure of DNA, Watson and Crick had started to devise a theory that genetic information is somehow transmitted from DNA to RNA and then to proteins. Part of the supporting evidence for this sequence of events lay in experiments that demonstrated that DNA never leaves the nucleus of a eukaryotic cell. These experiments also showed that most of the structures and processes involved in protein synthesis are found only in the cytoplasm of the cell. RNA, however, is found in both the nucleus and the cytoplasm.

Not long after Watson and Crick had established their model for the structure of DNA and its process of replication, Francis Crick coined the phrase "central dogma" to describe the two-step process by which he believed genes were expressed and proteins built. According to this dogma, a term he used to suggest a powerful theory that had little experimental support, a strand of DNA first serves as the template for the construction of a complementary strand of RNA. This strand of RNA then moves from the nucleus to the cytoplasm. Here, it guides the synthesis of a polypeptide chain. Over the next few years, experimental data confirmed that genetic information is indeed passed from DNA to RNA to protein.

Figure 8.3 summarizes the two main steps involved in the process of gene expression. The synthesis of an RNA molecule from a strand of DNA is called **transcription**. To transcribe means to make a copy of or otherwise transfer information from one medium to another in a way that preserves the original language of the information. During genetic transcription, the nucleotide sequence (the language) of the DNA strand is copied into the nucleotide sequence of the synthesized RNA molecule. In contrast, during the second step, called **translation**, the language preserved in the RNA must be converted to the amino acid sequence of a polypeptide in order for a protein to be synthesized. You will learn more about each of these steps later in this chapter.

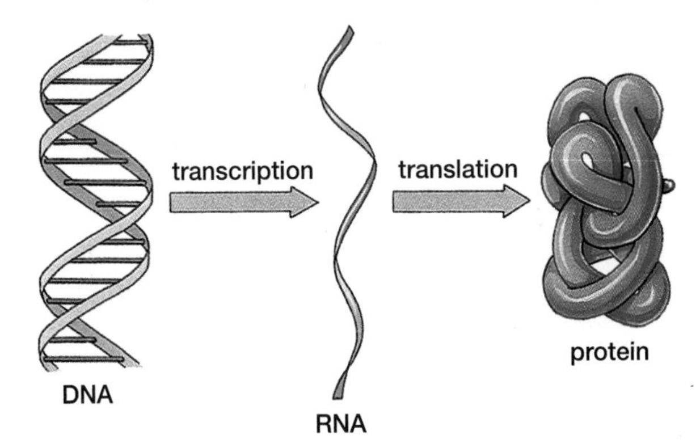

Figure 8.3 Crick's "central dogma" proposes a two-step process of gene expression. In this process, genetic information is first transcribed from DNA to RNA, and then translated from RNA to protein.

Why does DNA not code directly for a polypeptide product? At first glance, the two-step process seems like an unnecessary waste of cellular energy. However, the additional energy involved in producing an RNA copy of the DNA information is an investment that earns the cell a number of advantages. Consider, for example, the enormous amount of hereditary information that is contained on even a single chromosome. If the entire chromosome had to be transported out of the nucleus to take part in protein synthesis, the cell would expend a great deal of energy. The repeated transport of DNA back and forth between the nucleus and the cytoplasm would also place the hereditary material at risk of damage. With the two-step process, the cell copies a small portion of the DNA molecule and then carries this copy out of the nucleus while the DNA remains protected inside.

A single stretch of DNA can also be copied many times over to produce a large number of RNA molecules. Multiple copies help to speed up the process of protein synthesis. Finally, the fact that more than one step is involved in the process of protein synthesis means that there are several opportunities along the way to regulate gene expression. As you will see later in this chapter, different mechanisms operating at different stages of the process allow for very sophisticated regulatory functions.

The Genetic Code and mRNA Codons

The next step for researchers was to determine exactly which codons correlated with each amino acid. Late in 1961, Marshall Nirenberg and Heinrich Matthaei reported the first success in breaking the genetic code. First they synthesized an artificial RNA molecule made up only of uracil nucleotides. They then cultured this molecule (which they called poly-U) in a medium that contained the 20 amino acids and the various substances required to catalyze the formation of a protein. The polypeptide that was produced in this medium was made up only of the amino acid phenylalanine. As a result, Nirenberg and Matthaei concluded that the codon UUU must code for phenylalanine. Through a series of similar experiments, researchers worked out the entire genetic code by 1965. Table 8.1 shows the full set of RNA codons and their corresponding amino acids. Since information is passed from DNA to RNA, codons are always written in the form of the RNA transcript from the original DNA molecule.

By convention, the genetic code is always presented in terms of the RNA codon rather than the nucleotide sequence of the original DNA strand. The RNA codons are written in the 5′ to 3′ direction. To read the table, find the first letter of the RNA codon in the column titled "First letter." Then read across the rows in the column titled "Second letter" to find the second letter of the codon. This will take you to a set of four possible amino acids. Finally, read down the column titled "Third letter" to find the last letter of the codon. This will indicate the amino acid that corresponds to that codon. For example, the RNA codon GAG codes for glutamate. What amino acid corresponds to the codon CAU?

Table 8.1
The genetic code

First letter	Second letter: U	C	A	G	Third letter
U	phenylalanine	serine	tyrosine	cysteine	U
	phenylalanine	serine	tyrosine	cysteine	C
	leucine	serine	**stop**	**stop**	A
	leucine	serine	**stop**	tryptophan	G
C	leucine	proline	histidine	arginine	U
	leucine	proline	histidine	arginine	C
	leucine	proline	glutamine	arginine	A
	leucine	proline	glutamine	arginine	G
A	isoleucine	threonine	asparagine	serine	U
	isoleucine	threonine	asparagine	serine	C
	isoleucine	threonine	lysine	arginine	A
	start/ methionine	threonine	lysine	arginine	G
G	valine	alanine	aspartate	glycine	U
	valine	alanine	aspartate	glycine	C
	valine	alanine	glutamate	glycine	A
	valine	alanine	glutamate	glycine	G

Characteristics of the Code

The genetic code has a number of important characteristics. Three key features are its continuity, its redundancy, and its universality.

Continuity The genetic code reads as a long series of three-letter codons that have no spaces or punctuation and never overlap. This means that knowing exactly where to start transcription and translation is essential. Each sequence of nucleotides has a correct **reading frame**, or grouping of codons. Experiments such as Crick's show that if the reading frame on an RNA molecule is shifted by the insertion or deletion of an additional nucleotide, there is no mechanism in the cell that can reset the translation process back to the correct frame.

Redundancy There are a total of 64 possible codons but only 20 amino acids. As you can see from Table 8.1, only three RNA codons do not code for any amino acid. These three codons make protein synthesis stop. This means that each amino acid is associated with, on average, about three possible codons. This pattern offers a significant biological advantage. If there were a one-to-one correlation between codons and amino acids, 44 codons would code for nothing and therefore be read as "stop" signals to end protein synthesis. This would mean that a random mutation would be more than twice as likely to terminate protein synthesis than to code for an altered protein. Termination of protein synthesis too early is likely to result in a more severe mutation than simply building in a wrong amino acid. So, having a limited number of "stop" signals offers some protection against serious mutations.

Furthermore, the redundancy is not random — it follows a particular pattern. For example, Table 8.1 shows that four different codons code for the amino acid proline, and they all have the nucleotide sequence CC_. Similarly, of the six codons for arginine, four have the sequence CG_. The third position in a codon is often referred to as the "wobble" position, since in many cases it can accommodate a number of different nucleotides without changing the resulting amino acid. The wobble feature serves as an additional guard against harmful mutations. It also contributes to the efficiency of protein synthesis, as you will see later in this chapter. Redundancy and continuity are compared in Figure 8.4.

It is important not to confuse redundancy with ambiguity. The code is redundant in that several different codons can code for the same amino acid, but it is not ambiguous. No codon ever has more than one amino acid counterpart. While you cannot always read backwards from an amino acid to determine the precise nucleotide sequence of the codon, you can always read from the codon to its one correct amino acid.

Universality The genetic code shown in Table 8.1 is the same in almost all living organisms, from bacteria to mammals. That is, the same RNA codons correspond to the same amino acids in almost all organisms that have genetic material. The only known exceptions are found in a few types of unicellular eukaryotes and in the mitochondria and chloroplasts in eukaryotic cells. The universality of the genetic code provides evidence that all these organisms share a common ancestor, and that the code was established very near the outset of life on Earth (see Figure 8.5).

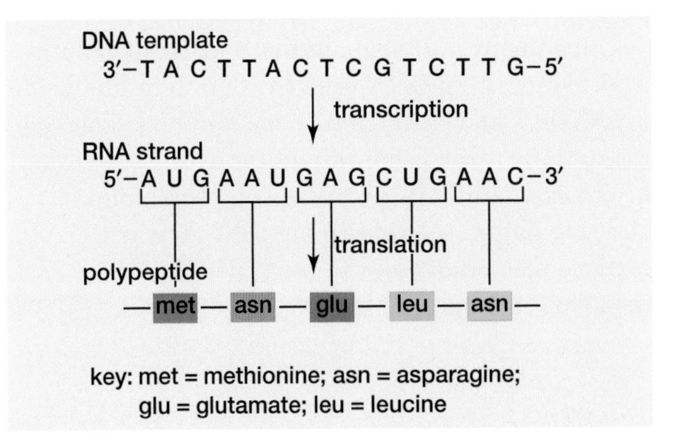

Figure 8.4 Two characteristics of the genetic code are continuity and redundancy. Here, a portion of a DNA molecule serves as a template for the synthesis of a strand of RNA. The genetic code is made up of three-letter codons along the RNA strand. Each codon correlates with one amino acid. Note that the codons are read as an unbroken series of non-overlapping words (continuity). Note also that two different codons can code for the same amino acid (redundancy).

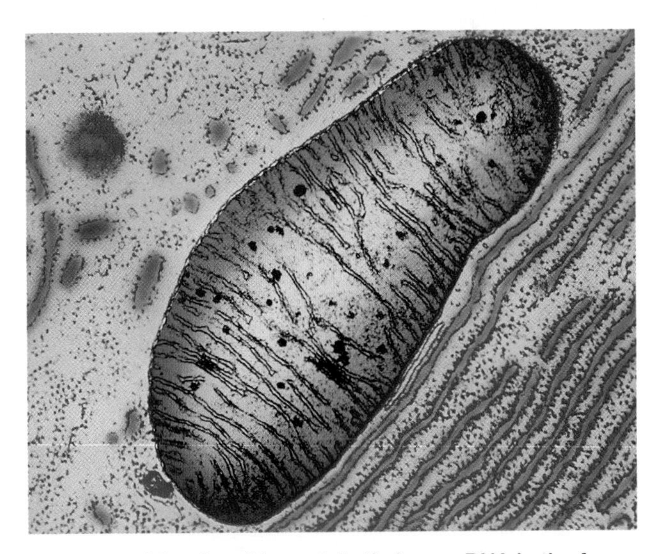

Figure 8.5 Mitochondria contain their own DNA in the form of a closed, circular molecule like the one found in bacteria. This DNA has a slightly different genetic code than the DNA found in most living cells, including the cells in which the mitochondria are found. This distinct code suggests that mitochondria might once have been separate organisms that became incorporated into eukaryotic cells early in the evolution of life forms. The fact that mitochondria have other similarities with bacteria lends support to this idea.

In addition to providing some hints about the origins of life on Earth, the universality of the genetic code also has important implications for society today. Since the same code is used by all cells, a gene that is taken from one kind of organism and inserted into another can still express the same polypeptide. For example, a corn plant can express genetic material from a bacterium. Later in this unit, you will see how this principle is applied in genetic technologies.

Once the overall steps in the transfer of genetic information from DNA to RNA to protein had been established and the genetic code had been cracked, researchers turned their attention to the question of precisely how the processes of transcription and translation work in living cells. You will explore these processes in more detail in the next two sections.

WEB LINK

www.mcgrawhill.ca/links/biology12

Between 1950 and 1965, genetics researchers went from puzzling over the molecular structure and function of DNA to establishing the complete genetic code and the first models of protein synthesis. Textbooks usually present the process of discovery as a series of separate steps, but in reality scientific research involves ongoing conversations among research teams at different institutions and in different countries. Use the Internet to identify some of the personal and institutional connections among the different teams involved in genetic research between 1950 and 1965. Go to the web site above, and click on **Web Links**.

SECTION REVIEW

1. **K/U** Explain how Sanger's work on the structure of proteins contributed to an understanding of gene expression.
2. **K/U** What is the triplet hypothesis? Draw a stretch of DNA that illustrates this hypothesis.
3. **C** Imagine you are Francis Crick, and you have just coined the term "central dogma" to describe your theory of gene expression. Some scientists do not agree with this theory. What evidence and arguments can you present to support your claim?
4. **K/U** Explain how the terms "transcription" and "translation" relate to the central dogma.
5. **K/U** Use Table 8.1 on page 254 to answer the following questions.
   - **(a)** What amino acids are coded for by each of the following codons?
     - **(i)** UUC
     - **(ii)** ACU
     - **(iii)** GCG
     - **(iv)** UAA
   - **(b)** What codons could code for the amino acid serine? for the amino acid aspartate?
6. **K/U** Write all the possible codon sequences that code for the polypeptide serine-methionine-glutamine.
7. **C** Use analogies from another field (such as music or sports) to describe the following features of the genetic code: a) redundancy; b) universality; c) continuity.
8. **K/U** What are the biological advantages to a eukaryotic cell of separating the processes of transcription and translation?
9. **K/U** What characteristics of the genetic code help protect the cell against the effect of mutations? Do any characteristics increase the potential damage from mutations? Explain.
10. **MC** The fact that the genetic code of mitochondria differs from that of most living cells has been used as evidence that mitochondria were once independent organisms. Working with a partner or in a small group, look for articles from recent newspapers or science magazines to find more examples of how information about the genetic code can be applied to research on other topics. Write a brief report summarizing your findings.
11. **I** A team of researchers has created a series of polypeptide chains for use in an experiment. Most of these polypeptides are made up of long chains of lysine with a small amount of arginine and glutamate.
    - **(a)** What process might the researchers have used to prepare these polypeptides?
    - **(b)** What would they have put into the reaction medium to obtain these results?
    - **(c)** What other amino acids might be found in trace amounts in the polypeptides produced through this process?

UNIT PROJECT PREP

As you learn about protein synthesis and gene expression in this chapter, consider how alterations in the transfer of genetic information might contribute to cancer. Specifically, how might they contribute to the form of cancer you have chosen to investigate?

8.2 From DNA to RNA: Transcription

EXPECTATIONS

- **Explain the steps involved in transcription.**
- **Explain the functions of the major enzymes and nucleic acids involved in the transfer of information from DNA to RNA.**
- **Illustrate the genetic code by examining and analyzing a segment of DNA.**

Imagine you have nearly finished writing an illustrated article about some science projects. You want to submit the finished piece to a magazine for publication. But before doing so, you want to get comments on some of the pages from a friend who lives across town. You could send her the full text and all the art, but then you would not be able to work on them while she records her comments. You also know there is a risk that the article and illustrations could be damaged or lost in transit. Consequently, you decide to photocopy just the text and art on which you want comments and send this copy to your friend. This process involves four key steps: (1) you have to decide where to begin the copying process; (2) you must copy the right number of pages; (3) you have to end the copy at the right place; and (4) you need to add some notes to the copy before packaging and mailing it. (For example, you might add notes to draw your friend's attention to particular passages or cross out paragraphs you intend to rewrite.)

The objective of transcription is to make an accurate copy of a small piece of an organism's genome. The process can be compared to someone using a photocopier, as shown in Figure 8.6. The same four key steps are involved: initiation, elongation (copying), termination, and processing. The **initiation** step locates the correct spot on the original DNA template where transcription is to begin. The **elongation** phase copies the correct number of nucleotides from the DNA template onto a particular type of RNA molecule called messenger RNA. **Messenger RNA** (or **mRNA**) is a strand of RNA that carries genetic information from DNA to the protein synthesis machinery of the cell. The **termination** step signals the right place to stop the copying process to make sure the mRNA molecule contains the complete set of instructions from the gene. Finally, the mRNA transcript undergoes some additional **processing** or final changes before it is transported from the nucleus to the cytoplasm of a eukaryotic cell. The following pages describe each of these steps in more detail.

Figure 8.6 DNA transcription works much like a photocopier in that it transcribes information from one medium to another while preserving the form and language of the information.

Initiation of Transcription

In a stretch of DNA that includes a gene, one strand of nucleotides known as the **sense strand** contains the instructions that direct protein synthesis. The other strand, or **anti-sense strand**, contains the complementary nucleotide sequence. An mRNA molecule synthesized from this anti-sense strand rarely codes for a functional protein. Therefore, in order for genetic information to be copied correctly from DNA to RNA, the initiation process must first select the correct strand of DNA from which to synthesize the mRNA molecule. Transcription must also begin at precisely the correct nucleotide, and it must proceed in the correct direction along the sense strand.

Specialized promoter sequences satisfy all three of these requirements. A **promoter sequence** is a particular nucleotide sequence on the DNA molecule. These nucleotide sequences provide a binding site for **RNA polymerase**, the main enzyme that catalyzes the synthesis of mRNA.

BIO FACT

On any given molecule of DNA, either strand can serve as the sense strand for different genes. In some viruses, the same stretch of DNA can contain two different genes — one transcribed in each direction along the same section of base pairs.

The promoter sequence on a DNA strand is usually rich in T and A nucleotides, and is therefore often referred to as a TATA box. Figure 8.7 shows the location of the two promoter sequences found in *Escherichia coli*. For transcription to be initiated, both of these promoter sequences must be present in their correct locations. For any given gene, the complete promoter signal (here comprised of the two promoter sequences) is found on only one strand of the DNA molecule. However, either strand can house the sense strand for different genes. At the same time, the nucleotide sequences in the promoters are slightly different from one another, which means that RNA polymerase will bind in only one orientation. The enzyme can thus face only one way along the DNA molecule, meaning that transcription will proceed in only one direction. Once the polymerase molecule has bound to the DNA molecule, it opens a section of the double helix and begins synthesizing a strand of mRNA that is complementary to the template DNA strand.

Promoter sequences ensure that transcription begins at the right place along a DNA molecule. These nucleotide sequences also keep the cell from expending energy on transcribing DNA that does not have a coding function. As you saw in the previous chapter, much of the genome of a eukaryote may be made up of stretches of DNA that do not code for protein products. The existence of promoter sequences means that RNA polymerase will only bind to and transcribe those small portions of the genome that contain genes.

Elongation of mRNA Transcript

The elongation of an mRNA transcript works in much the same way as the elongation of a new DNA strand during the process of DNA replication (which was discussed in Chapter 7). Like DNA polymerase, RNA polymerase works in the 5′ to 3′ direction, adding each new nucleotide to the 3′-OH group of the previous nucleotide. RNA transcribes only one strand of the DNA template, however. Thus, there is no need for Okazaki fragments. The process of elongation is illustrated in Figure 8.8 on the next page.

RNA polymerase moves along the double helix, opening it one section at a time. As the polymerase molecule passes, the DNA helix re-forms and the mRNA strand separates from its template DNA strand. As soon as this RNA polymerase begins tracking along the DNA molecule after leaving the promoter region, a new RNA polymerase can bind there to begin a new transcript. This means that dozens or even hundreds of copies of the same gene can be made in a very short time. Figure 8.9 on the next page shows elongating mRNA strands trailing out from their DNA template.

Unlike DNA polymerase, RNA polymerase has no proofreading function. As a result, transcription is considerably less accurate than DNA replication. Any errors in transcription are contained only in a single protein molecule, however. Therefore, they do not become part of the permanent genetic material of the organism. Consequently, the lower

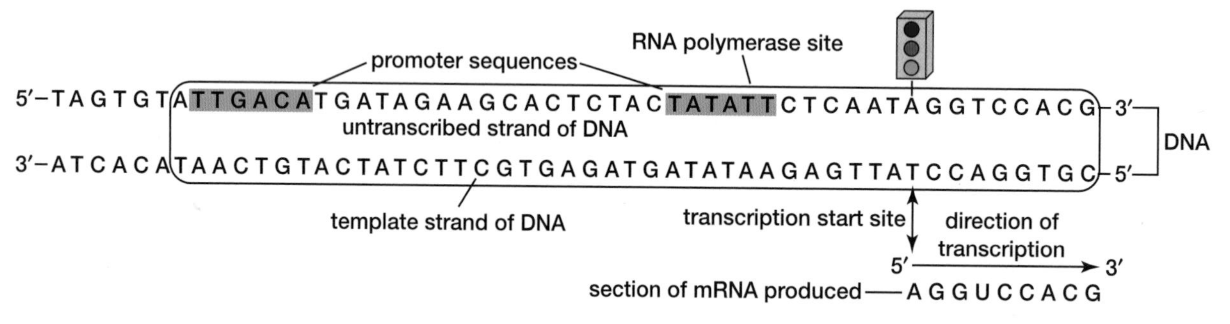

Figure 8.7 If the two different promoter sequences found in *E. coli* DNA are present at the correct locations, RNA polymerase can bind to the DNA molecule. By convention, the promoters are written according to their nucleotide sequence along the non-transcribed DNA strand. The combination of two different promoter sequences means that RNA polymerase will bind in only one way to the DNA molecule and, therefore, that transcription will only take place in one direction.

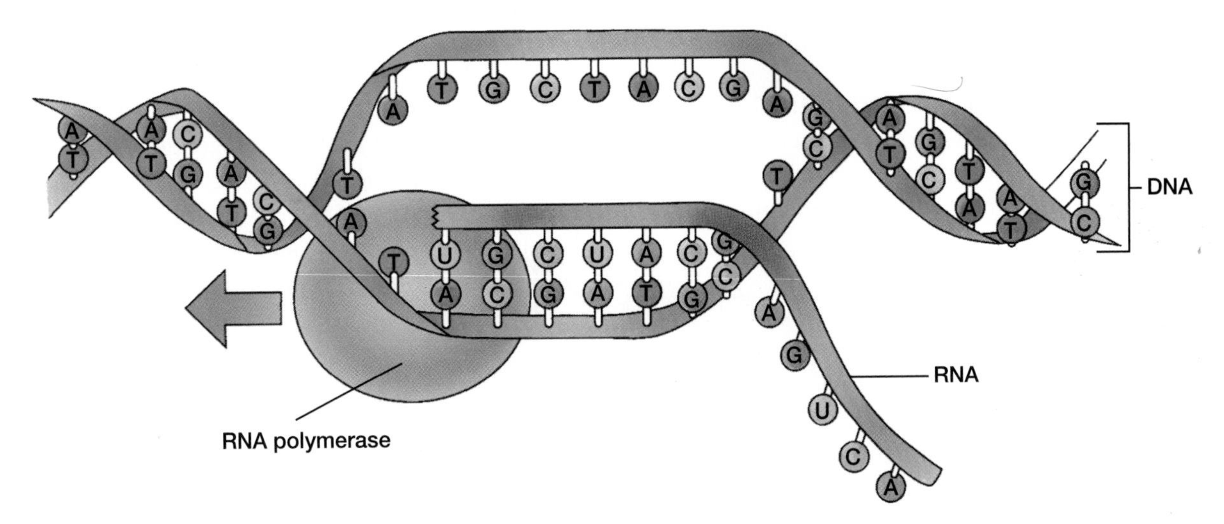

Figure 8.8 During elongation, RNA polymerase tracks along the sense strand of DNA, synthesizing a strand of mRNA by adding nucleotides in the 5′ to 3′ direction. In contrast to the synthesis of a new strand of DNA, no primer is needed. Only a short strand of the DNA helix is opened at any one time. As the helix re-forms, the elongating mRNA strand separates and trails out behind the RNA polymerase.

degree of accuracy does not pose a significant problem for the cell. At the same time, the absence of proofreading speeds up the process of protein synthesis. In eukaryotes, transcription occurs at the rate of 40 to 60 nucleotides per second.

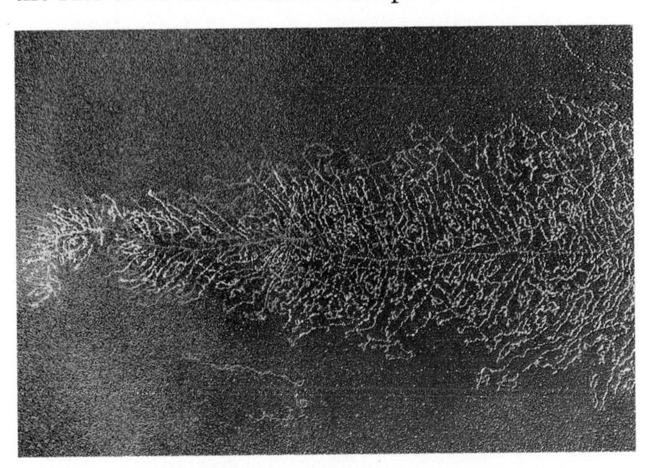

Figure 8.9 Many molecules of RNA polymerase are working along this DNA strand to transcribe mRNA molecules from a gene. The increasing length of the mRNA strands indicates the direction of transcription.

Termination of Transcription

Once transcription has been successfully initiated, the RNA polymerase continues along the DNA molecule until it encounters terminator sequences on the non-transcribed DNA strand. Like the promoter sequences, the terminator nucleotide sequences are highly specific. Figure 8.10 shows the terminator sequences found in *E. coli*.

Once transcription is terminated, the mRNA molecule separates from the RNA polymerase. In turn, the RNA polymerase molecule dissociates from the DNA molecule. The RNA polymerase can then bind to another promoter sequence and begin a new transcription process, either on the same gene or on a different gene. In prokaryotes, the mRNA molecule is now ready to begin directing the process of protein synthesis. In fact, since both transcription and translation take place in the cytoplasm of the cell in prokaryotes, the mRNA molecule may begin translation even before transcription is finished.

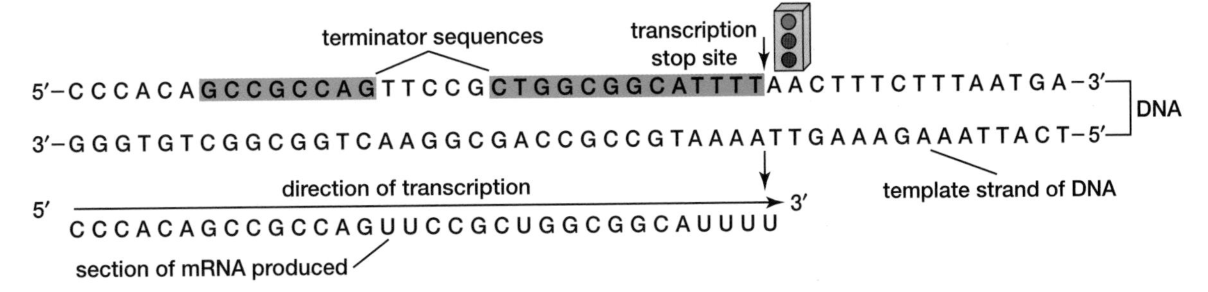

Figure 8.10 The terminator sequences, illustrated here on *E. coli* DNA, are transcribed onto the mRNA molecule. Transcription ends immediately after the final terminator has been transcribed. As with the promoter sequences, the terminator sequences are, by convention, written according to their nucleotide sequence along the non-transcribed DNA strand.

In eukaryotes, however, the processes of transcription and translation are physically separated by the double membrane that surrounds the cell nucleus. These processes are also separated by some additional reactions that modify the RNA transcript before it leaves the cell nucleus.

ELECTRONIC LEARNING PARTNER

To view animation clips on DNA transcription, refer to your Electronic Learning Partner.

Processing of mRNA Transcript

In eukaryotes, the mRNA molecule that is released when transcription ends is called **precursor mRNA**, or **pre-mRNA**. Pre-mRNA undergoes several changes before it is exported out of the nucleus as mRNA.

Pre-mRNA Cap and Tail

The first changes to the pre-mRNA molecule occur at the ends of the nucleotide strand. The 5′ end is capped with a modified form of the G nucleotide known as a **5′ cap**. At the 3′ end, an enzyme in the nucleus adds a long series of A nucleotides, referred to as a **poly-A tail**. The cap and tail have several functions. They help to protect the finished mRNA molecule from enzymes in the cytoplasm that are designed to break down nucleic acid fragments into their component nucleotides. In fact, the greater the length of the poly-A tail, the greater the stability of the finished mRNA. The cap and tail also serve as signals that help to bind the molecules that synthesize proteins. It also appears that the poly-A tail plays a role in helping to transport the finished mRNA molecule from the nucleus to the cytoplasm.

mRNA Splicing

As you saw in the previous chapter, most eukaryote genes contain both expressed nucleotide sequences or exons (which form part of the instructions for protein synthesis) and intervening non-coding nucleotide sequences or introns. The RNA polymerase does not distinguish between introns and exons as it transcribes a gene. The initial mRNA transcript therefore contains long stretches of nucleotides that must be removed before the transcript is used to construct a polypeptide.

The molecule that accomplishes this splicing act is called a **spliceosome**. This large molecule is formed from two components: proteins that are joined to nucleic acids of a type of RNA called **small nuclear RNA** (or **snRNA**) and other proteins. The spliceosome cleaves the pre-mRNA at the ends of each intron and then splices the remaining exons. This process is illustrated in Figure 8.11.

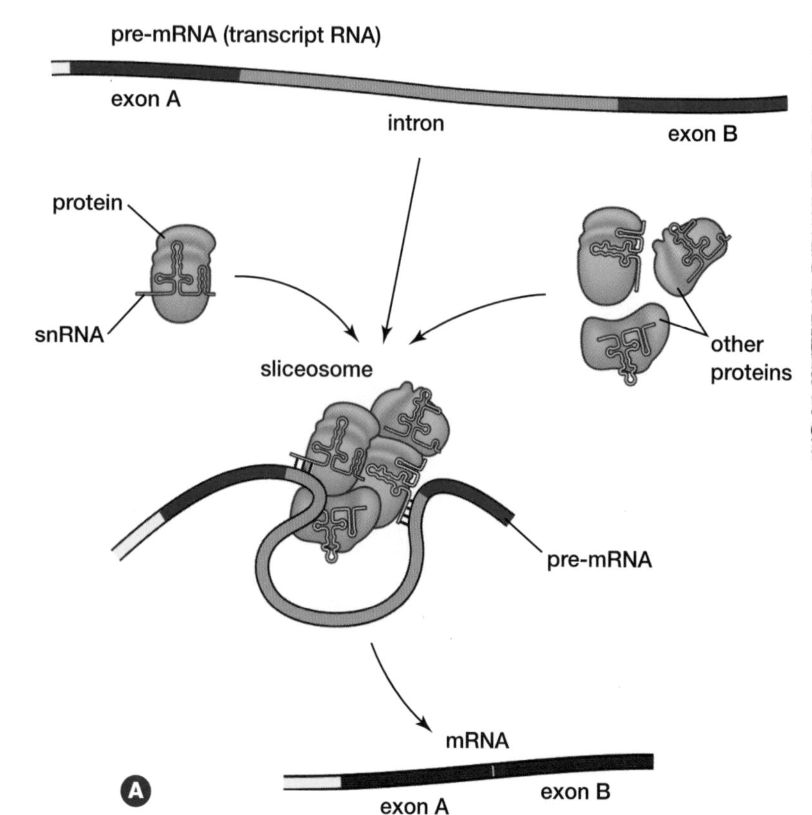

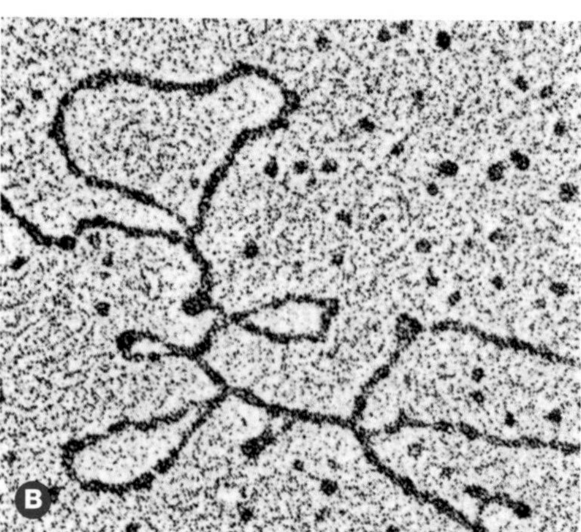

Figure 8.11 (A) During RNA splicing, the nucleotide sequence on the snRNA molecules in the spliceosome base-pairs with specific nucleotide sequences at the ends of pre-mRNA introns. (B) The spliceosome then excises the introns, as shown by the closed loops, and joins the exons. Studies show that it is the RNA component of the spliceosome that catalyzes the splicing. This is the first known example of enzymatic activity in a substance other than protein.

As you saw in Chapter 7, the removal of introns does not always follow the same pattern in any given gene. As shown in Figure 8.12, a stretch of nucleotides that is retained as an exon in one pre-mRNA transcript may be excised from another. As a result, a single eukaryotic gene can code for a variety of different protein products. This process helps to account for the fact that although your genome contains an estimated 30 000 to 35 000 genes, your body can produce well over 100 000 different proteins.

After the processing steps are complete, the finished mRNA molecule is transported from the nucleus to the cytoplasm, where it begins the process of translation. Figure 8.13 on the following page summarizes the main steps involved in transforming pre-mRNA into finished mRNA in a eukaryotic cell. After the processing steps are complete, the finished mRNA molecule is transported from the nucleus to the cytoplasm.

In both prokaryotes and eukaryotes (as you have learned in this section), the process of translation from mRNA to protein begins when a strand of mRNA reaches the cytoplasm of the cell. You will examine the events of translation in more detail in the next section.

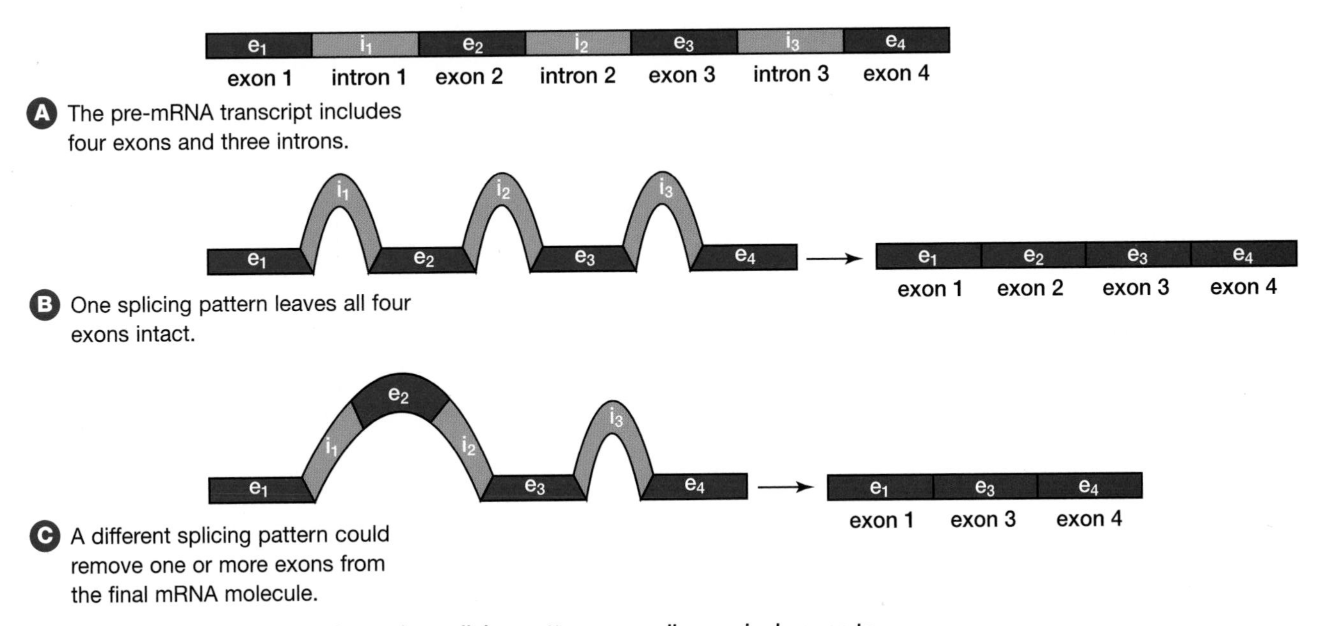

Figure 8.12 As shown here, alternative splicing patterns can allow a single gene to code for more than one polypeptide.

THINKING LAB

Transcription in Reverse

Background

Along with a gene, a coding stretch of DNA will contain other sequences that help guide the process of protein synthesis. In this activity, you will work backwards from a polypeptide chain to construct a stretch of DNA that might code for this product.

You Try It

1. The illustration at right shows an imaginary polypeptide produced by a bacterial cell. Using Table 8.1 (on page 254) and the information given about its amino acid sequence, draw one possible nucleotide sequence of the DNA molecule containing the gene for this polypeptide.
2. Label the template DNA strand, the promoter sequence, and the terminator sequence. How does the structure of the promoter sequence ensure that the right DNA nucleotide sequence is transcribed?
3. Write a paragraph or prepare a table contrasting the process of DNA replication with that of mRNA transcription.
4. Could DNA work as a messenger molecule instead of mRNA? What would be the effect on transcription? Write down some ideas and discuss them with a partner.

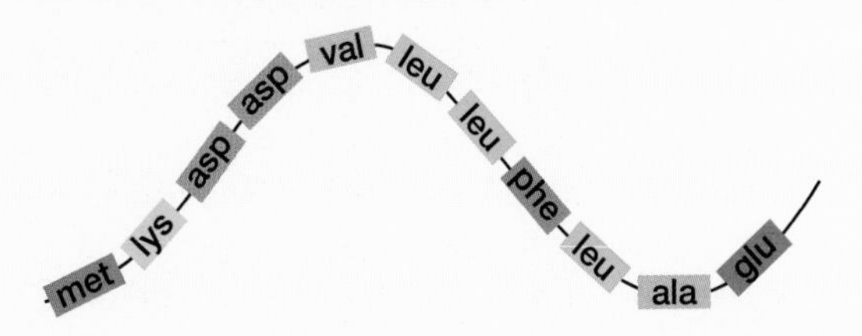

A The entire gene, including introns, exons, and leading and trailing segments, is transcribed from the DNA template.

B The ends of the pre-mRNA molecule are modified. One modified G nucleotide is added to the 5′ end, while a series of A nucleotides (anywhere from several dozen to a few hundred) is added to the 3′ end.

C The introns are removed and the exons are spliced together. The finished mRNA molecule is now ready for translation.

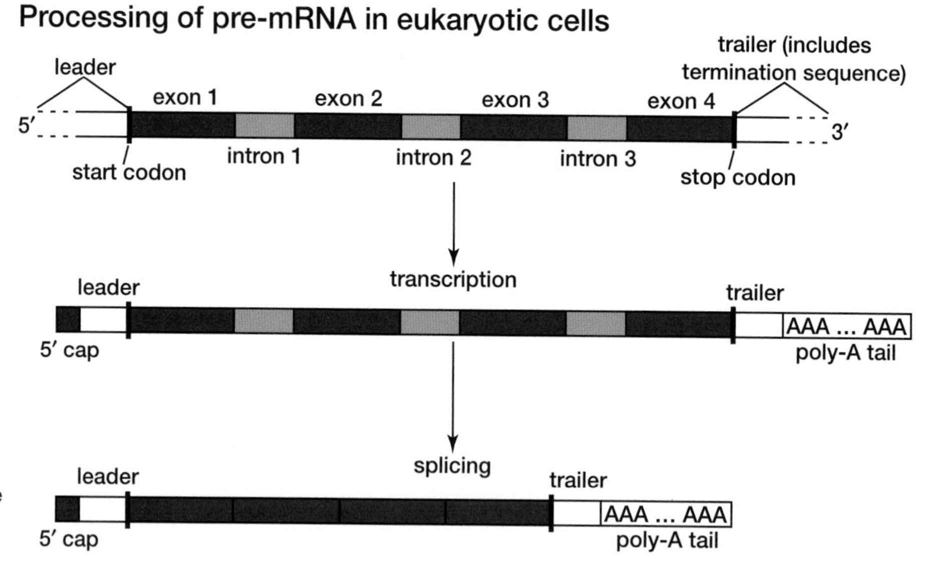

Figure 8.13 **During the processing of a pre-mRNA transcript in a eukaryotic cell, only the nucleotide sequence from the start codon to the stop codon is translated into a polypeptide. Both prokaryotic and eukaryotic mRNA contain non-coding leader and trailer sequences.**

SECTION REVIEW

1. K/U What is the objective of transcription?

2. K/U List the main steps in the transcription process in a eukaryotic cell.

3. K/U A portion of an mRNA molecule has the sequence CCUAGGCUA. What is the sequence of the anti-sense strand of the DNA?

4. MC Given that gene expression involves transcribing information from only one strand of the DNA molecule, what are the biological advantages of double-stranded DNA?

5. K/U Explain the difference between a pre-mRNA transcript and an mRNA molecule.

6. K/U Write one or two sentences to describe the function(s) of each of the following:
   (a) 5′ cap
   (b) poly-A tail
   (c) spliceosome
   (d) terminator sequence

7. K/U If a gene has four exons and three introns, what is the total number of different polypeptide products that the gene could code for? Explain your answer.

8. I Your research partner develops a form of RNA polymerase, poly-X, that recognizes the promoter sequence CCACC. She then decides to compare the action of this enzyme with that of the RNA polymerase normally synthesized by a bacterial cell. She transcribes the bacterial genome first with regular RNA polymerase, and then with poly-X.
   (a) What results do you think she will find?
   (b) Suppose the results do not conform to your expectations. What should her next steps be in trying to determine the properties of poly-X?

9. C You are to draw a flowchart for a magazine article that describes some evolutionary adaptations found in living cells. The flowchart will accompany a discussion about reasons why accuracy is less important in transcription than in DNA replication. Include a caption with your chart.

10. I In an experiment, you transcribe mRNA from *E. coli* DNA. You discover that when you put the mRNA in solution with a copy of the template DNA, the DNA and mRNA base-pair with each other. Then you transcribe all the *E. coli* mRNAs from every gene in the *E. coli* chromosome. When you put all these mRNA molecules together in solution with copies of the *E. coli* chromosome, how much of the DNA will remain unpaired? Explain your reasoning.

11. MC The action of the spliceosome is the first known example of enzymatic activity by a molecule other than a protein. For this reason, some RNA molecules are called ribozymes (ribo{some} + {en}zyme). Some scientists consider the properties of these RNA molecules to be important clues about the origins of life on Earth. In small groups, discuss the role such molecules might have played in this scenario. As a hint, consider the role of DNA and proteins in DNA replication and gene expression.

8.3 From RNA to Protein: Translation

EXPECTATIONS

- **Explain the steps involved in translation.**
- **Describe the structure and function of the main nucleic acids and enzymes involved in translation.**
- **Investigate, through simulation, the cell structures and molecules involved in transcription and translation.**

If you play a musical instrument, you know that it is possible to create a nearly infinite variety of tunes by experimenting with different combinations of notes. If you want to play a particular song accurately, however, you must read the symbols on a musical score and use these instructions to play a series of notes in precise sequence (as in Figure 8.14). Like any act of translation, turning the notes on a page into music requires a translator — in this case, someone who knows how to read music and how to play the instrument. The musical instrument is also an important component, since the correct sounds could not be made without this complex piece of equipment.

Figure 8.14 To play a particular song, a musician translates symbols on a page with the aid of a musical instrument. In an analogous way, a cell creates proteins by translating mRNA codons with the aid of its protein synthesis "equipment."

In an analogous way, it is possible to generate any number of polypeptides using the amino acids available in a cell. For a cell to create the particular proteins it needs, however, it must translate mRNA codons into amino acid sequences. This process requires both a chemical "translator" and a set of cellular protein synthesis "equipment." Once mRNA reaches the cytoplasm, these elements work together to assemble the protein products. In the text that follows, you will learn about the key components of a cell's protein synthesis equipment.

Transfer RNA (tRNA)

Different **transfer RNA** or **tRNA** molecules link each codon on the mRNA with its specific amino acid. Like mRNA, tRNA molecules are transcribed from genes on the DNA template. Unlike mRNA, however, tRNA molecules do not remain as a linear strand. Instead, base pairing between complementary nucleotides on different regions of each tRNA molecule causes the molecule to fold into the characteristic three-lobed shape you see in Figure 8.15.

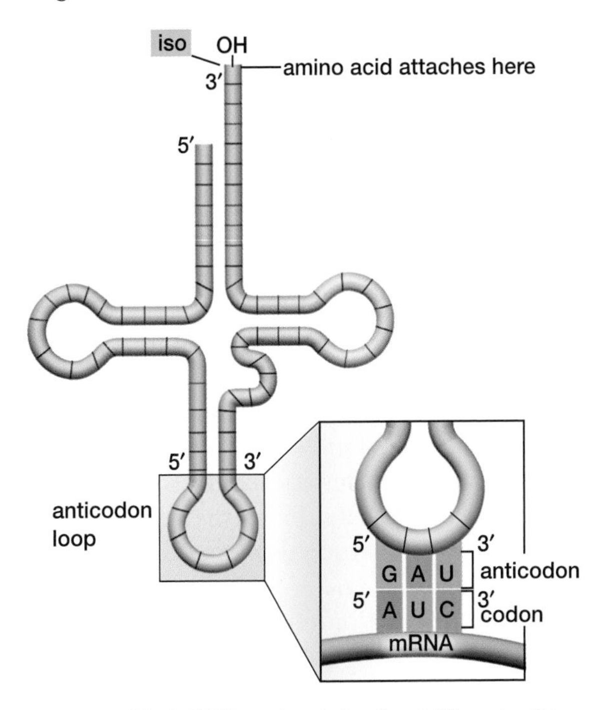

Figure 8.15 Each tRNA molecule is about 80 nucleotides long. By convention, the anticodon sequence is written in the 3′ to 5′ direction, opposite to the convention for DNA and mRNA. This practice makes it easier to match anticodons to their complementary codon sequence.

Two parts of the tRNA structure are particularly important to its role in translation. At the end of one lobe, an **anticodon** contains a nucleotide triplet with a sequence that is complementary to the codon of the mRNA molecule. At the 3′ end of the strand across from this lobe is an **amino acid attachment site**. This site binds the particular amino acid specified by the mRNA codon. A tRNA molecule bound to its particular amino acid is called an **amino-acyl tRNA**, or **aa-tRNA**.

As you saw in section 8.1, there are 64 possible combinations of nucleotide triplets that could form anticodon sequences. Of these combinations, 61 code for amino acids and three are stop signals. Therefore, you might expect to see 61 different kinds of tRNA molecules in each cell. However, a typical cell contains only about 30 to 45 different kinds of tRNA molecules, a result possible because of the wobble in the genetic code. In other words, the anticodons of some tRNAs can pair with more than one codon because the strict base-pairing rules are relaxed somewhat for the nucleotide in the third position of the anticodon triplet. For example, in many cases a base U in the third position of an anticodon can pair with either A or G in the third position of a codon. This versatility of the third nucleotide position helps to explain the pattern of redundancy you saw in Table 8.1, page 254. As a result, the cell can save energy by reducing the number of different tRNAs it must synthesize in order to manufacture a full range of proteins.

Activating Enzymes

Activating enzymes are the genetic code-breakers or chemical translators of the cell. These enzymes are responsible for attaching the correct amino acid to the tRNA molecule with the correct anticodon. Each activating enzyme has two binding sites. One recognizes the anticodon on the tRNA molecule, while the other recognizes the amino acid corresponding to that anticodon. When both the tRNA and the amino acid are bound to the enzyme, the enzyme catalyzes a reaction that attaches the amino acid to the hydroxyl group at the 3′ end of the tRNA molecule. The aa-tRNA molecule is then released. In the reaction illustrated in Figure 8.16, the enzyme that recognizes and binds the sequence UAC in its anticodon binding site will only bind methionine in its amino acid binding site. This means that the UAC tRNA will be attached only to methionine, and never to any other amino acid.

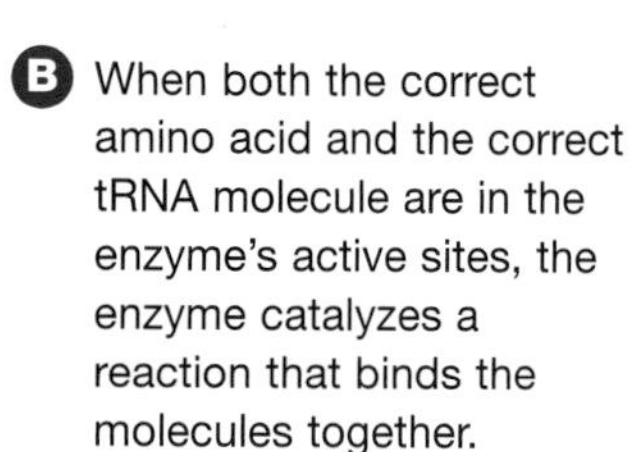

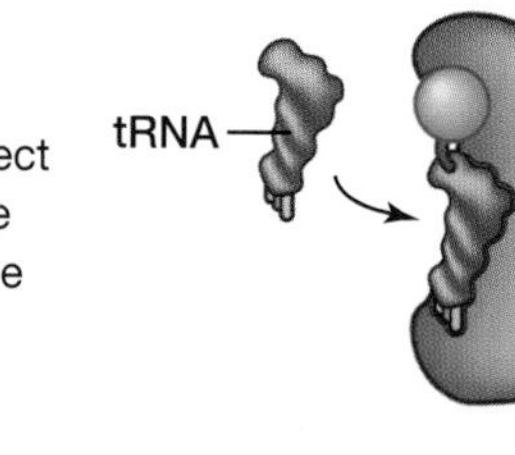

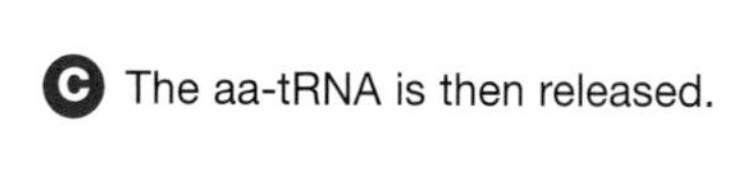

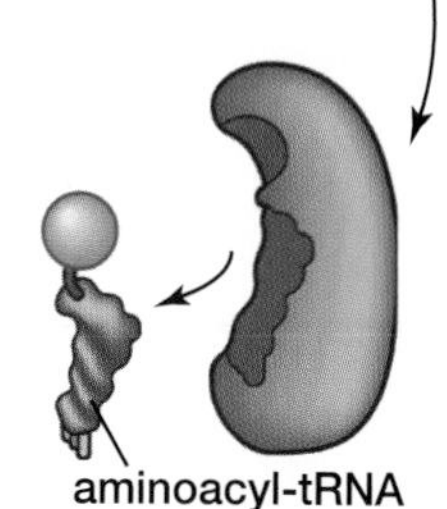

Figure 8.16 Activating enzymes link the correct amino acid to each tRNA molecule. The 3′ amino acid binding sites do not differ among different tRNA molecules. Instead, the match between a tRNA with a particular anticodon sequence and its one corresponding amino acid comes from the very specific nature of the binding sites on the activating enzyme.

Ribosomes

Along with a translator, the translation process also requires a structure that can bring together the mRNA strand, the aa-tRNAs, and the enzymes involved in building polypeptides. Structures called **ribosomes** perform this function and provide the site for protein synthesis within the cytoplasm. A ribosome is a complex that contains a cluster of different kinds of proteins together with a third type of RNA known as **ribosomal RNA** or **rRNA**. Ribosomal RNA is a linear strand of RNA that always stays bound to proteins in the ribosome assembly.

Each ribosome has two sub-units, one large and one small. The two sub-units fit together to produce an active ribosome, as shown in Figure 8.17. In a prokaryotic cell, the small unit contains one rRNA strand and about 20 different proteins, while the large sub-unit contains two rRNA strands interwoven with about 30 different proteins. Eukaryotic ribosomes tend to be larger and contain more rRNA and proteins than prokaryotic ribosomes.

Each active ribosome has a binding site for the mRNA transcript and three binding sites for tRNA molecules: the P site, which holds one aa-tRNA and the growing chain of amino acids; the A site, which holds the tRNA bringing the next amino acid to be added to the chain; and the E site, which releases the tRNA molecules back into the cytoplasm. These binding sites are provided by complementary nucleotide sequences on the different RNA molecules.

Like the transcription process, translation can be examined as a sequence of steps: initiation, elongation, and termination. These three steps are examined in more detail below.

BIO FACT

Many antibiotics work by paralyzing prokaryotic ribosomes. The structural differences between prokaryotic and eukaryotic ribosomes mean that these antibiotics can kill bacteria without harming the patient.

Initiation

Translation is initiated when an mRNA molecule reaches the cytoplasm of the cell. A sequence of nucleotides at the 5′ end of the mRNA molecule binds to a portion of the rRNA strand in a small ribosomal sub-unit. A special initiator tRNA molecule also binds to the ribosomal-mRNA complex. This molecule has the anticodon UAC (to complement the start codon AUG) and carries the amino acid methionine. The resulting combination binds to a large ribosomal sub-unit in order to complete the assembly of the translation initiation complex as illustrated in Figure 8.18.

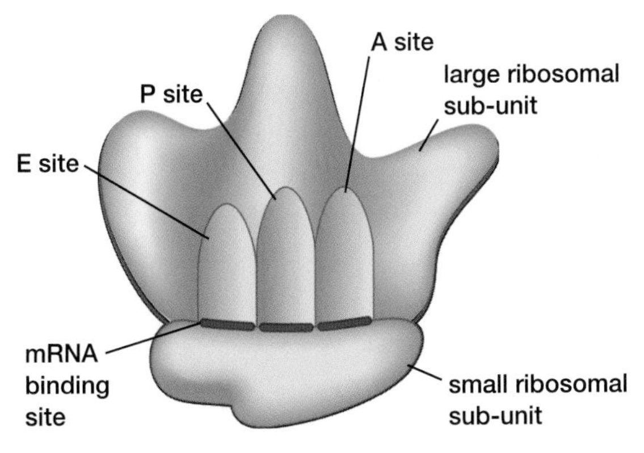

Figure 8.17 The two sub-units in an active ribosome form binding sites for the mRNA transcript and the aa-tRNA molecules involved in protein synthesis. Ribosomal RNA molecules are not specific to any polypeptide product, which means that the same ribosome can be used to synthesize any protein.

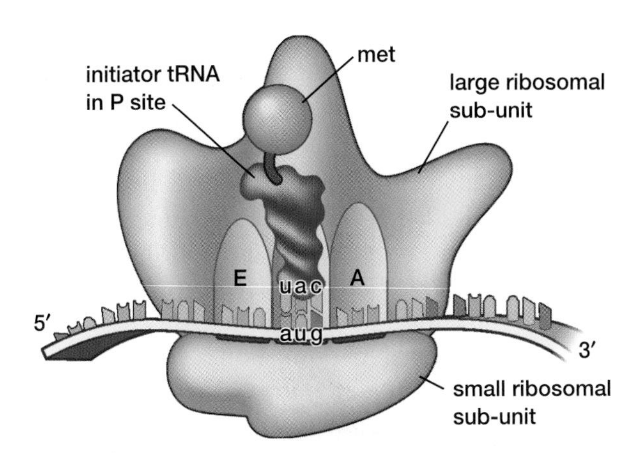

Figure 8.18 At the end of the initiation stage, the two ribosomal sub-units have come together to create an active ribosome bound to a strand of mRNA. An initiator tRNA sits at the P site. Both the A site and the E site are vacant.

As part of the initiation process, the ribosomal sub-units must bind at precisely the right place on the mRNA molecule. If the only signal for translation initiation was the start codon AUG, a tRNA anticodon could incorrectly bind to the sequence AUG at any point along the mRNA strand. Base pairing between the mRNA molecule and the nucleotide sequence in rRNA help resolve this problem.

To understand how this process works, remember that each mRNA molecule has a leader sequence ahead of the start codon. This leader sequence interacts with the rRNA in the small ribosomal sub-unit to help establish the correct starting place for translation. This interaction takes place before the large ribosomal sub-unit joins the initiation complex, with the result that the initial anticodon binding site is correctly established before translation begins.

Elongation

Once the initiation step is complete, the synthesis of a polypeptide involves repeating three general steps in a cycle. First the mRNA codon exposed in the A binding site forms a base pair with the anticodon of an incoming aa-tRNA molecule. Then enzymes and other molecules in the large sub-unit catalyze the formation of a peptide bond that joins the last amino acid in the growing peptide chain to the new amino acid. At the same time, the polypeptide chain is transferred from the tRNA in the P site to the tRNA in the A site.

In the third step, the ribosome moves a distance of three nucleotides along the mRNA molecule. This shift, which results in the ribosome assembly moving a distance of exactly one mRNA codon in the 5′– 3′ direction, is called **translocation**. Translocation brings the tRNA holding the polypeptide chain into the P site, and exposes a new A site for attachment by a new tRNA molecule. The tRNA molecule that was in the P site is now in the E or exit site, and it detaches from the ribosome. This released tRNA molecule will bind with a new amino acid to continue the process. Altogether, the three steps of the polypeptide elongation cycle, as shown in Figure 8.19 at the top of the next page, take a total of about 0.10 s to complete.

ELECTRONIC LEARNING PARTNER

To view an animation clip on tRNA and translation, consult your Electronic Learning Partner.

COURSE CHALLENGE

As you research the topic you have chosen for your Biology Course Challenge, consider how it relates to gene expression. Here are some ideas.

- If you are studying a disease or disorder, research how changes in gene expression can contribute to this condition. How do these changes arise at the molecular level?
- Create a list or flowchart of some of the ways in which interactions among DNA and proteins within cells can affect people and, more broadly, society and the environment. Use your results to help further your work.

Termination

The elongation cycle continues until a stop codon appears on the mRNA strand in the A binding site. Termination occurs when this codon is reached because there is no tRNA with the complementary anticodon. The previous tRNA, carrying the completed polypeptide chain, remains in the P site

DESIGN YOUR OWN
Investigation 8 • A

SKILL FOCUS

- Predicting
- Planning and initiating
- Performing and recording
- Communicating results
- Conducting research

Simulating Protein Synthesis

Throughout the 1950s and 1960s, scientists developed a number of models to explain the steps in protein synthesis even though they were not able to see most of the processes taking place at the cellular level. Today's researchers can now use electron microscopy to help them see and analyze molecular processes, but large-scale models are still an important tool in scientific research. In this investigation, you will work with a team to develop and present a simulation of protein synthesis.

Problem

How can you use materials available in your home or classroom to simulate the process of transcription and translation?

Prediction

Predict which aspects of protein synthesis will be relatively easy to simulate in the classroom, and which will be more difficult.

Materials

Select your own materials as determined by your experimental plan.

Experimental Plan

1. Working in a group, make a list of the steps involved in transcription and translation. Prepare a second list that includes the main structures, molecules, and processes involved at each step.
2. Discuss with your group how you might simulate the processes of transcription and translation in the classroom. Some possibilities include
   - assigning different locations in the classroom to different cellular structures, and the actions of different molecules to various students to role play; or
   - developing models using paper cutouts or other materials.

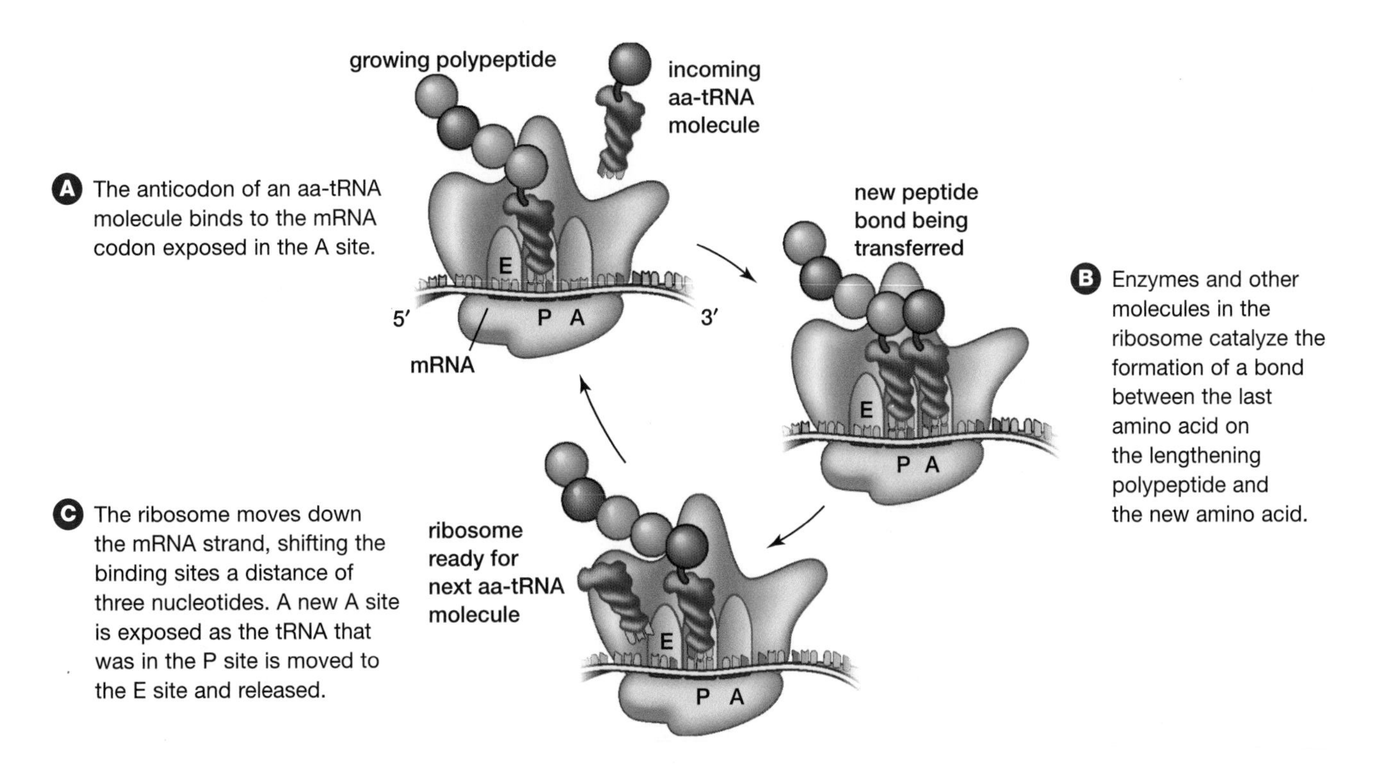

Figure 8.19 The three steps of the elongation cycle. Once initiated, this cycle will continue until the protein synthesis equipment encounters a stop signal.

3. Decide on one plan and assign responsibilities to each member of your group.
4. Assemble the materials you will need and prepare your presentation.

Checking the Plan

1. Check your plan against your initial list of the structures, molecules, and processes involved in protein synthesis. Have you included all the important steps?
2. Do the materials you are using help to explain the processes involved at each step?
3. Has your teacher approved your plan?

Data and Observations

Present your simulation of protein synthesis to the class. Make a note of any comments you receive from other groups. Identify which parts of your presentation seemed to be the most effective at simulating protein synthesis. Do these results match your initial hypothesis?

Analyze

1. Among the different approaches used by the groups in the class, which do you think was the most effective? Explain.
2. Now that you have seen how other groups approached the problem, how would you revise your own presentation?

Conclude and Apply

3. Identify three ways in which the two-step process of protein synthesis helps living cells conserve energy or reduce the risk of damage from mutations.
4. What are the adaptive advantages and disadvantages of the presence of introns in eukaryotic cells?

Exploring Further

5. Almost all living organisms on Earth use the same 20 amino acids. However, molecular biologists have been able to develop a number of artificial amino acids, which can be used to develop synthetic proteins that have many potential applications. In response, some research teams are exploring the possibility of expanding the genetic code to include new nucleotides that could be used to code for artificial amino acids. Use your library or the Internet to find out more about this research. Write a brief report explaining some of the hurdles that will have to be overcome as scientists try to expand the genetic code. What are some of the scientific and social implications of this research?

until a protein called a **release factor** binds to the A site. As shown in Figure 8.20, the release factor cleaves the completed polypeptide from the tRNA. The polypeptide then leaves the ribosome, and the ribosome assembly comes apart. The sub-units remain in the cytoplasm of the cell until they bind to a new molecule of mRNA and begin the process again.

Table 8.2
The main nucleic acids involved in transcription and translation

Nucleic Acid	Structure	Function
DNA	double helix	DNA stores genetic information.
messenger RNA (mRNA)	linear single strand	Messenger RNA carries genetic information from DNA to the protein assembly line. In eukaryotes, the precursor mRNA (pre-mRNA) must be processed before it moves to the cytoplasm for translation.
small nuclear RNA (snRNA)	linear single strand	Small nuclear RNA combines with proteins to create spliceosomes — molecular structures that cleave pre-mRNA at the ends of each intron and join the remaining exons.
transfer RNA (tRNA)	three-lobed "cloverleaf"	Transfer RNA carries a particular amino acid associated with a specific mRNA codon to the correct binding site in the protein assembly line.
ribosomal RNA (rRNA)	linear single strand	Ribosomal RNA combines with a complex of proteins to form a ribosome, the main structure in the protein assembly line. Separate rRNA strands in each ribosome sub-unit provide binding sites for the mRNA strand and tRNA molecules.

Molecular interactions among the amino acids along the completed polypeptide chain cause the chain to fold into its characteristic shape. In some cases, further modifications are required before the protein takes on its final form and becomes functional. The basic transfer of genetic information from DNA to RNA to protein, however, ends with the release of a completed polypeptide from the ribosome-mRNA complex. The main steps in the transfer of genetic information are reviewed in Figure 8.21 on the next page, Table 8.2 at left summarizes the structure and function of the nucleic acids involved in transcription and translation.

The average protein is about 100 amino acids long. It takes one ribosome about a minute to synthesize a protein of this length. At this rate, if mRNA strands were translated one at a time, it would take a long time for the cell to synthesize enough protein to serve its metabolic needs. The translation process is accelerated by having all three steps in the process take place simultaneously on any one strand of mRNA. As soon as the first ribosome has moved off the initiation sequence of an mRNA molecule, a new ribosome assembly can move in to bind to the same molecule. In this way, dozens or even hundreds of ribosomes can attach to a single mRNA molecule and thereby increase the rate of protein synthesis. An mRNA molecule bound to multiple ribosomes is called a **polyribosome**. A polyribosome complex is shown in Figure 8.22 at the bottom of the next page.

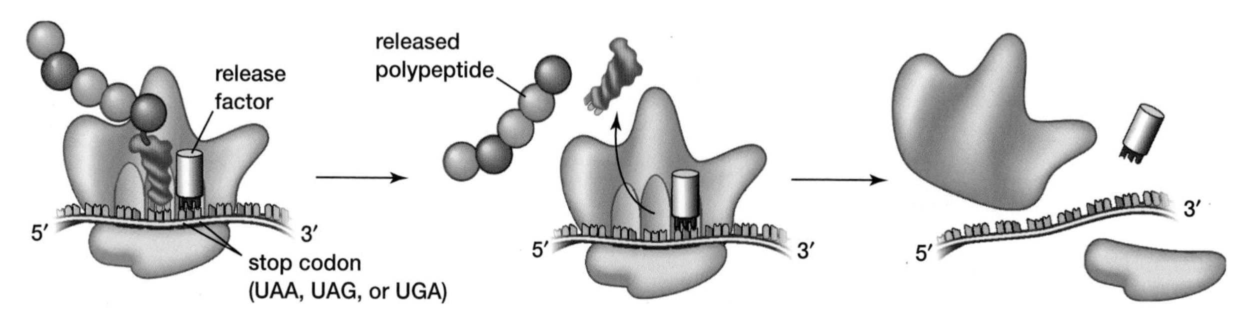

A Translocation exposes a stop codon in the A site. Instead of tRNA, a release factor binds here.

B The binding of the release factor causes the polypeptide to separate from the remaining tRNA molecule.

C The ribosome assembly now comes apart. Each element of the assembly can be used again.

Figure 8.20 Translation terminates when a stop codon is encountered on the mRNA.

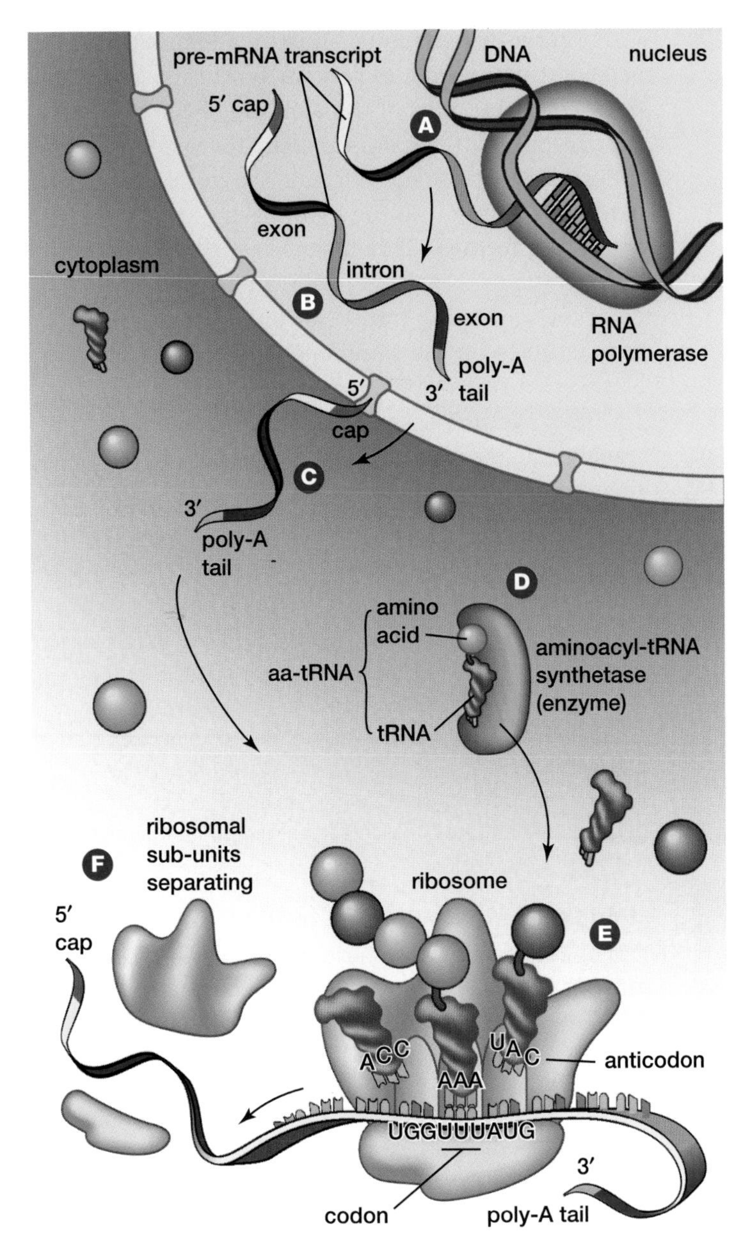

A A particular stretch of DNA, identified by a promoter sequence, serves as a template for a pre-mRNA transcript.

B Enzymes add a 5′ cap and poly-A tail to the pre-mRNA transcript, and spliceosomes splice out introns.

C The finished mRNA moves from the nucleus to the cytoplasm. There, the ribosome assembly and an initiator tRNA come together on the mRNA strand.

D Activating enzymes in the cytoplasm bind amino acids to their correct tRNA molecules to produce aa-tRNAs.

E These aa-tRNA molecules make their way to the ribosome assembly, where they attach to binding sites along the mRNA strand. The ribosome then moves along the mRNA strand, catalyzing the formation of peptide bonds along the growing chain of amino acids.

F The finished polypeptide is released and the ribosome assembly comes apart.

Figure 8.21 A summary of the main steps involved in transcription and translation in a eukaryotic cell. Which steps do not occur in prokaryotes?

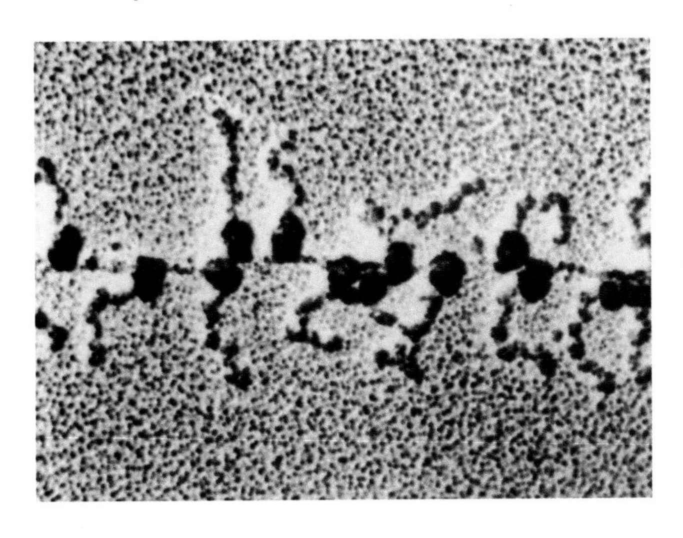

Figure 8.22 This electron microscope image of a polyribosome complex shows a strand of mRNA being translated by a number of ribosomes that are moving along the strand. A lengthening polypeptide chain trails out from behind each ribosome. Polyribosomes are found in both prokaryotes ~~and eukaryotes.~~ WRONG!

BIO FACT

A computer uses binary processing technology, so named because it stores information in the form of long sequences of just two digits, 0 and 1. These linear information sequences are translated into the text and graphics you see on your screen and the sounds you hear from your speakers. In comparison, the genetic code stores information using *four* "digits." Thus, the information-coding properties of DNA are much more powerful than those found in binary processing technology. A DNA-based computer the size of a lump of sugar could contain more information than a trillion CDs.

Prokaryotic cells have an extra advantage in accelerating the rate of protein synthesis. While the basic principles of translation are the same in prokaryotic and eukaryotic cells, there is one key difference. In a prokaryotic cell, the mRNA transcript extends directly into the cytoplasm of the cell as it is being formed. This means that ribosomes can bind to the mRNA even before transcription is complete, allowing transcription and translation to take place on the same mRNA strand at once, as shown in Figure 8.23. In a eukaryotic cell, in contrast, the processes of transcription and translation are physically separated by the membrane of the nucleus.

Biology At Work

Molecular Biologist

Alison Morrison is a molecular biologist with a mining company in Trail, British Columbia. When she thinks of mining, she considers an image of something best viewed through a microscope — bacteria. Surprising as her focus may seem, certain bacteria can be used to help recover metals.

Alison Morrison

New Technologies

The unique metabolisms of these bacteria are leading to the development of new mining technologies. Known as bio-leaching or bio-oxidation (see "bioremediation" on page 309), these technologies can both boost the recovery of metals from ores and remove water contaminants. Of particular value to mining firms is their potential to recover metals from sub-marginal ore.

The bacteria involved, primarily species of *Acidithiobacillus*, *Leptospirillum*, and *Sulfobacillus*, obtain their energy by using enzymes to oxidize iron and sulfur compounds in the presence of water and atmospheric gases. Part of Alison's research involves experiments designed to determine how much metal sulfide the bacteria can oxidize per unit of time. Mining companies would like to know more about these enzymes and the molecular genetics behind them. A logical first step is to identify the genes involved.

Unlocking Molecular Mysteries

One approach is to clone these genes and then use them to replace their counterparts in other bacteria whose genetics are well known. Plasmids of *Acidithiobacillus* have thus been introduced into *E. coli* to enable the expression and regulation of *Acidithiobacillus* genes to be studied. For example, the *rec* A gene in *E. coli* produces a protein that repairs DNA damage. (For more on DNA damage, see Chapter 9.) By replacing *rec* A with its counterpart from *Acidithiobacillus*, researchers discovered that the foreign gene could also carry out some DNA repairs. However, cloned genes may be expressed differently in the host bacterium than they are in the parent. This approach also cannot be used to study genes not represented in the host bacterium, such as those for iron or sulfur oxidation.

Alison's goal is to unlock some of the molecular mysteries behind the bio-leaching bacteria, thereby helping to make the new mining technologies they promise economically viable. Mining firms could then operate at lower costs while posing a significantly reduced environmental risk.

Career Tips

To become a molecular biologist, you will need to excel in courses in cellular biology and biochemistry in your undergraduate studies. You will then have to complete a master's degree or doctorate in the field.

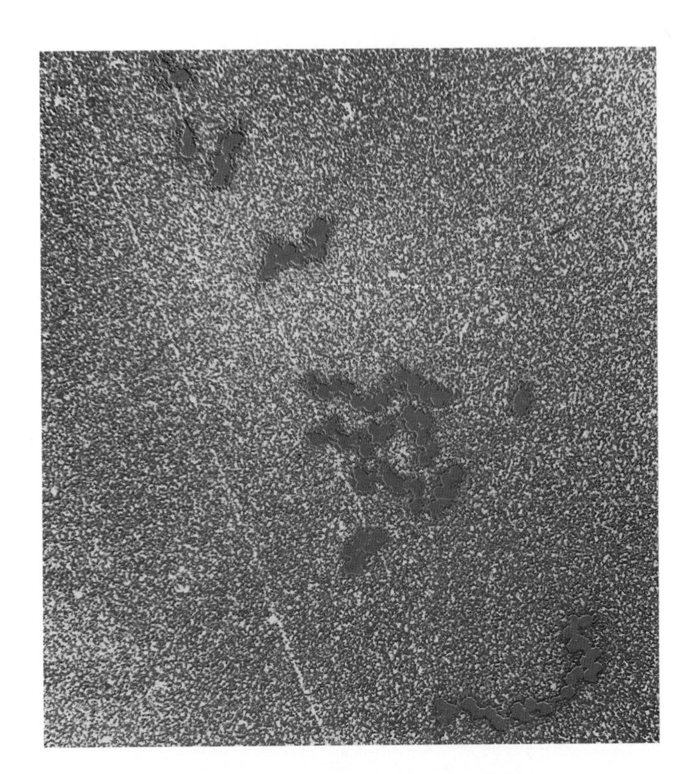

Figure 8.23 In this bacterial cell, mRNA is being transcribed from a DNA template (the pale linear strand). You can see the ribosomes (in red) already attached to the elongating mRNA molecules.

A prokaryotic cell typically has a very short life span. Consequently, it must be able to synthesize proteins very quickly in response to its rapidly changing needs. By way of contrast, a eukaryotic cell tends to live longer and have more complex metabolic requirements. By separating the processes of transcription and translation, the eukaryotic cell sacrifices some of the protein synthesis speed of a prokaryotic cell. As a result, however, it gains an additional opportunity to regulate the rate of its protein synthesis. In the next section, you will examine some of the mechanisms that regulate gene expression in both eukaryotic and prokaryotic cells.

SECTION REVIEW

1. K/U Explain how the wobble feature of the genetic code helps a cell conserve energy. What other advantage(s) does this feature give the cell?
2. K/U List the components of the protein synthesis equipment that a cell uses during translation. Identify whether each component is a polypeptide, a nucleic acid, or a combination of the two.
3. K/U Explain how a "stop" codon triggers the termination of the translation cycle.
4. K/U Arrange the following events into the order in which they occur during the elongation cycle of translation. Start from the binding of an aa-tRNA complex at the A site of the ribosome assembly:
   - translocation
   - transfer of the polypeptide chain from the tRNA in the P site to the tRNA in the A site
   - departure of tRNA from the E site
   - formation of a peptide bond
   - binding of tRNA to an amino acid to create an aa-tRNA complex
5. C You are working at a research facility that prides itself on developing creative ways to explain science to children. Your latest assignment is to develop a short story about protein synthesis using a dinosaur character called Toby the tRNA tyrannosaurus. Write a short chapter that follows Toby's path through the elongation cycle of translation.
6. MC A student researcher has developed a new form of activating enzyme. This enzyme works much faster than the activating enzyme normally found in mammalian cells because its amino acid binding site can recognize and bind any free amino acid found in the cell cytoplasm. "This enzyme will speed up the rate of protein synthesis and could have important medical benefits," claims the researcher. Is the researcher right? Explain your reasoning.
7. K/U Draw a polyribosome complex that identifies the following structures or features: 5′ end of the mRNA molecule; large and small ribosome sub-units; polypeptide chains; direction of translation.
8. MC Some antibiotics work by paralyzing bacterial ribosomes. With a partner or as part of a small group, identify other substances, such as medications or environmental toxins, that can alter the activity of eukaryotic ribosomes. Write a brief summary of your findings, including a description of the effects these substances can have on people.
9. C Discuss the implications of this statement: "The process of translation is very similar in prokaryotic and eukaryotic cells."
10. I Among the three types of RNA molecules involved in translation, which is/are responsible for determining the polypeptide sequence of a particular protein? How would you demonstrate this in an experiment?

8.4 The Regulation of Gene Expression

EXPECTATIONS

- **Explain how regulatory proteins act as control mechanisms for genetic expression.**
- **Demonstrate how more than one form of control can operate on a single gene at any one time.**
- **Discuss some of the reasons for the differences in control mechanisms in eukaryotes and prokaryotes.**
- **Interpret micrographs that demonstrate the cellular structures involved in protein synthesis.**

Every living cell has the ability to respond to its environment by changing the kinds and amounts of polypeptides it produces. Whether these polypeptides are the pigments that give fur a particular colour, the hormones that govern the development of sex organs, or the enzymes that metabolize nutrients, they all originate from the same processes of transcription and translation that you have studied so far in this chapter. By exercising control over these processes, a cell can regulate its gene expression.

For example, the white colour of arctic foxes (*Alopex lagopus*) in winter, as seen in Figure 8.24, helps to conceal the foxes from both prey and predators. As the temperature warms and the snow melts, the ground will take on the browns of the arctic tundra, and the foxes will also turn brown. The expression of the genes that govern the coat colour of these animals varies with the temperature, thus helping them to stay camouflaged in a changing landscape. Some animals show even more dramatic changes in gene expression during their lifetimes. For example, if a male slipper limpet (*Crepidola fornicata*) is surrounded by other males, as shown in Figure 8.25, it will turn into a female.

Figure 8.24 Seasonal changes in the coat colour of the arctic fox are the result of changes in gene expression.

Figure 8.25 Slipper limpets live in stacked colonies. If no female limpets are present, some males will turn into females.

Gene Expression in Prokaryotes

Figure 8.26 on the next page summarizes the main steps in the protein synthesis pathway in prokaryotes. Each step or control point (listed below) offers the cell an opportunity to prevent or otherwise influence the synthesis of an enzyme that is not needed.

- **transcriptional control** By exercising control here, the cell can speed up or slow down the transcription of mRNA from the gene that codes for the polypeptide.
- **post-transcriptional control** The cell may transcribe the mRNA but break it down before translation. Alternatively, it may lengthen or shorten the molecule's poly-A tail to control the time in which the mRNA remains stable and active.
- **post-translational control** After synthesizing the polypeptide, the cell may modify it chemically or vary the rate at which the polypeptide becomes a functional protein. Alternatively, the cell may break the polypeptide down before it becomes a functional protein.

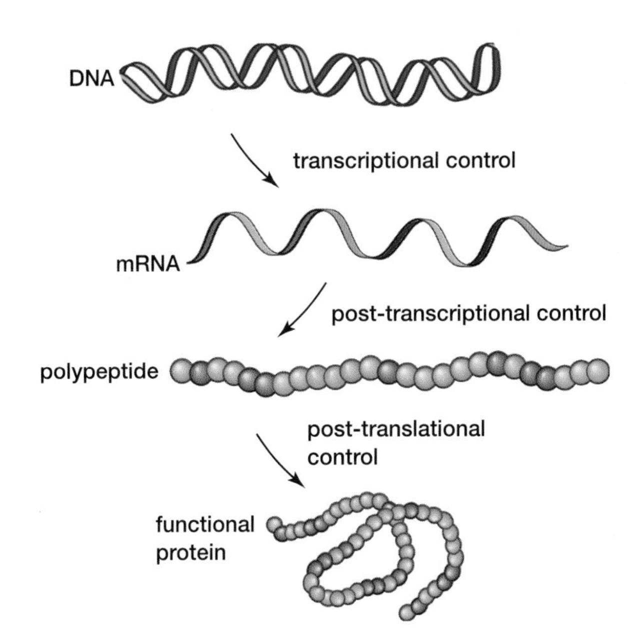

Figure 8.26 A summary of the main steps in the protein synthesis pathway in a prokaryotic cell. Each step in the pathway provides an opportunity for turning off the expression of a gene.

A cell can conserve more energy and resources if the control point is earlier rather than later in the sequence of protein synthesis steps. As you have seen, the processes of transcription and translation can take place simultaneously on the same mRNA molecule in prokaryotes, which means that post-transcriptional control is not likely to be very effective. Also, while prokaryotic cells contain mechanisms for breaking down polypeptides, these mechanisms appear to play a very small role in regulating gene expression. Thus, the most important pathways for the regulation of gene expression in prokaryotes are those that affect the rate of transcription of mRNA from DNA.

The Operon Model

The bacterium *Escherichia coli* has a gene that codes for an enzyme known as beta-galactosidase (or beta-g), which *E. coli* needs to metabolize lactose. If a colony of *E. coli* is grown on a medium containing no lactose, the cells will not manufacture beta-g. Within minutes after lactose is added to the medium, however, the *E. coli* cells will produce large quantities of the enzyme.

In 1961, the research team of François Jacob and Jacques Monod proposed a model to describe how changes in the environment affect gene expression in prokaryotes. From their research, Jacob and Monod determined that the gene for the beta-g enzyme is located in a transcription unit they called an operon. An **operon** is a stretch of DNA that contains a set of one or more genes involved in a particular metabolic pathway, along with a regulatory sequence called an operator. The **operator** is a DNA sequence located within the promoter sequences. It functions as a control element, governing whether or not RNA polymerase can bind to the promoter sequences to begin transcription of the genes. Figure 8.27 illustrates the structure of the lactose utilization operon, or *lac* operon, in *E. coli.*

Negative Gene Regulation in the *lac* Operon

When no lactose is available for the *E. coli* cell, a protein called a **repressor** binds to the operator region. The presence of the repressor makes it impossible for RNA polymerase to bind to the promoter. As a result, the genes of the *lac* operon are not transcribed. This is an example of **negative gene regulation** — that is, a situation in which a protein molecule interacts directly with the genome to turn off gene expression.

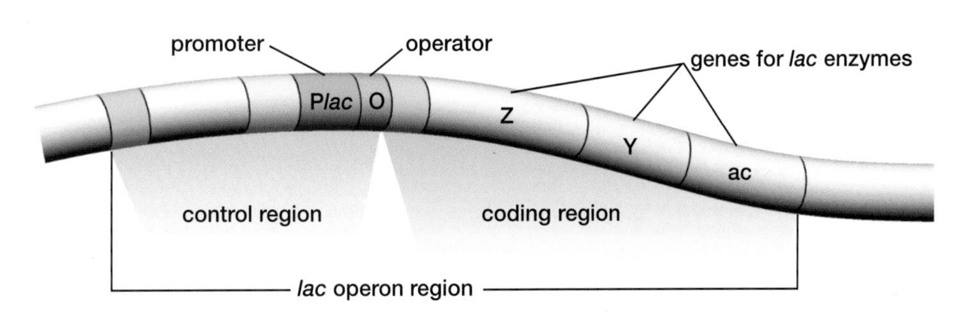

Figure 8.27 The *lac* operon of the *E. coli* chromosome contains a promoter sequence (known as P*lac*) that governs the transcription of all three genes involved in lactose metabolism. Transcription of the *lac* operon genes produces a single mRNA strand. During translation, internal stop and start codons within this mRNA strand break the polypeptide product into the three separate *lac* enzymes. The operator determines whether or not RNA polymerase can reach the genes in order to begin transcription.

If lactose molecules are present in the *E. coli* cell's environment, they will be taken up by the cell and some will bind to the repressor. When this happens, the repressor detaches from the operator site. In this case, the lactose molecule serves as an **inducer** by stopping the action of the repressor. As shown in Figure 8.28, RNA polymerase can then bind to the promoter sequence and begin transcription. The result is that the enzymes that allow the cell to break down lactose are synthesized only in the presence of lactose.

Positive Gene Regulation in the *lac* Operon

The *lac* genes code for enzymes that break lactose down into its component sugars, glucose and galactose. The *E. coli* cell then uses the glucose as a source of energy. But what if lactose and glucose are both available to the cell? In this case, it makes more sense for the cell to use the glucose that is already present than to expend energy on breaking down lactose. A second mechanism in the *lac* operon ensures that the *lac* genes are expressed at significant levels only when there is no glucose available to the cell.

If there is no glucose in its medium, an *E. coli* cell will be low on energy. Under these conditions, the molecule cyclic AMP (or cAMP) will accumulate in the cell (for a review of cell energy cycle, see Chapter 3). The accumulation of cAMP triggers the following series of events, as illustrated in Figure 8.29 on the following page:

- cAMP binds to a protein called an **activator** (also called a catabolite activator protein, or CAP).
- When cAMP binds to the activator, the activator then binds to a site close to the P*lac* promoter.
- The attachment of the activator makes it easier for RNA polymerase to bind to the promoter. As more RNA polymerase binds to the promoter, the *lac* genes are transcribed at a higher rate.

This is an example of **positive gene regulation**. In this situation, the direct interaction of a protein molecule with the genome increases the rate of gene expression.

The combination of the two types of gene regulation works rather like the combination of the ignition switch and the accelerator pedal in a car. Like the ignition switch, the negative control function determines whether the gene is "on" or "off." The positive control function regulates how fast transcription occurs once the gene is on, much like an accelerator pedal. If the ignition is off, pressing the accelerator pedal has no effect. Similarly, the *E. coli* cell will produce *lac* enzymes only if lactose is available. When lactose is available, the cell raises or lowers the amount of *lac* enzymes produced in accordance with its actual need for lactose as a nutrient source. The interaction between these two forms of control is illustrated in Figure 8.30 on the following page.

Co-repression in the *tryp* Operon

In the *lac* operon you saw two different examples of gene regulation. In one, a repressor protein binds to the genome unless it is removed through the action of an inducer molecule. In the other, an activator protein does not bind to the genome unless

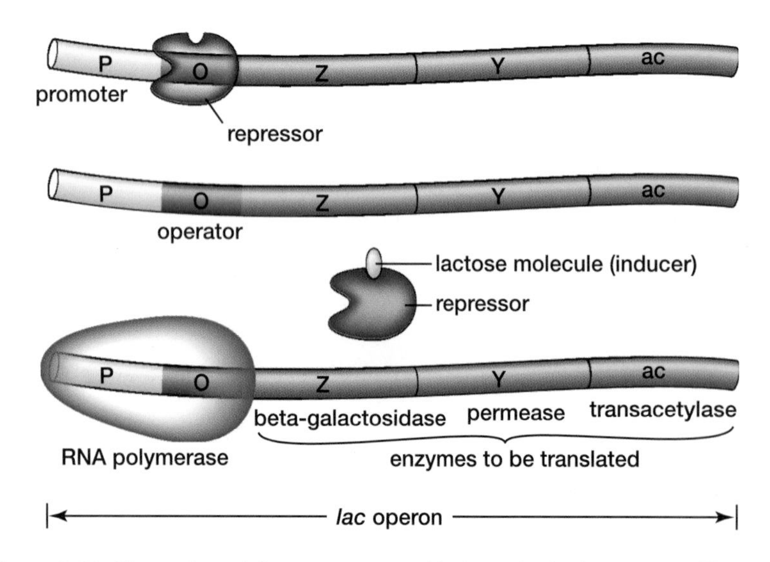

A When no lactose is present, the repressor binds to the operator and prevents transcription.

B If a lactose molecule binds to the repressor, the repressor dissociates from the operator.

C When the operator region is vacant, RNA polymerase binds to the promoter and begins transcription.

Figure 8.28 The action of the repressor and inducer in the *lac* operon. The gene that codes for the repressor is not part of the *lac* operon, so this protein is synthesized regardless of whether the *lac* system is active.

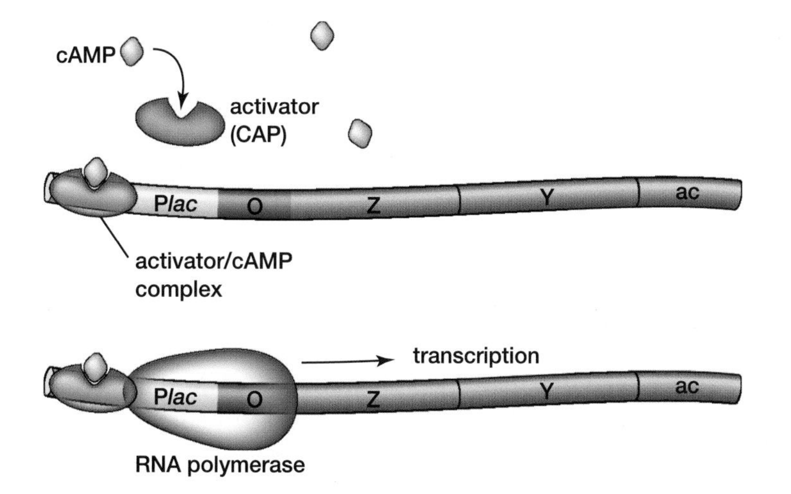

A When there is no glucose in the medium, cAMP accumulates in the cell and binds to the activator protein, which in turn binds to the DNA close to the promoter.

B The presence of the activator causes RNA polymerase to bind more readily to the promoter. As more RNA polymerase binds, the rate of transcription increases.

Figure 8.29 Positive gene regulation occurs in the *lac* operon via cAMP–CAP action.

it is activated by a separate inducer molecule (cAMP). Because an inducer plays a role in both forms of control, the *lac* operon is known as an inducible operon. Another operon in the *E. coli* genome functions as a repressible operon.

Under normal conditions, an *E. coli* cell produces the enzymes required to synthesize the amino acid tryptophan. But if the cell's medium already contains tryptophan, it does not make sense for the cell to expend energy synthesizing the amino acid from its precursor molecules.

Figure 8.31 on the following page shows the structure of the *tryp* operon in *E. coli.* This operon contains five different genes as well as a promoter and an operator sequence. Outside the operon, a separate gene codes for a repressor protein. Under normal conditions, the repressor protein does not bind to the operator region, so transcription usually takes place. When tryptophan levels in the cell are very high, however, some tryptophan molecules will bind to the repressor proteins. Once this reaction has occurred, the repressor protein is

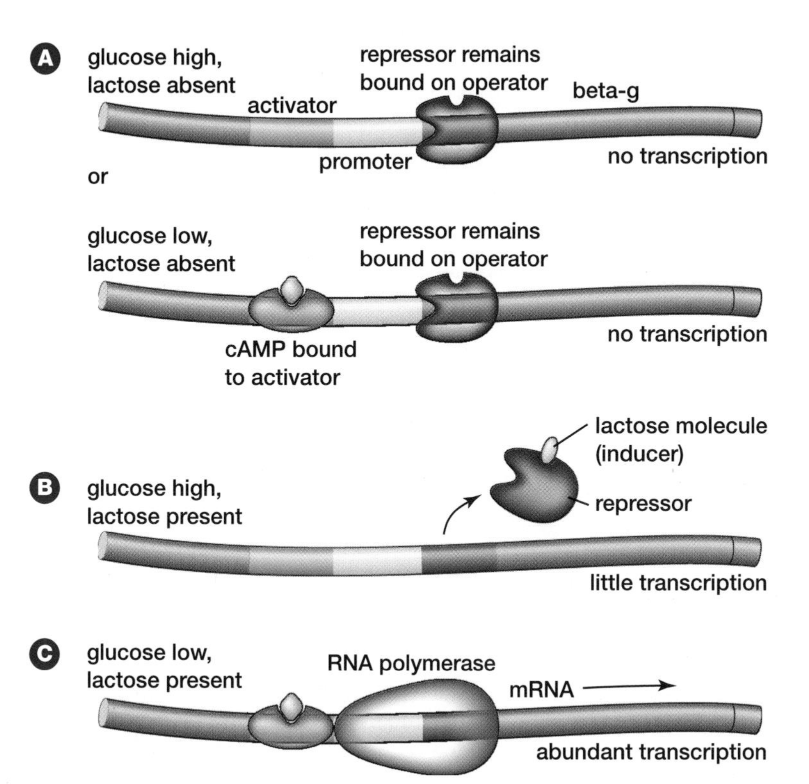

A If there is no lactose in the medium, the repressor will remain bound to the operator and prevent transcription of the *lac* genes regardless of the level of glucose present.

B If the medium contains lactose and abundant glucose, a lactose molecule (inducer) will bind to the repressor. When this happens, the repressor will dissociate from the promoter to allow transcription to proceed, although at a relatively low rate.

C If the medium contains lactose but very little glucose, a lactose molecule (inducer) will likewise bind to the repressor. The repressor then dissociates from the promoter sequence to allow transcription to take place. Under these low glucose conditions, however, a cAMP–activator complex will bind to the DNA. This action increases the rate at which RNA polymerase binds to the promoter and thus boosts the rate of transcription by 50 to 100 times.

Figure 8.30 Dual control of gene expression in the *E. coli lac* operon

more likely to bind to the operator and block transcription. The tryptophan molecule thus acts as a **co-repressor** to increase the affinity of the repressor for the operator.

A prokaryote is a single-celled organism with a relatively small genome and a short life span. Almost all of its hereditary material is likely to be expressed at some point during its life cycle. Therefore, gene regulation in prokaryotes relies primarily on mechanisms that can selectively turn off genes that are not required under particular circumstances. In contrast, eukaryotes may be multi-cellular organisms with many specialized tissues. Most of the hereditary material contained in any given eukaryotic cell is therefore not expressed during the life of that cell. As a result, eukaryotic cells are more dependent on mechanisms that keep gene expression turned off most of the time, and that turn on selected genes only as they are needed.

Gene Expression in Eukaryotes

Figure 8.32 on the next page summarizes the main steps in the protein synthesis pathway in eukaryotes. These steps or control points (listed below) offer more opportunities for control than those available to prokaryotes.

- **pre-transcriptional control** The cell controls the extent to which DNA is exposed to transcription enzymes, thus regulating the DNA's availability for transcription.
- **transcriptional control** The cell controls whether or not exposed DNA is transcribed into pre-mRNA.
- **post-transcriptional control** The cell controls the rate of processing of pre-mRNA into finished mRNA.
- **translational control** The cell manufactures the mRNA, but then controls its transport to ribosomes in the cytoplasm.
- **post-translational control** The cell manufactures the polypeptide product, but then modifies it chemically or varies the rate at which it becomes a functional protein. Another possibility is that the cell may break the polypeptide product down before it becomes a functional protein.

Like a prokaryotic cell, a eukaryotic cell can save energy if regulation of gene expression occurs sooner rather than later in the sequence of events leading to protein synthesis. Nevertheless, a eukaryotic cell is more likely to combine regulation at the pre-transcriptional or transcriptional steps with regulation at later steps. One reason for this approach lies with the greater stability of mRNA in eukaryotes. In prokaryotes, mRNA is typically degraded within a few minutes after it is synthesized. This is necessary because a gene that is expressed will continue to be transcribed until it is turned off by a regulatory mechanism. In contrast, mRNA in eukaryotes tends to remain intact for hours or even days, and it can be stored in the cell to be translated at a later point. This means that a eukaryotic cell can manufacture mRNA and then prevent it from being translated until its protein product is required. The following pages cover eukaryotic cell regulatory mechanisms through the protein synthesis pathway.

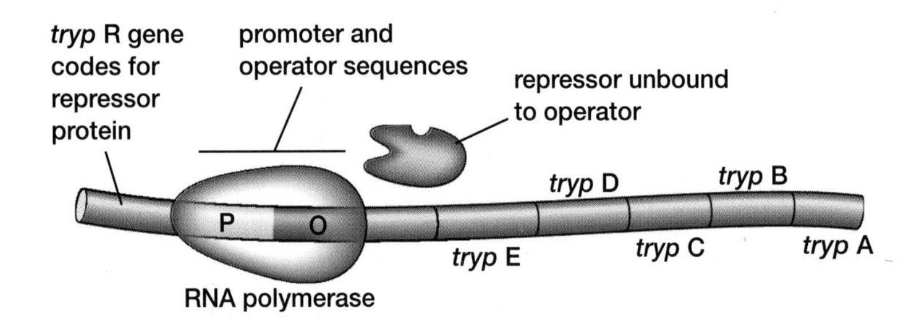

A When tryptophan levels are low, the repressor protein is present but does not usually bind to the operator. RNA polymerase is thus able to bind to the promoter, so transcription takes place.

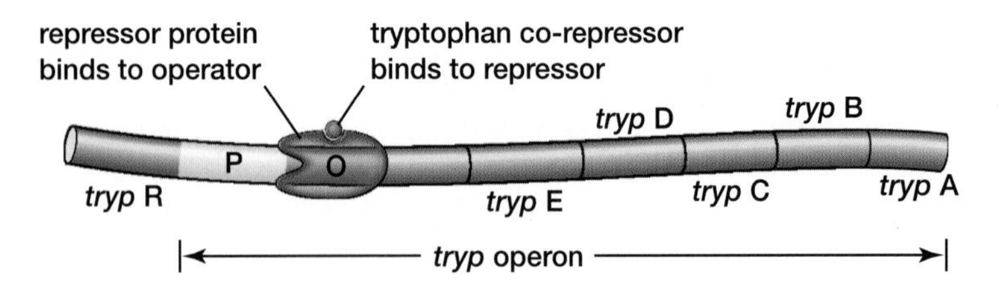

B When tryptophan levels are high, tryptophan binds to the repressor protein. As a result, the repressor binds more readily to the operator. When the repressor is bound to the operator site, RNA polymerase cannot bind and transcription will not proceed.

Figure 8.31 Co-repression in the *E. coli tryp* operon

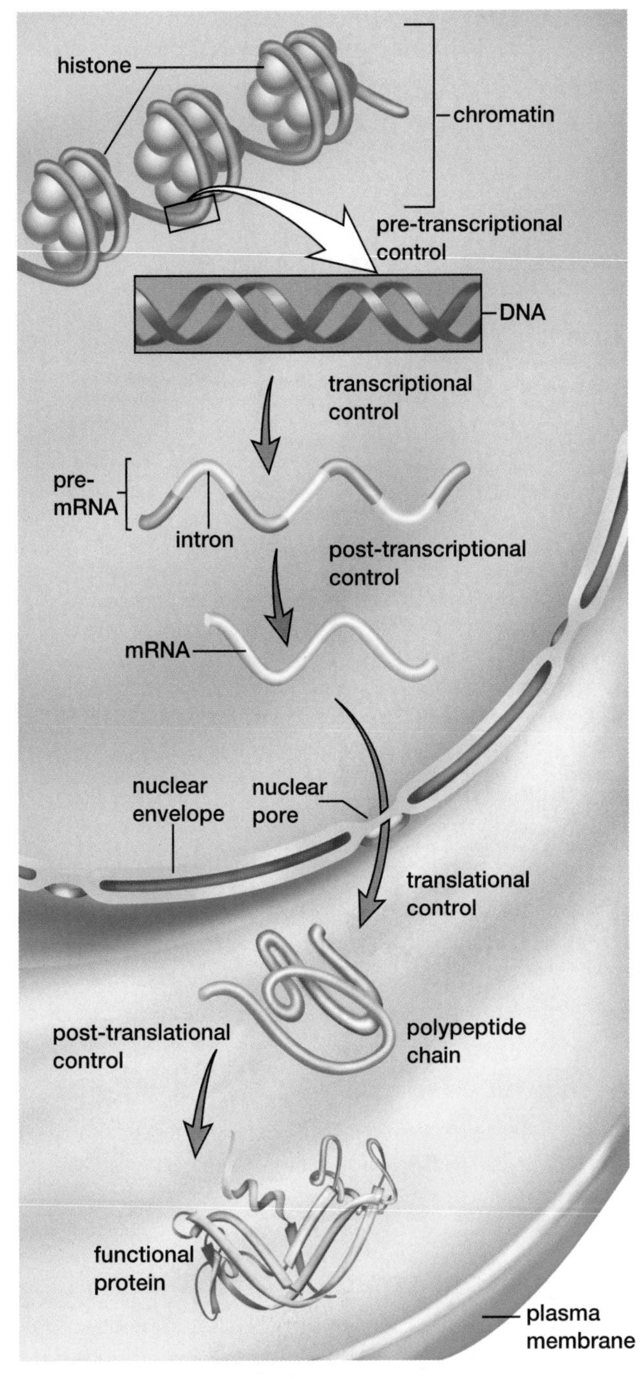

Figure 8.32 The main control points in the protein synthesis pathway in eukaryotic cells. The packing of DNA in the eukaryotic genome, the processing of mRNA, and the separation of transcription (in the nucleus) from translation (in the cytoplasm) offer opportunities for regulation not found in prokaryotic cells.

Pre-transcriptional Control

In Chapter 7, you saw that eukaryotic DNA undergoes several levels of packaging, and that different stretches of the genome may be more condensed than others during most of the cell's life cycle (see Figure 8.33). In general, the more condensed regions of DNA are not transcribed. Their more complex structure presents a physical barrier that keeps regulatory proteins and RNA polymerase from reaching their gene sequences to transcribe pre-mRNA. On the other hand, genes that are found on the more loosely packaged regions of DNA in various cells are more available for transcription, meaning that their sequences are the most likely to be expressed.

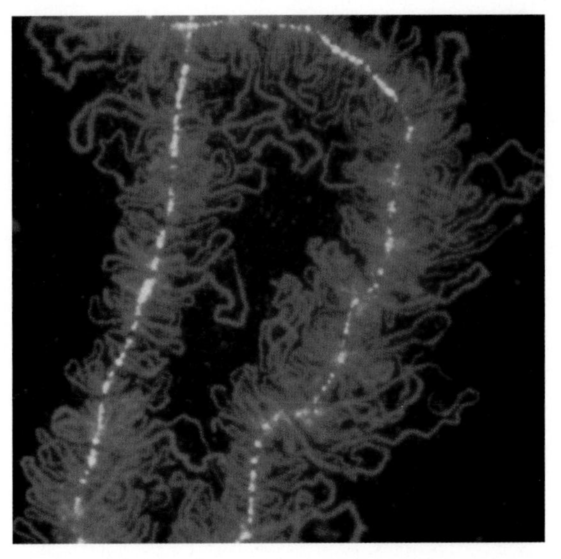

Figure 8.33 Sections of this chromosome from an amphibian egg cell have de-condensed to allow transcription to take place. The visible loops are stretches of DNA that have been released from the proteins that normally hold them in their tightly packed array. The mRNA transcripts will be stored in the egg cell until it is fertilized, at which point translation may take place.

Transcriptional Control

Researchers have not discovered any examples of operons in eukaryotic cells. Genes that code for different enzymes involved in the same metabolic pathway are much more likely to be found some distance from one another — even on different chromosomes — than to be grouped together in operons. As a result, transcription in eukaryotes relies on both the presence of the promoter DNA sequence and the action of regulator proteins called **transcription factors**. Before transcription begins, one transcription factor binds to the promoter. This action helps to bring together additional transcription factors. The resulting protein-protein interactions establish an initiation complex that allows RNA polymerase to bind to the promoter sequence of each gene.

As in prokaryotes, the rate of transcription in eukaryotes may rely on the action of operator

CONCEPT ORGANIZER Factors Influencing the Rate of Gene Expression

Structure of DNA (Chapter 7, section 7.2)

Protein–DNA interactions (Chapter 8, section 8.4)

Protein–protein interactions (Chapter 8, section 8.4)

Rate of gene expression

Gene expression is the process in which the molecular information stored in DNA is converted into structural and functional polypeptides by a living cell. The rate of gene expression is influenced by a number of factors. The packaging of DNA within a cell can make some stretches of DNA unavailable for transcription. Chemical modifications to the nitrogenous bases can also serve to keep specific portions of DNA from being expressed. In addition to these structural features of DNA, proteins play an important role in gene expression. Some proteins bind to DNA promoter or enhancer sequences to turn gene expression on or off. In other cases, proteins bind to other proteins to speed up or slow down the rate of gene expression. All of these factors work in combination with one another to determine the overall rate of gene expression. While genetic information still flows from DNA to RNA to proteins, the central dogma of gene expression also acknowledges that proteins play an equal role in determining when and how genetic information is expressed.

sequences (sometimes called control elements). These DNA sequences can bind to additional transcription factors in order to increase the rate of transcription. These operator sequences may form part of the promoter sequence, or may be located some distance away from the gene to be transcribed. Figure 8.34 on the next page shows one example of how one kind of transcription factor can bind to a distant operator sequence to enhance gene expression. The same transcription factor can influence transcription on a number of different genes in order to allow genes coding for related enzymes to be expressed at the same time.

Post-transcriptional Control

The processing of pre-mRNA into mRNA provides another opportunity to regulate gene expression. Sometimes the cell will not add the 3′ poly-A tail or 5′ cap to a pre-mRNA strand. This blocks translation in two main ways. First, an mRNA strand that lacks a poly-A tail will usually not leave the nucleus. This indicates that the tail plays some role in transporting the mRNA transcript to the cytoplasm. Second, a strand of pre-mRNA that lacks its cap and tail will usually be broken down very quickly by nuclease enzymes in the cell that reduce nucleic acid fragments to their component nucleotides. (These nucleotides are often recycled by the cell during DNA replication or repair, or during transcription). In either case, the mRNA will not reach the translation machinery of the cell.

BIO FACT

Sex hormones are one example of molecules that work as a kind of transcription factor to activate the transcription of a number of different genes at once.

Translational and Post-translational Control

Once a strand of processed mRNA has reached the cytoplasm, other regulatory proteins in the cell can bind to the leader region near the 5′ cap of the molecule. If the leader region is occupied by a regulatory protein, the small ribosomal sub-unit will not be able to attach to the mRNA. This means that the ribosome assembly will not form, and translation will not take place.

As in prokaryotes, enzymes in eukaryotes may provide some additional processing of a polypeptide product before it becomes a functional protein. If certain regulatory proteins are present in the cell, this processing might not take place. Even if it does, other regulatory proteins can interfere with the transport of polypeptides or functional proteins

A The operator sequence is located some distance from the promoter sequence that pertains to the gene to be transcribed. One group of transcription factors binds to this operator sequence.

B The interaction between the transcription factors and the operator sequence causes the DNA molecule to bend, bringing the transcription factors into contact with the promoter sequence. At the same time, their interaction causes other transcription factors to collect near the promoter sequence.

C The transcription factors bind with one another and with the promoter sequence to produce an initiation complex. This complex increases the affinity of RNA polymerase for the promoter, thereby increasing the rate of transcription.

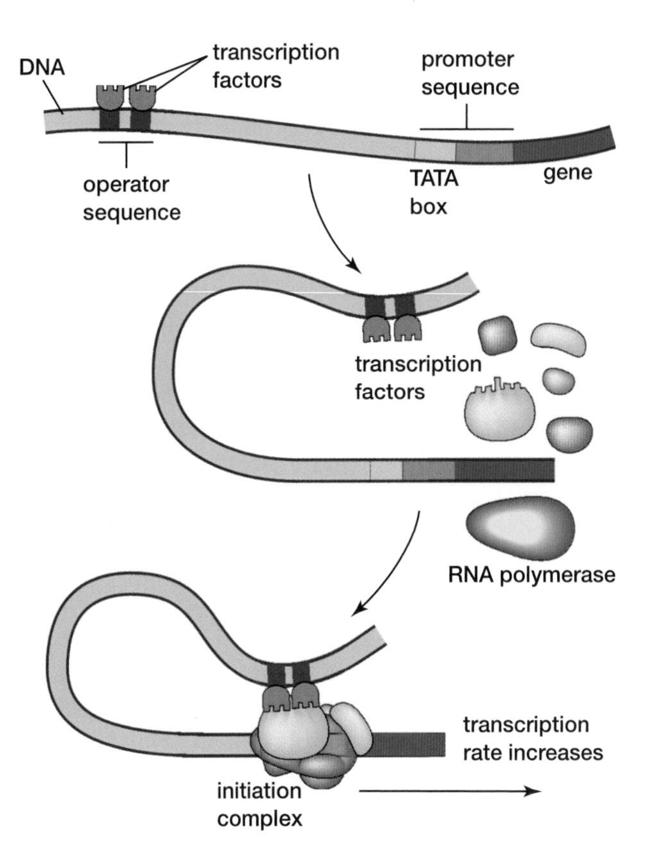

Figure 8.34 In a eukaryotic cell, a particular sequence of events (such as the one shown in this model) is needed to turn on gene expression. Most of the cell's genome remains unexpressed most of the time.

THINKING LAB

Interpreting Micrographs

Background

Transcription and translation have been understood largely through experimental analysis and deduction. However, micrographs (photographs taken through electron microscopes) can sometimes be used to confirm, clarify, or expand on hypotheses about such molecular events. In this lab, you will interpret and describe the events taking place in the micrographs shown.

You Try It

1. For each micrograph, write a title that briefly states what is taking place.
2. Write a short caption for each micrograph that identifies the visible structures and explains how the events shown should be interpreted.
3. In a few paragraphs of text, describe the events captured in each micrograph, the events that preceded those captured, and the events that will likely follow. Include information about structures or molecules that are involved but not pictured.
4. Draw and label diagrams of each micrograph that show in greater detail the events taking place.
5. Exchange your material with a partner. In what ways did your two interpretations differ? How would you change your interpretation as a result?

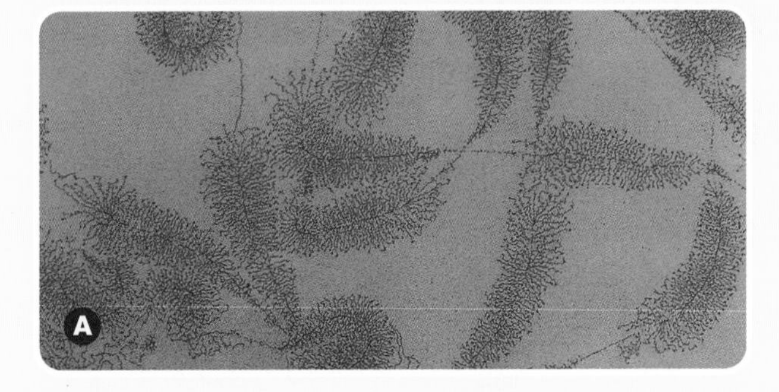

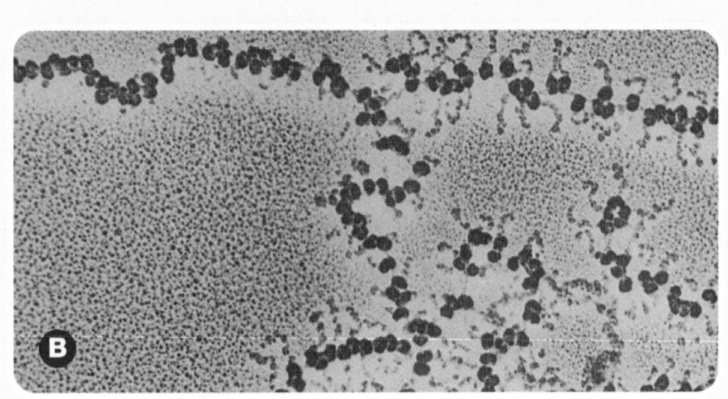

to their destinations outside the cell (that is, to elsewhere in the organism). The result in both cases is that post-translational control is achieved — the polypeptide product is broken down without ever becoming a functional protein.

In Chapter 7, you saw that the earliest theories of genetics held that DNA was not sufficiently complex to encode hereditary information. Scientists such as Levene concluded that this role must belong to proteins, and that DNA could be just a structural component of chromosomes. In the middle of the twentieth century, this theory was reversed. In other words, researchers concluded that DNA was the genetic material and proteins played only a secondary role. Since then, scientists have learned more about the role of proteins in regulating gene expression. The evidence available today suggests that the truth lies somewhere in between these two earlier models. DNA contains the information of heredity. However, by determining when and how that information is expressed, proteins play an equally important role in deciding the nature of life.

The more scientists learn about the interaction between DNA and proteins, the more they are able to manipulate living cells. In the next chapter, you will learn about some of the new genetic technologies that have given researchers the ability to change the expression of genes in living organisms, and even to design entirely new organisms.

WEB LINK

www.mcgrawhill.ca/links/biology 12

DNA is required to synthesize proteins, but DNA itself would not exist without the proteins needed to synthesize nucleic acids, replicate DNA, transcribe mRNA, and translate genetic information into polypeptides. Which came first? The more scientists learn about the interaction between DNA and proteins, the more questions they raise about how life may have originated. Use the Internet to research some of the theories on how nucleic acids first came into being. What are some of the practical applications of this research? To discover more, go to the web site above, and click on **Web Links**.

SECTION REVIEW

1. K/U Explain why gene regulation in bacteria is
   (a) more likely to take place at the level of mRNA transcription than after translation; and
   (b) unlikely to involve stopping protein synthesis between transcription and translation.

2. K/U Write a short sentence to distinguish between the following pairs.
   (a) gene and operon
   (b) transcriptional control and translational control
   (c) promoter and operator

3. K/U List three ways in which the process of gene regulation differs between prokaryotic and eukaryotic cells.

4. K/U Explain how the arrangement of DNA in a eukaryotic cell affects
   (a) the availability of DNA for transcription;
   (b) the rate of transcription of DNA to pre-mRNA; and
   (c) the number of different mRNA transcripts that can be produced from a single stretch of DNA.

5. MC A red blood cell contains no nucleus, but it can synthesize new hemoglobin molecules during its lifetime.
   (a) How is this possible?
   (b) Considering that the mRNA coding for hemoglobin in a red blood cell has an unusually long poly-A tail, what might be this feature's significance?
   (c) How might this principle be applied in other medical settings? List up to three possibilities and explain your choices.

6. C Positive gene regulation, negative gene regulation, inducer, repressor, and co-repressor are terms that can be confusing. Prepare a series of flash cards that can be used to help understand these terms.

7. I While working with a particular strain of *E. coli*, you discover that it transcribes the *lac* genes at a high level when no lactose is present in the cell's medium. List the possible causes of this activity. Then design an experiment to determine which of these possible causes is the case with your bacterial colony.

UNIT PROJECT PREP

All cancers arise because of changes to the process of protein synthesis in living cells. As part of your research, look for articles or other information sources that describe the genes involved in these changes and the polypeptides for which they code. Note any ways in which the cancer you are studying relates to changes in transcription or translation.

Summary of Expectations

Briefly explain each of the following points.

- The four different nucleotides found in DNA make up the genetic "words" that code for all 20 different amino acids. (8.1)
- The two-step process of gene expression has both costs and benefits for cells. (8.1, 8.4)
- The genetic code is redundant but not ambiguous. (8.1)
- Many transcripts can be simultaneously produced from a single gene. (8.2)
- Several types of processing are required before a pre-mRNA transcript can meet the needs of a eukaryotic cell. (8.2, 8.4)
- Four different types of RNA are involved in the main steps of protein synthesis. (8.2, 8.3)
- In prokaryotes, translation can begin before transcription is done. (8.2)
- The process of translation involves enzymes both inside and outside the ribosomes. (8.3)
- An active ribosome complex has at least four RNA binding sites. (8.3)
- An *E. coli* cell will only produce lactose-metabolizing enzymes if lactose is present in the cell's environment. (8.4)
- Crick's "central dogma" does not fully capture the process of gene expression. (8.1, 8.4)

Language of Biology

Write a sentence including each of the following words or terms. Use any six terms in a concept map to show your understanding of how they are related.

- codon
- triplet hypothesis
- transcription
- translation
- reading frame
- initiation
- elongation
- messenger RNA (mRNA)
- termination
- processing
- sense strand
- anti-sense strand
- promoter sequence
- RNA polymerase
- precursor mRNA (pre-mRNA)
- 5′ cap
- poly-A tail
- spliceosome
- small nuclear RNA (snRNA)
- transfer RNA (tRNA)
- anticodon
- amino acid attachment site
- amino-acyl tRNA (aa-tRNA)
- activating enzyme
- ribosome
- ribosomal RNA (rRNA)
- translocation
- release factor
- polyribosome
- operon
- operator
- repressor
- negative gene regulation
- inducer
- activator
- positive gene regulation
- co-repressor
- transcription factors

UNDERSTANDING CONCEPTS

1. A given organism has many different tissues, yet its cells all carry the same genetic information. Explain how this is possible.
2. Identify the significance of the discovery that the specific arrangement of amino acids is directly related to the structure of specific proteins.
3. Describe how Crick tested the triplet hypothesis.
4. What is the codon concept?
5. What was the term Crick coined to describe the process by which he believed genes were expressed? Why did he choose this term?
6. How is it possible for information on DNA that is confined to the nucleus of a eukaryotic cell to be expressed as protein products outside the nucleus?
7. How did Nirenberg and Matthaei begin working out the genetic code?
8. Identify the amino acids coded for by the following codons: AGC, GUU, UAU, AUG.
9. Name three characteristics of the genetic code and explain why they are important.
10. One species of bacteria manufactures 37 different tRNA molecules. Explain how these bacteria can still match anticodons to all of the 64 different potential mRNA codons.
11. RNA transcription in eukaryotes consists of four steps. Identify and describe each.
12. Define the term "sense strand." What is its counterpart called, and why?
13. Explain how activating enzymes work.
14. What causes RNA polymerase to cease transcription?
15. Draw a diagram that illustrates the main steps in the elongation cycle of translation. What enzymes play a role in maintaining this cycle?
16. Identify two different mechanisms that help a single gene produce very high levels of a polypeptide product in a short time.

17. Our bodies often benefit from antibiotics when fighting bacterial infections. Explain what this might have to do with ribosomes.

18. What functions are performed
   (a) by the polypeptide component of a ribosome?
   (b) by the rRNA component of a ribosome?

19. What is an operator, and how does it relate to operons? In what kind of cells are operons found?

20. Prokaryotic promoter sequences are sometimes described as being asymmetrical.
   (a) To what property of the promoter sequences does this term refer?
   (b) What is the significance of this feature?

21. List the main control steps in the protein synthesis pathway of prokaryotes, and briefly describe why each step is important.

22. Name the five control points available to eukaryotic cells and briefly describe each one.

23. How does the regulation of gene expression enable a cell to conserve energy? Give an example of a case in which another advantage might outweigh the need to conserve energy.

24. Explain what will happen to a pre-mRNA strand that lacks a 3′ poly-A tail or 5′ cap. How will this affect translation?

25. Choose the correct completion for the following statement: In a eukaryotic cell, most of the genome
   (a) remains unexpressed most of the time.
   (b) is expressed most of the time.
   (c) is repressed by a repressor most of the time.
   (d) is expressed by a promoter most of the time.

INQUIRY

26. Suppose you are provided with a sample of DNA. After you have analyzed its base composition, you divide the sample into three separate reaction mixtures in order to transcribe mRNA. Once the transcription is complete, you analyze the base composition of the mRNA from each reaction mixture. You obtain the following results. Based on these findings, answer the questions below.

	A	G	C	T	U
DNA strand I	19.1	26.0	31.0	23.9	0
DNA strand II	24.2	30.8	25.7	19.3	0
mRNA type A (reaction mixture A)	19.0	25.9	30.8	0	24.3
mRNA type B (reaction mixture B)	23.2	27.6	22.9	0	26.3
mRNA type C (reaction mixture C)	36.0	23.0	19.1	0	21.9

   (a) Which strand of the DNA is the sense strand that serves as the template for the synthesis of mRNA? Explain your reasoning.
   (b) How can you explain the composition of mRNA type B?
   (c) What substance(s) might have been added to reaction mixture B in order to obtain mRNA type B?
   (d) What might have happened in reaction mixture C in order to produce mRNA type C?

27. A researcher studying protein synthesis in *E. coli* uses an electron microscope to record the image below.

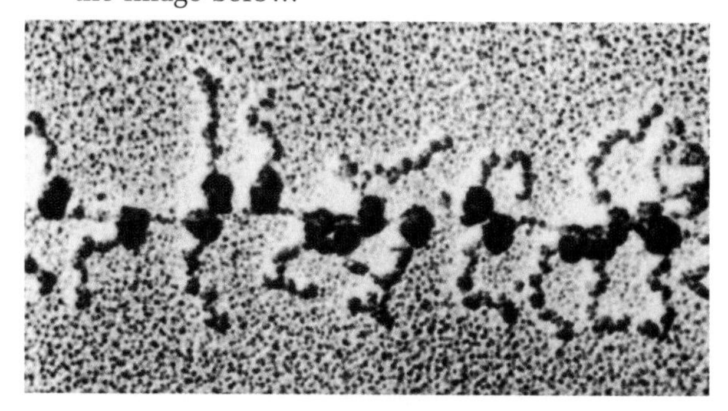

   (a) What structure is shown in this image? What is its significance?
   (b) The same researcher puts a new sample of *E. coli* into a medium containing a substance that destroys activating enzymes. She then records another image using the electron microscope. Assuming that this substance has no other effect on the cell, how would you expect the image to differ? Explain.

28. In a fictional species of bacterium, the *caf* operon contains genes that code for an enzyme that breaks down caffeine into its component molecules. The cell uses one of these molecules as a source of energy.

(a) Would you expect this operon to be repressible or inducible? Explain.
(b) Draw a diagram that illustrates how the presence of caffeine in the medium in which the cell is kept regulates the expression of the *caf* genes.

29. A particular yeast cell reacts to an increase in glucose in its medium by producing more of enzyme X.
(a) Design an experiment to determine whether the main control point for the synthesis of enzyme X is found during transcription or translation.
(b) Would the same experiment work if you were studying a bacterial cell? Explain.

COMMUNICATING

30. "Proteins, not DNA, are the key to cell specialization." In a report, explain what is meant by this statement.
31. Write a short essay that compares and contrasts the roles of DNA polymerase and RNA polymerase.
32. Create a table that could be used by others to distinguish between the various types of RNA and their roles in transcription and translation.
33. Assume for a moment that you want to author a paper on transfer RNA. With pen and paper, illustrate the basic structure of a tRNA molecule and label it accordingly. Write a caption to go with it that explains its functions.
34. With an opponent, debate ways in which the mechanisms involved in the control of the *lac* operon can be compared to traffic signals, and ways in which this analogy is not helpful.
35. When Crick first described the "central dogma", he had little experimental evidence to support his theory. In groups of three or four, brainstorm ways in which more recent evidence both supports and challenges the central dogma.
36. Form a debate team to discuss whether prokaryotes or eukaryotes, based on their methods of gene expression, have the evolutionary advantage.
37. Gene regulation is an important part of the growth and development of living organisms. Explain this statement in a short essay.

MAKING CONNECTIONS

38. Researchers have the ability to add a number of A nucleotides to the end of a strand of mRNA. What could be some practical applications of this procedure?
39. A molecular biologist creates a form of RNA polymerase that has the same proofreading ability as DNA polymerase. Explain what some of the advantages and disadvantages of this form of RNA polymerase could be:
(a) for researchers in a laboratory setting
(b) for living organisms
40. In small groups, interview people in your community who work in science, medicine, technology, or the environment to discover their thoughts on dogma (either current or historical) in their field. Do they believe this dogma has been largely helpful or misleading? Based on their answers, evaluate what role dogma should play in scientific discovery.
41. How could you use the information from this chapter to find ways to fight bacterial infections in humans? Write a short report (up to one page) identifying some processes that might be significant in the development of treatments. Could the same processes be applied to fight infections by eukaryotic cells such as yeast? Why or why not?
42. The study of the structure of genomes, including projects such as the sequencing of the human genome, is often referred to as "genomics." Many researchers claim that genomics is not nearly as significant as "proteonomics," the study of protein structure and function. Based on the information in this chapter, which field would you argue holds the greatest promise for advances in medicine and in understanding human development?

CHAPTER 9

DNA Mutations and Genetic Engineering

Reflecting Questions

- In what ways can substances in the external environment alter the information contained in DNA within a cell?
- How can scientists change the structure of DNA to produce new kinds of organisms?
- What are some of the benefits and risks of genetic engineering?

Prerequisite Concepts and Skills

Before you begin this chapter, review the following concepts and skills:

- comparing nucleotide sequences to the polypeptide product of a gene (Chapter 8, section 8.1),
- describing the process of complementary base pairing in the DNA molecule (Chapter 7, section 7.2),
- summarizing the processes and enzymes involved in DNA replication and proofreading (Chapter 7, section 7.3), and
- outlining the main pathways for gene expression and regulation in living cells (Chapter 8, section 8.4).

Much of what you have learned about genetics so far treats hereditary information as being relatively stable. You have probably learned, for example, that brown eyes are dominant and blue eyes are recessive, and that these traits are passed from parents to their offspring in a statistically predictable way. But if this is always so, how do you explain the appearance of this adult dog? Why doesn't the adult dog have either brown eyes or blue eyes?

In reality, the genome of any organism is far from stable. In the dynamic environment of the cell, the structure of DNA is constantly changing. Single nucleotide bases and entire genes are copied, shuffled, lost, and regained. Modifications in regulatory sequences activate some genes and silence others. Stretches of nucleotides "jump" from one chromosome to another. Within a single organism, changes like these produce such diverse results as an unusual eye colour or a cancerous tumour. As genetic changes are passed down over generations, they give rise to all the diversity of living organisms.

In the natural world, molecular changes in DNA are largely unpredictable, usually harmful if inherited, and infrequently passed from one species to another. In the laboratory, these natural barriers are breached. What happens, for example, when you mix bacterial, plant, and human DNA? If you follow the right recipe, you might create the corn plant shown here, which was engineered to produce a human antibody that fights cancer. Such "plantibodies" are only one example of the uses of a growing number of genetically engineered organisms. These organisms offer society benefits in such fields as medicine, agriculture, and the environment, but they may also come with significant risks.

In this chapter, you will learn about the kinds of changes that can occur in genetic material. You will examine the processes involved in genetic technologies to see how genetic material can be deliberately manipulated by researchers, and you will tackle some of the difficult moral and ethical questions that arise from activities such as DNA sequencing, genetic engineering, gene therapy, and cloning.

Genetic engineering can transform a corn plant into a source of human antibodies. Even in the natural world, the structure of DNA is constantly in flux.

Chapter Contents

Mutations and Mutagens

EXPECTATIONS

- Describe how mutagens such as radiation and chemicals can change the genetic material in cells by causing mutations.
- Describe the main DNA repair pathways in living cells.

When Gregor Mendel established his landmark theory of heredity in 1865, he described genetic traits as being determined by a particular combination of discrete factors of inheritance — factors we now call genes. Mendel found that these factors of inheritance passed unchanged from one generation to the next. The implication of this finding is that even when a gene is not expressed in one generation, the gene itself is not changed and it may be expressed in a future generation. This rule is a useful guide for determining general patterns of inheritance. However, if this rule always held true there would be no opportunity for genetic change and diversity.

In reality, the changes that take place at the molecular level within genes are an important source of genetic variation. A permanent change in the genetic material of an organism is called a **mutation** (see Figure 9.1). All mutations are heritable in that they will be copied during DNA replication. Not all mutations will be passed on to future generations, however. Only changes that affect the genetic information contained in the reproductive cells of an organism, called **germ cell mutations**, will be passed on to offspring. Mutations that arise in the other cells of an organism during its lifetime are called **somatic cell mutations**. Somatic mutations are not inherited by future generations, but they are passed on to daughter cells within the body of that organism during the process of mitotic cell division.

Figure 9.1 The northern leopard frog (*Rana pipiens*) is deformed as the result of mutations.

Mutations happen constantly in the DNA of any living organism. More than one trillion mutations occurred in your own DNA in the time it took you to read this sentence. Most of these are changes at the level of individual nucleotides, but mutations can also involve larger-scale re-organizations of genetic material.

Types of Mutations

Many mutations involve small changes in the nucleotide sequence within individual genes. A chemical change that affects just one or a few nucleotides is called a **point mutation**. Point mutations may involve the substitution of one nucleotide for another, or the insertion or deletion of one or more nucleotides.

Nucleotide Substitutions

A nucleotide substitution is the replacement of one nucleotide by another — a change from the DNA sequence CATCAT to CATTAT, for example. Such substitutions may have a relatively minor effect on the metabolism of the cell. One reason for this minimal effect is the redundancy of the genetic code. This redundancy means that a change in the coding sequence of a gene does not always result in a change to the polypeptide product of that gene. Even in a case where the point mutation results in the substitution of one amino acid for another, this substitution may not have a significant effect on the final structure or function of the protein product. A mutation that has no effect on the cell's metabolism is called a **silent mutation**. This kind of mutation, along with others that may result from nucleotide substitutions, is illustrated in Figure 9.2.

In other cases, a substitution may lead to a slightly altered but still functional protein product. Mutations that result in such altered proteins are known as **mis-sense mutations**. Mis-sense mutations can be harmful; for example, a change in a single amino acid in one of the proteins that makes up hemoglobin is responsible for the genetic blood disorder known as sickle cell disease. On the other hand, mis-sense mutations may help organisms develop new forms of proteins that can meet different requirements. For example, researchers have evidence that mis-sense mutations may play an important role in generating the enormous variety of antibodies that your body requires to fight infections.

Some substitutions can have severe consequences for a cell. A change in a gene's coding sequence that erases a "start" signal or that results in a premature "stop" signal can mean that the gene is unable to produce a functional protein. In the same way, a nucleotide substitution that affects a regulatory sequence may result in the cell being unable to respond properly to metabolic signals. Any mutation that renders the gene unable to code for any functional polypeptide product is called a **nonsense mutation**.

GUU–CAU–UUG–ACU–CCC–GAA–GAA

val – his – leu – thr – pro – glu – glu

A The normal coding sequence, with the codons in the top row and the resulting amino acids below them.

GUU–CAU–UUG–ACC–CCC–GAA–GAA

val – his – leu – thr – pro – glu – glu

B This mutation is silent, since the change in nucleotide sequence has no effect on the polypeptide product.

GUU–CAU–UUG–ACU–CCC–GUA–GAA

val – his – leu – thr – pro – val – glu

C This is a mis-sense mutation, since it causes the amino acid valine to be inserted in the place of glutamate within the polypeptide chain. The resulting protein is unable to transport oxygen effectively and produces a disorder known as sickle cell disease.

GUU–CAU–UAG

val – his –stop

D This substitution causes a nonsense mutation by changing the codon for the amino acid leucine (UUG) into a premature stop codon. No functional polypeptide will be produced from this gene.

Figure 9.2 A nucleotide substitution can result in different types of mutations, as shown here on a portion of the gene that codes for human beta-globulin, one of two polypeptides in the blood protein hemoglobin.

Nucleotide Insertions or Deletions

The insertion or deletion of one or two nucleotides within a sequence of codons produces a second type of point mutation known as a **frameshift mutation**. Unlike nucleotide substitutions (which do not affect neighbouring codons of DNA), nucleotide insertions or deletions cause the entire reading frame of the gene to be altered, as illustrated in Figure 9.3. It is possible for two frameshift mutations to cancel each other out — that is, the addition of one nucleotide at one location on a gene can be compensated for by the deletion of another nucleotide further along the coding sequence. In such a case, the result may be a mis-sense mutation. In most cases, however, a frame shift will result in a nonsense mutation.

GUU–CAU–UUG–ACU–CCC–GAA–GAA

val – his – leu – thr – pro – glu – glu

A The normal coding sequence, with the codons in the top row and the resulting amino acids below them.

GUU–CAU–GUU–GAC–UCC–CGA–AGA A

val – his – val – ala – ser – arg – arg

B The insertion of a single nucleotide, in this case guanine, results in a frameshift mutation.

A

GUU–CAU–UUG–CUC–CCG–AAG–AA

val – his – leu – leu – pro – lys

C Similarly, a deletion of even a single nucleotide, in this case adenine, also results in a frameshift mutation.

Figure 9.3 Frameshift mutations usually result in nonsense mutations.

Chromosomal Mutations

Substitutions and frameshift mutations typically affect only a single gene. Other mutations may involve rearrangements of genetic material that affect multiple genes, including genes located on separate chromosomes. One example of such a mutation is the exchange of portions of chromosomes that may take place between sister chromatids during the process of meiosis. (For a review of the somatic cell cycle, see Appendix 4.) Portions of chromosomes can also become lost or duplicated during DNA replication; this can result in changes to structural or regulatory DNA sequences.

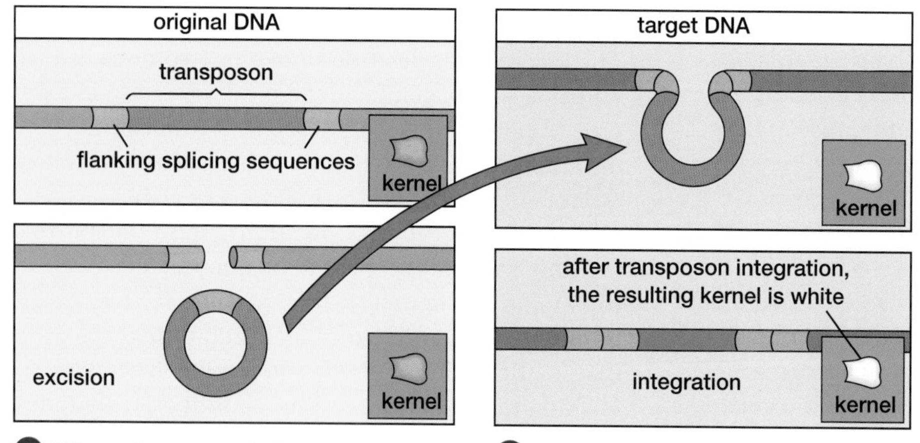

A When a transposon is located within a strand of DNA that does not code for kernel colour (top), the default kernel colour will be purple.

B When the transposon is excised and re-inserted into a gene that codes for kernel colour (top), it can disrupt the coding sequence and block the action of that gene.

Figure 9.4 Transposons were first discovered by American researcher Barbara McClintock in 1957. McClintock found that the random pattern of colours in the kernels of Indian corn could be explained by the movement of short strands of DNA from one position within a cell's chromosomes to another. In 1983, McClintock was awarded a Nobel prize for her work.

Another factor that can rearrange genetic material is the activity of transposable elements, also known as jumping genes or transposons. These are short strands of DNA capable of moving from one location to another within a cell's genetic material. A transposon is flanked by nucleotide splicing sequences that are recognized by an enzyme called transposase. Transposase excises the transposon out of one location and splices it into another. One effect of transposons is illustrated in Figure 9.4.

Within these flanking splicing sequences, a transposon may contain one or more regulatory or structural genes. Thus, the excision and splicing of a transposon from one place to another can interrupt the function of a gene or separate a promoter sequence from the gene it controls. The effects can be simultaneous yet different at the transposon's original location and at its new location. Mutations of this nature help to explain some of the rapid genetic changes that may lead to

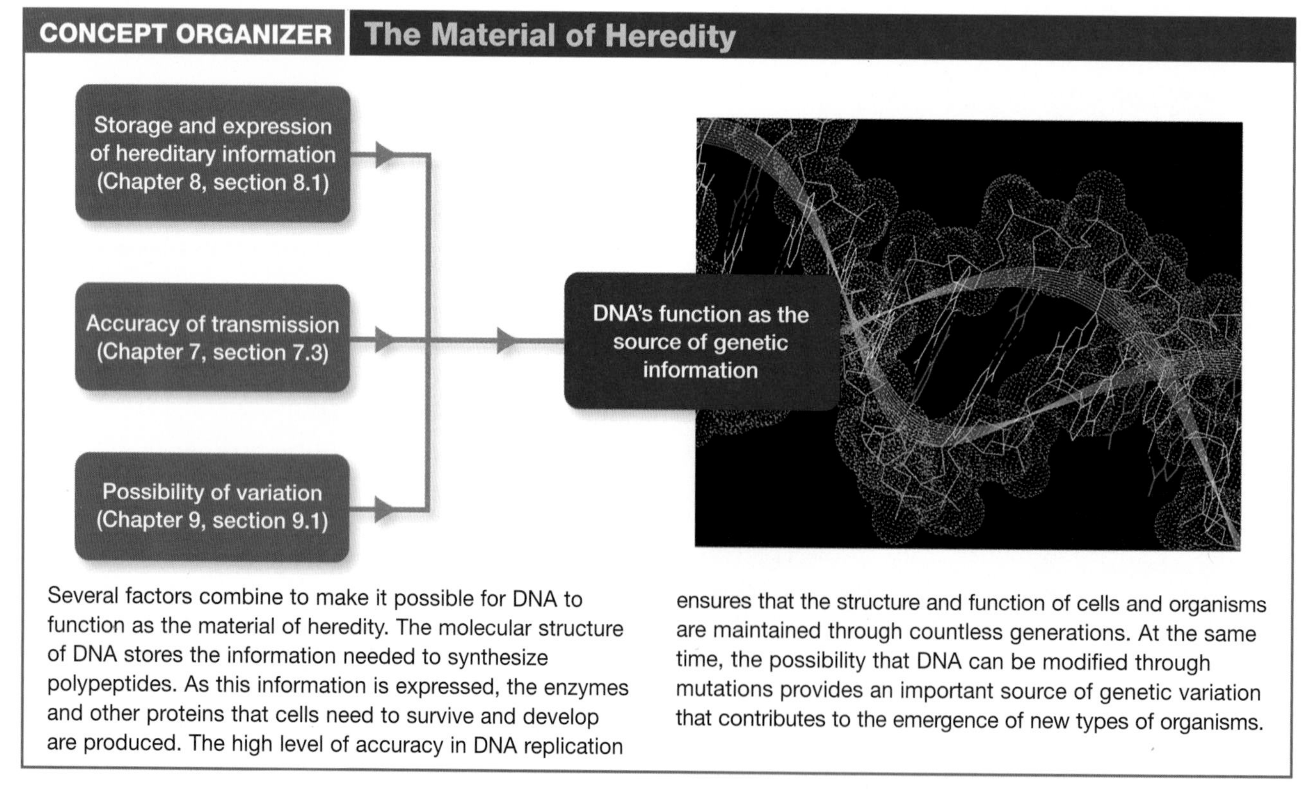

Figure 9.5 Three key factors make it possible for DNA to function as the material of heredity.

the development of new species of organisms. As Figure 9.5 indicates, stability and the variability that can result from mutations are both important features of the material of heredity

Causes of Mutations

Many mutations arise as a result of the molecular interactions that take place naturally within the cell. These mutations are known as **spontaneous mutations**. One source of spontaneous mutations is incorrect base pairing by DNA polymerase during the process of DNA replication. The rate of spontaneous mutations varies among organisms and even among different genes within a single cell.

While every cell undergoes spontaneous mutations, exposure to certain factors in the environment can increase the rate of mutation. Mutations that are caused by agents outside the cell are said to be **induced**. A substance or event that increases the rate of mutation in an organism is called a **mutagen**. Mutagens fall into two general categories, as described below.

BIO FACT

In mammals, spontaneous mutations are more than twice as likely to arise in the DNA of males than in females.

Physical Mutagens

Over a period of more than 20 years in the early 1900s, American researcher Thomas Morgan observed about 400 visible mutations in the tens of millions of fruit flies with which he worked. In 1926, one of Morgan's students (American Herman Muller) bombarded a population of fruit flies with X rays and produced several hundred mutants in a single day. Muller went on to study the mutagenic properties of X-ray radiation, and was awarded the Nobel prize in 1946.

High-energy radiation, like that from X rays and gamma rays, is known as a **physical mutagen** because it literally tears through a DNA strand, causing random changes in nucleotide sequences. It may break one or both strands of the DNA molecule, causing mutations ranging from the deletion of just a few nucleotides to the loss of large portions of chromosomes. High-energy radiation is the most damaging form of mutagen known.

Ultraviolet (UV) radiation, which is present in ordinary sunshine, has a lower energy level range than X rays but is still a powerful mutagen. It is most likely to affect the pyrimidine bases (C and T) within the DNA molecule. Where two pyrimidines are adjacent to each other, UV radiation can cause a chemical change in the bases that bonds them covalently to form a larger molecule called a dimer. Figure 9.6 shows how the resulting dimer distorts the DNA molecule. The distortion then interferes with DNA replication. UV radiation damage as a result of exposure to the Sun is one of the most common causes of cancer (specifically melanoma, a form of skin cancer) among light-skinned people. The skin pigment melanin helps to absorb UV radiation and offers some protection to the DNA within skin cells. For this reason, people with dark skin have a lower risk of developing cancer as a result of exposure.

BIO FACT

For light-skinned people, a single sunburn can double their risk of developing skin cancer.

Chemical Mutagens

A molecule that can enter the cell nucleus and induce mutations is called a **chemical mutagen**. These mutagens react chemically with the DNA molecule. Some chemicals, known as base analogues, have molecular structures that are similar to those

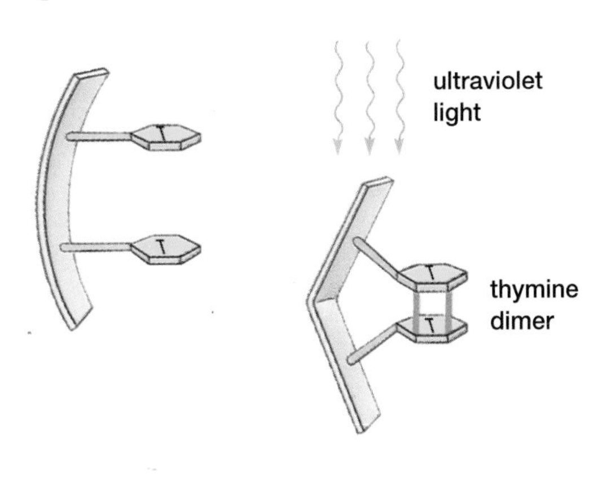

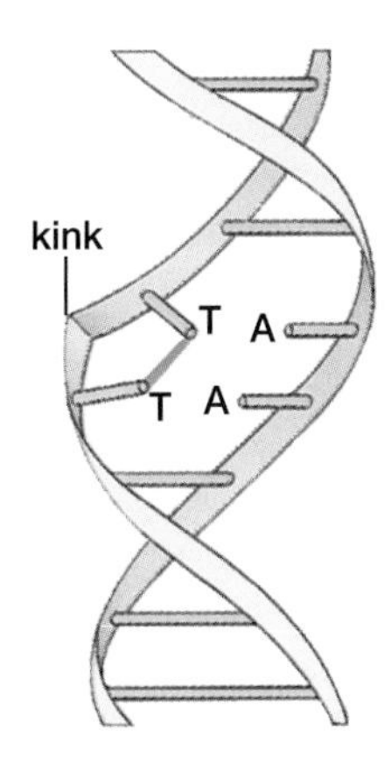

Figure 9.6 Exposure to UV radiation can cause a new bond to form between adjacent thymine bases. In turn, the resulting dimer distorts the DNA double helix and interferes with replication.

of nucleotides. Base analogues can be incorporated into DNA, but they will pair incorrectly during replication and cause substitution mutations. Other chemical mutagens react with nucleotides with the result that the shapes and bonding of the nucleotides are changed. For example, some mutagens can cause a C nucleotide to become a U nucleotide. Still other chemicals can insert themselves into the DNA double helix, causing distortions that stall replication or result in frameshift mutations. Examples of chemical mutagens include nitrites (which are sometimes used as a food preservative), gasoline fumes, and the benzene compounds found in cigarette smoke.

The **Ames test** is a simple test that measures the potential for a chemical to be mutagenic. As illustrated in Figure 9.7, this test uses a mutant strain of bacterium that is unable to synthesize the amino acid histidine. A suspension of the mutant bacteria is treated with the chemical and then spread onto a medium that lacks histidine. Only those bacteria that have undergone a **reverse mutation** — that is, a second mutation that restores their ability to synthesize histidine — will be able to grow. The more mutations induced by the chemical, the more likely it is that a reverse mutation will occur. Therefore, the number of colonies of bacteria that appear on the histidine-free medium indicates how strong a mutagen the chemical is.

Most chemical mutagens are **carcinogenic** — that is, they are associated with one or more forms of cancer. For this reason, the Ames test is used by many industries and government agencies as part of the process of screening new products for potential cancer-causing agents.

Cumulative Mutations

As you saw earlier, a single mutation often has little or no effect on a living cell. Over the life of a cell, however, a series of spontaneous and induced mutations can add up to more serious damage. For example, most cancers are the result of combinations of mutations. Some of these may be inherited, while others may arise as a result of exposure to mutagens in the environment. The fact that mutations accumulate within a cell helps explain why exposure to chemicals that are known to be carcinogens does not always result in cancer, and why cancer can occur without exposure to any known carcinogens. In the Thinking Lab on the following page, you will have the opportunity to identify different kinds of mutations and predict their effects.

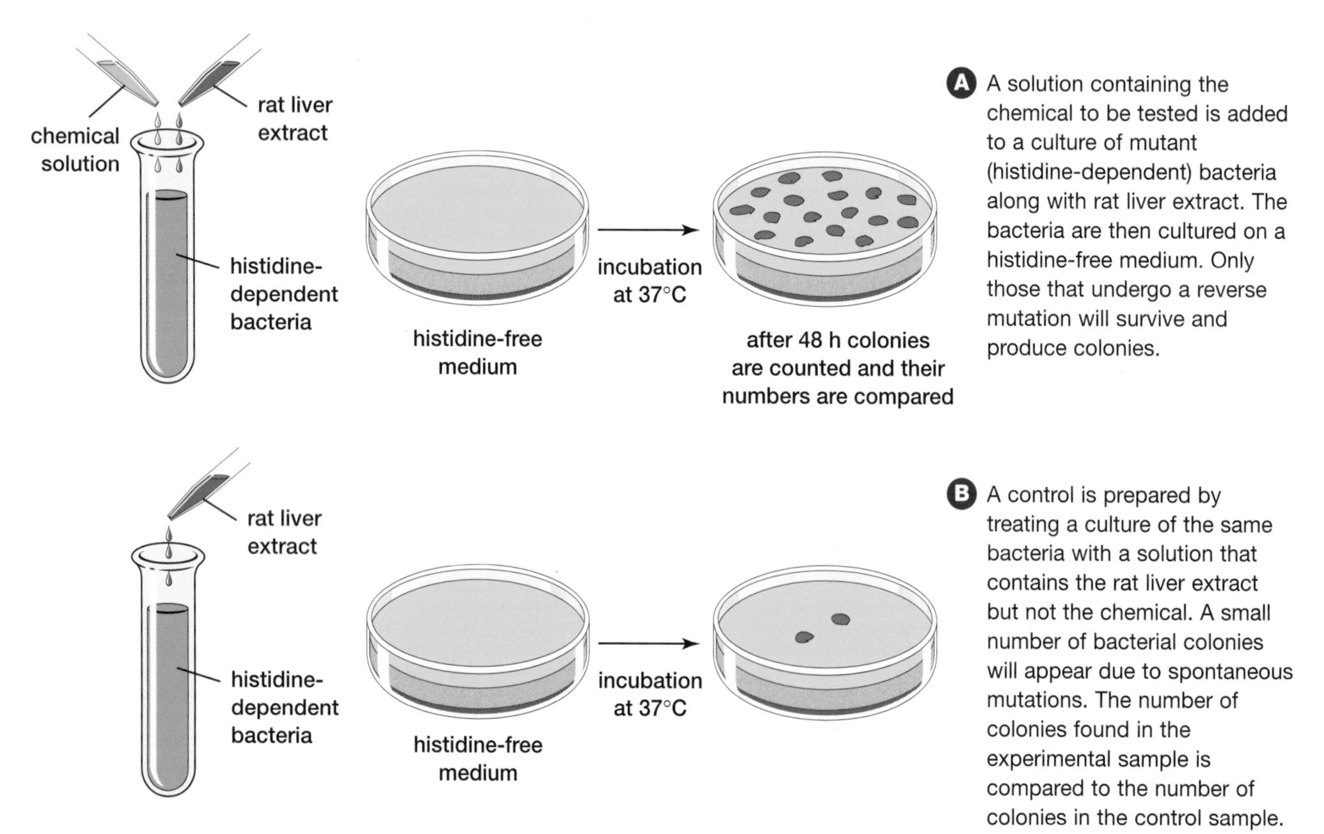

Figure 9.7 The Ames test was developed as a simple way to test for mutagens. An extract of rat liver is added to the chemical solution in order to simulate conditions inside a mammalian body. Most mutagens are associated with cancer, so the Ames test is an important part of consumer safety screening processes.

Mutation Repair Mechanisms

Even without exposure to mutagens, each of your genes undergoes thousands of mutations during your life. If these mutations all remained intact, the cumulative effect on your body would be disastrous. Fortunately, the cells of your body produce hundreds of types of enzymes that constantly work to repair damage to your DNA. Several repair pathways allow a cell to recognize and act on different types of mutations.

Direct Repair

In some cases, a cell is able to reverse damage to its DNA. Consider, for example, the proofreading function of DNA polymerase during DNA replication. As you saw in Chapter 7, DNA polymerase is able to recognize an incorrectly paired nucleotide and correct this error immediately. This kind of repair is called **direct repair** because it undoes the damage to the DNA molecule (as opposed to a repair process that excises and replaces the damaged section). Another example of direct repair is found only in prokaryotic cells. Bacteria such as *E. coli*, for example, produce an enzyme that can break pyrimidine dimers quickly.

Excision Repair

An **excision repair** occurs when a damaged section of DNA is recognized and replaced by a newly synthesized correct copy, as shown in Figure 9.8.

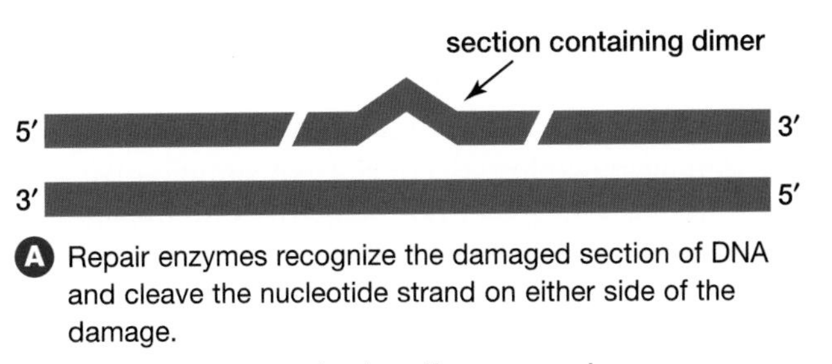

A Repair enzymes recognize the damaged section of DNA and cleave the nucleotide strand on either side of the damage.

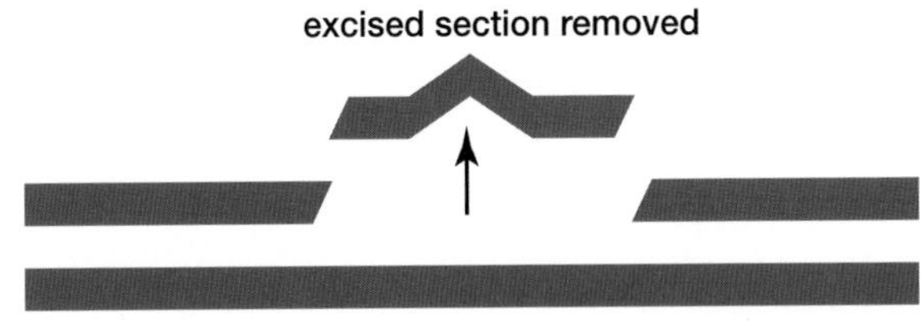

B The damaged section is removed from the DNA molecule. DNA polymerase builds a new DNA strand by adding nucleotides in the 5′ to 3′ direction.

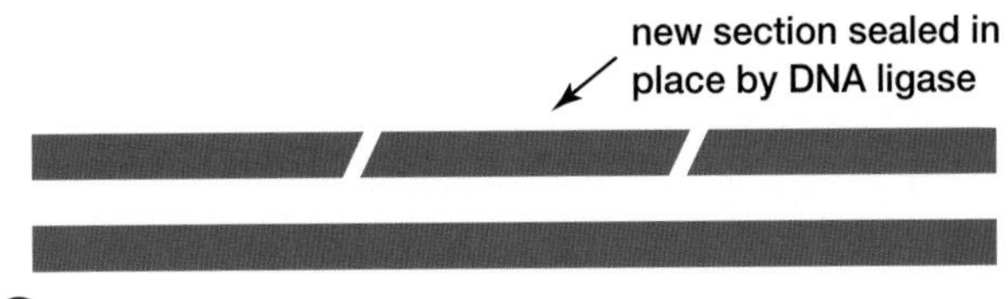

C DNA ligase seals the new stretch of nucleotides into the DNA molecule.

Figure 9.8 Eukaryotic cells contain over 50 different types of excision repair enzymes, each recognizing a particular type of DNA damage. DNA is the only macromolecule within living cells that can be repaired.

THINKING LAB

Identifying Mutations and Predicting Their Consequences

Background

As you have seen, a change in a single nucleotide makes the difference between a normal hemoglobin protein and the altered protein associated with sickle cell disease. This table shows nucleotide sequences from different forms of an imaginary gene. In this lab, you will examine these sequences to identify different types of mutations and predict their consequences.

Gene type	Nucleotide sequence (from template DNA strand)
wild (normal)	TACCTCTTTCGGGTAACAATAACT
Mutant 1	TACCTCTTTCGGGTAACTAATAACT
Mutant 2	TACCTCTTTCGGGTGACAATAACT
Mutant 3	TACCTCTTTCGGGTAAAAATAACT

You Try It

1. Translate the DNA sequence into the mRNA codon sequence, and then into the amino acid sequence of the polypeptide product.
2. For each of the mutant varieties of the gene, describe the type of mutation involved.
3. For each of the mutant varieties of the gene, predict the likely outcome of the mutation on the organism's metabolism. What variables would affect your prediction?
4. Write a nucleotide sequence for the template DNA strand of the wild gene that includes a point mutation. Trade mutual sequences with a partner to see how they compare. What would the cumulative effect of the two mutations be if they both took place in the same gene? Within your class, how many pairs of mutations cancelled each other out? What does this tell you about the likelihood of reverse mutations occurring in living organisms?

Excision repair involves three basic steps. First, a repair enzyme recognizes an incorrectly paired base or a distortion in the double helix and cuts out the damaged section. Second, DNA polymerase synthesizes a new section, using the undamaged DNA strand as a template. In the final step, DNA ligase seals the new section in place along the molecule.

Recombination Repair

When a mutation causes damage to both strands of a DNA molecule, there is no template from which to build a replacement section of DNA. When this type of damage occurs in a somatic cell, enzymes in the cell may use the homologous portion of a sister chromatid as a template to construct new DNA — a process called a **recombination repair**. While a repair made this way is likely to contain some errors, the extent of damage to the cell will, nevertheless, be much less than if no repair took place at all.

In cases where DNA damage is severe, the mutation may trigger the action of so-called suicide genes rather than DNA repair mechanisms. These suicide genes cause the cell to die, thereby preventing the mutation from being passed on to daughter cells.

In the natural world, mutations generally occur randomly. Even though mutagens do not always produce mutations, the probability that certain types of mutations will occur increases in the presence of specific mutagens. In the laboratory, however, scientists have developed a number of techniques by which they can alter the genetic make-up of living organisms in very specific and predictable ways. You will learn about these processes in the next section.

ELECTRONIC LEARNING PARTNER

Your Electronic Learning Partner has an interactive exploration on mutations.

SECTION REVIEW

1. K/U Write one sentence that explains the relationship between each of the following pairs of terms:
   (a) point mutation and substitution mutation
   (b) nucleotide insertion and mis-sense mutation
   (c) physical mutagen and thymine dimer
   (d) chemical mutagen and insertion mutation
2. K/U Describe the difference between a substitution mutation and a frameshift mutation. Which is likely to cause the greatest damage to a cell?
3. C Write a short paragraph explaining how a base analogue can act as a mutagen.
4. K/U Explain how the action of a transposon can affect the expression of two separate genes at once, even when these genes are located on different chromosomes.
5. MC A new food colouring is being tested in a lab. Using the Ames test, researchers find that a compound in the dye is associated with a rate of reverse mutation twice as high as that in the control sample.
   (a) What conclusions can the researchers draw from this information?
   (b) Many substances known to be mutagenic are, nevertheless, used in manufacturing and in food processing. In a small group, brainstorm some of the social and ethical issues that are associated with the deliberate use of mutagens. Write a short report that explains the circumstances under which you believe the risks of exposure to mutagens can be outweighed by the benefits of using these substances.
6. K/U Older people are at a higher risk of developing most cancers than young people. Why is this?
7. C Draw a diagram or a flowchart that could be used to teach the main steps involved in excision repair.
8. I One mutation results in the replacement of a G nucleotide with a T nucleotide in the sense strand of a DNA molecule. Under what circumstances will this substitution result in
   (a) a silent mutation?
   (b) a mis-sense mutation?
   (c) a nonsense mutation?

UNIT PROJECT PREP

Identify any mutagens that may be associated with the cancer you have chosen to study for your Unit Project. Explain how these mutagens work. Why can DNA repair mechanisms not correct the damage?

9.2 The Sequence of Life

EXPECTATIONS

- **Describe the functions of the cell components used in genetic engineering.**
- **Explain how eukaryotic DNA can be cloned within a bacterial cell.**
- **Describe the process and technique used to sequence DNA, and the major findings that have arisen from the Human Genome Project.**

In 1977, a new era in genetic engineering was launched by English biochemist Frederick Sanger (shown in Figure 9.9) and his colleagues. Sanger and his team worked out the complete nucleotide sequence in the DNA of the virus known as phage θX174, the unique shape of which is illustrated in Figure 9.10. This breakthrough enabled researchers to compare the exact sequence of the 5386 nucleotide bases in the virus with the polypeptide products of the virus's nine genes. As they studied the DNA sequence, the researchers made some new discoveries about the organization of genetic material. For example, from the fact that one of the genes of this virus is located entirely within the coding sequence of another, longer gene, they learned that genes can overlap. On a broader level, the work of Sanger and his colleagues opened the door to genome sequencing as a way to better understand the genetics of living cells.

The work of Sanger's team relied on three important developments. The first was the discovery of a way to break a strand of DNA at specific sites along its nucleotide sequence. The second was the development of a process for copying or amplifying DNA, which made it possible to prepare large samples of identical DNA fragments for analysis. The third development was the improvement of methods for sorting and analyzing DNA molecules. Although the techniques involved have been refined dramatically in the years since Sanger's discovery, these processes remain the basis of much of our genetic technology today. You will learn more about these processes in the following pages.

Figure 9.9 Frederick Sanger is one of only four people who have twice been awarded the Nobel prize in chemistry. He won the award in 1958 for his work on identifying the structure of proteins, and again in 1980 for his development of a technique for sequencing DNA.

Restriction Endonucleases

In order to defend themselves against infection by foreign DNA, most prokaryotic organisms manufacture a family of enzymes known as **restriction endonucleases**. Restriction endonucleases recognize a specific short sequence of nucleotides (the target sequence) on a strand of DNA and cut the strand at a particular point within that sequence. This point is known as a **restriction site**. Restriction sites occur by chance in one or more locations in almost any fragment of DNA.

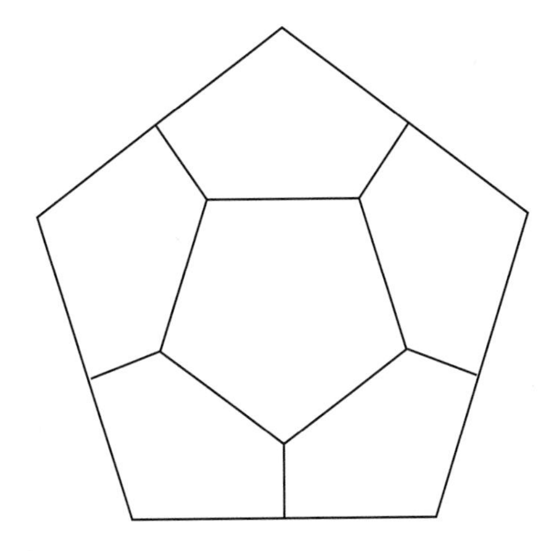

Figure 9.10 The genome of the virus known as phage θX174 was the first entire DNA molecule to be sequenced. This simplified illustration depicts the unique shape of phage θX174.

Figure 9.11 illustrates the results of a typical restriction endonuclease reaction. Many different endonucleases have been isolated, each recognizing a different sequence. Two key characteristics have made them useful to genetic researchers.

- **Specificity** The cuts made by an endonuclease are specific and predictable — that is, the same enzyme will cut a particular strand of DNA (such as a plasmid or chromosome) the same way each time, producing an identical set of smaller pieces. These smaller pieces are called **restriction fragments**.
- **Staggered cuts** Most restriction endonucleases produce a staggered cut that leaves a few unpaired nucleotides remaining on a single strand at each end of the restriction fragment. These short sequences, often referred to as **sticky ends**, can then form base pairs with other short strands having a complementary sequence. For example, they can form a base pair with another restriction fragment produced by the action of the same enzyme on a different strand of DNA. DNA ligase can then seal the gap in each strand in the new DNA molecule. In this way, researchers can produce **recombinant DNA** by joining DNA from two different sources.

Not all endonucleases produce sticky ends. Sticky ends can make binding and recombination easier, but they can also limit the uses to which endonucleases can be put. For some purposes, researchers use endonucleases that cleave the DNA molecule in a way that produces a blunt cut.

DNA Amplification

The process of generating a large sample of a target DNA sequence from a single gene or DNA fragment is called **DNA amplification**. Two different methods, as discussed below, are used by researchers.

Cloning Using a Bacterial Vector

Cloning using a bacterial vector relies on the action of restriction endonucleases. When a target sample of DNA is treated with an endonuclease, it is broken into a specific pattern of restriction fragments based on the location of the nucleotide sequences recognized by the enzyme. These fragments are then spliced (via their complementary sticky ends) into bacterial plasmids that have been cleaved by the same endonuclease. The result is a molecule of recombinant DNA.

The first recombinant DNA was created in 1973 by the American team of Stanley Cohen and Herbert Boyer. They used the process illustrated in Figure 9.12 to splice a gene from a toad into a bacterial plasmid.

The recombinant plasmid can then be returned to a bacterial cell. As the cell multiplies, it replicates the plasmids containing the foreign DNA. In this way, millions of copies of the DNA fragment can be produced. The plasmid here serves as a **cloning vector**, the term used to describe a molecule that replicates foreign DNA within a cell.

This cloning method is still in use today as a means of amplifying larger DNA sequences. For short fragments of up to about 1000 base pairs, however, a second and much faster method has since been developed.

A Most endonucleases recognize DNA sequences that have the same sequence of nucleotides running in opposite directions along the complementary strands. Pictured here is the restriction site of the endonuclease known as EcoR1.

B EcoR1 cleaves DNA in a specific way, producing the sticky ends shown here. The unpaired nucleotide bases along each staggered cut can then form hydrogen bonds with a complementary sequence of bases. DNA ligase can then seal the recombinant DNA.

C Different bacteria produce different endonucleases. The nucleotides within the target sequences of each bacterium are chemically modified such that the bacterium's own endonucleases cannot bind to them. Here, asterisks show which bases are modified in the bacterium that produces the EcoR1 endonucleases.

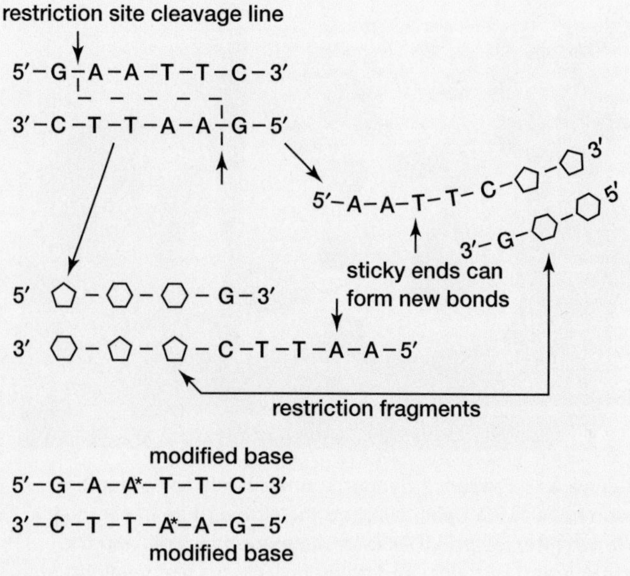

restriction site cleavage line
5′-G-A-A-T-T-C-3′
3′-C-T-T-A-A-G-5′
5′-A-A-T-T-C-3′
3′-G-5′
sticky ends can form new bonds
5′-G-3′
3′-C-T-T-A-A-5′
restriction fragments
modified base
5′-G-A-A*-T-T-C-3′
3′-C-T-T-A*-A-G-5′
modified base

Figure 9.11 Typical restriction endonuclease reactions result in sticky ends.

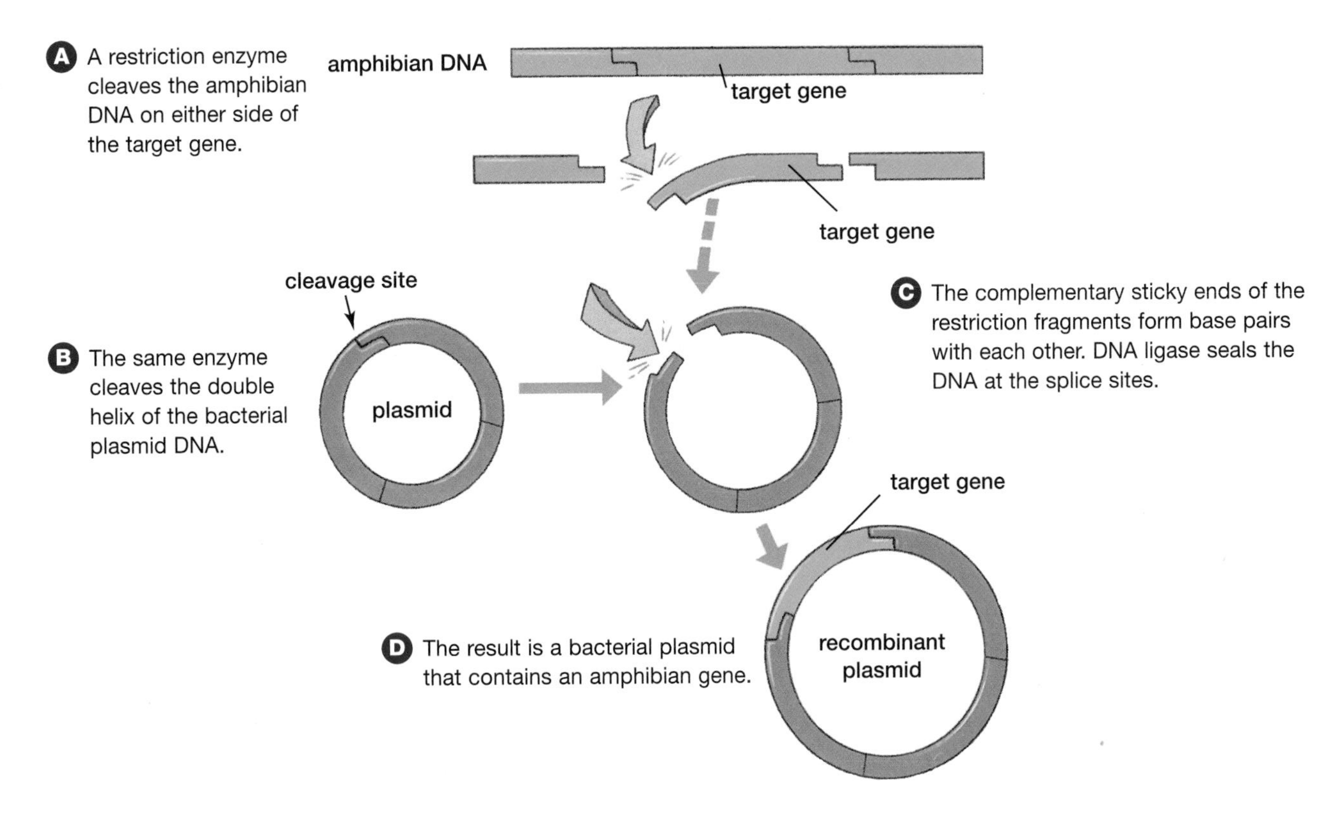

Figure 9.12 Using the bacterial vector cloning process, Stanley Cohen and Herbert Boyer developed a bacterial plasmid containing recombinant DNA.

Polymerase Chain Reaction

The **polymerase chain reaction (PCR)** is an almost entirely automated method of replicating DNA that allows researchers to target and amplify a very specific sequence within a DNA sample. It was developed by American researcher Kary Mullis in 1986 and earned him the Nobel prize.

The PCR process relies on the action of DNA polymerase, the enzyme responsible for replicating DNA. Remember that DNA polymerase cannot synthesize a new strand of DNA; rather, it can only attach nucleotides to an existing primer. This means that researchers must first prepare two primer sequences. These primers are each made up of about 20 nucleotides. The primer nucleotides have a sequence that is complementary to the 3′ end of each strand of the sample DNA molecule on either side of the DNA target sequence.

Once the primers are ready, the process (illustrated in Figure 9.13) begins. First, the sample DNA fragment is placed in a solution along with nucleotides and primers. The solution is heated to break the hydrogen bonds between base pairs, which causes the DNA double helix to open. Next, the solution is cooled. Heat-resistant DNA polymerase is then added. The DNA polymerase now starts adding nucleotides, in the 5′ to 3′ direction, to the daughter DNA strands. In just over

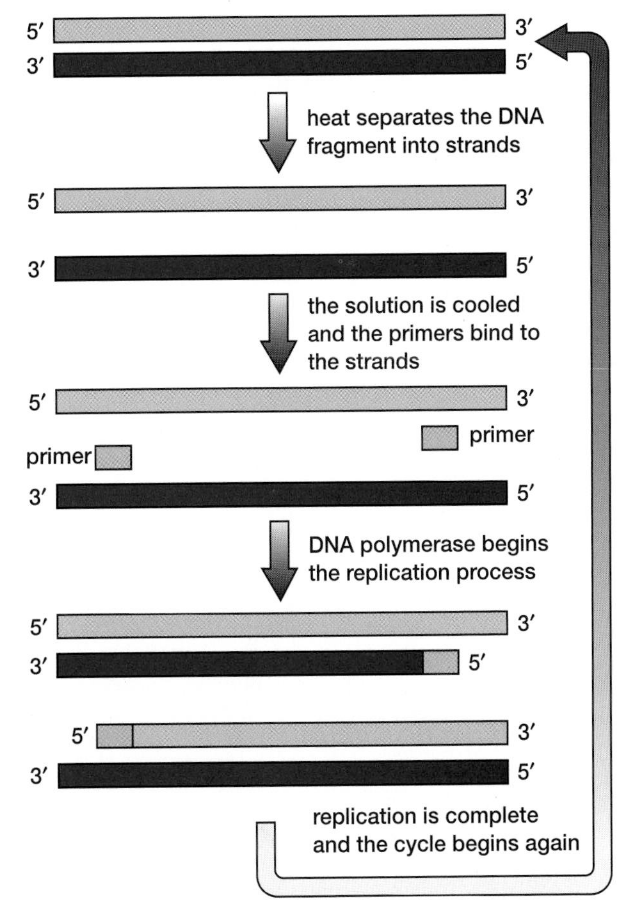

Figure 9.13 The polymerase chain reaction process is now almost entirely automated.

a minute, both DNA strands are replicated, resulting in two copies of the original target sequence. The cycle then repeats itself. Because the process uses a special heat-resistant form of DNA polymerase, it is not necessary to add new enzymes after each heating stage. Each cycle doubles the amount of target DNA in the sample, so the polymerase chain reaction can quickly generate billions of copies of a DNA sequence for analysis.

Sorting DNA Fragments

The third breakthrough that made Sanger's work possible was the development of a process called gel electrophoresis. **Gel electrophoresis** is used to separate molecules according to their mass and electrical charge. This process enables fragments of DNA to be separated so they can be analyzed.

In this process, which is illustrated in Figure 9.14, a solution containing DNA fragments is applied at one end of a gel. The gel is then subjected to an electric current, which causes the ends of the gel to become polarized. Being acidic, DNA has a negative charge. Therefore, the fragments tend to move toward the gel's positive end, with the smaller fragments moving more quickly. After a period of time, the fragments separate into a pattern of bands. This pattern is called a **DNA fingerprint**. One of the developments that made Sanger's work possible in 1977 was the refinement of electrophoresis to the point that DNA fragments could be separated if they differed in length by even a single nucleotide.

Analyzing DNA

The three processes described above — the use of restriction enzymes, DNA amplification, and gel electrophoresis — can be used in a number of ways to help researchers analyze and compare DNA samples. For example, investigators at a crime scene might find a single hair attached to a hair follicle. The DNA from this hair follicle can be amplified using DNA cloning or PCR to produce billions of copies of the sample DNA molecules. When a sample of the DNA is then cut with a restriction enzyme and run on a gel, the pattern of bands can be compared with the DNA fingerprint of the suspect. Since no two people (other than identical twins) have the same DNA pattern, a DNA fingerprint match is very strong evidence that the suspect was present at the crime scene.

In the same way, DNA fingerprint evidence can be used to solve disputes over parentage. Because a child's DNA is inherited equally from both parents, the child's DNA fingerprint will show some matches with the DNA fingerprint of each parent. As shown in Figure 9.15, a comparison of the DNA fingerprints of different people can help researchers identify the relationships among them.

ELECTRONIC LEARNING PARTNER

Your Electronic Learning Partner has an animation on restriction endonucleases and an interactive exploration on DNA electrophoresis gel results.

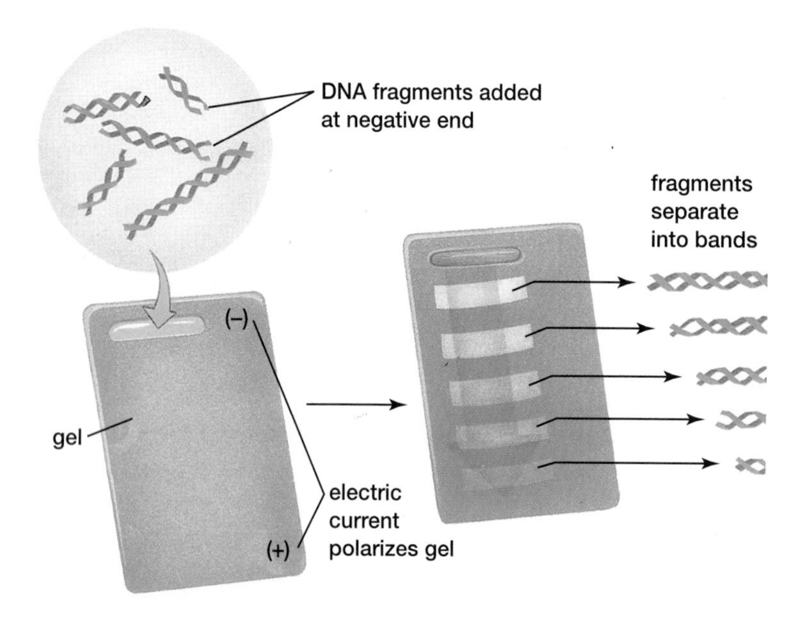

Figure 9.14 During gel electrophoresis, DNA fragments are added to a polarized gel at its negative end. Because the fragments carry a negative charge, they migrate toward the gel's positive end and separate into bands according to their mass. The resulting DNA fingerprint is an important tool in DNA analysis.

woman	man	child 1	child 2

Figure 9.15 This DNA fingerprint evidence shows the results of DNA analysis performed on a man, a woman, and two children. Based on this evidence, what can you conclude about the relationship between the children and each of the two adults?

Sequencing DNA

The same processes used to prepare DNA from different samples for comparison and analysis also play a role in determining the nucleotide sequence of a single DNA fragment. The process used to sequence DNA is known as **chain termination sequencing**. It relies on a modified form of the polymerase chain reaction.

Chain Termination Sequencing

Along with the nucleotides and polymerase used in the standard PCR process, the medium prepared for the chain termination reaction contains variants of each of the four DNA nucleotides that are known as dideoxynucleotides. As shown in Figure 9.16, these **dideoxynucleotides** resemble regular DNA nucleotides, but lack the 3′ hydroxyl group. Once a dideoxynucleotide has been added to an elongating DNA strand, DNA polymerase cannot add any more nucleotides. The replication process thus ceases, and the resulting DNA fragment breaks off.

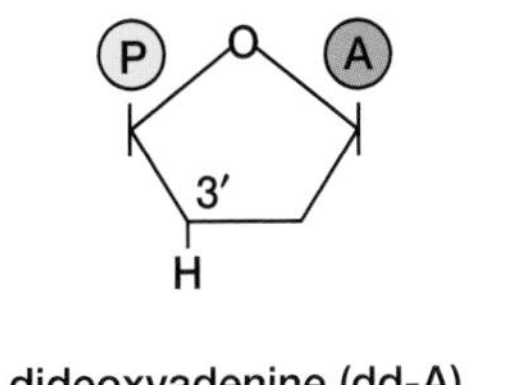

dideoxyadenine (dd-A) nucleotide

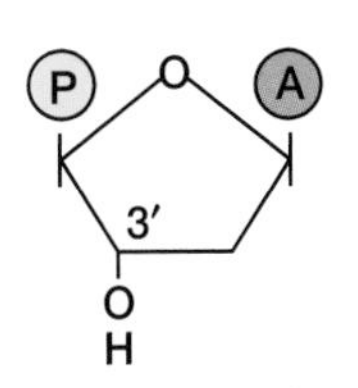

normal adenine nucleotide

Figure 9.16 Unlike the normal adenine nucleotide, the dideoxyadenine or dd-A variant lacks the 3′ hydroxyl group, which DNA polymerase needs to add another nucleotide.

The replication medium contains only a small quantity of the dideoxynucleotide variants of each of the four DNA nucleotides. As the polymerase chain reaction proceeds, there is a high probability that the polymerase enzyme will add a regular nucleotide to the growing chain and that the replication process will continue. But occasionally (using the case of adenine as an example) the polymerase will bind a dd-A to the chain instead, and the reaction will terminate. In a suspension that contains billions of elongating fragments, the end result is a series of fragments ending with a dd-A nucleotide — this is shown in Figure 9.18 on the following page. Together, these fragments represent all the possible A nucleotide locations on the elongating strand. The locations of C, G, and T nucleotides are identified in a similar way.

As shown in Figure 9.17, each of the four dideoxynucleotide variants can also be tagged with a different marker (for example, a dye that fluoresces a particular colour under ultraviolet light) to make the different nucleotides easily identifiable. As the fragments separate by length and mass during gel electrophoresis, the markers indicate which nucleotide ends each fragment. The gel can then be read from bottom to top to identify the nucleotide sequence. This step usually involves the use of an automated DNA sequencer, which speeds up the reading process.

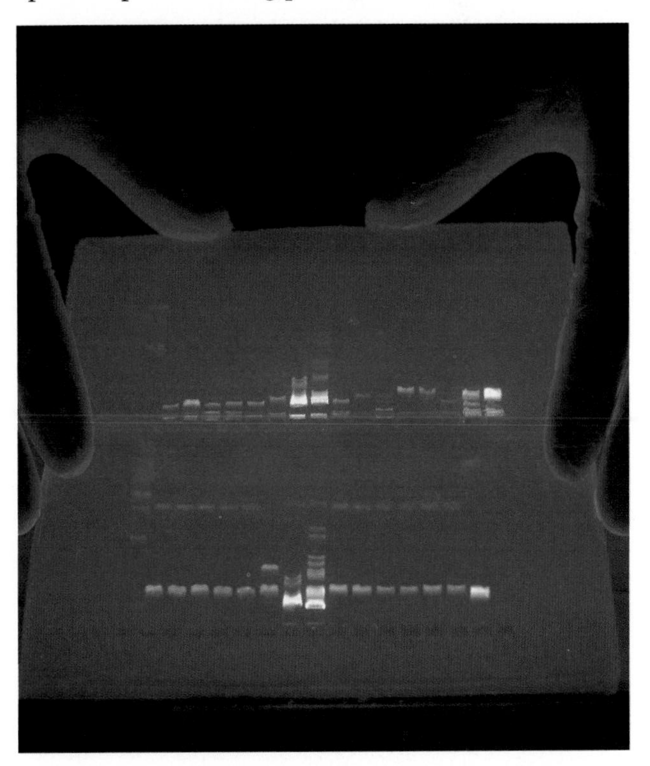

Figure 9.17 Fluorescent dyes are used to make different nucleotides easily identifiable.

The chain termination reaction can be used to sequence DNA samples of up to about one thousand base pairs in a single reaction. But even the small genome of a virus contains many thousands of nucleotides, and the genome of a mammal is

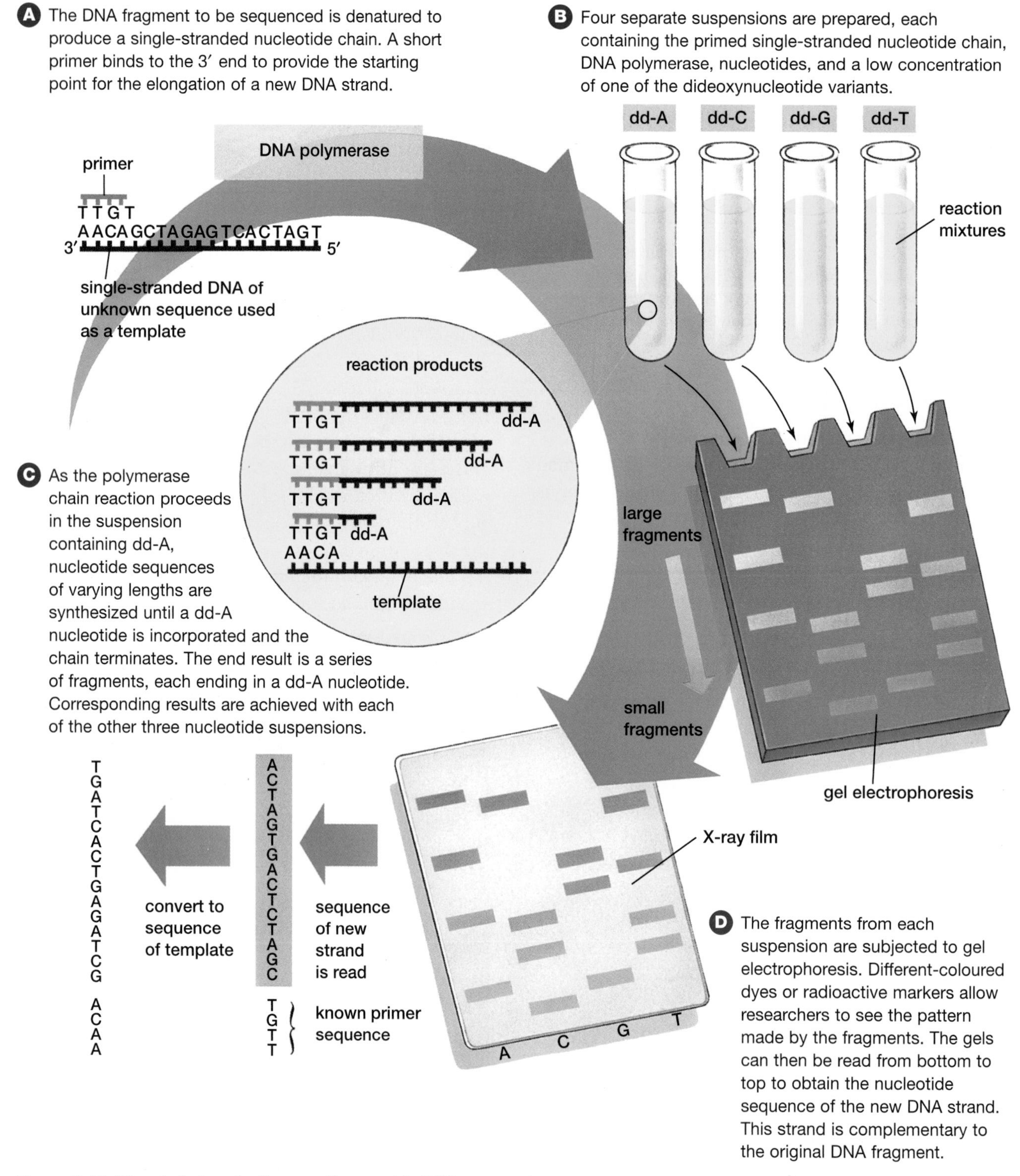

Figure 9.18 The chain termination reaction used in DNA sequencing

billions of nucleotides long. Because of this, for many years the main barrier to sequencing the DNA of eukaryotic organisms was the sheer size of the genomes involved. It was not until the late 1990s that advances in technology and computing software finally made it possible to sample and analyze enormous amounts of DNA relatively quickly. As a result, the sequencing of a large genome now involves the following three basic steps.

Genome mapping The entire genome is first randomly broken into smaller pieces of about 100 000 to 300 000 base pairs. These sections of DNA are then cloned in a bacterial vector called a **bacterial artificial chromosome** or **BAC**. By repeating this cycle several times, researchers obtain a series of overlapping BACs. These BACs are then run through gel electrophoresis to determine their individual DNA fingerprints. By studying the pattern of these fingerprints, researchers can determine the original order of the BACs within the genome.

Sequencing DNA Once the original order of the BACs has been mapped, each BAC is broken by restriction endonucleases into much smaller fragments that can be sequenced using the chain termination reaction. This sequencing step is sometimes referred to as **BAC-to-BAC sequencing**.

Analyzing the results The pattern among the resulting overlapping DNA sequences is used to determine the order of the fragments within each BAC. This procedure uses a number of different computer programs that can analyze DNA sequences. One of the biggest challenges is to distinguish true overlaps from the apparent overlaps that result from the many repetitive DNA sequences found in eukaryotic DNA.

Whole Genome Shotgun Sequencing

A second method of sequencing large genomes was developed in 1996 by American researcher Craig Venter. Called **whole genome shotgun sequencing**, this method skips the genome mapping stage entirely. Instead, it breaks the entire genome into random fragments of first about 2000 and then about 1000 base pairs. (Having fragments of different lengths helps make the nucleotide sequence assembly that follows more accurate.) These fragments, which number in the millions, are then sequenced and analyzed, after which nucleotide sequences corresponding to chromosomes are assembled. All of this is done with the aid of powerful computers and sophisticated software programs. Whole genome shotgun sequencing is faster than BAC-to-BAC sequencing, but can be less accurate. Both methods have contributed to the results of several genome sequencing projects.

The Human Genome Project

A complete draft of the human genome was first published in February 2001, making it the first mammalian genome to be sequenced. This landmark achievement, announced around the world at press conferences like the one shown in Figure 9.19, was the culmination of the work of thousands of researchers from laboratories around the world in a joint effort known as the Human Genome Project.

The **Human Genome Project (HGP)** determined the sequence of the three billion base pairs that make up the human genome. Among the project's immediate findings was the discovery that the DNA of all humans (*Homo sapiens*) is more than 99.9% identical. Put another way, this means that all the differences among individuals across humanity result from variations in fewer than one in 1000 nucleotides in each individual's genome.

Figure 9.19 In February 2001, lead Human Genome Project researchers announced the release of the first draft of the complete human genome.

Over the long term, the information from the sequencing of the human genome will provide geneticists with a better understanding of the relationship between the molecular structure of human genes and the biological mechanisms of gene function. Among other things, human genome sequencing will allow researchers to pinpoint specific nucleotide sequences that are involved in gene expression. Some of the potential benefits of these discoveries include better ways to assess an individual's risk of developing a disease, better ways to prevent disorders, and the development of new drugs and other treatments that are precisely tailored to an individual's personal genetic make-up.

In addition, comparing the human genome with the genomes of other species offers opportunities to learn more about the processes of development in living organisms. Such studies provide a significant foundation for further research.

At the same time, techniques such as DNA fingerprinting and DNA sequencing raise a number of legal and ethical concerns. For example, should a company that has worked out the sequence of a gene associated with a particular disease be allowed to prevent other companies from using this information to develop genetic screening treatments? Should an insurance agency be allowed to deny life insurance coverage to an individual who carries a gene that is linked to heart disease? In short, who should have access to an individual's genetic information, and for what purposes? These are just some of the questions that society must resolve as genetic technologies increasingly become a part of daily life.

In addition to their implications for, and potential applications in, human health and medicine, the genome sequencing techniques discussed in this

Investigation 9 • A

SKILL FOCUS

- Hypothesizing
- Performing and recording
- Analyzing and interpreting

Gel Electrophoresis

Gel electrophoresis is one of the most widely used means of analyzing DNA samples. Applications such as DNA sequencing and DNA fingerprinting rely on gel electrophoresis. In this activity, you will load and run a DNA sample and analyze the results. Alternatively, your teacher may wish to demonstrate this process to your class or ask you to conduct the activity using mock equipment and supplies.

Pre-lab Questions

- What properties of DNA allow it to be analyzed using gel electrophoresis?
- What variables might affect the movement of DNA fragments through a gel? Does this investigation control for these variables?

Problem

How can you use gel electrophoresis to compare two different samples of DNA?

Prediction

Predict what you will see if the DNA sample you load and run contains

- many copies of entire DNA molecules,
- many copies of DNA molecules broken into random fragments, or
- many copies of DNA molecules broken by a restriction endonuclease.

Explain your reasoning in each case.

CAUTION: **Active gel electrophoresis equipment generates enough electrical charge to cause serious harm. Always turn off the power and unplug the electrophoresis chamber before touching the gel box. Handle the micropipette with care. Some dyes used to stain DNA can pose serious health hazards if mishandled. Wear gloves and safety glasses at all times, and handle the dyes with care. Even non-toxic dyes will stain skin and clothing. If contact occurs, rinse the area thoroughly with water and inform your teacher. Wash your hands thoroughly after completing this lab, and dispose of the materials as instructed by your teacher.**

Materials

- prepared agarose gel slab
- electrophoresis equipment
- electrophoresis buffer solution
- prepared DNA sample
- micropipette
- prepared dye solution

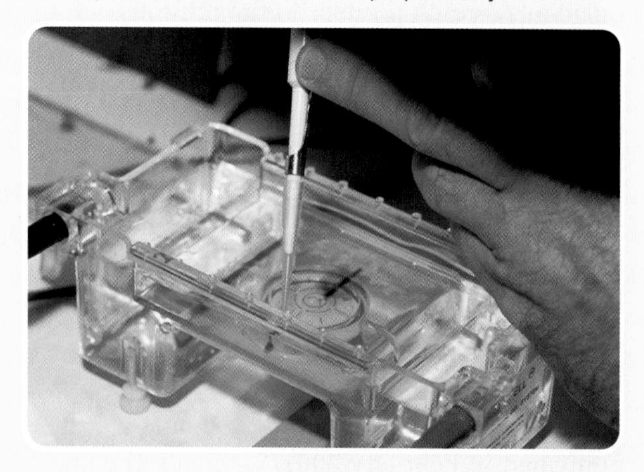

Procedure

1. Place the prepared gel slab into the electrophoresis chamber. The loading wells should be on the same side as the negative electrode.
2. Gently pour in the prepared buffer solution. The solution should cover the gel to a depth of about 2 mm.

section are now being applied to the study of the genomes of other living organisms. As you will discover in the sections that follow, these techniques are finding new applications in agriculture, industry, and environmental protection. You will also learn how researchers are using genetic engineering technology to develop new kinds of organisms, and consider what these new organisms might mean for society.

COURSE CHALLENGE

As you prepare for the Biology Course Challenge, think about how you could use each of the tools discussed in this section to analyze a sample of DNA. For example, how would you determine whether or not a certain individual carried the gene associated with porphyria?

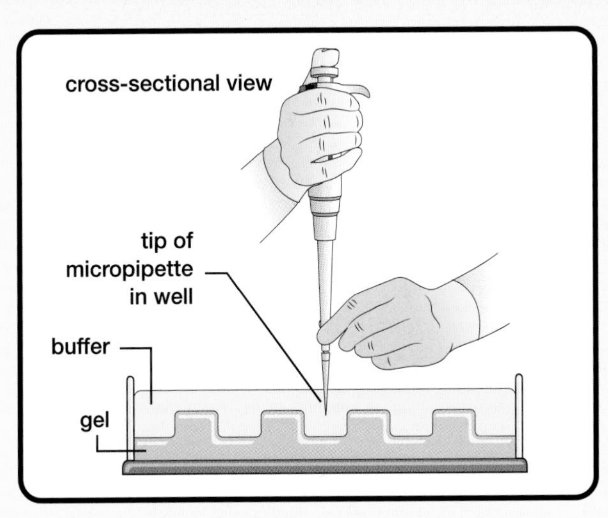

3. Mix the DNA sample with the dye as directed by your teacher. (Note: Rinse the micropipette clean between samples.)

4. Use the micropipette to gently place about 0.1 mL of each DNA sample into a different well. As you load your samples, keep the tip of the micropipette below the surface of the buffer but above the gel. One of your thumbs should be on the dispensing button, while the thumb and forefinger of your other hand steadies the other end of the micropipette just above the replaceable tip. Be very careful not to touch the agarose gel with the micropipette (if you puncture the gel, the sample will drain out). It is a good idea to practise your micropipetting technique using water before loading your samples.

5. Close the electrophoresis chamber. Make sure the power is off; then connect the cables from the power supply to the electrodes. Once the cables are connected, turn on the power.

6. Leave the power on for about 45 min, or until your teacher tells you to turn it off. Turn off the electricity and disconnect the power cables. Remove the gel slab and analyze your results. You may wish to place your slab on an overhead projector to see the bands more clearly.

Post-lab Questions

1. Draw a diagram showing your results. What do these results tell you about each of the DNA samples?

2. Compare your results with those of other members of your class. What differences do you see? How can you explain these differences?

Conclude and Apply

3. Gel electrophoresis can be used to analyze proteins as well as DNA. If you were using this technique to study samples of histone proteins, what would you do differently? Explain your reasoning. (To review the function of histone proteins in genetic material, see Chapter 7, section 7.2.)

4. The buffer solution contains sodium bicarbonate, boric acid, and salts dissolved in water. Why is this buffer solution necessary for the proper functioning of the gel?

Exploring Further

5. Use your library or the Internet to conduct research into a criminal trial in which genetic fingerprinting evidence was used. Write a report that describes how restriction endonucleases, polymerase chain reactions, and gel electrophoresis were used to compare DNA samples. Be sure to explain how the results were used in the trial, and what effect they had on the outcome of the case.

6. Use the Internet to research new discoveries that have arisen out of the Human Genome Project. Prepare a report that describes two recent findings about the structure of the human genome. Include a time line and concept map that show how these discoveries build on scientific research over the past 50 years. What do you think are some of the most important implications of these new discoveries?

SECTION REVIEW

1. K/U Compare and contrast cloning using a bacterial vector and polymerase chain reaction as methods of amplifying samples of DNA. When would you choose the cloning method as opposed to PCR?

2. K/U Explain how the addition of dideoxynucleotides to a replication medium stops DNA replication at specific points along the growing daughter DNA strand.

3. K/U To develop the first recombinant plasmid, Cohen and Boyer needed which of the following? Explain your reasoning.
   (a) a single restriction endonuclease, amphibian DNA with a single restriction site for this endonuclease, and plasmid DNA with at least two restriction sites for this endonuclease
   (b) a single restriction endonuclease, amphibian DNA with at least two restriction sites for this endonuclease, and plasmid DNA with a single restriction site for this endonuclease
   (c) two different restriction endonucleases plus amphibian DNA and plasmid DNA, each having one restriction site for each endonuclease
   (d) amphibian DNA and plasmid DNA with complementary restriction sites
   (e) a restriction endonuclease from the same bacterial species that produced the plasmid DNA, plus amphibian DNA and plasmid DNA each having one restriction site for this endonuclease

4. C At a party, you tell a new acquaintance that you intend to become a molecular geneticist and study the structure of viral genomes. He says, "Don't you think it would be more useful to study human DNA?" Considering the breakthrough work of Frederick Sanger and his team, how would you respond?

5. K/U Why is the enzyme DNA ligase necessary to produce recombinant DNA? Is this enzyme required for the polymerase chain reaction? Explain.

6. K/U DNA fingerprinting can be done using a very small amount of DNA because
   (a) new methods of gel electrophoresis are very sensitive.
   (b) the polymerase chain reaction allows researchers to amplify samples of DNA.
   (c) even a very small amount of DNA is likely to contain some restriction sites.
   (d) no two people have the same set of restriction sites on their DNA.
   (e) new DNA sequencing technologies make it possible to map a short section of DNA onto a larger chromosome.

7. C Develop a flowchart that illustrates how a bacterial artificial chromosome is used in DNA sequencing.

8. I This illustration shows the electrophoretic gel pattern that resulted from a chain termination sequencing process. What is the nucleotide sequence of the original DNA sample?

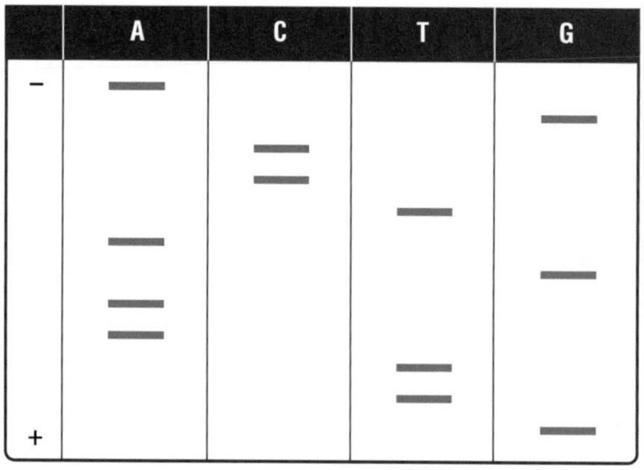

A C T G

–

+

9. MC Consider the kinds of arguments that a defence lawyer could advance to prevent DNA evidence from being used in court. Suppose that you are a forensic scientist called to testify in the case. Detail some of the points you would expect to use in your rebuttal, and why the jury would be wise to consider them.

10. MC In small groups, investigate the benefits and risks to the citizens of a nation of having the DNA fingerprints of all its permanent residents and visitors on file. Recall that DNA fingerprinting has in recent years been used to exonerate several people previously convicted of crimes.

The Chimera: From Legend to Lab

EXPECTATIONS

- Demonstrate an understanding of genetic manipulation, and of its industrial and agricultural applications.
- Outline contributions of genetic engineers, molecular biologists, and biochemists that have led to the further development of the field of genetics.
- Discuss social and ethical issues associated with genetic engineering.

The chimera is described in Greek mythology as a fire-breathing monster with the head and shoulders of a lion, the body of a goat, and a serpent for a tail (see Figure 9.20). Today, geneticists often use the term "chimera" to describe genetically engineered organisms that contain genes from unrelated species. The name may prove very fitting. The mythical chimera combined the strengths of many different animals to produce one creature that was exceptionally powerful. But this chimera was also frightening — it breathed fire and could be ferocious. In the same way, modern genetic chimeras bring together elements of different genomes in ways that can produce important social benefits. These new chimeras and the genetic technologies that create them also pose some disturbing risks; consequently, many people consider them to be dangerous.

Figure 9.20 The chimera is a mythical beast that combines parts of a lion, goat, serpent, and dragon.

As you saw in the last section, the first chimeric organism was created in 1973 when Stanley Cohen and Herbert Boyer successfully developed a bacterial plasmid that could express an amphibian gene. The work initiated by Cohen and Boyer remains the foundation of much of the genetic engineering done today.

Recombinant DNA Technology

All mammals produce a growth hormone called **somatotropin**. When cows are treated with high levels of this hormone they grow bigger, develop larger udders, and produce more milk than they normally would. In 1990, the gene coding for this hormone in cattle (bovine somatotropin, or BST) was successfully cloned and inserted into a bacterial vector using recombinant DNA technology. Produced on a commercial scale, the resulting hormone became the first **transgenic**, or genetically engineered, product approved for agricultural use in North America.

To insert a mammalian gene into a prokaryotic cell, two basic requirements must be met. First, researchers must isolate the target mammalian gene from the genome as a whole. Second, the researchers must find a way to ensure that the prokaryotic cell can express the mammalian gene correctly.

Creating and Isolating the Target Gene

In the previous section, you learned how restriction endonucleases can be used to break DNA into fragments, and how these fragments can then be inserted into bacterial plasmids for cloning. Recombinant DNA technology relies on a similar technique that employs an additional step to isolate plasmids containing the target gene.

To begin, the eukaryotic chromosome and selected bacterial plasmids to be used as a cloning vector are treated with a restriction endonuclease. When the eukaryotic DNA fragments are combined with the broken plasmids, some of the plasmids recombine with eukaryotic DNA. The plasmids are then returned to the host bacteria by simply culturing both in solution so that some of the bacteria will take up the plasmids. However, many of the plasmids will not contain recombinant DNA; of those that do, only a small portion will contain the target mammalian gene. Therefore, the next step is to isolate bacterial colonies that contain the recombinant plasmids incorporating the target gene. As illustrated in Figure 9.21, this step involves two stages of screening.

Stage 1: Identify the bacterial colonies that contain recombinant plasmids. Only a portion of the bacteria will take up recombinant plasmids. To identify those that do, researchers typically use plasmids carrying a particular genetic marker — that is, a trait that is easily identified. A common marker is the *E. coli* gene called amp^R, which confers resistance to the antibiotic ampicillin. When bacteria are plated onto a medium that contains ampicillin, only those bacteria that contain plasmids having the amp^R gene will survive and produce colonies.

Stage 2: Identify the bacteria containing the desired gene. When the mammalian DNA is broken with an endonuclease, the result is likely to be hundreds or thousands of fragments. Of these, only a small fraction will contain the target gene. As a result, another step is required to find those bacteria that contain a plasmid that includes the right gene. Identifying these bacteria involves the use of a nucleic acid probe in a technique called **nucleic acid hybridization**. If at least part of the nucleic acid sequence of the gene is known, this information can be used to construct a probe made of RNA or single-stranded DNA. The probe consists of a nucleic acid sequence complementary to the known gene sequence, along with a radioactive or fluorescent tag. As you learned in Chapter 7, the fluorescent tagging technique is also known as Fluorescence in situ Hybridization, or FISH.

To employ the probe, DNA from each bacterial colony is first heated to separate its two strands and then mixed with a solution containing the nucleic acid probe. The probe forms a base pair with its complementary sequence, making it possible for researchers to locate the tag to determine which bacterial colony contains the desired gene. Once the colony has been identified, it can be cultured to produce the gene product.

Expressing Eukaryotic Genes in Prokaryote Vectors

In Chapter 8, you examined some of the differences in transcription and translation between prokaryotic and eukaryotic cells. These differences complicate the process of expressing eukaryotic genes in

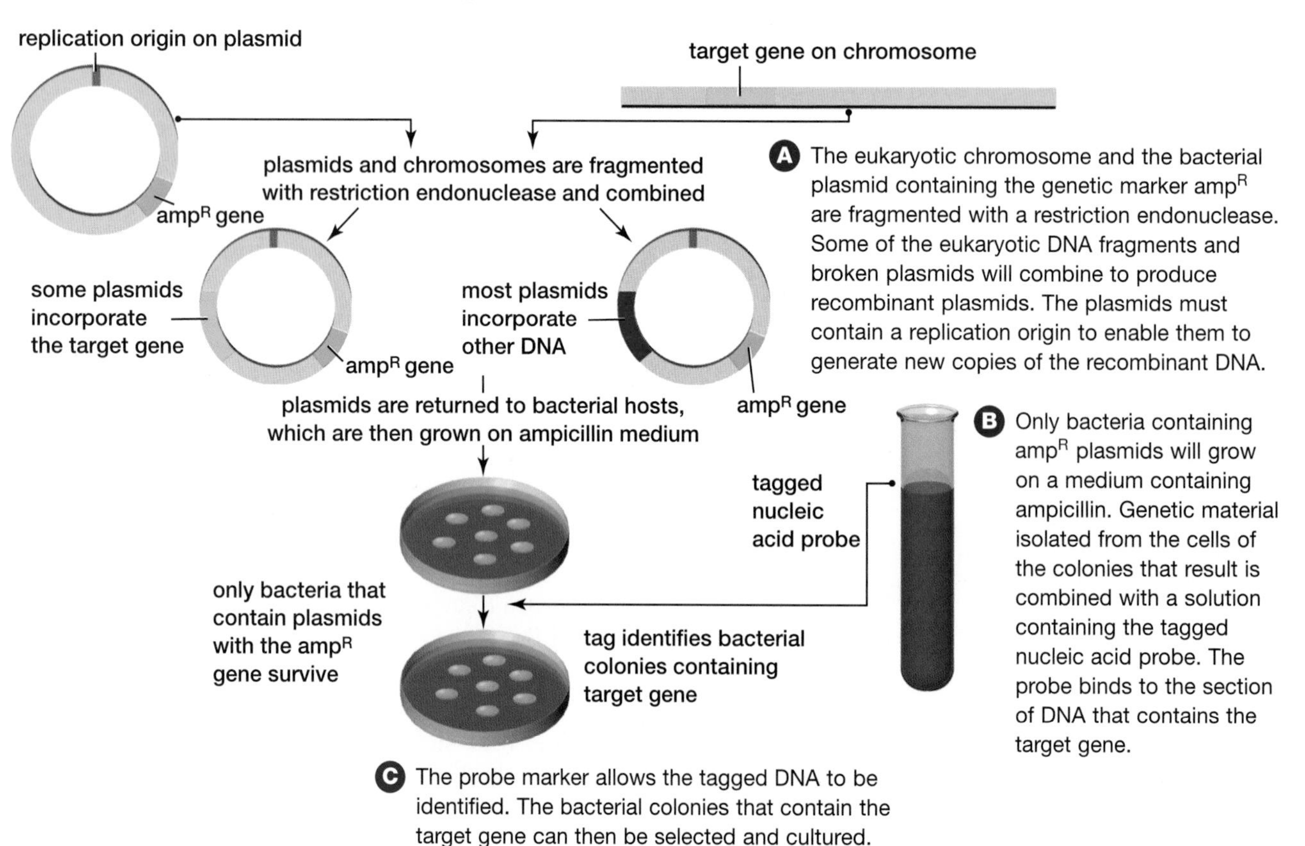

Figure 9.21 This screening process is used to identify bacterial colonies containing plasmids with the target gene. In addition to their application shown here, nucleic acid probes made up of an entire gene from one species have been used to locate similar genes in the genomes of other species.

prokaryotic cloning vectors. First, the promoter sequence of a eukaryotic gene will not be recognized by the prokaryotic form of RNA polymerase. To overcome this problem, researchers have developed a particular type of plasmid called an expression vector. An **expression vector** is a plasmid that contains a prokaryotic promoter sequence just ahead of a restriction enzyme target site. Thus, when recombination occurs, the inserted DNA sequence will lie close to the bacterial promoter. The host cell then recognizes the promoter and transcribes the gene.

Second, as discussed in Chapter 8, a prokaryote does not contain the snRNA or spliceosomes necessary to remove introns from a eukaryotic pre-mRNA transcript. This means that the mRNA transcript in a prokaryote will contain both coding and non-coding sequences, both of which will be translated by the cell. The solution to this problem has been to develop artificial eukaryotic genes that do not contain introns. Figure 9.22 shows how this is done. Researchers first isolate finished mRNA from the cytoplasm of a eukaryotic cell. The mRNA is then placed in a solution with an enzyme called **reverse transcriptase**, which creates a DNA strand complementary to the mRNA strand. This DNA strand is then isolated and added to a solution containing DNA polymerase, which synthesizes another complementary DNA strand. The result is a double-stranded molecule of DNA containing only the coding portions of the eukaryotic gene. This synthetic molecule is called **copy DNA** or **cDNA**.

Another solution to both of these problems is to use eukaryotic cells as cloning vectors. Yeast cells are often used for this purpose, since they can be cultured easily. Some yeast cells also contain plasmids, so similar techniques can be used to insert recombinant DNA into the cloning vector.

Inserting DNA into Plant or Animal Vectors

In some cases, only plant or animal cells will contain all the enzymes necessary to correctly manufacture a desired protein. Such cells can be grown in cultures to serve as cloning vectors. However, because these cells are more difficult to culture, it is harder to insert foreign DNA into them. To get around this apparent barrier and place foreign genes into eukaryotic genomes, biologists have developed several methods.

In 1983, for example, American microbiologist Mary Dell Chilton developed a process for inserting a foreign gene into a plant chromosome. For this purpose she employed the bacterium *Agrobacterium tumefacieins*, which causes tumor-like growths called galls to form on certain types of plants. This bacterium carries a plasmid known as

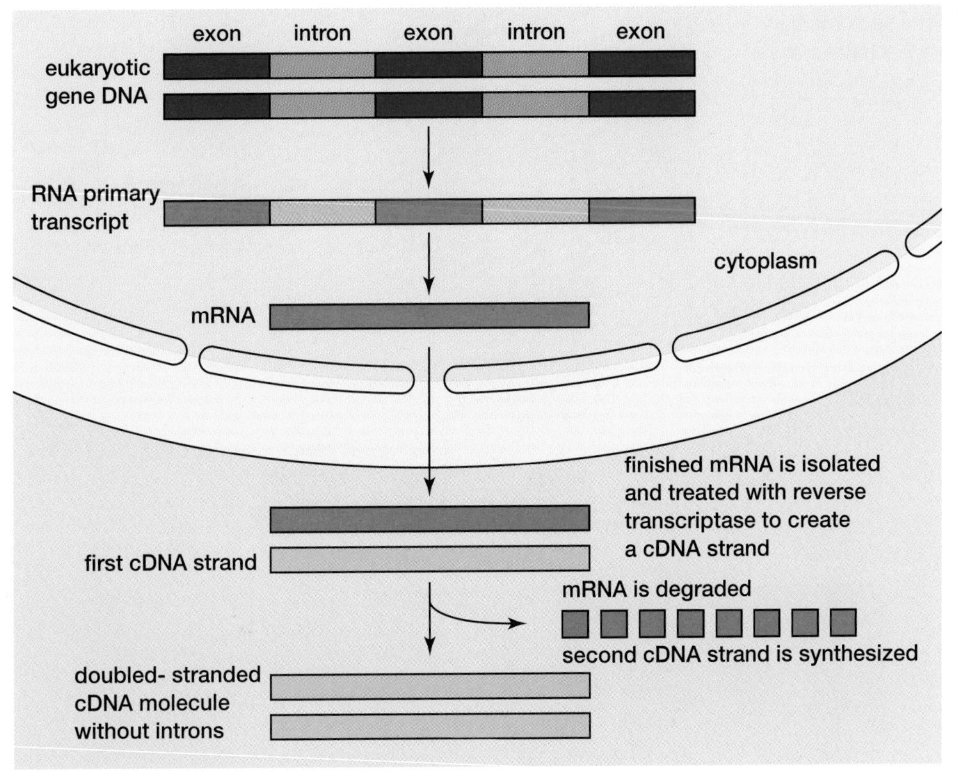

A When a eukaryotic gene is transcribed, the initial transcript contains both exons and introns. The introns are spliced out from the pre-mRNA before the transcript leaves the cell nucleus.

B The finished mRNA can be used as a template to synthesize a new strand of DNA. Reverse transcriptase, an enzyme found in some viruses, is used to create a single cDNA strand from an RNA template.

C The single strand of DNA then becomes a template for the synthesis of its own complementary strand. The result is a double-stranded cDNA molecule containing only the coding sequences of the gene.

Figure 9.22 A molecule of cDNA contains no introns. Therefore, this molecule can be correctly expressed by a bacterial host.

the **Ti** (for tumour-inducing) **plasmid**, which infects the host plant by integrating a segment of its DNA into the plant's DNA. Chilton used this plasmid's ability as a way to splice other genes into the plant genome. The process is illustrated in Figure 9.23.

This method of inserting foreign DNA into plant cells has been used to produce plants that carry many new genes. However, the Ti plasmid only infects dicots (plants having two seed leaves) as opposed to monocots (plants having a single seed leaf). As a result, researchers have had to develop other ways to bring new DNA into monocots, which include the agriculturally important cereal grains.

One key challenge has been to find a way to bring DNA from the cytoplasm into the cell nucleus. One means by which this can be accomplished is to pass a brief electric current through a solution containing a culture of eukaryotic cells. The current creates temporary pores in the cells' nuclear membranes, allowing fragments of DNA to cross into the nuclei. A second means was developed by American researcher John Sanford in 1988. Sanford invented a **DNA particle gun** (shown in Figure 9.24). This device can fire DNA-coated microscopic metal particles directly into plant cells and their nuclei. Once the foreign DNA is in a nucleus, there is a chance that it will be taken up by the host cell genome as part of the recombination process during cell division.

The new strains of organisms being developed through genetic technologies are examined by government agencies to determine their benefits and risks before they are approved for commercial use. Different countries often take different

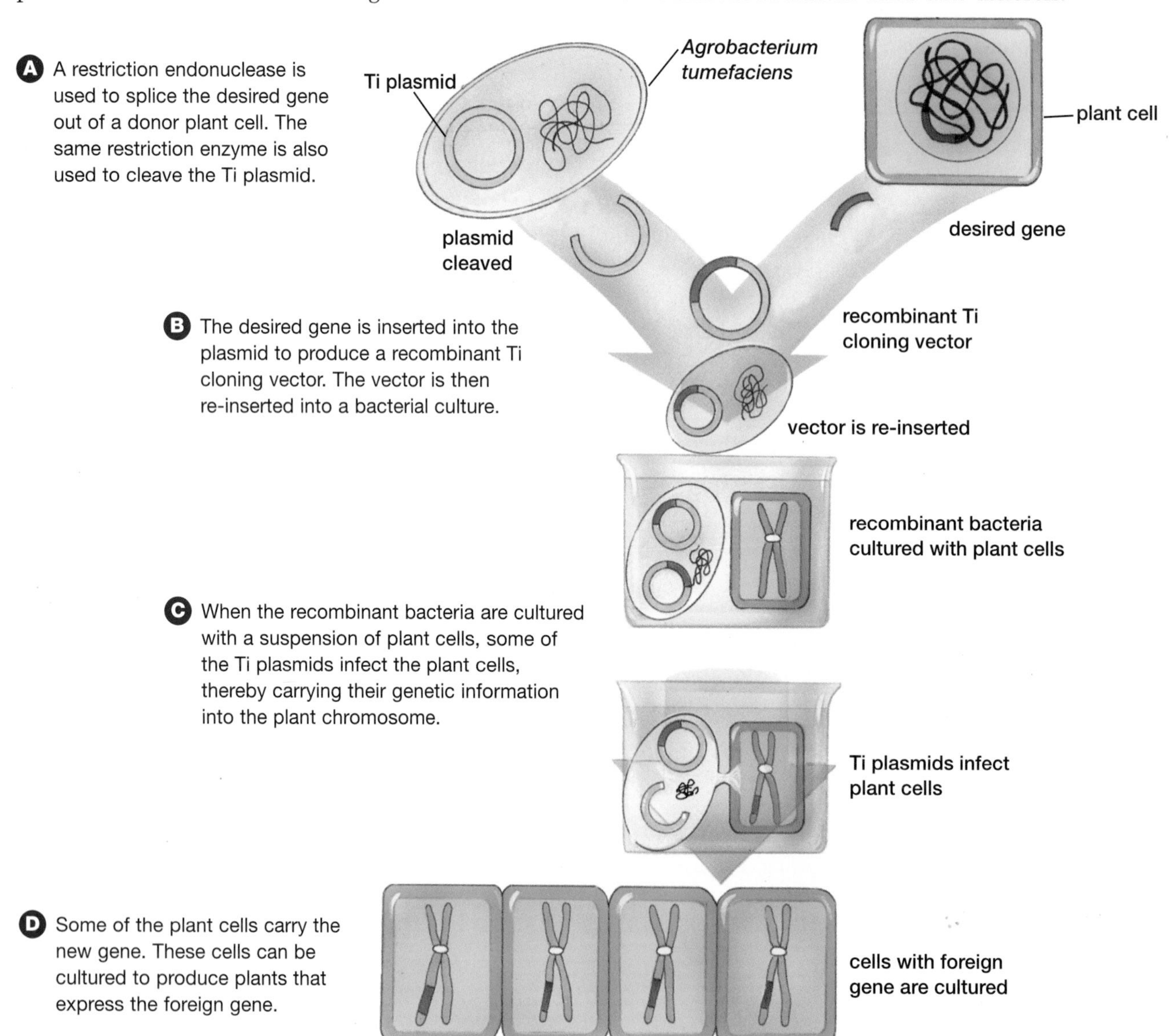

Figure 9.23 Chilton's process for inserting foreign DNA into a plant genome relies on the ability of the Ti plasmid to insert itself into plant DNA.

approaches to these decisions. For example, after studies found no evidence that milk from bovine somatotropin-treated cows posed any risk to human consumers, the commercial use of genetically engineered BST was approved in the United States in 1994. Canada, in contrast, decided in 2000 not to approve the drug for use. This decision reflected concern about the effects such hormone treatments might have on the health of cows. As this example illustrates, the potential benefits of the application of genetic engineering technologies and transgenic organisms must always be considered against the risks they may present to human health, the well-being of plant or animal stocks, and the environment. The need for standards and criteria in this area will only increase as transgenic options become increasingly available in various fields.

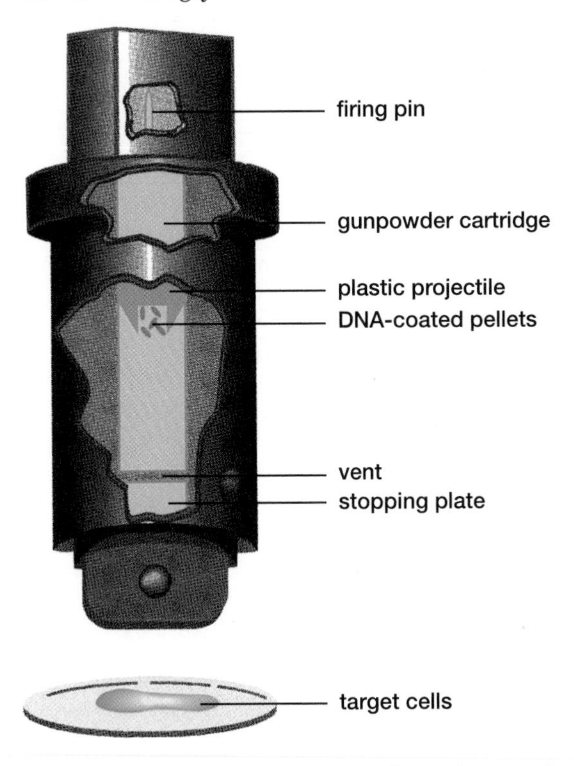

Figure 9.24 Sanford's DNA particle gun fires microscopic metal pellets coated with foreign DNA. The pellets are attached to a plastic projectile that is shot through the gun chamber. When the projectile hits the stopping plate, the pellets are torn off and carried into the plant cells and some of their nuclei. The foreign DNA may then be incorporated into the plant DNA.

Applications of Recombinant DNA Technology

Herbicide-resistant Corn

Crop plants containing recombinant DNA now account for over half of the production of corn and Canola™ in North America. Over 50 types of genetically modified crop plants have been approved for use in Canada. One example is herbicide-resistant corn. Geneticists working with a private corporation isolated and cloned a bacterial gene that provides resistance to the chemical glyphosate, an active ingredient in certain herbicides. They coated fine particles of gold with DNA fragments containing the bacterial gene and then fired the particles at a suspension of corn germ cells. Some of the cells took up the DNA. After screening for the right recombinants, the geneticists were able to grow corn that expressed the bacterial gene.

When these corn plants are grown as crops, as shown in Figure 9.25, farmers can apply herbicides to control weeds without damaging the corn. Fewer weeds means increased corn production. Corn containing this bacterial gene was approved for use in Canada in 2001, after the Canadian government concluded that the use of the transgenic corn did not present a risk to human health.

Figure 9.25 Genetically engineered plants can result in increased crop productivity.

Human Insulin

In 1982, a human insulin synthesized by transgenic bacteria was approved for medical use in the United States. This was the first example of a genetically engineered pharmaceutical product. Until that date, insulin (which is important for the treatment of diabetes) was extracted from cows and pigs that had been brought to market. Although the insulin of these livestock animals is very similar to human insulin, many patients suffered from allergic reactions. Consequently, an American pharmaceutical company developed a process for inserting the human gene for insulin into bacteria. The resulting transgenic bacteria provided a ready source of high volumes of human insulin. This ready source, in turn, lowered the cost of treatment and also reduced the number of side effects. Since that time, bacteria have been used as vectors for producing other pharmaceutical products.

Bioremediation: PCB-eating Bacteria

Among the by-products of a number of industrial processes, particularly those in the electronics industry, are polychlorinated biphenyls or PCBs. These highly toxic, environmentally persistent compounds can build up in soil and accumulate in the food chain, thereby presenting a risk to animal and human populations around the world. Cleaning up PCB-contaminated sites is difficult and costly. In response to this problem, a number of biotechnology companies have been experimenting with the development of recombinant bacteria that can degrade PCBs into harmless compounds. One technique is to locate bacteria that naturally contain genes coding for enzymes that break down PCBs and then transfer these genes into micro-organisms that can survive and reproduce well in soil. Multiple copies of these genes can be inserted into the host genome in order to increase the rate of PCB-degrading enzyme production.

Biology Magazine TECHNOLOGY • SOCIETY • ENVIRONMENT

Using Genes to Clean Up the Environment

How would you clean up millions of litres of oil spilled onto a gravel beach, or locate explosives buried in a minefield? Such difficult, dangerous, and costly tasks can now be made easier and safer with the help of genetically modified organisms.

Oil and explosives such as trinitrotoluene (TNT) are only two of the unlikely materials that can be used as raw materials by one organism or another. Bacteria in particular are able to use a huge variety of chemicals as a source of energy. Just as some insects can feed on leaves that are toxic to other insects, some bacteria can thrive on chemicals that would poison most organisms. For example, there are bacteria that can absorb phenol, cyanide, sulfur, and polychlorinated biphenyls (PCBs). These bacteria use enzymes to break down the chemical bonds in these molecules, much as we use enzymes in our stomachs to break down complex carbohydrates in our food into simpler glucose molecules.

Bacteria have lived on Earth for billions of years, and some are adapted to survive in Earth's most extreme natural environments — places that are acidic, radioactive, or that contain heavy metals. It is not so surprising, then, to find that certain of their species live in polluted areas around landfill sites, chemical factories, oil refineries, and mines. After collecting and culturing these organisms, scientists can identify their pollutant-breaking enzymes and the genes that encode them. They can then isolate the best clones and attempt either to improve the efficiency of their enzymes or to transfer their genes into other organisms in order to use them in bioremediation efforts.

Plants That Fight Soil Pollution

You may have heard that spinach is good for you because it contains iron. In fact, many plants contain metals, which they selectively absorb from the soil through their roots. For example, various members of the cabbage family (*Cruciferae*) absorb a long list of metals ranging from arsenic to mercury and zinc.

The key to plants' abilities to absorb metals is found in a group of proteins called metallothioneins. These proteins contain large numbers of atoms that readily bond onto metals. Depending on its particular structure, a metallothionein molecule may selectively bond to one particular metal. This bonding enables the species which express the protein to accumulate that metal in concentrations 30 to 1000 times greater than its concentration in the surrounding soil. In some examples, the plant may have a metal content of as much as 30 percent of the total dry mass of the plant's roots.

To be practical as a method of removing toxic metals from soil, plants must not only absorb these metals but also grow quickly in a range of different conditions and be easy to harvest. (If the plants are not removed, the metals will simply return to the soil when the plants die and decompose.) Through genetic engineering, scientists can add genes coding for metallothionein production to any plants that have these other desired properties. The result is a cheap, non-polluting way to remove or stabilize toxic metals that might otherwise be leached out of the soil into watercourses.

Can Transgenic Organisms Locate Toxins and Land Mines?

Genetically modified organisms called biosensors may soon play a role in the detection and monitoring of dangerous materials that cannot be discovered easily,

The use of living cells to perform environmental remediation tasks has become known as **bioremediation**. Other examples of bioremediation include bacteria that have been designed to clean up oil spills, filter air from factory smokestacks, or remove heavy metals from water.

Improved Nutrition

Millions of people worldwide suffer from malnutrition because they lack access to both sufficient food and balanced diets. Malnutrition can, in turn, lead to disease. Inadequate vitamin A, for example, is associated with vision problems, a weakened immune system, fatigue, dry skin, and joint pain. These symptoms affect hundreds of thousands of people in many Asian countries, where the diet consists chiefly of rice. To the potential benefit of these people, a Swiss company has recently developed a genetically modified strain of rice known as golden rice. This rice has been genetically engineered to produce beta-carotene, a vitamin A precursor. It also contains higher amounts of iron than regular rice. Golden rice is now being offered as a staple part of the food aid delivered to many developing countries, in the hope that its higher nutrient levels will help reduce the incidence of disorders linked to vitamin A and iron deficiencies.

safely, or economically by other means. For example, the standard procedure for determining if harmful toxins are leaching into the ground or water supplies is to periodically take and analyze soil and water samples. Similarly, the standard procedure for locating buried, plastic-housed land mines is for very brave individuals to search the area equipped with what is, sometimes, nothing more sophisticated than a long stick. The former procedure is an expensive, time-consuming, and labour-intensive task. The latter is a high-stakes gamble that can quickly maim its practitioner or curtail his or her life. How much more convenient would it be to have organisms living on dangerous materials sites that would unwittingly signal the location of pollutants or land mines?

Land mines buried by the millions kill or maim hundreds of civilians each year in countries where wars once raged. Plant biosensors may someday reveal their locations, allowing them to be safely detonated or removed.

How can you get an organism to send such a signal? One method is to add a gene for light production. Bioluminescence is the light produced by fireflies, glow-worms, some fungi, and many marine organisms. For example, a protein from a species of Pacific jellyfish (*Aequorea victoria*) fluoresces green when excited by blue or UV light. Genetic engineers can splice the gene that codes for this protein into a bacterium, linking it to a bacterial gene that responds to the presence of a certain toxin. When the genetically modified bacteria grow in an area containing this toxin, they will glow.

The glow of these bacteria may help identify toxins on an exposed surface, but what about those buried underground? For these applications, plants have the decided advantage of being larger and more easily tracked. One possible application of plant biosensors is to detect buried land mines containing TNT or other explosives. Historical areas of conflict such as Afghanistan, Angola, Cambodia, and the Falkland Islands are littered with millions of land mines, essentially making it impossible to farm large areas of otherwise arable land. Many land-mine casings are also plastic and thus cannot be located by metal detectors. However, some bacteria have gene promoters that are activated by TNT. These promoters can be linked to the green fluorescing genes described above and added to small, rapidly growing plants, the seeds of which could then be spread from the air over land mine sites. In time, the plant roots would spread out in the soil, absorb traces of explosives, and transport them to the leaves, which would fluoresce. Scientists could then map the location of buried explosives and decide on the best means of deactivating them.

Follow-up

Based on what you know or can find out about the characteristics of different organisms, suggest some possible combinations of genes that could be used for bioremediation or biomonitoring (using organisms to detect pollution).

Weighing the Risks

Products such as golden rice, shown in Figure 9.26, have been marketed as demonstrating the benefits of genetic engineering. However, many organizations and consumer groups argue that these benefits are outweighed by a variety of risks. In the case of golden rice, recent studies have shown that the amount of additional vitamin A offered by a regular daily serving may be as little as eight percent of an individual's daily requirement. Thus, these servings may not contribute much toward the goal of reducing the incidence of vitamin deficiency in those countries receiving food aid. The work undertaken to develop the rice has also consumed many millions of dollars that could have been spent on other, perhaps more meaningful forms of aid to developing countries. Given these questionable results, will the investment in genetically engineered foods prove worthwhile? Answers to questions such as this are part of the challenge involved in determining the advantages and disadvantages of genetic engineering technologies.

Figure 9.26 Genetically engineered foods such as golden rice can provide extra nutrition to people in countries receiving food aid, but at what long-term cost?

In Canada, proposals for the use of transgenic products are reviewed by government agencies such as Health Canada and the Canadian Food Inspection Agency. In deciding whether or not to approve such products for use in Canada, these agencies consider a number of criteria. These criteria include the potential social, economic, and environmental costs and benefits; the process by which the product was developed (including the source of genetic material); the biological characteristics of the transgenic product as compared with the natural variety; and the potential health effects, including the possibility that the product might contain toxins or allergens.

WEB LINK

www.mcgrawhill.ca/links/biology12

Each year, the Canadian government receives many applications for permission to market transgenic organisms. Use the Internet to research one application that was submitted to the government this year. Go to the web site above, and click on **Web Links**. How was this product developed? What issues does the Canadian government consider in deciding whether or not to approve a transgenic product? Prepare a brief report describing the product, its potential benefits and risks, and how you think the Canadian government should respond to the application.

Despite this review process, many organizations (including consumer advocacy groups and environmental groups) have opposed the use of a number of transgenic organisms. Among their concerns have been the following potential risks.

- **Environmental threats** The use of herbicide-resistant crops may encourage farmers to use higher levels of herbicides. This practice can lead to greater leaching of herbicides into water supplies and neighbouring ecosystems. Additionally, recent evidence suggests that genes can spread accidentally from genetically engineered organisms to wild organisms, thus posing a threat to biodiversity. In 2001, for example, a study of wild corn in Mexico found that some populations contained several different genes from transgenic corn plants. There is also a risk that herbicide-resistant crop plants could crossbreed with related natural plants, thereby producing "superweeds" that would be very difficult to control. In the same way, the development of pest-resistant plants could eventually lead to the development of "superbugs" that would be immune to certain pesticides.
- **Health effects** Many consumer groups argue that simply not enough is known about the long-term effects of transgenic products. Some believe that the consumption of transgenic products may be having effects that do not show up in the studies conducted by researchers to date. Others point to the problem of ensuring that the use of genetically modified crop plants complies with

government regulations. In the fall of 2000, for example, one biotechnology company was forced to recall stocks of a pesticide-resistant corn that had been approved only for use as animal feed. It was later found in human food products, including taco shells. This incident prompted many consumer groups to argue for tighter controls on the approval of transgenic agricultural products.

- **Social and economic issues** Advocates of genetically modified crops argue that these crops will help to alleviate world hunger. Their opponents argue that world hunger is the result of unequal food distribution, not food shortages, and that harvests of transgenic crops will not address this issue. Also, since transgenic organisms are primarily developed by private companies, many people fear the control of world food supplies could become concentrated in corporate hands. Other groups argue that the cultivation of transgenic crops favours large farms over small-scale or family farms, and that it increases farmer dependence on the corporations holding the patents on the organisms.

In addition to these issues, the treatment of living organisms as commodities to be manipulated, patented, and sold raises questions about how humans view their role in the world and their relationship to their environment. The potential benefits of genetic engineering must constantly be judged against a background of numerous social concerns, economic concerns, potential health risks, mores, spiritual beliefs, and potential environmental risks. The need to judge correctly becomes more pressing each year, as scientists apply genetic engineering technologies to various fields of medicine and develop genetically modified animals. In the next chapter, you will learn about some of the particular challenges involved in manipulating animal and human DNA, and you will examine some of the complex issues associated with this work.

ELECTRONIC LEARNING PARTNER

Your Electronic Learning Partner has animations on recombinant plasmids in bacteria.

SECTION REVIEW

1. K/U List the key steps involved in the development of bacteria that produce bovine somatotropin. In terms of the application of genetic technologies, what is the significance of the hormone produced by these bacteria?

2. K/U Describe how cDNA is used to overcome one of the challenges involved in expressing eukaryotic genes in bacterial vectors.

3. K/U Distinguish between the roles of ampicillin-resistant bacteria and nucleic acid probes as tools for preparing samples of recombinant DNA.

4. C Draw a bacterial plasmid that contains all the structures and features necessary to express a eukaryotic gene. Label each structure or feature with a brief note that explains its significance.

5. K/U Under what conditions would you use a eukaryotic vector rather than a prokaryotic vector to express a mammalian gene? What are some of the disadvantages of eukaryotic vectors?

6. C Using a computer, develop a table or flowchart that compares the main features of three different processes that can be used to insert bacterial genes into a plant cell.

7. I You are a member of a genetics research team searching for a cure for male pattern baldness. A member of your research team hands you a culture of bacteria and says, "These bacteria are supposed to express human keratin (the main protein component of human hair), but so far they have not produced any."
   (a) Make a list of the possible problems that could be preventing the bacteria from expressing the human protein.
   (b) What steps could you take to identify the problem?
   (c) What steps could you take to correct the problem?

8. MC Some companies that produce transgenic crop plants forbid farmers from saving the seeds from their crops in order to replant the transgenic organisms. Instead, the farmers must purchase new seeds each year. Write a brief report explaining some of the advantages and disadvantages of this policy. If you were a researcher working for one of these companies, what policy would you recommend?

9. C Interview one or more farmers in your community or a nearby community to discover their views on genetically engineered crops. Write a report that details their expectations and concerns.

9.4 Transforming Animal DNA

EXPECTATIONS

- **Describe the functions of the cell components used in genetic engineering.**
- **Describe the steps involved in inserting new genes into an animal genome and in cloning a mammal.**
- **Outline three different methods for inserting new genetic material into the cells of a human patient.**
- **Discuss medical and ethical issues associated with gene therapy.**

The quagga (*Equus quagga quagga*) shown in Figure 9.27 was a horse-like creature that lived in southern Africa until it was hunted to extinction in the late 1800s. More than 100 years after the last known quagga died, geneticists isolated quagga DNA from dried blood samples found in preserved quagga skins. Using DNA fingerprinting and sequencing techniques, they discovered that the quagga was a subspecies of the African zebra. Today, selective breeding projects are under way to try to recreate a quagga-like animal using zebras that carry genetic traits similar to those of the extinct quagga.

Why are the researchers working on this project relying on selective breeding rather than turning to genetic engineering techniques? Why do they not simply insert quagga DNA into a zebra genome, in the same way that DNA from one plant can be transferred to another?

Figure 9.27 Can genetic engineering help to bring back extinct species such as this quagga?

The researchers' choice of procedure hints at some of the difficulties involved in manipulating the genomes of animals. It is much more difficult to insert foreign DNA into an animal cell than into a plant cell. One of the main reasons for this is the dissimilar process of differentiation that takes place in plant and animal cells. **Differentiation** is the process by which certain portions of a genome are activated or silenced to enable a cell to take on the specific structure and function of a given tissue. As shown in Figure 9.28, differentiation is not permanent in most plant cells. This means, for example, that root cells taken from a fully grown plant can be cultured to produce an entirely new plant. In most animals, in contrast, once a cell has differentiated into a specialized tissue, it usually will be unable to give rise to other types of cells. As this cell differentiates, some portions of its DNA become permanently activated or repressed, making it very difficult to insert foreign DNA into the cell in a way that will allow that DNA to be expressed. In the following pages, you will explore two different fields of research that involve inserting and expressing foreign DNA in animal cells.

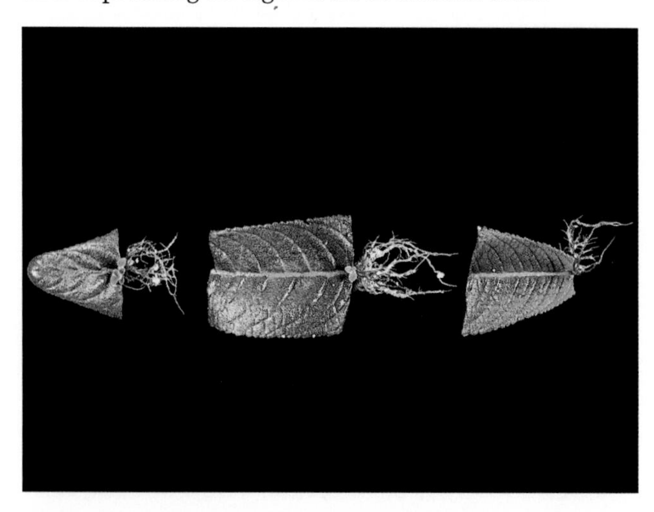

Figure 9.28 When a leaf from a succulent plant is planted in moist soil, it will grow a new set of roots. The cuttings shown here will eventually grow into complete plants. Why can an animal not produce new tissues or organs in the same way?

Cloning Animals

Organisms that are genetically identical are said to be clones of one another. A group of plants that have arisen through asexual reproduction from a single parent are clones. Identical twins, which form when a single zygote develops into two fetuses, are clones that arise naturally in animal populations. In recent years, researchers have developed laboratory techniques for cloning animals.

The first experiments in this regard date back to the 1950s. At that time, American biologists Robert Briggs and Thomas King transplanted nuclei from frog embryos and tadpoles into frog egg cells whose nuclei had been removed. When Briggs and King took the transplanted nuclei from the cells of very early embryos, they found that many of the eggs developed into tadpoles. When they took the nuclei from the cells of tadpoles, however, they discovered that very few of the eggs developed. Further, even when the eggs did develop into apparently normal tadpoles, the tadpoles never developed into adult frogs. These results gave support to the idea that differentiated cells could not be used to create clones. Many researchers concluded that the process of differentiation in animal cells meant that animal cloning from adult tissue would always be impossible.

Just some 40 years later, however, researchers began achieving that seemingly impossible goal. In the early 1990s, for example, mice were cloned by using the nuclei of cells taken from mouse embryos. More recently, an even more remarkable achievement signalled the discovery that differentiation in adult animal cells was not always irreversible.

In a country known for its extensive sheep herds, why would the birth of a lamb be headline news? If the country was Scotland and the year 1997, it would be because the lamb was Dolly, the first mammal to apparently be successfully cloned using cells taken from an adult donor. Ian Wilmut and his colleagues produced Dolly (shown in Figure 9.29) using genetic information taken from the udder cells of an adult sheep.

In order for these differentiated cells to be cultured to produce a viable embryo, the process of cellular differentiation had to be reversed. Figure 9.30 on the following page illustrates the main steps by which this was accomplished. First, Wilmut collected unfertilized egg cells from a donor sheep and removed the nuclei from these cells. Then, from a second donor animal, he removed a sample of udder cells. The udder cells were cultured in a special medium that stopped the cell cycle during the G phase. (For a review of the cell cycle, see Appendix 4.) The nuclei from these cells were then transplanted into the egg cells. When the resulting cells were cultured, a few began to divide. These early embryos were then implanted into the uterus of a third sheep that acted as a surrogate mother. One of these embryos developed into a lamb. After the birth of the lamb, now named Dolly, DNA tests confirmed that this animal was genetically identical to the sheep from which the udder cells were taken.

Figure 9.29 Dolly, the first mammal cloned using cells from an adult donor, was born in Scotland in 1997.

Since the birth of Dolly, other teams have cloned a number of other animals using cells from adult donors. As scientists study how these cloned animals develop, however, evidence is mounting that a number of problems may be associated with animal cloning. Dolly, for example, has shown signs of premature aging. Researchers working with other cloned animals have reported problems associated with gene expression.

Human Cloning

In late 2001, a team of scientists at an American research facility announced the first success at cloning human cells. The research team, led by Jose Cibelli, Robert Lanza, and Michael West, used two different techniques to clone human cells. First, using the cloning process developed by Wilmut to produce Dolly, the team obtained cloned human embryonic cells that survived long enough to divide several times. In a separate procedure, they induced human egg cells to divide, and were thus successful in producing a multicellular human blastula.

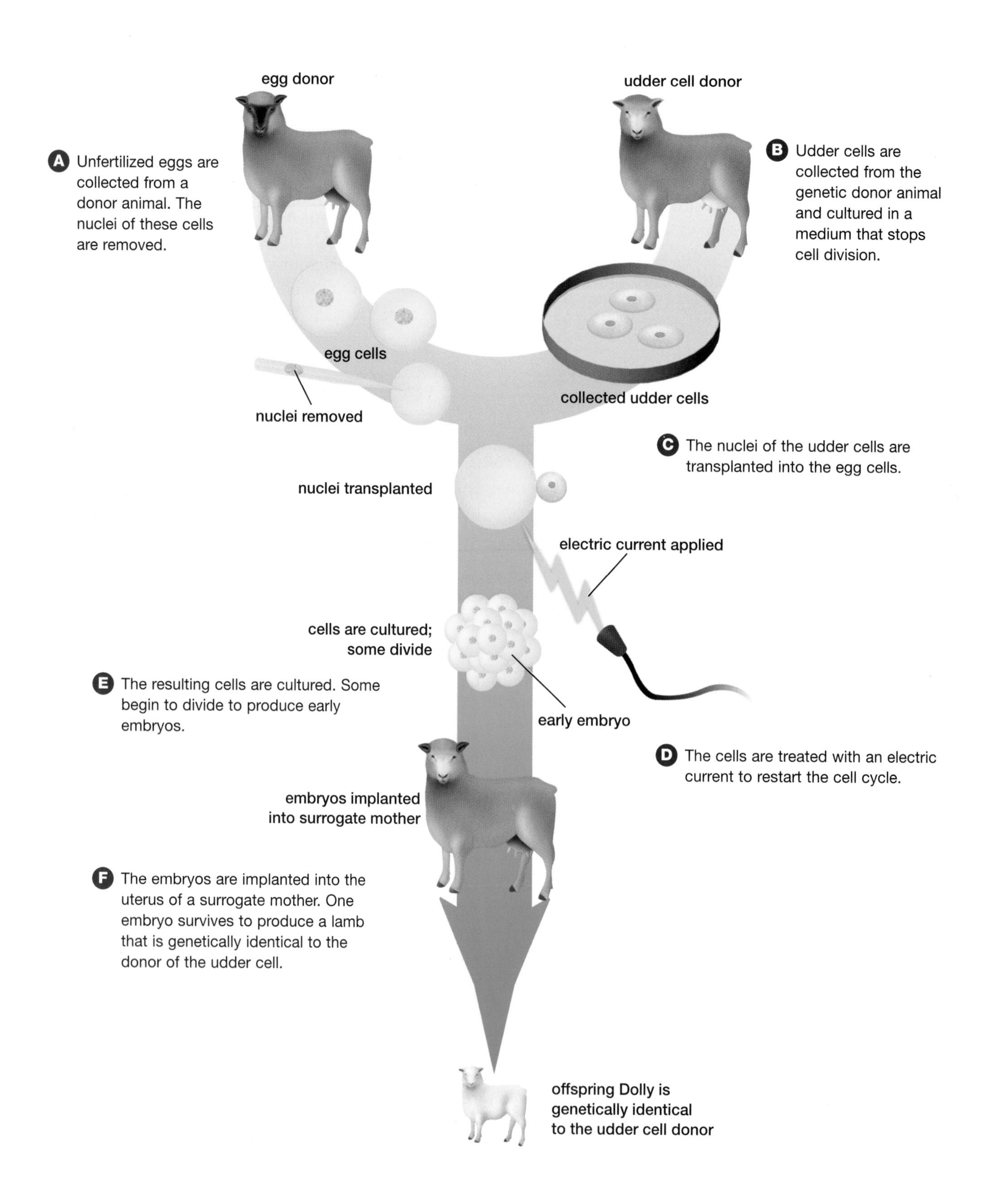

Figure 9.30 In order to produce a clone from an adult animal, geneticists must first find a way to reverse the process of cellular differentiation. In the case of Dolly, researchers stopped the cell cycle in the udder cells before inserting their nuclei into the egg cells. This procedure allowed the DNA in the differentiated udder cell nuclei to regain their potential to generate other types of cells.

Researchers involved in human cloning distinguish between therapeutic cloning and reproductive cloning. **Therapeutic cloning** is the culturing of human cells for use in treating medical disorders. **Reproductive cloning** is the development of a cloned human embryo for the purpose of creating a cloned human being. In either case, the potential benefits of these processes must be weighed against significant legal, moral, and ethical issues. Proponents of therapeutic cloning argue that this field holds the promise of eventually eliminating all human disease. On the other hand, all means of cloning animals and humans known to date involve the artificial creation and deliberate destruction of hundreds of embryos. These cloning technologies have the potential to change society's definitions of life and individuality, and as such will continue to be hotly debated in the years ahead.

Gene Therapy

Geneticists have already identified genes associated with more than 2000 human disorders, ranging from dwarfism to insomnia. In some cases, a single gene is associated with a disorder. In others, a certain gene might put an individual at a higher risk for developing a disorder. In both situations, genetic technologies have raised the possibility that cures might someday be found by correcting the function of the defective gene. The process of changing the function of genes in order to treat or prevent genetic disorders is called **gene therapy**.

The first successful human trial of gene therapy took place in 1990, when a four-year-old girl received an injection of genetically modified cells to help combat a severe immune deficiency disorder. The modified cells contained a working version of a gene that the girl lacked. While this treatment was not a cure — the girl continues to need regular infusions of the modified cells — the results indicated for the first time that it was possible to combat disorders by targeting genetic causes rather than by treating symptoms alone.

To date, gene therapy has neither produced any cures for genetic disorders nor been approved for general medical use on humans. Many clinical trials involving both animals and human patients are under way, however. As part of these experiments, genetic researchers developing gene therapy techniques must address two separate challenges. First, they must find a way to bring a working copy of the gene into a patient's body. Second, once the gene is inside the body, they need to ensure it will be expressed properly in the cell.

Transferring Genes into the Body

In gene therapy, as in recombinant DNA technology, the vehicle used to carry and replicate foreign DNA is called a vector. The two general types of vectors in use in gene therapy trials are known as viral and non-viral. Researchers are also exploring a number of other avenues, including the development of entirely artificial chromosomes.

- **Viral vector** Many viruses have the ability to target certain types of living cells and insert their own DNA into the genome of these cells. Using restriction enzymes, viruses can be genetically altered to carry a desired gene. As shown in Figure 9.31, these characteristics make viruses good candidates as vectors to deliver new genes into human patients. There are some risks associated with using viruses as vectors, however. Even though disease-causing genes are first spliced out of the viral genome, the remaining viral protein coat can trigger an immune response, including a very high fever. Several deaths in clinical trials have been attributed to such reactions in patients.

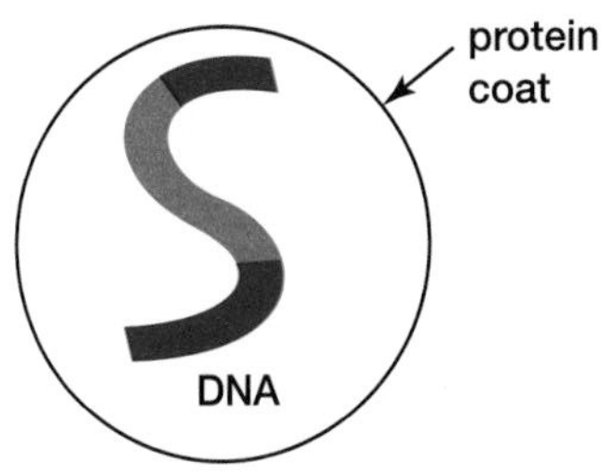

A The intact virus is made up of a protein coat containing a strand of DNA.

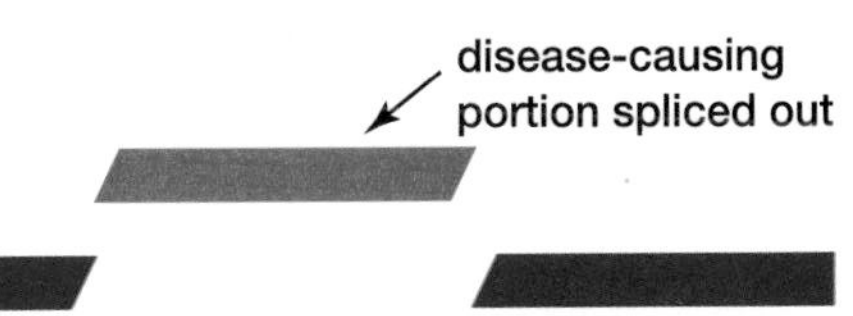

B The viral DNA is isolated and the disease-causing portion of the viral genome (red) is spliced out. Genes coding for the enzymes that allow the virus to insert its DNA into the genome (blue) of its host cell are left intact.

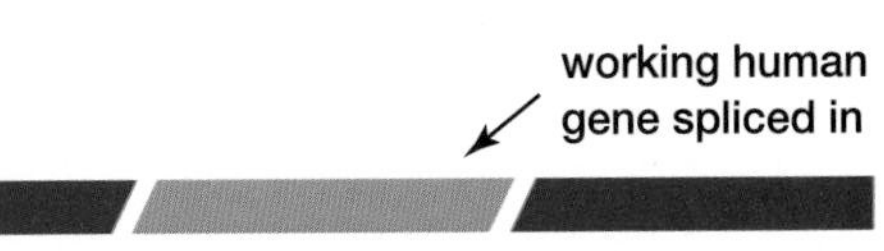

C A working human gene (green) is inserted into the viral genome. The modified viruses are then cultured with human cells. Some of the viruses will transfer the new gene into the cells' genome.

Figure 9.31 Some viruses can be modified and used as vectors to carry new genes into a human cell. The human immunodeficiency virus (HIV) — one of the deadliest viruses known — has the potential to be a very powerful viral vector because of its ability to infect many different types of cells.

- **Non-viral vector** Because of the risks associated with using viruses, scientists are exploring the possibility of developing other types of vectors. One approach being tested involves the insertion of DNA into fatty envelopes called liposomes. The liposomes can be transported across a cell's membrane and into its nucleus. Once in the nucleus, the liposome breaks down and releases the DNA. Since the liposome lacks a means of incorporating this new DNA into the cell's genome, the technique relies on the new DNA being taken up through recombination during cell division. Thus, while they are somewhat safer to use, liposomes are thousands of times less efficient than viruses at delivering new genes into a living cell.

Once a vector has been selected, there are several ways of bringing it into contact with a patient's cells. One way, known as ***ex vivo* therapy**, involves removing some of the target cells from the patient's body, culturing these cells with the vector, and then re-implanting the genetically modified cells. Blood cells are often used as the target cells for *ex vivo* therapy because they are easily removed from and replaced into the body. A second method, called ***in situ* therapy**, involves injecting the vector directly into the tissue containing the target cells. For example, a vector containing a gene that combats cancer can be injected directly into a tumour. Yet a third method being studied, called ***in vivo* therapy**, relies on a vector that can be taken up by the body through injection or inhalation. Once inside the body, the vector targets specific cells and inserts new genes directly into these cells. So far, treatments using *in vivo* therapy have not been successful, although this method continues to hold promise for future work.

Expressing New Genes in the Body

In order for gene therapy to be effective, the new genetic information must be inserted into cells that will continue to replicate this information over the life of the patient. Bone marrow cells, which produce the body's blood cells, are good candidates for lasting treatment of some conditions.

Part of the challenge facing scientists involved in gene therapy lies in the complexity of gene expression in human cells. For instance, a gene that is inserted incorrectly into a cell's genome might not be expressed correctly. Even worse, it might interfere with the expression of other genes. The effects of interactions among different genes, or between genes and the external environment, also make it very difficult for gene therapy researchers to develop reliable ways of bringing new genes into a human patient.

As work continues to try to make gene therapy an effective medical tool, some individuals and organizations are raising concerns about the ethical and moral aspects of manipulating the human genome. So far, all gene therapy trials in humans have focussed on **somatic gene therapy** — that is, therapy aimed at correcting genetic disorders in somatic cells. While such treatments can have an impact on the health of the patient, they do not prevent the disorder from being passed on to the patient's children. Other research has contemplated the possibility of **germ-line therapy**, or gene therapy that would alter the genetic information contained in egg and sperm cells. In theory, this kind of therapy could eliminate inherited genetic disorders. At the same time, however, germ-line therapy could have many unforeseen effects on future generations. So far, the Canadian government has only approved the use of somatic cell therapies for certain diseases for which no other treatments are known. Human germ-line therapy research is currently banned in Canada, and in many other countries as well.

The speed at which new genetic engineering technologies are developed and applied sometimes makes it seem as though scientists have a very good understanding of how genes work. In reality, as researchers learn more about the structure and function of genetic information in living organisms, the complexity of the interactions among DNA, proteins, and the environment is only starting to become apparent. More than a century after the discovery of DNA, exploration into the field of molecular genetics is only just beginning.

WEB LINK

www.mcgrawhill.ca/links/biology12

Gene therapy techniques may make it possible to alter the function of particular genes to either treat disease or change people in other ways. Should parents be allowed to select the eye colour of their children? Should a healthy person who is naturally short be treated to increase his or her height? Go to the web site above, and click on **Web Links** to find out more about some of the current legal and ethical issues raised by gene therapy research. With a partner, select one issue for discussion and prepare a five-minute classroom debate in which you each argue a different position on the best way to resolve the issue.

Biology Magazine TECHNOLOGY • SOCIETY • ENVIRONMENT

Medicinal Pigs and Other Clones

Every year, the lives of thousands of Canadians are saved by organ transplants. Thousands more, however, wait months or years before a suitable organ is found. As a result, each year hundreds of people on organ transplant waiting lists die before they can be treated. Even among those who receive transplanted organs, hundreds die due to organ rejection. Most of those who survive must take drugs to suppress their immune systems. Although this treatment enables their bodies to tolerate the transplanted organs, it also leaves them vulnerable to infections. Because of genetic engineering, there are now several potential avenues through which these problems may eventually be overcome.

Organs from Pigs

Pigs grow quickly and are easy to raise, and their organs are similar in size and structure to human organs. As you learned in Chapter 4, these facts have led researchers to consider pigs as a potential ready supply of organs for transplantation into human patients. So far, however, cross-species transplantation (also known as xenotransplantation) has had very limited success, because an antigen produced by animal cells usually triggers a serious immune response in humans that leads to organ rejection. Genetic researchers are exploring two different ways to avoid this result.

Genetic engineering can increase the chance of successful organ transplants from pigs to humans, but the prospect remains controversial.

In January 2002, researchers announced they had successfully developed and cloned "knock-out" piglets. These piglets were genetically modified to "knock out," or deactivate, the antigen gene in their cells. To create a knock-out pig, researchers isolate the antigen gene from the pig's genome and deactivate it by inserting a mutation into its DNA sequence. Next, they insert the modified DNA into a vector and culture the vector with the pig's cells. In some cases, the vector will insert the deactivated gene into the cultured cells. A screening process then identifies the knock-out cells, some of which are cultured into viable embryos.

Another possible way to avoid cross-species rejection is to insert a human gene into the pig genome to make the pig cells produce a human antigen instead of a pig antigen. The resulting transgenic pig embryos, along with the knock-out pig embryos described above, could then be cloned to produce a stock of pigs whose organs could be harvested.

The prospect of using pigs as a source of organs for transplantation has led to considerable debate. Many scientists are concerned about the risk of transferring diseases from pigs to humans. Others maintain it is unethical to create new kinds of animals purely for the purpose of harvesting their organs. Concerns such as these have led some researchers to look to human therapeutic cloning instead.

Become Your Own Organ Donor

Imagine you need a new heart. Your surgeon asks you for a DNA sample. From this sample a heart is grown that is genetically identical to your own, thereby eliminating both the wait for an appropriate donor and the risk of organ rejection. How could this be accomplished? The procedure would involve inserting your DNA into an enucleated cell to create an embryo — your own identical twin. This embryo would be cultured for about two weeks, then destroyed in order to collect the stem cells needed to grow your new heart.

In spite of the benefits, the practice of culturing human embryos with the object of destroying them is viewed by many as immoral and unethical. It also violates the tenets of many religions. Opponents of human therapeutic cloning also point to potential risks. Cloned animals have a higher than average rate of mortality and deformity, while those cloned using DNA from adult donors appear to age prematurely. Dolly the sheep, for example, became arthritic at the very early age of five. Such evidence suggests that organs produced through human therapeutic cloning may not function properly.

Successful therapeutic cloning for medical purposes may be years away. In the interim, governments around the world will be faced with the challenge of developing laws to guide genetic research and its applications in this rapidly changing field.

Follow-up

Different countries have enacted very different laws relating to cross-species transplantation and therapeutic cloning for transplantation purposes. Research the laws in Canada and one other country to see how they differ.

SECTION REVIEW

1. K/U Explain how the process of cellular differentiation affects methods of developing transgenic animals.

2. MC Decide whether Dolly could be described as the identical twin of the genetic donor sheep and give your reasons. What implications might your answer have on society if human cloning is someday allowed?

3. K/U In the cloning process that was used to create Dolly, what was the purpose of treating the egg cells containing the transplanted udder cell nuclei with an electric current?

4. K/U Describe the two main challenges that any gene therapy process must overcome.

5. K/U Compare the advantages and disadvantages of using viral and non-viral vectors to carry new genes into a human patient.

6. C With a partner or in a small group, brainstorm the benefits and risks of human cloning. Then write a report describing your own thoughts on whether human cloning is a good idea, and what restrictions should apply.

7. K/U What features of a genetic disorder would make it a good candidate for treatment using gene therapy?

8. C Imagine you have been asked to give a short presentation to a Grade 7 Science class describing the difference between somatic gene therapy and germ-line therapy. How would you explain the different terms? What examples could you use to support your presentation? Assume that the students are used to discussing social and ethical issues, but they have no background in genetics.

9. MC In addition to its role in the sequencing of the human genome, what are some of the potential uses of a human DNA library? What risks are associated with maintaining such a library?

10. MC Is it likely that researchers might someday develop a form of gene therapy that involves "gene pills" that you could swallow? Explain.

UNIT PROJECT PREP

Are genetic engineering technologies being used to treat the cancer you are studying for your Unit 3 Project? Find out how successful these treatments have been, and what challenges researchers are facing.

Summary of Expectations

Briefly explain each of the following points.

- All mutations are heritable, but not all mutations contribute to genetic variation within a species. (9.1)
- A number of factors contribute to the severity of a mutation. (9.1)
- The base-pairing properties of DNA are an important element in generating recombinant DNA. (9.2)
- A restriction endonuclease breaks a DNA strand in a predictable fashion. (9.2)
- Gel electrophoresis can be used to generate a DNA fingerprint. (9.2)
- The development of heat-resistant DNA polymerase has made automated DNA amplification much more efficient. (9.2)
- Two different sequencing techniques contributed to the results of the Human Genome Project. (9.2)
- Expressing a plant gene in a bacterial host presents a different set of challenges than those posed by expressing a bacterial gene in a plant cell host. (9.3)
- Genetically engineered organisms could have both positive and negative effects on the environment. (9.3)
- It is easier to clone a plant than an animal. (9.4)
- Germ-line therapy raises a number of social and ethical issues. (9.4)
- The same characteristics that make a virus deadly can make it a good candidate for use in gene therapy. (9.4)

Language of Biology

Write a sentence including each of the following words or terms. Use any six terms in a concept map to show your understanding of how they are related.

- mutation
- germ cell mutation
- somatic cell mutation
- point mutation
- silent mutation
- mis-sense mutation
- nonsense mutation
- frameshift mutation
- spontaneous mutation
- induced
- mutagen
- physical mutagen
- chemical mutagen
- Ames test
- reverse mutation
- carcinogenic
- direct repair
- excision repair
- recombination repair
- restriction endonucleases
- restriction site
- restriction fragments
- sticky ends
- recombinant DNA
- DNA amplification
- cloning vector
- polymerase chain reaction (PCR)
- gel electrophoresis
- DNA fingerprint
- chain termination sequencing
- dideoxynucleotide
- bacterial artificial chromosome (BAC)
- BAC-to-BAC sequencing
- whole genome shotgun sequencing
- Human Genome Project (HGP)
- somatotropin
- transgenic
- nucleic acid hybridization
- expression vector
- reverse transcriptase
- copy DNA (cDNA)
- Ti plasmid
- DNA particle gun
- bioremediation
- differentiation
- therapeutic cloning
- reproductive cloning
- gene therapy
- *ex vivo* therapy
- *in situ* therapy
- *in vivo* therapy
- somatic gene therapy
- germ-line therapy

UNDERSTANDING CONCEPTS

1. Contrast the likely effect of a frameshift mutation that occurs at the beginning of a gene with the likely effect of one that occurs near the end of a gene.
2. One mutagen acts by inserting itself into the double helix of the DNA molecule, causing the DNA strand to change its shape. Would this substance be considered a physical mutagen or a chemical mutagen? Explain.
3. What is a silent mutation? Explain your answer in terms of events at the molecular level.
4. Explain how a thymine dimer could arise within a bacterial cell, and how the cell might rectify this problem.
5. Complete the following table illustrating four different types of mutations.

	A	B	C	D
original DNA (sense strand)	TC*TAAG ATCTAC	C* changes to T before replication	C* lost during replication	A inserted after C* during replication
replication (sense strand)				
mRNA transcribed				
amino acid sequence				

6. Pseudogenes are slightly altered duplicates of genes that exist within an organism's genome. One organism may have hundreds of pseudogenes that correspond to a single active gene. What mechanisms could give rise to such pseudogenes over the course of time?
7. A research team conducts an Ames test on two different chemicals. The plate of bacterial cells exposed to chemical A is found to have 53 viable colonies. The plate of cells exposed to chemical B is found to have 22 viable colonies.
   (a) Which of the two chemicals is the stronger mutagen?
   (b) The control plate is found to have 12 viable colonies. Is this the result you would expect? Explain.
8. Describe the process of excision repair. In what ways is this process similar to DNA replication? In what ways does it differ?
9. Describe the contributions of the following researchers to the field of genetic engineering:
   (a) Frederick Sanger
   (b) Stanley Cohen and Herbert Boyer
   (c) John Sanford
   (d) Kary Mullis
10. What key features make a restriction endonuclease a useful tool in genetic engineering?
11. Describe how a bacterial vector can be used to amplify a sample of DNA. When would you choose to use this method rather than the polymerase chain reaction?
12. A PCR chamber in your lab develops a slight malfunction such that it is unable to heat the reaction mixture at the start of the reaction cycle. What would be the effect of this on the amplification process? Explain.
13. Explain how restriction enzymes, DNA amplification, and gel electrophoresis can be used to generate a DNA fingerprint.
14. The reaction medium used in the chain termination reaction normally contains only a very low concentration of dideoxynucleotides. What would be the effect of increasing the concentration of these molecules? Explain.
15. Describe how plasmids containing genes for antibiotic resistance can be used to screen bacterial colonies for recombinant DNA.
16. What problem did Mary Dell Chilton address with her research using Ti plasmids?
17. An electric current is sometimes used as part of the process to insert new DNA into a eukaryotic genome. Explain what the current does and why this is necessary.
18. Explain why the dissimilar processes of cellular differentiation in plant and animal cells make it more difficult to clone animals than plants.
19. Three different adult sheep were involved in the cloning process that led to the birth of the lamb Dolly. What were their roles? Which one was Dolly's clone?
20. What role can viruses play in gene therapy? Which characteristics of viruses make them good candidates for this role?

INQUIRY

21. You are given three different substances that are known mutagens. Using a variety of techniques, you analyze the results of exposure to these substances. Your findings are shown in the table at right. Using this information, link each of the three substances with one of the following molecular properties:
   (a) a molecule that can insert itself between a purine and a pyrimidine in an intact DNA strand;
   (b) a molecule similar in structure to thymine but capable of forming a hydrogen bond with guanine;
   (c) a molecule that converts cytosine to a form that can base-pair with adenine.

Substance in medium	Result of exposure
A	increase in the number of mutant colonies that synthesize mRNA in which the codon AAA is replaced by AGA
B	significantly fewer viable colonies
C	increase in the number of mutant colonies that produce modified proteins in which arginine is replaced by a "stop" codon

22. You are given two plasmids, one from each of two bacterial species. One contains gene A and the other contains gene B. You wish to create a single plasmid containing both genes.
   (a) Draw a diagram to illustrate the steps you would take to produce this recombinant plasmid.

(b) What other plasmids will result from this procedure? Which of them will be recombinant?

23. You are studying a plant cell that produces a protein with the amino acid sequence Met-Phe-Pro-Arg-Trp. Design a procedure that would enable you to locate the gene for this protein within the genome of the plant cell.

24. You successfully insert a plant gene coding for a growth hormone into a bacterial cell. When you extract and test the protein product, however, you find that it does not function as it should.
   (a) Identify and explain three problems that might lead the bacterial cell to produce a non-functioning protein.
   (b) How would you determine which of these problems has in fact occurred?

25. A molecular biologist involved in gene therapy research develops a viral vector by inserting a human gene into a viral genome. In clinical trials the vector successfully binds to target cells, but later analysis shows that none of the target cells have incorporated the new human gene into their DNA.
   (a) What might have gone wrong?
   (b) How could the biologist confirm the problem?
   (c) What steps could the biologist take to correct the problem?

COMMUNICATING

26. In a short essay, explain what is meant by the statement, "The genome of any organism is far from stable." What evidence can you offer to support this statement? What evidence can you offer to refute it?

27. Your cousin tells you that "whether or not you get cancer from smoking is just a matter of luck." Challenge your cousin to a duel of facts and list a series of points you would make in rebuttal. Will your cousin be able to identify any weak points in your argument?

28. Working with a partner, debate the statement, "The benefits of farming genetically engineered crops outweigh the risks." Then write a short report identifying what you thought were the three best points made by each side in the debate.

29. Imagine that scientists discover a gene associated with high IQ. They then develop a form of gene therapy that can insert this gene into the genome of a fetus.
   (a) Working with a partner or in a small group, brainstorm what you think might be some of the social effects of this discovery. What would happen, for example, if the treatment were very expensive? What if the treatment had the side effect of increasing the risk of mental disorders? Record your ideas in a list or concept map.
   (b) Write a brief report to describe what you think are some steps that could be taken now to prepare society for this kind of discovery in the future.

30. In a short report, discuss the risks and benefits associated with human germ-line therapy. Conclude with a recommendation on whether or not this kind of therapy should continue to be banned in Canada.

MAKING CONNECTIONS

31. A researcher develops a new form of polymerase chain reaction that can amplify long sequences of DNA. Is this technique likely to replace entirely the use of bacteria as cloning vectors? Why or why not?

32. In 1997, the United Nations declared that "Practices which are contrary to human dignity, such as reproductive cloning of human beings, shall not be permitted." In a short report, discuss why the United Nations considers human cloning to be a practice contrary to human dignity. In your summary, add your own reasons for why you agree or disagree with the United Nations' position.

33. A researcher develops a form of nucleotide variant whose 5′ phosphate group is unable to form a bond with the 3′ hydroxyl group of a neighbouring nucleotide. Could this molecule be used in a chain termination reaction in place of a dideoxynucleotide? Explain.

Cancer: Facts Versus Fiction

Background

In the fall of 2001, the sports world was stunned to learn that the captain of the Montréal Canadiens hockey team had been diagnosed with stomach cancer. International news agencies broadcast details of the planned course of treatment while fans, celebrities, and cancer survivors from around the world offered their support and encouragement to the 26-year-old athlete, Saku Koivu, shown below.

Today, cancer is recognized as one of the leading causes of death for people of all ages across North America. The term actually encompasses a set of diseases, each of which involves a breakdown in the molecular mechanisms that ordinarily control the rate of cell division. Affected cells grow uncontrollably, resulting in cancerous tumours that interfere with the normal function of body tissues and organs.

Each year, hundreds of millions of dollars are spent in Canada to support cancer research. Advances in our understanding of the molecular structure and function of the genetic material of the cell have made it possible to design new and increasingly effective treatments. Even so, it is not always easy to separate cancer myths from cancer realities. Is it true, for example, that eating blueberries or tomatoes can reduce your risk of developing cancer, or that men can get breast cancer? Is it true that inhaling the steam from a cup of coffee or drinking green tea can help to prevent cancer? In this project, you will sort fact from fiction in relation to one particular form of cancer.

ASSESSMENT

After you complete this project,

- **Assess your presentation and report based on how clearly your information was conveyed;**
- **Assess your presentation based on the responses from your classmates and the evaluation criteria that your class developed;**
- **Assess your research and communication skills as they developed over the course of the project. In what ways did your skills improve?**

Challenge

Working in a small group, research, design, and prepare a presentation on a particular form of cancer. Be sure to describe the molecular basis of the cancer and the means by which it is usually diagnosed and treated. Discuss the role of suspected risk factors, such as exposure to certain types of environments or mutagens. As part of your research, interview people who have different perspectives on the disease. You might, for example, interview people who have cancer, medical professionals or researchers, and/or people who have different attitudes about some of the ethical or social issues involved in prevention or treatment. Your presentation may include a variety of media and take any number of forms, such as the following:

- an audio-visual presentation to a simulated medical conference;
- a virtual tour of a research or medical facility on a web site;
- a drama you write and perform, either live or on video;
- a prepared speech; or
- a simulated TV panel discussion in which one team member plays the host and the others function as experts.

Materials

You will need to choose appropriate materials with which to present your data. If your presentation includes a web site, for instance, make sure it can be viewed on the equipment available at your school. If your plans call for an audio-visual presentation or speech, make sure any props or aids you use will be visible to the entire class. If you want to include recorded interviews, make sure you have access to good recording and playback equipment.

Design Criteria

A. Your presentation should combine information you obtain through your research with material you have learned in each chapter of this unit.

B. Include visuals to support your presentation. Some examples might be graphs showing the incidence of the cancer among different demographic groups or in different parts of the country, models illustrating the difference between normal and cancerous cells, or illustrations showing how environmental factors can act as mutagens to trigger the cancer.

C. Explain how changes in treatment and preventive measures are related to advances in our understanding of molecular genetics. This might include a description of key experiments that have helped researchers learn more about this form of cancer.

D. In addition to your presentation, you should prepare and submit the following written materials:

- an introduction that identifies your team members and explains how and why you selected your topic;
- a research log that records your progress, including any difficulties you encountered and how you dealt with them; and
- a bibliography that lists the research sources you used, along with the names of people you interviewed as part of your research.

Action Plan

1. As a class, decide how you will evaluate the presentations.
2. Decide on the particular form of cancer to study. Then develop a plan to find, collect, and organize the information you will need. Remember to keep a list of all your references and interview subjects for your bibliography.
3. Identify the materials you will need to prepare your report and presentation, including any props, visual aids, and equipment.
4. As you collect information, think about how your presentation might address issues such as the following:
   - What genetic, environmental, or other factors are associated with this cancer?
   - Is there a main group that is at risk of developing this cancer, such as people with a certain diet or lifestyle? Has this group changed over time? If so, how?
   - What is the state of current research on the causes, prevention, and treatment of this cancer? What have initiatives like the Human Genome Project contributed to this work?
   - What are some of the social, ethical, and legal issues associated with this research?
   - What are some common beliefs about the causes, prevention, and treatment of this cancer? Which of these beliefs can be supported by scientific evidence?
   - What do you think are some of the most promising avenues for research into the future treatment or prevention of the cancer you are studying? What are some of the technical or ethical hurdles that need to be overcome?
   - Have your teacher check and approve your action plan. As you carry out your plan, note any changes that you make along the way.
   - Once you are satisfied that you are properly prepared, make your presentation to the class.

Evaluate

1. Evaluate your own work and presentation using the evaluation criteria developed by your class. How effective do you think your presentation was?
2. Evaluate the presentations made by other members of your class. Were there any common themes or areas of significant difference among the reports presented by different members of your class?
3. After seeing all the presentations, what changes would you make to your own report if you were to present it again? Explain your reasons.
4. How did working on this project help you think about what you have learned in this unit?

UNDERSTANDING CONCEPTS

Multiple Choice

In your notebook, write the letter of the best answer for each of the following questions.

1. Friedrich Miescher is known for being the first scientist to
   - **(a)** isolate nuclein from white blood cells
   - **(b)** show the link between nuclein and Mendel's "factors of inheritance"
   - **(c)** argue that protein, not DNA, is the material of heredity
   - **(d)** distinguish between DNA and RNA
   - **(e)** propose that DNA is the material of heredity
2. The experiments of Avery, MacLeod, and McCarthy were significant because they demonstrated that
   - **(a)** bacteria can remain pathogenic even after they have been killed
   - **(b)** transformation takes place when the pathogenic bacteria are treated with a protein-destroying enzyme, but not when they are treated with an RNA-destroying enzyme
   - **(c)** transformation takes place when pathogenic bacteria treated with a DNA-destroying enzyme are cultured with non-pathogenic bacteria
   - **(d)** transformation takes place only when the DNA of the pathogenic cells is intact
   - **(e)** treating pathogenic bacteria with a DNA-destroying enzyme has the same effect as heat-killing the cells
3. Which of the following did not contribute to Watson and Crick's work to determine the structure of the DNA molecule?
   - **(a)** Chargaff's findings about the nucleotide composition of DNA
   - **(b)** Franklin's study of X-ray diffraction images of DNA
   - **(c)** Levene's findings about the arrangement of nucleotides in nucleic acids
   - **(d)** knowledge of the molecular properties of nitrogenous bases
   - **(e)** Franklin's observations about the way DNA reacts to water
4. In a strand of DNA, each nitrogenous base is
   - **(a)** bound to a phosphate group and to a deoxyribose molecule
   - **(b)** connected to the next nitrogenous base through a phosphate bridge
   - **(c)** connected to a deoxyribose molecule through a phosphate bridge
   - **(d)** capable of forming a hydrogen bond with another nitrogenous base
   - **(e)** (a) and (d)
5. Which of the following statements is not correct?
   - **(a)** Proteins play an important role in packing DNA in eukaryotes, but not in prokaryotes.
   - **(b)** Prokaryotic cells may contain more than one molecule of DNA.
   - **(c)** All living cells use DNA as their genetic material.
   - **(d)** DNA may be packed to different degrees within the same eukaryotic cell.
   - **(e)** The packing of DNA plays a role in regulating gene expression.
6. Which of the following statements about the eukaryotic genome is true?
   - **(a)** The best functional definition of a gene is that one gene codes for one polypeptide.
   - **(b)** Most of the DNA is usually found in a supercoiled form, although it will condense further during replication.
   - **(c)** The size of the genome is loosely related to the organizational complexity of the organism.
   - **(d)** The removal of exons determines the polypeptide product of a gene.
   - **(e)** Over 90 percent of the genome may be made up of non-coding sequences.
7. Which of the following is not evidence in support of the triplet hypothesis?
   - **(a)** Four nucleotides can code for 20 nucleic acids.
   - **(b)** The insertion of three nucleotides into a strand of DNA causes a slight change to the protein product.
   - **(c)** More than one codon can code for a particular amino acid.
   - **(d)** The insertion of a single nucleotide into a strand of DNA causes a major disruption to the coding sequence.
   - **(e)** Three codons do not code for any amino acid.
8. Which of the following statements about gene expression is incorrect?
   - **(a)** A molecule of mRNA will have a nucleotide sequence similar to that on the antisense strand of the DNA molecule, except that it will have U nucleotides in the place of the antisense strand's T nucleotides.

(b) In eukaryotes, mRNA must move to the cytoplasm before transcription can occur.
(c) Transcription and translation can take place simultaneously along a single molecule of mRNA.
(d) Proteins may determine when transcription takes place.
(e) Proteins may determine when translation takes place.

9. The characteristics of the genetic code include
(a) redundancy, which means that one codon can code for several amino acids
(b) universality, which means that the deletion of a nucleotide triplet will have the same effect in prokaryotic and eukaryotic cells
(c) continuity, which means that the addition of a single nucleotide will change the reading frame for all subsequent codons
(d) a "wobble" in the third nucleotide position, which creates ambiguity in the code
(e) both redundancy and continuity, which help to protect the cell against harmful mutations

10. Which of the following is not true of promoter sequences along a stretch of DNA?
(a) They are usually rich in T and A nucleotides.
(b) They can bind RNA polymerase in more than one direction.
(c) They can be blocked by polypeptides.
(d) They can be found on both strands of a DNA molecule.
(e) Their affinity for RNA polymerase can be changed.

11. Gene expression differs in prokaryotes and eukaryotes in that
(a) in eukaryotes, mRNA is processed so that exons are removed
(b) in prokaryotes, only a small portion of the genome is likely to be expressed during the life of the cell
(c) in eukaryotes, operons may include genes on more than one chromosome
(d) in eukaryotes, transcription must be completed before translation can begin
(e) in prokaryotes alone, many ribosomes can attach to a single mRNA strand to form a polyribosome complex

12. Which statement about the *lac* operon in *E. coli* is not correct?
(a) It is an example of transcriptional regulation.
(b) An activator can bind directly to the DNA molecule to promote transcription of the *lac* genes.
(c) A repressor can bind directly to the DNA molecule to inhibit transcription of the *lac* genes.
(d) Lactose does not bind directly to the DNA molecule.
(e) The rate of synthesis of the repressor protein drops when lactose is available to the cell.

13. Some mutations may be silent because
(a) the redundancy of the genetic code means that a change in the nucleotide sequence of a gene will not always change the amino acid sequence of the gene product
(b) they occur in somatic cells rather than germ cells
(c) they may not affect the transport of the polypeptide from the cytoplasm to the target tissue
(d) the continuity of the genetic code means that a change in the reading frame might not affect the structure of the final protein product
(e) cells contain many enzymes that can correct damage to the DNA molecule

14. A molecule that can enter a cell and cause random damage to the DNA molecule is known as a
(a) physical mutagen
(b) chemical mutagen
(c) base analogue
(d) dimer
(e) somatic mutagen

15. Which of the following statements about restriction endonucleases is not correct?
(a) They can be used to prepare DNA samples for fingerprinting.
(b) They can help protect bacteria from infection by viruses.
(c) They target a specific nucleotide sequence.
(d) They can be used in the polymerase chain reaction to amplify small amounts of DNA.
(e) They can be used to map sections of DNA along a chromosome.

16. Stanley Cohen and Herbert Boyer are known for being the first research team to
(a) sequence a bacterial plasmid
(b) sequence the genome of a virus
(c) clone a mammalian gene in a bacterial vector
(d) create a recombinant bacterial plasmid
(e) use a fluorescent probe to locate a gene on a chromosome

17. Which of the following is not used in the chain termination sequencing reaction?
    (a) DNA polymerase
    (b) dideoxynucleotides
    (c) a buffer solution
    (d) a primer strand of nucleotides
    (e) regular nucleotides

18. The process of cloning a mammal using genetic material from an adult donor involves a number of steps. Select the answer below that shows three of these steps in the correct order.
    (a) remove nuclei from egg cells; fragment DNA from genetic donor cells with endonuclease; culture egg cells and DNA from genetic donor cells
    (b) remove nuclei from donated egg cells; treat cell culture with an electric current; implant clone embryos into surrogate parent
    (c) collect cells from genetic donor; transplant nuclei from genetic donor cells into egg cells; culture cells to stop cell division
    (d) culture genetic donor cells to stop cell division; insert new genetic material; culture cells to produce early embryos
    (e) collect cells from genetic donor; culture egg cells to stop cell division; transplant nuclei from genetic donor cells into egg cells

Short Answers

In your notebook, write a sentence or a short paragraph to answer each of the following questions.

19. Explain how Chargaff's findings with respect to the composition of DNA provided evidence for the role of DNA in heredity.

20. Draw a DNA molecule that has the nucleotide sequence ATTCTGGC along one strand. Label the 5′ and 3′ ends.

21. Explain the role played by each of the following forces in maintaining the structure of DNA within eukaryotes.
    (a) hydrogen bonds
    (b) hydrophilic and hydrophobic interactions
    (c) phosphate bonds
    (d) histone–histone interactions
    (e) DNA–histone interactions

22. Explain how elongation of a daughter DNA strand takes place
    (a) in the direction of the movement of the replication fork
    (b) in the direction opposite to the movement of the replication fork

23. The genome of a eukaryote is made up of genes along with various kinds of non-coding DNA sequences. Give one example of how non-coding DNA could play a role in each of the following functions.
    (a) regulating gene expression
    (b) increasing the organizational complexity of an organism
    (c) governing the life span of a cell

24. Copy and complete the following table.

Polypeptide	Location in eukaryotic cells (cytoplasm or nucleus)	Function(s)
DNA polymerase		
activating enzymes		
excision repair enzymes		
RNA polymerase		
release factors		
helicase		
RNA primase		

Use the genetic code table on page 254 to answer questions 25–27.

25. What amino acid is associated with each of the following codons?
    (a) GCC
    (b) UUU
    (c) ACU

26. What anticodon sequences are associated with each of the following amino acids?
    (a) leucine
    (b) tyrosine
    (c) glycine

27. What features of the genetic code help protect the cell from the effects of mutations? Give specific examples using information from Table 8.1.

28. Answer the following questions in regard to the electron micrograph pictured on page 327.
    (a) What cellular process or structure does this figure show? Describe its significance in relation to the metabolism of a cell.
    (b) Where in the cell is this process taking place?
    (c) Identify each of the labelled structures.
    (d) Identify the 5′ end of the nucleic acid(s) shown.

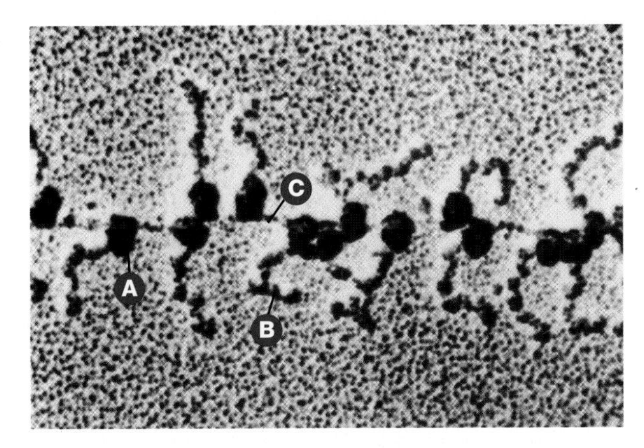

29. Prepare a diagram illustrating the elongation cycle of translation. Include a brief written description of each step in the cycle. How does a stop codon work to end the cycle?
30. Given the fact that complexes known as spliceosomes are found in eukaryotic cells, answer the following questions.
    (a) Where in a eukaryotic cell would you expect to find spliceosomes? Explain.
    (b) You are able to remove all the spliceosomes from a living eukaryotic cell. What effect would you expect this to have on the metabolism of the cell? Explain.
    (c) You are able to insert spliceosomes into a living bacterial cell. What effect would you expect this to have on the metabolism of the cell? Explain.
31. Explain how the action of the *tryp* operon helps a bacterial cell conserve energy and resources. Why is the *tryp* operon considered a repressible operon?
32. An *E. coli* cell undergoes a frameshift mutation in the gene that codes for *lac* repressor protein.
    (a) Would you expect this mutation to affect the *lac* genes? Explain.
    (b) Assume this mutation makes the repressor protein non-functional. In what way would this cell react differently than a normal *E. coli* cell to a sudden increase in the concentration of lactose in its environment?
33. A plant cell is subjected to a high level of UV radiation. What effect could this exposure produce in the DNA of this cell? What might the effect be on its daughter cells?
34. Create a labelled diagram that illustrates the steps in excision repair.
35. Describe two different methods that can be used to amplify a sample of DNA.
36. A researcher wishes to create a transgenic banana that will stay firm and yellow longer than a wild banana does. As a potential aid, she has isolated a strain of bacteria that secretes a protein that slows the ripening process in bananas.
    (a) What main steps might she follow to develop the new strain of banana?
    (b) What are the main challenges she will have to overcome?
37. List three features of bacterial plasmids that make them useful tools in genetic engineering.
38. Did the work of Briggs and King in cloning frogs support or challenge the theory that cloning could not be successful once the donor cells had differentiated? Explain.
39. Draw a replication machine and label it. Then draw a second diagram showing how a chemical mutagen can disrupt the replication process.
40. Describe the process that is used to create a viral vector for gene therapy. What are some of the benefits of using viruses as vectors? What are some of the risks?

INQUIRY

41. You are studying DNA replication in bacterial cells. In one culture, you note that replication is not producing viable daughter cells. Your analysis shows that about 50 percent of the daughter DNA appears normal, while about 50 percent is very unstable and does not form a double helix.
    (a) Draw and label a diagram that could account for your observations.
    (b) If you centrifuged the DNA from this culture, what would you expect to see?
42. When a suspension of heat-killed pathogenic bacteria is treated with enzyme X and then cultured with a strain of non-pathogenic bacteria, transformation does not occur.
    (a) How could you determine whether or not enzyme X is a restriction endonuclease?
    (b) If enzyme X is a restriction endonuclease, what steps could you take to determine its restriction site?
43. A lab studying protein synthesis discovers that three samples of mRNA have been contaminated by the addition of the nucleotide triplet UAA.

Ordinarily, the mRNA strand begins with the triplet AUG occupying nucleotide positions 1, 2, and 3 along the strand and codes for a protein 12 amino acids long. The table shows the results obtained when each of the tainted samples is placed in a suspension containing amino acids and the enzymes necessary for protein synthesis.

Sample	Observations
A	no functional protein product
B	fully functional protein product
C	slightly altered but generally functional protein product

After considering these results, answer the following questions.

(a) Which of the samples could contain the following mutation? Explain your reasoning. (You may wish to refer to the genetic code table on page 254.)

(i) UAA in position 6–8

(ii) UAA in position 16–18

(iii) UAA in position 26–28

(b) Further testing shows that your reasoning was incorrect. For each sample, suggest what additional factors might affect the results you expected.

44. In humans, brown eyes occur when cells in the iris produce the protein melanin, while blue eyes occur when melanin is not produced.

(a) Write a hypothesis that could explain, in terms of gene expression, why brown eyes are dominant and blue eyes are recessive.

(b) Design an experiment you could conduct to test your hypothesis. What would you use as a control?

(c) An individual who carries one gene for blue eyes and one gene for brown eyes may still have eyes that are as dark as those of an individual who carries two genes for brown eyes. Explain, in terms of the regulation of gene expression, how this is possible. Use a diagram to illustrate your explanation.

45. A bacterial cell that has been exposed to high levels of X rays soon afterward begins to produce enormous quantities of many different types of polypeptides, very few of which are the normal, functional proteins it usually produces.

(a) What kind of mutation could account for this effect?

(b) Assume the mutation has no other effect on gene expression. If your hypothesis about the cause of the mutation is correct, what other observation(s) about the polypeptide products would you expect to make?

COMMUNICATING

46. In 1943, 10 years before Watson and Crick published their paper on the structure of DNA, the atomic physicist Erwin Schrödinger predicted that the secret to life would be traced to the interactions among molecules within a larger, crystal-like compound. With a partner, discuss this prediction. Then prepare a debate for presentation to your class in which you take opposing sides to argue the ways in which you believe Schrödinger was right on the one hand and wrong on the other. You should each conclude your arguments with a single sentence that reflects your most important point.

47. Your community is hosting a series of public information meetings about health issues. The objective of the series is to teach people about the science behind healthy lifestyle choices. You are asked to make a 10-minute presentation on the topic "DNA and Mutations."

(a) Write a one- or two-sentence key message you would want your audience to remember.

(b) Outline your presentation under five main headings, beginning with "Introduction" and ending with "Conclusion."

(c) Under each heading list three points you would want to cover and describe each in a few sentences.

48. During a conversation with your neighbour, he emphatically states, "The cloning of human cells is a barbaric practice that should be banned immediately." Assuming your neighbour has no background in genetics, how would you respond? What information in support of your point of view would you provide?

49. As a teaching tool, imagine that a prokaryotic cell and a eukaryotic cell meet on a microscope slide and have an argument about which of them does a better job of regulating gene expression. Write a short exchange of dialogue between them that covers three points each cell might make to your class.

50. You are part of a team involved in creating genetically modified crops that can grow in very dry conditions. You intend to market these crops to countries that are suffering the effects of prolonged drought. Pitted against you are some consumer groups that are campaigning to have the governments of these same countries reject transgenic products.

(a) What arguments do you expect the consumer groups to make?

(b) What points would you include in a brochure to counter some of these arguments and promote your products?

MAKING CONNECTIONS

51. Explain how the study of other species' genomes can contribute to an understanding of human genetics. What can the study of a distantly related organism, such as a roundworm or yeast, contribute to this understanding that the study of another mammalian species could not?

52. When the results of the sequencing of the entire human genome were first released in 2001, many scientists were surprised to learn that the human genome contains far fewer genes than initially expected. What are some of the implications of this finding for the study of how human genetics relates to human health?

53. In the movie *Jurassic Park*, scientists found prehistoric mosquitoes preserved in amber. These mosquitoes had fed on dinosaur blood. The scientists extracted dinosaur DNA from the blood found inside these mosquitoes and then inserted this DNA into amphibian eggs in order to create living dinosaurs. Do you believe genetic engineering technology could make a project of this sort possible in the near future? Explain.

54. Advances in gene therapy research could someday make it possible for parents to select many of their children's genetic traits — including physical features such as hair or eye colour, IQ, and even certain personality traits. What effects might the creation of such "designer babies" have on society? Explain what laws, if any, you think the government should enact to regulate this area of genetic research.

55. You are a medical doctor with two young female patients who are hoping to be treated with a form of gene therapy that can build muscle mass. One has a genetic disorder that causes a breakdown in muscle tissue. The disorder is not life-threatening, although the treatment may help her live an active life and take part in sports with her friends. The other has the skill to be an exceptional hockey player, but is disadvantaged by having a very slight build. The treatment might help her develop the musculature needed to compete in professional hockey. The parents of each girl argue that the treatment is necessary if their child is to be able to live a fulfilling life and achieve her full potential. How would you respond to each family's request for treatment?

56. The Ontario government has decided to fund a five-year research study to determine whether or not chemicals contained in processed foods contribute to a higher incidence of cancer among children and youth. Your community has agreed to participate, and you are in charge of designing the study. Prepare a plan for the study that describes your hypothesis, the data you will collect, and the means you will use to collect them. Give reasons for your choices. What might be some of the implications of your findings?

57. Some groups argue that all food products containing genetically modified ingredients should be clearly labelled. Others argue that this labelling will harm producers, and that genetically modified foods have been demonstrated to be safe. What labelling policy would you propose? Explain your reasons.

58. You are a senior official in a government health department. You must decide how to allocate $100 million in genetics research funding among the following three areas: development of transgenic crops; somatic cell gene therapy; therapeutic cloning of human cells. How much funding will you allocate to each area? Justify your decision.

COURSE CHALLENGE

Consider the following as you continue to prepare for the Biology Course Challenge.

- Review the information you have gathered so far. Make a note of any possible links between your project and the processes by which mutations occur.
- Add important concepts and ideas from this unit. Do any of the new ideas or issues discussed here make you want to change the focus of your project?
- Research science magazines and sites on the Internet for new information related to genetic research. How could you use this information in your project?

UNIT

Evolution

Unit Preview

In this Unit, you will discover

- the scientific evidence that supports the theory of evolution,
- the mechanisms that result in evolution, and
- how the science of evolution is related to current biological research.

Unit Contents

UNIT PROJECT PREP

Read pages 420–421 before beginning this unit.

- Choose the types of technologies that are related to the unit project on Searching for the Common Ancestor.
- Set up files to organize your information on Searching for the Common Ancestor.
- As you read through this unit, collect information that would help you work on your project.

Dive into the water off one of Canada's coasts and you will be surrounded by a myriad of life. Life in the ocean, and in all habitats on Earth, is rich and varied. Species differ from one another, but individual members of a single species also differ. These sea stars are all the same species but they exhibit different colours. The diversity within a species can be obvious (such as variations in size or colour) or "hidden" (such as differences in DNA sequencing). This diversity is a cornerstone of evolutionary biology.

When *Hallucigenia* (inset photo) swam in the sea over 500 million years ago, there was a *greater* diversity of life forms than there is today. *Hallucigenia* was found in the Burgess Shale — a rich fossil bed in the Rocky Mountains in Yoho National Park, British Columbia. The fossils there provide a piece of the puzzle that has helped shape our current ideas about evolution — the process by which organisms alive today descended from ancient forms of life and have been modified over time. Using fossil evidence, observation of species over time, and modern techniques such as genetic analysis, scientists are adding to our understanding of life on Earth. How has life on Earth changed over the millennia? What are the mechanisms that give rise to new species? In this unit, you will explore the fascinating science of evolution. You will learn how early studies and observations of life shaped early theories of evolution. As well, you will see how new discoveries and scientific techniques and technologies contribute to our understanding of evolution.

Why did some species, such as *Hallucigenia*, become extinct, while others survived?

CHAPTER 10

Introducing Evolution

Reflecting Questions

- What ideas and observations helped develop the current theory of evolution?
- What is some of the evidence that helps explain evolution?
- What are the roles of genetics and the environment in evolution?

Prerequisite Concepts and Skills

Before you begin this chapter, review the following concepts and skills:

- describing how heritable traits are passed from one generation to the next (Chapter 6, section 6.4; Chapter 8, section 8.4).

Living organisms are constantly faced with challenges in their environment. Severe weather, drought, famine, and competition for food and space are all struggles living organisms may or may not overcome. Severe weather such as snowstorms and freezing temperatures are some of the challenges that animals, such as the wolf, face in northern environments. Animals that survive have the opportunity to reproduce and pass along to their offspring the traits that helped them survive. The diversity within species and the interactions of organisms with their environment help explain how populations can change over time and why some organisms survive while others become extinct.

The millions of species on Earth today are only a small fraction of the species that have ever lived. In fact, it is estimated that 99 percent of all species that have ever lived are now extinct. While some of the fossilized animals are ancestors of animals that are common today, others have long been extinct and are unlike anything in our modern oceans. Fossils help to show that there was a *greater* diversity of basic animal forms half a billion years ago than there is today. The animals like the fossils unearthed in the Burgess Shale lived during the Cambrian Explosion (over 500 million years ago), a time when there was a stunning burst of biodiversity that is now recorded in the fossil record.

What factors affect which organisms survive and pass on their genes to the next generation? How do environmental conditions affect survival and reproductive ability of organisms? Species diversity and environmental conditions are crucial factors when discussing evolution.

In this chapter, you will learn how changes in the environment and diversity within a *species* can result in changes in *populations* of particular species, and even the formation of new species. You will also learn how the early observations and ideas of naturalists and biologists helped provide a foundation for our current understanding of evolution.

Why is variation within a species necessary for evolution to occur?

Chapter Contents

Diversity of Life

EXPECTATIONS

- **Explain the process of adaptation of individual organisms to their environment.**
- **Analyze evolutionary mechanisms and their effects on biodiversity and extinction.**

The word "evolution" is commonly used in English but its meaning is often misunderstood or misused. In biology, **evolution** refers to the relative change in the characteristics of populations that occurs over successive generations. As you read through this unit, you will begin to better understand evolution and this definition will become clearer.

The grouse in Figure 10.1 lives in the boreal forest of northern Canada. Its brown, white, and black mottled feathers help it blend in with its environment. How do scientists explain that the grouse, and so many other living things, are so well suited to where they live? Recall what you have learned previously about adaptations and heredity. An **adaptation** is a particular structure, physiology, or behaviour that helps an organism survive and reproduce in a particular environment. Camouflage is one adaptation. The superb sense of smell of a shark or the bill shape of a shore bird are also adaptations.

Figure 10.1 This grouse is well-camouflaged in its forest environment. How could the coloration of *individuals* help the survival of a *population*?

Since adaptations help an organism survive, that organism will have a better chance of passing on to its offspring the particular characteristics that were advantageous to its survival. It is important to remember that although some differences between individuals are not outwardly evident, they do exist. For example, slight variations in bill size or shape, or mutations in a gene, are not immediately visible.

Meanwhile, environments can change: climates change over time, and droughts, floods, and famines occur. Thus, a characteristic that may not give an individual organism a particular advantage *now* may become critical for survival *later* if the environment inhabited by that species changes. This was demonstrated beautifully in the story of the English peppered moth.

The Peppered Moth Story

The story of the English peppered moth, *Biston betularia*, is often cited as an example of how the proportions of some inherited characteristics in a population change in response to changes in the environment. The peppered moth has two colour variations: greyish-white flecked with black dots (that resemble pepper) and black (see Figure 10.2 on page 335). In the past, the black variety was extremely rare. The first known black moth was caught in 1848 by lepidopterist R.S. Edleston. At that time, it was estimated that black moths made up less than two percent of the peppered moth population near Manchester, England. Yet 50 years later, in 1898, 95 percent of the moths in Manchester were of the black type. In rural areas, however, black moths were less frequent. What caused the sudden increase of black moths in Manchester? The answer lies in the behaviour and genetic makeup of the moths *and* the environment in which they lived.

Peppered moths are active at night. During the day, when they rest on the trunks of trees, they are

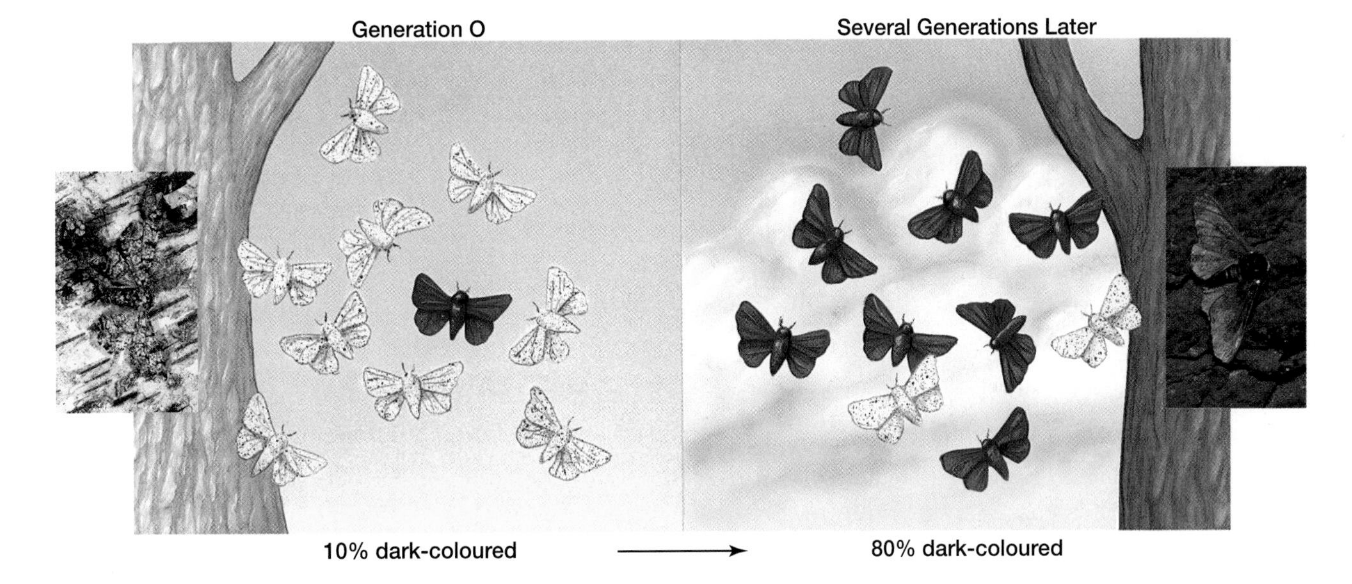

Figure 10.2 The frequency of flecked and black moths in this population of peppered moths changed in response to changes in the environment.

potential prey for birds. Until the mid-nineteenth century, the flecked moths in Manchester were camouflaged when they rested against the light-coloured lichens on tree trunks. The black moths, however, were easily seen and therefore easily preyed upon. The 50 years in which the black moths gradually became much more common in Manchester coincided with the Industrial Revolution in England. The air pollution from all the new factories killed the lichens, and soot began to cover Manchester's trees. As a result, the flecked moths were seen and eaten by birds and more black moths survived long enough to reproduce and pass on their genes to their offspring. Figure 10.2 shows how the peppered moth population evolved — that is, changed over the course of several generations.

The difference between the flecked and black forms of the peppered moth is caused by a single

THINKING LAB

Changes in Peppered Moth Populations

Background

The peppered moth story shows how genetic variety within a species can result in changes in the characteristics of a population when the environment changes. Since insects have a relatively short life cycle, this shift can happen quite quickly. In the 1950s, English biologist H.B. Kettlewell studied the camouflage adaptations in a population of flecked and black peppered moths.

Kettlewell raised over 3000 caterpillars to provide the adult moths. In a series of trials in the country and the city, he released and recaptured the moths. The number of moths recaptured in each trial indicated how well the moths survived in the environment.

You Try It

1. Examine the table on the right. Calculate the percentage of moths recaptured in each experiment.
2. Explain the differences in survival rates in the unpolluted and polluted environments.
3. Based on this lab, the information that has been presented so far in this chapter, and your understanding of genetics and evolution, discuss the following statements in a small group: Genes mutate. Individuals are selected. Populations evolve.
4. Discuss any other factors that may have influenced data in this study and the conclusions based on these data.

Numbers of peppered moths released and recaptured in polluted and unpolluted areas in England

Location		Number of flecked moths	Number of dark moths
Dorset (unpolluted)	released	496	488
	recaptured	62	34
Birmingham (polluted)	released	137	493
	recaptured	18	136

gene. Before the Industrial Revolution, more flecked moths survived and therefore passed on the gene for flecked-colouring in the **gene pool**. A gene pool is the total of all the genes in a population at any one time. However, when air pollution increased, more black moths survived with each successive generation and the ratio of flecked and black moths in the population essentially reversed. It is important to understand that the ratio of flecked to black moths in the *population* changed over successive generations. Individual moths were not transformed from flecked to black. A population is the smallest unit that can evolve.

In the 1950s, England started to enact clean-air legislation and lichens began to grow on trees again. As you might predict, the frequency of flecked moths increased in industrial areas such as Manchester. In these areas, nine out of 10 peppered moths were black in 1959. By 1985, five out of 10 were black, and the number dropped to three out of 10 by 1989. It is estimated that by 2010 the black peppered moth will again be as rare in Manchester as it was before the Industrial Revolution.

BIO FACT

The records and collections of lepidopterists were used by biologists to trace the spread of the black variety of peppered moth across England. Since the black moths were initially extremely rare, they created a frenzy among collectors in the mid-nineteenth century. The subsequent increase in the number of black moths collected from certain areas allowed historians and biologists to track colour changes in the populations of peppered moths.

Revisiting the Definition of Evolution

The peppered moth story demonstrates how a gene pool shifted from a population in which a particular gene coded for one expression of the characteristic (light, flecked colouring) in most individuals to a population in which the gene coded for a different expression of the characteristic (black colouring) in most individuals. Although no new species was formed, this *is* an example of evolution because there was a change in the gene pool of the population over successive generations. *Any shift in a gene pool is also used to*

Investigation 10 • A

SKILL FOCUS

- Predicting
- Performing and recording
- Communicating results

Diversity Within a Species

You have learned how diversity (also called variability) within a species can help populations survive environmental changes. Diversity within a species can be monitored genetically, or it can be demonstrated by measuring individuals within a population. In this investigation, you will measure a particular characteristic in each of three populations to determine variability within each population.

Pre-lab Question

- What would be the evolutionary advantage to a plant of having a larger seed?

Problem

How can variability among individuals be measured?

Prediction

Predict whether measurements of a particular characteristic (for example, length of bean seed) in a population would be evenly distributed throughout a population or whether most individuals would be the same length, with only a few individuals being longer or shorter than the norm.

Materials

10 kidney beans
callipers
string
10 lima beans
ruler
graph paper

Procedure

Part A

1. Use the callipers to measure the length of each of the 10 kidney beans.
2. Record your measurements.
3. Pool your measurements with other students in the class so you have between 50 and 100 measurements.
4. Calculate the average length of this population of kidney beans and prepare a bar graph of the class data.

define evolution. In fact, many scientists consider such a shift to be the most accurate and specific definition of evolution. This idea will be discussed in more detail in Chapter 12.

Natural Selection

The story of the peppered moths is an example of **natural selection**. Natural selection is a process whereby the characteristics of a population of organisms change because individuals with certain heritable traits survive specific local environmental conditions and pass on their traits to their offspring. You will learn more about natural selection later in this unit. For natural selection to occur there must be diversity *within* a species. Look around your classroom. You are all the same species but clearly there is a great deal of variety among you and your classmates. Without the extensive variability within a population, there would be no possibility for selection to occur. In the populations of peppered moths, the moths that survived were *selected*. In other words, *they survived the change in the environment around them*, and thus could reproduce and pass on the genes that coded for black. *Individuals* did not change colours during their lifetime; rather, the *populations* shifted in colour over time. The environment exerts a **selective pressure** on a population. In other words, an environmental condition can *select for* certain characteristics of individuals and *select against* those of others.

Artificial Selection

In the peppered moth story, change occurred naturally in the population in response to changes in the environment. However, people have been artificially selecting organisms for particular traits for centuries. Artificial selection for desirable traits has resulted in plants that are disease-resistant, cows that produce more milk, and racehorses that run faster. In **artificial selection**, a plant or animal breeder selects individuals to breed for the desired characteristics he or she wishes to see in the next generation. Figure 10.3 on page 338 shows some of the varieties of dogs that have been produced by artificial selection. As another example, a rose

Part B

1. Repeat the steps from Part A using lima beans.

Part C

1. Use the string to measure the length of your partner's forearm, from the crease inside the elbow to the wrist. Use the ruler to determine forearm length.
2. Record your data and pool your data with that of the rest of the class.

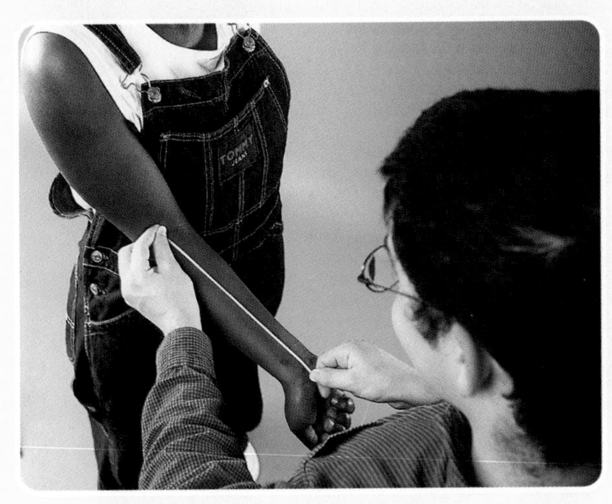

3. Calculate the average forearm length of students in your class and prepare a bar graph of the class data.

Post-lab Questions

1. How are your three graphs similar?
2. From your graphs, what can you conclude about the variability within a population? For example, is there a "typical" size, or is the distribution of individuals spread evenly from small to large?

Conclude and Apply

3. What advantage would large size have to a newly germinated seed? (Recall that a seed is stored food.)
4. What environmental pressures might favour small seeds?
5. Predict a situation (actual or imagined) in the environment in which having a longer forearm might be advantageous to a person's survival.

Exploring Further

6. Create a breeding strategy to favour the production of large seeds.

breeder could select the seeds from roses with a strong scent to produce generations of roses with an equally strong fragrance.

Artificial selection can also perpetuate characteristics that are not particularly desirable. For example, Pekinese and British bulldogs are bred for their flat faces, but this characteristic also results in severe respiratory problems. Hip dysplasia, a type of arthritis common in German shepherds, is also an unfortunate consequence of artificial selection.

Overhunting can result in future generations that have a higher proportion of individuals *without* the favourable trait. For example, in the 1970s and 1980s, between 10 and 20 percent of all wild elephants in Africa were being killed by ivory poachers each year. Since poachers preferred elephants with large tusks, elephants with smaller tusks were less likely to be killed. Elephants with no tusks were not shot at all. Since that period, elephant watchers and biologists have noticed more and more tuskless elephants in the areas that experienced the most intense poaching pressure.

The key difference between natural and artificial selection is that in natural selection, the environment plays the role that humans play in artificial selection. In natural selection, the environmental conditions determine which individuals in a population are most fit to survive in the current conditions. This, in turn, affects the proportion of genes among individuals of future populations because the genes from the surviving individuals are passed on to their offspring. When discussing natural selection and evolution, the word "fit" or "fitness" is often used. **Fitness** in this sense refers to how well an organism fits with its environment. A high degree of fitness means that an organism will survive and reproduce, thereby passing on its advantageous genes to its offspring.

Natural Selection Is Situational

It is important to note that natural selection does not anticipate change in the environment. Instead, natural selection is situational. It is essentially by chance that a trait that might at one time have no particular relevance to survival (for example, black coloration in moths) becomes the trait that helps a population survive. This trait then persists within a population in response to changes in the environment via subsequent inheritance of the trait by the offspring of survivors. Adaptations that are beneficial in one situation may be useless or detrimental in another. This has been demonstrated in the work of Peter and Rosemary Grant in their study of finches in the Galápagos Islands.

For over 20 years, the Grants have been studying medium ground finches (*Geospiza fortis*), one of the 13 species of finches in the Galápagos Islands.

Figure 10.3 All dogs are members of the same species, *Canis familiaris*, yet artificial selection has resulted in a wide variety of breeds.

These birds use their strong beaks to crush seeds, and tend to prefer small seeds that are produced in profusion during the wet years in the islands. Fewer small seeds are produced during dry years, and the Grants found that during these times the finches also have to eat larger seeds, which are harder to crush. As part of their study, the Grants measured the depth (dimension from top to bottom) of the finches' beaks. They found that the average beak depth in the population changes over the years. During droughts, the population's average beak depth increases. During wet periods, the average beak depth in the population decreases again. The Grants' study demonstrates a change in the finch population in response to the environmental conditions. During dry periods, birds with stronger (that is, slightly larger) beaks have an advantage because they are better able to crack large seeds. Since these birds have a feeding advantage, they survive in greater numbers and have greater potential to pass the gene for a larger beak on to their offspring.

This difference in ability to crack larger seeds within the finch population can only happen because there is variety within the population. As you found when you did the investigation on page 336, not all kidney beans are identical in length, nor are the forearms of all Grade 12 students. *It is the variety that is already present within a population that allows change to occur in response to local environmental conditions.* Natural selection acts like an editor; it only works with what is already present in a population.

The finches reproduce once a year. The Grants have been able to monitor morphological changes in the population only by measuring and monitoring the birds year after year as part of their multi-year study. In other organisms that reproduce more quickly, such as insects and bacteria, the change in a population in response to local environmental conditions can be observed in a much shorter time. How do you think natural selection is involved in insects becoming resistant to pesticides or in bacteria becoming resistant to antibiotics? This idea will be discussed further in subsequent chapters.

BIO FACT

As part of their study, the Grants needed to measure the force required to open seeds. Peter Grant designed a unique device with the help of an engineer from McGill University in Montréal. The "McGill nutcracker" looks like pliers with a scale attached. When a seed is squeezed with the pliers, a scale measures the force required to crack the seed.

SECTION REVIEW

1. **K/U** Can individuals evolve? Explain your answer.
2. **K/U** Give two definitions of evolution.
3. **K/U** How are adaptations and evolution related to each other?
4. **K/U** Describe how the study of peppered moths by Kettlewell demonstrates evolution in action.
5. **K/U** Define the term "gene pool."
6. **C** Explain the term "selective pressure" as it relates to the study of evolution.
7. **C** In a population of sparrows, most birds have a bill that is about 10 mm long. Some birds, however, have bills that are slightly longer or slightly shorter than the average. Explain why this variation within the population is important when discussing evolution.
8. **K/U** Give one example of artificial selection and one example of natural selection. What is the major difference between the two types of selection?
9. **MC** Give some examples of how people have used artificial selection to create new varieties of plants or animals. Describe the possible economic and environmental impacts of these new varieties.
10. **I** How would you test the hypothesis that larger finches on the Galápagos Islands had a greater survival rate in wet years than in dry years? What factors would you measure?
11. **C** With a partner, discuss what your understanding of evolution was before you read this section. Has your understanding changed in any way now that you have completed this section? If so, how has your definition of evolution changed?
12. **C** Explain what is meant by the statement "Natural selection is situational."

10.2 Developing the Theory of Evolution

EXPECTATIONS

- **Describe, and put in historical context, some scientists' contributions that have changed evolutionary concepts.**
- **Evaluate the scientific evidence that supports the theory of evolution.**
- **Identify questions to investigate that arise from concepts of evolution and natural selection.**

Ideas about natural selection and evolution began to be discussed in earnest in the early nineteenth century. Although the name Charles Darwin is often mentioned synonymously with the theory of evolution, in fact the work and ideas of many others helped to shape our current understanding of evolution. Indeed, as our technological and scientific techniques improve and our knowledge of the principles of evolution grows, our understanding of the processes of evolution also improves.

A Historical Context

The English naturalist Charles Darwin was by no means the first (or only) person to influence thought on what is commonly referred to as the theory of evolution. Several Greek philosophers believed that life gradually evolved. However, two of the most influential philosophers in Western culture, Plato and Aristotle, did not support ideas that organisms could change. For example, Aristotle thought that all organisms that ever would exist were already created. He also believed that these organisms were permanent and perfect and would not change. Religious beliefs of Darwin's time said that all organisms and their individual structures resulted from the direct actions of a Creator who formed the entire universe. It was thought that all species were created during a single week and that they remained unchanged over the course of time. The predominant belief that Earth was only a few thousand years old fortified the idea of a single act of creation.

In the nineteenth century, however, some scholars began to present new ideas. Some thinkers proposed that living things did change during the course of the history of Earth, and that the organisms that exist now might be different from the organisms that existed previously in history. Others said that populations of organisms perhaps even changed from one generation to the next. They observed variations in populations and saw that populations could adapt to particular situations. Although these ideas were discussed, especially in scholarly circles, they were contrary to the religious teachings of the time and as such were often dismissed as heresy. As well, no one could propose a plausible mechanism that explained these phenomena. Darwin's name is so closely associated with the theory of evolution because he linked all of the prevailing knowledge from paleontology, geology, geography, and biology with his own observations. In doing so, he developed a theory describing a mechanism that showed convincingly that life could change over time. (Another English naturalist, Alfred Russel Wallace, also came to the same conclusion as Darwin.)

Cuvier's Fossils

The science of **paleontology**, which is the study of fossils, provided important clues that helped to develop the theory of evolution. French scientist Georges Cuvier (1769–1832) is largely credited with developing the science of paleontology. Cuvier realized that the history of life was recorded in Earth's layers of rocks, which contained fossils. Cuvier found that each of the layers, or strata, of rock is characterized by a unique group of fossil species and that the deeper (older) the stratum, the more dissimilar the plant and animal life are from modern life (see Figure 10.4 on page 341). Cuvier also recognized that extinction of species was a fairly common occurrence in the history of life on Earth. As he worked from stratum to stratum, he found evidence that new species appeared and others disappeared over the course of millions of years.

Cuvier's work showed that something was causing species to appear and disappear, but he was strongly opposed to the ideas of evolution being suggested at the time. Instead, he proposed the idea of **catastrophism**. According to this idea,

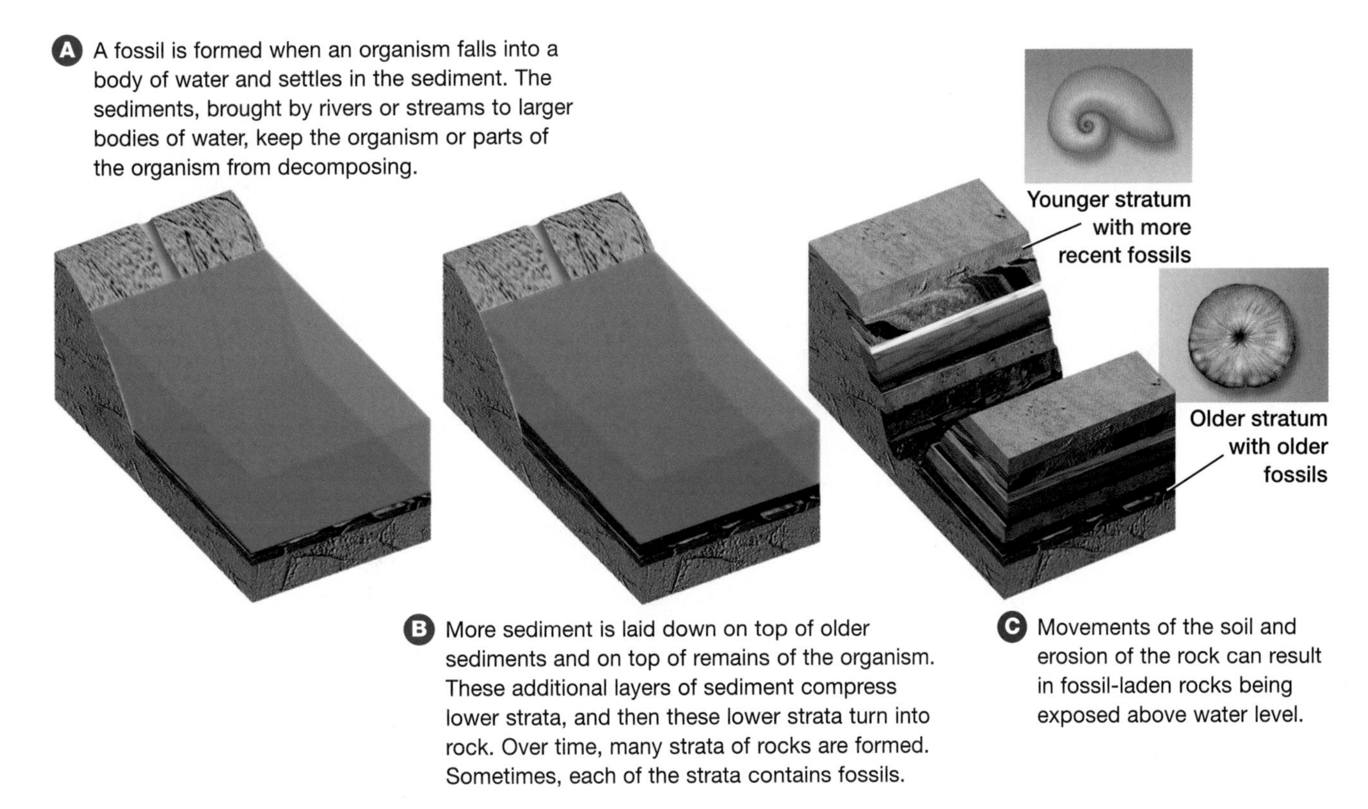

Figure 10.4 Layers of sedimentary rocks are of different ages and contain different groups of fossils.

catastrophes (such as floods, diseases, or droughts) had periodically destroyed species that were living in a particular region. He hypothesized that these catastrophes corresponded to the boundaries between each stratum in his studies. Cuvier proposed that these catastrophes were limited to local geographical regions, and that the area would be repopulated by species from nearby unaffected areas. This is how he explained the appearance of "new" species in the fossil record.

Lamarck's Theory of Inheritance of Acquired Characteristics

French naturalist Jean-Baptiste Lamarck published a theory of evolution in 1809, the year Charles Darwin was born and 50 years before Darwin would finally publish his own ideas on evolution. While working at the Natural History Museum in Paris, where he was in charge of the invertebrate collections, Lamarck compared current species of animals with fossil forms. He could see that there appeared to be a "line of descent" where the fossil record showed a series of fossils (from older to more recent) that led to a modern species. Lamarck proposed that microscopic organisms arose continually and spontaneously from non-living sources. He thought that species were initially very primitive, and that they increased in complexity over time until they achieved a sort of perfection. Lamarck believed that the organisms would become progressively better and better adapted to their environments. It was thought at the time that body parts that were used extensively to cope with conditions in the environment would become larger and stronger (the idea of "use and disuse"). Lamarck's idea fit with this line of reasoning. For example, he proposed that a blacksmith would develop a larger biceps in the arm in which he holds his hammer.

Lamarck further proposed that characteristics acquired during an organism's lifetime, such as large size, short hair, or large muscles, could then be passed on to its offspring. Following this reasoning, Lamarck claimed that the large biceps of a blacksmith would then be passed on to his offspring. He called this the **inheritance of acquired characteristics**. Lamarck's proposed mechanism of evolution is now known to be incorrect, but his ideas provoked thought and discussion. They also influenced the thoughts of others, including Charles Darwin. Although controversial for the time, Lamarck's thinking was visionary, especially his idea that adaptations to the environment result in the evolution of species.

BIO FACT

Recent understanding of the immune system has shown that, in some instances, characteristics acquired throughout one's lifetime *may* be passed on to offspring. For example, antibodies acquired during a mother's lifetime can be passed from mother to child during breastfeeding. This boosts the infant's immune system. Scientists Edward Steele and Reginald Gorczynski conducted an experiment that supported Lamarck's basic tenet when they were working at the Ontario Cancer Institute in Toronto in the 1970s. The researchers injected infant male mice with cells from different groups of mice and found that the infants' immune systems developed a tolerance to the foreign cells. They then observed that the offspring of these mice had the same tolerance. Other scientists who have tried and failed to replicate the experiment of Steele and Gorczynski refute the scientists' findings. Nevertheless, this work has sparked interest and critical scientific debate.

Darwin's Evidence

In 1831, a young man left England on the HMS *Beagle*, a British survey ship used for voyages of scientific discovery. No one, including 22-year-old Charles Darwin himself, knew what the voyage would mean to Darwin and the study of biology as he stepped aboard. The expedition had a primary mission to survey the coast of South America, yet it provided Darwin with an opportunity to travel much of the world with ample time to explore the natural history in various locations. Figure 10.5 shows the voyage of the *Beagle*. While the crew surveyed the coastline, Darwin spent hours on shore observing and collecting thousands of specimens in the diverse environments that the ship visited, from the towering Andes Mountains to the Brazilian jungle.

Darwin gathered evidence and made many important observations that led him to realize how life forms change over time and vary from place to place. First, he noted that the flora and fauna of the different regions he visited were distinct from those he had explored in England, Europe, and elsewhere. For example, the rodents in South America were structurally similar to one another but were quite different from rodents he had observed on other continents. Of particular importance was Darwin's observation that species living in the cooler, temperate regions of South America were more closely related to species living in the tropical regions of that continent than to the species in the temperate regions of Europe or elsewhere in the world. He noted that lands that have similar climates seemed to have unrelated plants and animals. Darwin and many others in his time wondered why it was that if all organisms originated from a single act of creation, there existed this distinctive

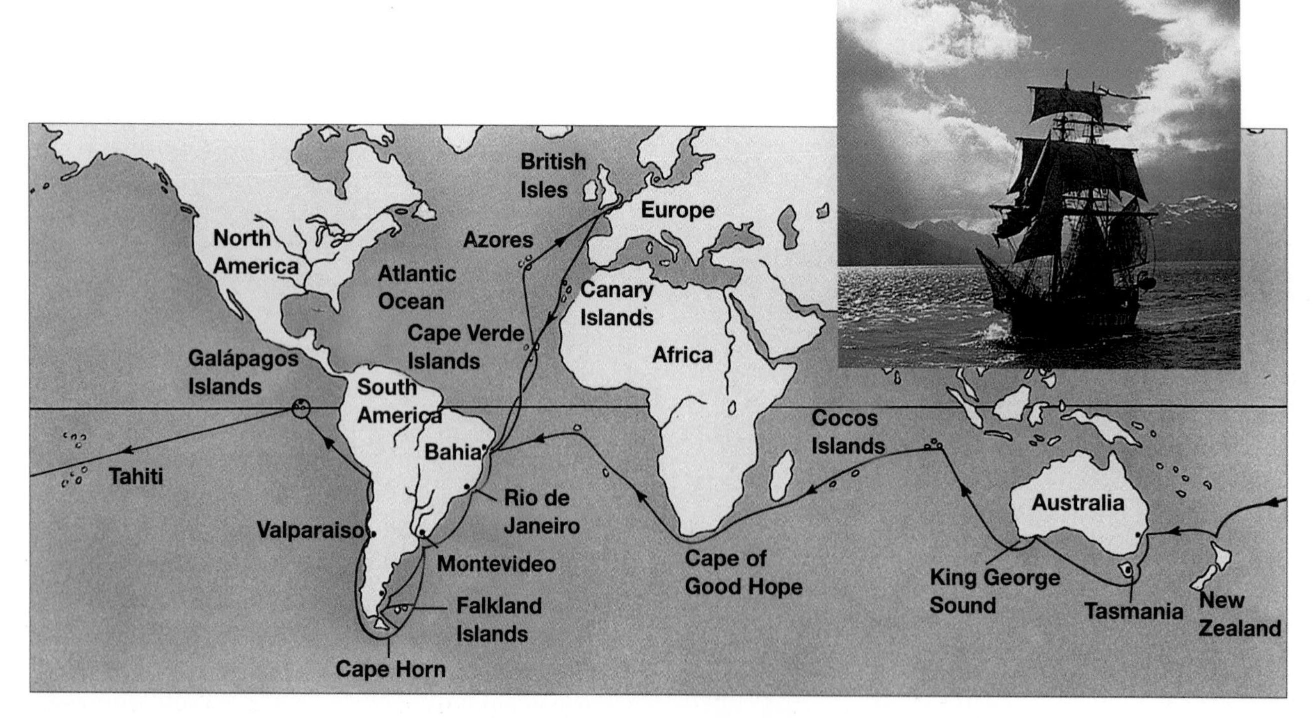

Figure 10.5 The five-year voyage of the HMS *Beagle* took Darwin around the world. Most of his time, however, was spent exploring the coast and coastal islands of South America.

clustering of similar organisms in different regions of the world. Why weren't organisms randomly distributed across Earth?

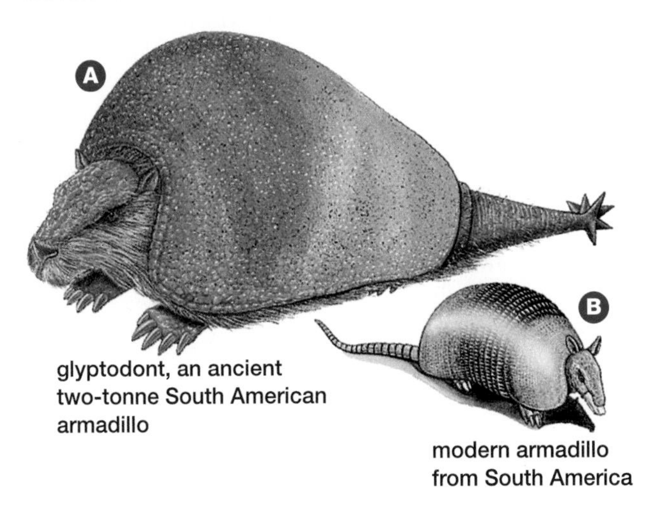

Figure 10.6 Comparison of the extinct glyptodont and a modern armadillo

Darwin also found several important fossil remains, including that of a glyptodont, an extinct armadillo-like animal. He wondered if this fossil was somehow related to the living forms of armadillos that lived in the same region (see Figure 10.6). Why would there be living and fossilized organisms that were directly related to one another in the same region? Could one have risen from the other?

Although it was not entirely evident to Darwin at the time, the *Beagle*'s five-week stop in the Galápagos Islands was particularly important in helping Darwin formulate his ideas on evolution. The Galápagos Islands are a group of over 20 small volcanic islands located in the Pacific Ocean approximately 1000 km off the coast of Ecuador. Darwin noted that the islands in the Galápagos supported relatively few animal species. (There was only one land mammal, for instance, and no frogs or other amphibians.) The species that were there, however, closely resembled animals of the west coast of South America, the nearest continental land mass. Darwin wondered: if these organisms had been created independently and placed in the Galápagos Islands (as the prevailing ideas of the time suggested), why did they so closely resemble organisms on the adjacent South American coastline? A single act of creation did not seem to support the trend Darwin was observing.

In the Galápagos, Darwin observed many new species, including huge land tortoises and giant cactus trees like those shown in Figure 10.7. These species were unique to the Galápagos, and were fairly common in the islands. Some of the species, such as the Galápagos tortoise, were slightly different from island to island. Darwin did not

BIO FACT

Although Darwin is often identified as being the naturalist on the *Beagle*, in fact he was not. Rather, he was welcomed aboard as a "gentleman's companion" to Captain Robert FitzRoy. At that time, captains did not socialize with their crew. Since the voyage was a long one, FitzRoy decided that he needed a companion and selected Darwin. The *Beagle*'s "official" naturalist was the ship's surgeon, Robert McKormick. At that time it was very common for the job of ship's surgeon to be combined with ship's naturalist.

Ⓐ Galápagos tortoise

Ⓑ Cactus trees

Figure 10.7 Unique species of the Galápagos

observe this himself. Rather, he was told by the vice-governor of the Galápagos that local residents could tell which island captured tortoises came from, just by looking at them. At the time Darwin dismissed this, later writing, "I did not for some time pay attention to this statement … I have never dreamed that islands, about fifty or sixty miles apart, and most of them in sight of each other, formed of precisely the same rocks, placed under a quite similar climate, rising to a nearly equal height, would have been differently tenanted." As it turns out, this fact became a critical piece of information that helped Darwin develop his theory.

Darwin also collected a variety of birds while in the Galápagos Islands, including 13 species of finches (as shown in Figure 10.8). "Darwin's finches" have since become well known in the history of evolutionary thought and, like the information on tortoises, they also became a key to the formulation of Darwin's final theory. While he was in the Galápagos Islands, Darwin scarcely gave the finches much thought, however. He collected several dozen birds but assumed they were similar to birds on the coast of South America or on other Galápagos islands.

Figure 10.8 Galápagos finches are adapted to gathering and eating different types of food.

Darwin *did* wonder why there was such a diversity of species in such a small area. Each type of Galápagos finch (see Figure 10.8) is adapted to gathering and eating a different type of food based on the size and shape of its beak. Tree finches, for example, have beaks largely adapted to eating insects and, at times, plants. Ground finches have beaks adapted to eating cactus or different-sized seeds. The woodpecker-type finch uses a tool, a cactus spine or twig, to probe in the bark of trees for insects.

In Britain, a colleague catalogued the birds for Darwin and became particularly excited about Darwin's finches. All of the birds were new species that had never been described before. On reflection, Darwin could now see that although the finches were somewhat similar to finches on the coast of South America, they were clearly distinct species. This suggested that they had been modified from an ancestral form of the bird that was blown by chance into the newly formed Galápagos Islands. In the *Voyage of the Beagle* Darwin wrote, "...in the thirteen species of ground-finches, a nearly perfect gradation may be traced, from a beak extraordinarily thick, to one so fine, that it may be compared to that of a warbler. I very much suspect, that certain members of the series are confined to different islands. ..." He continued, "Seeing this gradation and diversity of structure in one small, intimately related group of birds, one might really fancy that, from an original scarcity of birds in this archipelago, one species had been taken and modified for different ends."

In summary, Darwin's experience in the Galápagos Islands, particularly the information gathered on tortoises and finches, demonstrated a mechanism for how new species could arise from ancestral ones in response to the local environment.

While on the voyage, Darwin also read *Principles of Geology* by the geologist Charles Lyell. Lyell expanded on ideas first proposed in 1795 by another geologist, James Hutton. Hutton said that Earth's geological features were in a slow, continuous cycle of change. For example, the slow action of rivers eroding through rocks eventually forms canyons. This is called **gradualism**. Lyell expanded on Hutton's ideas to develop a theory known as **uniformitarianism**. Lyell said that geological processes operated at the same rates in the past as they do today. He rejected the idea of irregular, unpredictable, catastrophic events shaping Earth's history.

Lyell's work was significant and a strong influence on Darwin. If geological changes were indeed slow and continuous rather than catastrophic, then Earth was certainly older than the 6000 or so years espoused by biblical scholars. As well, Lyell's work showed that slow, subtle processes happening over a long period of time could result in substantial changes. Darwin, and others searching to explain the changes they saw in the organisms around them, applied Hutton's and Lyell's ideas to biology. Darwin hypothesized that slow, subtle changes in populations of organisms could translate into substantial changes over time.

Summarizing Darwin's Evidence

1. Plants and animals observed in the temperate regions of South America were more similar to plants and animals in the South American tropics than to plants and animals in other temperate regions in the world.
2. Darwin found fossils of extinct animals (such as the glyptodont) that looked very similar to animals presently living in the same region (for example, the armadillo).
3. Plants and animals living in the Galápagos Islands closely resembled plants and animals living on the nearest continental coast (the west coast of South America).
4. Species of animals (such as tortoises) that at first looked identical actually varied slightly from island to island in the Galápagos.
5. Finches collected in the Galápagos looked similar to finches from South America but were, in fact, different species. Finch species also varied from island to island.
6. After reading Lyell's work, Darwin understood that geological processes that are slow and subtle can result in substantial changes. As well, forces that affect change are the same now as in the past.

WEB LINK

www.mcgrawhill.ca/links/biology12

Upon his return to England, Charles Darwin wrote the memoirs of his journey and published *The Voyage of the Beagle* in 1839. This book, along with Darwin's other works, is still widely available today. To read some of the original text, go to the web site above, and click on **Web Links**. Read an entry that Darwin made about his time in the Galápagos Islands and the observations he made there.

WEB LINK

www.mcgrawhill.ca/links/biology12
Today, the Galápagos Islands continue to be an important site for scientific research in many subject areas, including evolutionary biology. The islands have been recognized by the United Nations and have been designated a Biosphere Reserve, World Heritage Site, and national park. To find out more about current scientific research in the Galápagos Islands, go to the web site above, and click on **Web Links**.

Darwin's Theory of Evolution by Natural Selection

After returning to England, Darwin compiled his memoirs of the voyage. He then devoted eight years to a study of barnacles, in which he filled four volumes on their classification and natural history. Darwin continued to develop his ideas and collect evidence to support his conclusion that species could and did change over time. He investigated variations in species by breeding pigeons and studying breeds of dogs and varieties of flowers. From this work he knew it was possible for traits to be passed on from parent to offspring, so it was clear that species could change over time. He could not explain, however, exactly *how* it happened.

In 1838, Darwin read *Essay on the Principles of Population*, which was written by English economist Thomas Malthus in 1798. In Malthus's paper Darwin found the key idea he had been searching for to explain his observation of changes in species over time. This idea was that plant and animal populations grew faster than their food supply and eventually a population is reduced by starvation, disease, or (as in the case of humans) war. How did this idea help Darwin's thinking? Malthus's idea helped Darwin refine his thoughts. Darwin knew that many species produce large numbers of offspring, but he also knew that population levels tended to remain unchanged. Malthus's vision of struggle and crowding helped Darwin realize that individuals had to struggle somehow to survive. This struggle was the force that constantly prevents a population explosion. A struggle could be competition for food, shelter, or a mate, for example. Only some individuals survive the struggle and produce offspring. Darwin recognized that the struggle between individuals of the same species competing for limited resources *selected for* individuals with the traits that would increase their chances of surviving. Then, the survivors could potentially pass this favourable trait on to their offspring. He realized this was similar to humans selecting for favourable traits when breeding dogs, horses, or plants.

BIO FACT

Erasmus Darwin (1731–1802), Charles Darwin's grandfather, also proposed that competition between individuals could result in changes in species. Erasmus Darwin was a physician, naturalist, and influential intellectual in eighteenth century England. He formulated one of the first formal theories on evolution, and published his ideas in papers and in a poem, *The Temple of Nature*.

ELECTRONIC LEARNING PARTNER

Refer to your Electronic Learning Partner for more information on the diversity of species in the Galápagos.

THINKING LAB

Could Pumpkins Rule Earth?

Background

Charles Darwin applied Malthus's ideas to various organisms. For example, he calculated that a single pair of elephants could have 19 million descendents in 750 years. He knew, of course, this could not be true and began to think about the mechanism that must be controlling populations of all species on Earth. The largest number of offspring produced by the members of a population is known as the **biotic potential** of a species.

You Try It

1. Assume there are 70 seeds in one pumpkin. These 70 seeds are planted and each seed grows into a plant that produces two pumpkins. Calculate the number of seeds produced by this generation.
2. If you plant all of the seeds from step 1, how many seeds are available at the end of the next generation?
3. Why is the maximum biotic potential never actually reached in nature?

WEB LINK

www.mcgrawhill.ca/links/biology12
Many scientists have contributed to our current understanding of evolutionary biology, and exciting work continues today. To learn more about the various contributions of scientists and philosophers, go to the web site above, and click on **Web Links**. Choose one of the individuals on the list and summarize his/her contribution to evolutionary biology.

Darwin's thinking was catalyzed by Malthus's ideas, his experience with pigeon breeding and artificial selection, and the observations he made during and after the voyage of the *Beagle*. He gradually synthesized his ideas to show that individuals that possess physical, behavioural, or other traits that help them to survive in the local environment are more likely to pass these traits on to offspring than those that do not have such advantageous traits. These favourable characteristics then begin to increase in the population and, over time, the nature of the population as a whole changes. Darwin called this process natural selection. Darwin drafted his initial ideas in two manuscripts shown only to trusted friends in 1842 and 1844. We know he realized their importance because he asked his wife to ensure they would be published in the event of his untimely death. Curiously, however, Darwin did not present his ideas publicly until 1859, when he released *On the Origin of Species by Means of Natural Selection*. (In this text we will refer to this book as *The Origin of Species*.)

Why did Darwin wait so long to publish his ideas? Thinking and discussions about evolutionary theory were becoming more and more commonplace in the mid-nineteenth century, but the discussions were inevitably heated. The subject was controversial, since it was perceived as being contrary to the religious teachings of the time. Perhaps Darwin was reluctant to publish because he anticipated the response and possible uproar it would cause. His friend Lyell, whose book on fossils had influenced Darwin, encouraged him to publish on the subject before someone else did, even though Lyell himself was not convinced of evolution.

Lyell's prediction came true in June 1858, when Darwin received a paper from British naturalist Alfred Russel Wallace. As a result of his studies in a group of islands near Indonesia, Wallace had reached a conclusion similar to Darwin's. In the paper, Wallace outlined an essentially identical theory of evolution by natural selection. With Wallace's paper was a letter asking Darwin to evaluate the paper and pass it on to Lyell if he thought it should be published. Darwin did as Wallace asked and in a letter to Lyell he wrote, "Your words have come true with a vengeance... I never saw a more striking coincidence... so all my originality, whatever it may amount to, will be smashed." Lyell presented Wallace's paper and parts of Darwin's unpublished 1844 essay to the scientific community on July 1, 1858. Darwin quickly went to work and wrote *The Origin of Species*, which was published in 1859. With *The Origin of Species*, Darwin was the first to gather an array of facts related to evolution and present them cohesively.

Descent with Modification

Darwin did not use the word "evolution" in the original edition of *The Origin of Species*. ("Evolved" is used once — it is the final word in the book.) Instead, Darwin spoke of **descent with modification**. One reason he did not use the word "evolution" is that he felt it implied progress — that each generation was somehow getting better (that is, was improving in some way). Natural selection does *not* demonstrate progress; it has no set direction. It results purely from an ability to survive local environmental conditions, thereby giving the survivors the opportunity to pass on the trait that helped them survive in the first place.

Darwin proposed two main ideas in *The Origin of Species*: present forms of life have arisen by descent and modification from an ancestral species; and the mechanism for modification is natural selection working continuously for long periods of time. Darwin said that all organisms descended from some unknown organism. As descendants of that organism spread out over different habitats over the millennia, they developed modifications, or adaptations, that helped them fit in their local environment. Darwin's theory of natural selection showed how populations of individual species became better adapted to their local environments. These ideas are summarized in the text box on the following page.

As Darwin anticipated, *The Origin of Species* created a sensation, since the ideas outlined in the work were deeply disturbing to many. Within a few years, however, his view was widely accepted by most scholars. This was partly because the gap between religious viewpoints and the idea of natural selection narrowed, and because Darwin supported his ideas logically with a great deal of evidence. *The Origin of Species* continues to be one of the most famous and influential books of all time.

Summary of Darwin's Ideas

Natural selection means that organisms with traits best suited to their environment are more likely to survive and reproduce. The factors Darwin identified that govern natural selection are:

1. Organisms produce more offspring than can survive, and therefore organisms compete for limited resources.
2. Individuals of a population vary extensively, and much of this variation is heritable.
3. Those individuals that are better suited to local conditions survive to produce offspring.
4. Processes for change are slow and gradual.

The work of Darwin, Lyell, Lamarck, and Cuvier helped shape the understanding of evolution. Many other people also helped advance these ideas. For example, Darwin was influenced by reading a work by Lyell on geology. Darwin supported his ideas with evidence of natural selection. In the next section, you will study some of the scientific evidence that supports the theory of evolution.

BIO FACT

Unfortunately, Alfred Russel Wallace, co-discoverer with Darwin of the idea of natural selection, is not well known by the general public. Wallace was an accomplished naturalist and contributed a great deal of knowledge to biological sciences, geography, and other disciplines. During his long life Wallace published over 150 works (including essays, books, and letters) and travelled and lectured widely. He did not, however, agree with all of the contents of Darwin's *The Origin of Species*. In fact, Wallace eventually became a "spiritualist" and could not extend the idea of natural selection to apply fully to humans. He believed that while natural selection worked at a biological level, there was a spiritual process that operated at the level of human consciousness. Humanity, he felt, had a special connection with God.

SECTION REVIEW

1. **K/U** An athlete breaks her leg. Years later she has a child who walks with a limp. Is this an example of evolution? Explain your answer.
2. **K/U** Describe the contributions of the following people to the understanding of evolution:
   (a) Cuvier
   (b) Malthus
   (c) Wallace
   (d) Lyell
3. **K/U** Charles Darwin was not the only person to discuss the idea of evolution. Why is his name most often mentioned synonymously with the idea of evolution?
4. **C** Write a brief presentation that explains the difference between catastrophism and uniformitarianism and how these ideas related to the development of the theory of evolution.
5. **K/U** Explain the idea of use and disuse as it relates to the theory of evolution by the inheritance of acquired characteristics.
6. **K/U** Summarize some of the observations Darwin made while on the voyage of the *Beagle* that he later incorporated into his theory of evolution by natural selection.
7. **MC** Nature writer Wallace Stegner once wrote of a population of trout in a mountain lake that were in a "Malthusian dilemma." Explain what Stegner meant.
8. **K/U** Describe what is meant by the term "biotic potential."
9. **K/U** Explain why Darwin referred to "descent with modification" rather than "evolution."
10. **I** At the site of a fossil bed, you come across fossils in a number of layers in the sediment. Which layers would have the oldest fossils and which would have the youngest fossils?

UNIT PROJECT PREP

For your Unit Project on Searching for a Common Ancestor, consider how evolution and fossils help scientists find examples of early forms of life.

Evidence of Evolution

EXPECTATIONS

- **Evaluate the scientific evidence that supports the theory of evolution.**
- **Analyze how technological development has extended or modified knowledge in the field of evolution.**

Charles Darwin assembled a group of facts that had previously seemed unrelated in *The Origin of Species*. However, before and after publication of this book, biologists, geologists, geographers, paleontologists, and other scientists provided a wealth of information that supported and strengthened the theory of evolution. Evidence in support of evolution has come from the fossil record, the sciences of genetics and molecular biology, the geographic distribution of organisms on Earth, and studies comparing the anatomy of adult and embryonic animals.

The Fossil Record

Fossils are made when organisms become buried in sediment that is eventually converted into rock. Sedimentary rocks with fossils reveal a **fossil record** of the history of life on Earth and show the kind of organisms that were alive in the past. While some fossils look similar to species we see today, most are very different. For example, the animals alive during the Cambrian period that were preserved in the Burgess Shale fossil beds in British Columbia had never been seen in the fossil record before. The animals unearthed in the Burgess Shale lived during the Cambrian Explosion (about 500 million years ago), a time during which a stunning burst of biodiversity occurred, much of which is now preserved as fossils. While some of the animals found in the Burgess Shale are ancestors of animals that are common today, others have long been extinct and are unlike anything in our modern oceans. An artist's representation of how the ocean might have looked when the Burgess Shale animals were alive is shown in Figure 10.9.

Fossils from more recent geological periods are much more similar to species alive today. This also supports the idea that life has evolved over time. Those species that were alive long ago have had a longer time to change, whereas those living only a few million years ago would have had comparatively little time to change. The geological time scale (Figure 10.10 on the following page) shows when organisms first appear in the fossil record.

Figure 10.9 An artist's representation of the habitat and animals now fossilized in the Burgess Shale

Another way in which the fossil record supports the idea of evolution is that fossils appear in chronological order — that is, probable ancestors appear earlier (in older rock strata) in the fossil record. The oldest fossils discovered thus far are of stromatolites that lived over 3.8 billion years ago. Stromatolites are unusual rings formed by cyanobacteria (blue-green algae). The stromatolite formation on the shore of Lake Superior (as shown in Figure 10.11) is approximately 1.9 billion years old. As Figure 10.10 shows, other organisms, from simple invertebrates to mammals, then appear sequentially in the fossil record through time.

The fact that organisms do not all appear in the fossil record simultaneously supports the idea that organisms have slowly evolved from ancestral forms. As an example, the fossil history of vertebrates shows that fossilized fishes are the oldest vertebrate fossils. Next to appear in the vertebrate fossil record are amphibians, followed by reptiles, and then birds and mammals (see Figure 10.10). Biologists and paleontologists have gathered evidence that shows that amphibians evolved from ancestral fish, reptiles from ancestral amphibians, and so on, up through the vertebrate groups.

It is important to keep in mind the vast amounts of time that the history of life covers. Changes are slow and can take millions of years, yet the fossil record gives us a "snapshot" of ancestral forms. Figure 10.12 on page 351 shows the evolution of oyster shells.

Figure 10.11 These stromatolites are among the earliest known organisms preserved in the fossil record.

About 200 million years ago, oyster shells were small and curved. The fossil record shows that the shells of later generations evolved into a larger, flatter shape over a period of about 12 million years. Oysters live on the ocean floor, and the larger, flatter shell shape may have proved a more stable shape to prevent shifting as water moved over the oysters.

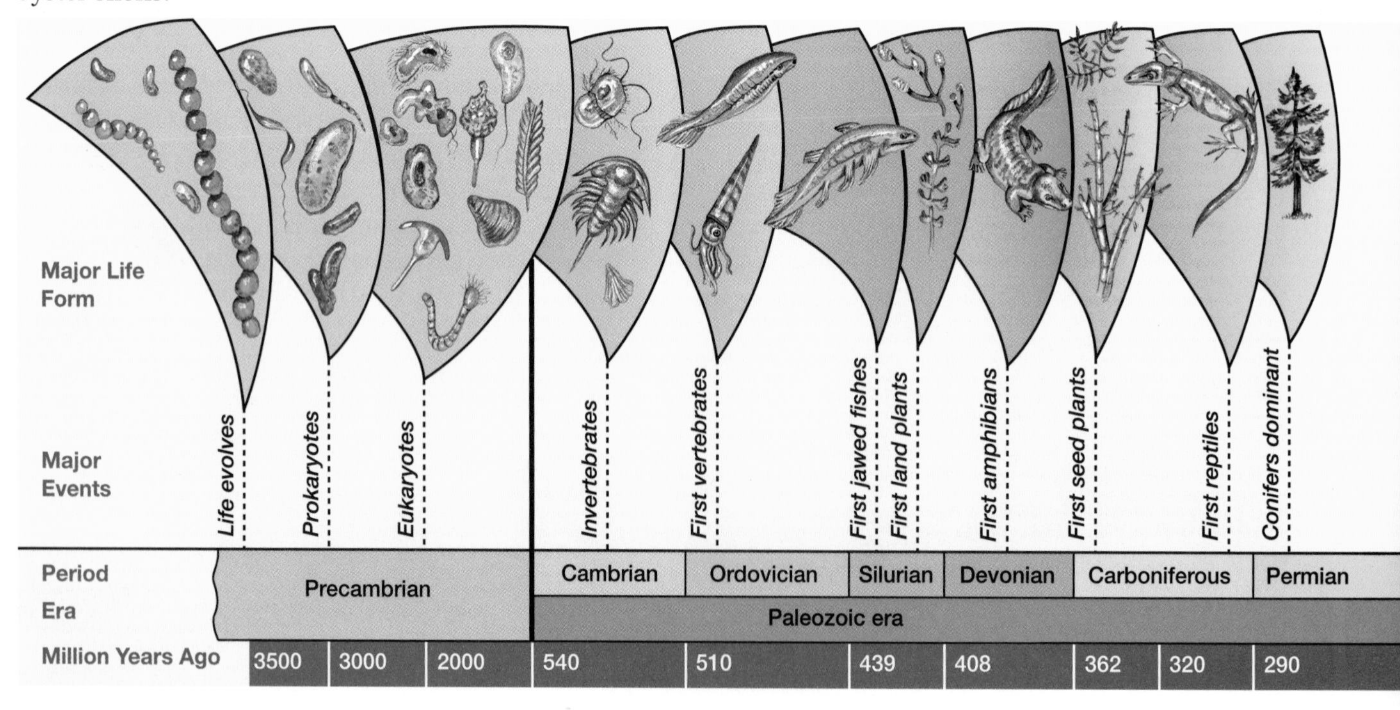

Figure 10.10 The geological time scale shows when organisms first appear in the fossil record.

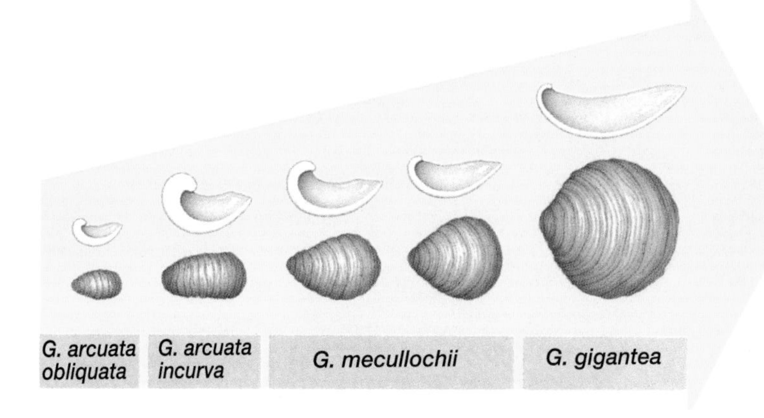

Figure 10.12 Evolution of the oyster shell

How did scientists begin to understand the links between fishes and reptiles or between reptiles and amphibians? What evidence was there to support the main ideas of natural selection — that organisms could slowly adapt and change (even into new species) given vast amounts of time? This idea has been supported by the discovery of hundreds of **transitional fossils**. These fossils show intermediary links between groups of organisms, and share characteristics common to two separate groups.

Archaeopteryx (see Figure 10.13A), for example, lived about 150 million years ago. Fossils of this species reveal characteristics of both reptiles and birds. This creature had feathers, but unlike any modern bird, *Archaeopteryx* had teeth, claws on its wings, and a bony tail. *Archaeopteryx* resembles certain dinosaurs more than any modern bird. This fossil, along with other types of evidence, supports the hypothesis that birds evolved from dinosaurs. (Indeed, if it had not been for the preservation of *Archaeopteryx* feathers, *Archaeopteryx* would have been placed in a group of small, carnivorous, bipedal (two-footed) dinosaurs called theropods.) Several other dinosaurs with feathers have since been unearthed, but *Archaeopteryx* is the first known true-flier and is considered to be the earliest bird.

Figure 10.13A Fossil of *Archaeopteryx*, which shows a link between birds and reptiles

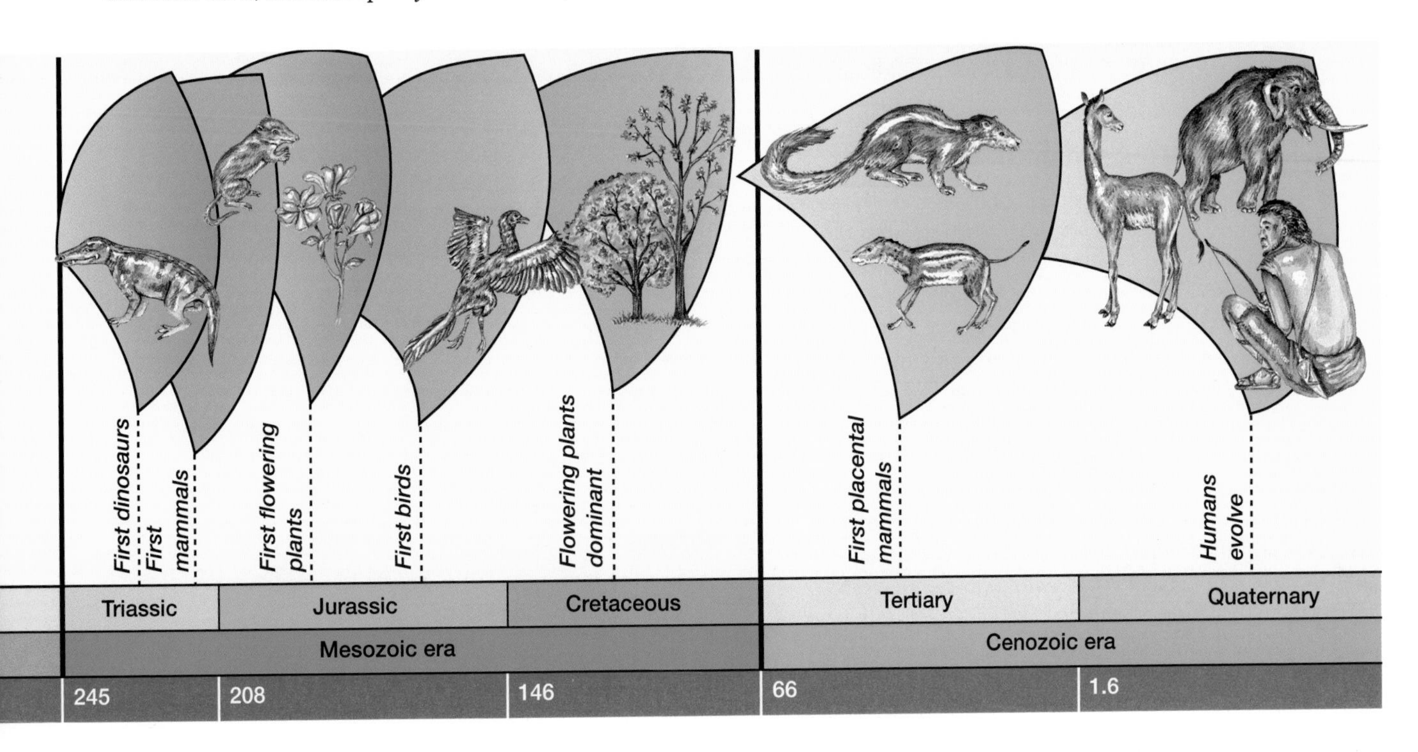

Acanthostega is a fossil that shows the link between fish and amphibians. *Acanthostega* lived about 360 million years ago (see Figure 10.13B). It had gills *and* lungs, stumpy legs, limbs and toes, a long tail almost the length of its body, a crocodile-like snout, and a jaw filled with teeth. Paleontologists do not think *Acanthostega* walked on land. Rather, they believe it used its limbs and toes to grab onto vegetation and pull itself through plant-choked swamps.

Historically, rocks were dated solely by their position relative to one another. Deeper rocks were considered to be older than shallower rocks. Today, new techniques and technologies provide a far more comprehensive understanding of the age of rocks and absolute rather than relative dates. Scientists can now date the rocks in which fossils are found by measuring the degree to which certain radioactive isotopes have decayed since the rock formed. The older the rock, the more its isotopes will have decayed. Radioactive isotopes in rock have been referred to as internal clocks. They measure the time since the rock was formed because they decay at a regular rate.

Paleontologists continue to add to our understanding of evolution. For example, researchers have recently found fossilized whales that link these aquatic mammals to their terrestrial ancestors. The *Basilosaurus* was an ancient whale that had hind limbs but led an entirely aquatic life (see Figure 10.13C on page 353). An earlier transitional form, *Ambolucetus*, had heavier leg bones and was thought to live both on land and in water.

Figure 10.13B *Acanthostega* lived in swamps and used limbs and toes to maneuver in swampy waters.

WEB LINK

www.mcgrawhill.ca/links/biology12

Canadian fossil sites, including the Burgess Shale in British Columbia, the rich fossil sites near Drumheller, Alberta, and the Joggins Fossil Cliffs in Nova Scotia, have revealed fascinating information on the evolution of life. To learn more about these sites, go to the web site above, and click on **Web Links**. Choose one fossil site and prepare a short oral presentation or one-page summary on the significance of the site and how it contributed to our understanding of the evolution of life.

MINI LAB

The Dinosaur-Bird Debate

Since 1996, fossils from six families of theropod dinosaurs have been found with preserved feathers or feather-like structures. While *Archaeopteryx* is the first known true-flier, paleontologists do not think that the feathers in the other species were used for flight. Instead, they might have been used for display or to cover eggs in the nests of brooding females. The dinosaur–bird debate actually began in the 1970s, when some paleontologists proposed that dinosaurs might have been warm-blooded and were the direct ancestors of birds. Scientists started to consider the possibility that some dinosaurs might have had feathers, hypothesizing that smaller, warm-blooded dinosaurs would need some sort of insulation to help regulate their body temperature. Dr. Phillip Currie, a paleontologist at the Tyrrell Museum in Alberta, has said that the theories on warm-blooded dinosaurs and the dinosaur origin of birds were "two of the biggest controversies in paleontology at the end of the twentieth century." At first, more people opposed the idea than supported it. However, this trend is now reversing, largely because of the discovery of a rich fossil find in northeastern China. Here, the first feathered dinosaur, *Sinosauropteryx prima*, a small chicken-sized animal that was covered in a downy coat of feather-like structures, was found.

In this activity, you will investigate the dinosaur–bird debate. Using print or Internet resources, gather evidence about the debate and answer the questions posed below. Dinosaurs with feathers include: *Sinosauropteryx*, *Sinornithosaurus*, *Beipiaosaurus*, *Caudipteryx*, *Protarchaeopteryx*, and *Microraptor*.

Analyze

1. List the evidence that supports the idea that birds evolved from dinosaurs.
2. What are the arguments against the idea that birds evolved from dinosaurs?
3. Describe the proposed origin of feathers. Is the use of feathers the same in birds as it is in dinosaurs?

Figure 10.13C Fossilized leg bones of *Basilosaurus,* an ancient whale that provides evidence that whales evolved from terrestrial animals

Geographical Distribution of Species

Biogeography is the study of the geographical distribution of species. Darwin's thinking was influenced by the distribution of animals. Recall that he wondered why the birds in the Galápagos Islands so closely resembled those on the closest continent, South America. This suggests that animals on islands have evolved from mainland migrants, with populations adapting over time to adjust to the environmental conditions of their new home. This idea has since been supported in many studies. Geographically close environments (for example, desert and forest habitats in South America) are more likely to be populated by related species than are locations that are geographically separate but environmentally similar (for example, a desert in Africa and a desert in Australia).

The biogeographical evidence for evolution also points to places such as Australia. Why would so many marsupials but relatively few placental animals live there? (Marsupials such as the kangaroo bear live young, but part of the offspring's development occurs outside the uterus in a pouch. Young develop in the uterus until birth in placental animals.) Australia can clearly support placental mammals; populations of introduced rabbits and mice have certainly increased! The unique marsupials of Australia evolved in isolation from places where the ancestors of placental mammals lived.

Neighbouring New Zealand also has a variety of animals found nowhere else, specifically a variety of flightless birds including the kiwi, the takahe, and the extinct moa, the largest bird to ever live (see Figure 10.14). New Zealand is a country comprised of two large and several small islands. Originally, New Zealand and Australia were part of the supercontinent Gondwana. As these countries drifted away from Gondwana, due to the shifting of the continental plates, they became isolated from other land masses. Once isolated, populations unique, or **endemic**, to these islands evolved.

Islands can have a volcanic origin (such as the Galápagos) or they may have broken off adjacent continental land masses (such as New Zealand). Islands can be colonized by species that swam, flew, or floated from the nearest mainland. Islands with nonvolcanic origins can also be populated by

Figure 10.14 Birds found only in New Zealand include the kiwi (A), the takahe (B), and the moa (C), which is now extinct.

species that remained on the island as it broke away from the mainland. Once isolated on the island, these species begin to evolve in different ways from their ancestors on the mainland. The degree of difference from their ancestors depends on the age of the islands. This can be demonstrated by looking at Madagascar and the Canary Islands, both of which are off the coast of Africa.

Madagascar is an island off the east coast of Africa that was originally connected to the African mainland. Madagascar is thought to have split from the African continent about 150 million years ago, although periodic fluctuations in ocean levels may have reconnected the two on a few occasions up until about 50 million years ago. Today, the channel between Africa and Madagascar is about 400 km wide, so species dispersing to the island during the last 50 million years would have had to cross this channel. Madagascar has 184 species of birds, 125 of which are endemic to Madagascar. Larger birds such as ducks, which can easily cross the water between the two countries, are found in both Africa and Madagascar. However, 90 percent of the land birds in Madagascar are found only there.

Madagascar is also the only place in the world where lemurs are found. However, the fossil record shows that lemurs were once widespread throughout Africa. Lemurs first appear in the fossil record about 65 million years ago; therefore, they were either present on Madagascar when it separated from the African continent or they floated to Madagascar when the channel was narrow. So why are lemurs no longer present in Africa? When Madagascar permanently separated from Africa 50 million years ago, monkeys had not yet evolved. Monkeys do not appear in the fossil record until about 35 million years ago, so they had no way of reaching the island of Madagascar (because the channel between Africa and Madagascar was too wide at that time). However, monkeys eventually took over the niche that lemurs had on the African continent and drove lemurs to extinction there.

The Canary Islands, off the northwest coast of Africa, are about 10 to 15 million years old. They were formed by volcanoes — they were never attached to the African continent. Therefore, unlike Madagascar, the Canary Islands have been colonized only by those animals and plants able to disperse from the adjacent coastline of Africa. Of the 53 bird species known to breed there, only two are endemic. As well, the Canary Islands have no snakes or land mammals (except bats). The eight species of lizards on the islands are thought to have drifted on pieces of wood from the adjacent coastline. They are similar to west African species, yet are sufficiently different to show that natural selection has created some change in the populations. In fact, some of the lizards are now recognized as new species.

Anatomy

When the anatomy of various animals is examined, more evidence for evolution of animals from common ancestors is revealed. Figure 10.15 shows the forelimbs and individual bones of five vertebrates. All of the limbs have the same basic arrangement of bones, yet they are modified into wings, arms, legs, and fins. The present arrangements of bones in the animals shown in Figure 10.15 are variations on a common structural theme. As these animals descended from common ancestors, the same bones were put to different uses. The bones have the same origin yet they now differ in structure and function. Such anatomical signs of evolution are called **homologous structures**. Homologous structures have not only similar numbers of bones but also similar numbers of muscles, ligaments, tendons, and blood vessels. They also have the same developmental origin.

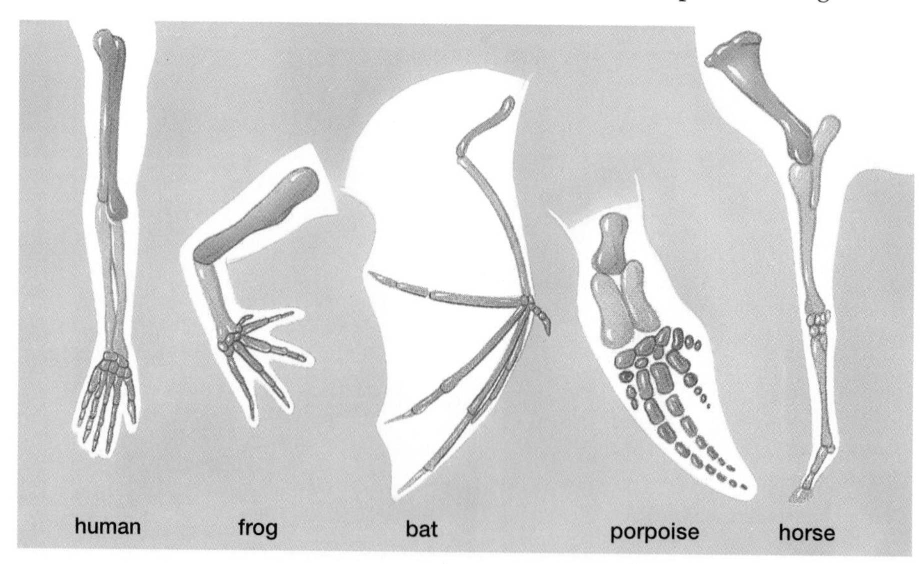

Figure 10.15 These vertebrates have the same basic arrangement of bones, but the bones have been put to different uses.

Homologous structures can be similar in structure, function, or both. For example, the limbs in Figure 10.15 are structurally similar. Also, the limbs of the human, frog, and horse perform the same function: they are designed for walking on land. Functional similarity, however, does not necessarily mean that species are closely related. For example, insect and bird wings are similar in function but not in structure. The wings of these types of animals evolved independently and have very different structures. Bird wings are supported by bones, whereas a tough material called chitin makes up insect wings. Body parts of organisms that do not have a common evolutionary origin but perform similar functions are called **analogous structures**.

Even though analogous structures do not show evolutionary relationships between animals, they do support the idea of natural selection. Bird and insect wings evolved separately when the ancestors of today's species adapted independently to a life that included flight.

Many organisms also possess **vestigial structures**. These are structures that were functional in the organism's ancestors yet have no current function. For example, the baleen whale in Figure 10.16 has vestigial pelvic bones. Pelvic bones perform no function in modern whales since they have no hind limbs. Their presence in modern whales points to the terrestrial origins of ancestral whales. The vestigial pelvic bones are artifacts from the whales' evolutionary history.

The forelimbs of the flightless ostrich are another example of a vestigial structure. The ancestors of modern ostriches were probably able to fly, but they likely foraged and nested on the ground. As a result, over time these animals became quite large and unable to fly, and the forelimbs became unnecessary.

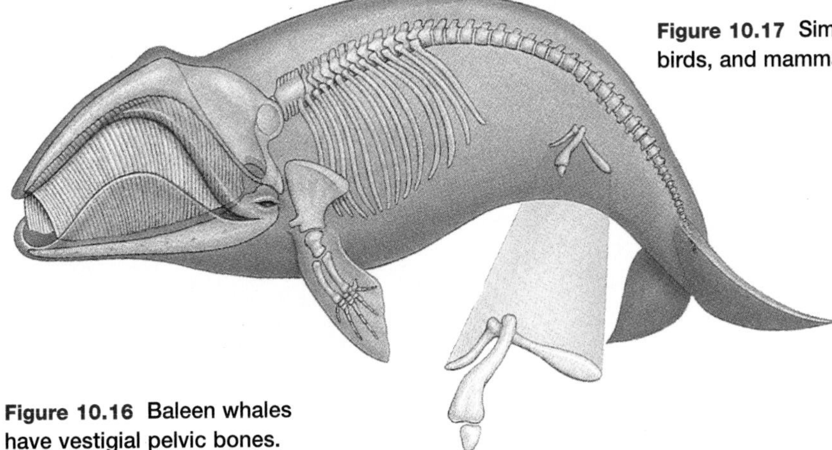

Figure 10.16 Baleen whales have vestigial pelvic bones.

Embryology

Embryology has also been used to determine evolutionary relationships among animals. When the embryos of organisms are examined, similar stages of embryonic development are evident. For example, all vertebrate embryos (including humans) go through a stage in which they have gill pouches (as shown in Figure 10.17). At certain stages in the development of the embryo, the similarities among fish, birds, humans, and all other vertebrates are more apparent than their differences. In Figure 10.17 for example, the early stages of development of fish, reptile, bird, and mammal embryos each have a tail and gill pouches. Gill pouches form gills in fish. In terrestrial vertebrates, the gill pouches are modified for other uses, such as the Eustachian tube in humans. The tail in a human embryo becomes the coccyx at the end of the spine.

These similarities between embryos in related groups (such as vertebrates) point to a common ancestral origin. It follows that related species would share both adult features (such as the number of arm bones, as discussed earlier) and embryonic features (such as the presence of gill pouches).

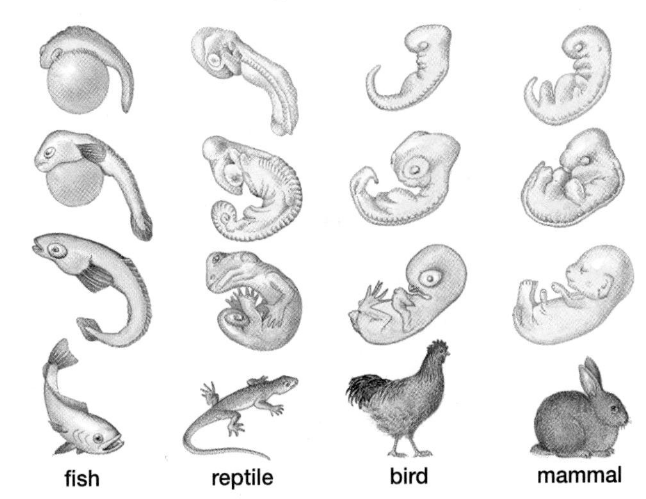

Figure 10.17 Similarities in the embryos of fish, reptiles, birds, and mammals show evidence of evolution.

Heredity

When Darwin published *The Origin of Species*, the science of genetics and an understanding of heredity was not yet established. This meant that Darwin could not completely explain the mechanism that drove natural selection. Today, since the laws of inheritance and the science of genetics are more clearly understood, the variations in organisms required for natural selection to occur can be explained. This will be discussed in further detail in Chapter 11.

Molecular Biology

The evolutionary relationships among species are reflected in their DNA and proteins. Since DNA carries genetic information, how closely related two organisms are can be determined by comparing their DNA. If two species have similar patterns in their DNA, this similarity indicates that these sequences must have been inherited from a common ancestor. For example, by studying gene sequences, scientists have determined that dogs are related to bears and that whales and dolphins are related to ungulates (hoofed animals such as cows and deer).

The degree to which DNA sequences are similar between species determines how closely related those species are. For example, humans and chimpanzees have an approximately 2.5 percent difference between their DNA sequences, while humans and lemurs have a 42 percent difference.

The science of molecular biology has also helped show that all forms of life are related to the earliest

Biology At Work

Paleontologist

In 1985, Wu Xiao-Chun was a student at the Chinese Academy of Sciences. He was studying paleontology. While searching for early crocodiles and dinosaurs, he unearthed a small fossil in southwestern China's Lefung Basin. Thinking it was a fragment of an unimportant lizard-like fossil, he paid little attention to it.

Three years later, Wu Xiao-Chun decided to study the little fossil. As he started chipping away the rock from around the fossil, he began to suspect this was not just a fragment, but rather a complete skull unlike any other reptile fossils found in the same area.

Wu's professor passed the skull on to Dr. Luo Zhe-Xi, a paleontologist at the Carnegie Museum of Natural History in Pittsburgh, Pennsylvania. Dr. Luo and others examined the skull, which was only about the size of a fingernail. Astonishingly, they found that the organism's ear was not encased in its jaw, as is the case with mammal-like reptiles. Instead, the creature's jaw was like that of a modern mammal, even though it had lived in the age of the dinosaurs, some 195 million years ago. The creature also had other modern-mammal features, including a brain cavity that was large in proportion to its skull. Dr. Luo and his colleagues concluded that this was the closest known relative of modern mammals, despite being 45 million years older than any previously discovered mammal. They named it *Hadrocodium wui*. *Hadrocodium* is Latin for "full head," and *wui* refers to its discoverer, Wu Xiao-Chun.

Today Wu Xiao-Chun, now Dr. Wu, is a paleontologist at the Canadian Museum of Nature in Ottawa. He is known for his study of fossils, including feathered dinosaurs, lizard-like animals, and crocodiles in North America, Europe, and Asia. Though Dr. Wu's main interest is in reptiles, he still fondly recalls the day when he unearthed the skull of the world's earliest known mammal.

"I like paleontology because it includes both indoor and outdoor work," he says. "I love hiking." To Dr. Wu, each sediment is "like a mysterious book and the fossils inside like inset pictures or illustrations. Each fossil contains a story."

Career Tips

1. Paleontologists work at universities and at museums such as the Canadian Museum of Nature and the Royal Tyrrell Museum of Paleontology in Drumheller, Alberta. Some also work for private companies.
2. What does this feature tell you about what paleontologists do?

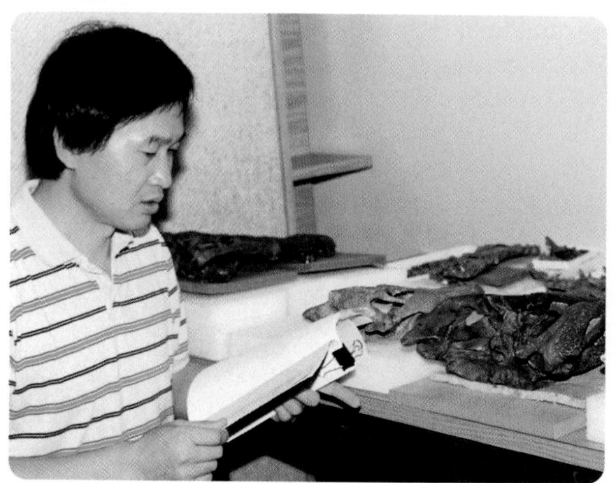

Dr. Wu Xiao-Chun

organisms to some extent. Even organisms that are only remotely related have some proteins in common. One example is **cytochrome c**, a protein involved in cellular respiration that is found in the mitochondria. The amino acid sequence of cytochrome c is so similar among organisms that it can be used to indicate relatedness. The length of the cytochrome c enzyme varies from 103 to 112 amino acids, depending on the organism. The amino acid sequence of the cytochrome c in chimpanzees and rhesus monkeys (both primates) differs by only one amino acid; the sequence in chimpanzees and horses (both mammals) differs by 11 amino acids; and the sequence of the chimpanzee and dogfish (both vertebrates) differs by 24 amino acids.

Scientists have also tracked the evolution of cytochrome c itself. Figure 10.18 shows that the longer the time since an organism evolved from a simple ancestor, the greater the number of differences in nucleotide sequences in the gene for cytochrome c. This also points to the evolutionary idea of organisms having common ancestors. While mutations have substituted amino acids in various places in the protein cytochrome c during the long period of evolution, cytochrome c still has a similar structure and function in all species.

Scientists can also study the evolutionary history of a gene using DNA sequencing. The gene for the protein hemoglobin has been well studied. The pattern of descent, or the **phylogenetic tree**, of the hemoglobin gene is shown in Figure 10.19 on the following page. (A phylogenetic tree shows the pattern of descent. A phylogenetic tree is similar to an evolutionary family tree for an organism.) The progressive changes in the hemoglobin molecule have produced a tree that shows the evolutionary relationships between organisms — the shorter the line in the tree, the more amino acids in common and the closer the evolutionary relationship.

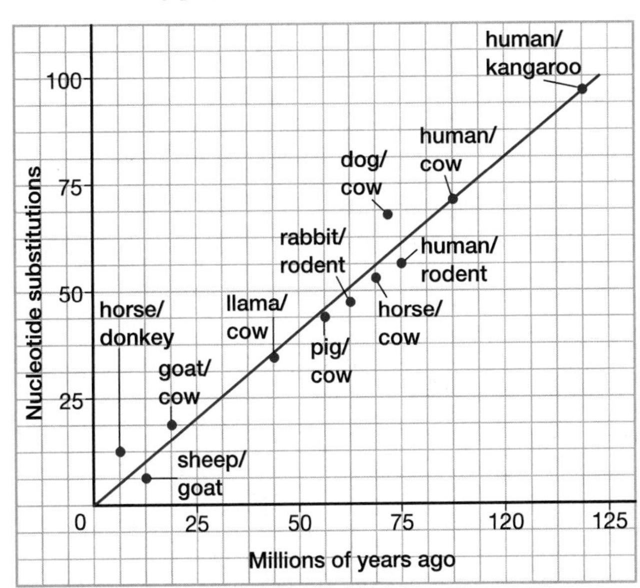

Figure 10.18 Evolution of cytochrome c. The longer an animal diverged from a common ancestor, the greater the difference in genetic sequence.

Defining a Theory

Current understanding of the theory of evolution is dismissed by some as being "just a theory." This implies that somehow a theory is just a guess and therefore is easy to refute. It is important to clarify the use of the words "theory" and "fact" in the realm of science. Scientific facts are the data that have been collected. For example, homologous structures, the fossil record, the DNA sequencing in individual organisms, and the other information presented in this chapter are scientific facts. Scientific theories attempt to explain facts and tie them together in a comprehensive way. For example, the facts gathered by Darwin and people before and after him show that evolution is happening. Darwin's theory of evolution by natural selection is the theory that attempts to tie these facts together.

This chapter has outlined the development of the theory of evolution and the facts from various disciplines, including geography, paleontology, and molecular biology, that all support this theory. The study of evolution continues to spark debate even today. Although there are still lively debates in the scientific community over specific details of exactly *how* life evolved, biologists do not refute the idea of evolution itself.

COURSE CHALLENGE

The evolution of rattlesnake venom or the evolution of sickle cell disease are two potential topics that can be examined in the Course Challenge. Start making notes on the links that evolution (in general) has with metabolic processes, homeostasis, and molecular genetics. In the next unit, think about how the population dynamics might, in turn, be affected by changes in populations.

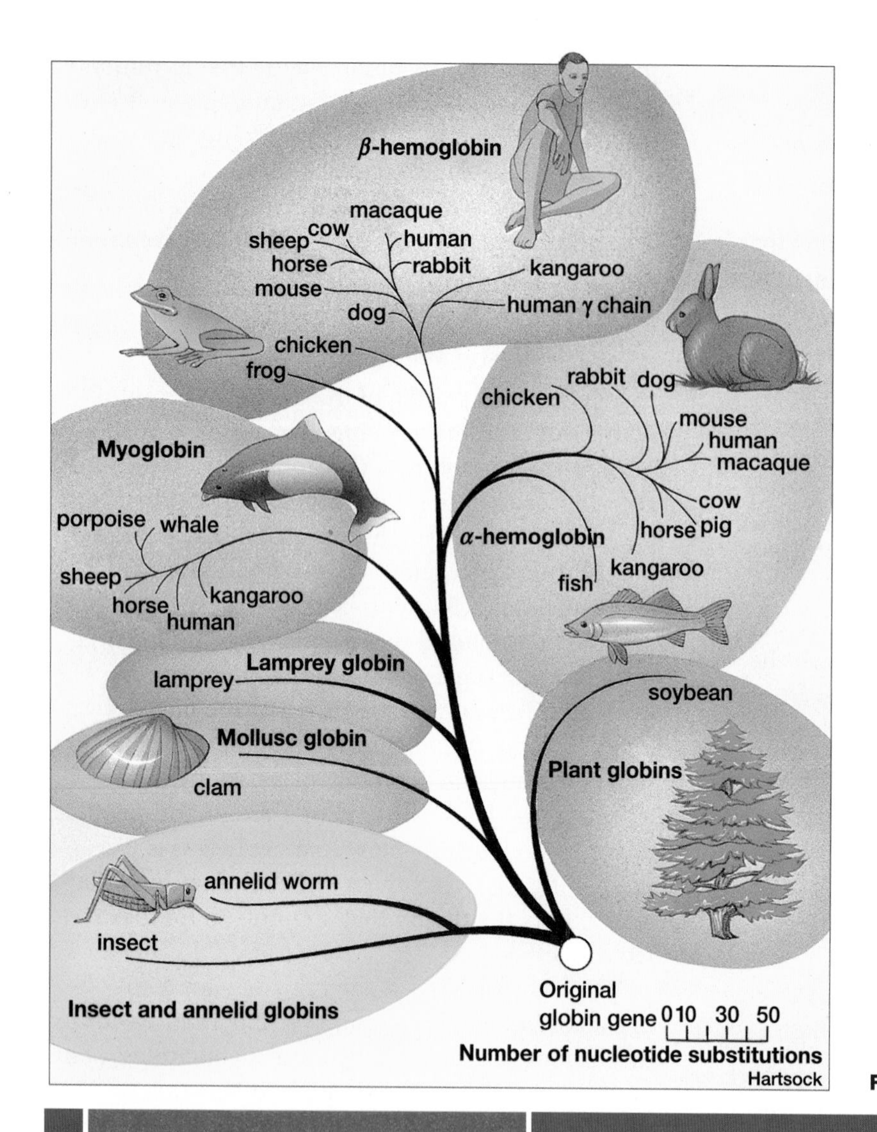

Figure 10.19 Evolution of the globin gene

SECTION REVIEW

1. K/U How does the discovery of so-called missing links in the fossil record help us to understand evolutionary events of the past?

2. C Choose a fossil (either one described in this text or another one you have researched) and describe what information this fossil provides that helps us understand evolution. Give reasons for your opinion.

3. K/U How do the number of endemic species differ between Madagascar and the Canary Islands? Explain why these differences exist.

4. K/U Describe how the anatomy of animals is used to explain evolution.

5. K/U Baleen whales, such as grey and humpback whales, have teeth and body hair while they are embryos, but they lack these features as adults. What does this tell us about the evolutionary history of these animals?

6. K/U When human organs are transplanted, the rate of success is higher in cases where the donor and recipient are close relatives. Why do you think this is so?

7. C Explain how the differences in the sequence of amino acids that make up cytochrome c in different kinds of organisms help us understand evolution.

8. MC Biologist Stephen Jay Gould wrote, "The fact of evolution is as well established as anything in science (as secure as the revolution of the earth around the sun) … Theories, or statements about the causes of documented evolutionary change, are now in a period of intense debate — a good mark of science in its healthiest state. Facts don't disappear while scientists debate theories." Explain the difference between fact and theory as they relate to science.

9. I Make a hypothesis concerning what species changes and what environmental changes you would expect to see on Madagascar if it were to become reconnected to the mainland of Africa. How might a scientist test your hypothesis?

Summary of Expectations

Briefly explain each of the following points.

- Variety within a population and the environment in which organisms live allow natural selection to happen. (10.1)
- Genetics and environment can affect evolution. (10.1)
- The ideas and observations of many people helped develop the current theory of evolution. (10.1)
- Natural selection is the process whereby a population of organisms changes because individuals who inherit certain traits can survive the local environmental conditions and pass on these traits to their offspring. (10.1)
- Artificial selection is the process whereby humans artificially select organisms with certain traits and these organisms pass on these traits to their offspring. (10.1)
- Observations of Charles Darwin led him to develop a theory of evolution. (10.2)
- Contributions of Cuvier, Lamarck, Malthus, Hutton, Lyell, and Wallace helped develop the theory of evolution. (10.2)
- Darwin's theory of evolution by natural selection is compared with Lamarck's theory of evolution by the inheritance of acquired characteristics. (10.2)
- Technology, such as DNA sequencing and amino acid sequencing, has provided more evidence for evolution. (10.3)
- Fossils, biogeography, anatomy, and molecular biology all provide evidence for evolution. (10.3)
- A scientific fact is data or information that have been collected, and a theory attempts to explain facts. (10.3)

Language of Biology

Write a sentence including each of the following words or terms. Use any six terms in a concept map to show your understanding of how they are related.

- evolution
- adaptation
- gene pool
- natural selection
- selective pressure
- artificial selection
- fitness
- paleontology
- catastrophism
- inheritance of acquired characteristics
- gradualism
- uniformitarianism
- biotic potential
- descent with modification
- fossil record
- transitional fossil
- biogeography
- endemic
- homologous structures
- analogous structures
- vestigial structure
- embryology
- cytochrome c
- phylogenetic tree

UNDERSTANDING CONCEPTS

1. If you teach children to look both ways before they cross the street, this action will help them survive. Is this an example of natural selection at work? Explain your answer.
2. Compare the controlling factors for artificial selection and natural selection.
3. How might the colour of a field mouse affect its survival?
4. Some caves contain fish that are blind. These fish have eye sockets and vestigial eyes. Explain how (a) Lamarck and (b) Darwin would account for the origin of sightlessness in these fish and other blind cave-dwellers.
5. How do vestigial structures provide evidence for evolution?
6. Summarize the main points of Darwin's theory of evolution by natural selection.
7. Explain, in the context of evolution and natural selection, common misunderstandings and misinterpretations about the following words: (a) evolution; (b) fitness; (c) theory.
8. Although Lamarck and Darwin explained evolution in different ways, their theories had some similarities. Describe these similarities.
9. Does evolution mean that organisms are becoming progressively better with each generation? Explain your answer.
10. Are a bird wing and an insect wing homologous structures? Explain your answer.
11. How does the study of embryology support evolution?
12. Why is it more accurate to speak of the evolution of populations rather than of individual organisms?
13. Explain why DNA is a useful tool for determining possible relationships among species of organisms. Give an example.

14. Distinguish between fact and theory.

15. Describe how the following items contributed to Darwin's thinking on evolution:
    (a) his experiences on the voyage of the *Beagle*;
    (b) Lyell's *Principles of Geology*;
    (c) the experience of plant and animal breeders; and
    (d) Malthus's essay on population.

16. Much of the theory of evolution has been developed by interpreting certain observations or making inferences about these observations. For each observation below, outline the inferences that Darwin, other scientists, and other naturalists made from this information.
    (a) Populations tend to remain stable in size.
    (b) No two individuals are exactly alike.
    (c) Resources such as food are limited.

INQUIRY

17. Outline a breeding program that would help you develop a cow that produces more milk. Is your cow a new species? Explain your answer.

18. Design an experiment that would demonstrate variation within a population.

19. Examine the fossils found in the sedimentary rocks shown below. The rocks are older as you go deeper into the rock strata. Explain what these rock strata and the fossils within them can tell you about evolution.

20. You are analyzing the amino acids in the hemoglobin of various species. You find that this protein in rhesus monkeys differs by about eight amino acids from the protein in humans. The difference in this protein between mice and humans is about 26 amino acids, and the difference between lampreys (a primitive fish) and humans is about 125 amino acids. Interpret these data and explain how they relate to our understanding of evolution.

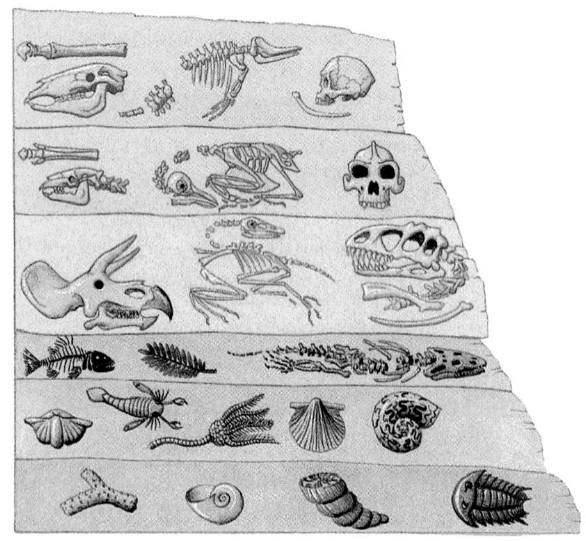

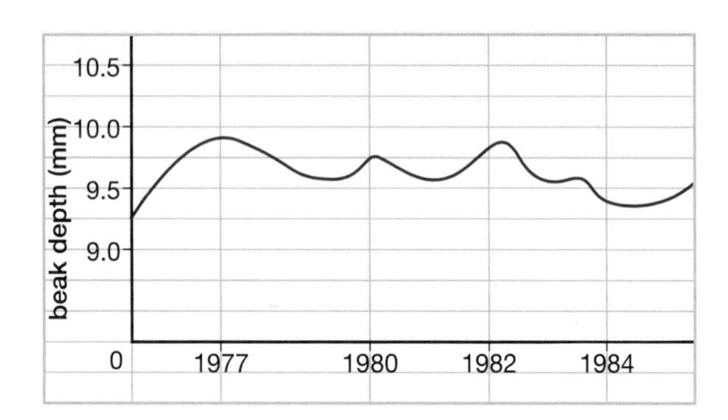

21. This graph shows how the average beak size (depth) in a population of ground finch shifted during particularly wet and dry years. 1977, 1980, and 1982 were all drought years; 1984 was a wet year.
    (a) Interpret these data and explain how they relate to natural selection and the definition of evolution.
    (b) An observer suggested that during drought years all the seeds were large and tough to open. This meant that birds exercised their beaks more, making the beaks stronger. Is this a plausible explanation for these data? Explain your answer.

22. Lamarck's idea of inheritance of acquired characteristics has recently gained support by some scientists. Search the Internet for information on renewed interest in Lamarck's idea as it relates to the immune system. The new ideas suggest that in some instances, characteristics acquired during one's lifetime may be passed on to offspring. Summarize these ideas.

COMMUNICATING

23. Red Island and Blue Island are hypothetical islands 500 km off the coast of South America. Red Island is volcanic in origin and is only five million years old. Blue Island separated from South America over 80 million years ago. Describe the origin of the animals on these islands and how they may be similar to or different from those of South America.

24. Create a time line showing the various individuals whose contributions eventually led to the development of the theory of evolution by natural selection. State their contributions.

25. Richard Dawkins, a popular writer and evolutionary biologist, refers to natural selection as the "blind watchmaker," meaning that natural selection is totally blind to the future. Explain what Dawkins means by this statement.

26. Explain how the examination of proteins can demonstrate relatedness among species.

27. A population of fish in which 95 percent of individuals are light-green and 5 percent are mottled grey lives primarily among kelp that grows on the ocean bottom. A disease kills the kelp, leaving the population without cover. Use a diagram or objects (such as poker chips) to describe how the population might change over several generations.

28. You are a biologist working with a student to make a collection of plants found in Hawaii. You notice that your assistant did not label the island that one of the plants was collected on. When asked, he explained that he did not think this was necessary as this particular plant was found on all of the Hawaiian islands. Write a memo clearly explaining why it is necessary to label the exact island and location where the plant was found.

MAKING CONNECTIONS

29. A tan-coloured insect lives in a sandy area. Some insects in the population show some green in their coloration. The climate begins to cool and become moister; slowly the habitat is covered by green plants. Use Darwin's theory of evolution by natural selection to explain how the insect population might evolve to be green. Use a diagram.

30. Darwin recognized that variation occurred within populations and that these variations could be inherited. He could see the results but could not explain the mechanism. Explain the advances in science and technology that would eventually make Darwin's theory of evolution even more convincing and would help fill in this missing piece of the puzzle.

31. Given your understanding of diversity within species and natural selection, explain why it is important to maintain biodiversity.

32. Two populations of flowers of the same species are found in nearby meadows. There are slight differences in the plants between the two populations, such as flower colour and leaf shape. How might Darwin have interpreted these facts?

33. A farmer sprays an insecticide on a field to combat a beetle that is destroying the crops. The spray works very well the first year it is used. However, after five years of spraying on an annual basis, the insecticide does not seem to be effective any longer and the beetles are still present. Explain how this illustrates natural selection.

34. Analyze the following data. The proteins present in four organisms are shown below. (Each letter represents a protein.) Determine which of the organisms are closely related. Explain your answer.

Organism 1 A, G, T, C, L, E, S, H
Organism 2 A, G, T, C, L, D, H
Organism 3 A, G, T, C, L, D, P, U, S, R, I, V
Organism 4 A, G, T, C, L, D, U

35. Evolutionary biologist Stephen Jay Gould said "Local environments change constantly. They get colder or hotter, wetter or drier, more grassy or more forested. Evolution by natural selection is no more than a tracking of these environments by differential preservation of organisms better designed to live in them: hair on a mammoth is not progressive in any cosmic sense." Explain what is meant by this statement.

36. Recommend ways that would help ensure that non-native plants and animals would not be accidentally brought to islands such as the Galápagos Islands or the Hawaiian Islands. What can be done once non-native plants and animals invade these types of islands?

CHAPTER 11

Mechanisms of Evolution

Reflecting Questions

- How does an understanding of genetics help explain how changes within a species can occur?
- How can we measure genetic variation in a population?
- What are the mechanisms that result in genetic variation?

Prerequisite Concepts and Skills

Before you begin this chapter, review the following concepts and skills:

- understanding how mutations occur (Chapter 9, section 9.1),
- understanding the mechanism of natural selection (Chapter 10, section 10.1),
- explaining the relationship between variation and natural selection (Chapter 10, section 10.1), and
- explaining how molecular biology contributes to the scientific study of evolution (Chapter 10, section 10.3).

Each fall, swarms of monarch butterflies leave southern Canada and begin their southward migration. Along the route of their incredible journey — which, for most, will end in the mountains of southern Mexico — populations of monarchs will be eagerly awaited and counted by keen observers. Monarch butterflies are one of the few migratory insects, and their migratory routes are well-studied. Every year, biologists and volunteer observers monitor monarch butterfly populations. These observers have found that populations shift dramatically from year to year. Numbers rise some years, then fall other years. During their migration and during their time in Canada, the United States, and Mexico, populations of monarchs (like populations of all organisms) are subjected to variable environmental conditions. There can be a drought one year, an early snowfall the next, or the loss of valuable habitat in another year. Populations peak following favourable conditions, while adverse conditions can result in precipitous declines. Populations continue to shift naturally from one generation to the next.

Monarch butterfly populations are estimated to total over 120 million individuals and, for now, their populations are large enough to ensure a sizeable gene pool that can withstand many challenges. But what would happen if a small population became isolated from the larger group? If the smaller population's gene pool changed, how might the smaller population itself change? Would the organisms start to look different? How does population size and the genetic variation within the population affect the evolution of the monarch butterfly population?

As you have learned, evolution is the shift in the frequency of genes in the gene pool of a population from one generation to the next. Genetic variation within a population allows changes to occur within a species. You will examine the relationship between genetics and evolution in greater detail in this chapter. You will also learn about some of the mechanisms that affect genetic variation and learn how to measure this variation.

How does genetic variety, such as the variety that exists in this population of bacteria, make evolution possible?

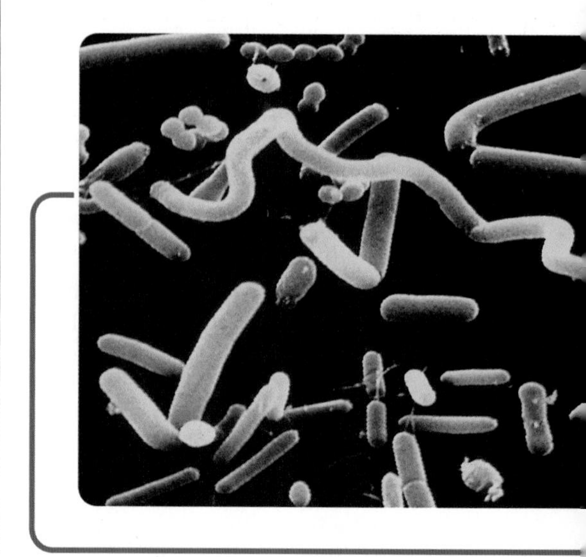

Chapter Contents

11.1 Population Genetics

EXPECTATIONS

- **Analyze evolutionary mechanisms and their effects on biodiversity and extinction.**
- **Understand how alleles are distributed in a population.**
- **Understand how allele ratios reflect selective pressure in a population.**

Evolution can be divided into macro-evolution and micro-evolution. **Macro-evolution** is evolution on a grand scale; it is large evolutionary change such as the evolution of new species from a common ancestor or the evolution of one species into two. (You will learn more about the evolution of new species in Chapter 12.) The modern camel, for example, evolved over 65 million years from a small ancestor that was not much larger than a rabbit. This long, visible sequence of changes and the categorization of organisms (extinct and living) in relation to one another are examples of macro-evolution. Figure 11.1 shows the sequence that paleontologists propose for the evolutionary path of the modern camel. (Ancestral camels actually evolved in North America, and then expanded their range to include parts of Asia and Africa.)

Micro-evolution is the change in the gene frequencies within a population over time. It is evolution *within* a species, or evolution on a small scale. For example, adaptation by natural selection is an example of evolutionary change within a species, or micro-evolution. As these changes accumulate, they can lead to the formation of a new species.

This chapter focusses on micro-evolution and the mechanisms that result in genetic variation within a population.

Heredity and Evolution

While Darwin's *The Origin of Species* convinced most biologists that species could change over time, Darwin's mechanism for change — natural selection — took longer to gain acceptance. Part of the difficulty in explaining the mechanism of natural selection was that there needed to be a plausible explanation for how traits could be inherited. If variation within a species was necessary for natural selection, what was the ultimate source of this variation?

In Darwin's theory of natural selection, new variants of species arise continually in populations. Some variants thrive and produce more offspring,

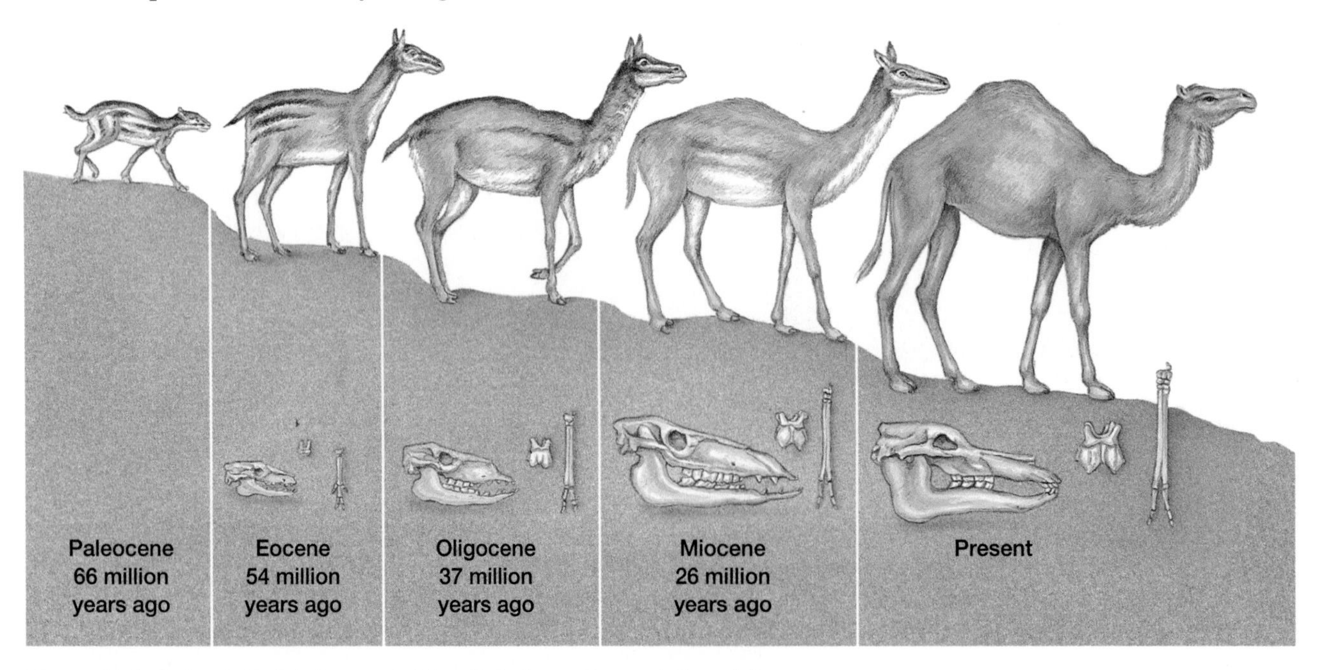

Figure 11.1 Paleontologists have used fossils to trace the evolution of the modern camel.

thus slowly leading to change in a population (this may even lead to new species over time), while other variants die off because they cannot thrive in the environment. This idea did not mesh with the ideas about inheritance at that time, which said that characteristics were blended and an offspring was an "average" of its parents. (For example, according to blended inheritance, the offspring of a plant with a red flower and a plant with a white flower would be pink. This offspring would then pass on the pink colour to its offspring. In reality, this is not always the case.) What was needed to support natural selection was an understanding of how chance variations arise in a population and how these variations are passed from parents to offspring.

Part of this missing information needed to explain inheritance and support the idea of evolution by natural selection was actually discovered during Darwin's lifetime. Gregor Mendel, an Austrian monk who is shown in Figure 11.2, conducted experiments with pea plants in the 1850s, and his work provided the basis for an explanation of inheritance. His experiments showed that, in contrast to the idea of blended inheritance, parents pass on discrete factors of inheritance, which he called genes. Mendel showed that genes do not blend in the offspring; genes retain their characteristics when they are passed to the offspring. Mendel's work and the subsequent work of others on inheritance would eventually help support the idea of natural selection by showing how the variation created through the mechanisms of heredity is the raw material on which natural selection acts.

Figure 11.2 Gregor Mendel (1822–1884) conducted experiments that explained the inheritance of characteristics.

Mendel's work was some of the first that helped to explain mechanisms of inheritance. But his work was not read by Darwin, and it would be several decades before ideas about inheritance were used to help explain natural selection. In the late nineteenth and early twentieth centuries, there was a growing interest in genetics. In the 1930s, a new field of science emerged — population genetics. As scientists began to broaden their understanding of genetics, they demonstrated that there is substantial genetic variation within populations. They showed that variations could arise in populations through changes, or **mutations**, in genes. A mutation is a permanent change in the genetic material of an organism. (Refer back to Chapter 9, section 9.1 to review mutations.) It was recognized that mutations provide the genetic variation within a population. Evolution, therefore, depends on both random genetic mutation (which provides variation) and mechanisms such as natural selection. (You will learn more about mutations in section 11.3.)

Scientists, including geneticist Theodosius Dobzhansky, biogeographer and taxonomist Ernst Mayr, paleontologist George Gaylord Simpson, and botanist G. Ledyard Stebbins, combined ideas from their fields of study with Darwin's ideas about natural selection and the current understanding of inheritance to develop a revised theory of evolution. This modification to evolutionary theory, and the meshing of Mendel's and Darwin's ideas, was called the **modern synthesis**.

Reviewing the Language of Genetics

To understand and discuss genetic variation, it is important to review certain terms. **Alleles** are alternate forms of a gene. In humans, for example, there are three alleles — I^A, I^B, and i — that determine whether an individual has A, B, AB, or O blood type. Since individuals generally have two sets of chromosomes — one received from the male parent and one received from the female parent — there are two alleles for every gene at every locus. (A **locus** [plural loci] is the location of a gene on a chromosome.) So, humans could be I^AI^A, I^AI^B, I^Ai, I^BI^B, I^Bi, or ii at the locus for blood group. If the two alleles at a locus are identical (for example, I^AI^A or *ii*), the individual is called **homozygous** for that characteristic. An individual with two different alleles at the locus (for example, I^AI^B) is called **heterozygous**. The three blood type alleles, I^A, I^B, or *i*, exist in the population, but no single person can have all three. In some populations, the allele possibilities are even greater, and far exceed the two possible alleles any human can have.

If the two alleles inherited from parents are different, one of them (the **dominant allele**) will be fully expressed in the organism's appearance and therefore will become the phenotype. (Note that dominant in this sense does not mean that this allele is somehow better. Trait for the dominant allele is simply the one that is always expressed in an individual.) The other allele, the **recessive allele**, has no noticeable effect on the organism's appearance, but it remains as part of the genotype of the organism. Figure 11.3 shows a cross between a pure purple-flowered pea plant and a pure white-flowered pea plant. The alleles for colour are W and w. Since W is the dominant allele, the flowers can only be white when the two alleles are both recessive (that is, ww).

The **genotype**, or genetic make-up, of an *individual* remains constant throughout its life. However, over time, the alleles within a *population* may change. New alleles may arise and may be recombined, thus producing individuals with novel **phenotypes**. Phenotypes are the physical and physiological traits of an organism. (Physical in this sense refers to how the organism with this trait appears.) A phenotype of an individual can be the product of both the environment and heredity. For example, environmental factors such as disease, crowding, injury, or the availability of food can all affect the appearance of an individual. But these acquired characteristics are not heritable; that is, they are not passed on to the next generation. Because of dominant and recessive alleles, an organism's appearance does not always reflect its genetic make-up. For example, Figure 11.4 shows a cross between two pea plants that have the alleles W or w at the locus for colour. The genotypes WW and Ww both result in a purple flower, while the genotype ww results in a white flower. Table 11.1 summarizes how genotype is related to phenotype.

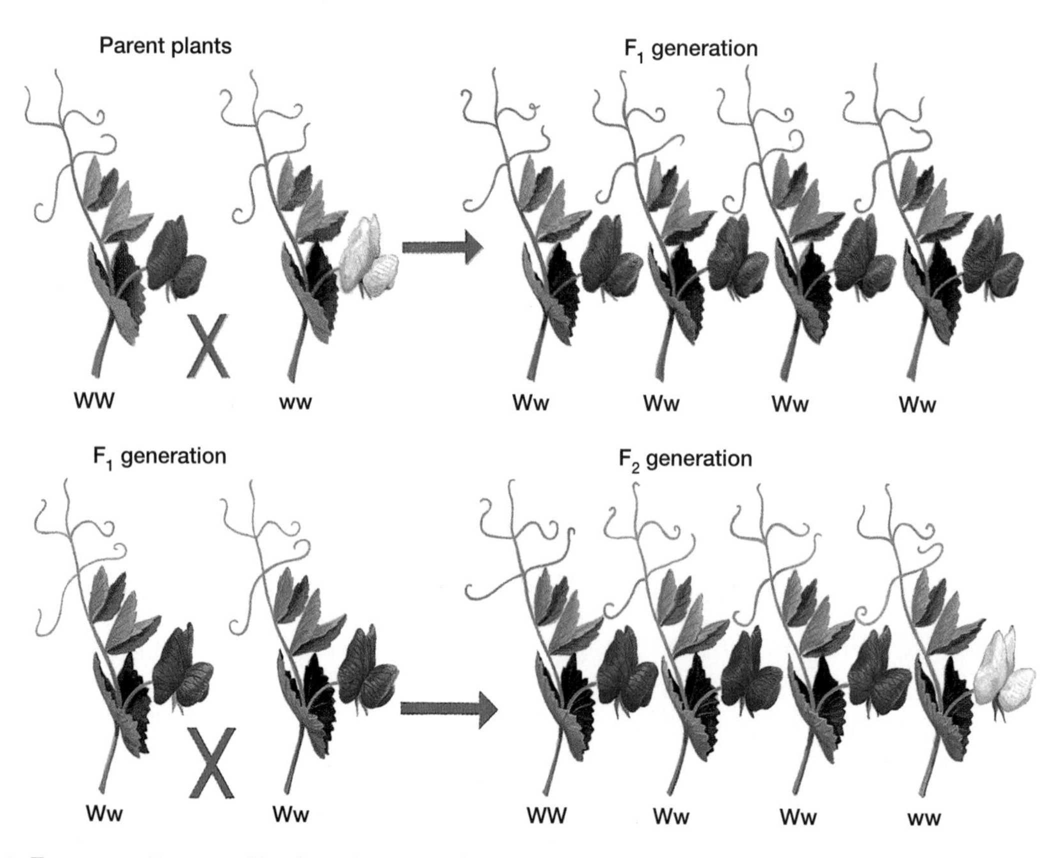

Figure 11.3 Two generations resulting from the cross of a purple-flowered pea plant and a white-flowered pea plant.

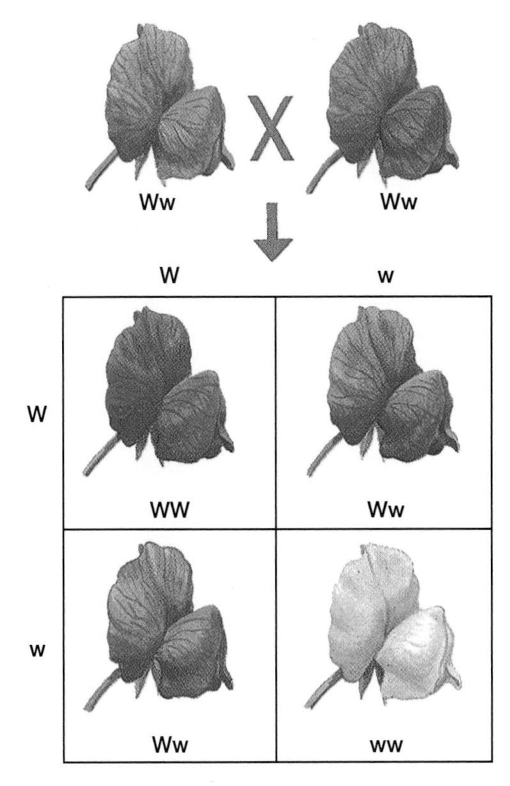

Figure 11.4 The result of a cross of two pea plants is shown in this Punnett square.

Not all traits are purely dominant or purely recessive. Sometimes neither of the alleles controlling a trait are dominant. In this case, blending of the two traits can occur. This is called **incomplete dominance**. Figure 11.5 gives an example of how incomplete dominance can occur. White or red flowers are homozygous, while pink flowers are heterozygous.

Occasionally both alleles for a trait may be dominant. These alleles are said to be **co-dominant** and both alleles are expressed in the heterozygous individual. In some varieties of chickens, for example, two alleles for a trait may be expressed equally. A black rooster crossed with a white hen produces offspring that have some black feathers and some white feathers.

Table 11.1
Genotype and phenotype in peas with alleles W and w

Genotype	Genotype	Phenotype
WW	homozygous dominant	purple flowers
Ww	heterozygous	purple flowers
ww	homozygous recessive	white flowers

Population Genetics

A **population** is a localized group of a single species occupying a particular area. For example, the field of lilies in Figure 11.6A on page 368 is a different population from a field of lilies in an adjacent valley. The two populations are not completely isolated (since pollinating insects may travel between them), but it is more likely that members of the same population will interbreed to produce the next generation. The same is true for the pond of frogs in Figure 11.6B. Although it is not out of the question that these frogs would mate with frogs from a nearby pond, it is more likely they would mate with individuals that live in the same pond.

Monarch butterflies (such as those shown on page 363) are all from the same species, yet there are distinct populations within this species. When the butterflies migrate in winter, they travel in huge flocks and become mixed. In their summer breeding grounds, however, they have strong family groupings. Genetic mixing during migration and in their winter habitat ensures that the species does not begin to diverge into two or more species. However,

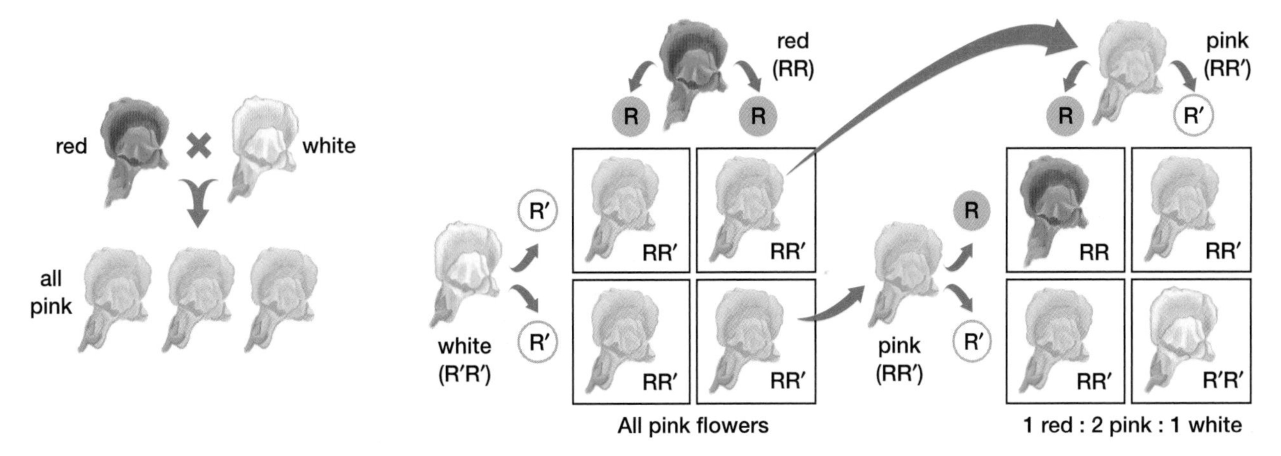

Figure 11.5 Flower colour in the snapdragon is an example of incomplete dominance. Pink flowers are heterozygous (*RR′*), where neither allele is dominant.

Figure 11.6 A field of glacier lilies and a pond of frogs are each considered populations.

the genetic similarity of the smaller summer groupings creates diversity among local populations.

All of the genes in a population or, more specifically, all of the alleles at all gene loci in all individuals of the population, make up the population's gene pool. There can be genetic variation both within individuals (when they are heterozygous for alleles) and within populations. For example, most plants within a population have more than one allele, or are **polymorphic**, at 45 percent of the loci. As well, individual plants are likely to be heterozygous at about 15 percent of their loci. A polymorphic population (with organisms exhibiting different phenotypes and genotypes) and heterozygous individuals contribute to the level of genetic variability within a population. When all members of a population are homozygous for the same allele, that allele is said to be **fixed** in the gene pool. In most circumstances, however, there are two or more alleles for a gene and each exists with a relative frequency within the population. The unique combination of alleles in individuals provides the variation within a population.

Scientists can use the technique of **electrophoresis** to help measure genetic variation within populations. Recall from Chapter 9 (section 9.2) that in the process of electrophoresis, samples of DNA from individuals are placed in a special gel that is then placed in a solution and connected to an electrical circuit. The DNA fragments move through the gel at varying speeds and the resulting pattern of bands — called the DNA fingerprint — is stained and analyzed. Biologists can use this technique to look at the variability of genes (and, consequently, genetic variation) in the population. To do this they compare the samples from different individuals within a population to calculate the percentage of loci that are polymorphic. The more sites that are polymorphic, the greater the genetic variety within the population.

Polymerase chain reactions (PCR) are also used by evolutionary biologists. (Recall that PCR was introduced on page 287.) PCR techniques are used to amplify (generate multiple copies of) DNA from small samples. For example, even minute samples of DNA gathered from mummified organisms or fossils can be copied using PCR techniques. Then, the DNA can be analyzed and compared with DNA sequences of other organisms to help determine evolutionary relationships. Electrophoresis and PCR techniques can be used to sequence and analyze DNA taken from long-dead, or even long-extinct, organisms. For example, DNA has been taken from a 76 000-year-old mummified human brain, fossilized bacteria, and a 40 000-year-old frozen woolly mammoth. This information will help determine the evolutionary history of organisms, because the relatedness of species can be reflected in DNA and proteins. Species that are closely related share a greater proportion of their DNA sequences and proteins.

BIO FACT

In the blood hemoglobin molecule of 146 amino acids, humans and gorillas differ by just one amino acid. Humans and frogs, however, differ by 67 amino acids.

Population geneticists study the frequencies of alleles and genotypes in populations. The study of population genetics is important to the study of micro-evolution because changes in the genetic variability within the population can be used to determine if a population is undergoing micro-evolution. To illustrate how frequencies of alleles and genotypes can be calculated, let's consider a hypothetical population of 400 field mice that are either white or black (see Figure 11.7). The allele for black, A, is dominant to the allele for white, a. (For this example we will assume there are only two alleles for this locus.) In this population of mice, 364 are black and 36 are white. Of the black mice, 196 are homozygous dominant (AA) and 168 are heterozygous (Aa). The 36 white mice are homozygous recessive, aa. Since these mice inherit one set of chromosomes from each parent (that is, they are **diploid**), there are a total of 800 copies of genes for fur colour in the population of 400 field mice. The dominant allele (A) accounts for 560 of these genes ($196 \times 2 = 392$ for AA mice and $168 \times 1 = 168$ for Aa mice). The recessive allele (a) accounts for 240 of these genes ($36 \times 2 = 72$ for aa mice and $168 \times 1 = 168$ for Aa mice). The overall frequency of the A allele in the gene pool of this population is $560/800 = 0.7 = 70\%$, and the frequency of the a allele is 0.3 or 30%. (**Frequency** is the number of occurrences of a particular allele in a population divided by the total number of alleles in the population.) The genotypic frequencies in this population are: AA = 0.49 (196 out of 400 mice), Aa = 0.42 (168/400), and aa = 0.09 (36/400). The frequencies of alleles and genotypes are called the population's **genetic structure**.

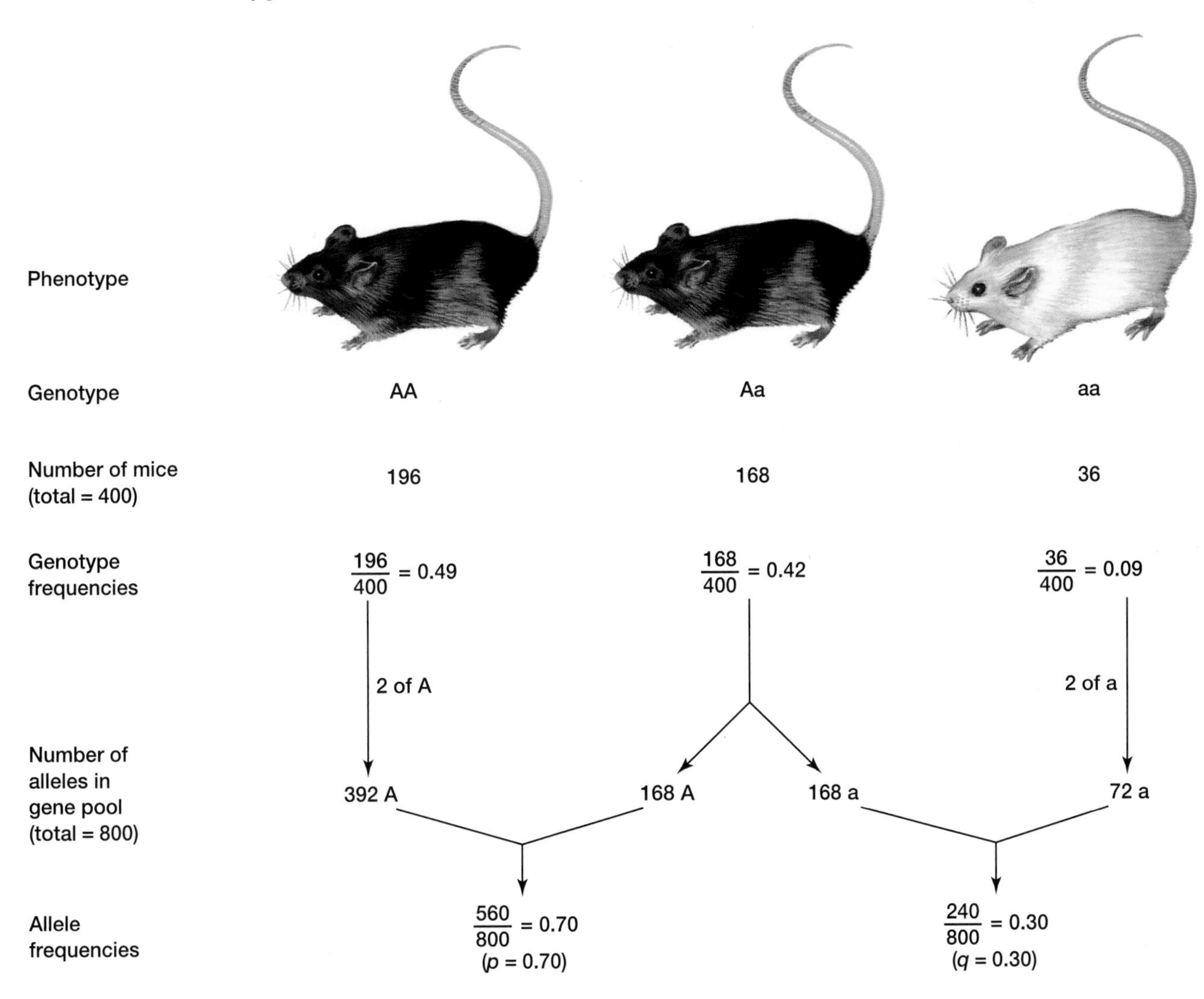

Figure 11.7 Genetic structure of a parent population of field mice

Practice Problems

1. Calculate the allele frequency and genotype frequency in a population of 500 flowers in which 275 are homozygous dominant, 137 are heterozygous, and 88 are homozygous recessive.

2. In a population of 500 plants with purple flowers and white flowers, the genotype frequency is AA = 0.64, Aa = 0.32, and aa = 0.04. Calculate the frequency of each allele in the population. If A is the allele for purple and a is the allele for white, calculate the numbers of each colour of plant. Copy the following chart into your notebook and fill in the data.

phenotype			
genotype			
number			
genotype frequencies			
number of alleles in gene pool			
allele frequencies			

SECTION REVIEW

1. K/U Distinguish between macro-evolution and micro-evolution. Give an example of each.

2. K/U Explain how genetic variation and micro-evolution are related.

3. K/U Explain how variations within a species are affected by natural selection.

4. K/U Use the word "population" to explain local differences in a species.

5. K/U Describe the similarities and differences between the following pairs:
   (a) allele and gene
   (b) phenotype and genotype
   (c) dominant and recessive
   (d) homozygous and heterozygous

6. I Assume that a white animal is crossed with three other animals of the same species, A, B, and C. (For this example we will assume there are only two alleles for this locus.) Animal A is brown and produces offspring A′, which is also brown. Animal B is white and produces offspring B′, which is brown. Animal C is brown and produces offspring C′, which is white. Give the genotypes and phenotypes of all seven animals. Show how you came to this conclusion.

7. K/U Explain how polymorphic populations and heterozygous individuals contribute to the level of genetic variety in a population.

8. C Describe the genetic structure (genotype and allele frequency) using a chart or table for a population of 300 frogs. The frogs are either spotted or spot-less. The allele for spots, A, is dominant to the allele for no spots, a. In this population, 240 frogs have spots and 60 are spot-less. Of the 240 frogs with spots, 200 are homozygous and 40 are heterozygous.

9. I If a group of scientists is trying to determine how a particular fossil of a woolly mammoth is related to an animal on Earth today, what techniques could they use? How could they determine relatedness?

The Hardy-Weinberg Principle

EXPECTATIONS

- **Solve problems related to evolution using the Hardy-Weinberg equation.**
- **Develop and use appropriate sampling procedures to conduct investigations into questions related to evolution.**

For a population to undergo change there must be genetic variation. If all members of a population were genetically identical, all of their offspring would be identical and the population would not change over time. One way to determine how a real population *does* change over time is to develop a model of a population that *does not* change genetically from one generation to the next. Then, actual populations can be compared with this hypothetical model. Such a model was developed independently and published almost simultaneously in 1908 by English mathematician G.H. Hardy and German physician G. Weinberg. These men noted that in a large population in which there is random mating, and in the absence of forces that change the proportions of the alleles at a given locus, the original genotype proportions will remain constant from generation to generation. Their theory is referred to as the **Hardy-Weinberg principle**. In the example shown in Figure 11.7 on page 369, this principle says that the genotypes of 0.49 AA, 0.42 Aa, and 0.09 aa would persist in the mouse population from generation to generation. Because their proportions do not change, the genotypes are said to be in **Hardy-Weinberg equilibrium**.

The Hardy-Weinberg principle is written as an equation. For a gene with two alternative alleles, say A and a, the frequency of allele A (the dominant and, usually, more common allele) is expressed as p, and the alternative allele a (the recessive and, usually, more rare allele) is expressed as q. Because there are only two alleles, $p + q$ must always equal one. The Hardy-Weinberg equation is:

$$p^2 + 2pq + q^2 = 1$$

where:

p = frequency of dominant allele
q = frequency of recessive allele
p^2 = frequency of individuals homozygous for allele A
$2pq$ = frequency of individuals heterozygous for alleles A and a
q^2 = frequency of individuals homozygous for allele a

Let's apply the Hardy-Weinberg principle to the population of field mice introduced in Figure 11.7. In this population, 70 percent (0.7) of the fur-colour loci in the gene pool have the A allele and 30 percent (0.3) have the a allele. The equation can be applied to see how genetic recombination during sexual reproduction will affect the frequencies of the two alleles in the next generation of field mice. The Hardy-Weinberg principle assumes that mating is completely random and that all embryos will survive. The gametes — sperm and ova — each have one allele for fur colour, and the allele frequencies of the gametes will be the same as the allele frequencies in the parent. Every time a gamete is drawn from the pool at random, the chance that the gamete will bear an A allele is 0.7, and the chance that the gamete will have an a allele is 0.3 (see Figure 11.8). Using the Hardy-Weinberg equation, $p = 0.7$ and $q = 0.3$ ($p + q$ must equal 1).

Figure 11.8 shows the possible scenarios that can result when gametes combine their alleles to form zygotes. The Hardy-Weinberg equation states that the probability of generating an AA genotype is p^2. So, in our population of field mice, the probability of an A sperm fertilizing an A ovum to produce an AA zygote is 0.49 (which is 0.7×0.7).

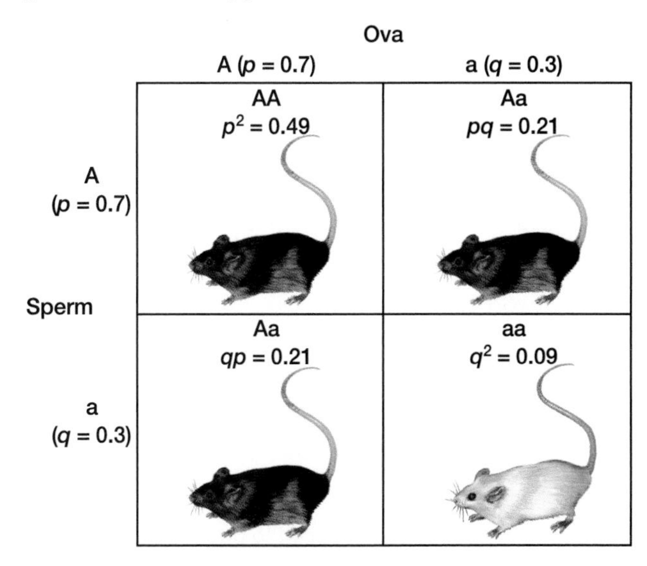

Figure 11.8 The genetic structure of the second generation of field mice

The frequency of individuals homozygous for the other allele (aa) is q^2, or $0.3 \times 0.3 = 0.09$. The genotype Aa can arise in two ways, depending on which parent contributes the dominant allele. Therefore, the frequency of heterozygous individuals in the population is $2pq$ ($2 \times 0.7 \times 0.3 = 0.42$ in our example). All of these possible genotypes add up to 1 ($0.49 + 0.09 + 0.42 = 1$).

The Hardy-Weinberg principle predicts the expected allele and genotype frequencies in idealized populations that are not subjected to selective pressure. Deviations from the frequencies that are expected by the principle indicate that natural selection is occurring. The five conditions required to maintain the Hardy-Weinberg equilibrium are:

- Random mating — Mating must be random with respect to genotype. For example, females cannot select males with a particular genotype or phenotype when they mate.

Biology Magazine TECHNOLOGY • SOCIETY • ENVIRONMENT

Biotechnology and Evolution

Can technology create new species? This question might bring to mind a scientist putting together bits and pieces from various organisms, like a child making new designs from the parts of a construction toy. Modern biotechnology includes techniques that do make novel combinations of species possible. In many ways, humans have been producing new animals, plants, and micro-organisms for thousands of years.

Charles Darwin himself explained the process of evolution by first describing how livestock, crops, ornamental plants, and pets are developed by artificial selection — breeding only those organisms that have certain desirable inherited characteristics. This early form of reproductive technology gave us cows, chickens, apples, roses, dogs, and many other types of animals and plants previously unknown in nature.

In the twentieth century, other technologies led to the unexpected evolution of new organisms with less desirable characteristics. As farmers added more pesticides to their fields, new forms of resistant weeds and insect pests appeared on farms. As physicians prescribed more antibiotics for their patients, new varieties of antibiotic-resistant bacteria sprang up in hospitals. By adding toxins to the environment, humans unintentionally selected organisms able to survive exposure to the chemicals meant to kill them.

Playing with Genes

The tools of genetic engineering now make it possible to bypass the selection of individual organisms and directly select particular genes with useful characteristics. For example, biotech companies add genes to organisms to give them new traits, such as the ability to produce beneficial enzymes or resist certain herbicides. Unlike

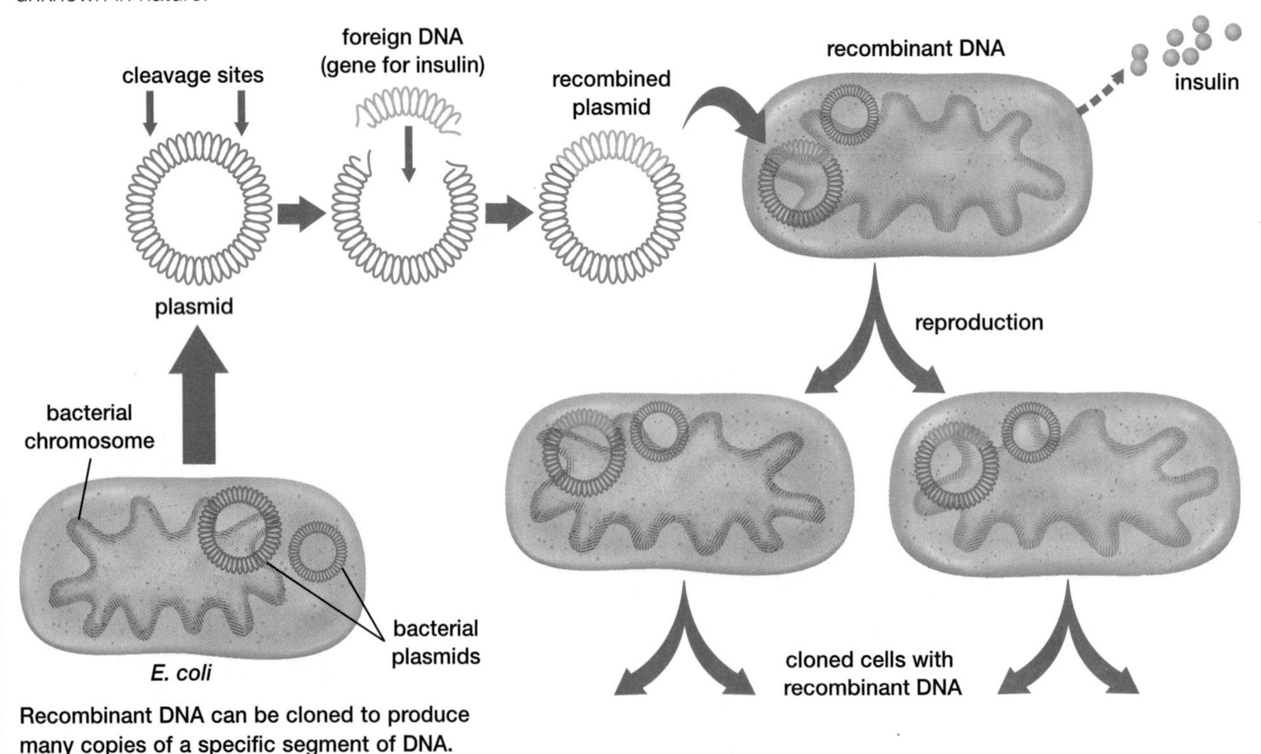

Recombinant DNA can be cloned to produce many copies of a specific segment of DNA.

- No mutations — Alleles must not mutate. In other words, allele choices in the gene pool (as defined in section 10.1) must remain unaltered.
- Isolation — There must be no exchange of genes among populations, since this would alter the gene pool.
- Large population size — The population must be very large.
- No natural selection — No genotype can have a reproductive advantage over another.

These conditions will be explored further in the next section.

In most natural populations, the allele and genotype frequencies *do* change from generation to generation — they are not in Hardy-Weinberg equilibrium since natural populations cannot meet all of the criteria listed above. Therefore, most natural populations are changing. You will explore what causes gene frequencies to deviate from Hardy-Weinberg equilibrium in the next section.

earlier forms of artificial selection, genes from one species can be introduced into other species. For example, one of the earliest applications of genetic engineering created populations of bacteria containing human genes that code for the production of the human insulin molecule. These transgenic bacteria are now the main source of insulin for diabetics.

Many people are concerned that the transfer of genes from one species to another is unnatural. However, gene swapping between species is not entirely a human invention. Different bacteria routinely exchange genetic material by the processes of transformation and conjugation, producing recombinant cells. Viruses also carry genetic material from species to species, even among widely different groups of organisms such as insects and mammals. Because of this ability, viruses are commonly used by molecular biologists.

Another misconception is that transgenic organisms are a type of hybrid, like the result of crossing a lion with a tiger (which has been done in zoos). The offspring of a lion and a tiger have equal genetic contributions from both parents. In contrast, the genetic contribution added to a transgenic organism by genetic engineering is only a tiny fraction of the organism's genome — far less than 1 percent. It is no more of a change in the genome than might be produced by normal random mutation. Even though the change is very small, a directed change can have large impacts on the phenotypes of organisms (for example, the changes made to the genes of people who have certain genetic disorders).

Directed Evolution

The characteristics of each species are determined by their genes. The genes that are of most interest to genetic engineers are those that code for the production of useful molecules such as enzymes and other proteins. These biological molecules have evolved within living organisms over billions of years to perform specific functions. But some of the properties we want enzymes to have for industrial or medical use are not found in any organisms we know of — perhaps because they would clash with the needs of the organism, or because they were never required. For decades, scientists have been applying the principles of evolution to explore a vast universe of novel protein designs that never evolved in nature.

Directed evolution can produce proteins that have capabilities not found in naturally occurring organisms. By speeding up rates of mutation and selection, researchers have created completely new enzymes from purely random pools of DNA sequences in only a few days. For example, one lab increased the catalytic efficiency of an enzyme more than 100-fold by applying random mutagenesis, gene recombination, and screening over a sequence of generations.

As we learn more about the relationships that genes, proteins, and organisms have with their environments, it may be possible to artificially evolve entire organisms. A study of the history of life on Earth shows us that millions of strange and remarkable species have evolved and disappeared. The future will bring new species and new diversity, some of it deliberately introduced by humans but most of it produced by the never-ending process of natural selection.

Follow-up

1. Biotechnology analyzes and manipulates genomes. This makes it seem like each type of organism is simply the product of various molecules working together in a co-ordinated way. Are organisms more than their genes? If so, what else helps define and separate one species from another?
2. The numbers of some rare and endangered animals have been increased by techniques such as cloning and implanting embryos in surrogate mothers of a different species. It is also possible that recently extinct species could be revived by using genetic material from well-preserved specimens. Do you think these techniques might affect the course of evolution?

Investigation 11 • A

SKILL FOCUS

- Predicting
- Performing and recording
- Communicating results

Population Genetics and the Hardy-Weinberg Principle

There are several characteristics that you can easily measure to determine some of the genetic variability that exists within your classroom. Eye and hair colour, the presence or absence of freckles, hair on your fingers, and even how you cross your hands are all genetic traits.

Studies like this investigation often involve interviewing or examining people. All those involved must respect the confidentiality of the subjects of the study. This confidentiality should be maintained in all studies unless permission has been obtained from the subjects.

Pre-lab Question

- Does your class meet all the requirements to maintain the Hardy-Weinberg equilibrium? (Consider your class as a population.)

Problem

How can genetic variety among the students in your classroom be measured?

Prediction

If there are three possible genotypes (and two possible phenotypes) for a particular characteristic, predict the frequency of each allele in your class.

Part A

Trait	Dominant	Recessive
hairline	pointed on forehead	straight across forehead
freckles	present	absent
thumb joint	last joint bends out	last joint is straight
finger hair	present	absent
folded hands	left thumb over right	right thumb over left
tongue rolling	can be rolled in U-shape	cannot be rolled

Procedure

1. Choose one of the traits from the table shown here. (Your entire class should test the same trait.) Copy the table below into your notebook, and fill in the genotype and phenotype for the three possible combinations of alleles. (Use A for the dominant allele, and a for the recessive allele.)

Trait	Possible genotypes	Possible phenotypes

2. Survey the class to determine the total number of students with each phenotype of the selected trait. Copy the following table into your notebook and record these results as percentages. Change each percentage into decimal form.

	Class Phenotypes				Allele Frequency	
	Dominant phenotype $p^2 + 2pq$		Recessive phenotype q^2			
	Number of students	Percentage of students	Number of students	Percentage of students	p	q
class population						
larger population						

3. Use the Hardy-Weinberg equation to calculate the frequency of each allele in your class. Record your calculated frequencies in the last two columns of your table.
4. (Optional) Repeat steps 2 and 3 for a larger population of two or more classes.

Post-lab Questions

1. What is the frequency of homozygous dominant students, p^2, in your classroom? What is the frequency of heterozygous students, $2pq$?
2. What is the frequency of homozygous recessive students, q^2, in your classroom?
3. What are the percentages of the three genotypes in your classroom?
4. (Optional) How did genotype and allele frequencies change when you sampled a larger population size?

Conclude and Apply

5. Explain the relationship between population size and genotype and allele frequency.

6. What does the answer to question 5 imply about the need for an appropriate sample size in order to obtain an accurate picture of what is occurring within a population?
7. Explain how the Hardy-Weinberg equation can be used to study genetic diversity in populations.

Part B
Materials

4 playing cards for each participant (2 from red suits and 2 from black suits)

Procedure

1. Each playing card represents an allele. Cards from red suits are recessive alleles, and cards from black suits are dominant alleles.
2. Find a partner. Place your four cards face down randomly on your desk, but do not mix your cards with your partner's.
3. Each person in this partnership, or random mating, should turn over one card. This is the offspring of the first generation. Copy this table into your notebook and record the genotype of the first offspring.

Generations		Class total for each phenotype		
		AA	Aa	aa
first generation mating	first offspring			
	second offspring			
second generation mating	first offspring			
	second offspring			
third generation mating	first offspring			
	second offspring			
fourth generation mating	first offspring			
	second offspring			
fifth generation mating	first offspring			
	second offspring			
	Class Totals			

4. Retrieve your card and shuffle your four original cards again. Repeat step 3. This is the second offspring of the first generation. Record the genotype of the second offspring in your table.
5. You and your partner must now each assume the genotype of one of your offspring. For example, if the first offspring was AA, one partner now begins with four black cards. The other partner should assume the genotype of the second offspring. If this offspring was Aa, for example, this person now begins with two red cards and two black cards.
6. Randomly select a different partner in your class. Repeat steps 3 and 4 and record the first and second offspring from the second generation.
7. Repeat step 5, selecting new cards if necessary to reflect the alleles of the offspring from the second generation.
8. Continue choosing a different partner at random to create third, fourth, and fifth generation mating, with two offspring from each generation.
9. Collect and record class totals for each genotype from each mating in each generation.

Post-lab Questions

1. What is the initial allele frequency in your class population? Express this as a percentage converted to a decimal.
2. Consider the data you collected over the five generations as a single large population, so your class totals in the last row of your table are the genotypes of an entire population. Calculate the frequency of each genotype as a percentage converted to decimal form. Calculate the allele frequency in the population.
3. Use the Hardy-Weinberg equation to determine the genotype frequencies of the beginning population where $p = 0.5$ and $q = 0.5$.
4. Calculate the genotype and allele frequencies of the class population for the fifth generation only.

Conclude and Apply

5. How do the allele frequencies change from generation to generation? Explain whether this population is in Hardy-Weinberg equilibrium.
6. Predict what would happen if you completed this activity with only half of your class.

Exploring Further

7. Repeat this activity with only half of your class. Compare the results between the two populations of different sizes.
8. What limitations does this simulation have in imitating what is actually occurring in the population?
9. Describe how this activity could be changed to replicate an actual, natural population that is evolving rather than a hypothetical, non-evolving population.

Practice Problems

1. An investigator has determined that 16 percent of a certain human population *cannot* roll their tongue. The ability to roll the tongue is controlled by a dominant allele. Calculate the genotype and allele frequencies for the population.

2. In a certain population, 30 percent are homozygous dominant, 49 percent are heterozygous, and 21 percent are homozygous recessive. What percentage of the next generation is predicted to be homozygous recessive, assuming a Hardy-Weinberg equilibrium?

3. In a population of pea plants, 1 percent are short, which means they are homozygous recessive. What are the frequencies of the recessive allele t and the dominant allele T? What are the genotypic frequencies in this population?

WEB LINK

www.mcgrawhill.ca/links/biology12

One application of the Hardy-Weinberg equation is to predict how many people in one generation of a human population are carriers of a particular recessive allele. If the number of babies born annually with a particular disease (such as phenylketonuria [PKU] or cystic fibrosis) are known, the number of adults that carry the allele can be predicted. This information can be used to track trends in the conditions, help medical researchers garner support for their work, and help public-health workers allocate their time and resources effectively. To learn more about these diseases and the frequency of these recessive alleles in our population, go to the web site above, and click on **Web Links**. Determine the frequency of the recessive traits for PKU and cystic fibrosis using the Hardy-Weinberg equation.

UNIT PROJECT PREP

For your Unit Project on Searching for a Common Ancestor, consider how new techniques in DNA analysis can help determine relatedness among fossils. How would evolving populations of possible ancestral fossils differ from the Hardy-Weinberg principle?

SECTION REVIEW

1. **K/U** The Hardy-Weinberg principle is a model that uses a hypothetical situation that would rarely, if ever, be replicated in nature. Explain why it is useful to use the Hardy-Weinberg principle to help understand population genetics.

2. **K/U** A biologist has found that 10 percent of a population of bats are hairless, which is a recessive trait. Assuming that the population is in Hardy-Weinberg equilibrium, determine the genetic structure (genotype and allele frequencies) of the population.

3. **K/U** List the conditions necessary to maintain the Hardy-Weinberg equilibrium.

4. **K/U** Select one condition that is necessary for the Hardy-Weinberg equation to work, and explain why this condition must be met for no change to occur.

5. **C** A population of flowers are in Hardy-Weinberg equilibrium with 32 white flowers and 168 yellow flowers. The white flowers are bb and the yellow flowers are Bb or BB, where b is recessive and B is dominant. Draw a table showing the phenotypes, genotypes, frequency of the genotypes in the population, and frequency of alleles B and b. Create a Punnett square that shows the potential crosses for this population of flowers.

6. **K/U** If a plant breeder started selecting either the white or the yellow flowers from Question 5 above, would the population be in Hardy-Weinberg equilibrium?

7. **MC** Is the human population of North America in Hardy-Weinberg equilibrium? Explain your answer.

8. **MC** Why would you expect the whooping crane population of North America to not be in Hardy-Weinberg equilibrium?

11.3 Mechanisms for Genetic Variation

EXPECTATIONS

- **Explain the role of mutations in micro-evolution.**
- **Analyze evolutionary mechanisms and their effects on biodiversity and extinction.**
- **Explain three ways in which natural selection can affect genetic variation.**

Even though rattlesnakes (such as the one in Figure 11.9) are found throughout much of North America, few humans have close encounters with them. Most rattlesnakes take cover in the underbrush if danger is present or, at least, give warning with their distinctive rattle. Nonetheless, thousands of people are bitten by rattlesnakes each year, although only 0.2 percent of victims will die. In recent decades, however, there have been several reports of unusual reactions to certain rattlesnake venoms. As well, doctors are reporting that they often have to use many more vials of antivenom to treat bites. In some cases, a species whose venom was previously not considered a threat to humans delivered bites that were deadly. In other cases, patients showed symptoms that were not usual for the venom of the snake that bit them. (Rattlesnake venom usually contains either neurotoxins that affect the nerve impulses to muscles and can restrict breathing, or hemotoxins that affect the tissue near the bite.) Victims showed signs of neurotoxin poisoning when they had been bitten by snakes that previously were thought to deliver only hemotoxins. Why do the toxins seem to be changing? Are venoms evolving?

Figure 11.9 What factors might have caused rattlesnake venom to become more potent in recent decades?

Scientists studying this phenomenon have presented several explanations. Some scientists suspect that closely related snakes with differing types and potencies of venom are interbreeding in places where their populations overlap. Others suggest that in some populations, snakes with more potent venoms are being naturally selected because their prey are developing increasingly powerful substances in their blood to block venoms. For example, studies have shown that populations of California ground squirrel that overlap the range of the northern Pacific rattlesnake have a factor in their blood that makes them better able to combat the snake's venom.

A third suggested explanation for the changes in rattlesnake venom relates to the change in age-structure of snake populations. Juvenile rattlesnakes have stronger venom than larger adult snakes. Because humans usually hunt, capture, or even run over larger snakes, the overall age of some snake populations may be shifting to favour young snakes with more potent venom.

All of these possible explanations for the increased toxicity in rattlesnake venom provide scenarios of micro-evolution in action. The gene pools of these populations are changing because of natural selection or because individual snakes are entering or leaving the population. These situations deviate from the Hardy-Weinberg equilibrium. In this section, you will investigate the five conditions that have the potential to result in micro-evolution: mutation, genetic drift, gene flow, non-random mating, and natural selection.

BIO FACT

Of the five causes of micro-evolution, only natural selection always adapts a population to its environment. The other agents of change — gene flow, genetic drift, non-random mating, and mutation — can affect populations in positive, negative, or neutral ways.

WEB LINK

www.mcgrawhill.ca/links/biology12
To learn more about the possible evolution of rattlesnake venom, go to the web site above, and click on **Web Links**. Read through the article and identify specific situations in which the gene pool in the population might be changing. Using resources from the Internet and library, find an example of how the gene pool of a Canadian species is changing.

Mutations

Mutations may provide new alleles in a population and, as a result, may provide the variation required for evolution to occur. (Recall that you learned about mutations in Chapter 9, section 9.1.) When DNA mutates, a cell may die, malfunction, or multiply rapidly into a tumour. Whatever the result, the mutation disappears when the organism dies. If, however, the mutation alters the DNA in a gamete, the mutation may be passed on to subsequent generations. Mutations can have effects that are favourable, unfavourable, or neutral, and the fate of the mutation depends on how it acts in the population.

Mutations alone are not likely to cause evolution. But if a mutation provides a selective advantage (such as the ability of the California ground squirrel to break down rattlesnake poison), it may result in certain individuals producing a disproportionate number of offspring as a result of natural selection. Eventually the favourable mutation will appear with increased frequency in a population.

The fate of particular mutations may also change. Neutral, or perhaps even harmful, mutations can be a source of variation that ultimately helps a population survive given the right circumstances, such as when environments change. For example, the water flea *Daphnia* (shown in Figure 11.10) normally lives in water that is around 20°C and cannot survive in water 27°C or warmer. However, there is a mutation that enables *Daphnia* to survive in temperatures between 25°C and 30°C. This mutation is only advantageous — and thus only perpetuated in the population — when water temperatures are so warm that the other *Daphnia* die off.

In situations in which the environment is changing extremely rapidly, mutant alleles that were previously insignificant in the population may, by chance, fit the new environmental conditions better. As a result, the organisms containing the mutant allele survive and the mutant allele is perpetuated because it provides a selective advantage. The once neutral, or even negative, mutation can in some cases mean the survival of a population. For instance, there are many examples of insects, bacteria, and viruses quickly becoming adapted to new environments because of mutations that prove to be beneficial. Populations of mosquitoes have rapidly developed resistance to certain ingredients in insecticides because of a mutation that resulted in alleles that could break down and withstand the chemical poisons. When the mosquitoes were first sprayed by the insecticide, most died. However, those with the mutant allele that withstood the chemicals were naturally selected for and thus were more likely to survive and reproduce. This scenario is repeated generation after generation until there is a mosquito population resistant to the insecticide. Another scenario is being played out today as strains of bacteria become increasingly resistant to once-effective antibiotics.

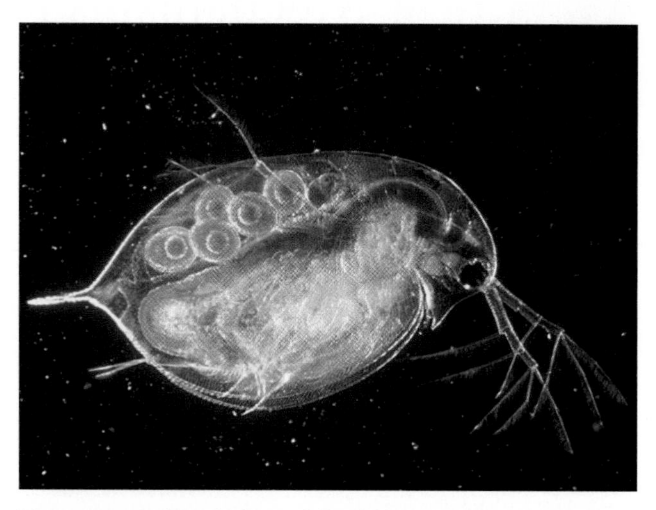

Figure 11.10 Populations of *Daphnia* can have a mutation that allows them to survive at higher-than-normal water temperatures.

Bacteria and other micro-organisms reproduce quickly, and mutations that affect the population's genetic variation can have a significant impact in a short period of time. Bacteria can reproduce asexually by dividing as frequently as every 20 minutes. This could result in a single cell having close to a billion descendants in about 10 hours. Because of these astounding reproductive rates, any new mutation that proves beneficial can increase its frequency in the population quickly. This phenomenally rapid asexual cloning of individuals resistant to the new environment (the "poison" of an antibiotic, for example) makes the development of new antibiotics increasingly challenging for biochemists.

COURSE CHALLENGE

You may choose to examine an example of micro-evolution in a particular species as your Biology Course Challenge topic. Start making notes on the links that this topic has with metabolic processes, homeostasis, and molecular genetics. In the next unit, think about how the population dynamics of this species are related to its predator or prey species.

Genetic Drift

In small populations, the frequencies of particular alleles can be changed drastically by chance alone. This is called **genetic drift**. As an example, imagine flipping a coin 1000 times. Every time you flip a coin you have a 50–50 chance of having an outcome of heads or of tails. In a large sample size (for example, 1000 flips), you would logically expect the number of outcomes of heads and tails to be fairly close. If, however, you flipped heads 700 times and tails 300 times, you might start to wonder about your coin. On the other hand, in a small sample size (for example, 10 flips), it would not be too unusual to flip heads seven times and tails three. The smaller the sample size, the greater the chance of sampling error. In population genetics, the sample size can greatly affect the gene pool of a population; the smaller the population, the less likely that the parent gene pool will be reflected in the next generation. In a large population, there is a better chance that the parent gene pool will be reflected in subsequent generations.

Figure 11.11 on the following page illustrates how genetic drift can happen in a small population and how these changes can be rapid and significant. In any population, not all of the individuals in each generation will necessarily reproduce. This further amplifies the effect of genetic drift. For example, in the first generation of flowers in Figure 11.11, only four plants produce seeds that give rise to fertile offspring. In such a small population size, the allele frequencies shift in the second generation. Allele frequencies again change in the third generation when only two of the plants in the second generation leave fertile offspring. In this example, genetic drift reduced variability because one allele was lost (it "drifted" out of the population) and the other allele became fixed in the population. By the third generation, only mutation or migration of new individuals into the population could re-introduce the lost allele.

THINKING LAB

An Evolving Disease: Tuberculosis

Background

Tuberculosis is an infectious lung disease caused by the bacterium *Mycobacterium tuberculosis*. Tuberculosis is a contagious disease that can be spread by the inhalation of the bacteria. Although anyone can get tuberculosis, people who are already in poor health and who live in crowded conditions are particularly susceptible. Tuberculosis was once fairly easy to treat — an antibiotic discovered about 60 years ago treated the disease effectively. At one time it was thought that tuberculosis could one day be eradicated. Today, however, new drug-resistant strains of tuberculosis are causing great concern to medical researchers.

In one recent study, researchers have been working with officials from the Russian prison system to help stave off the rapid evolution of drug-resistant forms of tuberculosis. Prisons in Russia are very crowded and it is thought that up to 100 000 prisoners carry strains of tuberculosis resistant to at least one antibiotic. Tuberculosis is readily spread in the prisons and bacteria move quickly from host to host. *Myobacterium* can be destroyed with a long course of antibiotics. However, since few prisoners get the full course of antibiotics (either because it cannot be provided or because patients are discharged before the treatment is completed), resistant bacteria spread easily through their bodies. When prisoners are released and left untreated, they can spread a new drug-resistant version of the bacteria to the general public. The tuberculosis rate increased five-fold in Russia between 1990 and 1996, and it is now one of the leading causes of death of young Russian men. Health officials who monitor tuberculosis are beginning to see drug-resistant strains of *Myobacterium* in places such as North America, where tuberculosis is relatively uncommon. As well, *Myobacterium* is just one of the many bacteria that are becoming resistant to antibiotics.

You Try It

1. Create a model showing how a population of *Myobacterium* could become resistant to antibiotics.
2. Is poorly supervised or incomplete treatment of antibiotics better than no treatment? Discuss this statement with a partner.
3. Using the Internet or library resources, investigate how researchers are treating drug-resistant tuberculosis or another disease that can be treated by antibiotics. Also, find out how they are trying to limit the spread of the disease.

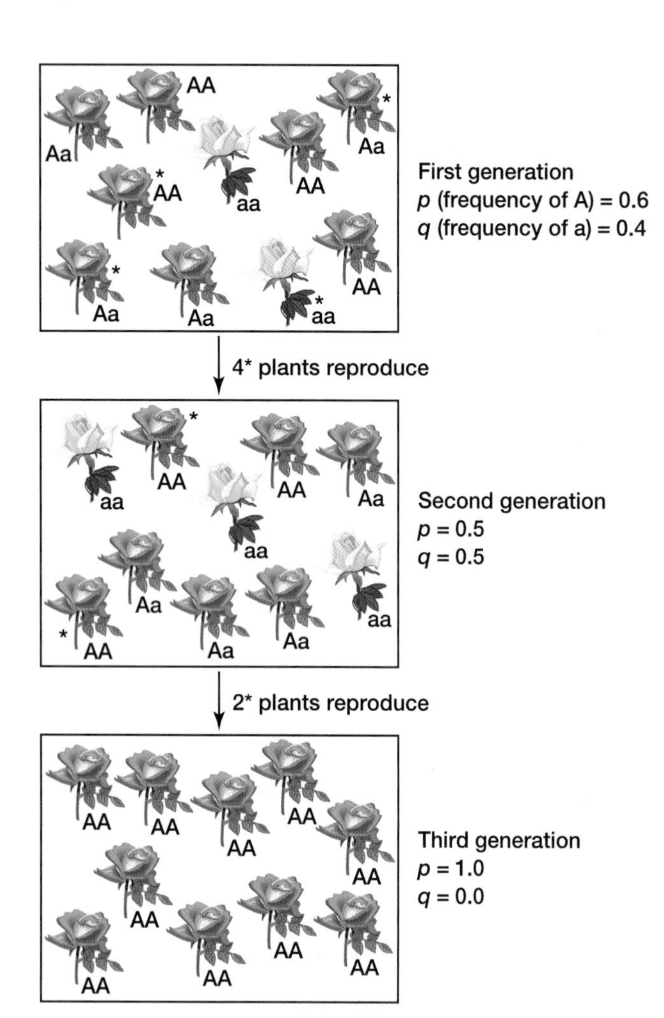

Figure 11.11 The frequency of alleles changes in this population over three generations because of genetic drift.

Most natural populations are large enough that the effects of genetic drift are negligible. However, two situations — population bottlenecks and the founding of new colonies by a few individuals (the founder effect) — lead to genetic drift.

The Bottleneck Effect

Populations can be subject to near extinction as a result of natural disasters such as earthquakes, floods, or fires, or of human interferences such as overhunting or habitat destruction. The surviving population is unlikely to represent the gene pool of the original population. The **bottleneck effect** is a situation in which, as a result of chance, certain alleles are overrepresented and others are underrepresented (or even absent) in the reduced population. Genetic drift then follows and the genetic variation in the surviving population is reduced.

The population of northern elephant seals (see Figure 11.12) passed through a bottleneck in the 1890s when overhunting reduced the population to, possibly, as few as 20 individuals. Since the species became protected, the population has increased to over 30 000 individuals. Biologists have studied 24 gene loci of several of the 30 000 individuals and have found no genetic variation; electrophoresis showed that at each of the 24 loci there is only one kind of allele. This is markedly different from what is found in populations of southern elephant seals that were not subject to the bottleneck effect, in which there is a high degree of genetic variation.

Whooping cranes, which breed in Wood Buffalo National Park in the Northwest Territories, also went through a genetic bottleneck. In 2001, the population at Wood Buffalo National Park was 177 whooping cranes. According to the data, scientists hypothesized that these birds were descendants of at most 12 (and more likely six or eight) founding birds. In addition to these 177 birds, another 86 whooping cranes are found in flocks that scientists are trying to establish in the Rocky Mountains (two individuals) and in Florida (84 individuals.) Biologists are working on strategies to limit loss of diversity due to genetic drift.

Figure 11.12 The reduced genetic variation in populations of northern elephant seals is the result of the bottleneck effect and genetic drift.

WEB LINK

www.mcgrawhill.ca/links/biology12

To learn what biologists in Canada and the United States are doing to help preserve genetic diversity in whooping cranes, go to the web site above, and click on **Web Links**. Read the essays on the web sites and summarize some of the problems associated with the small population size of whooping cranes. What strategies are being used to help preserve the species?

The Founder Effect

When a small number of individuals colonize a new area, chances are high that they do not contain all the genes represented in the parent population. The change in allele frequencies that result in this new population is called the **founder effect**. The particular alleles carried by these founders are dictated by chance only. As well, since the new population is in a new environment, its members will experience different selective pressures than the members of the parent population do. In practice, it is difficult to tell how much of the genetic difference between two populations is because of the founder effect and how much is a result of natural selection and the different selective pressures working on the populations. The ancestral population of Hawaiian honeycreepers shown in Figure 11.13 was thought to have migrated from North America about five million years ago. Individuals from the population became isolated on different islands and evolved into different species. Each population started with a small assortment of genes (founder effect), which were subjected to different selective pressures depending on the local environmental conditions.

While the founder effect is important on islands and other isolated habitats, other populations that have limited input of new genetic material also show the effects of a limited gene pool. Isolated human populations occasionally have high frequencies of inherited genetic disorders. An example of the founder effect is found in a small village in Venezuela, where the incidence of Huntington's disease is remarkably high. Huntington's disease is a debilitating, degenerative disease of the nervous system. It is caused by a lethal dominant allele that is not manifested in any particular phenotype. Because of this, individuals show no symptoms of the condition until they are about 30 years old (or older), when the deterioration of their nervous system begins. The presence of this condition in this particular village can be traced back to one woman who carried the dominant allele. Since the symptoms of the disease do not appear until later in life, carriers can reproduce (thereby potentially passing on the allele for this disease) before it is clear whether they carry the dominant allele or not.

In another example, in 1814 a population of 15 people founded a British colony on the Tristan da Cunha, a small group of islands in the Atlantic

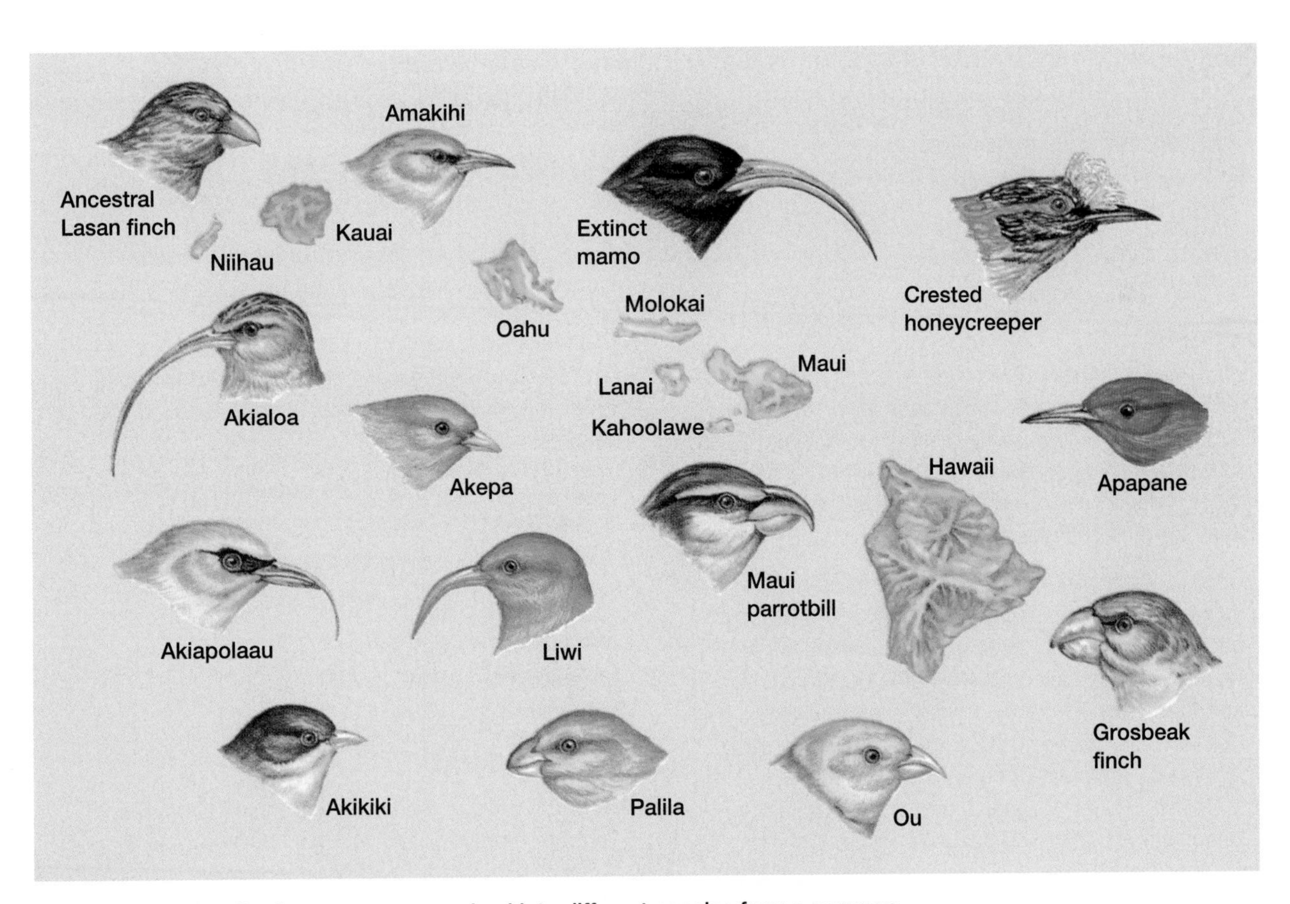

Figure 11.13 Hawaiian honeycreepers evolved into different species from a common ancestral population as a result of the founder effect and selective pressures.

Ocean. Apparently, one of these individuals carried the recessive allele for a type of blindness called retinitis pigmentosa. Studies have shown that the frequency of this allele in the current population on Tristan da Cunha is much higher than in populations from which the original founders came.

Gene Flow

To maintain genetic equilibrium, the gene pool of a population must be completely isolated. In practice, this is rarely the case. A windstorm or tornado can deliver new seeds or pollen to a population. This movement of new alleles into a gene pool, and the movement of genes out of a gene pool, is called **gene flow**.

Gene flow can reduce the genetic differences between populations that may have arisen because of natural selection or genetic drift. Previously isolated populations, including human populations, can accumulate differences over generations because of selective pressure or as a result of having a closed gene pool (no new alleles entering or leaving) in a small population. If the gene flow is extensive enough between two neighbouring populations, they can eventually become amalgamated into a single population with a common genetic structure. How do you think the relative ease of travel has contributed to micro-evolutionary change in human populations? What factors have limited the extent to which micro-evolutionary change can take place even with the accessibility of travel?

Non-random Mating

Genetic equilibrium can be maintained in a population only if that population mates on a random basis. However, not all organisms mate in such a way. Individuals will usually mate more often with neighbours than with more distant members of the population. **Inbreeding** (mating between closely related partners) is a type of **non-random mating** that causes frequencies of certain genotypes to change in the population. Inbreeding does not change allele frequencies; it results in a population with more homozygous individuals. Self-fertilization is particularly common in plants and is the most extreme case of inbreeding. Pea flowers, for example (as shown in Figure 11.14), include both the male and female reproductive structures. This ensures that self-fertilization will take place unless the flower is disturbed by an insect or other means.

Assortative mating is another type of non-random mating, in which individuals choose partners that have a similar phenotype such as size. For example, many animals (such as toads) select mates that are

THINKING LAB

Genetic Diversity and Fish Hatcheries

Background

Fish hatcheries have helped to preserve fish stocks for commercial and sport fishers. Hatcheries have stepped in to help make up for losses in the numbers of fish due to habitat destruction and overfishing. However, several genetic risk factors have the potential to affect the genetic diversity of both hatchery and wild stocks.

Hatcheries take the gametes from wild salmon, fertilize them, raise the salmon to a small size, and then release them back into their native stream. Fish in hatcheries are raised in controlled situations in which the objective is to maximize the output from the hatchery. In most cases, native salmon also spawn in the same areas. Each stream has a genetically unique salmon species. Therefore, hatcheries are careful to release their salmon only in the same streams from which the gametes were taken. The exception to this would be a situation in which the salmon that run in a river are extinct — hatcheries may release fish grown from gametes taken from other streams into a river that no longer has native salmon.

Analyze

Note: You may use the Internet, library, or other resources (such as interviews with fisheries biologists) to help you with this activity.

1. Speculate on how, or if, the following practices or situations might happen in hatcheries:
   - loss of variability within a population
   - loss of variability between populations
   - genetic drift
   - artificial selection
2. What are the advantages and disadvantages of hatchery-raised fish?
3. In many places, there is now a movement to save and/or improve spawning habitat in even the smallest creek that bears a population of salmon. How does this trend support improving the genetic diversity of wild salmon populations?

similar sized. Assortative mating is the basis of artificial selection, in which animals such as dogs are bred for particular characteristics. This inbreeding has led to a decrease in the genetic diversity in breeds of dogs and the perpetuation of certain diseases and conditions (such as hip dysplasia) in some breeds.

Natural Selection

The Hardy-Weinberg equilibrium says that all individuals are equal in their ability to survive and reproduce. In actual situations, however, this condition can rarely, if ever, be met. Populations have a range of phenotypes and genotypes, and some individuals in the population will leave more offspring than others. As you learned in Chapter 10, natural selection is the mechanism that results in this differential reproductive success. Selective forces such as predation and competition work on populations, and consequently some individuals are more likely to survive and reproduce than others. If having a single allele gives even a slight yet consistent selective advantage, the frequency of the allele in the population will increase from one generation to the next at the expense of the less favourable allele. There is a greater chance of the organisms with the slightly favourable allele living and reproducing and then passing this slightly favourable allele to their offspring. Therefore, selection causes changes in a population's gene frequencies that shift the population away from Hardy-Weinberg equilibrium.

There are three ways in which natural selection can affect the frequency of a heritable trait in a population: stabilizing selection, directional selection, and disruptive selection.

Stabilizing selection favours an intermediate phenotype and acts against extreme variants. This type of selection reduces variation and improves the adaptation of the population to aspects of the environment that remain relatively constant. Figure 11.15 shows how stabilizing selection keeps the majority of baby weights between 3 and 4 kg. Infant mortality is greater for babies who are smaller or larger than this size.

Directional selection favours the phenotypes at one extreme over the other, and results in the distribution curve of phenotypes shifting in that direction. This type of selection is common during times of environmental change or when a population migrates to a new habitat that has different environmental conditions. Figure 11.16 on page 384 shows the directional selection shift that took place as horses evolved from an ancestral form that was adapted to a forest habitat to the modern form, which is adapted to a grassland habitat. This shift took place in response to a changing environment. *Hyracotherium* was about the size of a dog and was well-adapted to the forest environment present during the Eocene epoch. During the Miocene and Pliocene epochs, however, grasslands began to replace the forests and the ancestral horses were selected for larger size, more durable teeth suitable for grinding grasses, and longer legs for increased speed in the more open habitats.

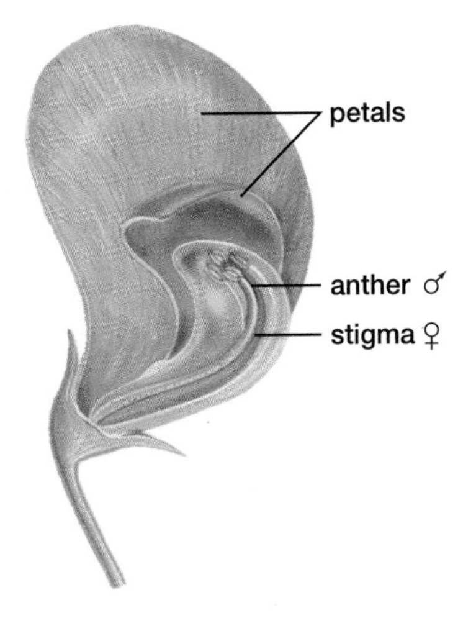

Figure 11.14 Pea flowers are designed so that self-fertilization is ensured if the flower is not disturbed.

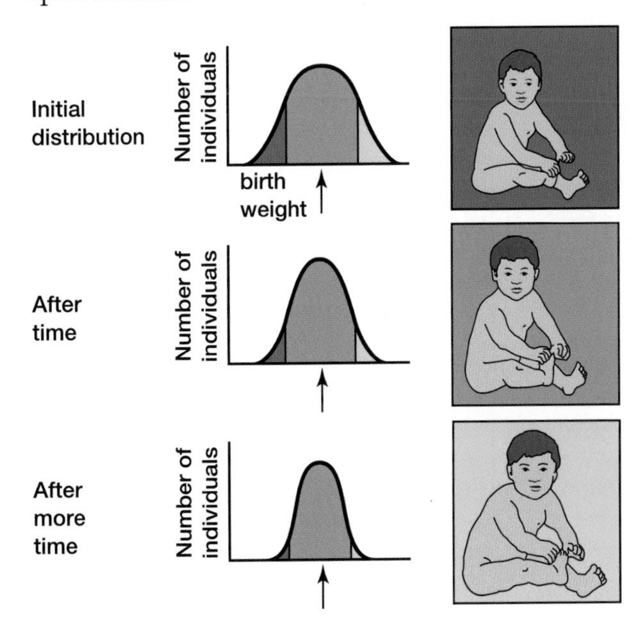

Figure 11.15 Natural selection favours the intermediate phenotype (for example, human baby weight) in stabilizing selection. Now, most babies are of intermediate weight.

Global climate change may also cause directional selection in some populations. Imagine that a hypothetical population of penguins lives near Pacific islands where the water temperatures have been moderate until recently. Global climate change has resulted in a shift in ocean currents such that the water is now consistently much colder. In the changing environment, birds with less body fat are less successful because they need to use more energy (food) to keep themselves warm. Birds with more body fat stay warmer in the water and can afford to use more energy for reproduction. As a result, the fatter penguins have more success raising young, so there is an overall increase in the number of alleles for increased body fat in the penguin population.

The shift in the population of peppered moths you learned about in Chapter 10 is an example of directional selection. The resistance of insects and bacteria to pesticides and antibiotics, respectively, are also examples of directional selection.

Disruptive (diversifying) selection takes place when the extremes of a phenotypic range are favoured relative to intermediate phenotypes (as shown in Figure 11.17). As a result, intermediate phenotypes can be eliminated from the population. In several salmon species (including coho salmon) there are two male phenotypes that are extremely different. When small "jack" males mature, they weigh about 500 g and are approximately 30 cm in length. In comparison, the "normal-sized" males average about 4.5 kg (and can be as large as 8.5 kg) when they are mature.

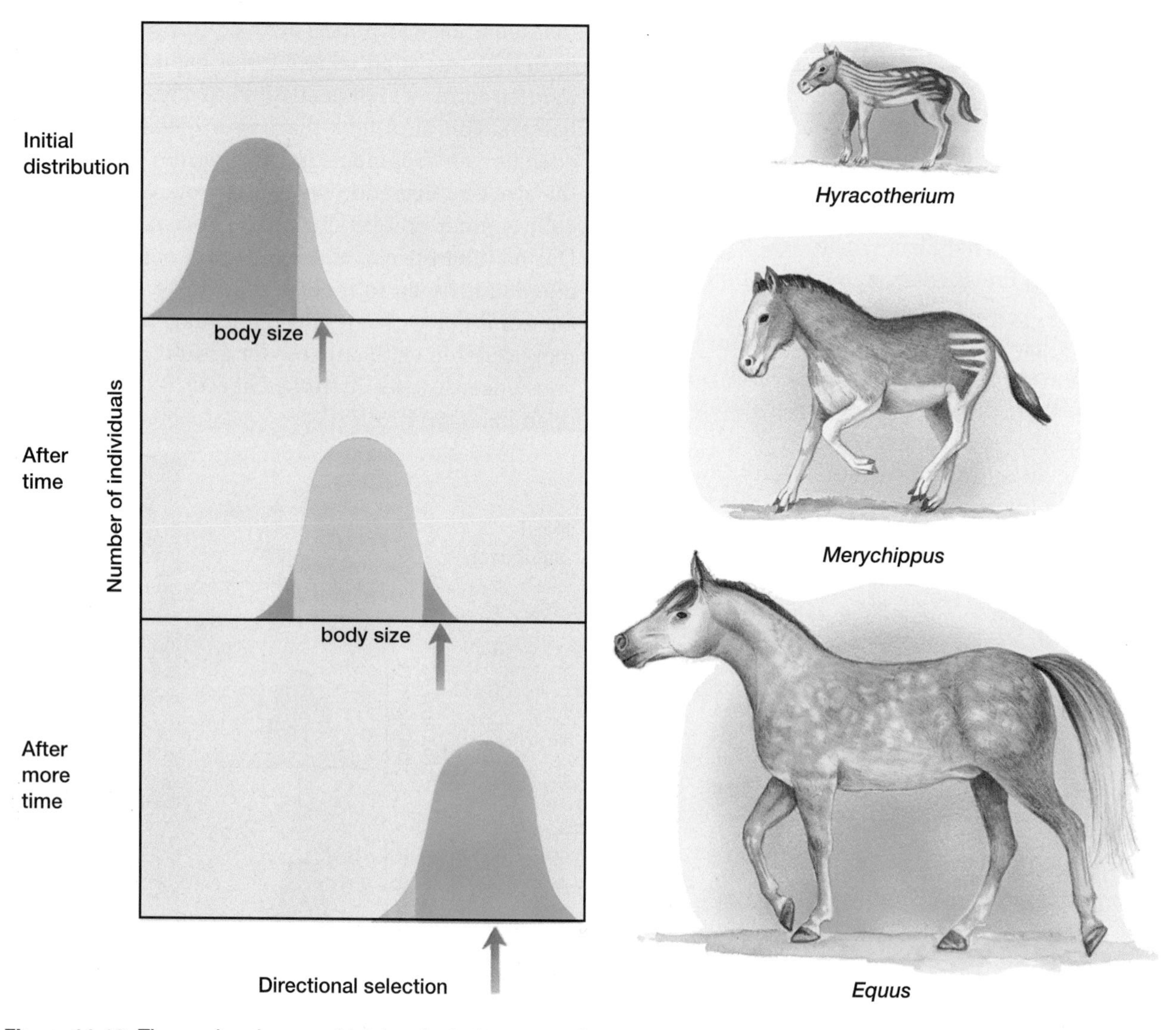

Figure 11.16 The modern horse, which is adapted to a grassland habitat, evolved from an ancestral horse that was adapted to a forest habitat. This shift in phenotype is called directional selection.

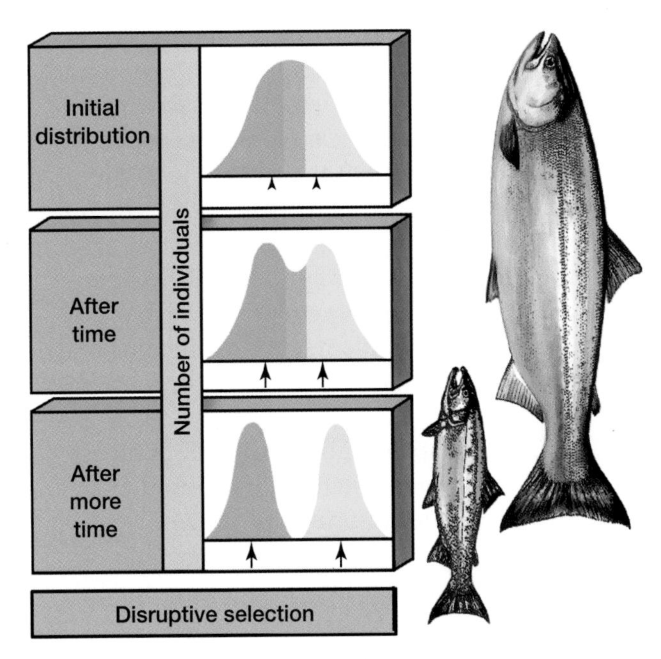

Figure 11.17 In disruptive selection, two extreme phenotypes are favoured over the intermediate form. Shown are two phenotypes of male coho salmon — the smaller "jack" salmon and the noticeably larger, regular-sized male.

Sexual Selection

Sexual reproduction has evolved independently several times throughout the course of evolutionary history. Remarkably, most forms of sexual reproduction share similar characteristics: the ova are large and immobile, while the sperm are small swimmers. Instead of both gametes wandering around searching for each other, it is more practical for one (the ova) to stay in place while the other searches for it. Two things occur to increase the probability of the gametes meeting — the ova releases powerful pheromones and males release millions of sperm.

Evolution has favoured mutations that make a species' sperm smaller and eggs larger. If sperm's only function is to carry genes (and not to carry the energy required for cell division), a species can have more, smaller-sized sperm. To complement this strategy, fewer, larger eggs that have the stored energy needed for cell division are required. This trend towards females with large eggs and males with more than enough sperm to fertilize the entire population of their species has produced a very competitive situation for sexually reproducing species. While one male may be able to fertilize all of his species, virtually every other male is in the same position. This has led to the evolution of a wide array of sexual behaviours and sexual attractants, such as plumage or scents, as males compete for the chance to mate with a female.

Males and females of many animal species often have markedly different physical characteristics, such as colourful plumage in male birds and antlers in male deer. This difference between males and females is called **sexual dimorphism**. Figure 11.18 shows the striking difference between male and female orioles. These obvious characteristics, as well as courtship displays and other mating rituals, result in another type of selection — **sexual selection**. Although the selection of mates has many facets, in general, competition between males (through actual combat or visual displays) and the choices made by females result in sexual selection and enhanced reproductive success.

Characteristics used in sexual selection may not be adaptive in the sense that they help an individual survive in the environment in any way. The larger mane of a lion or the antlers of a moose, for example, do not help these animals better withstand environmental conditions. However, if such characteristics give them the advantage of being chosen by a female, then their alleles can be perpetuated in the population and sexual selection has occurred.

Many organisms, including bacteria and many protozoa, reproduce without sex. Even some plants and animals can reproduce asexually. Sexual reproduction can be costly — it takes vast amounts of energy to grow new plumage, or a large rack of antlers. In many ways, sex seems to make no sense. A very well-adapted individual that could create clones of itself would create offspring that were equally well-adapted, so why has sexual reproduction evolved at all? Recently, scientists proposed a surprising idea: sex is a way of fighting off parasites and disease because sexual reproduction enhances genetic variability within the species.

Figure 11.18 Many birds, such as Baltimore orioles, show a high degree of sexual dimorphism.

Since sexual reproduction mixes the genes from both parents, any ability to fight off parasites or disease can be passed on to offspring. While a population may be perfectly adapted to the current environment, a population of clones will have little genetic variation to work with to survive changing conditions. Biologists tested their hypothesis by studying a type of fish — called topminnows — that live in Mexican ponds and streams. These fish sometimes mate with a closely related species, producing hybrids that are only female and that always reproduce by cloning rather than by mating. Curiously, in order to activate their eggs to grow, these hybrid females get sperm from male fish but they do not incorporate DNA from the sperm into their eggs.

In one pond the scientists studied, they found that the clones, rather than the sexually reproducing fish, were infected by parasitic cysts. Because the clones were exact replicas of one another, this strain of fish was an easy target for the parasites. Once the parasites became established in the population, they began to reproduce quickly.

In a second pond, where there were two different strains of clones, the researchers found that the more common strain of clone was subject to more infections. This also fit their hypothesis, which said that parasites able to attack the most common fish will thrive and spread throughout the population. (Meanwhile, the numbers of the other strain of fish remained low — at least temporarily — since they had less habitat available to them.) Eventually, however, the parasites became so successful that they killed their hosts and the more common strain of clone died out. This gave more habitat for the other strain of fish and their population rose. Of course, this also provided more opportunities for *another* parasite to infect *this* strain, and the cycle began all over again.

In a third pond, scientists saw something that seemed to contradict their hypothesis: they found that the fish that sexually reproduced were more infested than the clones. On closer inspection, however, it was clear that their hypothesis did fit. The pond had dried up years before, and when it refilled it had been recolonized by just a small population of fish. As a result, these fish were highly inbred and therefore deprived of the genetic variety that is the important advantage of sexual reproduction.

So, in many ways, sex is a compromise. A perfectly well-adapted individual will, in most cases, still mix genetic material with another individual to create offspring that are not exact clones. The questions is, who to mix genetic material with? Parasites may also play a role in determining which males are selected as mates by females. The displays demonstrate the fitness (because it takes energy and resources to produce the displays) and genetic potential of the males. Scientists speculate that a strong display — whether a loud song or a particularly bright display of feathers — shows that the male is healthy, strong, and not weakened by parasites or disease.

WEB LINK

www.mcgrawhill.ca/links/biology12

Our understanding of evolution grows daily as new ideas are presented and new information is gathered. To learn more about recent ideas and discoveries, go to the web site above, and click on **Web Links**. Investigate one new discovery or scientific study that advances our understanding of evolution. Create a short summary of this new information and post it on a bulletin board in your classroom.

SECTION REVIEW

1. MC Hunters often seek "trophy" animals — those that have large sets of antlers or horns. You are a wildlife biologist who recommends stopping trophy hunting in a certain area. Justify your reasoning and explain how this hunting behaviour might affect a population.
2. C Create a diagram that explains how genetic drift can shift the allele frequency in a population in just a few generations.
3. K/U Explain the difference between how natural selection changes phenotypes observed in populations and how the other four agents of micro-evolutionary change (genetic drift, gene flow, etc.) act on populations.
4. K/U Can the role of a particular mutation present in a population change over time? Explain your answer.
5. K/U Compare and contrast the founder effect and the bottleneck effect.

Summary of Expectations

- Macro-evolution is evolution on a grand scale; a large evolutionary change. (11.1)
- Micro-evolution is the change in the gene frequencies within a population over time. (11.1)
- The Hardy-Weinberg equation is $p^2 + 2pq + q^2 = 1$. (11.2)
- The five conditions required to maintain a population in Hardy-Weinberg equilibrium are random mating, no mutations, isolation, large population size, and no natural selection. (11.2)
- Mutations may provide new alleles in a population and, as a result, may provide the variation required for evolution to occur. (11.3)
- The five causes of micro-evolution are natural selection, gene flow, genetic drift, non-random mating, and mutation. (11.3)
- The bottleneck effect and founder effect lead to genetic drift. (11.3)
- Three ways in which natural selection can affect genetic variation are stabilizing selection, directional selection, and disruptive selection. (11.3)

Language of Biology

Write a sentence including each of the following words or terms. Use any six terms in a concept map to show your understanding of how they are related.

- macro-evolution
- micro-evolution
- mutation
- modern synthesis
- allele
- locus
- homozygous
- heterozygous
- dominant allele
- recessive allele
- genotype
- phenotype
- incomplete dominance
- co-dominant
- population
- polymorphic
- fixed
- electrophoresis
- polymerase chain reaction (PCR)
- diploid
- frequency
- genetic structure
- Hardy-Weinberg principle
- Hardy-Weinberg equilibrium
- genetic drift
- bottleneck effect
- founder effect
- gene flow
- inbreeding
- non-random mating
- assortative mating
- stabilizing selection
- directional selection
- disruptive (diversifying) selection
- sexual dimorphism
- sexual selection

UNDERSTANDING CONCEPTS

1. Differentiate between (a) dominant and recessive, and (b) gene and allele.
2. Explain the relationship between population size and the frequency of change in gene pools.
3. If a person gets his straight hair permed, explain whether this affects his (a) genotype; (b) phenotype.
4. Describe the possible fates of a mutation and the effects a mutation may have on a population.
5. A fly has a mutation that allows it to survive being sprayed by an insecticide. Is the mutation alone an example of micro-evolution? Explain your answer.
6. Are sex characteristics such as antlers adaptive in any way? Explain your answer and describe how sexual selection may affect the frequency of particular alleles in a population.
7. A species of toad commonly selects mates that are similar in size. How does this behaviour affect micro-evolution?
8. Give five examples of ways in which populations deviate from the Hardy-Weinberg equilibrium.
9. Choose an organism introduced in this unit and explain how two of the five situations that result in micro-evolution affect this population.
10. Describe how the work of Mendel and Darwin were blended to help develop the modern synthesis of the theory of evolution.
11. In pea plants, yellow peas are dominant over green peas. Predict the phenotypes and genotypes of the offspring of a cross between a plant heterozygous for yellow peas (Yy) and a plant homozygous for green peas (yy).
12. Describe the genotype of the parents and offspring in the following situations:
    (a) A black mouse is crossed with a white mouse. There are 16 offspring, of which 75 percent are black and 25 percent are white.
    (b) A bean with speckled seeds is crossed with a bean heterozygous for this characteristic. All offspring have speckled seeds.
    (c) A tall dog and a short dog have two tall pups.

13. If a population has two alleles for a particular locus, B and b, and if the allele frequency of B is 0.7, calculate the frequency of heterozygotes if the population is in Hardy-Weinberg equilibrium.
14. If 16 percent of individuals in a population have a recessive trait, calculate the frequency of the dominant allele in the population. Assume the population is in Hardy-Weinberg equilibrium.
15. Describe three situations that might result in a bottleneck effect in a population.
16. Describe four situations that might result in gene flow in a population.
17. Explain why most mating is *not* random. Give an example of non-random mating in plants and in animals.
18. Describe assortative mating and provide an example.
19. Identify whether each of the following is an example of stabilizing, directional, or disruptive selection.
    - **(a)** a population has only very large and very small snails
    - **(b)** a population of ducks lays eggs of intermediate weight
    - **(c)** in different parts of Africa, the colour pattern of the butterfly *Papilio dardanus* is dramatically different
    - **(d)** most individuals in a population of hummingbirds have long beaks
    - **(e)** a population has only medium-sized spiders
    - **(f)** a population shifts from being primarily black moths to being primarily flecked moths
20. Compare natural selection with sexual selection.

INQUIRY

21. Plan an experiment or model that explains gene flow.
22. Devise a demonstration using coins, poker chips, or another item of your choice to explain how the founder effect works.
23. Create a demonstration, game, or other activity that explains one way in which populations can change over time.
24. The diagrams on the right illustrate different types of natural selection. The red bell-shaped curves indicate a trait's variation in a population. The blue bell-shaped curves indicate the effect of natural selection. Determine the type of selection occurring in each illustration and provide an explanation for how and/or why that type of selection might be occurring.
25. Using the data given below on the peppered moths (which you read about in Chapter 10, section 10.1), create bell-shaped curves that illustrate the natural selection of peppered moths from a polluted environment (in 1959) to a less polluted environment (in 1985 and 1989). Explain what is happening from 1959 to 1989 in the peppered moth population.

Year	In the region near Manchester
1959	9 out of 10 peppered moths were black
1985	5 out of 10 peppered moths were black
1989	3 out of 10 peppered moths were black

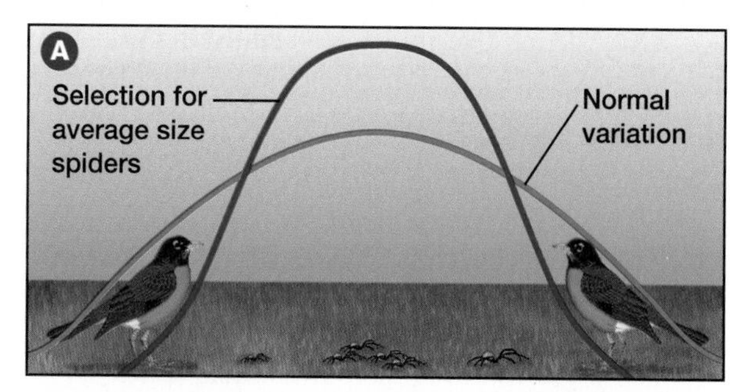

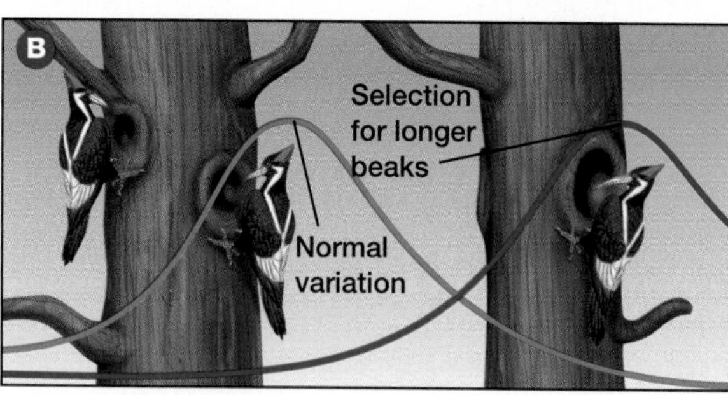

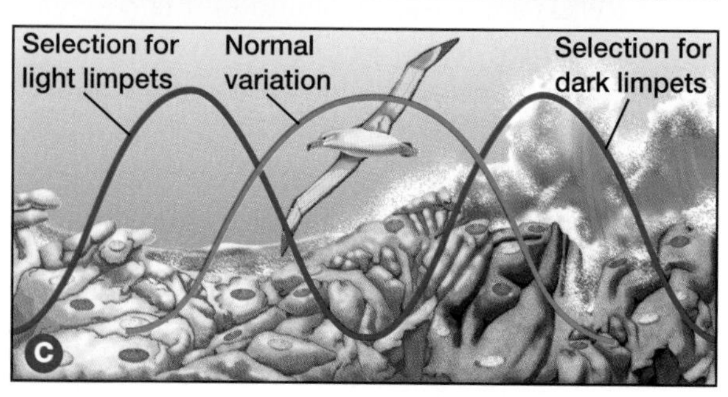

COMMUNICATING

26. Create a diagram that shows how non-random mating can increase the frequency of homozygous individuals in a population.
27. Explain the conditions in which a seemingly neutral mutation present in a small portion of a population may become quickly perpetuated in, and advantageous to, the population.
28. Theodosius Dobzhansky, a pioneer in the field of population genetics and one of the architects of the modern synthesis, said "Nothing in biology makes sense except in light of evolution." Explain your understanding of this statement.
29. You are a biologist who has been asked to explain evidence for micro-evolution to a class. Prepare your talk in point form. Provide examples of micro-evolution in action and of ways in which biologists study micro-evolution.
30. Summarize the three ways in which natural selection can shift the traits in a population's gene pool over time. Use diagrams to illustrate your summaries.
31. Scientists have used various types of biochemical and genetic analyses to determine the relatedness among the giant panda, the red panda, bears (such as polar bear, brown bear, black bear), and the raccoon. Results showed that the giant panda has DNA that more closely resembles the DNA of bears, and the red panda has DNA that more closely resembles the DNA of the raccoon. Draw a phylogenetic tree that shows these relationships.
32. You are a doctor who often prescribes antibiotics. Make a list of criteria for your patients explaining why they must take antibiotics only as prescribed.

MAKING CONNECTIONS

33. Why might a plant breeder be interested in knowing how certain traits are inherited?
34. Suppose paleontologists unearth a human skeleton that has been partially mummified and has had some of its hair preserved. What techniques could scientists use to gather more information from this discovery that would add to our understanding of evolutionary history? What are the limitations of the data and the techniques?
35. You are a biologist studying an endangered species of fox. Explain how you might use your understanding of population genetics in your work.
36. Describe different ways in which plant or animal biologists working with endangered species try to enhance genetic variation in populations.
37. What would happen to the conservation efforts if a number of alleles were eliminated from the current whooping crane population?
38. Sickle cell disease is caused by a recessive allele. Explain why the fact that we are diploid organisms keeps this allele at lower frequencies in the population. Imagine that a population of 20 individuals, three of whom carried the recessive allele for sickle cell disease, colonized a deserted island 200 years ago. The descendents of these individuals still live on this island. Predict the incidence of sickle cell disease on the island compared with the incidence of the disease in the human population at large.
39. Do you think that antibiotics should be available without a prescription? Give reasons for your opinion.

CHAPTER 12

Adaptation and Speciation

Reflecting Questions

- What determines a species?
- How do new species arise?
- What is the relationship among adaptation, natural selection, and the formation of new species?

Prerequisite Concepts and Skills

Before you begin this chapter, review the following concepts and skills:

- understanding the mechanisms that result in genetic variety, such as mutations and gene flow (Chapter 9, section 9.1), and
- understanding the process of natural selection (Chapter 10, section 10.1).

From the bacteria that thrive in your digestive tract, to a species of algae that survives on glaciers, to the elephants in the forests of Asia, there are millions of species on Earth inhabiting vastly different habitats. As well, there are innumerable species that once thrived and are now extinct. The formation of most new species takes thousands of years, but as you read this page, there are forces at play that are affecting populations. These forces may ultimately lead to the creation of a new species. The bacterial species shown here (*Staphylococcus aureus*), for example, is common in hospitals. For years, the antibiotic penicillin was highly effective in killing this bacteria and others. In fact, the discovery of penicillin meant that World War II was the first war in which fewer soldiers were killed by disease than by bullets or other shells. But today, just over 60 years since the discovery of penicillin, this wonder drug is virtually unable to fight off *S. aureus*. The ability of populations (such as bacteria) to adapt rapidly to changes in their environment is just part of the story of speciation in bacteria.

Defining a species is an ever-present challenge for biologists. For example, speciation differs in sexually reproducing species and in micro-organisms. In the past, biologists measured and recorded differences between individuals and noted their habitat and behaviour. However, this is not a practical approach for all species. With new advances and discoveries in microbiology and the unearthing of new fossils, we are learning more about how and when species form. Through experimentation and observation, a biologist can determine differences between populations and also determine the evolutionary lineage of a species. What criteria would you use to distinguish the eastern maple shown here from maple trees in western Canada or Europe?

In this chapter, you will investigate adaptation and speciation. What are the situations needed for new species to form? How quickly do species form? How do populations adapt to new environments? How do we distinguish one species from another? These are some of the questions that you will explore in this chapter.

What determines whether this Ontario maple is a different species from maple trees that grow in other parts of Canada?

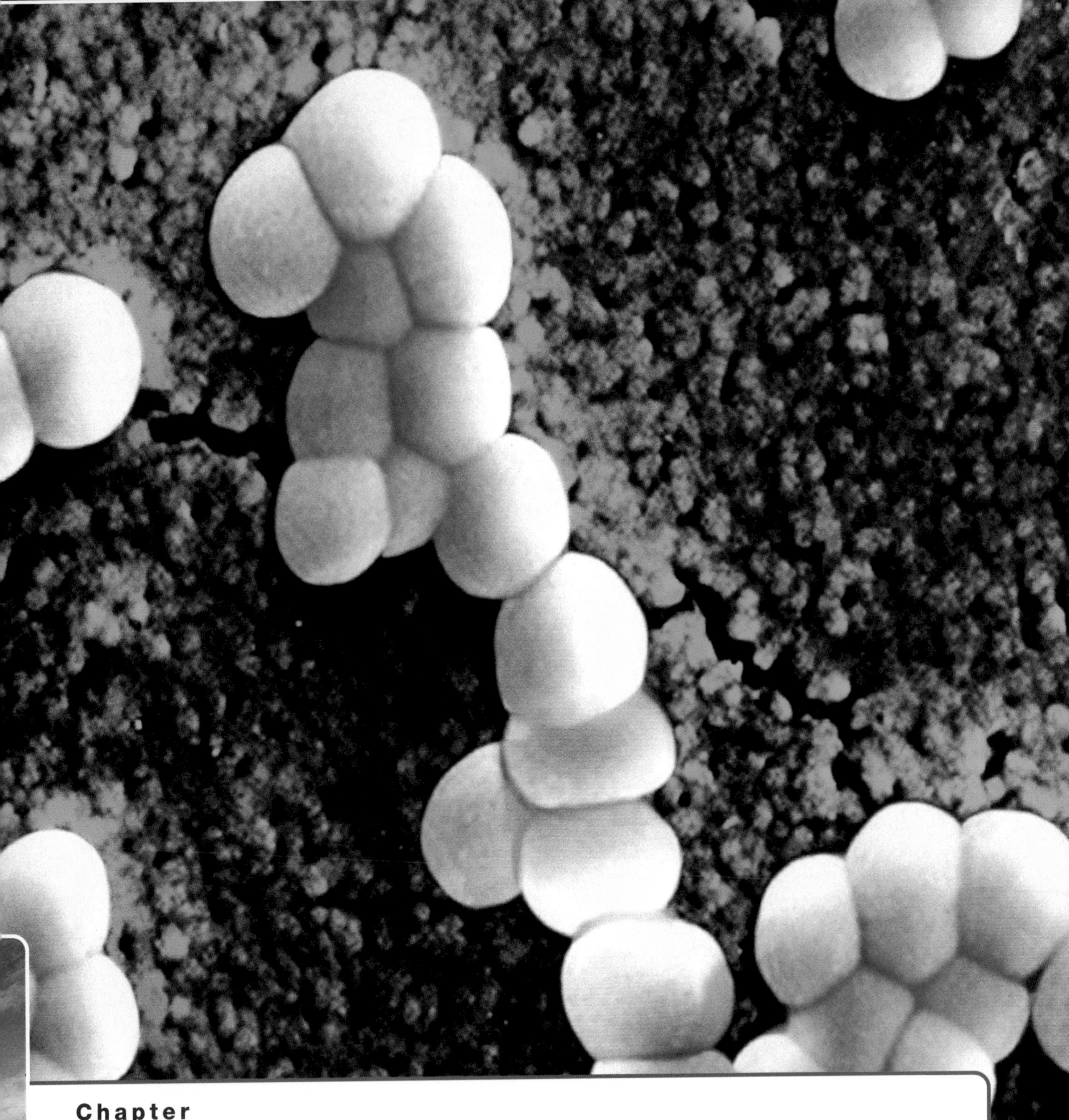

Chapter Contents

12.1 Adaptation

EXPECTATIONS

- Explain the process of adaptation of individual organisms to their environments.
- Describe the relationship between natural selection and adaptation.
- Describe different types of adaptations, explain how complex adaptations might have evolved, and describe exaptation.

The broad, flat leaves of a maple tree and the spines of a cactus are features that enable these plants to live in environments that have different conditions. A species of broad-leafed tree would not survive in the hot, dry desert or in the cold, dry tundra of northern Canada. In these environments, such trees would lose too much water across the large surface of their leaves. In contrast, the spines (which are modified leaves) of cacti, along with other characteristics, reduce water loss. With respect to absorption of light, the broad maple leaf provides a large area to absorb the moderate amounts of sunlight present in a temperate climate. In contrast, cacti live in an environment with an abundance of strong sunlight and a generally dry atmosphere, so they can absorb enough light through their small leaves or through their stems without losing moisture. Leaf shape is an important trait with respect to survival in plants. The sharp canine teeth of cougars and other carnivores; the agile, flexible hooves of mountain goats; and the ability of Arctic char to withstand near-freezing water temperatures are all traits that are important to survival (see Figure 12.1). Any trait that enhances an organism's fitness or that increases its chance of survival and probability of successful reproduction is called an **adaptation**. How exactly do adaptations arise?

Adaptation is essentially a product of natural selection. Organisms become adapted to their immediate environment over a period of time through natural selection. As populations are subjected to the vagaries of their environment, the genetic characteristics that are best adapted or well-suited to the environment are selected. For instance, populations living in cold areas will have a variety of features and behaviours that make them better adapted to withstand the cold. Those individuals that possess characteristics that enable them to survive in the cold will reproduce and may pass on these favourable adaptations to their offspring. Natural selection can, along with selective pressures, affect the number of individuals with particular traits. The result may be an adaptation of the population.

When discussing adaptations, it is important to note that the environment is more than just the immediate surroundings of an organism. Environment includes all the factors, other than genetic make-up, that can affect whether or not an organism lives through the embryo, juvenile, and adult stages to reproduce. For example, whether a plant successfully resists the selective pressure of its environment depends on many factors. These factors include the speed and normality of its germination, whether bacteria or fungi infect it as a seedling, and whether the soil in which it grows can support it. To complicate matters further, selective pressures can be contradictory. For example, warm or hot temperatures may increase the rate of plant growth, but they can also dry out the soil, thus impeding proper root growth.

Figure 12.1 The Arctic char (*Salvelinus alpihus*) has adapted to cold Arctic waters.

Adaptations may occur as particular variations increase in frequency within a population of organisms. However, variation and adaptation are not the same. A variation may improve fitness, but it may also have no effect on or even reduce fitness. Any variations that are the result of changes in the dominant alleles in the population and that may reduce fitness in the current environment will decrease in frequency in the population by natural selection. For example, the length and shape of dragonfly wings are adaptations for flight and, thus, for survival, since dragonflies prey on other insects in flight. Wing length within a population will vary slightly, but there is an optimal wing length that best suits the current environment in which the population lives. If a dragonfly has wings that are too short, it may not be able to generate enough lift to stay off the ground. If its wings are too long, they may become too heavy. So, there is a certain length of wing that results in the greatest fitness for dragonflies. Variations — such as the length of dragonfly wings, or the sharpness of eagle talons — that aid in survival and increase fitness will be preserved in a population by natural selection. If a variation is favourable for an individual, the chances are greater that the individual will survive to pass on its genes to its offspring. Over time, all surviving members of a population will have inherited that variation, at which point the original variation becomes an adaptation. In other words, the adaptation has become a general characteristic of the entire population, like a dragonfly's wing length or the sharpness of eagle talons. *In summary, while adaptations are products of natural selection, variations within a species are the raw material upon which natural selection acts.*

Evolution of Complex Adaptations

When you imagine the human eye, it seems impossible that all of its intricate parts (the lens, pupil, retina, muscles, vitreous humour, blood, nerves, and pigment), which work together to focus light into images, could have combined randomly to make such a complex organ. Adaptations, particularly ones such as the change from a simple to a complex eye, do not arise all at once. Rather, adaptations evolve over time as a result of a series of small adaptive changes. Each change is a slight modification of the traits of the previous generation.

The adaptations in the organisms living today are the result of natural selection acting on chance variations that arose at particular times in the evolutionary history of these organisms. For example, the eye has evolved in a series of steps, with each step providing organisms with vision that was slightly better for its given environmental conditions. Many marine invertebrates, such as the scallop in Figure 12.2, have ocelli — clusters of light-sensitive cells that allow the organisms to detect movement and luminosity (light level). Their eyes do not form an image. On the other hand, insects such as the fly in Figure 12.3 have compound eyes, which are excellent for detecting movement and which also form an image.

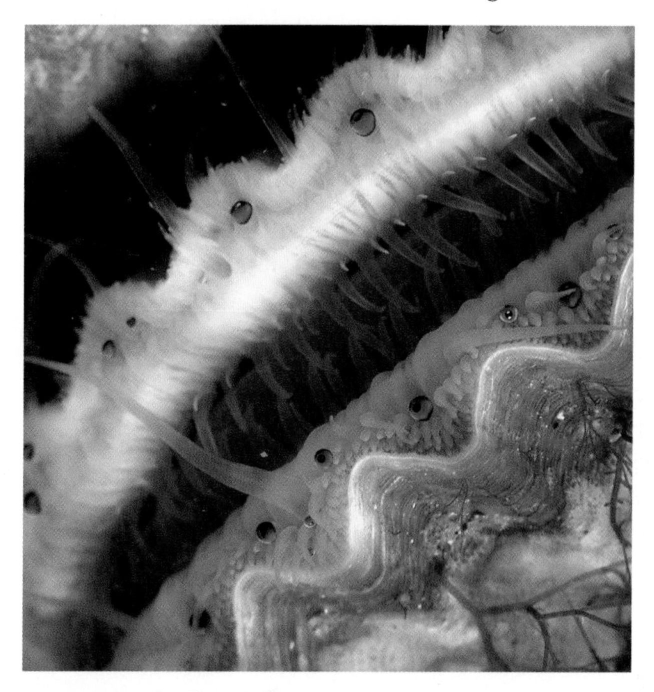

Figure 12.2 Scallops have simple eyes that are able to detect changes in light and movement, but they cannot form an image.

Figure 12.3 A compound eye enables a fly to see images.

Primitive eyes were simply a cluster of light-sensitive cells. These rudimentary "eyes" probably gave the ancient organisms an ability to see movement and to distinguish between light and dark. This gave them a selective advantage in their environment as they could detect movement of a potential predator. Over time, new variations of eyes arose in populations. For example, natural selection resulted in the formation of a simple lens that provided a blurry image. Since seeing even a blurred image is generally an advantage over seeing no image at all, this characteristic would be selected for in the population and would eventually become fixed in the population. Subsequent changes in some animals led to a sharpening of focus and, eventually, permitted colour vision. In other animals, there was no selective pressure for an advanced type of eye. In these cases, the genes for a simple lens would continue to be passed on to future generations. Each step in the evolution of eyes was due to random variations that arose in populations, and to the perpetuation of these variations within the population where the traits provided a selective advantage in a particular habitat. As a structure such as an eye becomes more adaptive for some animals and improves an animal's chances of survival, the chances of these genes being passed to offspring are increased.

WEB LINK

www.mcgrawhill.ca/links/biology12
While the eyespots of flatworms are not nearly as complex as the human eye, they still provide the flatworm with an advantage in its environment. To learn more about the evolution of a fish eye, and how long biologists think this might have taken, go to the web site above, and click on **Web Links**. Make a time line showing the changes that might have led from an eyespot to a fish eye.

MINI LAB

Small Changes, Large Gains

The adaptations that enable species to live within their environment are often difficult, or impossible, to see. Many adaptations are internal, such as changes in biochemical pathways responsible for metabolic processes. Other adaptations happen in very small steps. In the population of finches that you read about in Chapter 10 (on page 347), researchers found than even a millimetre in beak length could mean the difference between life and death in some situations. In this MiniLab, you will learn how small advantages can result in large gains for particularly well-adapted individuals.

You will need a number of different sizes, lengths, and styles of forceps and/or household tweezers. You will also need three types of small- to medium-sized seeds, such as sesame seeds, lentils, and rice. (These seeds are referred to as seeds A, B, and C here.) Mix about 30 to 40 of each of the three types of seeds together in one tray, making sure that there are an equal number of each type of seed at the beginning of the lab. Choose one style of forceps and attempt to gather seeds (any type) for 20 s. Record the number of seeds gathered by type, and record the particular characteristics of the forceps used to gather each seed. Repeat this trial three times and determine the average number of seeds gathered. Repeat this procedure using two other styles of forceps.

Now assume that there has been an environmental event (such as a drought or flood) that has reduced the availability of seed A. To simulate this, leave only 10 percent of seed A in the tray. Repeat the trials and compare the results.

Finally, assume there is an environmental event that has reduced the number of seeds B and C and doubled the number of seed A. Leave only 10 percent of seeds B and C in the tray and double the number of seed A. Repeat the trials and compare the results.

Analyze

1. Graph your results from these trials.
2. Describe any correlation between the characteristics of the forceps and their ability to pick up particular types of seeds.
3. Describe what happened after the first environmental event when the number of seed A available was reduced. How might this have affected the subsequent generations if the tweezers were actually a type of bird beak?
4. Describe what happened after the third trial. Were any of the effects of the first trial reversed? Explain how this might happen in natural situations.
5. Natural populations can have good years when the populations boom and poor years when the populations decline. Did your experiment demonstrate this phenomenon? How could you have adjusted your experiment to make it more realistic?

The Changing Function of Adaptations

Sometimes an adaptation that evolved for one function can be co-opted for another use. Originally this was called pre-adaptation, but since this term implies that there is a level of conscious planning in advance (which is not the case in evolution), a new term was coined — **exaptation**. As an example, the invertebrate ancestors of vertebrates may have stored phosphate in their skin to help them survive lean times. It turns out that the best way to store phosphate was in a matrix of calcium, which created a hard tissue. This hard tissue (for example, the shell in Figure 12.4) could also protect an animal from predators. Therefore, what originally evolved as an adaptation for metabolic processes was exapted and used for protection. Later, a calcium matrix of bone was used for muscle attachment and became the framework, or skeleton, of vertebrates.

The limbs and digits of terrestrial vertebrates did not evolve in response to a demand for walking on land. Instead, they evolved in fully aquatic tetrapods (four-legged creatures) such as *Acanthostega* that used legs and toes to move in coastal wetlands. (You were introduced to *Acanthostega* in section 10.3 on page 354.) These organisms used these limbs to crawl over logs, grip onto rocks, and clamber through marshy areas. When some of these tetrapods ventured onto land, the limbs proved useful. A living example is the lungfish of Africa, that uses its fleshy fins to move from pond to pond and to bury itself in the mud during dry periods. Paleontologists have discovered approximately 12 species of early tetrapods and all appear to have been aquatic. Thus, what evolved as an adaptation for an aquatic existence eventually became useful for a life on land. It is as though evolution borrowed something adapted for one function to perform a new function.

Types of Adaptations

Adaptations can be broadly classified as structural (or anatomical), physiological, or behavioural. The different arrangement of teeth in carnivores, herbivores, and omnivores; the tissues in vascular plants that allow transport of water and food; and the shape of fins or beaks are all **structural adaptations**. These adaptations can be anatomical (that is, dealing with the shape or arrangement of particular features), but structural adaptations can also include **mimicry** and **cryptic coloration**.

Mimicry enables one species to resemble another species or part of another species. Often, a harmless species will mimic a harmful species; the result is that predators that avoid the harmful species will also avoid the mimic. For example, the fly in Figure 12.5 is a harmless mimic of a yellow-jacket wasp. This fly, as well as other insects including some beetles, capitalizes on the fact that many predators will avoid anything with black and yellow patterning after being stung a few times by bees or wasps.

Figure 12.4 Shells are made of calcium carbonate. Calcium was originally stored by invertebrates as a way to stockpile phosphates, an energy source.

Figure 12.5 Mimics, such as this syrphid fly, copy the coloration or patterns of harmful species as a defence against predators.

Cryptic coloration makes potential prey difficult to spot. For example, animals may have colouring that blends well with their surroundings. Other animals can be camouflaged by shape and colouring, such as the bizarre sea horse (called a sea dragon) that lives among a particular type of seaweed. Figure 12.6 shows how a sea dragon can look like the algae in which it lives. Potential prey do not distinguish the sea dragon from its seaweed surroundings, and can be lured into the relative safety of the seaweed only to be consumed by the sea dragon.

Many structural adaptations are internal rather than external. For example, the strong muscle walls of the human heart are an adaptation than enables the heart to pump blood throughout the body. As another example, the digestive tracts of herbivores and omnivores are much longer relative to body size than the digestive tracts of carnivores. Vegetation is more difficult to digest than meat because of its tough cellular walls. A longer digestive tract permits more time for digestion and a greater surface area for the absorption of nutrients.

Physiological adaptations are those adaptations associated with functions in organisms. The enzymes needed for blood clotting, the proteins used in spiders' silk, the chemical defences of plants, and the ability of certain bacteria to withstand extreme heat are all examples of physiological adaptations.

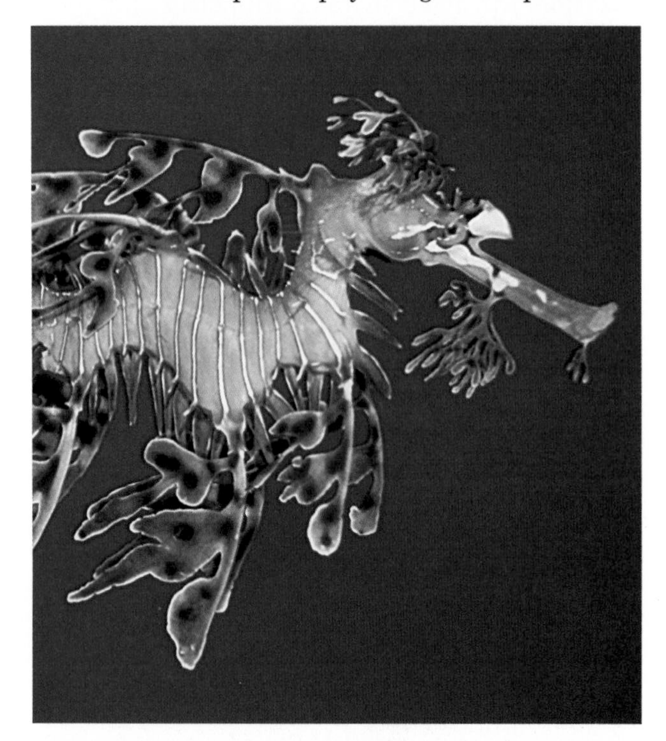

Figure 12.6 The coloration and leafy appearance of the sea dragon's (*Phycodurus sp.*) fanlike fins keep it well hidden among the seaweed in which it usually lives.

Organisms are also adapted in how they respond to the environment. These **behavioural adaptations** include migration, courtship displays, foraging behaviour, and the response of plants toward light and gravity. Animals have found different ways to avoid severe environmental conditions with adaptations such as the migration of monarch butterflies, hummingbirds, caribou, and wildebeests; the winter sleep of bears and skunks; and the hibernation of jumping mice, some turtles, and garter snakes. No doubt, some of these adaptations evolved in response to changes in environmental conditions as the continents formed and moved. All these behavioural changes are the result of natural selection — those individuals that survived passed on their genes to the next generation. For example, the monarch butterflies that moved to warmer climates survived and passed on the behavioural traits for migration.

In natural situations, it is unrealistic to isolate and classify adaptations in rigid categories because adaptations often depend upon one another. For example, bird migration is considered a behavioural adaptation. But migration would not be possible without a complex set of structural adaptations such as feathers, light bones, and strong wing muscles. As well, a variety of physiological adaptations, from nerve impulses to the release of hormones, enable flight and migration to happen.

BIO FACT

The feathers and lightweight, honeycombed bones of birds are examples of exaptation. The fossil record shows that light bones actually predated flight. This means that lightweight bones must have had some use on land. It is thought that the agile, bipedal (two-legged) dinosaurs that were the probable ancestors of birds benefited from a lightweight frame. The wing-like forelimbs and the feathers (which originally had other uses, perhaps in courtship displays or in providing warmth) were also co-opted for flight. The first flights may have been hops when pursuing prey or escaping predators. As this behaviour became advantageous in the environment, the structural adaptations that allowed it to happen were passed on.

ELECTRONIC LEARNING PARTNER

Refer to your Electronic Learning Partner for more information on mimicry.

Is Evolution Perfection?

It is sometimes assumed that the result of adaptation and natural selection is perfection in organisms. This is not the case, however, for a variety of reasons. As mentioned earlier in this section, selection can only edit variations that already exist in a population; evolution essentially has to "make do" with what is presented. As a result, designs are often awkward or less than optimal. An example is the human eye, since the neurons in our retina point backward. Although our eye works well, in many ways it is quite inefficient. In general, organisms are locked into the constraints of their evolutionary history; therefore perfection is not easily achieved. Since species have descended from a long line of ancestors, they are tied to their existing anatomy. It is not the case that old structures are scrapped and new structures are created with each step in evolution. Rather, existing structures are co-opted and adapted for the new environment. The result is designs that are sometimes less than perfect. The chronic back pain experienced by many humans is thought to result from the musculature and skeleton that have been modified from our four-legged ancestors, who were not adapted specifically for an upright posture.

Another reason that adaptations and natural selection do not achieve perfection is that adaptations are often compromises. A sea lion must swim, but it must also move about on land. In their present structure, sea lions can swim well but they are far less efficient at walking.

Finally, not all evolution is necessarily adaptive. Chance events such as tropical storms or volcanoes can also affect the composition of the gene pool. Some individuals survive this type of event randomly, and it is these individuals that remain to supply the variation upon which natural selection acts as future generations emerge.

The individuals that survive and reproduce will pass on their genes to their offspring. Over time, the populations of individuals change. In the next section, you will find out how new species can be formed from changing populations.

BIO FACT

Examples of ineffective adaptations include thumbs in pandas (which require the redirection of muscles from the hand to operate), hollow bones in flightless birds such as penguins (which do not need light bones since they do not fly anyway), as well as teeth in fetal baleen whales and tails in humans (both of which are re-absorbed before birth, and thus never used).

SECTION REVIEW

1. **K/U** Describe two mammal adaptations. Explain how each trait is adaptive.
2. **I** Stomata are openings on the surface of leaves that allow plants to release water. Analyze the following data showing the number of stomata on the leaves of one tree species. What might these data tell you about the rainfall in the areas where the data were collected? What is the relationship between rainfall and number of stomata?

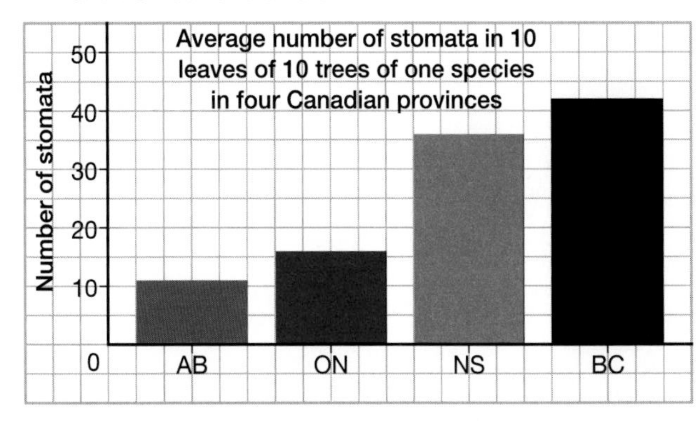

3. **C** Describe, with the aid of a sketch, a plausible pathway for the evolution of a complex adaptation such as the vertebrate eye.
4. **K/U** Was a primitive eye that was 95 percent less effective than a modern eye useless? Explain your answer.
5. **K/U** Are the following adaptations behavioural, structural, or physiological? Give reasons for your answers.
   (a) plant stems grow toward light
   (b) woodpeckers' bills are pointed and sharp
   (c) cacti have spines
   (d) spiders use proteins in their webs
   (e) flowers produce scent
6. **K/U** Describe the relationships among variations, adaptations, and natural selection.
7. **K/U** Give two examples of behavioural adaptations and explain how they may have evolved.
8. **K/U** Explain why adaptation and natural selection do not result in perfection.
9. **K/U** Evolutionary biologist Karel Liem said that "Evolution is like modifying a machine while it's still running." Explain what this statement means.

How Species Form

EXPECTATIONS

- Define the concept of speciation and explain the mechanisms of speciation.
- Explain biological barriers to reproduction.
- Describe alternative concepts of species.

The meadowlarks in Figure 12.7 look remarkably similar, yet they are different species. How is a species defined? Historically, species were described in terms of their physical form. But obviously physical similarity does not necessarily mean organisms are the same species. To this end, scientists now also consider physiology, biochemistry, behaviour, and genetics when distinguishing one species from another.

The most common definition of a species describes a **biological species** concept. In this context, a species consists of a reproductively compatible population; that is, a population that can interbreed and produce viable, fertile offspring. To accomplish this, the populations must be able to interbreed in the same time period. If one population breeds in the spring and one in the fall, the two populations would generally not interbreed. The concept of a biological species, therefore, centres on the inability of two species to hybridize. (A hybrid is the offspring of a cross between individuals of two species.) In those cases where hybrids can form, they are usually infertile or the gametes produced are not viable. A well-known example of two biological species are the horse and the donkey. According to the biological species concept, the horse and the donkey are two separate species. They may interbreed, but the offspring of a horse and donkey (called a mule or hinny) is almost always infertile.

Figure 12.7 Even though eastern and western meadowlarks (*Sturnella magna* and *S. neglecta*, respectively) overlap in parts of their range, they are separate species.

There are two general pathways that lead to the formation of new species: transformation and divergence. Figure 12.8 illustrates the two pathways. A species can be the result of accumulated changes over long periods of time such that one species is transformed into another (**transformation**). The alternative is **divergence**, in which one or more species arise from a parent species that continues to exist. Both pathways are the result of natural selection. Divergence promotes biological diversity because it increases the number of species. In transformation, however, a new species is gradually created while the old species is gradually lost. Identifying instances of transformation is subjective, since it is difficult to determine when the new species became reproductively isolated from the original species, which no longer exists.

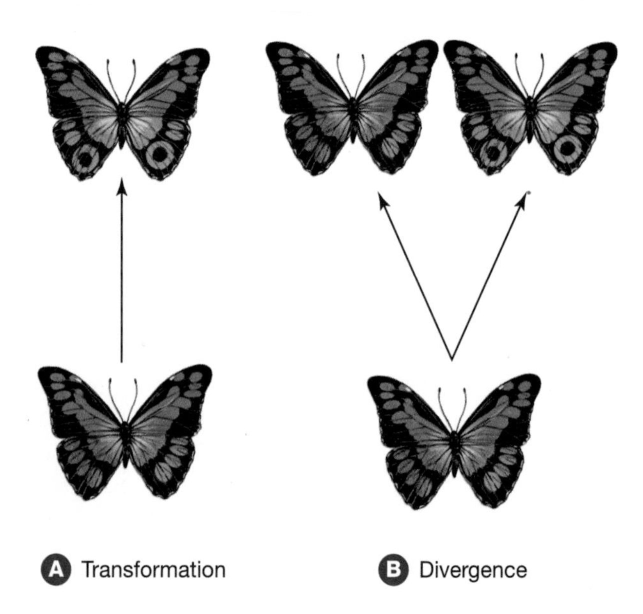

Figure 12.8 There are two patterns of speciation: (A) transformation (or anagenesis), in which one species evolves into another; and (B) divergence (cladogenesis), in which one or more species arise from a parent species.

Speciation, the formation of species, is a continuous process. Examples of speciation occurring today are often difficult to pinpoint because there are so many factors affecting the natural selection of individuals within a species. It is challenging to determine exactly when a species becomes a new species, or whether two different populations are the same species. For example, bird species such as the Baltimore oriole (*Icterus galbula*) and the Bullock's oriole (*I. bullockii*) in North America were once considered to be one species, called the northern oriole. After more research, the northern oriole (*Icterus galbula*) was again separated into the two species. Modern DNA analysis helps scientists to determine which populations may be a single species and which ones (such as the Baltimore oriole and the Bullock's oriole) may be two or more species.

Biological Barriers to Reproduction

In order for species to remain distinct, they must remain reproductively isolated. Various barriers prevent interbreeding and restrict genetic mixing between species, and species are generally separated by more than one type of barrier. **Geographical barriers** such as rivers prohibit interbreeding because they keep populations physically separated. However, there are many **biological barriers** that keep species reproductively isolated even when their ranges overlap.

It is clear that a bat will not mate with a squirrel, nor will a fern fertilize a rose, but what biological barriers keep species that are closely related to each other from interbreeding? Reproductive barriers are one example — they can act before or after fertilization to isolate gene pools.

Pre-zygotic Barriers

Pre-zygotic barriers (also known as pre-fertilization barriers) either impede mating between species or prevent fertilization of the ova if individuals from different species attempt to mate.

- **Behavioural isolation** The songs of birds, the courtship rituals of elk, and the chemical signals (called pheromones) of insects are all examples of behavioural barriers to reproduction. Any special signals or behaviours that are species-specific prevent interbreeding with closely related species. For example, even bird species that look virtually identical (such as the meadowlarks in Figure 12.7) and have overlapping ranges can remain separate biological species, largely because of differences in their songs. The songs allow them to recognize individuals of their own species. Another example of behavioural isolation is that females of some species release powerful, species-specific pheromones to attract males.
- **Habitat isolation** Although two species may live in the same general region, they may live in different habitats and therefore encounter each other rarely, if at all. For example, two species of North American garter snakes — the common garter snake and the northwest garter snake — live in the same area, but the northwest garter snake prefers open areas (such as meadows) and rarely enters water, while the common garter snake is most commonly found near water. These snakes are shown in Figure 12.9.

Figure 12.9 The northwest garter snake (*Thamnophis ordinoides*) (A) and the common garter snake (*T. sirtalis*) (B) occupy different habitats in a similar geographical area. This keeps the two species reproductively isolated.

The Patterns of Evolution

EXPECTATIONS

- **Explain the mechanisms of speciation.**
- **Describe the different patterns of evolution.**
- **Compare two models describing the pace of evolution.**

Types of Speciation

Speciation is the process by which a single species becomes two or more species. Biologists generally recognize two modes of speciation, the definitions of which are based on how gene flow is disrupted within a population.

Sympatric Speciation

When populations become reproductively isolated — even when they have not become geographically isolated — **sympatric speciation** occurs. In sympatric speciation, factors such as chromosomal changes (in plants) and non-random mating (in animals) alter gene flow. This type of speciation is far more common in plants than in animals.

Given the right set of conditions, a new species can be generated in a single generation if a genetic change results in a reproductive barrier between the offspring and the parent population. For example, errors in cell division that result in extra sets of chromosomes (a mutant condition called **polyploidy**) can lead to speciation.

A polyploid organism has three or more sets of chromosomes in the nucleus of each of its cells. Most animals are diploid — they have one set of chromosomes inherited from each parent. While it is quite rare for animals to be polyploid, this condition is relatively common in plants, particularly among flowering plants. (Polyploidy in plants is possible because many species are able to self-fertilize and reproduce vegetatively.)

Recall that during reproduction, a sequence of events must occur during meiosis in order for organisms to reproduce successfully. If errors occurred during meiosis and chromosomes did not separate, the gametes produced would have two sets of chromosomes (diploid, $2n$), instead of one set (haploid, $1n$). Then, if two diploid gametes fuse, the offspring would have four of each chromosome (tetraploid, $4n$). If tetraploid offspring survive, they could undergo normal meiosis and produce diploid gametes. The plant can now self-pollinate or reproduce with other tetraploids. However, it cannot produce viable seeds when crossed with diploid plants from the original population, since any offspring from this mating would be triploid ($3n$) and therefore sterile (because unpaired chromosomes result in abnormal meiosis). In just one generation, a reproductive barrier has been established in a population, because gene flow is interrupted between a small population (as small as one individual) of tetraploids and the parent population.

Figure 12.11 shows the stages in speciation by polyploidy. Many species, including cotton, apples, day lilies, and chrysanthemums, have been developed by plant breeders who artificially double

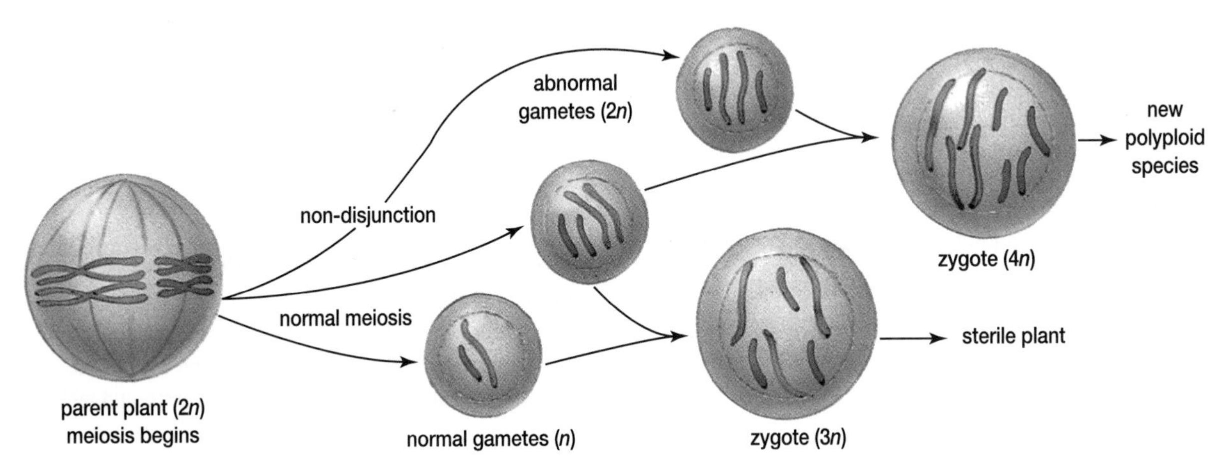

Figure 12.11 Polyploidy can lead to the formation of new species.

populations do not last long enoug
enough to become new species.

The population of finches being
Galápagos is an example of speciat
(You learned about the study of th
Peter and Rosemary Grant in Chap
of the ancestral species reached on
in the Galápagos, possibly as a resu
off course in a tropical storm. Unal
the mainland, the ancestral species
differently than their mainland rel
ancestral birds or their successive
since spread through the islands. N
developed as they evolved in respo
environments on individual island

By observing the finches now pr
islands, measuring features such as
and analyzing the DNA of the bird

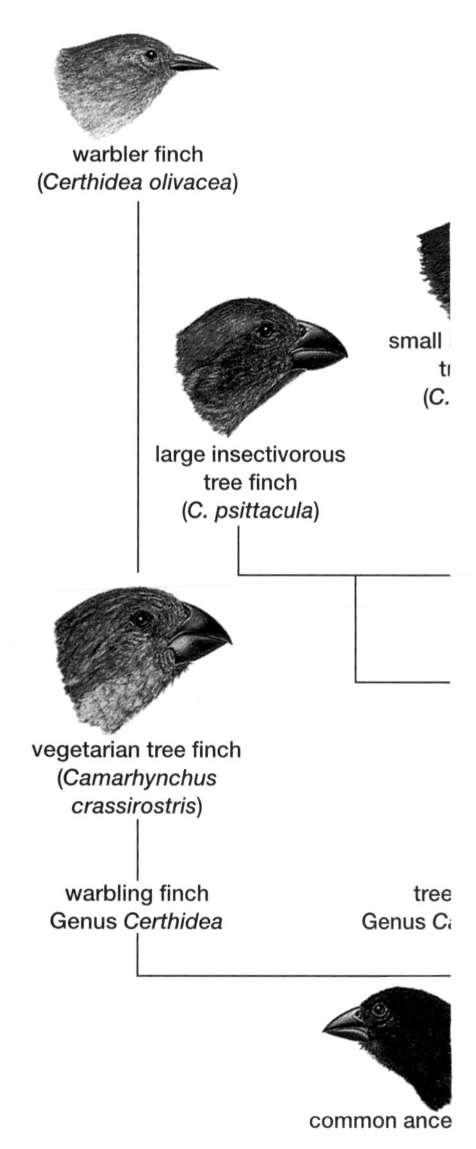

Figure 12.15 A phylogenetic tree for finc

the chromosomes in a plant and hybridize the resulting polyploids.

BIO FACT

Polyploidy is one way that plant breeders can create seedless fruit, such as watermelon. Plant breeders produce triploid watermelon seeds by crossing a normal diploid parent with a tetraploid parent. (The tetraploid plants are created by genetically manipulating diploid plants to double their chromosome number.) The resulting watermelons are triploid, and thus sterile — they do not have seeds, yet they still produce fruit.

In another model of sympatric speciation, two species can interbreed to produce a sterile offspring. Although the offspring is infertile, it can reproduce asexually — resulting in the formation of a separate population. Through mechanisms such as errors in meiosis, the sterile hybrids can be transformed into fertile polyploids in subsequent generations, thus forming a new, fertile species even though geographical isolation has not occurred. Figure 12.12, for example, shows the evolution of wheat. Chromosome analysis has shown that wheat is the result of two hybridizations of wheat with wild grasses, and two instances of meiotic error. As a result, a new species — the wheat that has been used to make bread for 8000 years — arose. Many other species grown for agriculture, including cotton, oats, and potatoes, are polyploids.

Sympatric speciation may also occur in the evolution of animals, but it is much less common. The mechanisms for sympatric speciation in animals are also different from those in plant populations. Generally, animals will become reproductively isolated within the geographical range of a parent population as they begin to use resources not used by the parent population. This, in turn, will lead to non-random mating and eventual speciation. For example, Lake Victoria in Africa holds almost 500 species of closely related fishes called cichlids (some cichlids are shown in Figure 12.13). Each species has a feature that makes it unique from other species in the lake, and none of these species are found anywhere else on Earth. It is thought that this incredible explosion of diversity happened as small groups of the parent population began to exploit different food sources and habitats in the lake. The speciation of cichlids has resulted in a remarkable variety of cichlids with a fascinating diversity of teeth, jaws, mating behaviours, and coloration. What makes this example even more astounding is that all of this diversity evolved from a single ancestor in less than 14 000 years — a relative blip in the geological time scale.

Figure 12.13 Several species of cichlids. Almost 500 species of cichlids live in Africa's Lake Victoria.

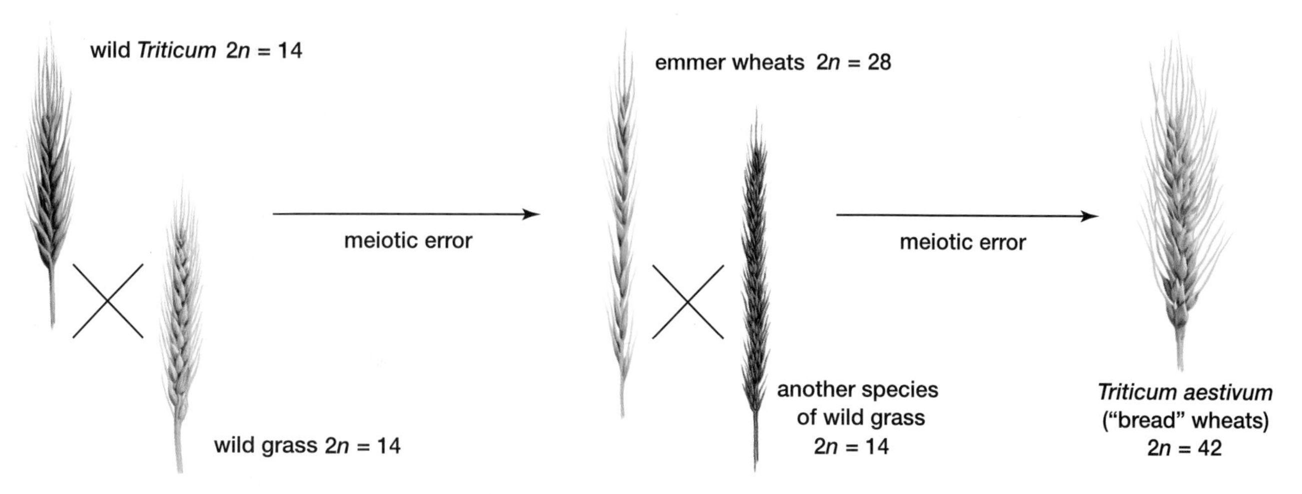

Figure 12.12 Wheat that is used to make bread evolved as a result of two hybridizations and two instances of meiotic error.

Biologists studying the speciatio
differences among cichlids are run
In the 1950s, the Nile perch, a fish
other east African lakes, was introc
of food for people living near Lake
huge fish can grow up to two metre
by preying on cichlids. As well, far
around the lake have resulted in m
erosion. The soil erodes into the lak
the once clear waters muddy. Sinc
the distinct markings of potential r
having difficulty clearly identifyin
and, as a result, have been mating
closely related species. This interb
eroding the reproductive isolation
these fishes into hundreds of new

Allopatric Speciation

Also called geographic speciation,
speciation occurs when a populati
two or more isolated groups by a g
barrier. (Figure 12.14 illustrates the
Eventually, the gene pool of the sp
becomes so distinct that the two gr
to interbreed even if they are broug
For example, a glacier or lava may
populations, fluctuations in ocean
a peninsula into an island, or a few
reach a geographically separate hal
populations are reproductively iso
frequencies in the two populations
diverge due to natural selection, m
drift, or gene flow. This geographic
population does not have to be ma
for transformation to occur. Howev
maintained long enough for the po
become reproductively incompatib
are rejoined.

The effect of a geographical barr
large part to an organism's ability t
mobility of animals or the ease wit
plant spores are dispersed limits g
the cohesive influence of a commo
affects the impact of a geographica
example, while birds easily cross th
it is impossible to rodents. As a res
bird species inhabit either side of t
different species of squirrels inhab
of the canyon.

Generally, small populations tha
isolated from the parent populatio
to change enough to become a new

UNIT 4 REVIEW

UNDERSTANDING CONCEPTS

Multiple Choice

In your notebook, write the letter of the best answer for each of the following questions.

1. If evolution occurs, we would expect different biogeographical regions with similar environments to
   - **(a)** all contain the same mix of plants and animals
   - **(b)** have land masses that are connected to each other
   - **(c)** each have its own specific mix of plants and animals
   - **(d)** have plants and animals that have similar adaptations
   - **(e)** both (c) and (d)
2. The fossil record provides direct evidence for common descent because you can
   - **(a)** see that types of fossils have changed over time
   - **(b)** sometimes find common ancestors
   - **(c)** trace the ancestry of a particular group
   - **(d)** sometimes find arrangements of bones similar in common ancestors
   - **(e)** all of the above
3. Assuming a Hardy-Weinberg equilibrium, 21% of a population is homozygous dominant, 50% is heterozygous, and 29% is homozygous recessive. What percentage of the next generation is predicted to be homozygous recessive?
   - **(a)** 21%
   - **(b)** 50%
   - **(c)** 29%
   - **(d)** 25%
   - **(e)** 42%
4. In a population of diploid individuals that is in Hardy-Weinberg equilibrium, the frequency of a dominant allele for a certain hereditary trait is 0.3. What percentage of individuals in the next generation would be expected to be homozygous for the dominant trait?
   - **(a)** 9%
   - **(b)** 14%
   - **(c)** 42%
   - **(d)** 49%
   - **(e)** 90%
5. From which of the following areas of study did Darwin and Wallace derive *most* of their evidence for evolution?
   - **(a)** mechanisms of heredity
   - **(b)** comparing the anatomy of different species
   - **(c)** geographic distribution of organisms
   - **(d)** embryology
   - **(e)** animal behaviour
6. Genetic equilibrium occurs when
   - **(a)** populations are small
   - **(b)** there is no immigration to or emigration from a population
   - **(c)** natural selection acts on particular phenotypes
   - **(d)** mutations arise in a population
   - **(e)** individuals that are related or live in close proximity to one another mate
7. A human population has a higher-than-usual percentage of individuals with a genetic disease. The most likely explanation is
   - **(a)** gene flow
   - **(b)** stabilizing selection
   - **(c)** directional selection
   - **(d)** genetic drift
   - **(e)** all of the above are possible
8. Which of these is/are necessary in order for natural selection to occur?
   - **(a)** variation
   - **(b)** differential success at reproduction
   - **(c)** inheritance of difference
   - **(d)** all of the above
   - **(e)** only (b) and (c)
9. Which of the following is a pre-zygotic barrier?
   - **(a)** habitat isolation
   - **(b)** temporal isolation
   - **(c)** hybrid inviability
   - **(d)** hybrid sterility
   - **(e)** (a) and (b)
10. The many species of Galápagos finches were each adapted to eating different foods. This is an example of
    - **(a)** gene flow
    - **(b)** adaptive radiation
    - **(c)** sympatric speciation
    - **(d)** all of the above
    - **(e)** (b) and (c)
11. Which of the following types of reproductive barriers are not pre-zygotic?
    - **(a)** mechanical isolation
    - **(b)** geographical isolation
    - **(c)** temporal isolation
    - **(d)** gametic isolation
    - **(e)** behavioural isolation

Short Answer

In your notebook, write a sentence or a short paragraph to answer each of the following questions.

12. Explain the difference between a fact and a theory. Give an example of each.
13. Explain the difference between analogous structures and homologous structures.
14. Distinguish between mutations and variations.
15. Give an example of each of the following types of mutations: one that would be beneficial to an individual; one that would be detrimental to an individual; and one that would have no effect on an individual.
16. "Evolution can occur without new species arising." Do you agree with this statement? Explain your answer.
17. Explain why diversity within a population is necessary for evolution.
18. Artificial selection can sometimes perpetuate traits that are not desired, such as respiratory problems in some breeds of dogs. Does the same thing happen in natural selection? Explain your answer.
19. Does the process of natural selection always improve the design of organisms? Explain your answer.
20. How might (a) Lamarck and (b) Darwin have explained the elephant's long trunk?
21. Insects reproduce fast enough that they could quickly populate and "overrun" Earth. Explain why this does not occur. How was this significant to Wallace and Darwin?
22. Explain how the ability to sequence DNA furthered the understanding of evolution.
23. Distinguish between macro-evolution and micro-evolution.
24. How do heterozygous individuals and polymorphic populations contribute to variation within a population?
25. In the past, ideas of natural selection have been used to justify injustice and prejudice. Explain why this is a scientifically incorrect use of the idea of natural selection.
26. Explain why the effects of genetic drift are more significant in small populations.
27. Outline the limitations to defining species purely on the basis of reproductive isolation.
28. Distinguish between allopatric and sympatric speciation.
29. Describe how adaptive radiation helps colonize volcanic islands.
30. Describe an example of (a) convergent evolution and (b) coevolution.
31. Give an example of (a) structural adaptation, (b) physiological adaptation, and (c) behavioural adaptation.
32. A doe tends to favour bucks with larger antlers. Is this an example of natural selection? Explain your answer.
33. Explain why most species would not be in Hardy-Weinberg equilibrium.
34. Explain why the evolution of resistance to antibiotics in bacteria is an example of directional selection.
35. If a human population has a higher-than-usual percentage of individuals with a genetic disease, is the most likely explanation gene flow or genetic drift? Explain your answer.

INQUIRY

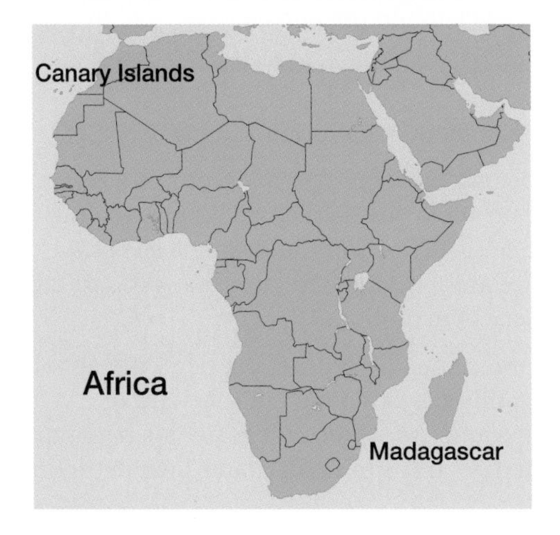

36. Madagascar separated from the African continent about 50 million years ago. The Canary Islands are volcanic in origin and are about 10 to 15 million years old. Discuss the types of organisms you would expect to find on these islands and why this information supports the theory of evolution.
37. Calculate the genetic structure of a population of flowers in which 150 individuals are homozygous dominant, 130 are heterozygous, and 58 are homozygous recessive. Assuming that the allele for pink is dominant to the allele for white, describe the population's phenotype as well.

38. The following diagrams represent a distribution of genotypes in a population. Copy the diagrams into your notebook and draw and label: (a) another line to show that disruptive selection has occurred; (b) another line to show that stabilizing selection has occurred; and (c) another line to show that directional selection has occurred.

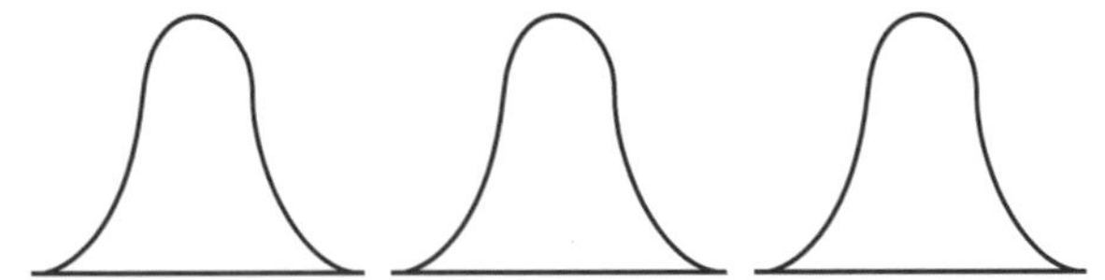

39. You are an investigator studying the frequency of certain traits in a population. You find that 73% of the individuals in the population has freckles. The presence of freckles is controlled by a dominant allele. Calculate the genotype and allele frequencies for the population.

40. Are pandas more closely related to bears or raccoons? This has been a long-standing question for biologists. A biologist determined in the 1950s that, based on behavioural traits, red pandas and giant pandas were closely related to each other and that both more closely resemble bears than raccoons. However, in the 1980s, molecule analysis (including DNA comparison) led to the determination of the evolutionary characteristics shown below. Use this illustration to answer the following questions.

(a) Is the raccoon or the bear more closely related to the red panda?

(b) Is the raccoon or the bear more closely related to the giant panda?

(c) Approximately how long ago did raccoons and bears split into two lineages?

Is the panda a bear or a raccoon?

COMMUNICATING

41. Complete this concept map by using the following vocabulary terms: frequency of alleles, speciation, gradualism, natural selection, geographical isolation, reproductive isolation, punctuated equilibrium.

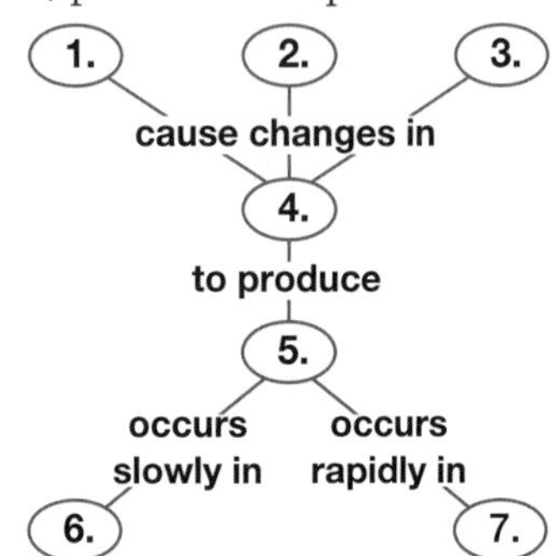

42. Create a time line or use another graphic organizer to outline the major events and ideas that have led to the current theory of evolution.

43. Use a labelled diagram to show natural selection at work in a population as environmental conditions change.

44. A person tells you that evolution is a hoax because it is "just a theory." Explain to the person what a theory means in a scientific sense and provide five facts that support the theory of evolution.

45. There is concern within a community about the outbreak of a dangerous species of bacterium. As a precaution, people begin to purchase and administer their own antibiotics, without the advice of physicians. Prepare a communications brief that explains why this practice could worsen the situation.

46. You are organizing a debate on gradualism vs. punctuated equilibrium. Develop an information brief for each debating team.

MAKING CONNECTIONS

47. You discover the remains of an extinct animal that has a small amount of brain tissue preserved in its skull. Outline the scientific techniques you might use to learn more about the evolutionary history of this animal.

48. You are a biologist heading a team of scientists trying to bring whooping cranes back from the brink of extinction. At its smallest, the population had six to eight individuals. Develop a brief presentation that explains to funding officials why this population is still in peril even though it now numbers over 200. Outline the steps you would take to help save this population.

49. A scientist observes that members of a particular plant species are shorter at the top of a mountain than at the bottom. Give an explanation based on natural selection.

50. Several articles published recently in a scientific journal call for the reduced use of antibiotics in the feed given to animals (such as chickens and cattle). Based on your understanding of coevolution, explain why scientists are calling for this change.

51. Explain why zoos exchange animals of one species. How does this benefit society? How does it benefit the environment? What are some of the economic issues?

52. You are a gardening expert who runs a local nursery. A gardener calls you and explains that she had an insect infestation in her garden. When she applied an insecticide, 99 percent of the insects were killed. However, when she applied the insecticide again six weeks later, only 50 percent of the insects were killed. How would you explain why the insecticide did not work as well the second time it was applied?

53. In Canada, Atlantic salmon are farmed on both the Pacific and Atlantic coasts. Some people are concerned about the introduction of domestic salmon to the oceans, fearing that Atlantic salmon that escape from fish farms might affect the genetics of wild salmon if they begin to interbreed and hybridize. Biologists point to the selectional forces that are at play in the two populations. Farmed salmon, for example, are artificially selected and bred for increased growth rate and larger size, among other characteristics. In populations of wild salmon, however, natural selection is at play. Describe the selectional forces that might affect wild salmon populations and note whether the type of selection in farmed and wild salmon populations is directional, stabilizing, or disruptive.

UNIT

Population Dynamics

Unit Preview

In this Unit, you will

- analyze the components of population growth and explain the factors that affect the growth of populations,
- investigate and evaluate populations, their interrelationships, and their effect on the sustainability of life on Earth, and
- evaluate the carrying capacity of Earth and relate it to the growth of populations, consumption of natural resources, and technological advances.

Unit Contents

UNIT ISSUE PREP

Read pages 538–539 before beginning this unit.

- Choose a country from the list or select one you would like to investigate.
- Begin research to identify the country's most pressing environmental issue.
- Collect information on the issue and its effect on human populations.

On Easter Sunday 1722, Jacob Roggeveen discovered a 650-km^2 island in the Pacific Ocean, 3700 km from the nearest continent. He described it as a dry wasteland with no trees or bushes over 3 m high or native animals larger than an insect.

Evidence suggests that humans arrived on Easter Island about A.D. 400, having come from Polynesia. With them came chickens and rats, which reproduced rapidly. Much of the island was forested; huge palms and a diversity of other trees provided food, wood for timber and dugout canoes, and fibre for rope. The human population grew rapidly and developed an elaborate culture, sufficiently organized to permit quarrying, transportation, and raising hundreds of enormous statues to honour ancestors.

Gradually, things changed. By 1400, trees were extinct on the island. Without large canoes, the islanders had to rely on near-shore fish, as well as land birds. As food supplies dwindled, society broke down. Warring tribes pulled down most of the statues. Eventually the human population fell to about 1000, one tenth its peak size. Perhaps the experience of the Easter Islanders can provide lessons for us currently living on the "island" known as Earth.

Human activity combined with increased drought can turn rich fertile land into deserts. The understanding of ecology and its effect on the environment are critical to preserving the biosphere.

CHAPTER 13

Ecological Principles

Reflecting Questions

- Why are species distributed as they are on Earth?
- Why are some species rare while others are abundant?
- What impact does one species have on other species in its environment?
- What impact do organisms have on Earth itself?

Prerequisite Concepts and Skills

Before you begin this chapter, review the following concepts and skills:

- understanding how organisms will only survive and reproduce in environments to which they have adapted, therefore maintaining their homeostasis (Chapter 6, section 6.3), and
- explaining why organisms need to obtain energy and specific nutrients from their environment to support normal metabolic functions (Chapter 3, section 3.4).

To avoid the fate of the Easter Islanders, we need to understand the impact we have on Earth. The study of Earth is challenging because the planet is so large and difficult to manipulate experimentally. One possible solution may lie in an enormous greenhouse, situated in the Arizona desert. Its massive steel and glass shell contains roughly 170 000 m^3 of atmosphere, 1 500 000 L of fresh water, 3 800 000 L of salt water, and 17 000 m^3 of soil, all organized into different zones that model natural Earth environments. There is, for example, a rain forest, a desert, and an ocean.

This greenhouse, called Biosphere 2, is one of the world's biggest laboratories. Since Earth is so complex and unpredictable, it is difficult to determine exactly how specific organisms are affected by the types of changes (in variables such as temperature and carbon dioxide concentration) we see occurring, or to predict what might occur in the future. By using Biosphere 2, scientists can double the amount of a variable in one area, while leaving it at a "normal" level in another similar area. Controlled experiments like this are vital to understanding how change affects organisms.

Originally built for another purpose (see question 10 in the Chapter Review section), Biosphere 2 was taken over by Columbia University in 1996 and is now a research and teaching laboratory. With the participation of many other universities, Columbia has built a campus adjacent to the lab, which provides unique educational opportunities for students from around the world. Conferences are also held at the site, for those involved in studying or managing Earth's resources. All of these activities are designed to help people understand global problems and consider the challenges faced by our planet.

In this chapter, you will examine concepts related to these challenges. How can you learn about the complexity of interactions among organisms so you can understand them better? How do conditions on Earth affect its inhabitants? How do living things affect their physical environment?

The portion of Earth inhabited by living things is called the biosphere. In a facility called Biosphere 2, scientists are attempting to mimic conditions on Earth (Biosphere 1) to study them in a controlled manner.

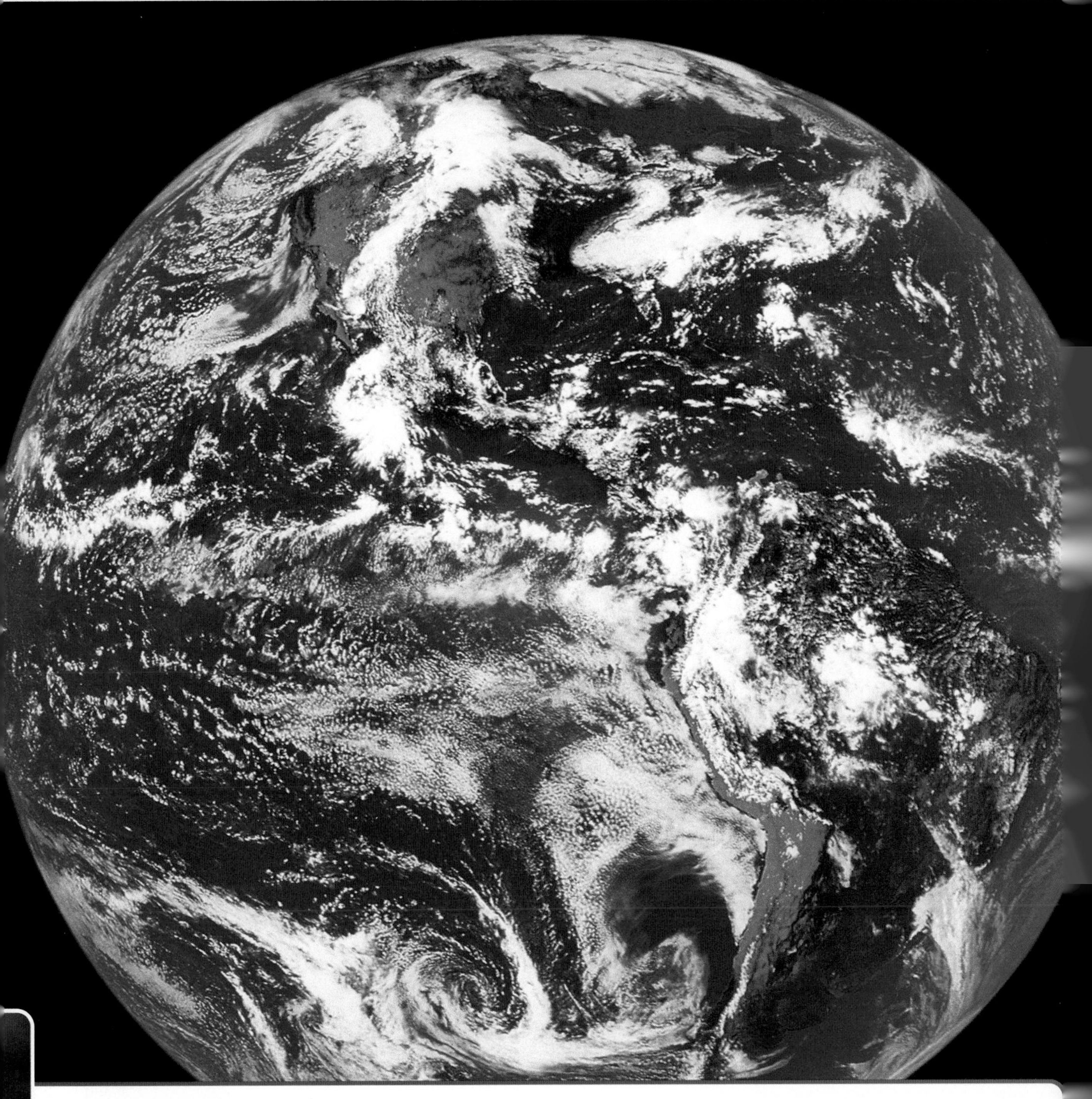

Chapter Contents

13.1 Where on Earth Do You Find Organisms?

EXPECTATIONS

- Distinguish between abiotic and biotic components of the environment.
- Describe populations and their interrelationships within communities and ecosystems.
- Describe the process of ecological succession.
- Differentiate between the habitat, range, and niche of a population.

Ecology is often described as the study of the patterns of distribution and abundance of plants, animals, and other types of organisms on Earth (see Figure 13.1). Since these patterns depend on interactions among organisms as well as between individuals and their environments, ecology also includes the study of these interactions.

It is important to realize that the environment of every organism includes other organisms as well as the air, soil, or water that surrounds it. In other words, an organism's environment includes biotic and abiotic components. As you learned in previous studies, the **biotic** components are the living things with which the organism interacts, such as those it consumes or is consumed by, competes with or is helped by, and parasitizes or is host to. The **abiotic** components are all non-living physical and chemical factors that influence an organism's survival, such as temperature, light, water, and nutrients.

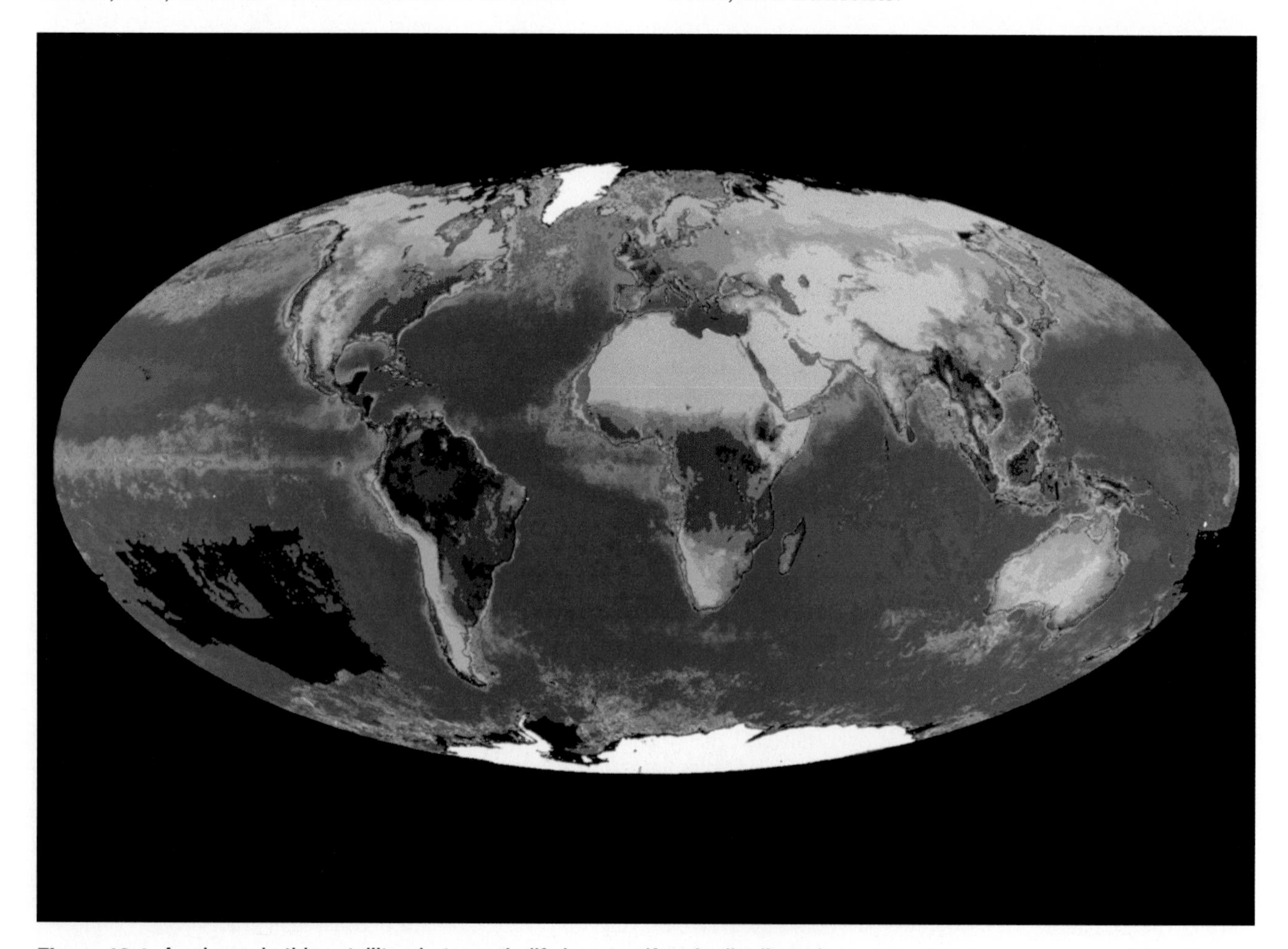

Figure 13.1 As shown in this satellite photograph, life is not uniformly distributed throughout the biosphere. The dark green areas are nutrient rich, while the yellow areas are relatively barren.

Organizing Interactions

Ecology as a whole is a very large area of study. Ecologists thus often specialize, focussing their research efforts on specific types of interactions. For the most part, these specialties relate to four increasingly general categories, which ecologists use to organize their thinking — individual organism, population, community, and ecosystem (see Figure 13.2).

For example, many ecologists study how individual organisms interact with the abiotic components of their environment. These researchers are interested in how the behaviour or physical features of an organism allow it to cope with such things as the temperature and moisture conditions typically found in its environment. This understanding helps explain why some types of organisms are only found in specific areas (that is, it helps us to understand the distribution of species), or why they are more abundant in some places than in others.

Populations

As you saw in Chapter 11, a **population** is a group of individuals of the same species living in the same geographic area. Figure 13.3 shows a population of monarch butterflies. The size of the geographical area defining the population can vary. The geographical area can depend on how fast or how far an organism can travel, or on how the abiotic conditions vary from place to place. Differing conditions may explain why different populations of the same species display variations in behaviour, physiology, and physical features. The variations evolve over time, as populations become adapted to their local environments.

Population ecologists focus mainly on factors that affect the size and composition of populations. They are interested in what causes populations to grow or decline, the rate at which populations change in size, and what determines the relative numbers of males and females (or young and old) in populations.

Figure 13.3 Each monarch butterfly is part of a population. How would you define this population — what are its edges or boundaries?

Communities

In nature, populations are rarely isolated. Instead, they interact with each other to form the next level of organization: a community. A **community** consists of all the organisms in all the interacting populations in a given area. Community ecologists typically study how interactions among the

Figure 13.2 The study of ecology occurs at several levels, from individual organisms, to populations and communities, to ecosystems.

members of different populations affect community structure. For example, they may be interested in why some communities are made up of many different species, while others contain only a few. Interactions that might influence community structure include competition between individuals in different populations and the relationship between who gets eaten and who does the eating (for example, prey and predator populations). Environmental factors (such as how much moisture is present or the number of hours of sunlight) also have a strong influence on which and how many organisms live in a community.

Communities are dynamic; they do not remain the same. The number and types of organisms that communities contain change over time. As the populations in a community interact with each other, they often change the abiotic environment such that it becomes more suitable for other species. These other species then gradually take over and form a new community. This process of successive change in species composition is called **ecological succession**.

Succession can be clearly witnessed after some type of disturbance (such as a fire) has removed some or all of the members of an existing community. After a fire like the one shown in Figure 13.4, an open area emerges. The first organisms to establish themselves in this open area are members of species that do well in open, disturbed habitats. Such species are referred to as **colonizers**, not just because they are the first to arrive after the disturbance, but also because their growth and reproduction processes change the habitat (just as the activities of the human colonists on Easter Island did). As colonizers fill up the area, they create shade, alter the soil, and in various other ways make the habitat less suitable for species like themselves and more suitable for other species. In some cases, this process may continue with new populations replacing old ones, until a climax community is formed. A **climax community** is a self-perpetuating community in which populations remain stable and exist in balance with each other and the environment, indicating that succession has come to an end.

Figure 13.4 Wildfires may be destructive in the short term, but because they occur frequently in the natural environment, some species have adapted to them. What effects might long-term suppression of fire have on ecosystems? What other types of disturbances could affect succession?

Figure 13.5 Mount St. Helens, located in the state of Washington, erupted violently on May 18, 1980. It sent ash into Canada and significantly disturbed an area of over 600 km^2 around the volcano. (A) shows the mountain four months after the eruption. Five years later, (B) shows various kinds of shrubs starting to grow alongside the smaller plants that had first colonized the bare areas.

The combination of species making up a climax community varies widely from one area to another, limited in part by climate and other abiotic features. The species living in a climax community tend to dominate the area — until the next disturbance occurs. In reality, most communities are routinely disturbed at various stages of succession and never achieve what might be considered a typical "climax" for the area. Because the size, frequency, and severity of disturbances vary, many communities actually consist of a mosaic of patches. These patches are at different stages of succession, and the particular species that replace others in a particular area may differ from one successional series to the next. Therefore, it is often difficult to predict the future composition of species in a particular community.

The process described above is known as **secondary succession**, and is the redevelopment of a previously existing community after a disturbance (see Figure 13.5). Since disturbances are common (over an area as large as Ontario, for example, many disturbance events occur in various locations each year), secondary succession happens repeatedly. Secondary succession can always be found occurring somewhere.

Although less common, succession can involve the development of a first-time community in an area. This process, called **primary succession**, occurs on newly formed volcanic islands, as shown in Figure 13.6, or in areas left bare by retreating glaciers. In these situations, soil has not yet formed and the only organisms present may be prokaryotes (bacteria and archebacteria). The first eukaryotic species (protists, fungi, plants, or animals) to arrive in these new areas are lichens, which are carried in by the wind and can grow on bare rock (see Figure 13.7). Their growth begins the breakdown of the rock, eventually leading to the production of soil.

Figure 13.6 When volcanoes erupt beneath the floor of an ocean, the result is often the formation of an island. Over time, the island may be colonized by organisms that arrive accidentally.

Figure 13.7 Several species of lichen have formed a community on this rock. What other types of organisms might be living on the rock as part of this community?

Organisms (such as lichens) that are the very first to arrive in a barren landscape, and whose activity changes the landscape dramatically, are referred to as **pioneer species**. Primary succession is generally a much slower process than secondary succession. Producing enough soil to support the first grasses, for example, can take up to 1000 years, depending upon abiotic conditions. Once soil is present, small plants are able to colonize the area. The process of succession then continues with other larger plants, animals, and other organisms arriving and replacing the pioneers. Later these are replaced in their turn, until either a climax community is reached or a disturbance occurs and the process starts all over again.

Ecosystems

A community of living organisms, together with the abiotic factors that surround and affect it, is called an **ecosystem**. An ecosystem includes all of the non-living parts of the environment and the living organisms in a particular area, as well as the interactions among them.

Although we often think of ecosystems as being quite large, they can be very small. In fact, the size of an ecosystem can depend on what you are studying. There are small ecosystems within large ones, which are within even larger ones. For example, the tree in Figure 13.8 could be considered an entire ecosystem. It contains a community of living things and abiotic components such as air, water, and nutrients. This ecosystem interacts with, and is part of, a larger ecosystem containing the community of organisms living in the small wooded area. Both of these ecosystems are part of a larger ecosystem, which extends beyond the boundaries of the photograph.

Figure 13.8 The ecosystem in this tree is part of the larger ecosystem that comprises ZooWoods, located outside the Ramsay Wright Zoological Laboratories in downtown Toronto. Many of the plants in ZooWoods were salvaged from ecosystems that were destroyed as a result of development in Toronto's suburban communities.

All of these ecosystems and their interactions make up the **biosphere** (as shown in Figure 13.1 on page 430). The biosphere can be thought of as the largest possible ecosystem. It includes all portions of Earth inhabitable by some type of life (that is, all the land surfaces and bodies of water on the planet) and extends several kilometres into the atmosphere and into the soil under Earth's crust. All living things that inhabit these environments are part of the biosphere.

A Species' Place in the Biosphere

The various types of organisms that live on Earth are not scattered randomly — each has its own "place" in the biosphere. The large-scale patterns we see in the distribution of species are ultimately the result of unequal heating of Earth's atmosphere, land masses, and oceans (see Figure 13.9).

This unequal heating gives rise to Earth's major climate zones, from the tropics near the equator through the temperate zones to the cold regions near the north and south poles. In addition, unequal heating sets up conditions that produce global air and water movements (trade winds and ocean currents) that interact with physical features (mountains, islands, and lakes) to produce the various patterns of rainfall we experience. The result is that some areas of the world are quite dry, while others are very wet.

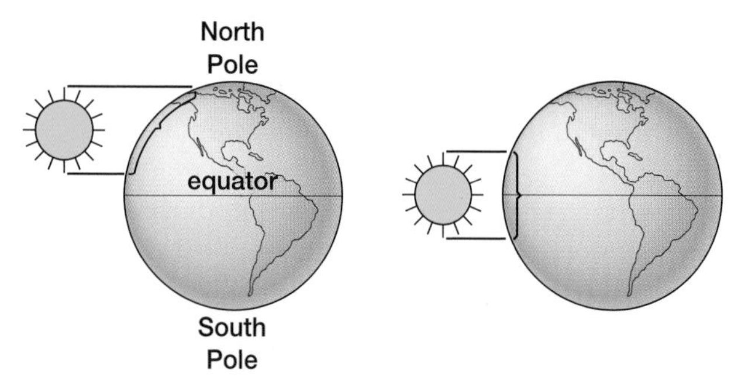

Figure 13.9 Earth is roughly spherical, so the solar energy that arrives near the poles is spread out over a greater area (and is therefore less intense at any given point) than the solar energy that arrives near the equator. The spreading out of solar energy results in the differential heating of the equator and the poles.

The pattern of rainfall influences the climate and the type of soils found in the different regions. These factors together determine the type and abundance of plants and other photosynthetic organisms that can survive. This, in turn, determines the size of the populations and the variety of species that will inhabit the area. It is possible to identify some very general types of large ecosystems, called **biomes**, or groups of ecosystems found in specific regions on Earth. These large regions (shown in Figure 13.10) are characterized by a particular combination of physical conditions and communities of organisms.

WEB LINK

www.mcgrawhill.ca/links/biology12

Biosphere 2 provides a variety of interesting experiences for students of all ages. University students can enroll in Earth Semester, a four-month course during which they participate in Biosphere 2 research projects, take field trips to other environments in the surrounding area, and learn about the interactions between Earth and its inhabitants. If you want to learn more about these courses or the research going on at Biosphere 2, go to the web site above, and click on **Web Links**.

Habitat and Range

The division of the world into large biomes is somewhat arbitrary. Each type of biome blends into the adjacent one gradually, making it difficult to identify exactly where the boundaries of the major biomes are located.

There is also a tremendous amount of variation within each biome. For example, the taiga biome (sometimes referred to as the northern coniferous forest or the boreal forest) covers a major part of central and northern Canada, Europe, and Asia. Taiga also varies widely from north to south and from east to west with respect to the density and type of trees it contains. Thus, each biome may contain a variety of different habitats, each with its own set of organisms and abiotic conditions. A **habitat** is a place or area with a particular set of characteristics, both biotic and abiotic. Each type of organism is found in the specific type of habitat in which its physical, physiological, and behavioural adaptations equip it to survive and reproduce optimally.

In some cases, all members of a certain species live in the same general type of habitat. The habitat may be spread over a single large area or be found in a number of separate locations. In other cases, species are divided into populations that prefer different types of habitats. The caribou, for example, is a species found in many parts of Canada. It is divided into a number of populations that live in different habitats, as shown in Figure 13.11. Some caribou live in forests, while others prefer the open tundra.

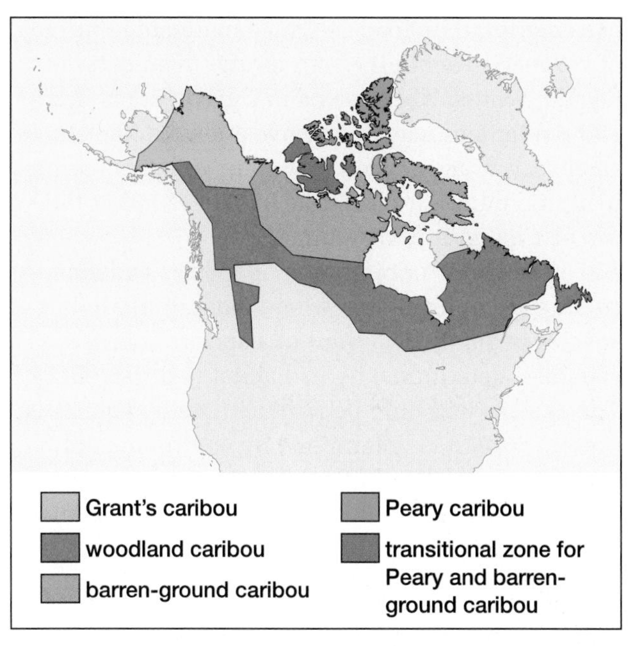

Figure 13.11 The distribution of caribou (*Rangifer tarandus*) in Canada. Why do you think caribou are found in these areas?

These different populations are sometimes referred to as subspecies or ecotypes, and they vary somewhat in behaviour and physical features. The barren-ground caribou is smaller and lighter in colour than many other types of caribou. Most

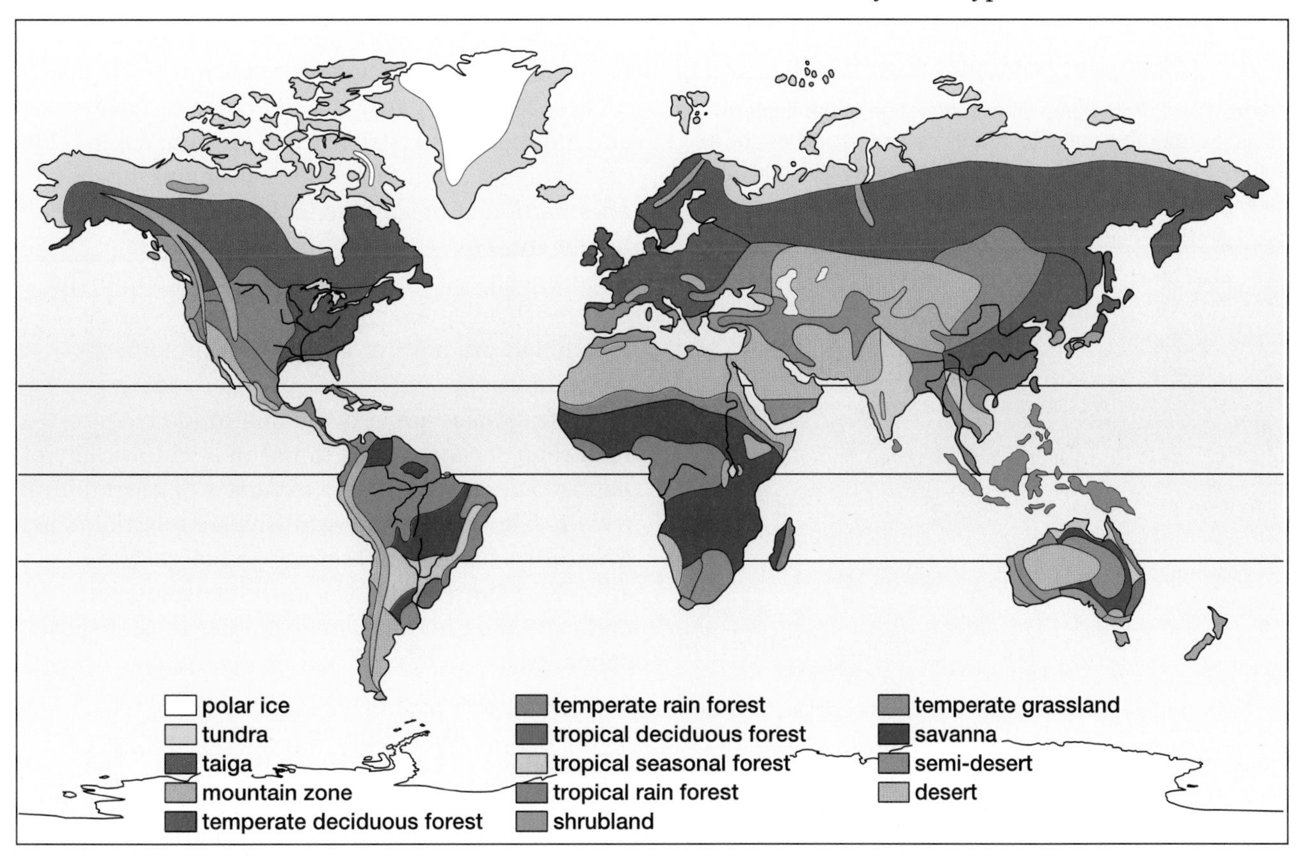

Figure 13.10 Distribution of Earth's major biomes. What do you think are the two most important factors determining the distribution of these vegetation zones?

barren-ground caribou in Canada migrate hundreds of kilometres annually from tundra in summer to taiga in winter. The woodland caribou are larger and darker and may only move a few kilometres each year, from forested mountain valleys in winter to alpine tundra (tundra-like habitat found at the top of a mountain) in summer.

The **range** of a population or species is defined as the geographical area where that population or species is found. The limit of a species' range is generally determined by its habitat requirements. The species will only be found where its habitat is present, which is determined by environmental variables, including both abiotic factors (such as temperature and rainfall) and biotic factors (such as type of food).

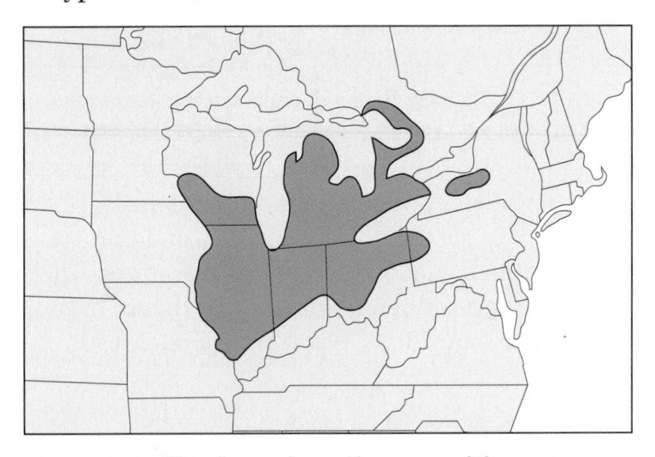

Figure 13.12 This figure shows the range of the eastern massasauga rattlesnake (*Sistrurus catenatus*). What are some of the reasons why this rattlesnake is not found outside this range?

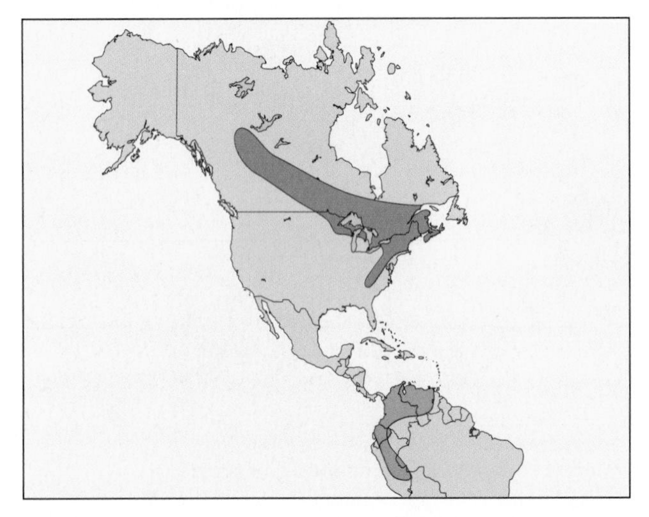

Figure 13.13 The Canada warbler (*Wilsonia canadensis*) has what are referred to as disjunct summer and winter ranges. Why do you think this species is called the Canada warbler? Why does this bird, like many others, fly such a long way to spend the winter elsewhere? Why does it fly back to Canada each summer?

Some species, such as the caribou, have large ranges. The species we call caribou also live in northern Europe and Asia, where they are known as reindeer. Other species, such as the eastern massasauga rattlesnake, have relatively small ranges (see Figure 13.12). Still others, including many types of birds, live in different locations in summer and winter and migrate long distances between these separate ranges each year (see Figure 13.13).

Ecological Niche

Members of different species can share the same range and even the same habitat, or at least show considerable overlap in the type of habitats they prefer. This is possible because they have different ecological niches. The **ecological niche** of a population is the role its members play in a community. This includes the resources (such as food and living space) they need and how they interact with the other members of the community (what they compete with, and perhaps what they get eaten by).

Many ecologists describe a population's habitat as its street address, and its ecological niche as its job in the community. It is common for two species to share the same habitat and be able to survive and reproduce because they have different ecological niches. Some species, such as the giant panda and the koala, have very narrow ecological niches. They eat a very limited range of food types, and are therefore restricted to a specific habitat. For other species, such factors as a strong preference for a particular breeding site, a low tolerance for temperature extremes, or specific moisture requirements may restrict their habitat and niche. Species like this are called **specialists** because they have very narrow, specialized preferences or tolerances.

Other species are very flexible in their requirements and can be found in a wide variety of habitats, often across large expanses of geographical terrain. These types of organisms are referred to as **generalists** because of their more general (less specific) requirements. Cockroaches, mice, rats, and especially humans are examples of ecological generalists.

In Investigation 13-A, you will perform a simulation investigation to determine the effectiveness of being a generalist or specialist when foraging for food.

In the next section, you will explore how communities are structured and what specific roles individuals have within a community.

Investigation 13 • A

SKILL FOCUS

Conducting research

Performing and recording

Analyzing and interpreting

Foraging Strategies — A Simulation Experiment

In this investigation, you will compare the foraging (feeding) strategies of specialists and generalists by doing a simulation experiment. You assume the role of a forager — perhaps a predator — and various types of beans will be your food or prey. A pan of sand will simulate the habitat in which you live and look for food.

Pre-lab Questions

- Identify at least one species that is considered a specialist.
- Identify at least one species that is considered a generalist.
- What are some of the advantages and disadvantages of being a specialist or a generalist?
- What criteria would you use to evaluate the success of the two methods of foraging for food?

Problem

Identify differences between the foraging strategies used by specialists and generalists.

Predictions

Predict differences in the foraging behaviour of generalists and specialists. Also predict which type will be the most successful.

CAUTION: **Wash your hands when you have completed the investigation.**

Materials

plastic dishpan (or other similarly sized container)
sand
dried beans (40 kidney beans, 30 pinto beans, 20 lima beans)
stopwatch
graph paper
sieve

Steps 1 and 2

Procedure

Part A: One Species of Prey

1. Fill the dishpan one-half to three-quarters full of sand.
2. Bury 40 kidney beans, 30 pinto beans, and 20 lima beans in the sand. None should be visible on the surface.
3. Design a data table that will allow you to record an identifying number for each forager, the number of prey items (beans) remaining buried before each forager begins searching, and the time taken by each forager to find enough beans to ensure survival.
4. Appoint one member of your group as the "referee" prior to the first turn. This person will time the first foraging trial and ensure that each forager abides by the foraging rules (step 5). The position of referee should be rotated so all members of the group have an opportunity to forage.
5. Foraging must be done with one hand only, and care must be taken not to remove any of the sand from the pan along with the beans. No one but the referee can watch the current forager (the other group members must turn their backs).
6. When the stopwatch starts, the first forager begins searching for beans. He or she has 10 seconds to find three beans — the number required for survival. Immediately after the beans are located, foraging stops. If the forager cannot find three beans, he or she dies of starvation and is eliminated. As each prey item is located, it must be placed in a container beside the pan immediately after capture. (This simulates the time it takes for a predator to handle — kill and/or eat — its prey).
7. The referee records in the data table the time taken to find the prey, smoothes over the sand after each turn, and ensures all remaining beans in the pan are buried before the next forager takes a turn.
8. Repeat steps 4 to 7 until all foragers in your group have died.
9. Produce a graph showing the relationship between the time taken to find three beans (dependent variable) and the number of beans remaining in the pan (independent variable).

Continued on page 438

10. Run the sand through a sieve to remove all remaining beans.

Part B: Multiple Prey Species

1. Bury 40 kidney beans, 30 pinto beans, and 20 lima beans in the sand.
2. Design a table that will allow you to record an identification number for each forager, the number of each type of bean remaining in the pan before each forager starts, the number of each type of bean captured by each forager, and the time taken by each forager to locate enough prey items to survive.
3. Repeat Steps 4 to 8 from Part A. This time, however, survival is ensured by finding any of the following five combinations of beans:
   4 pinto beans　　3 kidney beans
   2 lima beans　　3 pinto beans and 1 kidney bean
   3 pinto beans and 1 lima bean
   As before, the forager must find one of these combinations within 10 seconds or die of starvation. As soon as survival is ensured, foraging must stop. The time taken to find sufficient prey is recorded in the table.
4. Construct a graph showing the relationship between time taken to find sufficient prey to ensure survival and the number of prey items remaining in the pan.

Post-lab Questions

1. What type of foraging strategy — that of a specialist or a generalist — was represented by each of the situations in Parts A and B of this investigation?
2. Why were other group members required to turn their backs while the forager looked for beans? What does this prevent? Does this aspect of the simulation represent what occurs in nature or not? Explain.
3. Were there any beans left when all the foragers had died in Part A or Part B? Do you think there might be prey items left in a natural habitat? Why?
4. What happens as the food is depleted (that is, the number of beans decreases) in the habitat? Why?
5. In general, which of the two types of foraging strategies worked more quickly to allow individuals to find enough food to survive? Why do you think this was the case?
6. In Part B, did you use any particular strategy in deciding what combination of prey to search for? If so, what was it and why did you use this strategy?
7. Do you think this simulation truly represents natural foraging activity? Why?

Conclude and Apply

8. In nature, what features of both the prey and the habitat would contribute to the success of an individual forager? Were any of these features simulated in this experiment? Explain.
9. In nature, how might a forager increase its efficiency at finding prey? How might the simulation be changed to represent this? Repeat Part B using your changed simulation to test your new prediction.
10. How many foragers (represented by the number of turns taken before all foragers died) were able to survive in each of the two habitats (in Parts A and B)? Explain your results.
11. Is it possible for the same species to act as both a generalist and a specialist in its foraging behaviour? Explain your answer.

Exploring Further

12. Given sufficient quantities of all types of food, what factors in real life might influence the choice of food taken by a generalist?
13. **(a)** In nature, what do animals do when their food supply is depleted?
    (b) How might an organism try to ensure that there continues to be sufficient food in a habitat to support it and/or its young?
14. How might you change this simulation to incorporate reproduction of prey or predators?
15. Repeat the procedures after changing the substrate to simulate the effects of human interaction (for example, by mixing stones, pieces of plastic, or other substances into the sand). Compare your new results with the previous trials for undisturbed substrate. What did you find? Explain your findings in a one-page essay.

COMPUTER LINK

If spreadsheet software is available, input the data into the spreadsheet and use the computer to generate graphs that illustrate the data.

SECTION REVIEW

1. K/U In ecological terms, describe the difference between a population and a community.

2. K/U Many wolf families are found within the boundaries of Ontario's Algonquin Park. Should all these animals be considered members of a single population or more than one population? Explain your answer.

3. K/U Explain why it almost always takes longer to achieve climax stage through primary succession rather than through secondary succession in the same region.

4. K/U On May 18, 1980, the eruption of the Mount St. Helens volcano (in Washington State) caused widespread environmental destruction. Over the years, many plant and animal species have returned to the region. What type of succession is demonstrated by this chain of events? Why would you not expect to find exactly the same variety of plant and animal species inhabiting the region now as compared with the period before the eruption?

5. K/U Occasionally a volcano on the sea floor erupts and forms a new island above the water's surface. Over time, the bare lava often becomes colonized by various plant and animal species. Identify the kind of succession described in this example. Speculate about the origin of the flora and fauna that eventually make their homes on these new islands.

6. K/U Describe how simple organisms (such as lichens) can survive on bare rock substrate. How do they alter their environment to make it more suitable for higher organisms?

7. K/U **(a)** Explain how a fallen tree in a forest can be regarded as an entire ecosystem.
   (b) Is this or any other ecosystem completely separated from other ecosystems?

8. K/U Does Earth's biosphere ever change, or has it remained constant over time? Explain your answer.

9. K/U Would you expect that physical conditions on the other planets in our solar system vary as they do on Earth? Explain your answer.

10. I Identify the distinguishing features of two different climatic zones that are found in the Great Lakes region of Canada.

11. K/U Identify the features that ecologists use to distinguish populations of barren-ground caribou from populations of woodland caribou.

12. K/U In ecological terms, explain how a particular species of fish found in the Great Lakes (such as the northern pike) is not only part of a population but also part of a community, an ecosystem, and a biome.

13. I The whooping crane (*Grus americana*), an endangered species in North America, spends the summer months in northern Canada. Use references to locate and identify the biome or biomes that lie within the summer or breeding range of this species. Describe the general biotic and abiotic features of the biome(s) you have identified.

14. I The ranges of two distinct species of birds — the Baltimore oriole (*Icterus galbula*) and the grasshopper sparrow (*Ammodramus savannarum*) — seem to overlap in many parts of North America. Research and describe the differences between the ecological niches of the two species that make it possible for them to co-exist within the same ecosystem.

15. K/U Explain why ecologists might consider members of the species *Homo sapiens* to be generalists rather than specialists.

16. K/U (A) The tree swallow (*Iridioprocne bicolor*) and (B) the little brown bat (*Myotis lucifigus*) breed in similar habitats in Ontario, and both feed on insects they catch while in flight. What differentiates their ecological niches?

17. I The great-horned owl (*Bubo virginianus*) is a ubiquitous predator species that generally inhabits rural areas. Some populations survive on a rather limited diet, while others feed on an extremely varied number of prey species. Research and compare the feeding habits of different populations of great-horned owls (such as a population found in western Canada versus one in the southern United States).

UNIT ISSUE PREP

You have just learned that an organism's environment includes biotic and abiotic components. When researching your environmental issue, consider these components and the effect they have on human populations.

The Structure of Ecosystems

EXPECTATIONS

- **Describe what is meant by the trophic structure of a community.**
- **Describe the ecosystem roles of producers, consumers, and decomposers.**

As mentioned in section 13.1, ecosystems contain both biotic and abiotic components. Each ecosystem consists of all the organisms in one or more communities, as well as the physical and chemical factors affecting them. The boundaries of ecosystems are not distinct — one ecosystem may overlap or exist inside another. There are terrestrial (land-based) and aquatic (water-based) ecosystems, and ecosystems that contain both land and water. An ecosystem can be small (such as the one shown in Figure 13.14), or as large as the biosphere — a global ecosystem.

Figure 13.14 Small insects fall into the pitcher plant's cup-shaped leaves, drown, and decompose. This plant, which typically lives in nitrogen-poor environments, extracts the nutrients it requires from the bodies of insects. The pitcher plant and its surroundings comprise a small ecosystem.

Despite the wide range in sizes and types of ecosystems, the same basic processes take place in all of them. Two processes are particularly important: energy flow and chemical cycling. The proper functioning of these processes is vital to the survival of organisms in the ecosystem and to the integrity of the ecosystem itself.

Trophic Structure

When ecologists refer to the **trophic structure** of an ecosystem or community, they are describing the feeding relationships among its members. Each species is assigned to a specific **trophic level** in the structure, depending on its main source of nutrition. Most ecosystems have several trophic levels through which energy flows and chemicals (matter) cycle.

The first (or lowest) trophic level consists of autotrophic organisms. Autotrophs are organisms that can make energy-rich organic molecules (such as glucose) from the raw materials available in the environment. They then break these "homemade" organic molecules down during cellular respiration to provide the energy that fuels the rest of their life processes. Photosynthetic autotrophs use the energy of the Sun to drive this manufacturing process. Almost all plants, as well as some types of protists (algae) and bacteria (cyanobacteria), are photosynthetic autotrophs.

All organisms need energy to drive cellular processes. They must, therefore, have a source of organic molecules from which they can release this energy during cellular respiration. Autotrophs, which are at the first trophic level, produce organic molecules; this makes the first level the most important. Because the first trophic level supports all life at the higher levels, autotrophs are referred to as the **primary producers** in an ecosystem. This first trophic level provides all the potential energy required to drive the other levels in the ecosystem.

All organisms in the trophic levels above this one are heterotrophs. Heterotrophs are unable to make the energy-rich molecules they need to fuel their life processes. Instead, they must obtain these molecules by consuming other organisms, either autotrophs or other heterotrophs. Therefore, they are referred to as consumers.

Herbivores that eat autotrophs are termed **primary consumers**, since they are the first eaters in the chain. On land, insects, snails, grazing mammals, and birds and mammals that eat seeds and fruit are the most common herbivores. In aquatic ecosystems, this role is often taken by heterotrophic protists, various types of small invertebrate animals, and some species of fish.

Carnivores that eat mainly herbivores are **secondary consumers**. Spiders, frogs, and insect-eating birds are examples of secondary consumers. In most ecosystems, these secondary consumers are themselves eaten by other carnivores, which are known as **tertiary consumers** (the third set of eaters). There may also be higher levels of consumers above these.

The members of another consumer group, often referred to as **decomposers**, obtain their energy-rich molecules by eating leftover or waste material derived from all the trophic levels, including the feces of living organisms, dead bodies, or body parts (for example, fallen leaves). Decomposers are very important to every ecosystem. Their role is to break large molecules (that were once part of living organisms) down into smaller ones and return them to the abiotic environment. Thus, decomposers return vitally important nutrient elements such as carbon and nitrogen to the soil and air. These materials can then be used again by autotrophs to make new energy-rich organic molecules. Decomposers are an ecosystem's recyclers, ensuring that the biosphere's limited supply of required nutrients is not lost.

THINKING LAB

Shrinking Polar Bears and Expanding Snow Geese

Background

An incredible variety of living things inhabits the biosphere. This biotic diversity reflects the abiotic diversity of Earth. Through the process of evolution, populations have evolved adaptations that enhance their survival and reproductive ability in diverse habitats. The result is the diversity of life.

During the last 100 years, scientists have reported significant changes in the world's ecosystems. While some of these changes seem to have been beneficial for some species, they seem to have had negative effects on others. For example, dramatic changes have been observed in populations of polar bears (*Ursus maritimus*) and lesser snow geese (*Chen caerulescens*) that breed in Canada's arctic. Field surveys of polar bears have revealed that populations are declining, fewer cubs are born each year, and individual bears are smaller and weigh less than what has been considered typical for members of this species. In contrast, snow goose populations have tripled since 1968. Their numbers are so high that the feeding activity of these geese is causing substantial damage to the habitats where they and other species breed and overwinter.

Why do you think these changes are happening? In completing this assignment, you will draw on your knowledge of the nature of ecology and evolution. You might choose to work on one of these two species, and then compare your findings with those of students who worked on the other species.

You Try It

1. Describe the ecological niches of the polar bear and the snow goose. Include diet, preferred habitat, time of peak activity during a day, yearly activities such as migration, and other relevant information.
2. Would you consider the polar bear to be a generalist or a specialist? What about the snow goose? Explain your answer.
3. To which trophic levels do the polar bear and snow goose belong?
4. Describe the feeding, sensory, and locomotory adaptations that improve the ability of each species to survive and reproduce in its habitat and niche. What features allow each species to cope with environmental stresses (for example, temperature or moisture extremes) in its habitat?
5. Draw a map illustrating the approximate range of each species.
6. Find out as much as you can about the evolutionary history of these species. What other species are they most closely related to? When might they have appeared as a species?
7. How might changes that currently appear to be occurring in Earth's climate affect the future evolution of each of these species?
8. Estimate what chance each of these species has for surviving for the next 100 years. Does either face possible extinction? Why or why not?

Some types of bacteria use energy derived from breaking the chemical bonds in hydrogen sulfide molecules to form the organic molecules they use as food. This type of autotrophic food production is called chemosynthesis. It is quite rare and occurs in some very unusual environments, including ocean depths of 2500 m. At these depths there is no light and very little oxygen. In addition, hot magma from Earth's molten core escapes to superheat the surrounding water.

Some unusual marine organisms live near undersea vents off the coast of British Columbia. These organisms include tube worms, limpets, and palm worms.

ELECTRONIC LEARNING PARTNER

Your Electronic Learning Partner has animation clips that will enhance your understanding of deep-sea vent communities.

Food Chains and Webs

The trophic structure determines the route taken by the energy and matter (chemical elements) contained in food as it moves through an ecosystem. Food is transferred from primary producers to primary consumers and then to secondary consumers along a pathway referred to as a **food chain** (see Figure 13.15).

In reality, few ecosystems are so simple that they consist of only a single, unbranched food chain. More commonly, many species may feed on a single species below themselves, while at the same time a single upper-level consumer species may eat many different species below itself. In addition, organisms may eat individuals from two or more levels. The hawk shown in Figure 13.15 may eat mice, grasshoppers, and snakes, while the mouse may eat plant parts (fruits and seeds) and grasshoppers. In fact, many organisms, including humans, are referred to as omnivores because they eat plants, animals, and other types of organisms. In other words, they are both primary and higher-level consumers. The result is that the feeding relationships in most ecosystems form complex **food webs**, rather than simple food chains. Figure 13.16 illustrates a food web.

In the Thinking Lab on page 441, you explored how communities are structured and the specific roles individuals can take within a community. The next section will show you how energy flows through ecosystems.

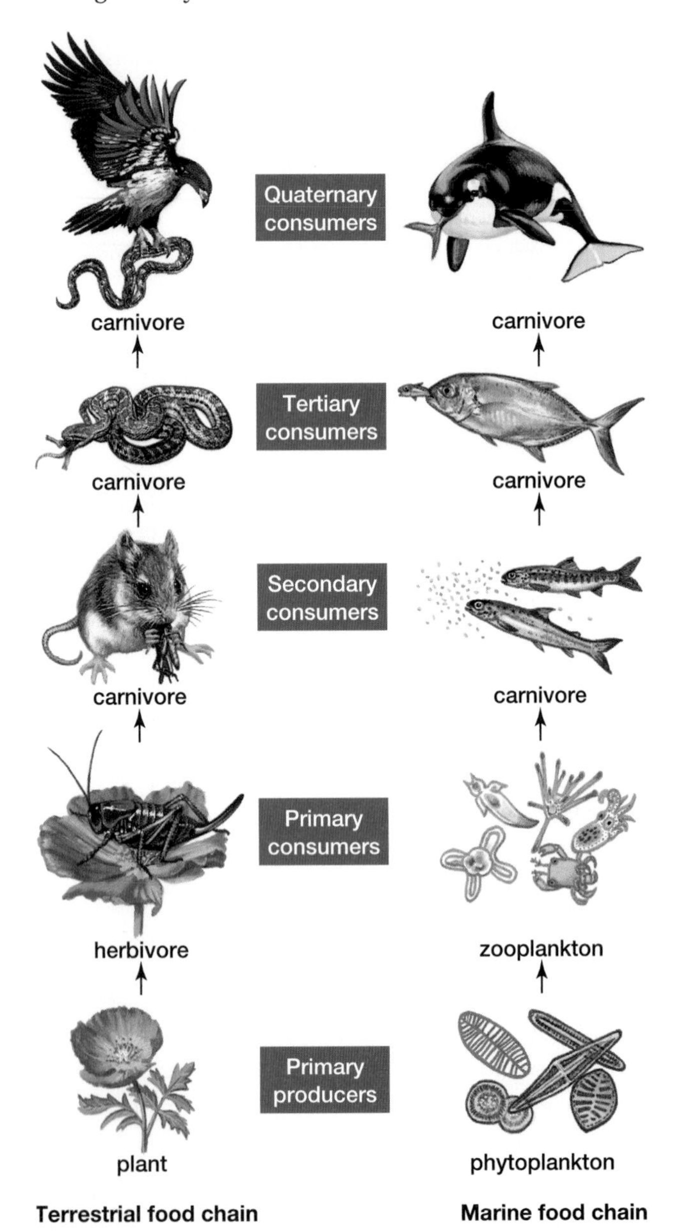

Figure 13.15 Terrestrial and aquatic communities contain different species, but can have the same overall trophic structure. "Plankton" is a general term referring to small aquatic animals and protists. What distinguishes the two types of plankton?

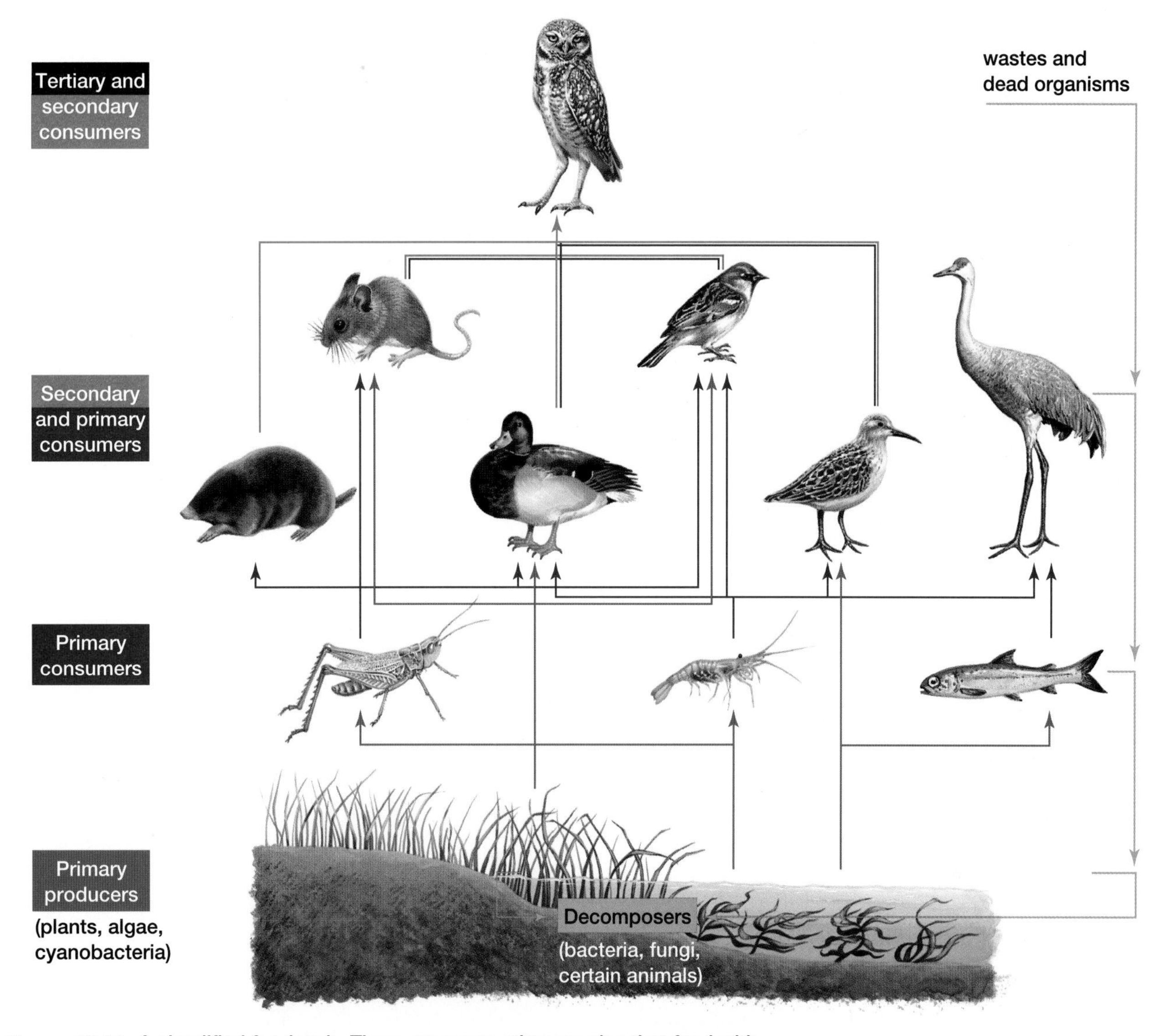

Figure 13.16 A simplified food web. There are many other species that feed with or on the ones shown in the diagram. Which species eat organisms on more than one trophic level?

SECTION REVIEW

1. K/U Explain why some ecosystems can support highly complex food webs while others can support only relatively simple ones.

2. C Draw a chart to illustrate a food web that could be found in a typical pond ecosystem (show at least three distinct trophic levels). Identify the organisms and label each trophic level.

3. K/U Explain how the same species can occupy more than one trophic level within the same food web. Explain how this type of ecological interaction can enhance the stability of a food web.

4. K/U Describe the types of biotic and abiotic factors that can lead to the collapse of a food web in an ecosystem.

5. K/U Explain why autotrophs rather than decomposers occupy the lowest level of a food chain.

6. K/U Could photosynthetic producers exist in the absence of consumers in an ecosystem? Explain your answer.

7. K/U Describe the typical energy sources that decomposers rely on in an aquatic ecosystem (such as a pond or lake).

8. K/U Describe the difference between a food chain and a food web. Which is more realistic in its depiction of what actually exists in nature?

9. K/U Explain why producer organisms that live deep below the surface of Earth's oceans rely on chemosynthesis rather than photosynthesis to manufacture high-energy food molecules.

Energy Flow in Ecosystems

EXPECTATIONS

- **Explain how autotrophs sustain ecosystems by supporting higher trophic levels.**
- **Describe the concept of primary productivity and explain why it varies among ecosystems.**
- **Use the energy pyramid to explain production, distribution, and use of food resources.**
- **Using the ecological hierarchy of living things, evaluate how a change in one population can affect the entire hierarchy.**
- **Relate the pyramid of primary productivity to the pyramids of biomass and numbers.**

All organisms require energy for growth, body maintenance (such as repairing damage to body parts), and reproduction, and many species require energy for locomotion. Energy to support these activities is released from large, energy-rich organic molecules during the process of cellular respiration (or, in a few species, fermentation) and stored in the form of ATP.

Most primary producers use energy from the Sun to drive the production of energy-rich molecules in their own bodies. Consumers subsequently obtain these molecules by eating either the bodies of primary producers or the bodies of other consumers. Therefore, the amount of energy available to an ecosystem is determined by the amount of energy captured by the autotrophs.

Every day the Sun bombards Earth with about 10^{22} joules of solar radiation. The majority of this radiation is absorbed, scattered, or reflected by the atmosphere. Much of the radiation that does reach the biosphere hits bare ground or water, with only a small fraction landing on photosynthetic organisms. Of this, only a portion is of a wavelength suitable for photosynthesis. The result is that only one to two percent of the total radiation emitted by the Sun is converted into chemical energy. The amount varies somewhat from place to place, depending on the type of organisms found in each region, the intensity of the light (recall Figure 13.9 on page 434), and many other factors (see Figure 13.17).

Even though the fraction of solar radiation actually captured by primary producers is very small, these organisms produce between 150 and 200 billion tonnes of organic material each year. This amount supports the majority of life on Earth.

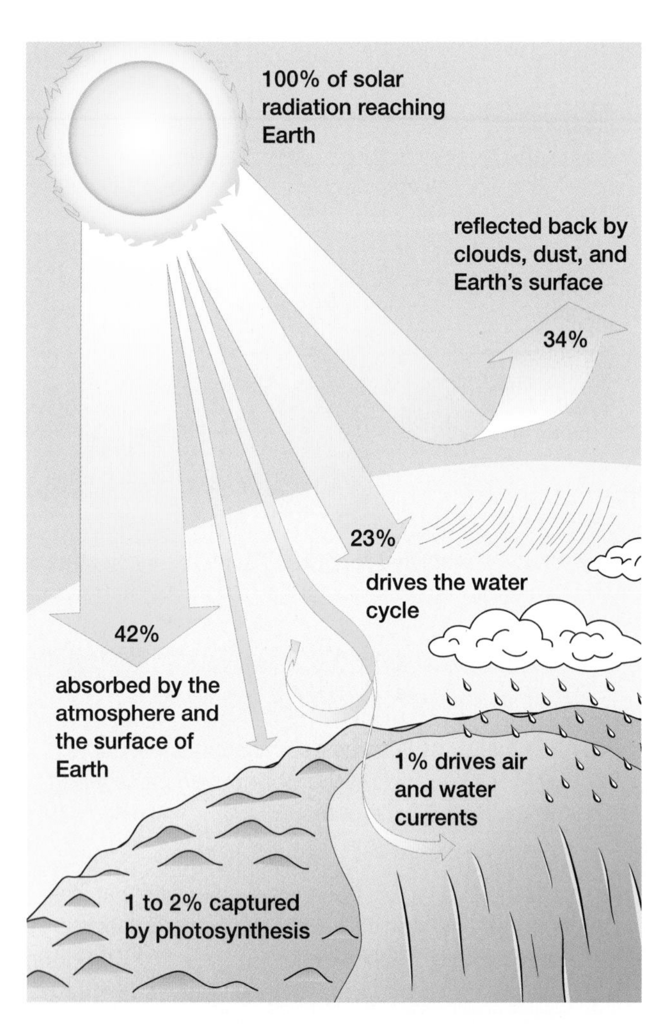

Figure 13.17 Very little of the total amount of radiation leaving the Sun is converted to chemical energy during photosynthesis.

Primary Productivity

Primary productivity is the amount of light energy that autotrophs in an ecosystem convert to chemical energy (and store in organic compounds) during a specific period of time. It is commonly measured in terms of energy per area, per year ($J/m^2/a$). It can also be expressed as **biomass** (mass) of vegetation added to an ecosystem per area, per year ($g/m^2/a$). It is important to remember that primary productivity is the *rate* at which organisms produce *new* biomass, which is not the same as the total mass of all photosynthetic autotrophs present in an area at one time. For example, a forest has a very large biomass — the mass of its vegetation is greater than that of a grassland of equal size. But primary productivity of the grassland may actually be higher because animals are constantly eating the plants and new ones are being produced. Thus, new mass is being accumulated in the grassland at a higher rate than in the forest.

The amount of primary productivity (as shown in Figure 13.18) can vary significantly, both among ecosystems and within an ecosystem over time. The rate of productivity depends on many factors, including the number of autotrophs present in the ecosystem, the amount of light and heat present (the process slows during winter), and the amount of rainfall the system receives (since water is a raw material of photosynthesis). In addition, photosynthetic organisms need certain nutrients, such as nitrogen, potassium, and phosphorus, to grow. Even though many of these nutrients are not directly involved in photosynthesis, they contribute to limiting the rate of primary productivity by affecting plant growth.

Energy Transfer at Higher Trophic Levels

Not all solar energy captured by primary producers is passed on to higher trophic levels. A substantial portion of the energy captured by producers is used in their own cellular respiration reactions. Another large portion is simply never eaten by consumers — which is why many ecosystems look green.

Only some of the total biomass eaten by consumers is converted into the body tissues of the organism that ate it. Figure 13.19 on the following page shows what happens to the energy a herbivore (caterpillar) obtains from the plant material it eats. Approximately half the plant tissue is indigestible, and the energy from this portion is expelled with the caterpillar's feces. Although some of this energy will be consumed by decomposers and will continue to be part of the ecosystem, it will not

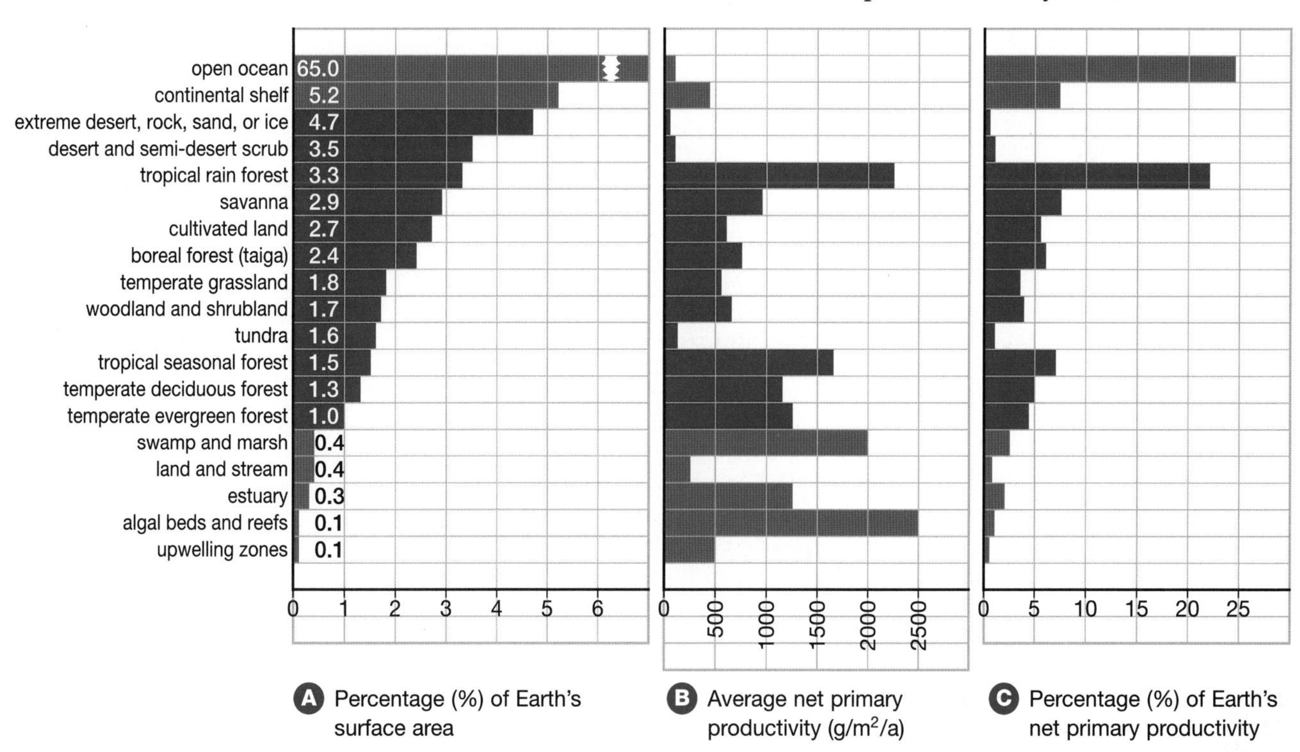

Figure 13.18 Why do the oceans contribute such a high proportion of Earth's total primary productivity when their average productivity is low compared with that of algal beds and reefs? Net primary productivity is the total amount of solar energy transformed to chemical energy by autotrophs (called gross primary productivity), minus the amount used by the autotrophs during cellular respiration.

be available to the secondary consumers that eat caterpillars. Approximately one third of the energy the caterpillar obtains is used in its own cellular respiration (providing energy for locomotion, maintaining body temperature, and other body processes), and therefore lost to the ecosystem. In fact, only about one sixth of the energy is incorporated into new caterpillar tissue — tissue that can be eaten by secondary consumers.

This energy loss (and its related unusable heat) occurs between all trophic levels in a food web (see Figure 13.20). Although the efficiency with which energy is transferred from one level to the next varies among different types of organisms, it usually ranges between 5 and 20 percent. In other words, roughly 80 to 95 percent of the potential energy available at one trophic level is not transferred to the next one. This pattern of energy loss is often illustrated as a **pyramid of productivity** (see Figure 13.21 on page 448).

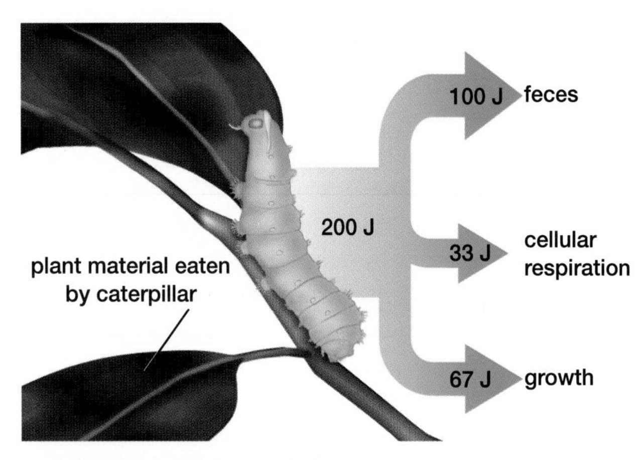

Figure 13.19 Why is so much energy lost as waste from a caterpillar?

The next MiniLab will help you understand pyramids of productivity by examining three typical food chains.

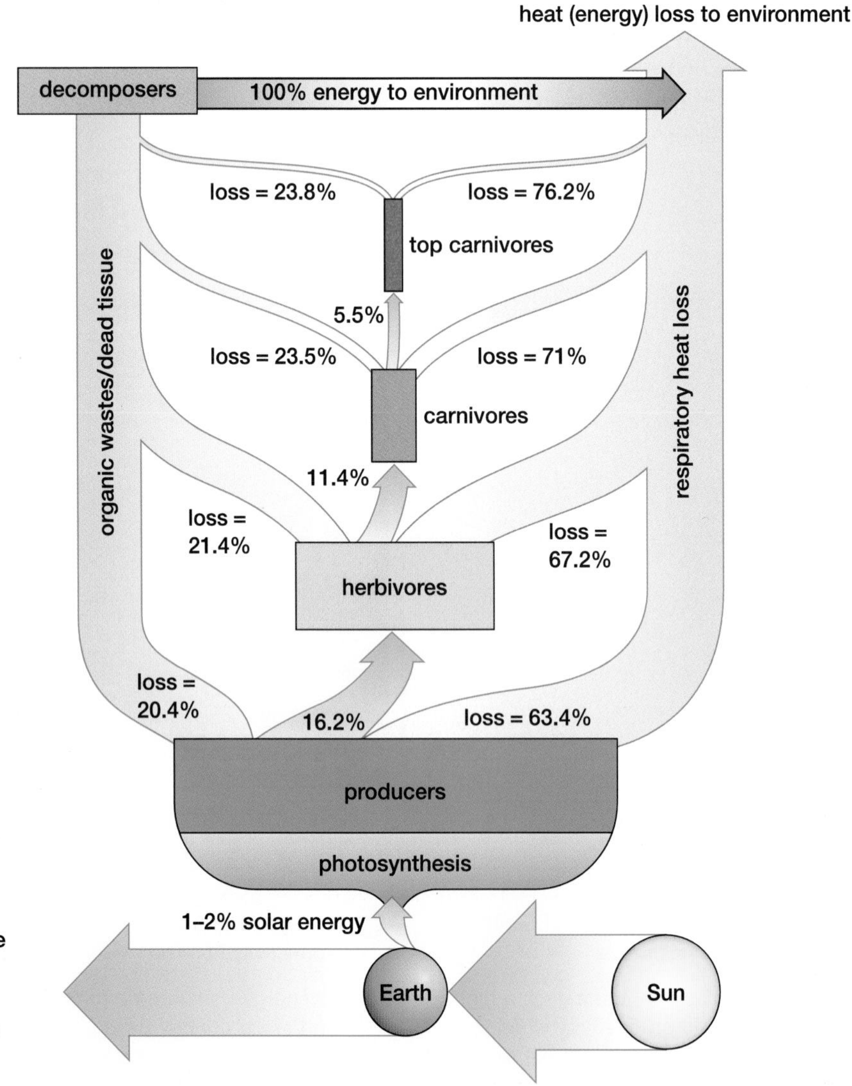

Figure 13.20 Why is the respiratory heat loss of consumers higher than that of producers? Note that it is highest for top-level carnivores.

BIO FACT

Meat is more easily digested than most plant materials, so carnivores are slightly more efficient at converting food into biomass. However, carnivores typically use up more of this biomass during their own cellular respiration because their energy needs are higher than the energy needs of herbivores. Carnivores tend to move around more to find food, and many of them are endothermic.

Since progressively less energy is transferred from lower to higher levels in a food web, less biomass can be produced at the higher trophic levels. This concept can be represented in a **biomass pyramid**, in which each tier represents the biomass of that trophic level (see Figure 13.22A on the following page). Typically, the shape of a biomass pyramid is similar to that of a pyramid of productivity. However, in some aquatic ecosystems, a relatively low biomass of primary producers (called phytoplankton) supports a higher biomass of primary consumers (zooplankton), as shown in Figure 13.22B. This

MINI LAB

Where Do You Fit in the Food Chain?

How might your knowledge of pyramids of productivity influence your decision about the type of foods you include in your diet? Recall that only a very small fraction of the energy released by the Sun is assimilated into plant material (see Figure 13.17 on page 444). For ease of calculation, assume that the amount of energy captured by plants and contained in their tissues is two percent of the total energy available from sunlight. Additionally (although it is a simplification), assume that 10 percent of the energy at one trophic level is transferred to the next level. Study the diagram and determine the percentage of the Sun's energy available to humans (as shown at the top of each of the three food chains).

Analyze

1. About 80 percent of the world's population eat mostly grain-based foods. Why do you think this is the case?
2. How might diet influence the number of humans Earth can ultimately support?
3. A square metre of land planted with rice produces about 5200 kJ of energy per year. A chicken farm produces about 800 kJ/m^2 of potential food energy per year. Assume that a human must consume 2400 kJ per day to survive. Although it is an oversimplification to imply that a person could survive by eating only one type of food, calculate the total area of land needed to support the student population of your school for one year on a diet of:

   (a) rice **(b)** chicken
4. Research the differences between the food used to feed chickens and other poultry in small family-run farms and the type of feed used in large, commercial agribusiness operations. Which do you think is more environmentally friendly?

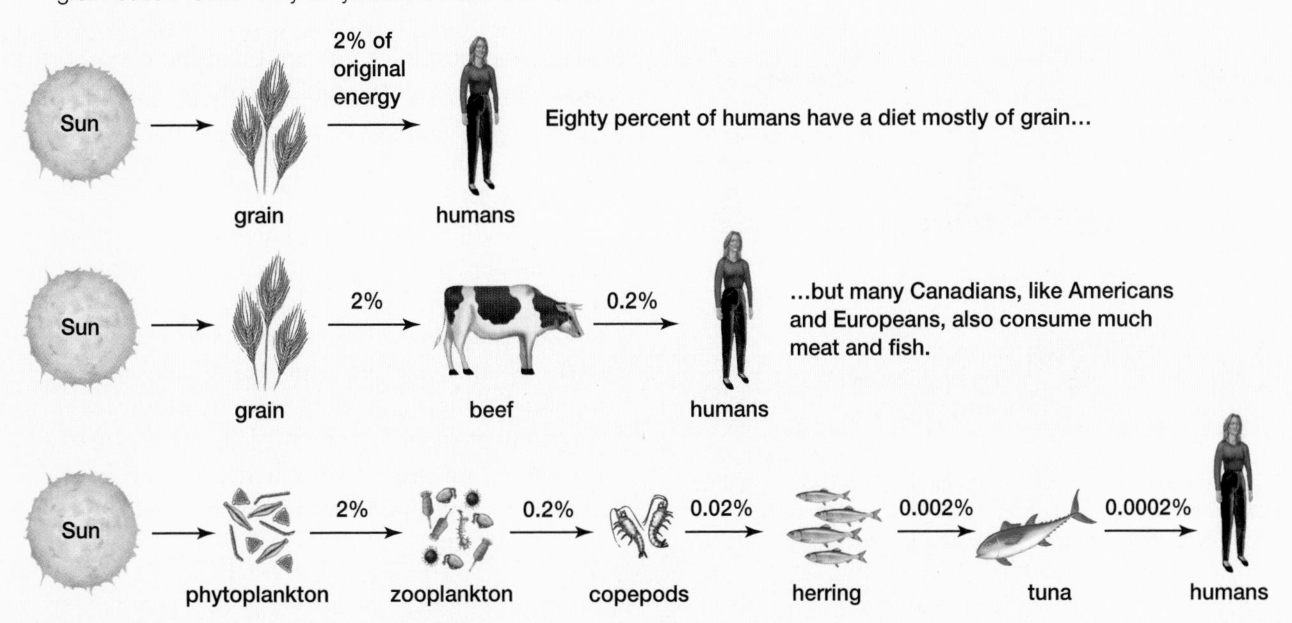

Three typical food chains for humans with different diets.

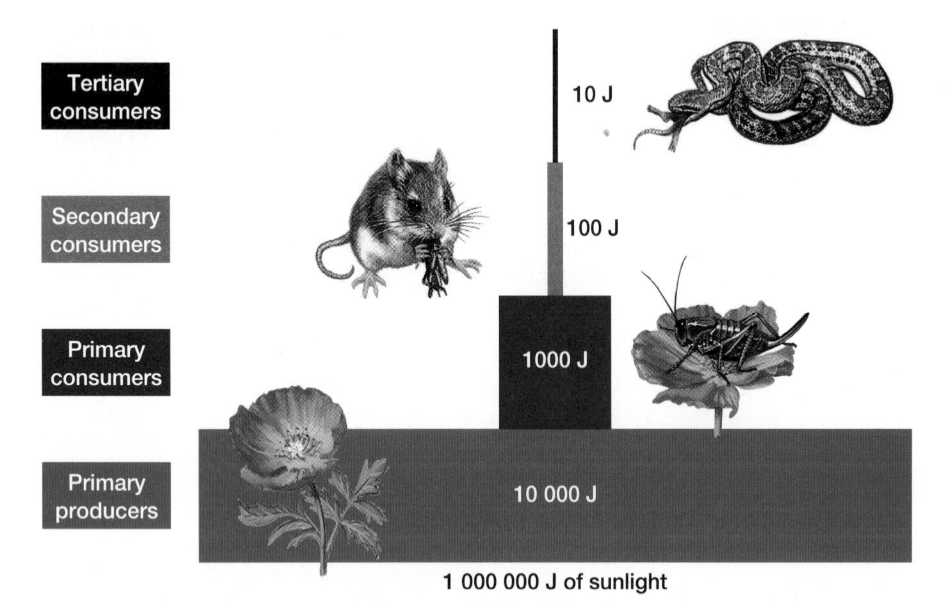

Figure 13.21 This figure is drawn to show a 10 percent efficiency of energy transfer from one trophic level to the next. Although the rate of efficiency varies (from 5 to 20 percent), 10 percent is a commonly used average figure. This gives rise to what is sometimes called "the rule of 10" when describing the shape of this pyramid.

occurs because the phytoplankton are eaten so quickly that there is no time for a large population to develop. The population of zooplankton can only exist because the phytoplankton have an extremely high reproductive rate — new organisms appear as fast as others are eaten. This means that the productivity of phytoplankton is very high, and the pyramid of productivity for this ecosystem is therefore wide at the top and narrow at the bottom.

In every ecosystem, the biomass of carnivores at the highest trophic level is very limited. Only a tiny fraction of the chemical energy captured by photosynthesis flows all the way through a food web to a tertiary or higher-level consumer. Thus, most food webs are limited to five or fewer trophic levels — there is just not enough energy left to support more levels.

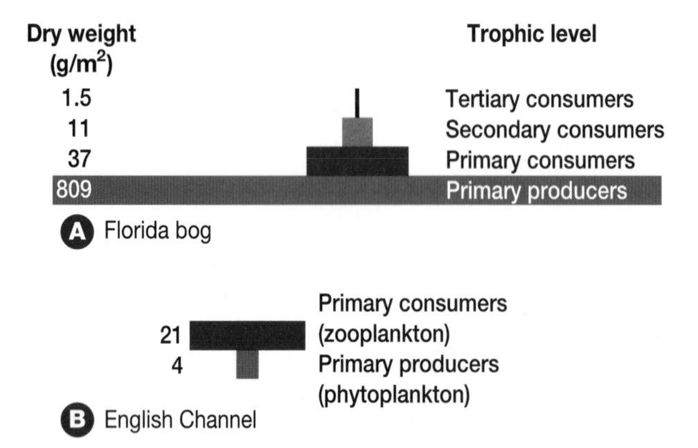

Figure 13.22 The biomass pyramid in (A) is based on data collected from a Florida bog. In the English Channel ecosystem, the pyramid of biomass is inverted (B).

Animals that make up the highest trophic level in an ecosystem tend to be large, predatory species such as lions, whales, hawks, and eagles. Since biomass is limited at the top of the pyramid, there can only be a few of these large animals in any ecosystem at one time. In fact, when you compare the number of individual organisms at each trophic level, you will find that the same pyramidal shape appears. The **pyramid of numbers** in Figure 13.23A shows the effect of the decreasing energy supply on the number of individuals at each level. Although the supply of solar energy is almost limitless, almost all of the energy is eventually lost from ecosystems as a result of inefficient transfers between trophic levels.

Not all pyramids of numbers have this shape. In a forest, for example, a few individual primary producers (trees) have enough biomass to support a large population of herbivores. As is true for the pyramid of biomass in some aquatic ecosystems, this could result in a pyramid of a different shape (see Figure 13.23B).

Completing the Thinking Lab on the following page will demonstrate that it is not only energy that can be passed through a food web. Certain toxic compounds can cause serious damage to species in an ecosystem when those compounds are passed from one trophic level to another.

In the next section, you will learn that nutrients are essential for proper growth and the repair of body tissue, and that they also cycle in ecosystems.

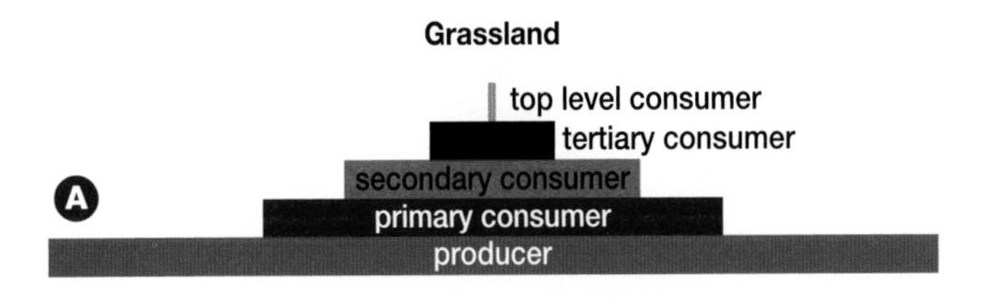

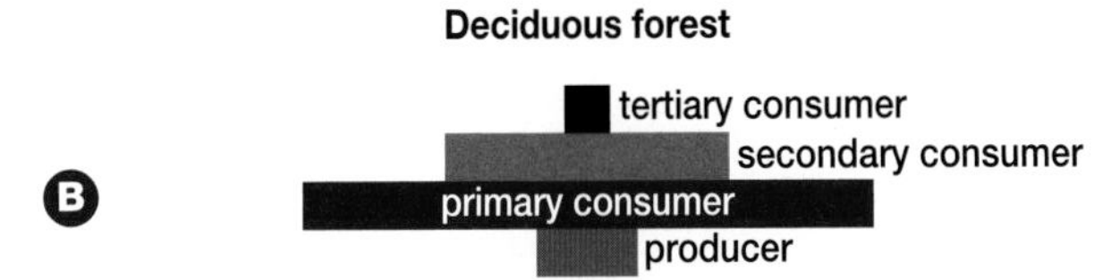

Figure 13.23 (A) Many ecosystems have a trophic structure that produces a pyramid of numbers with a broad base. (B) In some ecosystems, however, there can be fewer producers than primary consumers.

BIO FACT

The pyramid of productivity shows that primary consumers harvest more of the energy trapped during the process of photosynthesis (by eating photosynthetic organisms directly) than secondary consumers (who eat the organisms who ate the photosynthetic organisms). For humans, it is far more efficient in terms of obtaining energy to eat grain directly rather than to eat grain-fed beef. The pyramid of numbers shows that the biosphere could successfully feed far more humans if they were herbivores.

THINKING LAB

Biological Magnification

Background

Certain toxic compounds are not easily broken down by decomposers, so they remain in the water or soil for long periods of time. This increases the probability that these compounds will be ingested by small organisms — which are then eaten by increasingly larger organisms and passed through food webs. Since each animal on a trophic level tends to eat many organisms from the level below, the amount of toxins taken in becomes magnified with each step up the food chain. These compounds, which accumulate in the fatty tissues of animals, have been observed to have diverse (and often harmful) effects on these organisms. These effects include damage to the nervous and reproductive systems as well as the production of genetic mutations leading to various forms of cancer.

One of the first cases in which the phenomenon of biological magnification was observed involved DDT. DDT is a powerful insecticide that was used widely during World War II to control mosquitoes and other insects that transmitted diseases (such as malaria) to humans. Signs of damage caused by this chemical started to turn up about 10 years after it was first used. Mounting evidence of its deleterious effects caused the Canadian government to restrict the use of DDT after 1969. Study the the table on the right and the diagram on the following page, and answer the following questions.

You Try It

1. What is the relationship between an organism's trophic level and the concentration of DDT in its body?
2. How might an animal that lives a long distance from an area sprayed with DDT accumulate the chemical in its body?
3. Describe the general patterns you find in the data in the table. Speculate on the possible reasons for the differences in concentrations of DDT measured among species and localities and over time.
4. Research the specific types of prey consumed by one of the bird species listed in the table. Draw a biomass pyramid involving this species, incorporating trophic levels. Include the correct quantity of DDT (in ppb) at each level of your chart.

Species	DDT concentration in eggs (ppb)		
	Year	Bay of Fundy	Atlantic Ocean
Leach's storm-petrel (*Oceanodroma leucorhoa*) (feeds on small organisms near the water surface)	1968	no data	1460
	1972	6810	2480
	1976	1750	750
	1980	1130	460
	1984	1050	400
Atlantic puffin (*Fratercula arctica*) (feeds on small fish)	1968	no data	890
	1972	2570	760
	1976	1270	590
	1980	1030	550
	1984	740	300
double-crested cormorant (*Phalacrocorax auritus*) (feeds on larger fish)	1972	6510	2850
	1976	1490	2180
	1980	1910	1340
	1984	1070	1880

DDT concentration in the eggs of three species of sea birds breeding along Canada's east coast

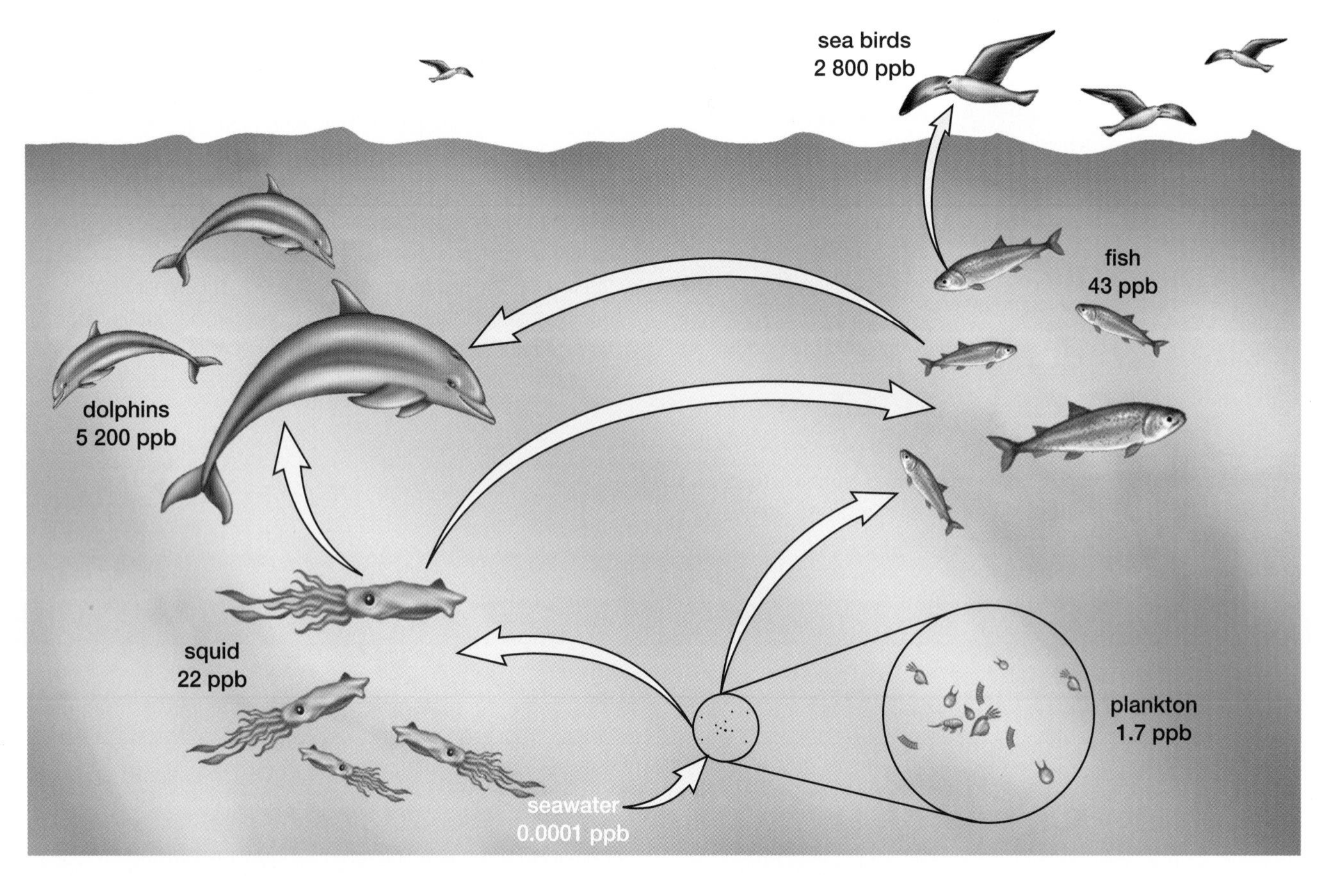

Figure 13.24 An example of data collected on DDT in a food web of organisms in the north Pacific Ocean. Concentration of DDT is measured in parts per billion (ppb). The arrows indicate the flow of energy from one trophic level to another.

SECTION REVIEW

1. K/U Explain why the primary productivity of a grassland usually exceeds the primary productivity of a forest ecosystem covering an area of the same size.
2. K/U Identify the factors that might cause annual fluctuations in the primary productivity of a specific grassland ecosystem.
3. K/U On average, less than 20 percent of the food energy consumed by a grasshopper is converted into new grasshopper tissue. What happens to the rest of the food energy that was contained in the tissues of the producer species consumed by the grasshopper?
4. K/U Explain why many models showing the relationships among trophic levels in an ecosystem are shown in the shape of pyramids. Describe the difference between the information depicted in a biomass pyramid, a pyramid of productivity, and a pyramid of numbers.
5. K/U Explain why the primary productivity levels of equatorial ecosystems generally exceed those of ecosystems situated farther north.
6. K/U Explain why the energy transfer from herbivores to carnivores is more efficient than the energy transfer from producers to herbivores within the same food chain.
7. K/U Suppose that the biomass for primary producers is less than the biomass for primary consumers. Can this ecosystem survive? Explain your answer.
8. K/U Describe the factors that determine the shape of a pyramid of numbers diagram for a specific type of ecosystem. Explain why a grassland pyramid would have a wider base than a forest pyramid.
9. MC We have seen that there are fewer carnivores than herbivores in ecosystems because of the inefficiency of energy transfer between trophic levels, and that the world could support more people if we ate only plant material. Some people feel this means that humans should switch to a vegetarian diet; others disagree. There are, in fact, a variety of issues to consider in addition to the relatively simple one of energy transfer. Take a stand. Prepare your arguments carefully and be prepared to debate the issue in class. You might want to prepare a pamphlet that could be used to educate the public (or the rest of the class) about your point of view.

Nutrient Cycling in Ecosystems

EXPECTATIONS

- **Distinguish between how energy and nutrients move in ecosystems.**
- **Describe the general features shared by all biogeochemical cycles.**
- **Differentiate between biogeochemical cycles of materials that exist in the environment in gaseous form and those that do not.**
- **Illustrate some of the processes that move nutrients through ecosystems, using one nutrient cycle as an example.**
- **Recognize the impact that humans and plants have on biogeochemical cycles.**

In addition to energy, organisms must also obtain 17 chemical elements (termed essential nutrients) for proper growth and repair of body tissue. Several of these elements, including carbon, hydrogen, oxygen, and nitrogen, are required in large amounts, since they make up 95 percent of the mass of all living organisms. Calcium, iron, and other elements are needed in smaller amounts, while only trace (very small) amounts are required of the remaining nutrients.

Energy and nutrients are vital for the survival of organisms. Although both are carried as part of the same complex molecules from one trophic level to another, their overall paths through ecosystems are different. Energy is continuously supplied to Earth by the Sun, and while some is captured and flows through the ecosystem, much is lost to the environment along the way.

In contrast, the supply of nutrients is not constantly replenished. The existence of life in the biosphere depends on the recycling of chemical elements. Recycling begins when organisms die, parts are lost (such as dead skin cells or broken branches), or wastes are eliminated.

Decomposers then go to work, releasing the nutrients that were contained in these bodies into the atmosphere or soil. From these "pools," nutrients are picked up by various types of organisms and re-used to build new bodies — which can later be eaten by other organisms.

Since the routes these chemicals travel involve both biotic and abiotic components of the environment (including rocks and soil that are *geo*logical features), they are referred to as **biogeochemical cycles**. Read on to learn about the general features of biogeochemical cycles.

Biogeochemical Cycles

The route a specific element takes in its biogeochemical cycle depends on the element and on the trophic structure of the ecosystem in which it is travelling. However, there are two general types of cycles. The first type includes the cycles of carbon, oxygen, nitrogen, and sulfur — elements that can occur as gases in the atmosphere. These nutrients have global cycles because individual nutrient atoms may travel long distances. For example, a plant living in one location may take up carbon (in a carbon dioxide molecule) that was released by an animal living far away.

The second type of cycle involves nutrients that are more static and tend not to move around (including phosphorus, potassium, calcium, and trace elements). These elements are typically found in the soil rather than the atmosphere. They are absorbed by plant roots and return to the same soil "pool" when the plant dies.

In general, all biogeochemical cycles involve the movement of elements between four reservoirs — which can be thought of as nutrient "banks" (see Figure 13.25 on the following page). Nutrients can be easily withdrawn by organisms from two of these banks. The first of these consists of living organisms and the bodies of recently deceased organisms; therefore it is a biotic bank. The second storehouse consists of abiotic parts of the environment from which nutrients can be easily accessed — atmosphere, soil, and water.

In the other two reservoirs, nutrients are held more tightly and cannot be accessed by living organisms. Again, one is biotic and the other is abiotic. The biotic reservoir is formed from the compressed (fossilized) remains of organisms that died long ago. Over time, the nutrients contained in the bodies of these organisms became incorporated into oil, peat, or coal deposits, where they are unavailable to living organisms. When we burn these fossilized remains as fuel (or when they are eroded), the elements move from this biotic "vault" into the accessible abiotic bank formed by soil, water, and atmosphere.

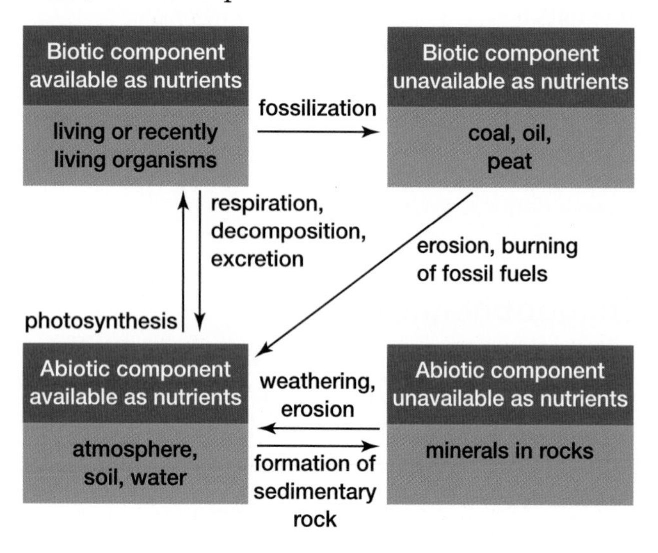

Figure 13.25 The four nutrient reservoirs are categorized with respect to (a) whether they involve biotic or abiotic components of the ecosystem and (b) whether the nutrients they contain are directly available to living things or not.

The abiotic vault in which nutrients are locked consists of rock. When soil is converted to rock, the chemicals contained in the rock are held very tightly and become inaccessible to living things. However, these chemicals can be released back into the soil, water, or atmosphere by weathering or erosion.

Having looked at some of the features shared by all biogeochemical cycles, it is useful to examine a few specific cycles in more detail. Because ecosystems are complex and can exchange materials with each other, it is difficult to trace the route taken by any of the cycling elements. For example, an ecosystem with finite boundaries (such as a pond) exchanges nutrients with many other ecosystems. Water running off adjacent land introduces chemicals to the pond from distant sources, and visiting birds carry away nutrients in the bodies of the fish or insects they have eaten. Following the inputs and outputs of ecosystems is challenging, but researchers have worked out the basic routes followed by materials in several biogeochemical cycles. One of the most familiar is the water (hydrologic) cycle, in which oxygen and hydrogen cycle together, as shown in Figure 13.26.

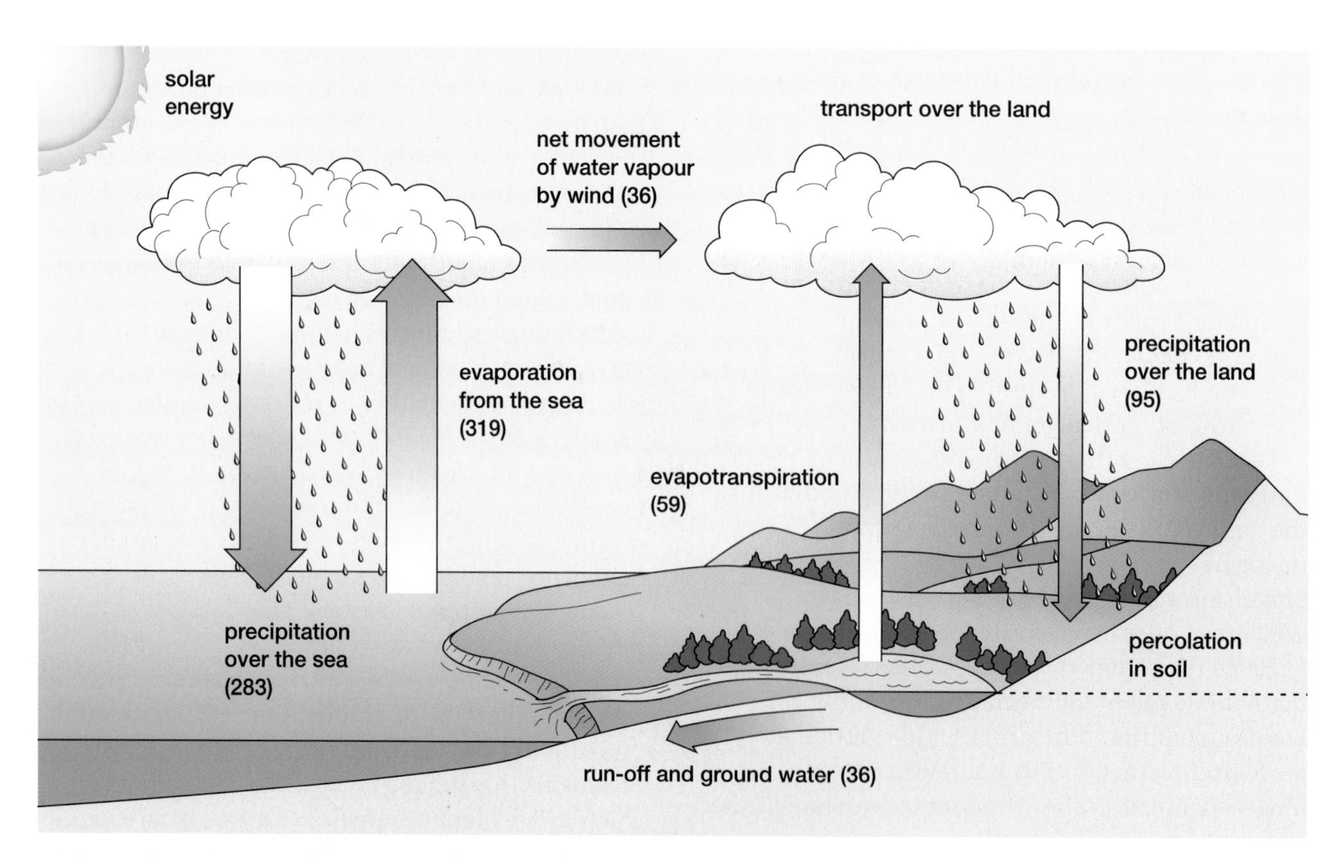

Figure 13.26 The water or hydrologic cycle. The width of the arrows indicates the relative amounts of water being carried in each route. The numbers in parentheses represent estimates of global water flow in billion billion (10^{18}) grams per year.

The Carbon Cycle

Another important biogeochemical cycle is the carbon cycle (shown in Figure 13.27). Other than oxygen and hydrogen, which are mostly locked up in water molecules, carbon is the most abundant element in living things. Plants and other autotrophs take in carbon dioxide as a raw material for photosynthesis, and almost all organisms (including autotrophs) give it off as a waste product of cellular respiration. The carbon that accumulates in long-lived, durable plant material (such as the wood of trees) does not cycle as rapidly, unless the trees are burned.

In some environments, living things die faster than decomposers can break them down. This was particularly evident in the warm, moist conditions of the Mesozoic era, roughly 150 to 250 million years ago. During that time, giant ferns and a multitude of other plants grew thickly in vast forests. Under these conditions, so much detritus built up that it became compressed, resulting in the formation of coal or oil deposits. Today, we refer to these substances as **fossil fuels**. Since these deposits were formed from what were once living things, they contain substantial amounts of carbon. As described above, these carbon supplies are not accessible to living things unless the carbon is released by burning — either naturally or in a coal stove or oil furnace.

In one sense, the burning of fossil fuels simply returns to the atmosphere the carbon that was removed by photosynthetic activity in Mesozoic forests. However, during the time these carbon supplies were locked up underground, a new balance developed in the global carbon cycle. With the addition of substantial amounts of so-called new carbon to the cycle, the environment is changing. What effects do you think this might have on the biosphere? This is an issue you will consider further in Chapter 15.

The Nitrogen Cycle

Nitrogen is another element that is relatively abundant in ecosystems, making up almost 80 percent of Earth's atmosphere. But since nitrogen gas (N_2) cannot be taken up by photosynthetic organisms, it is not available to heterotrophs. In fact, plants can only use nitrogen when it is a part of ammonium (NH_4^+) or nitrate (NO_3^-) molecules. Nitrogen gas is occasionally

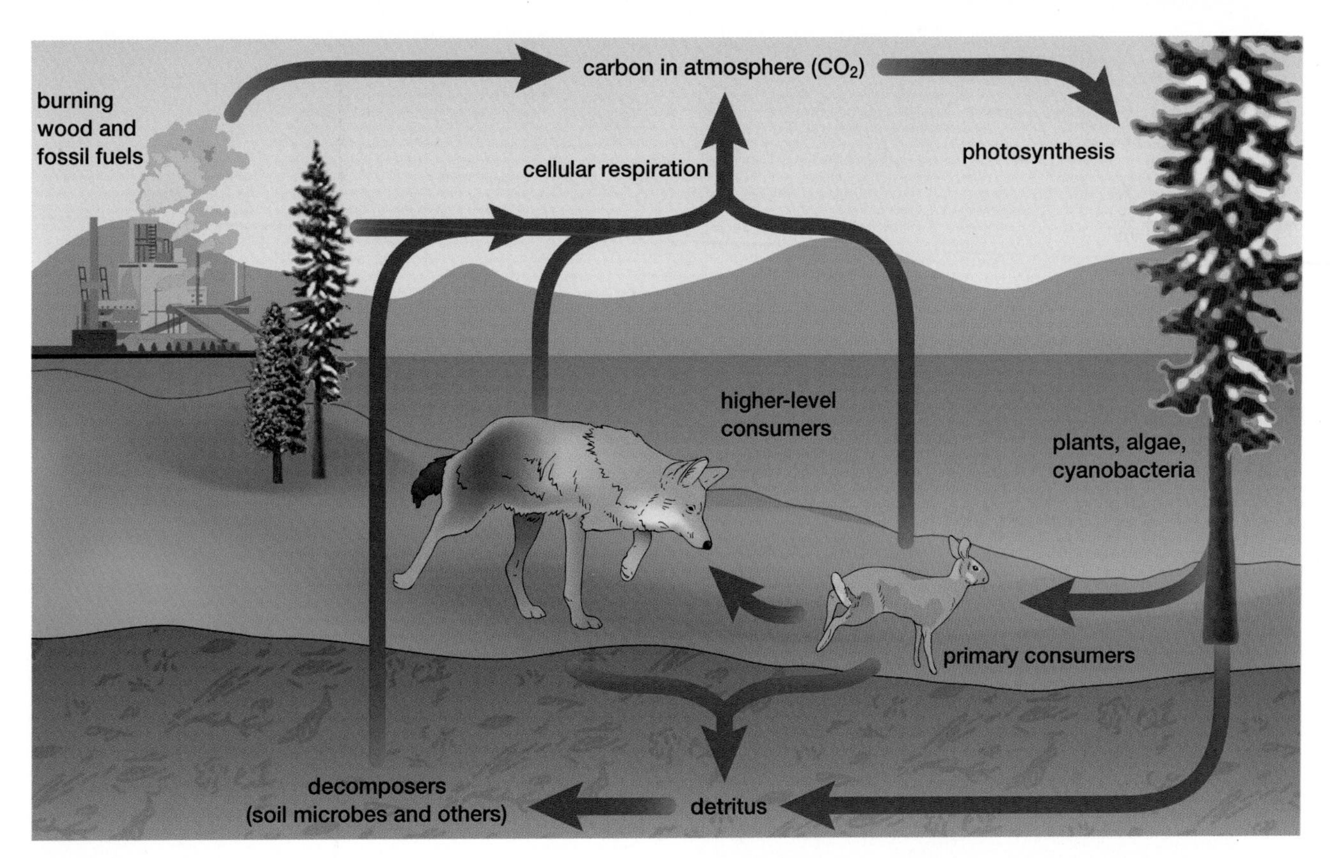

Figure 13.27 On a global scale, the amount of carbon entering the atmosphere as a result of decomposition and respiration balances the carbon removed by photosynthesis. The burning of wood and fossil fuels unbalances the equation. What might result from this practice?

converted to these materials in the atmosphere if it becomes dissolved in rain or attached to dust particles. It can then enter the soil and be taken up by plants (see Figure 13.28).

More commonly, the conversion of nitrogen to a form useful to plants occurs as a result of the activity of bacteria. Certain bacterial species are capable of a process known as nitrogen fixation, which converts nitrogen gas to ammonium. Other types of bacteria can convert ammonium to nitrite (NO_2^-), in processes referred to as nitrification. Bacteria and cyanobacteria are the only organisms that carry out these processes. Without them, very little nitrogen would be available for living things in ecosystems.

Nitrogen is a vital component of amino acids, which are the building blocks of which proteins are made. Since proteins make up the majority of the structural material of living organisms, act as enzymes, and perform many other functions, it is clear that without the bacteria involved in the nitrogen cycle the biosphere as we know it today would be devoid of life.

The Phosphorus Cycle

Water, carbon, and nitrogen exist as gases in the atmosphere and therefore cycle over large areas. In contrast, phosphorus is an example of a nutrient that cycles on a more localized scale (see Figure 13.29). Most phosphorus is cycled through food webs in both aquatic and terrestrial ecosystems. Some phosphorus escapes from the land and enters Earth's supply of ground water (the reserve of underground water), drains into rivers, and eventually flows into the sea. However, an almost equal amount re-enters the soil from the weathering of rocks.

Over long periods of time, phosphates that reach the ocean accumulate in sediments and eventually become incorporated into rock. These rocks may re-enter terrestrial environments as a result of geological processes that either raise the sea floor or lower seawater levels at a particular location. Most phosphorus cycles locally through soil, plants, and consumers, but some will cycle over a longer time and larger scale. This is true for most elements that do not exist in the atmosphere as a gas.

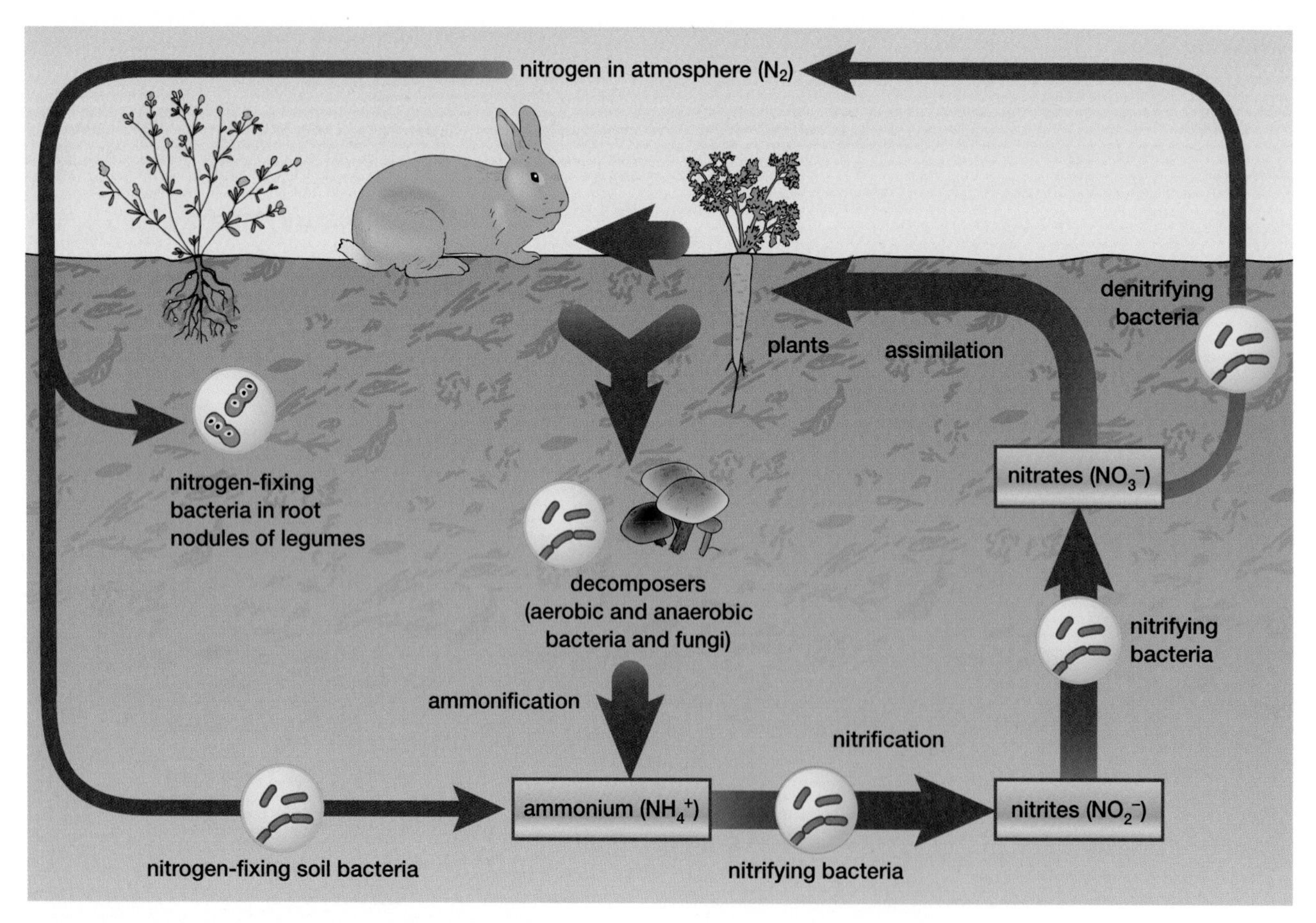

Figure 13.28 The width of the arrows in this diagram represents the relative amount of nitrogen being moved by each process. Why do you think some gardeners inoculate some types of plants with a specially prepared mix of bacteria before planting them?

The Impact of Humans on Biogeochemical Cycles: A First Look

Could the destruction of the Easter Island ecosystem have been due to the upset of one or more biogeochemical cycles? Can humans actually have a large enough impact to disrupt natural cycles? These questions will be examined in more detail in Chapter 15; for now, you need only look at some general aspects.

All over the world, researchers have realized that because of weather variations, the rates of biogeochemical cycling and the routes taken by nutrient molecules differ from time to time, even within the same ecosystem. As a result, it is necessary to study ecosystems for long periods of time to fully understand the movement of nutrients and the time required to complete certain cycles or parts of cycles. Many scientists are therefore conducting what is often referred to as long-term ecological research (LTER), either by monitoring change in certain variables over a broad area or by studying specific ecosystems in detail.

WEB LINK

www.mcgrawhill.ca/links/biology12

In some cases, LTER involves members of the public assisting in the collection of data that are later analyzed by scientists. In Canada (as well as elsewhere in North America), programs like Frogwatch, Feeder Watch, Tree Watch, The Breeding Bird Survey, and The Christmas Bird Count encourage the participation of individuals and groups. Various web sites show you how to get involved, obtain data, analyze populations, or learn more about various LTER projects. To get started, go to the web site above, and click on **Web Links**.

One very productive LTER project has been in existence since 1963 at the Hubbard Brook Experimental Forest in New Hampshire (see Figure 13.30A on the following page). At this location, scientists began by monitoring the amount of important nutrients that entered the ecosystem (contained in rain and snow) and the amount that left it (flowing out in streams or being given off to the atmosphere by plants). The results indicated that the amounts were almost balanced — cycles operating within the forest ecosystem conserved the nutrients the ecosystem contained.

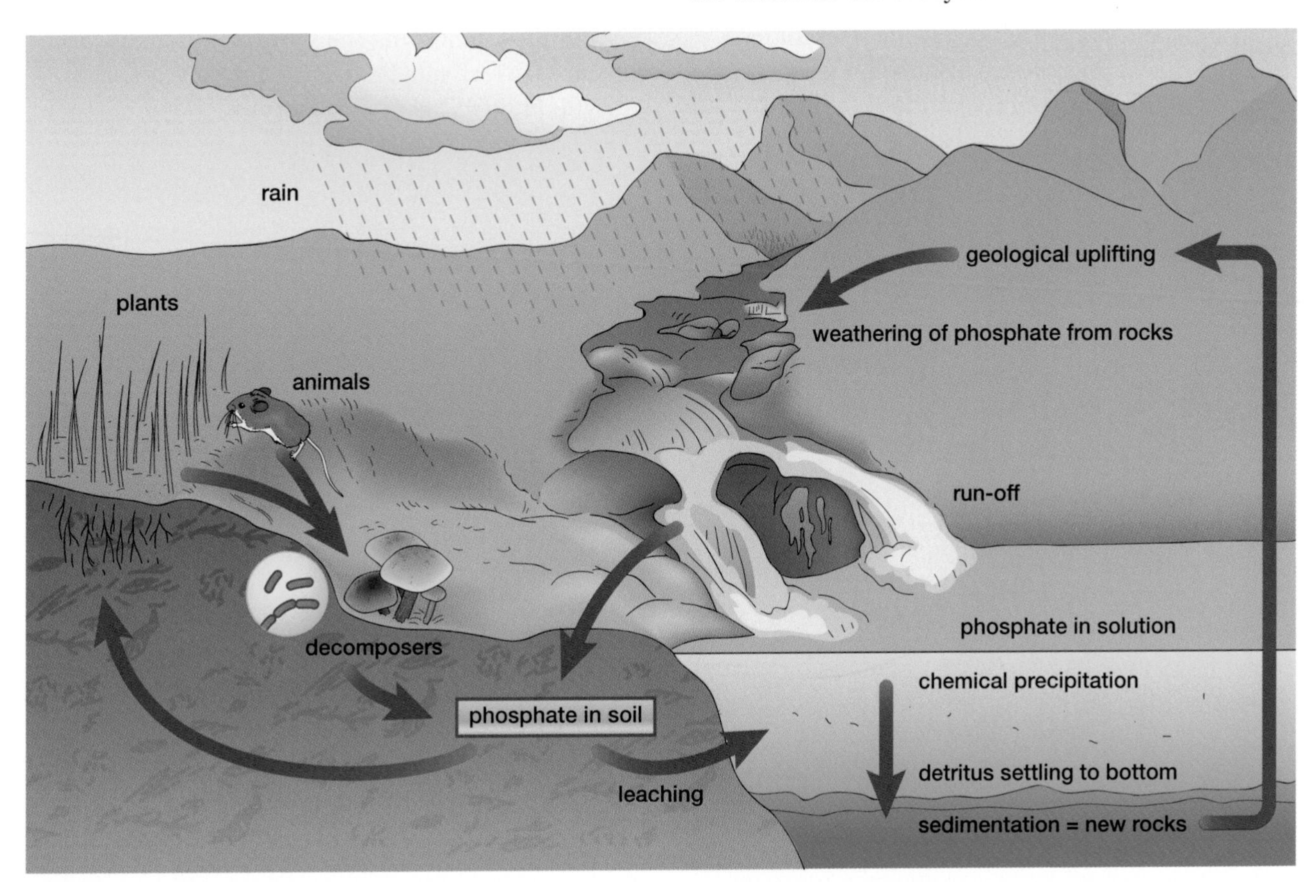

Figure 13.29 Manufacturers have responded to the demands of various environmental groups and removed most of the phosphates from laundry and dishwashing detergents. Why do you think these groups campaigned for the removal of phosphates?

In 1966, one of the valleys in the forest area was logged (as in Figure 13.30B). All the trees were cut down and left to decay, and the area was sprayed with herbicides to prevent new plant growth. The inflow and outflow of nutrients in this valley was then compared with that of an untouched experimental forest. The amount of water that ran through and out of the logged valley increased by 35 percent, presumably because there were no plants to absorb the water from the soil. Substantial amounts of nutrients were lost from the ecosystem, with most flowing into the stream that ran through the valley. The amounts of calcium and potassium in the stream increased by 4 and 15 times respectively, and the nitrite concentration was 60 times higher after logging. Not only were these nutrients being lost from the forest ecosystem, but in some cases they were making the water from the stream unsafe to drink.

These results, as well as other data collected over the last 35 years, have demonstrated that plants control the amount of nutrients leaving a forest ecosystem. When plants are removed, nutrients are lost. The loss begins immediately and continues for as long as the plants are absent. After succession starts, nutrient loss is reduced. It may take a long time to achieve the balance of loss and gain seen in an intact ecosystem — if it ever does return.

Environmental factors can have a significant influence on the variety, distribution, and productivity of autotrophs in an ecosystem. To enable you to understand the potential effect these factors can have, you will design and test your own model ecosystem in the investigation on page 458. Using controlled testing, you will observe and make comparisons between a control ecosystem set-up and one in which you have altered the environmental factors.

In this chapter, you recognized the challenge of learning about the complexity of the interactions among organisms so they can be better understood. You also learned about trophic structure, including various ecological principles involving abiotic and biotic components; the complex relationships that exist among individuals, populations, communities, and ecosystems; ecological succession; and the differentiation between the habitat, range, and niche of a population. You also learned about trophic structure, including the ecosystem role of producers, consumers, and decomposers, and saw how energy flows and nutrients cycle through an ecosystem. In Chapter 14, you will learn about population ecology, and examine features of populations, including their size and density. You will also investigate the factors that influence the growth and decline of populations.

A

Figure 13.30 Hubbard Brook Experimental Forest (A) Dams built along streams running through the area allowed researchers to collect data on the usual outflow of nutrients. Rain and snow samples were collected and analyzed to measure nutrient inflow. (B) After this area had been logged, the loss of nutrients increased dramatically. Do you think the loss would have stopped if secondary succession had been allowed to occur naturally? Defend your answer.

Biology At Work

Ecological Policy Maker

"Aboriginal wisdom teaches that all species should be protected, and all species interrelate," says Dr. Deborah McGregor, Head of Aboriginal Policy and Intergovernmental Co-ordination for Environment Canada, Ontario Region. Similarly, Dr. McGregor stresses that society should safeguard all natural habitats, not merely those deemed environmentally sensitive. "You have to protect the whole system," she says. "It's not enough to just create a park."

"Certain responsibilities come with being in Creation," she adds. "One key principle is that everything has life, while another is the ethics of non-interference. For example, we shouldn't interfere with water and prevent it from carrying out its responsibilities by polluting it. If we do, we create imbalances. The water can no longer quench our thirst." A member of the Whitefish First Nation on Manitoulin Island, Dr. McGregor believes that Aboriginal wisdom is spiritually derived — it comes from the Creator.

According to Dr. McGregor, all Aboriginal people are scientists in the sense that they have ecological wisdom passed down to them by their elders. "Aboriginal science is different from traditional western (non-Aboriginal) science," she says. "It's important for western-trained scientists to work with Aboriginal people — to consider more bases of knowledge." Dr. McGregor provides advice to staff at Environment Canada on how to interact with Aboriginal communities. "Visitors shouldn't just come with a predetermined agenda, a clipboard, and tape recorder," she says. "The way to really learn is to build relationships."

Among traditional western professions, Aboriginal people have tended to enter teaching, law, social work, and health sciences. Dr. McGregor and her Environment Canada colleagues try to encourage more Aboriginal people to go into ecology and other natural resource fields. Dr. McGregor has a master's degree in environmental studies from York University, and a doctorate in forestry from the University of Toronto. She stresses, however, that academic achievement goes only so far in preparing a person for a career. "Community service is important too," she says. "You have to prove that you can deliver at the community level — that you're not just book learning and talk."

Career Tips

1. What community services could you perform to help prepare yourself for a career that interests you? Make a list.
2. "Traditional western science doesn't have a monopoly on truth," says Dr. McGregor. In 1997, she was among Aboriginal scientists featured in an Ontario Science Centre exhibit called Question of Truth. Use the Internet and/or other resources to learn more about Aboriginal scientists. Choose one scientist and write a report about his or her work.

Manitoulin Island

DESIGN YOUR OWN
Investigation 13 • B

SKILL FOCUS

- Initiating and planning
- Hypothesizing
- Identifying variables
- Performing and recording
- Analyzing and interpreting

Ecosystem Productivity

Various vital environmental factors (such as sunlight, water quality, nutrient availability, and temperature) influence the variety, distribution, and productivity of autotrophs within an ecosystem. In this investigation, you will devise model ecosystems and use them to study the impact of specific environmental factors on the productivity of various trophic levels of a food web. You will also investigate how energy flows through the various trophic levels in these ecosystems.

Problem

How are model ecosystems used to investigate the impact of various environmental factors on the production of biomass at various trophic levels of an energy pyramid?

Hypothesis

Devise a hypothesis about how one or more specific environmental factors may affect the productivity and biomass of each trophic level within a specific type of ecosystem.

CAUTION: **Treat all of the organisms you have selected with care and respect. Wash your hands after handling the organisms.**

Materials

From your procedure, develop a list of required materials. Be sure to include an appropriate number of equal-sized containers for your ecosystem simulations.

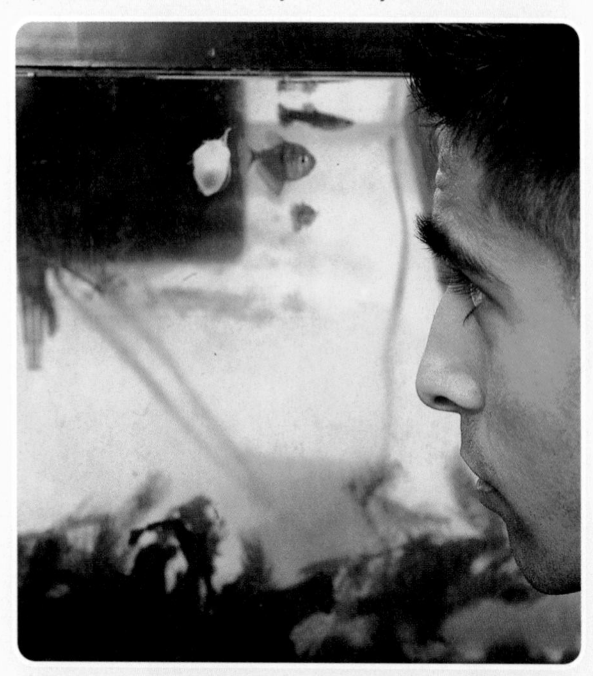

The ecosystem in this aquarium shows three trophic levels (primary producer, consumer, and detrivore). There is also an invisible fourth trophic level. Can you identify it?

Experimental Plan

1. For this investigation, organize yourselves into small work groups. Each group will be responsible for investigating how a specific variable (such as light intensity) affects some aspect of its model ecosystem. Groups may wish to co-ordinate their efforts by testing different ranges or values of the same variable. This approach will generate a wide range of data for analysis regarding the impact of a single variable. Alternatively, work groups may choose to investigate different variables — generating a variety of data about the impact of different variables. Whichever organizational strategy you select, each group will prepare one control set-up and a minimum of two experimental set-ups for the investigation.
2. Use containers to set up terrariums or aquariums that will serve as simulated ecosystems for your investigation (each set-up should be identical). Select organisms (such as phytoplankton, zooplankton, snails, aquarium plants, and small fish species) that will inhabit the ecosystems. A typical ecosystem consists of three to four trophic levels.
3. Identify the organism(s) in each trophic level and draw a food web of your ecosystems. Include less obvious organisms (such as bacteria and fungi) that form important links in the food web.
4. Identify a specific limiting factor or variable, such as the intensity of light, temperature, fertilizer, or specific nutrient that will impact the productivity of your model ecosystems. Make this your independent (manipulated) variable.
5. Identify the other factors or variables to be controlled; that is, those that must remain the same in each of your ecosystem set-ups.
6. Design a procedure to test the effect of the selected experimental variable on productivity and biomass. This procedure should allow for quantitative measurement of changes in biomass production in each ecosystem. Select an appropriate parameter to be measured, such as transparency of water, pH, CO_2 concentration in aquarium water, population samples of micro-organisms, and so on.

7. Devise a sampling procedure (be sure to include sample size, procedure, timetable and frequency of sampling, and sampling instruments) to measure changes in productivity and biomass of the target organism(s) at each trophic level in each model ecosystem. Ensure that your sampling techniques will provide an accurate estimation without unduly disturbing the integrity of the ecosystems.
8. You may wish to set up and run the ecosystems for a period of time to ensure they are stable, and to give you an opportunity to sample and identify the various species inhabiting your systems. You may also wish to monitor biotic and abiotic conditions such as temperature, water transparency, nutrient content, population size of various species, and the pH of the water. The data derived from this initial sampling could provide you with a baseline or standard for your subsequent experimental observations.
9. Design a procedure for evaluating the effects of changing the selected variable and measuring ecosystem response.
10. Your procedure should provide quantitative as well as qualitative data. Determine how your sample data could be translated into specific biomass values at each trophic level for the food webs in your model ecosystems.
11. Determine your method of data analysis. Select the format for graphs and charts that will illustrate most effectively how key parameters of your ecosystems responded to the variables manipulated in your experimental procedures.
12. Establish the amount of time you will require. Ask your teacher to approve your experimental design and your safety plans and arrange for the equipment and materials you require.

Checking the Plan

1. Describe the dependent and independent variables in this investigation.
2. What will you measure and how?
3. Design a table to record your data and observations.
4. Select the factors to be graphed, such as sunlight versus biomass.

Data and Observations

Conduct your investigation and record the observations and measurements. Enter your data on a summary table and graph your results.

Analyze

1. Describe any changes in biomass production and other factors in your ecosystem models over the duration of the experimental procedure.
2. What did the results indicate about how the variable tested for in this procedure affected biomass at each trophic level?
3. Compare the biomass of heterotrophs with the biomass of autotrophs.
4. How did your results relate to your original hypothesis?
5. Discuss possible sources of error in your procedure (relating to such factors as the sampling procedures used and the ability to control extraneous variables) that may have influenced your results.
6. How do the results of your group compare with the data provided by the other groups working on similar variables? Did dividing the class into groups provide you with a more complete understanding of the effects of certain variables on the quality of functioning ecosystems?

Conclude and Apply

7. What can you conclude about the accuracy of your original hypothesis?
8. How could your original procedure be improved?
9. Discuss the benefits and drawbacks of studying model ecosystems rather than studying actual ecosystems in the field.

Probeware 

If you have access to probeware, do the activity "Temperature and Dissolved Oxygen Use by Goldfish."

COURSE CHALLENGE

As you continue to prepare for the Biology Course Challenge, consider the ecological principles presented in this chapter. These principles offer many opportunities for selecting an issue to analyze from various perspectives. For example, look again at the carbon cycle presented in Figure 13.27 on page 453. This figure shows the effects of carbon on the following (keep in mind that there are also effects on living things you cannot see, such as insects and bacteria):

- plants
- animals
- atmosphere
- land surfaces
- industries

SECTION REVIEW

1. K/U Describe the similarities and differences you would expect to see between biogeochemical cycles that occur in aquatic ecosystems (such as a pond) and those that occur in terrestrial ecosystems (such as a grassland).
2. K/U What are the essential differences between the biogeochemical cycles for elements like nitrogen or oxygen and for elements like phosphorous or calcium?
3. C Draw a flowchart that illustrates the carbon biogeochemical cycle.
4. K/U Explain why oil and coal are often referred to as fossil fuels.
5. K/U Describe the role of nitrification in supplying essential nutrients to the species found at higher levels in a food web.
6. K/U Explain the similarities and differences between terrestrial and aquatic ecosystems with respect to carbon cycling.
7. K/U Describe the role of bacteria in terrestrial biogeochemical cycles.
8. C Make a diagram that illustrates each phase of the hydrologic cycle. Identify the biotic and abiotic elements of the cycle depicted in your diagram.
9. K/U Identify the types of green plants that play a critical role in the perpetuation of the nitrogen cycle.
10. MC Speculate about the possible impact of global warming on the carbon cycle in the northern boreal forest, which is one of Canada's more predominant biomes.
11. MC Some concerns now exist about the purity of fruits and vegetables that are labelled organic. Some academics suggest that it is not possible to grow food that is completely free of pesticides, fertilizers, and other commercial additives that are commonly used by today's highly industrialized agricultural operations. Discuss these concerns within the context of your knowledge of how biotic and abiotic factors interact in natural ecosystems.
12. K/U In Earth's northern hemisphere, the amount of carbon dioxide in the atmosphere is less during the summer than during the rest of the year. Examine the carbon cycle and suggest why this is so. (Note: there is more land area in the northern than the southern hemisphere, and therefore more vegetation).
13. I Design an investigation to compare the ecological efficiency of a small terrestrial ecosystem (such as a small terrarium) with an aquatic ecosystem (such as a small aquarium). Specifically, you may want to address the following generalization: aquatic ecosystems are more efficient than terrestrial ecosystems because the bodies of aquatic producers are made up of a smaller percentage of indigestible material than the bodies of terrestrial producers.

UNIT ISSUE PREP

LTER projects are used to study ecosystems for long periods of time to fully understand the movement and time required for nutrients to complete certain cycles. Are LTER projects being undertaken in your country? What impact might changed biogeochemical cycles have on humans?

Summary of Expectations

Briefly explain each of the following points.

- An organism's environment consists of both abiotic and biotic features. (13.1)
- Individuals are part of populations, communities, and ecosystems. (13.1)
- Ecosystems and communities change over time as a result of ecological succession. (13.1)
- Range, habitat, ecological niche, and biome are all terms describing different aspects of where a species lives or what role it plays in a community. (13.1)
- The movement of energy and nutrients in ecosystems is affected by the trophic structure of communities. (13.2)
- Autotrophs (primary producers) form the first trophic level, which sustains all of the higher levels. (13.2)
- The number of consumers and the biomass at each level is limited by the inefficiency of energy transfer between levels. (13.3)
- Primary productivity, biomass, and the abundance of organisms are usually highest at the first trophic level and decrease at each successive level. With respect to these three features, the trophic structure of an ecosystem can be represented by a pyramid. (13.3)
- Energy flows through, whereas nutrients cycle in, ecosystems. (13.4)
- The biogeochemical cycles of elements that exist in gaseous form differ from those elements that exist in another form. (13.4)
- Nutrients held in some environmental reservoirs are unavailable to living things, whereas in other reservoirs they are easily accessed by organisms. (13.4)
- Plants act to conserve nutrients within ecosystems. (13.4)

Language of Biology

State the biological significance of each of the following terms. Use these terms in concept maps to show your understanding of how they are related. Be prepared to explain your rationale.

- ecology
- biotic
- abiotic
- population
- community
- ecological succession
- colonizer
- climax community
- secondary succession
- primary succession
- pioneer species
- ecosystem
- biosphere
- biome
- habitat
- range
- ecological niche
- specialist
- generalist
- trophic structure
- trophic level
- primary producer
- primary consumer
- secondary consumer
- tertiary consumer
- decomposer
- food chain
- food web
- primary productivity
- biomass
- pyramid of productivity
- biomass pyramid
- pyramid of numbers
- biogeochemical cycle
- fossil fuels

UNDERSTANDING CONCEPTS

1. If the habitat of a species can be compared to an address, and its ecological niche can be compared to a job or occupation, what metaphor might you use to describe its range?
2. In your notebook, state whether each of the following statements is true or false. Correct each false statement and explain your reasoning.
   (a) All members of a particular population belong to the same species.
   (b) All members of a particular species belong to the same community.
   (c) A species with a diverse diet is probably a habitat specialist.
   (d) Species living in the same habitat can have different ecological niches.
   (e) Species with different ecological niches can live in different habitats.
   (f) All members of the same species live in the same type of habitat.
   (g) Cellular respiration does not occur in the cells of autotrophs.
   (h) Heterotrophs are not capable of photosynthesis.
   (i) There are generally more tertiary consumers than secondary consumers in an ecosystem.
   (j) There are generally more primary producers than primary consumers in an ecosystem.
3. Why are there typically only three to five trophic levels in an ecosystem?

4. Give three reasons why solar energy is not completely converted to chemical energy by primary producers.
5. Which are more common in nature, food chains or food webs? Why?
6. How do the biogeochemical cycles of elements that can exist as gases in the atmosphere differ from the cycles of those elements that do not have a gaseous form?
7. Describe the roles of plants in an ecosystem with respect to:
   (a) energy flow
   (b) nutrient cycling

INQUIRY

8. Calculate the Sun's overall efficiency in sustaining life using the following information.
   (a) 98 percent of the Sun's energy is unused
   (b) of the two percent used by plants, only 0.6 percent ends up in glucose (assume glucose is the only useable energy form created by photosynthesis)
   (c) half of the glucose is used by plants for their own life processes
   (d) 60 percent of the glucose reaching your cells is used to make ATP
   (e) your usage of energy is 55 percent efficient

   Why do you think these processes are not 100 percent efficient? What would happen if the efficiencies of these processes doubled? tripled?
9. Assume that a human lives entirely on the meat of a single species of bird, and that the bird only eats herbivorous insects. Estimate how much plant material this person would need to consume directly in order to gain 1 kg of weight.

COMMUNICATING

10. Construction of Biosphere 2 was completed in 1991. It was designed so a crew of eight people could live for two years completely isolated (except for communication contact) from the rest of the world. They were to care for and survive off resources (including plants and animals) placed inside the building. During this process, they were to learn how this could be done on a larger scale, perhaps even on another planet. Unfortunately, not everything went as planned. Oxygen levels inside the building dropped so low that more had to be pumped in, breaking the rule of isolation. Crops did not grow as well as expected and nearly all the animals placed in the building died (only cockroaches and ants reproduced abundantly). What are some of the things that might have happened to produce these results? If you have any ideas, suggest how the problems you described might have been corrected.
11. Some ecologists like to emphasize ecosystem processes more than ecosystem structure. Rather than labelling organisms as producers, consumers, and decomposers, they note that most organisms can play all these roles. These ecologists identify three basic ecosystem processes: production (the incorporation of energy and materials into bodies), consumption (the use of organic molecules for growth and reproduction), and decomposition (the breaking down of large organic molecules with the release of nutrients they contain into the atmosphere, soil, or water). With this in mind, explain why plants could be considered consumers as well as producers, and why animals might be called producers. Various kinds of organisms that are commonly referred to as decomposers can each be thought of as taking a different role in the breakdown process. Look up definitions for "scavenger" and "detrivore," and give examples of each type of decomposer. In what way can these organisms also be thought of as consumers — and even producers?
12. Design a flowchart to illustrate what happens to solar energy as it enters the biosphere, is converted by producers to chemical energy, and moves up to tertiary consumers. Be sure to show all the things that happen to the various amounts of energy along the way.
13. There are two schools of thought about the formation of ecological communities. While some scientists think that biotic factors are most important in determining what species make up a community, others hold the view that abiotic factors are more significant. Research the concepts to discover more about this debate, and come to your own conclusions

about the issue. Explain your position in a one-page essay.

14. Over a two-week period, collect newspaper items that describe or cover what might be called environmental issues. (Alternatively, do a web search and research a single topic that spans a two-year period.) For each article or topic, list the concepts from this chapter (refer to the Language of Biology list) that relate to the story or issue described.

MAKING CONNECTIONS

15. A famous ecologist once wrote a book called *The Ecological Theater and the Evolutionary Play*. What do you think he meant by this title? Write a paragraph explaining this title and discussing the relationship between evolutionary and ecological change.

16. Nature's most unusual organisms are generally found in extreme environments (such as deserts, the Arctic, or at the bottom of oceans). To many people, these organisms are so anatomically different that they appear grotesque. They are also often physiologically different from organisms we are more familiar with. What reasons can you give for this phenomenon?

17. Why do some species have small ranges? Part of the answer to this question lies in this chapter. Can you think of any other factors that might contribute to limiting the geographic distribution of a species to a very small area — usually on one continent?

18. When asked to describe the difference between plants and animals, many students answer that photosynthesis only occurs in plants and cellular respiration only occurs in animals. What is wrong with this statement? Why do you think this is a source of confusion for students? What might you do to reduce the problem (that is, to make the facts clearer and more memorable)?

19. Scientists have discovered that the rates of decomposition in ecosystems vary from an average of four to six years in tropical forests to 50 years in the tundra. This means that there is little decaying matter (leaf litter) containing nutrients on the ground in tropical forests. Instead, the nutrients are taken up very quickly by rapidly growing trees, with only about 10 percent of the nutrients existing in the soil. Seventy-five percent of the nutrients in a tropical forest ecosystem are contained in the woody trunks of the trees. On the other hand, 50 percent of the nutrients in a temperate forest ecosystem may be present in the soil and leaf litter. How do you think these two types of ecosystems differ with respect to their ability to withstand logging? Explain your answer in full, addressing any differences in the process of secondary succession that you might see between the two.

20. Many of the endangered species in the world are specialist species, with narrow ecological niches. Why is this?

21. Many of the world's top carnivores are also endangered. Why do you think this is so?

22. Grizzly bears (*Ursus horribilus*) and Gray wolves (*Canis lupus*) are top predators in various ecosystems, and were extirpated (that is, they went extinct locally) between 60 and 150 years ago in parts of their range. As a result, moose populations in these areas grew rapidly. Researchers have found that in some locations, human hunters replaced grizzlies and wolves as a top predator and reduced moose densities. In areas where hunting was not permitted, moose numbers remained so high that the pressure of their grazing (particularly on shrubs along the shores of lakes and streams) reduced the density of the plants in these habitats. This in turn had an impact on birds, which had previously used these shorelines for nesting or resting spots during migration, and certain bird species disappeared entirely. To some ecologists, these results showed that the structure of certain communities is determined from the "top down" by predators, rather than from the "bottom up" by the type and density of plants. They argue that this shows how vital it is to re-introduce grizzlies and wolves into areas where they have disappeared, and to ensure they are conserved where they still exist. Do you agree? Pretend that you are attending a public meeting discussing the possibility of re-introducing wolves to an area near your community. What will you say to the scientists proposing the re-introduction?

CHAPTER 14

Population Ecology

Reflecting Questions

- How do ecologists determine the size of a population?
- What factors determine how fast a population will grow?
- What makes some populations more vulnerable to decline than others?

Prerequisite Concepts and Skills

Before you begin this chapter, review the following concepts and skills:

- using examples, explain the process of adaptation of organisms to their environment (Chapter 11, section 11.1),
- using the ecological hierarchy of living things, evaluate how a change in one population can affect an entire hierarchy (Chapter 13, section 13.2), and
- using examples of the energy pyramid, explain production, distribution, and consumption of food resources (Chapter 13, section 13.3).

In 1988, a species of mollusc never seen before in North America was discovered in Lake St. Clair (located on the river connecting lakes Huron and Erie). This species was the zebra mussel (*Dreissena polymorpha*). A few individuals of this species had probably been carried into the lake a few years earlier, in the ballast of a ship from the Caspian Sea. Since then, the zebra mussel has spread throughout the Great Lakes and into various rivers and lakes in eastern Canada and the United States.

Populations of these tiny mussels become so dense that they can cause enormous damage to the ecosystems they invade. As larvae, zebra mussels attach themselves to hard surfaces, including the shells of various species of North American molluscs. The rapid growth of zebra mussels not only kills many individuals of these other mollusc species, but also fills up their habitat. Efficient consumers of plankton, zebra mussels have significantly reduced the food supply for fish, such as the lake whitefish (*Coregonus clupeaformis*). Since few animals feed on zebra mussels, their biomass contributes little to the next higher trophic level. But at the same time, zebra mussels use up a considerable amount of energy contained in the lower levels.

Zebra mussels have also had a substantial economic impact on humans whose living is dependent on healthy fish populations. In addition, mussels clog water pipes of cities and towns and encrust boats and docks.

What is it about the zebra mussel that made it possible for its population to grow so quickly in North America? Did the same rapid growth occur in the population of rats imported to Easter Island? In this chapter, you will learn about factors that contribute to the growth or decline of populations. Why do some grow faster than others? Why do some appear to remain stable, while others fluctuate wildly (growing very large and then "crashing" to a few individuals)? These questions interest population ecologists. The answers are vital to the conservation of endangered species and the sustainability of ecosystems.

Individual zebra mussels reach only 4 cm in size, but grow in such high density that they can completely blanket an area in a short period of time.

Chapter Contents

14.1 Characteristics of Populations

EXPECTATIONS

- Describe characteristics of a population, such as size and density.
- Estimate the size and density of populations using various sampling methods.

Demography is the study of populations — particularly the characteristics of populations that can be quantified, such as size, density, and growth rate. Demographers (those who study demography) are interested in what are sometimes called the vital statistics of a population. These include the characteristics mentioned above as well as others, such as the ratio of males to females and the ratio of old to young individuals in a population. These are among the factors that determine if, and by how much, the population is going to grow. Demographers generally start their study by defining the population of interest.

Recall from the previous chapter that a population consists of a group of individuals, all of the same species, simultaneously occupying a particular area. The boundaries of this area, which partly define the population, may be natural. For example, the shores of Lake Ontario define the limits of the population of lake whitefish that live in the lake. Similarly, the shores of Isle Royale in Lake Superior define that island's moose (*Alces alces*) population.

Many populations, however, have arbitrary boundaries set by a researcher or a government agency for management or study purposes (see the example in Figure 14.1). White-tailed deer (*Odocoileus virginianus*) in Algonquin Park often travel beyond the park's boundaries, and there is not much to distinguish between a maple tree inside the park and one just outside. Nonetheless, it is often useful to talk about the park's deer or tree populations, with the understanding that the geographic limits of these populations are somewhat arbitrary.

Regardless of how the boundaries of a population are determined, all populations can be described in terms of two important characteristics: density and dispersion. The **density** of a population is the number of individuals per unit of volume or area (for example, the number of oak trees per square kilometre in Prince Edward County). **Dispersion** relates to how the individuals in a population are spread out within its geographical boundaries. (Are they spaced uniformly throughout the habitat, or are they clumped together in groups?) Both of these features can affect how a population grows and what impact the population has on the environment.

Figure 14.1 Although we might refer to Ontario's blue jay (*Cyanositta cristata*) population, the birds themselves are not restricted by provincial boundaries. Individuals fly in and out of the other provinces and across the border into the United States daily.

Measuring Population Density

The density of a population depends on its size — that is, how many individuals it contains. In rare cases, it is possible to count all the members of a given population. Such a complete count is referred to as a **census**. The governments of many countries regularly conduct censuses of their human populations.

Often, however, a complete census is impossible due to time, energy, or financial constraints — a population may simply be too large to count. Therefore, ecologists must frequently estimate the size or density of the population of interest. There are a variety of ways to do this. In most cases, the number of individuals in a number of **samples** (small portions or subsets of the entire population) are counted or estimated and then averaged. The results are then extrapolated to the entire area occupied by the population, as shown in Figure 14.2.

Estimating Numbers by Using Transects or Quadrats

In some cases, organisms are sampled along a **transect**, which is essentially a very long rectangle. In transect sampling, a starting point and direction are randomly chosen and a line of a certain length (for example, 100 m) is marked out. The occurrence of any individual within a certain distance of this line is recorded. This distance might be 1 m if plants are being sampled, or perhaps 50 m if more mobile organisms (such as birds or mammals) are being studied. A sample transect is shown in Figure 14.3. Transects are particularly useful when the density of a species is low, or when organisms are very large (such as trees in a forest).

For plants and other types of organisms that tend to stay in one spot all their lives, ecologists generally use **quadrats** to sample a given population. First, several locations are randomly chosen within the area, and at each site a quadrat of known size (for example, 1 m^2) is marked out. Next, the number of individuals of a species within the quadrat is counted (as illustrated in Figure 14.4 on the next page), or the number of individuals of each of several species are counted if more than one population is being studied. The density of the population is determined by calculating the average number of individuals per quadrat, and dividing by the size of the quadrat (for example, 2.5 individuals per m^2). The size of the population can then be estimated by extrapolating from the density figure. For example, if the geographical area occupied by a population is 1000 m^2 and on average there are 2.5 individuals per m^2, then the population size is roughly 2500 individuals.

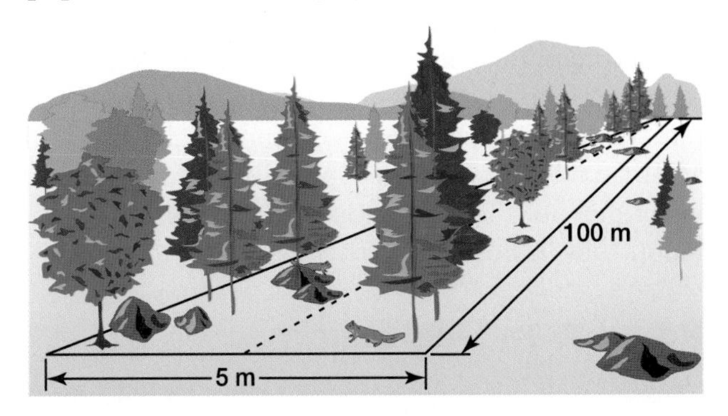

Figure 14.3 What is the area sampled along this transect?

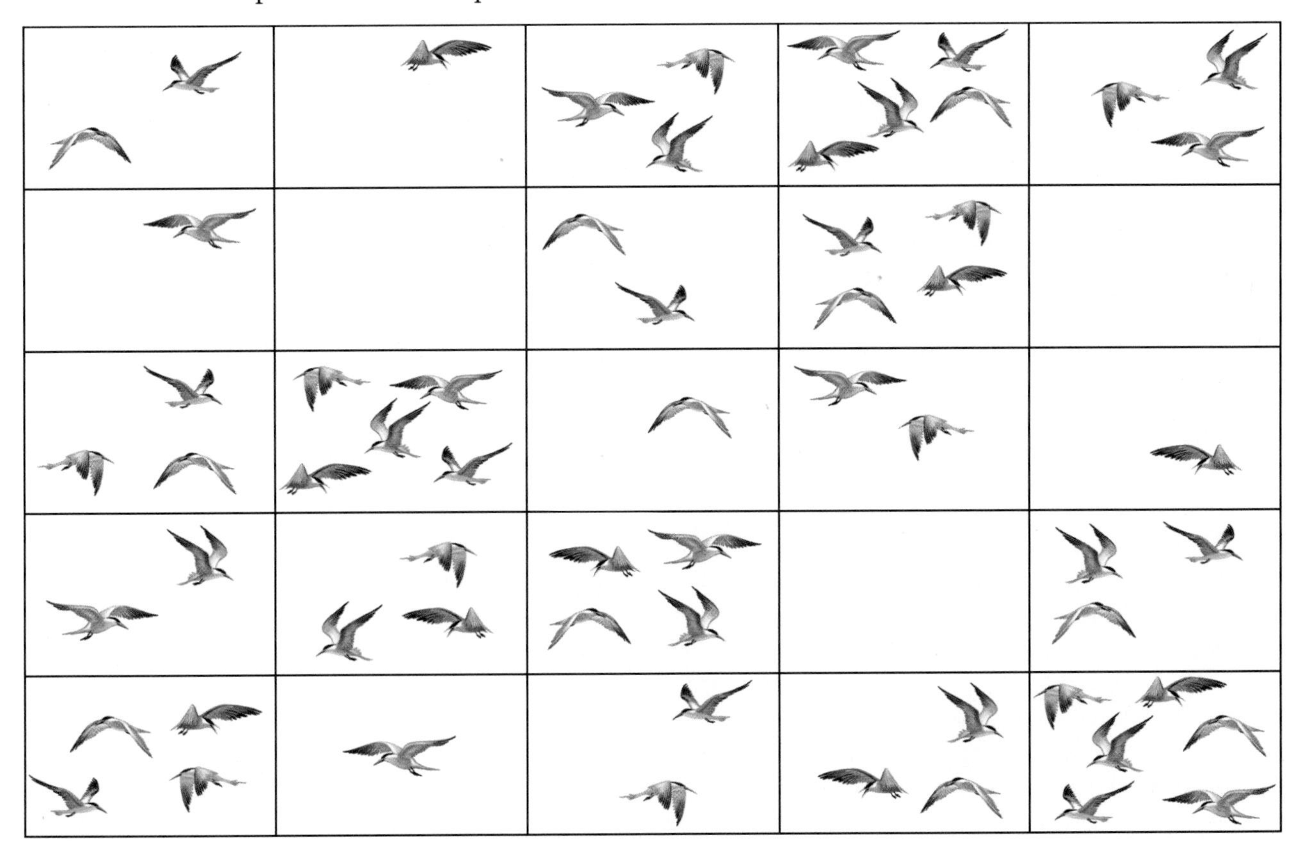

Figure 14.2 Examine a random sample of 5 of these 25 quadrats. Based on your sample, what is the average number of birds per quadrat? How many birds do you estimate there are in this flock?

Similar methods are used in studies of aquatic ecosystems to estimate the size of various populations. Often, samples of a known volume of water are collected. The water is then passed through a net or sieve, and the number of organisms in each sample is counted. As with quadrats, the average density (number of individuals per unit volume) can be used to estimate the size of the population contained in the entire body of water.

When using any sampling technique, it is important to take *random* samples — that is, samples in which all individuals in the population have an equal chance of being represented. To have the best chance of taking such a sample, it is important to know something about how the individuals in the population are distributed — in other words, to understand the population's dispersion. In the MiniLab on the next page, you will practise one technique used for estimating populations of mobile species.

Population Dispersion

Imagine walking along a transect and making a note on your data sheet each time you see a member of the population in which you are interested. Do you think you would be counting individuals at regular intervals — perhaps one every 10 m? Or do you think it more likely that you would come across several individuals at once, and then none, until you reach another group? This, of course, depends on how the members of the population are dispersed.

BIO FACT

Ecologists are often more interested in knowing the *relative* densities of two or more populations (that is, whether one is more or less dense than another) rather than the actual or *absolute* density of each one. Relative densities can be determined by looking for signs of activity along transects. For example, the relative density of bears in an area is often estimated by looking for footprints, claw marks on trees, clumps of hair caught on bushes, and piles of droppings (usually called "scats" by biologists). Since it is difficult to distinguish between grizzly (*Ursus horribilis*) and black bears (*U. americanus*) using some of these signs, samples of hair or scats are brought back to the lab for DNA analysis.

To simplify discussion, ecologists tend to recognize three general patterns of dispersion: uniform, random, and clumped (see Figure 14.5). It is important to realize, however, that there is actually a continuum of dispersion patterns. Many species typically display patterns that fall between two of these three types. Factors important in determining dispersion include the distribution of

Figure 14.4 When sampling the density of certain plants in a quadrat, it is often too difficult to count the number of individuals of a species. In this case, the proportion of the quadrat covered or shaded by one or more plant species is recorded. The area sampled is usually square or circular.

resources, the types of interactions that typically occur among members of a population, and the distance offspring generally disperse from their parents. In general, how the members of a population are spaced throughout an area depends on a complex interplay among these factors (each will be discussed in more detail below). As always, ecology and evolution are intertwined. The dispersion pattern typically seen in a population is produced by behaviours and other features that increase the ability of individuals to survive and reproduce. Dispersion patterns, which occur in ecological time, are therefore the result of characteristics that have been selected for over evolutionary time.

In some cases, it is primarily the distribution of food, water, or other needed resources that determines the dispersion of individuals. If these resources are not distributed evenly but rather are

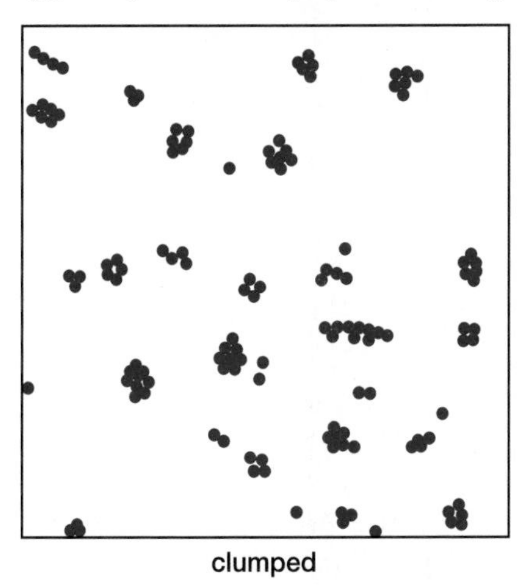
clumped

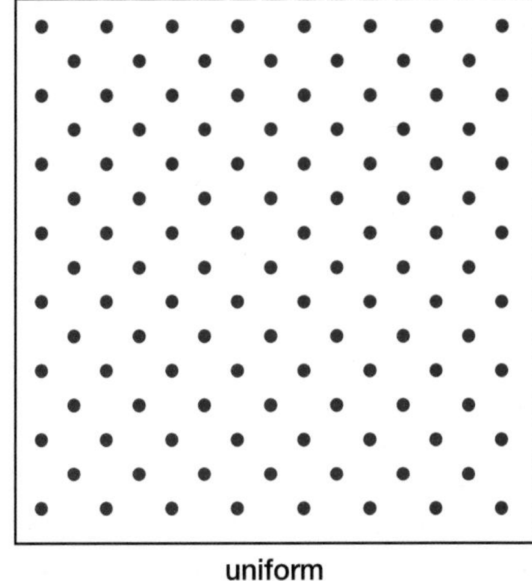
uniform

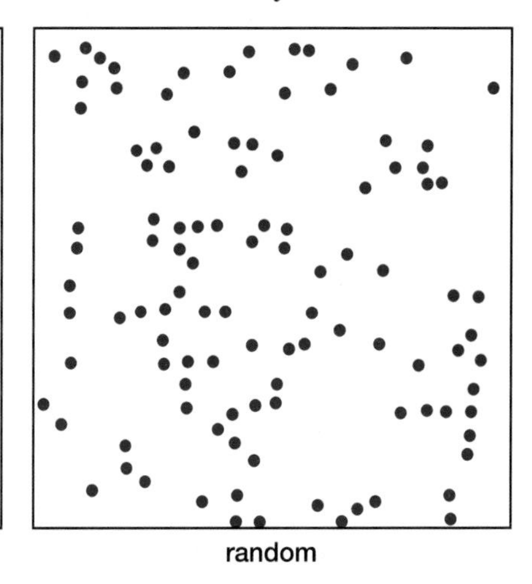
random

Figure 14.5 Think of examples of populations that fit each of these dispersion patterns.

MINI LAB

What Is the Population of Your School?

The size of a population of mobile organisms is often estimated using one of a number of mark–recapture sampling techniques (sometimes referred to as capture–recapture sampling). Small, lightweight leg bands are often used to mark birds, fin or gill tags are used for fish, and ear tags are sometimes used to mark captured mammals. It is important that the markers do not interfere with the ability of the subject to forage and perform other activities related to survival or reproduction.

In this MiniLab, you will estimate the population of your school. To do so will require careful planning, your principal's permission, and co-operation from other students. "Capture" a random sample of 25 or 50 students (for example, capture the first 50 students passing you in the hallway of your school) and "mark" them in some way. You might use pieces of string or ribbon as armbands to mark your captured individuals. Release them and let them again become randomly mixed with the general population. You will have to decide in advance how long it might take them to become re-mixed (an hour? a day?).

Take a second random sample of students (perhaps the same size as your first sample) after re-mixing has occurred, and record how many of them are marked. The proportion of individuals in the second sample that are marked can be used to estimate the size of the entire population by using the formula:

$$\frac{\text{number of individuals caught and marked in the first sample}}{\text{total population size}} = \frac{\text{number of marked individuals recaptured}}{\text{total number in second sample}}$$

For example, if 100 students were caught initially, marked, and released, and 10 of these individuals were recaptured when a second random sample of 100 was taken, an estimate of the population size (N) could be obtained by re-arranging the formula given above:

$\frac{100}{N} = \frac{10}{100}$ therefore,

$N = \frac{100 \times 100}{10} = 1000$ students

Analyze

1. Compare your estimate of the size of the school's population with the true size. How close were you?
2. **(a)** What problems do you think might affect the accuracy of your estimate?
   (b) Do you think any of these problems might also be faced by ecologists studying non-human animal populations?
3. How might you improve your sampling design?

clumped in specific places, then members of the population also tend to be clumped (see Figure 14.6). For example, many types of plants tend to be clustered in locations where the moisture, temperature, and soil conditions are optimal for their growth and reproduction.

Figure 14.6 These animals show a clumped distribution because the lakes, rivers, and ponds they rely on are only found in certain locations in the environment.

Clumping is also typical of species in which individuals gather into groups for protection from predators or to increase hunting efficiency. For example, schools of fish and flocks of birds often form to protect individuals from predators. Humpback whales (*Megaptera novaeangliae*) gather to feed so they can make "bubble nets" around their prey, and wolves hunt in packs to increase the probability of catching food. Wolf packs, chimpanzee groups, and lion prides are also important in ensuring that young members of the population are cared for while some adults hunt or engage in other activities. In general, species in which the interactions among individuals in a population can be described as *positive* (that is, the interactions draw individuals together) tend to show clumped distribution patterns.

Finally, clumping is typical of species in which offspring grow in close proximity to their parents (Figure 14.7 shows one example). Some species of plants, for example, tend to be found growing in clumps even if there is suitable habitat elsewhere, simply because they produce seeds that cannot travel far from the plant that produced them.

Figure 14.7 Aspens reproduce asexually, with new individuals sprouting from the roots of older ones. They therefore tend to grow in groves of genetically identical individuals.

By contrast, if the resources needed by members of a population are evenly distributed across an area and are in short supply, a uniform dispersion pattern is often seen. Many species of birds and mammals, for example, are said to be territorial. They defend a territory that contains the shelter or food they need for survival, mating, or raising young. Defending this territory involves keeping other individuals (or members of other families) out of the area. Since the amount of the resource (and thus the area) an individual or pair needs is relatively constant within a species, the result is that members of the population are more or less evenly spread over the habitat.

Certain species of plants, for example the black walnut tree (*Juglans nigra*), achieve the same result by producing chemicals that discourage the growth of other trees nearby. Secreted into the soil, these chemicals ensure that an individual plant has its own territory, containing the water or soil nutrients it needs. Whether a territory is defended with behaviours or chemicals, this type of interaction among the individuals in a population can be described as *negative* (that is, the individuals repulse or repel each other). Such interactions lead to uniform dispersion patterns.

For a population to display a random dispersion pattern, the resources needed by its members have to be uniformly distributed. These resources also must be so abundant that there is no need for individuals, pairs, or families to defend their share. In addition, randomness requires that there be neither positive nor negative interactions among

THINKING LAB

Sampling a Moose Population

Background

Knowing about the dispersion pattern typical of a species and how this pattern can vary among populations can help us understand the behaviour and ecology of the species. It is also valuable to know about dispersion before taking samples to estimate population density, since the accuracy of your estimate can depend on the dispersion. In this lab, you will see how transects might be used to sample populations of a large mammal — the moose.

You Try It

1. Study the three diagrams showing hypothetical moose populations with two different dispersion patterns. Describe how the moose are dispersed in each diagram.
2. The shaded portions of the diagrams show the transects that were used to sample each population. What proportion of the total study area (defined by the box) was sampled in each of the three cases?
3. Count the number of moose encountered *within the transect areas* for each case. (That is, do not count the total number of moose in any of the diagrams.) From these counts, estimate the size of the three moose populations.
4. The actual numbers of moose in the three populations are 60, 133, and 133, respectively. How close were your estimates? What do you think affected the accuracy of the estimates in each case?
5. How would you design a sampling experiment on a wild population? Note that in real life, the time and expense involved usually restricts the proportion sampled to between 10 and 20 percent of the total area of interest.

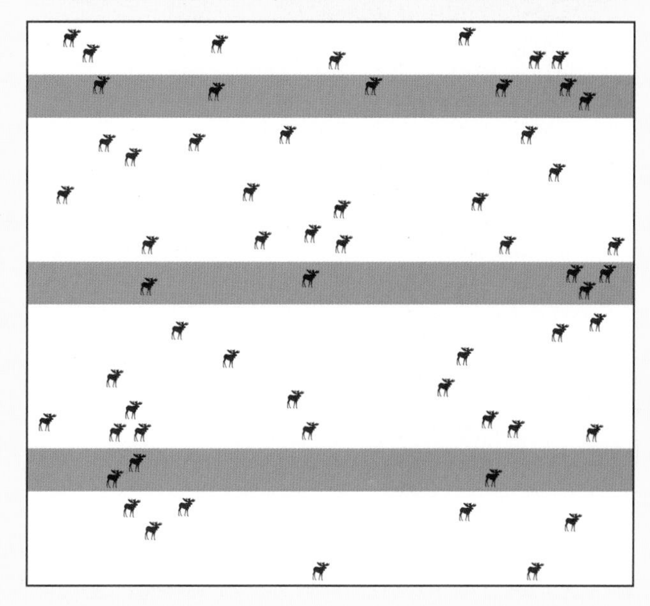

Dispersion pattern 1

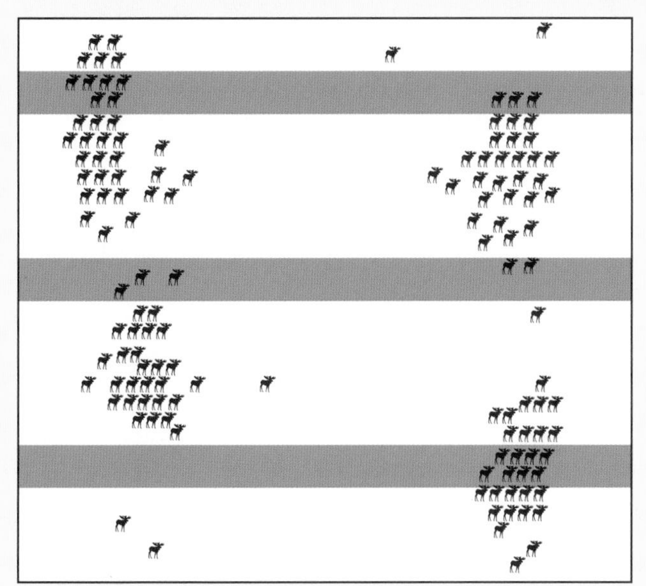

Dispersion pattern 2

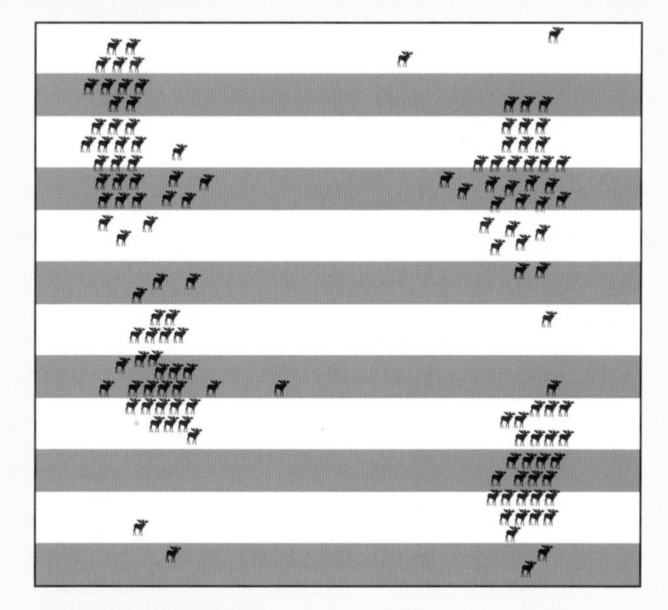

Dispersion pattern 3

the members of a population, and that offspring disperse more or less equally over all distances. Since these conditions are rarely met in nature, random dispersion patterns are unusual.

It is important to recognize that the dispersion pattern displayed by a population may vary between seasons, or even with the time of day (as with the bats shown in Figure 14.8). The scale at which a population is examined is also important. For example, the human population of Canada has a clumped dispersion. Looking at the country overall, most of its human population is concentrated in cities and towns. But if you focus on a smaller scale and look at the population of one city, you will often find a more uniform distribution (as illustrated in Figure 14.9).

Whether members are clumped, randomly scattered, or uniformly dispersed, many populations change in size. In the next section, you will learn how and why this happens and what factors affect the rate at which such change occurs.

Figure 14.8 Many species of bats tend to forage (find food) alone at night, but roost in colonies during the day.

Figure 14.9 Each homeowner in this development has his or her own territory.

SECTION REVIEW

1. K/U What is the biological definition of the term "population"?
2. MC Why is it important to be able to develop accurate estimates of natural populations of plants, animals, and other species of living organisms?
3. K/U Why is it often necessary to estimate population size, rather than counting each member of a population individually?
4. K/U Describe what is meant by the density of a population.
5. K/U When is it useful to use transects to sample populations?
6. I Develop a procedure in which transects are used to sample the species within a region. Select a local ecosystem for study.
7. I Suppose that in a study of a particular species of rodent, a sample of 200 was trapped, marked, and released. Later, a second sample of 200 was obtained and it was discovered that 20 percent of these animals were recaptured (that is, they were already marked). Based on these data, estimate the size of the total population.
8. K/U How might the abundance and distribution of food in a habitat influence how a species is dispersed in its environment?
9. I Research the types of sampling techniques typically used to estimate populations of migrating species, such as the caribou (*Rangifer tarandus*) in northern Canada. How do researchers attempt to improve the accuracy of their observations?

14.2 Describing the Growth of Populations

EXPECTATIONS

- **Analyze the components of population growth and explain the factors that affect the growth of various populations.**
- **Use conceptual and mathematical models to determine the growth of populations of various species (e.g., use the concepts of exponential and logistic (sigmoid) growth to describe and predict the future size of various populations.)**
- **Differentiate exponential and logistic growth curves.**
- **Predict the future size of an exponentially growing population, given its growth rate and starting size.**
- **Describe how limits on the supply of necessary resources can affect the growth of populations.**

In July 1996, Canada's human population was about 29 672 000. By July 2000, it was roughly 30 750 000. What processes accounted for this change in population size? **Birth** and **immigration** (the movement of individuals into a population) added to the population, while **death** and **emigration** (the movement of individuals out of a population) reduced it. These four basic processes cause change in the size of all populations. In Canada, birth and immigration rates are higher than death and emigration rates, so the population is growing. If the opposite were true, the population would be declining.

Growth of Populations in Unlimited Environments

Humans probably move among populations more than is typical of most species. For many populations, immigration and emigration rates are either very low or are roughly equal to each other, and thus can often be ignored. For this reason, and for the sake of simplicity, ecologists tend to focus on only two of the four processes — births and deaths — when considering how population size changes. Imagine a population growing under ideal conditions: with lots of food, space, shelter, and so forth. You could use the following word equation to describe how such a population changes in size:

Change in population size during a time interval	=	Number of births during the interval	–	Number of deaths during the interval

This word equation can be effectively summarized in a mathematical model. Using N to represent population size, t for time, and the Greek letter Δ (delta) for change, we can use ΔN to symbolize a change in population size and Δt to represent a change in time (that is, a time interval such as a day or a year). If B is the number of births during a time interval and D is the number of deaths, the word equation becomes:

$$\frac{\Delta N}{\Delta t} = B - D$$

The absolute number of births (B) that occurs during a given time interval depends on the size of the population at the beginning of the interval (which determines how many individuals are present to give birth). The same is true for deaths (D). A small population will probably experience fewer deaths during a time interval than a large one, simply because there are fewer individuals that can die.

It is useful to have a mathematical model that describes change in population size for all populations, regardless of their size. Therefore it is better to express B and D as rates (the average number of births or deaths that occur *per individual* during a specified time interval). Ecologists use b to symbolize per capita (which means per individual) birth rate — the average number of offspring produced by an individual during a given time interval. For example, if 50 offspring are born in a population of 1000 individuals during an interval, the birth rate b would be $50/1000 = 0.05$. Knowing the birth rate, we can then calculate the expected number of individuals (B) that would be added during an interval to a population of any size. For example, if $b = 0.05$ and $N = 500$, $0.05 \times 500 = 25$

small enough to be far below its carrying capacity, density-dependent factors have no effect and population growth is rapid. Eventually, however, the population reaches a density at which these factors start to have an effect; after this point the population growth slows, and it eventually stops when the carrying capacity is reached.

Later in this chapter, Investigation 14-A will give you an opportunity to examine the effect of both density-dependent and density-independent regulating factors on a species of micro-organisms. The remainder of this section describes some specific types of density-dependent factors in more detail.

Competition

In contrast to density-independent factors, density-dependent factors are typically biotic — they involve living things. As has been described, these living things are often members of the population itself. For example, the carrying capacity of a population's environment may depend chiefly on the availability of food. When the population reaches the inflection point in its logistic growth curve, there is no longer an abundance of food for each member of the population. Members must now compete with each other for the limited food supply, which becomes even more limited as the population size increases. The result is that the birth rate decreases or the death rate increases, or both (see Figure 14.19), and the population growth slows more and more as the density increases.

This type of competition among the members of a population is referred to as **intraspecific competition**. You have encountered its effects already, since it leads to evolutionary change as a result of natural selection. A higher proportion of successful competitors survive longer and have greater reproductive success; they are said to have higher fitness. This allows them to pass on more copies of their alleles to subsequent generations. The result is a change in allele frequencies within a population — which results in evolution.

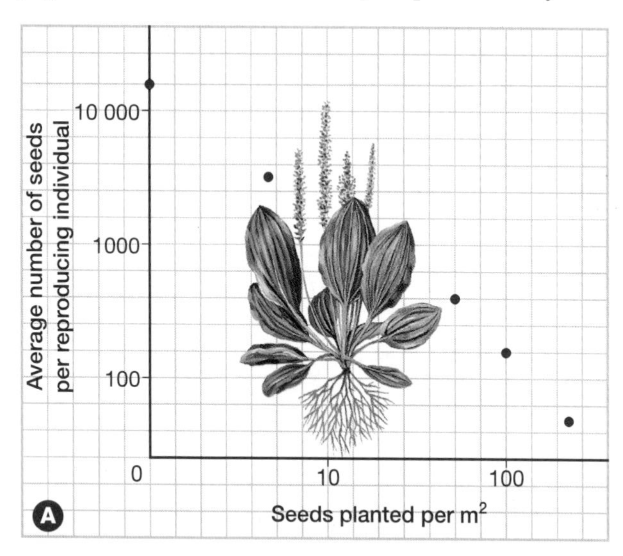

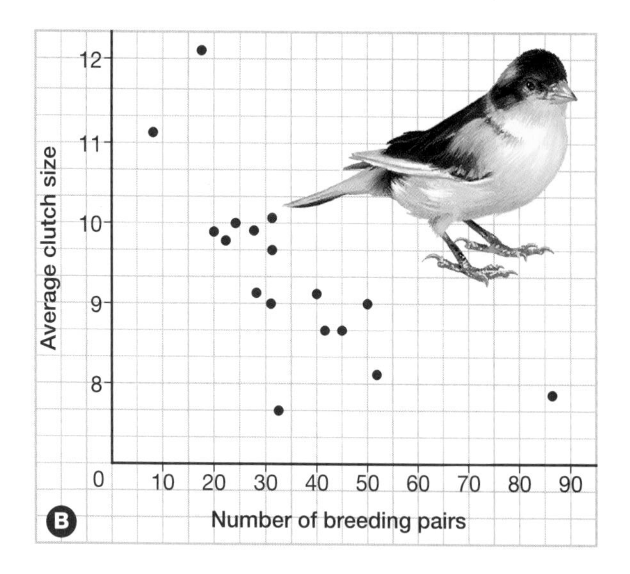

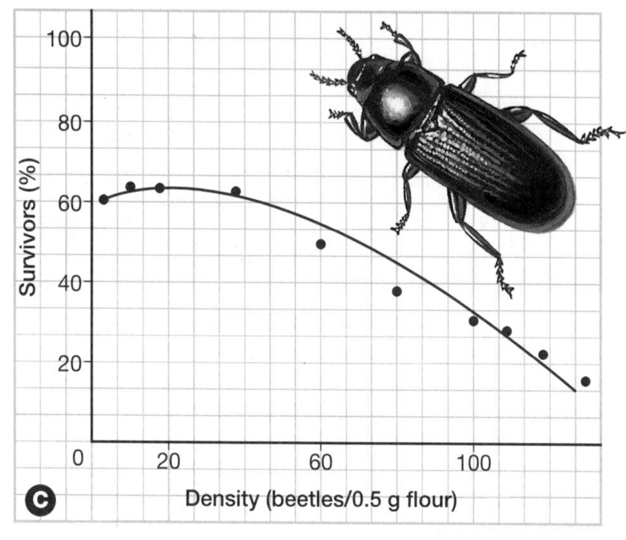

Figure 14.19 (A) A varying number of plantain seeds (a common Ontario weed) were planted in experimental plots. The average number of seeds produced by an adult plant decreased as the density of the plants in the plots increased. (B) Many songbird populations are limited by the amount of food available in their habitat. In some species, the average number of eggs laid by females declines as the density of breeding pairs increases. (C) In some cases, increased population density reduces the ability of individuals to survive rather than (or in addition to) reducing reproduction. This is true for laboratory populations of flour beetles. This graph shows the percentage of beetles that survive from egg to adult, in relation to the population density of the beetle.

In addition to its role in evolution, intraspecific competition is an important density-dependent factor limiting the growth of many populations. Members of a population may compete for a variety of needed resources, including food, water, sunlight, soil nutrients, shelter, or breeding sites (see Figure 14.20). The effect is always the same — a reduction in the population's growth rate.

Figure 14.20 For gannets, which nest on rocky islands such as this one off Cape St. Mary's in Newfoundland, the limited number of nesting sites determines the carrying capacity of the environment. As a result, only a certain number of pairs can nest and reproduce at any given time. When *N* is low, all birds can find a suitable nest site and reproduce, so the population grows. Above a certain *N*, however, many pairs fail to breed successfully and, as a result, population growth slows and eventually levels off.

BIO FACT

For some species, internal rather than external cues (such as a reduced food supply) seem to limit population size. Small rodents kept in experimental conditions with abundant food and shelter will reproduce quickly until their population reaches a certain density. At this point, even though the resources they need most are still unlimited, reproduction decreases. High density seems to produce a stress response in which hormonal changes delay sexual maturation, cause shrinkage of reproductive organs, reduce the effectiveness of the immune system, and often produce aggressive behaviour (sometimes including cannibalism). Exactly what cues trigger these hormonal changes is unclear, but similar effects of crowding have been noticed in some animal populations in nature.

Density-dependent limiting factors can involve interactions between two or more populations as well as interactions within a single population. For instance, two species with similar habitat requirements (for example, see Figure 14.21) may compete with each other for soil nutrients, food, or other resources found in that habitat. This type of interaction between two or more populations is referred to as **interspecific competition**. In some cases, one species may eventually out-compete the other or others, and the "losing" species disappears from the area (see Figure 14.22 on the next page). A result like this often indicates that the interacting species had very similar ecological niches. Ecologists explain the effects of interspecific competition by referring to the **competitive exclusion principle**. Essentially, this theory states that species with niches that are exactly the same cannot co-exist; if two species have completely overlapping niches, one will always exclude the other, as shown by the dashed line in Figure 14.22B. However, if the niches of competing species are sufficiently different, they can both live in a particular area, although the density of one or more of the populations may be lowered by the presence of the other (as shown by the dashed line in Figure 14.22A).

Figure 14.21 In Ontario, garlic mustard (*Alliaria petiolata*, a plant introduced from Europe) invades moist woodland habitats and crowds out many native species. It is a particular threat to some endangered plant species, such as the wood poppy (*Stylophorum diphyllum*).

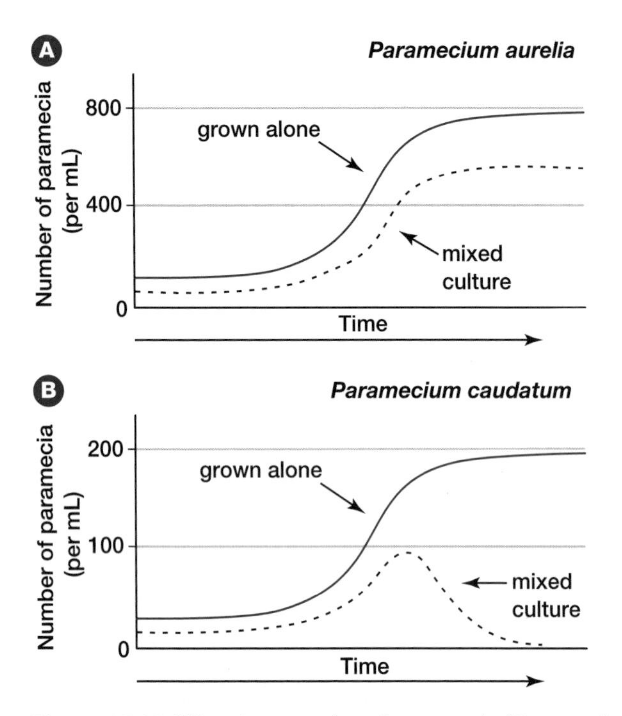

Figure 14.22 When two species of paramecia (*Paramecium aurelia* and *P. caudatum*) are grown in separate cultures in a laboratory, their populations follow a logistic curve to reach their own carrying capacities. This is shown by the solid lines. However, when they are grown together, *P. caudatum* goes extinct in the mixed culture, shown by the dashed line in (B), since *P. aurelia* is a much better competitor.

What this means is that differences between species can be increased as a result of natural selection, with interspecific competition as the driving force of evolutionary change. In each competing species, the individuals that are most different from their competitors will be best able to avoid competitive interactions and will therefore obtain the most resources. For example, if two species of birds compete for seeds of roughly equal sizes, those individuals of both species that can eat larger or smaller seeds will be able to find more food. They will, therefore, be more likely to survive and reproduce than will members of their own species that cannot avoid interspecific competition. As a result, their alleles — coding for the characteristics (such as beaks that can handle different seeds) that distinguish them from their competitors — will increase in frequency in subsequent generations. In this way, natural selection can produce increased divergence between competing species. In Figure 14.23, natural selection may take two different routes to lead to reduced competition. The total range of prey sizes taken by the two species may increase, with one species extending its preferences to smaller items than were taken previously, while the other may include larger foods than were eaten before. Alternatively, the total range of prey eaten may remain the same, but the niche of each species may shrink; one species may become a specialist on small prey while the other takes mainly large prey. Over time, this can increase the diversity of species living in a community.

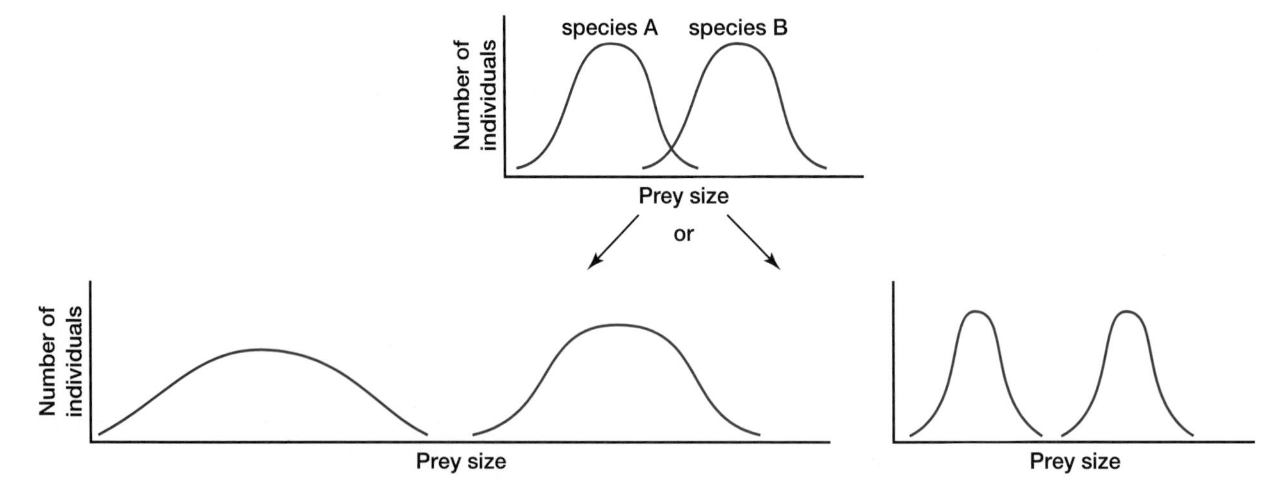

Figure 14.23 Within a species, individuals vary in many of their features. As a result, they use a range of whatever resources are necessary for their survival and reproduction. In this example, the members of two species, A and B, vary with respect to the size of their prey, but some individuals of both species eat intermediate-sized prey. This overlap produces interspecific competition, which may lead to evolutionary changes in both species.

Interaction of Predator and Prey Populations

Some populations, especially those of certain insects, birds, and mammals, fluctuate regularly in density. These alternating periods of high and low populations are often referred to as population cycles. While some small herbivorous mammals, such as voles, have 3- to 5-year cycles, larger herbivorous mammals, such as snowshoe hares (*Lepus americanus*) and muskrats (*Ondatra zibethica*), have 9- to 11-year cycles. These longer cycles are also typical of some birds, including ruffed grouse (*Bonasa umbellus*). The causes of cycles vary with the species and from population to population within a single species. Some may be due to a lag in response to density-dependent factors, as discussed in section 14.2. If such a lag is fairly constant, a more or less regular cycle of fluctuation above and below the population's carrying capacity could result.

Some of the mammal species that display fluctuations in population density are predators that have cycles overlapping those of their prey (such as some of the herbivores already described). One explanation for these cycles is the density-dependent effect of each population on the other. For example, some populations of Canada lynx prey almost exclusively on snowshoe hare (see Figure 14.24). An increase in the hare population would reduce competition for food among the lynx (*Lynx canadensis*), allowing them to increase their reproduction rate and survive longer. The result would be an increase in the population density of lynx. However, the presence of a large number of these predators would eventually cause the hare population to decrease. This, in turn, would increase competition among lynx for food, causing a decline in the predator population and permitting the prey population to expand once again.

BIO FACT

Some of the most remarkable population cycles in the world are those displayed by various species of cicadas (insects in the order Homoptera). The life cycle of these species takes place mostly underground, and requires 13 to 17 years to complete. At the end of this period, they emerge as adults in extremely high densities (as high as 600 per m^2). When the adults lay their eggs and die, the aboveground population shrinks to virtually nothing. The long life cycle may be an adaptation to reduce predation. Since these species are around for such a short time, few predators have learned how to prey on them efficiently.

However, there is more to this story than just the relationship between predator and prey. Hare populations on arctic islands where there are no lynx also undergo a cycle, indicating that it is not simply the effect of predators that causes hare populations to increase or decrease. An alternative hypothesis for fluctuations in the size of snowshoe hare populations is that the grazing activity of a large number of these herbivores causes serious damage to the plants (especially willows) that they eat. When the hare population is small, only a small portion of each plant is consumed. The plants can maintain high survival and reproduction rates in this situation, resulting in high plant density. This, in turn, allows the hare population to increase, perhaps to a point where their grazing damages the plants, thereby lowering plant survival and reproduction. The hare population will then decline because of a decrease in their food supply. A long-term research

Continued on page 488

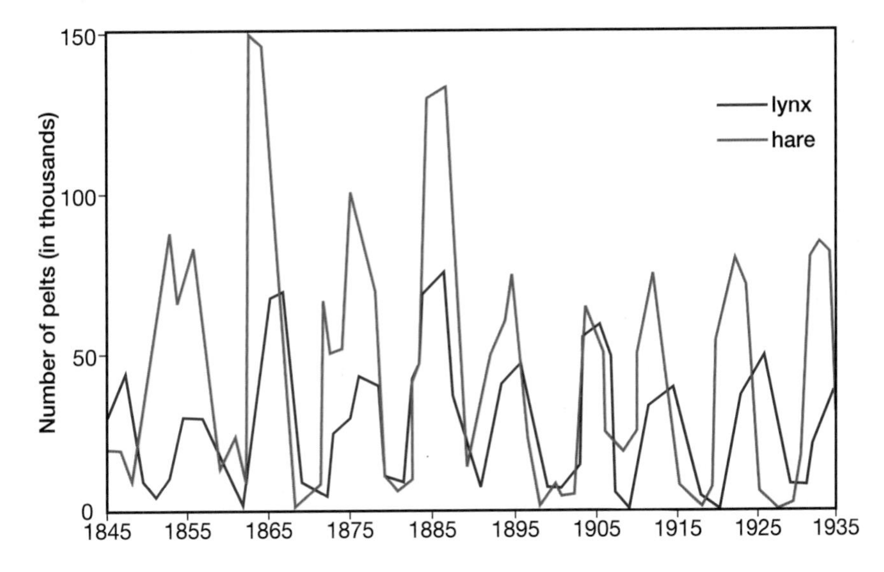

Figure 14.24 This graph shows the number of Canada lynx and snowshoe hare pelts traded annually to the Hudson's Bay Company over a 100-year period in Canada's arctic (data which allowed biologists to estimate the population sizes for both species). Note that approximately every 10 years, populations of both species become very large, decline, and then become large again. These cycles seem much too regular to be explained by randomly occurring abiotic factors.

Investigation 14 • A

SKILL FOCUS

- Initiating and planning
- Predicting
- Identifying variables
- Performing and recording
- Analyzing and interpreting
- Conducting research

Paramecium Populations

Cultures of small organisms, such as microscopic paramecia, can be used to study how various abiotic and biotic factors affect population growth patterns. These unicellular protozoa are ideal for studying changes in population size because they have a relatively short reproductive cycle and cultures can be easily maintained, measured, and manipulated in a laboratory environment. This investigation focusses on how density-dependent factors and density-independent factors affect the growth rate of paramecium populations. These observations will allow you to suggest how abnormal pH levels in natural ecosystems, resulting from acid precipitation or other types of contamination, might have negative effects on organisms living in aquatic environments.

Pre-lab Questions

- What is the natural food supply of micro-organisms such as paramecium?
- Describe the life cycle of paramecium.
- How would lower pH levels affect populations of organisms that inhabit aquatic ecosystems?
- How would changes in food availability affect populations of organisms that inhabit aquatic ecosystems?

Problem

How can you determine the impact of density-dependent factors (such as food supply) and density-independent factors (such as pH) on the growth rate of paramecium populations?

Prediction

Predict the effect of changes in pH level and food supply on the growth rate of paramecium populations.

CAUTION: **Handle hydrochloric acid with care. Wash your hands after each observation.**

Materials

glass jars of uniform size
culture medium (skim milk powder)
distilled water
paramecium cultures
droppers
heating sources
thermometers
dilute hydrochloric acid
pH hydrion papers (or other indicators that provide accurate pH readings)
microscopes
methyl cellulose

Procedure

1. For this investigation, your class should be split into several groups. One half of the class should focus on testing the effects of a density-dependent limiting factor (food supply), while the other should investigate the effects of a density-independent limiting factor (pH). Each student group will run one control and one or more experimental set-ups.
2. Identify the specific factor or variable (such as pH or food supply) to be investigated. This will be your independent (manipulated) variable. Select the range of factors to be tested. (For example, one group could maintain one control culture at neutral pH and an experimental culture at a slightly lower pH while the other groups maintain cultures at different pH levels. That way, a broad range of pH levels can be analyzed.)
3. Maintain all cultures at constant temperature and at uniform, medium light conditions. Avoid direct sunlight, drafts, and contamination of cultures. Ensure that all glassware is clean and uncontaminated by soap or other chemical residue. Leave each culture open to the air, but ensure that water levels remain constant by adding room-temperature distilled water as required.
4. To start, add about 5 g of skim milk powder to 250 mL of distilled water. Inoculate each set-up with an identical volume of paramecium culture (about three or four full pipettes or droppers). The control set-up for each group should be kept at neutral pH and be provided with the same food supply and starter paramecium culture. In fact, you may wish to maintain each set-up for three to four days with a standard food supply and neutral pH, to stabilize each paramecium population before you begin to manipulate the environments.
5. If you are investigating food supply as a variable, add different amounts of food to your experimental set-ups (the class should work with a range of 0–20 g of food per experimental culture).
6. If you are investigating pH as a variable, add varying quantities of dilute HCl (start with 0.1 mol/L HCl) to produce a range of pH, from neutral to about pH 3. Maintain a constant pH in each set-up throughout the experiment.

7. Use a microscope to view samples of your cultures each day. Estimate the paramecium population of each culture. Try the following procedure to view and count the paramecia in your samples:
   (a) Squeeze all the air out of the dropper bulb. Use the dropper to remove a few drops of your paramecium culture.
   (b) Place a drop of methylcellulose and one or two drops of water on the centre of a glass microscope slide. (You can also add a few grains of very fine, washed sand to slow down the paramecia and prevent the cover slip from crushing the specimens in your sample.)
   (c) Carefully squeeze the bulb to deposit your culture sample into this mixture.
   (d) Gently lower a cover slip over the sample.
   (e) View your sample with the low-power objective lens of your microscope. Scan the sample and count the number of paramecia in five different fields of view (view the centre and the area near each edge of the cover slip). Add up the total number of individuals counted in all of the samples and divide by five to determine an average estimate of the current population size. Enter your data in a summary data table.
   (f) Use the same procedure to observe samples daily from each set-up and record your estimates of population size in the appropriate data table.
8. Ensure that your sampling technique can provide an accurate estimate of the population size without unduly disturbing the integrity of the ecosystem (and, if possible, without causing any changes to the appearance of individual paramecia in each culture).
9. After the data are entered in your summary table, graph the results. If available, you may wish to use a computer spreadsheet program to record your experimental data and generate graphs. Combine data from other student groups to generate graphs depicting the class results for each experimental variable tested.

Post-lab Questions

1. Describe any observed changes in the paramecium population size. Graph the control and the experimental population sizes over the duration of your experimental procedure.
2. How do the size and growth rate of the control culture compare with the populations of your experimental paramecium cultures? Specifically, what do your results and the class results suggest about the impact of a density-dependent factor and a density-independent factor on the rate of growth of paramecium populations?
3. How did your results relate to your original prediction?
4. Discuss possible sources of error in your procedure (relating to such factors as the sampling procedures used and the ability to control any extraneous variables) that may influence your results.
5. Were your results consistent with the observations of other groups?

Conclude and Apply

6. What can you conclude about the accuracy of your original prediction?
7. How could your procedure be improved?

Exploring Further

8. Devise a supervised investigation to study populations of micro-organisms in actual aquatic ecosystems, such as ponds or streams. Have your teacher approve your experimental design and safety precautions in advance. Test the pH of each sample. Identify other significant density-dependent and density-independent regulating factors that may have an impact on certain species of microscopic organisms in local aquatic ecosystems (such as rivers or ponds). Indicate your sample locations on a local map.
9. Conduct research to assess population changes through the seasons in the ecosystems you have chosen.
10. What do you think is the mechanism by which changes in pH increase or decrease the rate of population growth? That is, why does the pH of its environment matter to a paramecium, or to any other organism? How would an effect on an individual organism translate into changes in population growth?
11. Which general type of regulating factor — density-dependent or density-independent — is most important in regulating the size of paramecium populations in nature? Or do you think both might be important, but under different circumstances? Explain your answer in full.

abiotic and biotic factors are part of the population's environment, the combination of their effects is sometimes referred to as **environmental resistance** to population growth.

The examples discussed under "Symbiotic Relationships" might suggest that populations of some kinds of organisms (such as insects) are regulated by abiotic, density-independent factors, whereas populations of other kinds of organisms (including birds and mammals) are regulated by biotic, density-dependent mechanisms. This apparent contrast led early ecologists to argue about which form of population regulation was more important, or at least more common. These discussions, along with more research, eventually led to the understanding that for most populations, both types of regulating factors play a role. For example, most of the time a population of birds may be regulated by density-dependent factors. However, occasional catastrophes may reduce the population to a point where density has little influence on the growth rate. Similarly, severe weather may regularly cause the size of an insect population to decline. If the environment contains a limited number of protective sites where members of this population can hide to avoid the weather,

THINKING LAB

Sustainable Harvesting

Background

Theoretically, populations regulated by density-dependent factors can be harvested. In other words, individuals can be removed for commercial and/or recreational purposes in such a way that the population is not depleted. This will be the case if the number of individuals being harvested (referred to as the yield of the population) is just equal to the number being added as a result of reproduction. One of the tasks of those who manage populations of organisms that humans use as resources (for example, forests; fish stocks; and deer, moose, and bear populations) is to try to determine the Maximum Sustainable Yield (MSY). MSY is the largest number of individuals that can be removed from the population each year (or other harvest period) without reducing the population's growth over the long term. In this lab, you will see how the MSY can be assessed — in theory.

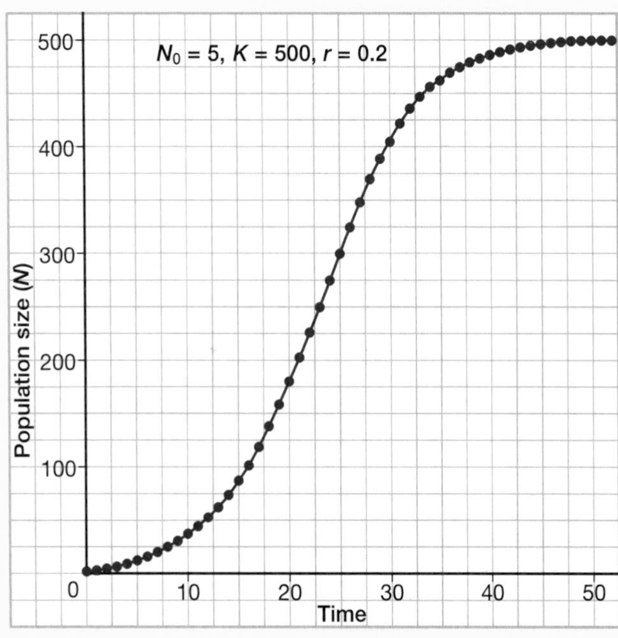

Growth curve for a theoretical, density-dependent population

You Try It

1. The graph shows the growth curve for a theoretical, density-dependent population living in an environment in which the carrying capacity is 500 ($K = 500$). The population started out with five individuals ($N_0 = 5$) and grew at a rate of 0.2 offspring per individual per time period ($r = 0.2$). Using the curve, estimate the number of individuals added to the population in the first five time units. Estimate the size of the population in the middle of this time interval (that is, at time = 2.5 units). Follow the same procedure for intervals 5–10, 10–15, 15–20, and so on up to 45–50. Record your data in a table, or use a spreadsheet program if available.
2. Plot the data you obtained in Step 1 for each of the 10 time intervals on a graph of "Number of individuals added to the population" (*y*-axis) versus "Population size" (*x*-axis). How big is the population during the interval when the maximum number of individuals is being added? Is N at this time near K, far below it, or halfway to K? Explain the shape of the curve you have plotted. Why does the curve increase to a maximum and then decrease?
3. Suppose you are the manager of this population and you are trying to determine its MSY. At what size would you want to maintain the population, and why? How would an increase in the population's size above the MSY point affect yield? How would a decrease affect yield?
4. Although MSY values have often been used to set harvest quotas (such as the number of fish that can be caught or deer that can be hunted), the system has not worked well in some situations. Many populations have been seriously depleted as a result of overfishing or excessive hunting, in part because quotas were set too high. Why do you think this might have happened? What problems can you see occurring as a result of using this method to set quotas?

density does play a role. The amount by which N exceeds the number of hiding places will affect the proportion of the population killed. In fact, most populations are probably regulated by a combination of density-dependent and density-independent factors.

The abiotic and biotic factors that regulate population size also influence other characteristics of populations. These characteristics will be described in the last section of this chapter.

SECTION REVIEW

1. MC What factors might eventually limit the exponential growth rates of a particular insect species found in such regions as the Canadian prairies?
2. K/U What might allow some animal populations to grow beyond the theoretical carrying capacity of their environment and then crash to abnormally low levels?
3. K/U What factors might limit the population size of animals (such as vultures) that live as scavengers?
4. K/U In a density-dependent population, why does population growth decline as population density increases?
5. MC Provide a real-life example of a situation in which interspecific competition limits the population size and growth of a particular species.
6. K/U What combination of factors might produce regular population cycles typical of small herbivore species such as mice and squirrels?
7. I Rat populations in many urban areas in Canada have increased dramatically in recent years. Use the Internet to research public health records and other web sites for data about the extent of this problem. Describe the factors that seem to be contributing to the growth in local rat populations and the factors that may ultimately limit their population size.
8. MC Biologists have observed increasing instances of coral bleaching in recent years. This phenomenon involves the death of small green algae that live in the tissues of coral polyps (the tiny animals whose bodies secrete much of the material that forms coral reefs). Following the death of the algae, the coral organisms that hosted them also die. Describe the probable type of relationship that exists between the algae and coral organisms.
9. K/U Some species of sharks are often seen with small fish called remoras attached to their bellies. The remoras seem to feed off the scraps of food dropped by the sharks when they are feeding on prey. The sharks seem to derive no benefit from the presence of the remoras. What type of symbiotic relationship might the association of the shark and remora illustrate? Do you think it likely that the shark is totally unaffected by the remora? In what way might it be affected? In what way might your answer to this last question change your interpretation of the type of relationship exhibited here?
10. K/U What is the difference between the way in which abiotic and biotic factors typically produce environmental resistance to population growth?
11. MC Roosevelt elk (*Cervus elaphus roosevelti*) were re-introduced to the Sunshine Coast region of British Columbia because overhunting eliminated the original elk population many years ago. Their population has now grown to about 250, and some local residents are complaining that the elk are devouring plants in gardens, parks, and golf courses. Limited hunting of this population is now permitted in an attempt to control the problem. Do you think hunting of this elk population should be permitted? How do you think the resumption of hunting might affect the long-term growth patterns of this population and other species that share the same environment as the elk?
12. MC In ecological terms, what measures would be most effective in controlling populations of pest species such as rats, mice, or certain types of insects?
13. I Having been recently appointed regional wildlife manager, you must set a quota on the number of moose that can be hunted during the next hunting season. What information do you need? Design a study that will allow you to obtain the necessary data.
14. C You have just been hired to teach Grade 6 at a local school. Design a lesson plan, including an activity, that you might use to teach your students about the factors that affect population growth in different environments.
15. C Although all populations eventually face environmental resistance to continued growth, the contribution of abiotic and biotic factors to this resistance may vary from species to species. Compare a micro-organism, such as *E. coli*, a plant (such as a type of tree), and a mammal, such as the snowshoe hare or the black bear, with respect to the type of factors that typically limit the growth of populations of each species.

14.4 Life History

EXPECTATIONS

- **Describe the features of an organism's life history, and discuss the evolution of the life history typical of a particular population.**
- **Distinguish between three general patterns of survivorship.**
- **Discuss the factors that relate to the fecundity of females in different populations.**
- **Describe what is meant by the age structure of a population and what produces it.**
- **Differentiate between equilibrial and opportunistic life histories, and describe the type of selection, *r* or *K*, that tends to produce each of these life history patterns.**

As described earlier, populations vary with respect to the factors that regulate their size. They also vary with what are called the life history features of their members. **Life history** refers to the schedule of certain important events that occur during the life of an organism. Important features of an individual's life history include the age at which it reaches sexual maturity, how often it reproduces, and how many offspring it has each time. Life history also includes features related to the life span of the organism.

Like most of its traits, the life history features of an organism are genetically controlled and are the result of natural selection operating over evolutionary time. That is, these features have been selected to maximize evolutionary fitness (the number of offspring that survive to reproduce) under certain environmental conditions. Because the members of a population all live in the same environment, they share similar life history features. Ecologists therefore tend to refer to the particular survival and reproductive patterns shown by the individuals in a population as the life history typical of that population.

Survivorship

As previously mentioned, one life history feature that varies among populations is the average life span of individuals. Ecologists often compare different populations in this respect by using a standard measure, called survivorship. **Survivorship** is the proportion (percentage) of individuals in the population that typically live to a given age. The most accurate way to determine survivorship is to monitor a large group of individuals all born at the same time (a group like this is termed a **cohort**) throughout their lives, recording the age of death for each organism until the last one dies. This information is then usually graphed as shown in Figure 14.29.

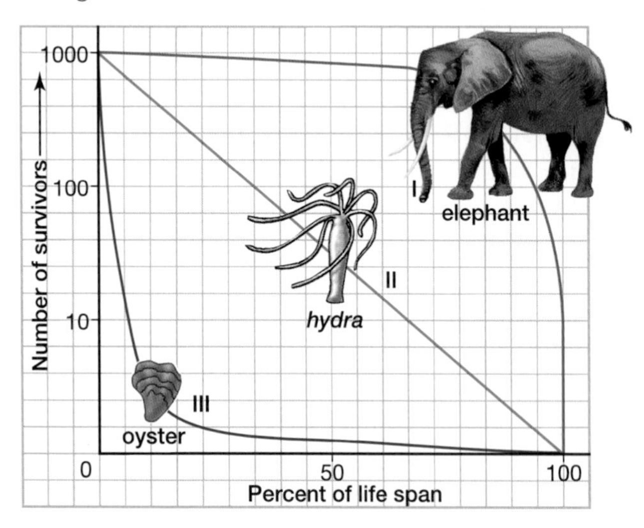

Figure 14.29 The three general patterns of survivorship are known as Types I, II, and III. Note that the number of individuals in the population is often standardized to 1000, regardless of the true size of the cohort. Which type of survivorship do you think is characteristic of human populations?

After studying many different populations, ecologists established that there were three general patterns of survivorship (as Figure 14.29 illustrates). In populations that have a Type III survivorship pattern, most individuals die at a young age (probably soon after birth or hatching) and very few survive long enough to reproduce. In contrast, populations with a Type I survivorship pattern have high juvenile survival. Most individuals in the population live past the age when sexual maturity is reached, and the percentage of survivors from a cohort does not start to decline rapidly until a relatively old age. The Type II survivorship curve lies in between these two. In populations with this pattern, the percentage of individuals dying is constant over all ages.

It is important to realize that not all populations fit one of these three survivorship patterns. There is, in fact, a continuum of survivorship curves and the survivorship pattern for a particular population may fall somewhere between Types I and II, or between Types II and III. As is often the case, ecologists start by simplifying the complexities of survival patterns in order to try to understand them. Nonetheless, many populations do fit one of these three survivorship curves fairly closely. Populations that share a common survivorship pattern also tend to be similar with respect to other life history features, as described below.

Fecundity

As mentioned, populations vary with respect to patterns of reproduction as well as survivorship. The average number of young produced by a female over her lifetime can be referred to as the **fecundity** of the population. In some populations, females produce their lifetime quota of young all at once and then die. This is typical of many species of insects and annual plants (those in which individuals live for only one growing season, such as sunflowers and petunias), as well as some vertebrates, including several species of salmon. Populations in which females only reproduce once and then die have non-overlapping generations.

In other populations, females typically survive their first reproductive event and go on to reproduce several more times. Many types of vertebrates, including humans, songbirds, and elephants, are like this. So are perennial plants (those that survive for more than one year), including many species of shrubs and trees. The fact that females reproduce more than once leads to populations with overlapping generations. In such cases, it is common for a female and her previous offspring to all be reproducing during the same time interval.

Populations also vary with respect to the number of offspring females typically produce each time they give birth. The females of some species (for example, humans, elephants, and grizzly bears) commonly produce only one offspring at each reproductive event. In contrast, some females have many young each time they reproduce. A single marigold plant may produce hundreds of seeds. In some populations of Chinook salmon (*Oncorhynchus tshawytscha*), each female may lay over 15 000 eggs.

Since the energy a female can accumulate and use for reproduction is limited, the size of individual offspring tends to be inversely related to the number produced. Newborn humans and elephants are much larger in relation to the size of their mothers than are newly hatched salmon or spiders. For the same reason, the number of offspring also tends to be inversely related to the amount of care parents give their young. A mother that produces hundreds of offspring cannot feed or protect each one for long, if at all. A single young, however, often receives (and requires) a substantial amount of care, in some cases staying with one or both parents for years.

Age Effects

As the survivorship curves show, in any given population an individual's probability of dying varies with its age. In populations with a Type I survivorship pattern, a very young individual has a low probability of dying, compared with an older one. As you might expect, the fecundity of an individual also varies with its age. Often very young and very old individuals have a zero probability of producing offspring. In many populations, individuals have the highest fecundity in the early to middle period of their lives (see Figure 14.30). Of course, in populations in which females reproduce only once, individuals do not survive after reproducing; therefore, there are no older, non-reproducing members.

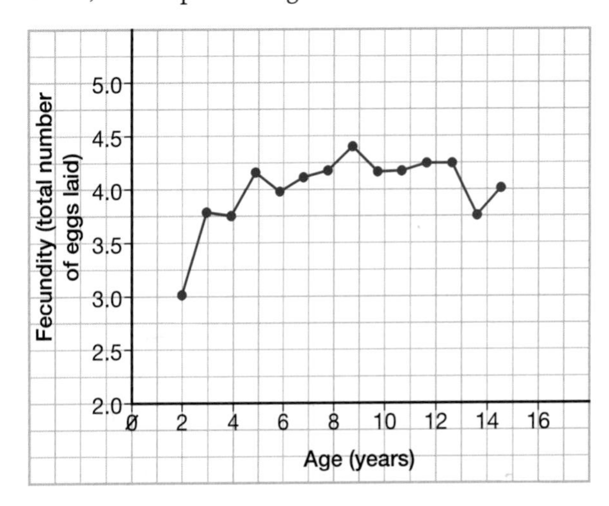

Figure 14.30 The fecundity (measured in terms of the number of eggs laid) of female snow geese (*Chen caerulescens*) increases until the birds are about nine years old and then gradually declines.

For a particular population, ecologists often summarize the relationship of survivorship (and its opposite, probability of death, which is often referred to as **mortality**) and fecundity to age in what is called a **life table**. The first column of a life table typically shows the age of individuals in the cohort being studied. In Table 14.2 on page 496, age is expressed in years, but in life tables for short-lived species, it may be in months or even days.

Life tables also vary with respect to the amount of detail they include. Although ecologists sometimes build tables with many columns expressing survivorship and fecundity in different ways, many look just like the one shown in Table 14.2 on page 496. The third column of the table records age-specific survivorship — the proportion of the cohort that survives to the start of the age class. For example, of the 530 grey squirrels forming this cohort, only 0.253 or 25.3 percent were alive one year later. Put another way, when one of these squirrels was born, it had a 25.3 percent chance of surviving for one year. Understanding survivorship is important for making accurate estimates of future population size.

Biology Magazine TECHNOLOGY • SOCIETY • ENVIRONMENT

Alien Invaders

What do rice, dogs, gypsy moths, and *Ophiostoma ulmi* (a fungus that causes Dutch elm disease) have in common? All originated in other parts of the world and were introduced to North America by people. They are non-indigenous, or alien, species. Many species were transported deliberately to provide food and other materials, or for purposes of sport, landscaping, or biological control. Some were introduced to this continent to be pets. Many others, however, arrived as unwanted stowaways — hidden passengers on ships and aircraft.

The rate of transportation of species from one part of the world to another has increased greatly over the past 40 years, largely due to an increase in international trade and travel. Some alien species are not suited to their new environment and do not survive. Others thrive and rapidly expand their populations and range. When uncontrolled, these invasive organisms produce catastrophic changes in local ecosystems. They compete with and prey on many indigenous species. They have become a major cause of species extinction and a loss of biodiversity.

An estimated 50 000 non-indigenous species live in North America today, and their number is increasing yearly. Over 40 percent of the plants and animals on the threatened or endangered species lists in North America (and as many as 80 percent of endangered species in other regions of the world) are at risk as a result of pressures created by alien invaders.

The Alien Advantage

Why do some introduced species undergo a rapid population growth? In their original home, their numbers are limited by the availability of resources (such as food, soil nutrients, and water) and by various predators, diseases, or competitors. When transported to another environment, they potentially have a double advantage. Their new home may lack some or all of the predators, diseases, and other factors that kept their numbers in check, and it may have more available resources because of less competition.

The process of natural selection ensures that species have continually adapted to their surroundings over time. Moving organisms from the ecosystem in which they evolved to a different ecosystem that has had no previous experience with them suddenly creates a new selective factor that may put native organisms at a disadvantage. For example, many species living on islands evolved without the predators and competitors faced by their relatives on the mainland. As a result, the island populations commonly lack important defences. Why put energy and materials into structures and behaviours that are not needed? Many island birds cannot fly and many island plants do not have the thorns or bitter tastes and poisons that deter herbivores. When predators such as rats, snakes, and cats are introduced to islands, flightless birds are helpless and are quickly reduced in numbers or exterminated. When herbivores such as goats and cattle are taken to islands, the native vegetation is overgrazed and destroyed.

Invading species have dramatically altered aquatic ecosystems as well as terrestrial ones. In the introduction to this chapter, you learned about the zebra mussel. It colonizes on surfaces such as docks, boat hulls, commercial fishing nets, water intake pipes and valves, native molluscs, and other zebra mussels. A rapid breeder, zebra mussels have been known to reach a density of 700 000/m^2 in some locations. These large mussel populations have completely covered and displaced native mussels, clams, and snails, and have reduced food and oxygen for native fauna. Their only known predators are some diving ducks, fresh-water drum, carp, and sturgeon, but these predators are not numerous enough to have a significant effect on them. As well, zebra mussels have caused billions of dollars in

The fourth column of Table 14.2 gives age-specific fecundity — the average number of offspring born to females of that age. For this grey squirrel population, fecundity is highest at ages three and four. Although some life tables record the average number of offspring for both males and females, most concentrate only on females (since in many species males play no role in parental care and it is thus difficult to determine how many offspring belong to them). Some ecologists prefer to include only data on survival and mortality in life tables. They refer to tables with data on the numbers of births as fecundity or fertility tables. Others, however, include both types of data in the same table. To make comparisons among populations easier, the size of the cohort is often standardized to 100 or 1000, regardless of how many individuals were studied.

WEB LINK

www.mcgrawhill.ca/links/biology12

To find current data on the status of introduced species (such as the zebra mussel) and other invaders (like the West Nile virus), visit the web site above, and click on **Web Links**.

damage by invading and clogging water intake pipes at water filtration and electrical generating plants (see the photographs on pages 464 and 465).

Nearly all the crops and livestock found over thousands of square kilometres of Canadian farmland are alien organisms, brought here originally by immigrants. Wheat replaced indigenous grasses on the prairies and cattle replaced the bison.

Challenges Ahead

The patterns of plants, animals, micro-organisms, and diseases that have shaped the North American continent since the last ice age are changing. As invading species spread and transform the landscape, other changes caused by humans threaten to encourage the population growth of a few species while threatening the survival of many. A warming climate may give some introduced insect pests and weedy plants a greater survival rate over winter. The escape of genetically modified organisms, farmed fish, and other exotic species will add further pressures to indigenous organisms and ecosystems. The natural world of tomorrow may mirror the gradual narrowing of the cultural diversity being experienced in human society. Some people maintain that large multinational companies are contributing to the increasing homogeneity of products found in cities around the globe. So, too, a relatively few hardy, invasive species with high population growth rates may, by the end of the twenty-first century, replace the rich variety of organisms still living today.

Follow-up

1. Fish farms on Canada's west coast have been raising increasing numbers of Atlantic salmon, which originated from the east coast. How do Atlantic salmon differ from Pacific salmon? What threats are posed by the potential release of Atlantic fish into Pacific waters?
2. Carry out a case study of an invasive alien species living in your area or another part of Canada. Document its origin and the problems caused by its introduction. What can be done to reduce or eliminate these problems?
3. In 2000, Canadian authorities tested many dead birds for the presence of West Nile virus. First detected in the United States in 1999, it was feared that this virus would soon spread to Canada. It was not until August 2001 that this virus was first discovered in Ontario, and by November of that year there were 125 birds in which the presence of the virus had been confirmed. Although a virus is not typically considered a living organism (and therefore might not be described as an "introduced species" by some), this pattern of rapid spread is typical of many species introduced into an area. Do research to learn more about the West Nile virus. What is its effect on humans? How did it arrive in North America and why is it able to spread so quickly? Where in Canada is it currently found?

Table 14.2
Life table for a grey squirrel (*Sciurus carolinensis*) population

Age class (years of age)	Number of individuals in studied cohort alive at beginning of age class	Proportion of cohort alive at beginning of age class	Fecundity (average number of offspring born to females in that age class)
0–1	530	1.000	0.00
1–2	134	0.253	1.28
2–3	56	0.106	2.28
3–4	39	0.074	3.24
4–5	23	0.043	3.24
5–6	12	0.022	2.48
6–7	5	0.009	2.28
7–8	2	0.003	2.28

Because individuals of different ages have different probabilities of surviving, every population has a different **age structure** — a different distribution of individuals through the age classes. The age structure of a population can be represented by an **age pyramid**, which shows the proportion of individuals in the population in each age class. Populations like the one shown in Figure 14.31A have by far the largest number of individuals in the youngest age class. In other populations, individuals are spread more evenly throughout all age classes (see Figure 14.31B). You will look more closely at the effect of age structure on population growth in Chapter 15.

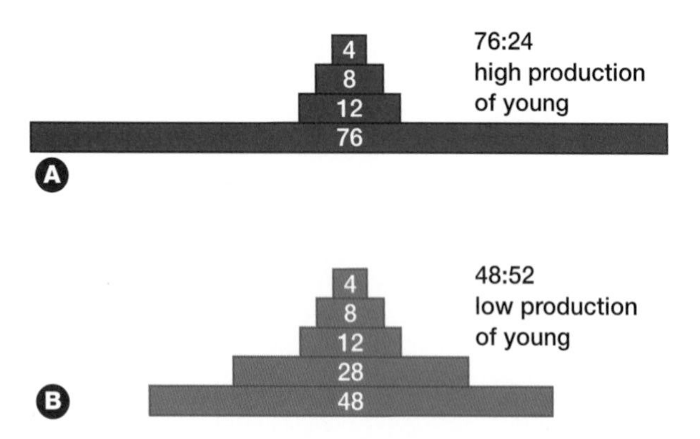

Figure 14.31 In both of these age pyramids, the bottom level represents the youngest members of the population, which are not yet of reproductive age. The other levels represent older age classes containing sexually mature adults that are reproducing or past reproductive age. Note that in (A) three quarters of the population is very young, while in (B) less than half of the population is in the youngest age class.

Life History and Environmental Stability

Recall that life history traits are genetically controlled and are the result of natural selection acting to increase the evolutionary fitness of the individuals in a population. How might an organism maximize the number of its offspring that survive to reproductive age? Theoretically, the best way to do this would be to start reproducing at a young age, live for a long time, reproduce many times over that long life, produce a large number of offspring each time, and give each of them enough parental care that they survive to reproductive age.

Unfortunately, this is not possible in real life. The amount of nutrients, energy, and time that can be gathered or used by each individual is finite, so not all of these things can be maximized simultaneously — trade-offs must be made. Many studies have shown the existence of such trade-offs in nature. For example, female beetles that produce fewer offspring during one reproductive event have a higher chance of surviving to reproduce again.

Given that trade-offs must occur, in which aspect of its life history should an individual invest the most? What combination of survival and reproduction will maximize its fitness? The optimal combination of life history features is determined by the environment in which the organism lives. One factor that ecologists have recognized as particularly important in selecting for specific life history features is the stability or predictability of the environment.

In many habitats (such as the shorelines of certain oceans and mountain highland areas), conditions are relatively variable and unpredictable. In such situations, an individual may have only a small chance of surviving from one breeding season to the next, regardless of how large and healthy it is. Here, an organism that expends all its available energy on producing as many young as possible while conditions are good will probably leave behind more offspring than one that produces fewer young and invests some of its energy in survival. Chances are high that the latter individual will lose this energy investment, since conditions are likely to become unsuitable for reproduction before it can try again.

Environments of this type tend to be unpredictable because of abiotic factors, such as variable weather conditions. Populations in these environments are regulated by density-independent factors and can grow rapidly and reach a large size when conditions are good, but may suddenly crash (perhaps due to a storm, flood, or drought). Individuals making up such populations are selected to take advantage of whatever opportunity they can to reproduce, since conditions may become unfavourable at any moment and their chance of surviving to reproduce again is small.

An organism like this, which typically produces many young in a single burst of reproduction, displays what is often referred to as an **opportunistic life history**. Because populations made up of such organisms have a high *r* value (a high growth rate), this type of life history is also referred to as *r*-selected. Natural selection that produces such a life history is referred to as ***r*-selection**.

In contrast, populations living in relatively stable environments generally fluctuate much less in size. In fact, they usually remain near an equilibrium point set by the carrying capacity (*K*) of the environment. Because they live in high-density situations, individuals in such populations must compete with one another for limited resources. The winners of these interactions are the individuals whose adaptations give them a competitive advantage. These adaptations tend to make them highly specialized for survival and reproduction in

THINKING LAB

Life History — Putting It All Together

An equilibrial life history is the result of selective forces acting in a relatively stable environment, just as an opportunistic life history is the result of the forces typical of unpredictable environments. Because these forces act simultaneously on more than one of an organism's traits, each of these idealized life history categories is characterized by a whole set of features, most of which have been discussed in this chapter. This Thinking Lab enables you to summarize your understanding of the features typical of *r*- and *K*-selected life histories, which are explained on pages 496–498.

You Try It

1. Construct a table like the one shown here. Fill it in by answering the following questions. Your answers should describe characteristics of the environments in which *r*- versus *K*-selection occurs, as well as some of the life history features that result from these two types of selection.
   - **(a)** Is the climate or environment relatively stable or is it unpredictable?
   - **(b)** Is the population regulated mainly by density-independent or density-dependent factors?
   - **(c)** What is the form of the survivorship curve: Type I (low death rate for young individuals) or Type III (high mortality early in life)?
   - **(d)** Does the population vary in size or does it tend to remain stable near the carrying capacity of the environment?
   - **(e)** Is the level of competition in the population high or low?
   - **(f)** Do organisms grow quickly and reach sexual maturity at a young age, or do they grow slowly and delay reproduction?
   - **(g)** Does each organism in the population typically reproduce only once or several times over its lifetime?
   - **(h)** Do the organisms usually produce many small offspring or a few larger ones?
   - **(i)** How extensive is parental care?
   - **(j)** What is the relative body size of the organisms?
   - **(k)** Do the organisms have relatively long or short lives?
2. Study Table 14.2. Construct two life tables, one for an extremely *r*-selected species (one at the "*r*" end of the *r*–*K* selection continuum), and the other for an extremely *K*-selected species. Your tables should reflect what you know about the differences between the fecundity and survivorship of organisms with these two types of life histories.

Comparison of opportunistic and equilibrial life histories

	r-selection (opportunistic life history)	*K*-selection (equilibrial life history)
climate/environment		
type of factors regulating the population		
survivorship		
population size/density		
level of competition		
growth rate/age of sexual maturity		
number of reproductive events during lifetime		
number and size of offspring		
parental care		
body size (small/large)		
life span		

a particular habitat. For example, they might be specialists at finding, hunting, or eating a particular type of food, or at physiologically coping with a particular temperature and moisture regime.

In a stable environment, individuals that produce a few large offspring that are cared for to ensure their survival are favoured. They are more likely to pass their genes along to subsequent generations. An organism like this is said to have an **equilibrial life history**. This type of life history is also called K-selected, since it is seen in populations that are near K in size. In these environments, ***K*-selection** is at work.

As always, it is important to realize that the distinction between these two general types of life history is not absolute. Real environments are highly complex. So, too, are the selective forces that act within those environments. Most organisms therefore, experience both r- and K-selection to a greater or lesser extent, and many display features of both types of life histories. It is probably useful to think of purely opportunistic (r-selected) and purely equilibrial (K-selected) life histories as forming opposite ends of a theoretical continuum. Most populations lie somewhere between these two extremes. In the Thinking Lab on page 497, you will illustrate your understanding of the differences between an opportunistic life history and an equilibrial life history.

Looking Ahead

Human life history features are closer to what ecologists would consider the equilibrial or K-selected end of the scale. This suggests that for most of its evolutionary history, the human population has been at or near its carrying capacity. Several lines of evidence from archeologists, anthropologists, and historians support this viewpoint.

However, as will be discussed in the next chapter, about 8000 years ago humans managed to change their environment's carrying capacity. The human population then entered into a period of rapid growth that is still underway. Much of the biosphere has been affected by this growth, and the ecology of many species has been altered as a result. The future survival of an array of species, including *Homo sapiens*, may depend on whether the growth of the human population and its impact on Earth can be controlled.

SECTION REVIEW

1. **K/U** Identify each of the elements in the $(K-N)/N$ portion of the logistic growth equation (that is, what does each of the letters represent?). How does this portion of the equation affect the overall rate of growth of a population?

2. **K/U** Distinguish between opportunistic and equilibrial life histories.

3. **K/U** Distinguish between r- and K-selection.

4. **K/U** What factors typically limit the population size of species with very high birth rates?

5. **C** Draw a survivorship curve for a population in which most of the mortality occurs among the youngest members of the species. Provide examples of two species that demonstrate such a survivorship pattern.

6. **K/U** What stage of an organism's life cycle typically has the highest fecundity rate? Explain your answer.

7. **K/U** Do humans typically have only one reproductive event or several over their lifetimes? Explain your answer.

8. **MC** What is the survivorship pattern for Canada's human population?

9. **K/U** Compare the typical age structures seen in populations with Type I, Type II, and Type III survivorship patterns. Draw a diagram comparing the three types of survivorship patterns.

10. **K/U** How might generation time affect r, the population growth rate?

11. **I** Design a study that would allow you to determine whether a particular species had a more opportunistic or equilibrial life history.

12. **I** Think of a habitat near your school or home. List 10 species that this habitat contains. Do these species have equilibrial or opportunistic life histories? Design a study that would allow you to find out.

13. **C** Compare two species of plants, the dandelion (*Taraxacum officinale*) and the red maple (*Acer rubrum*) and two species of mammal, such a mouse and a black bear (*Ursus americanus*), with respect to whether they are more r- or K- selected. Specifically, compare them with respect to their fecundity, the shape of their survivorship curves, life span, population density, and whether the environment each species typically inhabits is more or less stable or unpredictable.

Summary of Expectations

Briefly explain each of the following points.

- The size and density of populations can be estimated using mathematical formulas. (14.1)
- Individuals in a population can be dispersed randomly, uniformly, or in a clumped pattern. (14.1)
- Four basic processes cause change in population size. (14.2)
- Population growth can be described using either an exponential or logistic (sigmoid) growth curve. (14.2)
- The rate of change in the size of populations growing exponentially or logistically can be determined mathematically. (14.2)
- The future size of exponentially growing populations with overlapping or non-overlapping generations can be predicted. (14.2)
- For populations growing in limited environments, the effect of population density on the rate of change in population size can be determined. (14.2)
- Density-dependent and density-independent regulating factors affect a given population. (14.3, 14.4)
- Various types of relationships exist among the members of different populations. (14.3)
- Different populations have characteristic life history features; certain features are typical of equilibrial life histories, while others are characteristic of opportunistic life histories. (14.4)
- Populations in unpredictable environments tends to be *r*-selected, whereas those in stable environments tend to be *k*-selected. (14.4)
- Populations vary with respect to their age structure. (14.4)
- Ecologists sometimes simplify explanations of complex phenomena. (14.2, 14.3, 14.4)

Language of Biology

State the biological significance of each of the following terms. Create concept maps to show your understanding of how they are related. Be prepared to explain your rationale.

- demography
- density
- dispersion
- census
- sample
- transect
- quadrat
- birth
- immigration
- death
- emigration
- growth rate (*r*)
- biotic potential
- exponential growth
- J-shaped curve
- non-overlapping generations
- replacement rate (*R*)
- overlapping generations
- logistic growth
- S-shaped curve
- carrying capacity (*K*)
- density-independent factor
- density-dependent factor
- intraspecific competition
- interspecific competition
- competitive exclusion principle
- symbiotic relationship
- host
- symbiont
- parasitism
- mutualism
- commensalism
- environmental resistance
- life history
- survivorship
- cohort
- fecundity
- mortality
- life table
- age structure
- age pyramid
- opportunistic life history
- *r*-selection
- equilibrial life history
- *K*-selection

UNDERSTANDING CONCEPTS

1. What factors contribute to determining how individuals are dispersed in an area?
2. For the rate of population growth (*r*) to be greater than zero, what has to be true of the birth and death rates (*b* and *d*) in the population?
3. In general terms, describe the rate of change in the size of a logistically growing population in each of the following situations, and explain why the population size is changing at this rate.
   (a) when the population is very small relative to the carrying capacity of the environment
   (b) when the population size is roughly one half the carrying capacity (that is, near the inflection point of the logistic growth curve)
   (c) in the shoulder of the curve, when the population is near the carrying capacity
4. Why might the carrying capacity of the environment for a particular species change as its population size increases?
5. Select an organism living in an environment with which you are familiar, and list the things that might limit the growth of its population.
6. In general terms, how does competition with other species affect the growth of populations?

7. How might predation or parasitism explain why a population's size does not increase indefinitely?

8. In what way is the relationship between a population of herbivores and the plants they eat similar to the relationship between predator and prey populations?

9. Compare competition and mutualism with respect to their effect on the growth of two interacting populations.

10. (a) Explain why two species with very similar niches have a high probability of competing with each other. How can they co-exist?
    (b) What process or processes might occur to reduce the competition?

11. Is a virus a predator or a parasite?

12. Female grizzly bears typically have only one offspring each year. Black bears, however, often produce twins or even triplets if conditions are favourable. Female grizzlies do not begin breeding until they are three to four years old, whereas black bear females reach sexual maturity at two years of age. Grizzly bears have a more restricted diet and are more particular about what type of sites they will use for dens. Grizzly bears are also larger than black bears. Which of these species do you think is more *K*-selected? Justify your answer.

13. Give three examples of species that typically have Type III survivorship curves and three that usually have Type I survivorship curves.

14. Clumped dispersion patterns are often the result of what might be called positive interactions among members of a species; if one individual is present at a given spot, the chances are higher that more will be at the same spot. Uniform dispersions are the result of negative interactions. In this case, the presence of an individual (or pair) tends to lower or decrease the probability (chance) that another individual will be found nearby. With this in mind, how might you describe a random distribution in terms of the probability of finding individuals near each other?

15. For a population with overlapping generations to increase in size, the growth rate (r) must be greater than zero. In a population with non-overlapping generations, what must the replacement rate be for the population to increase in size?

16. A population of deer contains 10 000 individuals. During a year, 5000 individuals are born and 4500 die. What will be the size of this population 10 years from now?

INQUIRY

17. In order to estimate the size of a population of oak trees, a forester runs several 100 m transects through a 100 ha woodlot and counts the number of oaks within 5 m of the line. Five of the transects produced the following results: 15, 17, 25, 16, and 20.
    (a) What is the forester's estimate of the density of oaks in the woodlot?
    (b) Estimate the size of the oak population in this woodlot.
    (c) Design a sampling procedure that would allow the forester to obtain a more accurate estimate.

18. A researcher caught and tagged a sample of 50 deer mice in an isolated field. She waited two weeks and then captured a second sample, consisting of 63 mice. Of these 63 mice, 10 were tagged — they had been in the original sample and were captured a second time. How big was the population of mice in the field? What might cause this estimate to be biased? How might the researcher reduce such bias in future studies?

19. Brewer's yeast is a type of fungus that produces ethyl alcohol as a waste product of metabolism. Reproduction most often occurs asexually by a process called budding, in which one cell produces another identical to itself. Consider this to be an example of reproduction in which two offspring are produced and the "parent" dies. Under good conditions, yeast reproduces every 15 minutes.
    (a) If a small amount of yeast (say 500 cells) were used to start a culture, how many cells would be present after 24 h?
    (b) What do you think naturally limits the growth of yeast populations?
    (c) Design an experiment that would allow you to determine what naturally limits the growth of yeast populations.

20. Moose are not native to Newfoundland. The first moose — two males and two females —

were transported to the province in 1904. They spread over the entire island relatively quickly and the current moose population is 120 000 to 150 000 individuals, the densest in North America. The population is this large even though a moose hunt was started in 1930 and some estimates suggest that more than 180 000 moose have been taken legally by hunters since 1945. Do research to investigate the biology of the moose, the environment it encounters today, and the past environment to explain why this species has been so successful.

21. Do research to compare the life histories of timber rattlesnakes (*Crotalus horridus*) and massasauga rattlesnakes (*Sistrurus catenatus*). Alternatively, choose two Ontario mammals that interest you to compare. Which species is more *K*-selected? What is the status of each of these species in Ontario? That is, is either one threatened or endangered? How do you think the life history of each species has affected its ability to cope with environmental change?

22. You are part of a team that has been assigned to study the bald eagle (*Haliaeetus leucocephalus*), which is endangered in Ontario. The overall goal of the research is to prepare a recovery plan for this species in the province. Select one aspect that you would like to study and write a research proposal detailing what you need to know and how you plan to get the data.

COMMUNICATING

23. Make up a pamphlet, poster, or Powerpoint™ presentation that could be part of a public education campaign designed to limit the spread of zebra mussels. Explain why the biology of this organism makes it such a problem.

24. Recently, scientists have discovered that the presence of zebra mussels can be beneficial in at least one way. Do research on the zebra mussel to identify benefits that this species might provide. Prepare a poster showing what you have found and compare it with what you prepared for question 23.

25. The population of trout in a certain small lake has fluctuated around 2000 for the last 10 years. The owner of a fishing resort on the lake wants to increase the number of trout available for his clients. He plans to stock the lake with 1000 more individuals of the same species. Explain to him what you think will happen. Include a graph in your answer.

MAKING CONNECTIONS

26. Prepare an essay describing what you think was the cause of the decline of either the cod population off the coast of Newfoundland or the Pacific salmon population off the coast of British Columbia. What do you think should have been done to avoid this problem? If you think it is possible to restore these populations, suggest how it might be done. If you think it is not possible, explain your reasoning.

27. Farmers and gardeners refer to many types of plants as weeds and various species of insects as pests. What features do weeds and pests have in common? What is it about the typical life histories of these species that make them difficult to deal with? Why do you think they share these life history features?

28. A resource manager has data on the number of plants browsed by moose along a transect through a particular area. She wants to convert this measure of relative density to an estimate of absolute density. What would she need to know in order to do this?

29. Review the list of plant and animal species that are at risk (threatened, endangered, or otherwise at risk) in Ontario. Taking into account concepts you have learned in this chapter, do these species tend to have any features in common? Are there features that distinguish them from species that are not at risk in the province? How might you use this information to better manage or conserve Ontario's biological diversity?

30. Describe three examples of mutualism not mentioned in this chapter. In any of these cases, is one of the species at risk or is its population size declining? If so, what do you predict might happen to the other species in the relationship?

31. Many species considered to be at risk (threatened or endangered) in the world have equilibrial life histories (that is, they are *K*-selected). Give two reasons why such species might be more at risk, explaining your reasoning.

CHAPTER 15

Human Ecology

Reflecting Questions

- How do you determine the carrying capacity of Earth for humans?
- Is the human population already exceeding its carrying capacity, or will it do so soon?
- What effect does the human population have on Earth?

Prerequisite Concepts and Skills

Before you begin this chapter, review the following concepts and skills:

- evaluating how a change in one population can affect the entire hierarchy of living things (Chapter 13, section 13.3),
- explaining the concept of interaction among different species (Chapter 14, sections 14.3 and 14.4),
- describing characteristics of a population, such as growth, density, distribution, and carrying capacity (Chapter 14, sections 14.2 and 14.3), and
- analyzing the components of population growth and explaining the factors that affect the growth of various populations (Chapter 14, sections 14.3 and 14.4).

Corn has long been an important agricultural crop in Ontario. Archeological evidence shows that Aboriginal people grew corn near Campbellville, Ontario before A.D. 1200. In 2000, Ontario farmers planted 829 600 ha of corn (approximate value $5.2 billion), with the majority grown in monoculture — an agricultural system in which only one crop at a time is planted over a large area. In contrast, early Aboriginal people and European settlers typically grew a variety of crops in small fields. Why the difference?

The answer is related to the number and distribution of people who rely on this resource. Canada's population is much larger now, with the majority of it being concentrated in cities and towns. This results in fewer opportunities for individual families to grow crops. In industrial societies, jobs are specialized — only a few people grow produce while others transport, process, market, and research the product. Most of us simply eat this product, knowing little about how it got to the grocery store.

In the past, produce from a garden plot commonly fed a single family, who would then sell any excess to neighbours or local markets. Today, 70 percent of Canada's corn comes from Ontario. Large amounts of energy are expended to plant, harvest, and transport the crop, since much of it ends up far from where it was grown. Large amounts of fertilizer and pesticides are used to increase the yield. A recent decline in the rate of use for some pesticides may be due to the availability of new types of corn, including genetically modified forms. Some people express concern regarding use of these new forms, noting that they may cause as yet unknown problems.

About 60 percent of Ontario's corn feeds farm animals, with the remainder used to manufacture products including clothes and fuel. Very little is directly consumed by humans. Are these practices sustainable? Do they echo some of the mistakes made by the inhabitants of Easter Island? In this chapter, you will examine issues related to the demands placed on Earth by the global human population and the ecological consequences of human population growth.

In the past, a garden like this might have supplied the needs of one family. Probably, little was added to the soil other than manure for fertilizer.

Chapter Contents

15.1 The Human Population: Past Meets Present

EXPECTATIONS

- **Explain the demographic changes observed over the past ten thousand years, relating these to advances in medical care, technology, and food production.**
- **Describe what happens during a demographic transition.**
- **Relate the age structure of a population to its stage of demographic transition and its rate of increase.**
- **Describe the causes and effects of urbanization.**

Humans exhibit many of the characteristics of *K*-selected species. We delay sexual maturity and reproduction for many years, usually produce only one offspring at a time, and give those offspring intensive, long-term parental care. Compared with the vast majority of other species on Earth, we have relatively large bodies and long life spans. These life history traits are typical of species living in fairly predictable environments in which populations are controlled mainly by density-dependent factors such as disease, predation, and competition for resources (including food and shelter).

The History of Human Population Growth

As hunter-gatherers, early humans depended on plant foods (using different parts of many types of plants) that were naturally available in the environment, and on their ability to hunt and capture other animals. Although unpredictable density-independent events (such as floods and droughts) undoubtedly had some effect, for the most part the growth of human populations was probably controlled by density-dependent factors (especially competition for limited food resources). During this time, the worldwide human population grew slowly, as shown in Figure 15.1.

It is difficult to document the number of humans who lived on Earth hundreds or thousands of years ago. It is even more difficult to determine prehistoric numbers (from the period before people began to record descriptions of their daily lives and the events that influenced them). Nevertheless, many demographers estimate that 10 000 years ago, when the glaciers of the most recent ice age receded, the human population stood at roughly five million. With this many humans on Earth, the hunter-gather lifestyle became more difficult. There was not enough land available for new populations to forage and existing land began providing marginal resources. At about this time, what is often referred to as the **Agricultural Revolution** occurred — humans started to plant and harvest crops and domesticate animals. Now that people had some control over their food supply (instead of being controlled by it), a long period of increasing human population growth began.

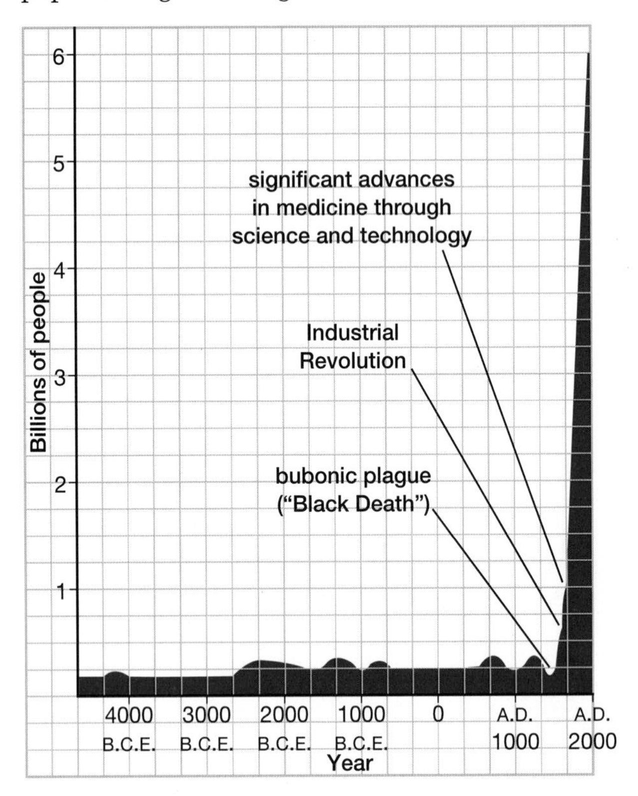

Figure 15.1 For most of its history, the human population was stable or grew only slowly. Explosive growth began following the Industrial Revolution in the eighteenth century, when the death rate dropped dramatically.

Some researchers suggest that by A.D. 0, this number had risen to about 250 million. Truly reliable estimates of the human population are not available until 1650, when there were about

500 million people on Earth. By 1850, many of the world's nations were conducting actual censuses of their citizens and the world's population was estimated to be about one billion. Human population growth has been very rapid since then; additional billions are being added in much less time than it took for the first billion to be reached. Figure 15.2 shows the time required to reach each additional billion.

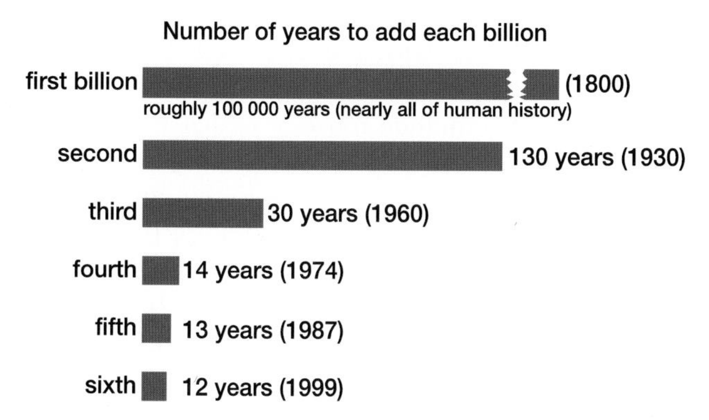

Figure 15.2 Since the development of agriculture, the rate of human population growth has dramatically increased. It took much longer for the population to reach one billion than it took for each additional billion to be added.

Demographic Transition

Recall from Chapter 14 that the growth rate of a population (r) depends on the difference between the birth and death rates of that population (that is, $r = b - d$). From studying many human populations and how they changed after the introduction of technology, demographers have developed a model of how shifts in birth and death rates alter patterns of population growth. The model suggests that increases and decreases in r occur as a population goes through a series of stages known as a **demographic transition**.

In all early human populations and in some modern societies, death rates (especially of infants and children) were and are high. As a result, these populations developed customs that encouraged high birth rates so that the birth and death rates came close to balancing. This corresponds to the first stage of the demographic transition, in which the overall growth rate of a population is more or less stable. This is illustrated in Figure 15.3.

Demographers have estimated that until 1700, mortality rates were constant or declined very slowly and were just about matched by birth rates. Temporary increases in mortality occurred as a result of wars or epidemics, including that of the black death (bubonic plague) that killed millions of humans in Europe and Asia during the fourteenth century. Although it sometimes took years for a local population to recover from these periods of high mortality, high birth rates eventually compensated and the world population continued to grow slowly.

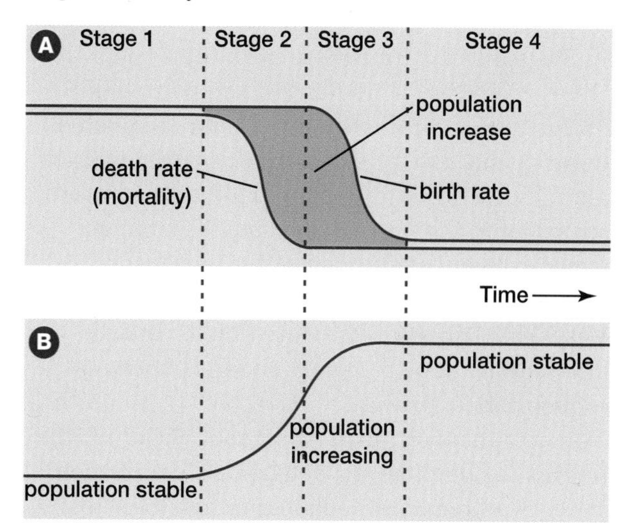

Figure 15.3 Diagram (A) shows how birth and death rates change as a population passes through the stages of a demographic transition. Diagram (B) shows the corresponding changes in the rate of population growth.

In the early years of the eighteenth century in England, and slightly later in Europe and North America, the **Industrial Revolution** took place. During this time, people shifted from working in traditional agricultural jobs and making goods by hand to working in factories and making mass-produced goods. As industries expanded over the next 150 years, the standard of living for many people increased. At about this time, scientists such as Louis Pasteur and Joseph Lister provided convincing evidence that bacteria and other small organisms caused many of the diseases that had previously ravaged populations. The resulting improvements in sanitation and medical care lowered the death rate in some countries, as did improvements in nutrition. The invention of the steam engine made it possible for a single farmer with a tractor to do the work of dozens, thus increasing food production. Trains and ships could distribute food quickly and reduce the impact of local famines. Later, near the middle of the twentieth century, the widespread use of antibiotics, vaccines, and other medicines lowered the death rate even more.

Since it generally takes several generations for cultural values and customs to shift in response to lowered death rates, birth rates in these industrialized countries remained high (and may have increased with improvements in the health of mothers). This corresponds to Stage 2 of the demographic transition shown in Figure 15.3(A). Because birth rates were so much higher than death rates, the result was a period of rapid population growth, as shown in Figure 15.3(B).

In Stage 3 of the transition, birth rates begin to decline since there is no need to compensate for high infant and child mortality — and there is no cultural memory of a time when this was necessary. Eventually the birth rate again matches the lowered death rate and the population stabilizes, but at a higher carrying capacity than before (due to improvements in food production, among other things). This condition, in which the birth and death rates are approximately equal, is known as **zero population growth**.

Demographers estimate that it took Britain 250 to 300 years to complete the demographic transition process. The transition took less time (closer to 200 years) in other industrialized countries in Europe, North America, and Japan, probably because Britain shared its information and technology. Many of the remaining nations in the world (those in Africa, Latin America, and parts of Asia) only began their demographic transition during the twentieth century. They experienced a much more rapid drop in death rates, mostly as a result of the availability of knowledge and technology from developed nations. In many cases, these remaining countries have gone from a traditional, high mortality rate to a low one in a single generation and are still in Stage 2 of their demographic transition. However, evidence suggests that many of these populations are already in Stage 3 — birth rates have begun to decline, as shown in Table 15.1.

Table 15.1 shows birth rates for a few countries in two general groups. As the name suggests, **more industrialized nations** (sometimes called more developed nations) tend to have large industries requiring skilled workers and a higher standard of living. These include the countries in Europe, as well as Canada, the United States, New Zealand, Australia, and Japan. The **less industrialized nations** (less developed) include those in Africa and most of Asia. Since countries were not grouped in this same way in 1974, the averages for more and less developed countries are not available for that year.

The demographic transition process can be analyzed for populations ranging in size from the global population to populations of individual countries, and even down to those of small regions (such as a city or a county). In Investigation 15-A on page 508, you will determine the stage of demographic transition that characterizes a particular region. Note that you should conduct this investigation with your teacher during a class excursion.

WEB LINK

www.mcgrawhill.ca/links/biology12

If your class is unable to go out into the field to collect data for Investigation 15-A, several data sets suitable for doing this investigation are available on the Internet. Go to the web site above, and click on **Web Links**.

Table 15.1
Births and deaths for selected countries and regions (from the Population Reference Bureau)

Country or region	Births per 1000 individuals in 1974	Deaths per 1000 individuals in 1974	Births per 1000 individuals in 1995	Deaths per 1000 individuals in 1995	Births per 1000 individuals in 2001	Deaths per 1000 individuals in 2001
World	30	12	24	9	22	9
less industrialized nations (average)			28	9	25	8
Uganda	45	16	52	19	48	19
Costa Rica	28	5	26	4	22	4
Turkey	39	12	23	11	22	7
more industrialized nations (average)			12	10	11	10
Canada	15	7	14	7	11	8
United States	15	9	15	9	15	9
Norway	15	10	14	11	13	10

Demographic Transition and Population Age Structure

Human demographers tend to refer to the average number of children born alive to each woman in a population during her lifetime as the **fertility rate** rather than as the rate of fecundity. The latter term is more commonly used by ecologists working with other types of organisms (see Chapter 14, section 14.4). The fertility rate required to achieve long-term zero population growth is called the **replacement fertility rate** and is 2.1 children per woman. This rate is just enough to replace the parents who produced them and to compensate for children who die before reaching reproductive age.

Countries that have completed their demographic transition (or that are in Stage 3) have declining birth rates. In fact, they have fertility rates at replacement level or below. Both of these rates have been dropping substantially in recent years, causing demographers to re-adjust earlier projections of the future size of the human population. According to the United Nations Population Division, 44 percent of the world's population lived in countries with fertility rates at or below replacement level in 1995. Only a few of the less developed nations had shown no signs of fertility reduction, and the fertility rates of 49 nations (including China, which comprises about one fifth of the world's population) had been below replacement level for over a decade.

Some countries, however, still have fertility rates above replacement level. As a result, the global average fertility rate in 2001 was approximately 2.8 children per woman per year. If it stays at this level, world population is projected to reach 7.82 billion in 2025 (see Table 15.2). Many demographers predict that the worldwide average fertility rate may decline further to reach 2.1 by 2050 as more countries progress through the demographic transition. If this happens, they estimate that the human population will stabilize at just under 10 billion by 2150. However, if the average fertility rate is higher or lower than this in 2050, the projected population could be quite different, as shown in Figure 15.4 on page 510. Notice that although the difference between the high and low fertility rates in Figure 15.4 is only about one child per couple, it can alter the projected population sizes quite dramatically.

Note that just because the fertility rate of a country or region has reached replacement level, it does not mean that the population will stop growing (that is, that it has reached zero population growth). There are other factors to consider as well. For example, infant mortality varies among countries and will alter the rate of growth. Notice also that there is variation in the proportion of young individuals in the populations listed. In Chapter 14, you saw that the birth rate not only depends on the number of offspring each female has, but also on the number of females available to have young or the number of females who will be

Table 15.2
Approximate fertility rates, projected population sizes, and other parameters for selected countries and regions (from the Population Reference Bureau)

Country or region	Projected population 2025 (millions)	Projected population 2050 (millions)	Infant mortality rate in 2001 (deaths per 1000 live births)	Total fertility rate in 2001	Percentage of population less than 15 years of age in 2001
World	7818	9036	56.0	2.8	30
less industrialized nations (average)	6570	7794	61.0	3.2	33
Uganda	48	84	97.0	6.9	51
Costa Rica	5	5	12.0	2.6	32
Turkey	85	97	35.0	2.5	30
more industrialized nations (average)	1224	1242	8.0	1.6	18
Canada	36	36	5.5	1.4	19
United States	346	413	7.1	2.1	21
Norway	5	5	3.9	1.8	20

Investigation 15 • A

SKILL FOCUS

Predicting

Performing and recording

Analyzing and interpreting

Communicating results

Cemetery Studies of Human Demography

Cultural changes often have profound effects on the birth and death rates of a population. In this investigation, you will have an opportunity to collect (or use) data to construct survivorship curves (Chapter 14, section 14.4) for two or more human populations. You will also examine the process of demographic transition.

Pre-lab Questions

- What is a population? Are its boundaries defined only in space?
- What data do you need to construct a survivorship curve for a population?

Problem

Does survivorship change over time in human populations?

Prediction

Make a prediction about how the survivorship curves for past and present human populations might differ.

CAUTION: **When exploring the cemetery, take care not to disturb the landscape or injure yourself.**

Materials

pencil
clipboard
field data sheet
appropriate footwear and clothes (possibly including rain gear)

Procedure

1. Design a field data sheet to facilitate data collection. The sheet should be standardized for all members of your class and for all teams collecting data. It should have columns or blanks in which you can record gender, age at death, year of death, and, if possible, the cause of death for each individual studied.
2. Divide yourselves into groups and designate which ones will collect data on certain segments of the population. The data collected will depend partly on the number of gravestones or death records available and the age of your community. Ideally, collect data on at least 200 individuals who died during each of these three intervals: 1980 to 2000, 1930 to 1950, and 1880 to 1900. For each of these intervals, have one group collect data on females, while another group collects data on males. If your opportunities for data collection are limited, a division of deaths into two time intervals, pre- and post-1950, will show some of the same patterns. Ensure all team members are clear on their respective roles regarding data collection before proceeding to the cemetery.
3. Visit one or more local cemeteries (the older the cemetery the better) and record the information described in step 1. (If unable to visit a cemetery, use the Web Link preceding this investigation to gather your data.) Ensure that members of the same group avoid collecting data on the same individual. Behave in a way appropriate in cemeteries by not stepping on burial plots. Data can also be obtained (with permission) from death records, which may be held at city hall or at a records office elsewhere in your community. In some cases, death records for individuals buried in a cemetery are kept in a central office at the cemetery. In some ways, death records are better to work with than gravestones; they are not subject to weathering or the effects of acid precipitation (which can make inscriptions impossible to read). Also, the cause of death is regularly recorded (it is seldom found on a gravestone).

4. Return to the lab and sort your data so you can collate the numbers of individuals of each gender in your time period who died during each of the following age intervals: zero to just under five years of age, five to just under 10, 10 to just under 15, 15 to just under 20 and so on, up to a final interval of 95 years of age and over. Again, all groups should use similar data sheets.

5. Calculate the mortality rate (the proportion of the population that died) for the first age interval. Then standardize the data from each group to a cohort of 1000. For example, if records for 200 deaths were obtained and 90 of these involved children less than five years of age, then the mortality rate equals the number of deaths per cohort in that age group. Multiplying this number by 1000 then gives you the number that would have died if you had a cohort of 1000, thus standardizing the data to this cohort size. For this example, the calculations would be:

$$\frac{90}{200} = 0.450$$

$$0.450 \times 1000 = 450$$

6. Survivorship is described as the proportion of the total population surviving through an age interval and therefore alive at the start of the next interval. The survivorship at the start of the first age interval is 100 percent, since all members of the cohort are assumed to be alive at the start of the exercise. The proportion surviving through this interval (and therefore alive at the start of the second age interval) is one minus the mortality rate for the first interval. For the example given above, the survivorship for individuals in the zero to just under five interval would be:

$$1 - \frac{450}{1000} = 1 - 0.450$$
$$= 0.550$$

The number surviving out of the original standardized cohort (of 1000) would be $0.550 \times 1000 = 550$.

7. Calculate the mortality through the next age interval as in step 5. For example, if 50 of the 200 people in your sample died between the ages of five and just under 10 years of age, the mortality rate for this interval is:

$$\frac{50}{200} = 0.250$$

$$0.250 \times 1000 = 250$$

To calculate the survivorship through this interval, this interval's mortality rate must be subtracted from the proportion of the population alive at the beginning of this age interval. This proportion is the same as the survivorship through the previous age interval as determined in step 6. The survivorship for this interval would then be:

$$\frac{550}{1000} - \frac{250}{1000} = 0.550 - 0.250$$
$$= 0.300$$

$$0.300 \times 1000 = 300$$

8. Calculate the mortality and survivorship rates for each subsequent age interval using the method described in steps 5 to 7.

9. Each group should construct a curve showing the pattern of survivorship for the males or females who died during a given time interval. This graph should have survivorship on the *y*-axis and age intervals on the *x*-axis. Survivorship can be expressed either as proportions or actual numbers, so the *y*-axis can extend from 0 to 1 or from 0 to 1000.

Post-lab Question

1. Describe the shape of your group's survivorship curve. Does it seem most similar to a Type I, II, or III survivorship curve (refer to Chapter 14, section 14.4)?

Conclude and Apply

2. Compare the curves for different sexes and time periods. Were your predictions about potential differences in the survivorship patterns at different historical times correct? Describe any differences between your predictions and what you actually observed.

3. Explain the differences you see between historical periods with respect to survivorship patterns.

4. Were there any differences between the sexes with respect to their survivorship patterns in a particular time period and/or age interval? If so, explain why these differences might exist.

5. Do you think there might be differences among cemeteries in different regions (for example, urban versus rural) or between different parts of a city with respect to the survivorship patterns of people who died during certain eras? Explain your answer.

6. Do you think your data provided an accurate picture of survivorship in your area? What might you do to get a clearer picture?

Exploring Further

7. If the data are available, calculate the proportion of individuals in each time interval who died from as many of the following general categories of causes as possible.

   (a) stillborn

   (b) accident

   (c) infectious diseases (that is, contagious diseases caused by a bacteria, virus, or other type of parasitic organism)

   (d) degenerative diseases (for example, cancer, heart disease, stroke)

   (e) other

   Use a bar graph or pie chart to illustrate your results. Is there a difference among the time intervals with respect to the proportion of individuals in each of these categories? Relate any differences you see to historical developments within your community (for example, wars, the installation of water treatment plants, hospital construction, and so forth).

available in subsequent years. In other words, the age structure of a population is also very important in determining its rate of growth.

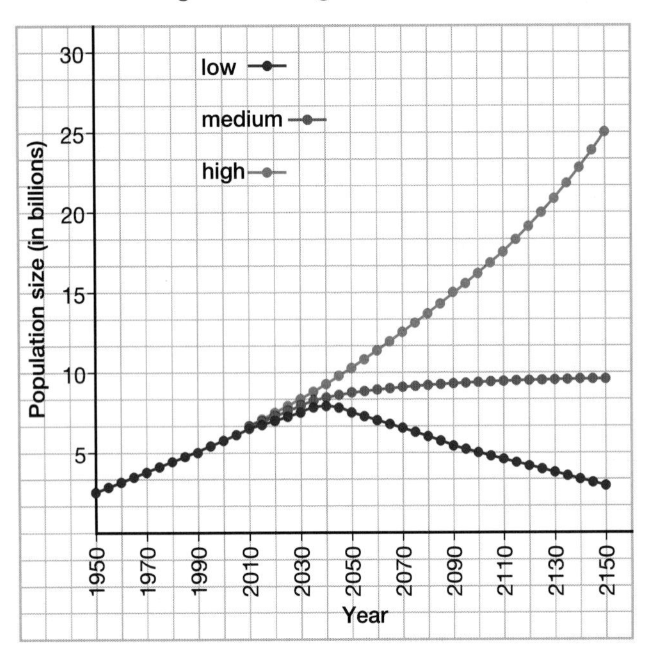

Figure 15.4 The medium projection on this graph depends on all regions of the world reaching the replacement fertility rate by 2050. The high and low projections assume that the fertility rates are somewhere within the ranges shown on the legend. Projections were calculated in 1998 by staff of the United Nations Population Division.

As a country or region passes through a demographic transition, its age structure changes. This can be seen by looking at a series of age pyramids for a nation like Sweden, as shown in Figure 15.5. Sweden's population was growing in 1900 ($r = 0.01$), whereas in 2000 its r value was zero and its fertility rate was 1.5. Notice that in Figure 15.5(B), a large number of children under the age of five are shown in the bar at the base of the diagram, which had already started to become less pyramid-shaped. In most countries involved in World War II, birth rates increased substantially in the years following the war (from about 1948 to 1960 in many countries). This period was referred to as the *baby boom*. In many countries, when this group of children reached reproductive age (roughly 25 to 30 years later), they produced their own baby boom (referred to as the *echo*). Because fertility rates had already fallen in most of these countries, the number of births resulting from the echo tended to be smaller than the original boom. In general, as a country completes its demographic transition, its age structure becomes less pyramidal and, in some cases, starts to taper at the bottom as well as at the top.

Figure 15.6 on page 511 shows the contrast between the age pyramids of three countries, drawn using population statistics for the year 2000. The Democratic Republic of Congo was still showing rapid growth ($r = 3.1$), whereas the United States displayed slow growth ($r = 0.6$) and Germany negative growth ($r = -0.1$). Some countries with age pyramids currently shaped like the Democratic

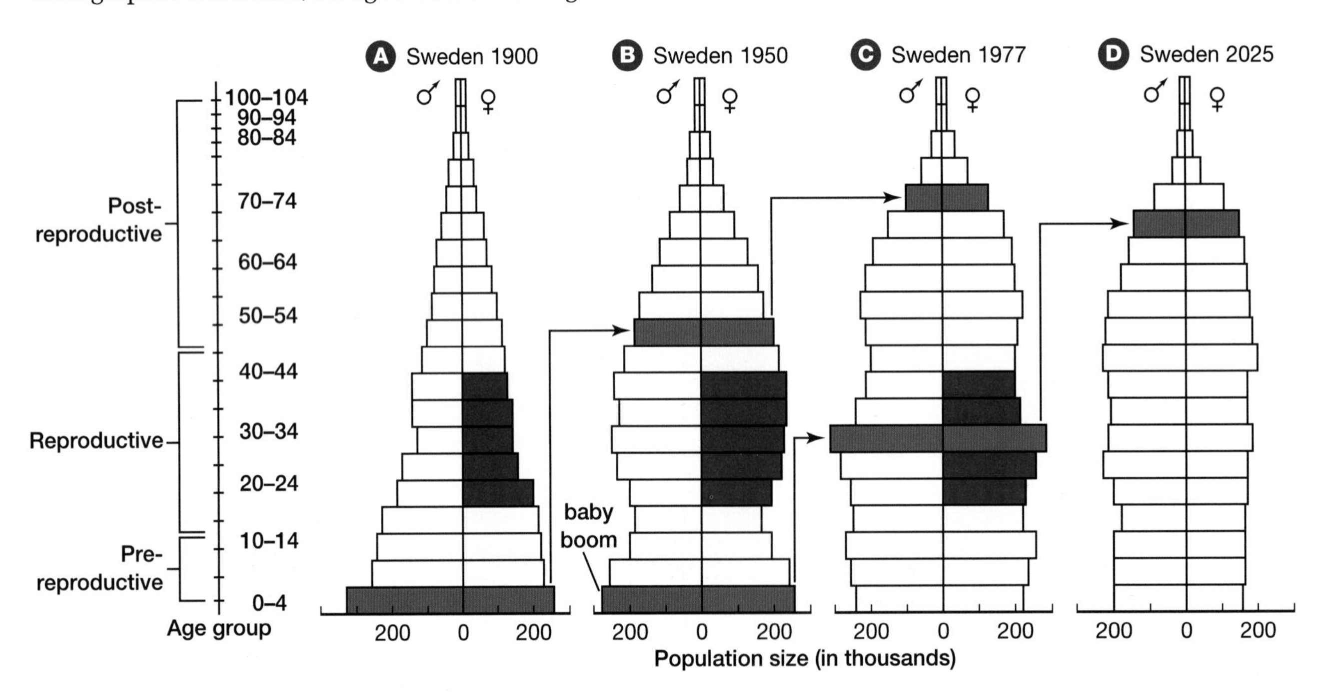

Figure 15.5 When a country is in the early stages of demographic transition, a large proportion of its population is young, and therefore the population grows quickly. As the population passes through Stages 2 and 3, the shape of the age pyramid changes to become less pyramidal.

Republic of Congo's actually have fertility rates that are declining to near replacement level. However, these populations will continue to grow rapidly for a period because so many females are in or nearing their reproductive years. Populations like this, with a large proportion of young individuals, are often said to have a *momentum for growth.*

As you have seen, age pyramids are one tool demographers use to help them examine a population's potential for growth and determine its stage of demographic transition. The next Thinking Lab will help test your understanding of how age pyramids are created and how they can be used.

Urbanization

From artifacts, historical records, and observation of similar human groups existing today, we know that hunter-gatherer societies typically exploited (or exploit, in the case of the few remaining modern societies) many different resources lightly, rather than relying heavily on only a few. They often moved seasonally, using different food sources as they became available. Their need for mobility and their typically subsistence-level lifestyle provided no incentive for large families and their birth rate was therefore generally low. Because their populations were small and nomadic, their accumulation of waste (material and human) was minimal and easily handled by Earth's natural processes. Therefore, for many reasons, hunter-gatherer societies had relatively little impact on the environment.

Around 10 000 years ago a new lifestyle emerged. It may have been spurred by the retreat of the last ice age, which left behind a warmer climate and fertile soils. Agriculture, the sowing and harvesting of crops, and the domestication of animals paved the way for a more sedentary lifestyle. It also set the stage for the development of what is referred to as civilization — the appearance of a highly organized society, characterized by the use of written language and advances in government, the arts, and the sciences.

The earliest agricultural sites in the world have been found in the Near East (in what is now Turkey), the Nile Valley, and Mesopotamia (the region east of the Mediterranean Sea between the Tigris and Euphrates rivers). The fertile soil and hot climate in these areas made it possible to harvest several times each year. Farmers planted fields of wheat and barley, which grew naturally on forested hills above the plain. Because these cereals did not provide all the proteins needed by humans, it was also necessary to eat meat — a commodity supplied by domesticated animals. Farming provided a relatively predictable and abundant food supply, but it was also labour intensive, requiring many hands to bring in a crop. These factors favoured large families, and the human population began its inexorable increase.

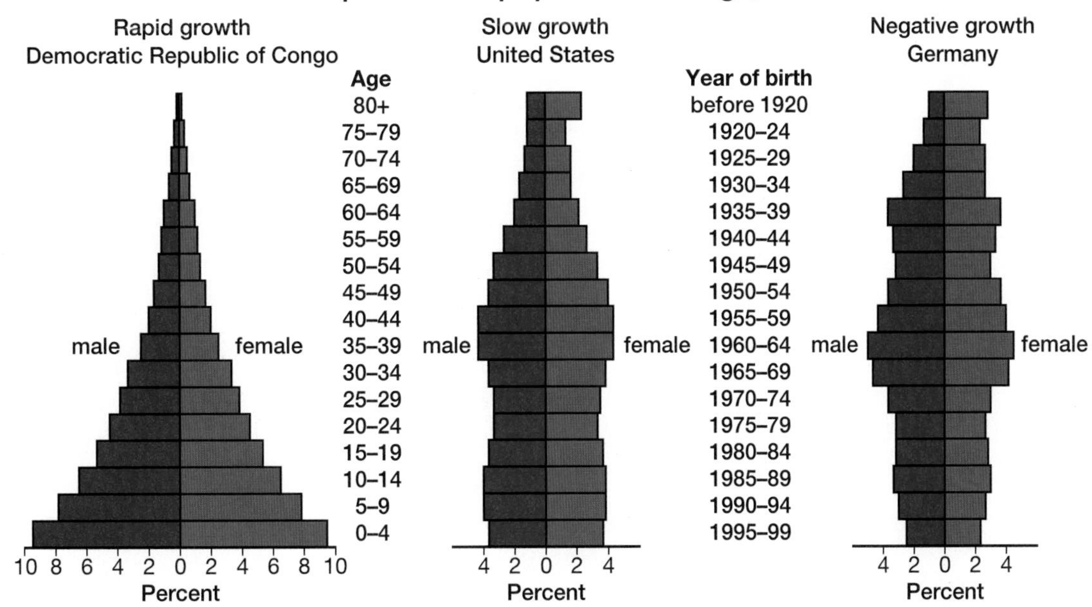

Figure 15.6 Age pyramids for three countries with different age structures. What is the ratio of post-reproductive individuals (those past their reproductive years) to pre-reproductive and currently reproducing individuals in each country? How will this affect the future growth of the population?

Six thousand years ago, Mesopotamian farmers dug ditches from the rivers that surrounded the plain on which they grew their crops. These ditches not only controlled seasonal floods, but also provided water when it was desperately needed during dry periods. These ancient irrigation ditches were very successful for a long time. High crop yields supported a large population in which trade and social roles became highly diversified, as illustrated in Figure 15.7 on page 513. Knowledge increased and systems of government developed to organize the large number of people in new cities and states.

Over time, however, the soil became less and less fertile. Surplus irrigation water was not removed from the land, and it left behind salts and other toxins as it evaporated. Farmers switched from wheat to the more salt-tolerant barley, but eventually it also became difficult to grow. The previously forested hills above the flood plain became denuded as larger numbers of people sought fuel, building materials, and land on which to grow crops and graze animals. Some of the ancient cities of Mesopotamia seemed to fade away from lack of food, others apparently succumbed to massive floods or mudslides caused by soil erosion on the cleared hillsides.

THINKING LAB

Building Age Pyramids

Background

The International Programs Center (IPC) of the United States Bureau of the Census has created an international data base to meet the needs of organizations that require its assistance, and for its own research. The data it contains come from censuses and surveys done by individual countries, as well as estimates and projections provided by the IPC. One of the data sets is extremely useful for building age pyramids. Demographers use such pyramids as a tool to help them project population trends.

You Try It

1. Using the data in the table, construct age pyramids for both countries. The age pyramids can be drawn by hand or by using a spreadsheet program.
2. What do the pyramids indicate about the stage of demographic transition each country is experiencing?
3. Which country, Canada or Mexico, will grow most in the next decade? Explain why.
4. Are there any differences in the proportion of males versus females in each age category? That is, is the sex ratio (the ratio of the number of males to the number of females) the same in all age categories? If so, which age categories differ and is the pattern the same for both countries? What factors might explain any differences you see between countries and between genders? Would these differences have any effect on population growth? Why or why not?

Age structures for the Canadian and Mexican populations for the year 2000 (from the United States Bureau of the Census)

Age Category	Canada		Mexico	
	Number of males (x1000)	Number of females (x1000)	Number of males (x1000)	Number of females (x1000)
0–4	945.9	901.5	5815.6	5580.2
5–9	1069.7	1015.6	5799.8	5573.9
10–14	1062.1	1013.1	5690.1	5478.6
15–19	1065.6	1019.8	5429.8	5269.5
20–24	1051.8	1009.1	5060.5	4999.2
25–29	1087.6	1059.1	4569.5	4620.3
30–34	1187.8	1171.0	3676.1	3988.2
35–39	1396.9	1369.3	3012.8	3401.6
40–44	1332.3	1318.5	2382.6	2741.6
45–49	1160.8	1168.8	2039.9	2311.1
50–54	1034.1	1035.6	1651.6	1845.2
55–59	782.2	795.7	1333.6	1491.5
60–64	615.2	644.8	1067.0	1199.7
65–69	547.9	598.7	800.6	921.4
70–74	458.9	553.1	545.2	652.2
75–79	339.8	481.9	329.8	427.3
80+	337.1	647.1	252.3	390.1

WEB LINK

www.mcgrawhill.ca/links/biology12

If you would like to build and compare age pyramids for other countries or regions, go to the web site above, and click on **Web Links**.

Figure 15.7 How many different occupations or activities can you recognize in this representation of an ancient clay tablet?

The same events seem to have repeated themselves many times throughout history. In A.D. 100, for example, Rome was a city of roughly one million people and the centre of a great empire. The Romans did not develop agricultural systems sufficient to support their city. As their empire expanded, they supported their growing population by importing food from farther and farther afield. Eventually, challengers to the empire were able to cut off these supply lines and Rome collapsed. This was due, at least in part, many historians think, to poor environmental practices (including loss of soil fertility in a wide area around the city).

This is one of the main ecological problems posed by cities. Typically, cities disrupt the biogeochemical cycles that sustain ecosystems. Large numbers of people concentrated in urban areas require vast amounts of food, but have little space in which to grow it. Therefore, they must import food, often from long distances, thus requiring the input of extra energy (usually from fossil fuels) to transport it. The nutrients in this food are usually not returned to the soil in which it was grown. They are commonly dumped in landfills and/or pumped through sewage lines into oceans or rivers. Unconnected to the land and knowing little about what is involved in getting food to the store where they buy it, most city residents are unaware of the true ecological cost of this lifestyle.

Cities also demand many trees for fuel, paper, and building materials, but rarely involve themselves in reforestation activities. They require large amounts of water, but do not typically return the same amount of usable water to the environment. In other words, most cities have what can be described as a linear metabolism, in which no attempt is made to balance inputs and outputs. Most cities do not return nutrients to the soil, plant new forests, or maintain the hydrologic cycle, as shown in Figure 15.8(A).

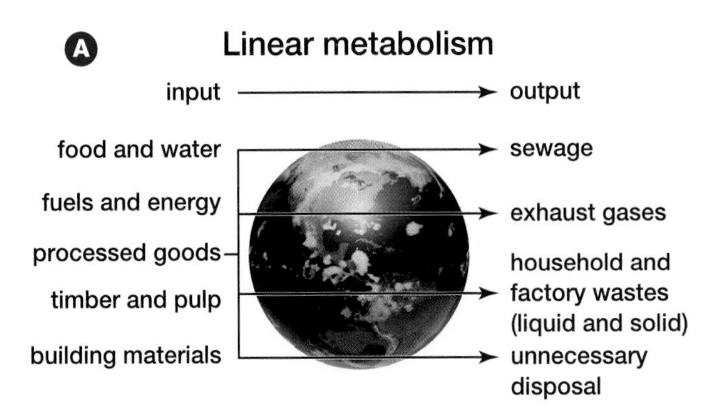

Figure 15.8 (A) Is linear metabolism sustainable?

What is needed is a rethinking of how cities work so that they have a more circular metabolism, in which every output can also be used as an input. Figure 15.8(B) shows a circular metabolism. This pattern reduces the demand for "new" resources, thus affecting a much smaller area.

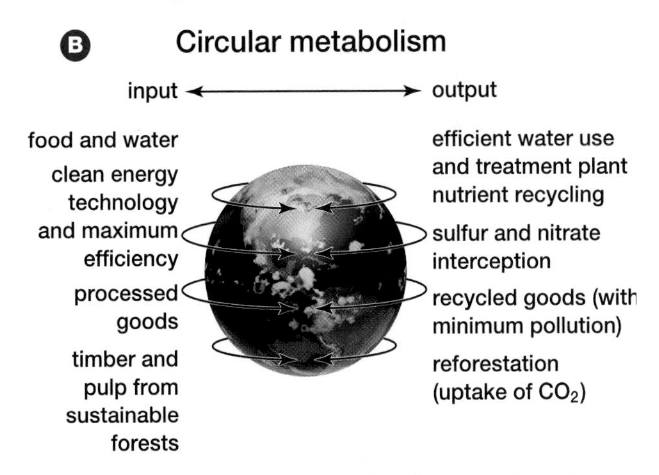

Figure 15.8 (B) Does the city you live in or are closest to have linear metabolism or circular metabolism?

Historically, many Chinese cities demonstrated circular metabolism. They took great care to replenish the soil from which they got their food. Both animal and human wastes were returned to the belt of farmland that surrounded the city. Forests were replanted and water was used carefully. This is true of many Chinese cities even today (for example, Beijing, Shanghai, Wuhan).

The effect of urbanization is an issue that needs urgent attention, as the size and number of cities — and their impact on Earth — increase daily (see Table 15.3). In 1950, New York was the world's only megacity (urban area with over five million inhabitants). By the year 2000, there were 19 such cities. Tokyo alone had a population of 26.4 million and New York, with 16.6 million was only the fifth largest city. In less developed countries, many cities are swelling rapidly as a result of migration and high birth rates. Some rural inhabitants move to cities because their land is degraded, because they are forced to look for new opportunities as large companies take over agricultural production, or because they are unable to find sufficient firewood or water. Others come seeking jobs or a new life. Many find it difficult to locate or afford housing and end up in crowded settlements where conditions (such as lack of clean water or sanitation) make them vulnerable to fire, disease, and bad weather, as shown in Figure 15.9. In some of the world's cities, densities are very high. In Lagos, Nigeria for example, there are an average of 5.8 people per room, compared to the North American average of one or two.

Table 15.3
Changes in urbanization occurring throughout the world (from the United Nations Population Division)

	Percentage Urban (living in cities)			
	1950	1975	2000	2030
world	29.7	37.9	47.0	60.3
Canada and United States	64.0	74.0	77.0	84.0
more industrialized nations	54.9	70.0	76.0	83.5
less industrialized nations	17.8	26.8	39.9	56.2

Lower density (as shown in Figure 15.10) — that is, a less clumped, more uniform dispersion pattern — does not necessarily solve the problems caused by cities, however. In fact, one of the biggest issues for North American cities is **urban sprawl** — a pattern of scattered, uncontrolled development in which working, living, and recreational areas are widely separated. This results in very large, diffuse cities in which it is necessary to drive from place to place. Sprawl increases pollution and destroys farmland, parks, and other open spaces. Today, a new trend toward *smart growth* is finding enthusiastic support among city planners. Communities that bring working and living spaces into close proximity and provide a variety of housing options (including some high-density housing) are preferred. The principles of smart growth encourage cities to be pedestrian- and cyclist-friendly, have efficient rapid transportation systems, and set aside land for parks. Food security is also viewed as important, and consideration is given to renewing or ensuring the health of agricultural land near cities.

Figure 15.9 Contrast this with a Canadian city such as Toronto.

Figure 15.10 A residential area of Toronto. How do living conditions compare with those in Figure 15.9?

In June 2001, the city of Ottawa held a Smart Growth Summit to explore these ideas and make them more accessible to its citizens. Like many Ontario cities, Ottawa grew rapidly during the last half of the twentieth century. The city leapt over a greenbelt (a reserve of open space) that had been created around Ottawa in the 1950s and 1960s to limit its growth. In 2001, Ottawa was at a crossroads, attempting to plan a city that would ensure quality of life while still encouraging population growth (see Figure 15.11). In other words, it was attempting to ensure ecologically and economically sustainable growth — a difficult balancing act faced by cities all over the world. For population growth to be sustainable, it must not take a population past the carrying capacity of its environment. The first step in ensuring that this does not happen is to determine the carrying capacity for a particular population. This challenging task will be discussed in the next section.

Figure 15.11 Why do you think so many cities are built on waterways — rivers, lakes, or oceans?

SECTION REVIEW

1. K/U (a) What features of human life history suggest that *Homo sapiens* is a *K*-selected species?
   (b) Why did humans evolve these particular features?
2. K/U What effect did the Agricultural Revolution have on the rate of growth of the human population?
3. I How might you determine the positive and negative effects of monoculture farming on the human population? What factors would you have to consider and what variables would you measure? Design a study that will allow you to prepare such a balance sheet for a crop of your choice.
4. K/U Compare the birth and death rates during each of the four stages of a demographic transition.
5. MC How did the Industrial Revolution have an impact on the rate of population growth?
6. K/U Define the term "zero population growth."
7. K/U If the fertility rate is at replacement level, can a population still be growing? Explain your answer.
8. K/U What is the replacement fertility rate for humans? Explain why it is called this.
9. MC Compare more industrialized nations with less industrialized nations with respect to typical age structure, fertility rate, and stage of demographic transition. What aspects of being more industrialized do you think have contributed to differences in these demographic features?
10. K/U How does the age structure of a population influence its rate of growth?
11. I How might you measure the impact of large urban areas on surrounding ecosystems?
12. MC If it is not there already, how might your school or community come closer to the circular metabolism model for water use? In your answer, consider the volume of water used (could it be less?), where the water ends up, how the used water is treated (is the method effective?), and the various uses to which water is put (should it all be treated to the same extent?).
13. MC Describe some of the impacts of urban sprawl on the ecosystems surrounding a city. What would you consider to be the most serious of these?

UNIT ISSUE PREP

As you have seen, the Agricultural and Industrial Revolutions significantly impacted human population growth. This growth, along with the effects of urbanization, had a number of environmental impacts on Earth. When choosing your environmental issue, consider these events and their potential impact.

What Is Earth's Carrying Capacity?

EXPECTATIONS

- **Evaluate the carrying capacity of Earth and relate the carrying capacity to the growth of populations and their consumption of natural resources and advances in technology.**
- **Explain why it is difficult to determine human carrying capacity.**

In 1679, Anton van Leeuwenhoek, the Dutch scientist whose main contribution to science involved improving the microscope and studying microscopic organisms, estimated that the maximum number of people the world could support was 13.4 billion. He calculated this figure by multiplying the population of the Netherlands by the ratio of Earth's total inhabited land area to the area of the Netherlands. These numbers were, of course, estimates. At that time, there were no reliable census data available for the Netherlands or any other country. In addition, parts of the world were poorly mapped, so the amount of inhabitable land was uncertain.

Many more estimates of Earth's carrying capacity for humans have followed van Leeuwenhoek's. These have ranged from less than one billion people to over one trillion, and they are getting more scattered over time, as illustrated in Figure 15.12. There has been no clear trend since 1679 for the estimates to get increasingly larger or smaller, or even to move closer together. In fact, Figure 15.12 shows that the spread between the highest and lowest estimates has increased with time. This indicates that trying to determine the value of K (the carrying capacity) for humans is a difficult task. It also suggests that we may be aiming at a moving target — perhaps the carrying capacity is not constant but changes or fluctuates over time. This should not be surprising given that the carrying capacity depends on many variables, including the condition of the environment, which itself changes over time.

Methods of Estimating Carrying Capacity

A few theorists who have estimated human population limits have based their assertions on little or no evidence or logic. However, most estimates have been based on some sort of data and have been calculated using one of a variety of methods. Some researchers analyzing this question have divided the terrestrial portions of Earth into regions and specified a maximum density of population that each of these regions could support. They have then multiplied these maximum densities by the area of each region and taken the sum to obtain the total number of people that could be supported on Earth. The difficulty with this method lies in determining the densities for each region. Few researchers used objective, repeatable methods to do this. In addition, they assumed that these densities would not change over time, which is unlikely given that environmental conditions change.

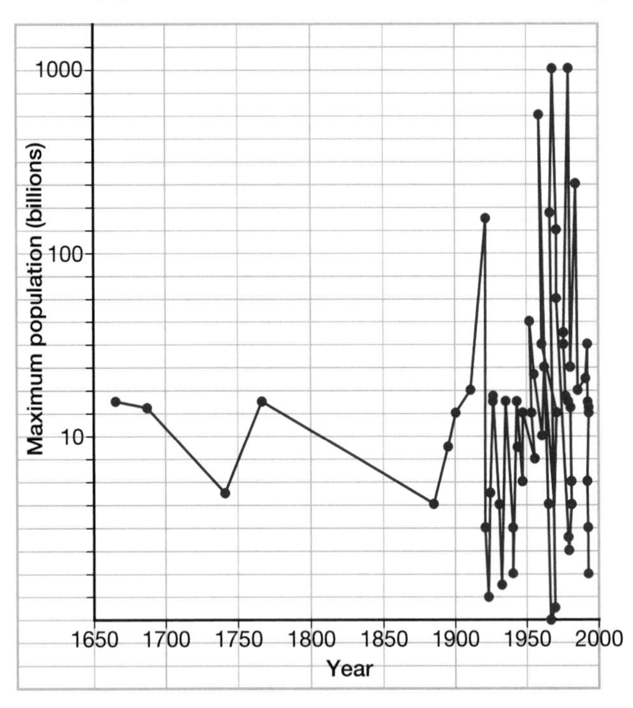

Figure 15.12 This graph gives an indication of the extent of variation in estimates of the human carrying capacity. It also illustrates how the scatter in estimates has increased over time. Notice how the estimates do not seem to be converging on a single true value. In fact, the largest estimate ever made (published in 1964) would be off the scale of this diagram and therefore has been omitted.

A second method has involved fitting logistic curves to historical data on human population growth and extrapolating into the future, to determine at what population growth could be expected to level off. In 1924, researchers using this method estimated that the maximum size of the human population would be two billion. The number of people on Earth reached this around 1930 and kept increasing. In 1936, another group estimated that the maximum would be 2.65 billion and that this would not be achieved for another 150 years. In fact, this number was surpassed less than 20 years later. Clearly, it is difficult to fit curves to the pattern of human population growth, perhaps because for awhile the human population grew faster than any model would predict.

Limiting Factors

Many studies of the human carrying capacity have looked at the impact of a single constraining factor on growth. Food has been commonly thought to be the single most important limiting factor. Thus, a widely used formula for determining the maximum population size is:

$$\text{Population that can be fed} = \frac{\text{food supply}}{\text{individual food requirement}}$$

Although this formula seems objective, results obtained by using it can vary depending on estimates of the amount of food available on Earth and estimates of the amount of food required by each person. The food supply varies with the choice of crops and the variety of each crop planted; these will change as weather and soil conditions change or as advancements in technology produce new varieties. The supply will also depend on the land area available for planting and amount of watering that can be done. Year-to-year variation in food availability will occur as a result of loss due to pests, weather fluctuations, and waste due to spoiling if food must be transported long distances. The denominator of this equation can also vary, depending on the trophic level of the humans involved (whether they are primary or secondary consumers) and on the amount of energy (measured in calories or kilojoules) that they are assumed to require.

Some ecologists have identified variables other than food as being the most important for setting a limit on population size. These other factors have included things like energy, biologically accessible nitrogen or phosphorus, fresh water, light, and space for waste disposal. Regardless of the factor believed to be most important, all the researchers who have used an equation similar to the one above have failed to examine whether some other variable or variables might take effect even before the factor they consider to be the most important has begun to have an impact.

Multiple Factors

In contrast, some studies have considered population size to be limited by the action of multiple factors acting independently. These are combined into equations in a variety of ways. For example, if both water and food are considered to act as constraints on population growth, then:

$$\text{Population that can be fed and watered} = \text{minimum of} \left(\frac{\text{food supply}}{\text{individual food requirement}}, \frac{\text{water supply}}{\text{individual water requirement}} \right)$$

A problem with even these more complex equations is that they do not take into account that different parts of the human population have different food and water requirements. Also, these requirements may fluctuate over time. For example, the water requirements of people in tropical countries may increase during the dry season, while the food requirements of people in temperate countries may be higher in winter. Therefore, using such equations to estimate the carrying capacity for Earth will likely produce inaccuracies.

Chance Events

All of the methods described above are more or less static. That is, they do not take into account chance events such as volcanic eruptions, epidemics, El Niño effects, and genetic changes in viruses and bacteria, among other things. Any of these chance events can produce random fluctuations in populations on a local or regional scale. Although some very complex models have been developed to estimate carrying capacity while taking such random events into account, they are in the earliest stages of development and have not been well tested. Further study of such models is important. They could help us determine at what level the human population could be maintained most of the time (perhaps 95 years out every 100), given a certain anticipated amount of random variation. The concept organizer shown in Figure 15.13 on the following page provides a summary of some of the factors influencing the carrying capacity of Earth for humans.

In the Thinking Lab on page 519, you can use your own food intake for one day as an estimate of the requirements of a typical person. Extrapolating

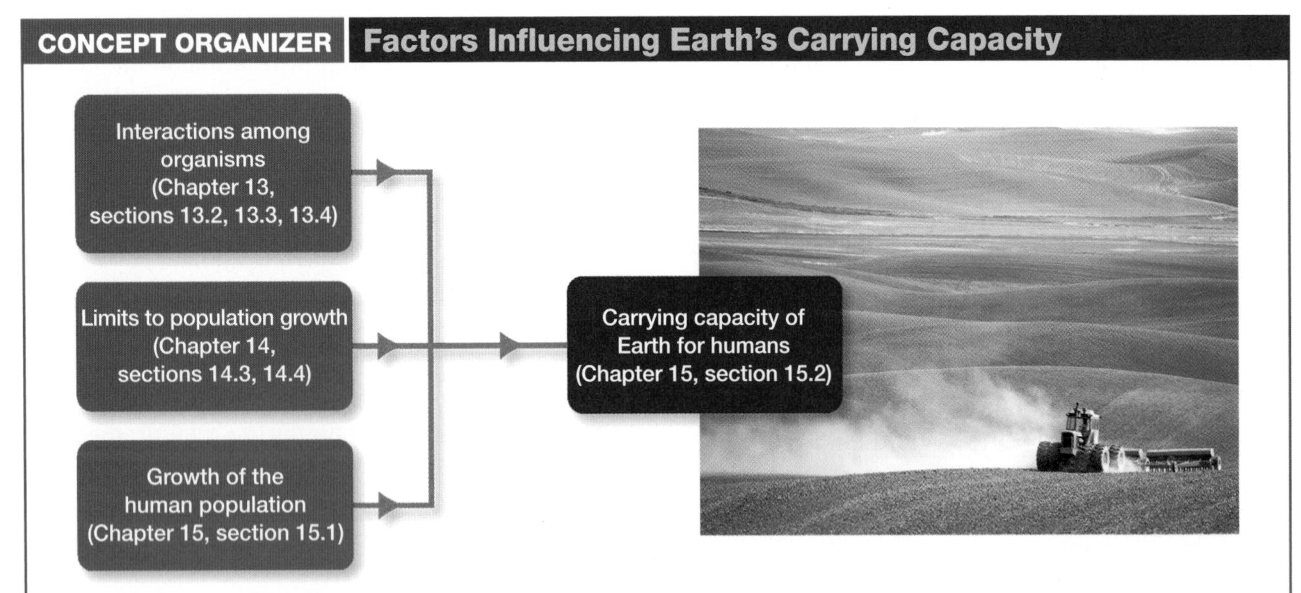

The survival and reproduction of individuals depends on interactions among species and between living things and the abiotic components of their environment. All sorts of interactions play a role in determining whether a population's size will grow, decline, or remain stable. These interactions occur between organisms at different trophic levels (such as between a predator and its prey), those at the same trophic level (for example, competitive interactions or mating), and those between organisms and their physical world. For example, do climatic conditions promote survival? Are there toxins present? As is true for all species, human populations face both abiotic (storms, floods) and biotic (limited food supplies, disease outbreaks) limits to their potential growth. Determining the ultimate limit to the growth of the global human population — the carrying capacity of Earth for *Homo sapiens* — is difficult, if not impossible. This is true for many reasons. One of the most important relates to the fact that humans, perhaps more than any other species, constantly change their environment in ways that both increase (for example, improving food crops and the efficiency of resource use) and decrease (polluting the air and water) this carrying capacity.

Figure 15.13 Factors that influence the carrying capacity of Earth for humans.

this to a much larger population should give you some insight into the difficulties involved in determining carrying capacity, even when only one variable is considered as a limiting factor.

Carrying Capacity: A Challenging But Useful Concept

We have not yet come up with a foolproof method of estimating the carrying capacity of Earth for humans. This is partly because it depends not only on environmental variables (in ways we do not yet fully understand), but also on humans themselves — technology we develop or accept, political institutions, cultural values, and many other factors (including individual and group choices about how resources can and should be distributed). It depends, for example, on how many of us eat meat, eat grains, drive cars, ride bicycles, and a multitude of other things.

Some people have suggested that rather than trying to determine the global carrying capacity, we should concentrate on estimating local, national, or regional limits to growth. Although it may be easier to define the extent of certain resources (such as minerals or water) on a smaller scale, global events or forces may still affect these resources. Human carrying capacity cannot be defined for one country alone if that country trades with other nations and shares global resources such as the atmosphere, oceans, and biodiversity.

Many of those who have attempted to determine the human carrying capacity have provided an upper and a lower estimate. Taking into consideration only the upper estimates in these cases and all single estimates developed by other researchers, the median (middle value) of 65 separate estimates of the maximum size of the human population made before 1996 is roughly 12 billion people. If the lower conservative values are considered along with all single values, the median of these 65 estimates is 7.7 billion. The Population Division of the United Nations predicts that the world's population will be roughly 7.8 billion — just above this lower value — in 2025 and about nine billion in 2050. Although calculating the median value of a wide range of historical estimates does not give us the actual carrying capacity, this analysis does provide some indication that the human population may be coming close to what many people perceive as its limit.

In summary, we cannot currently predict (with any level of confidence) how many people Earth can support. We do know that it depends on more than simply the availability of resources. We also know that regardless of what the actual carrying capacity may be, we are already at the point where the human population has had a substantial impact on environmental quality. Researchers in various fields have suggested three general approaches to reduce this impact: the *bigger pie* theory, the *fewer forks* approach, and the *better manners* approach.

The Bigger Pie Theory

First, some suggest that the carrying capacity can be increased to whatever is needed by improving technology. This has been called the *bigger pie theory*. This theory suggests that Earth can be made to sustain any number of people — we just need to grow more food, improve energy technology, and so on.

The discovery of how to grow crops and domesticate animals is certainly an example of increasing the carrying capacity by increasing the food supply. Many more inventions followed, from the hand plough to the steam engine, commercially

THINKING LAB

The Carrying Capacity of the Food Supply

Background

Imagine that food is the most important factor limiting the growth of the human population. If you knew how much land was required to support an individual, you could estimate how many people Earth could sustain — as long as they all ate the same amount as this *standard* individual. In this lab, you will take the first steps toward making such an estimate.

You Try It

1. The table shows the average amount of energy contained in human food that could be produced per square metre of land per year. For example, 1 m^2 of a field of wheat yields a number of kernels that, when converted to bread, will provide approximately 2720 kJ of energy per year. The same area would yield enough corn to produce 540 kJ of energy in beef when fed to cattle.
2. Many experts recommend that active teenage males should each consume about 11 720 kJ of energy per day. For an active teenage female, this figure is 9 200 kJ per day. From the list of food types in the table, select the foods that you might eat in one day. Using the labels shown on food packages or other nutritional information, determine the amount of energy you might obtain from each type of food in a day. Many food labels give the amount of energy contained in a single serving in calories (Cal or Kcal). To convert this to kJ, multiply by 4.184.
3. Multiply your results by 365 to obtain the amount of food energy you would consume in a year.
4. Determine the area of land required to support your yearly requirement of each food type by dividing consumption (step 2) by yield (from the table). Sum these values over all food types to obtain the amount of land that would be needed to support you for a year.
5. Assuming that you represent the typical male or female in your class, determine how much land it would take to grow enough food to feed all of your class for a year. How much would it take to feed the population of the entire school?
6. How much land would it take to support six billion people for a year assuming this consumption pattern? Do you think this is a reasonable assumption? Explain your answer.
7. What else would you need to know in order to determine the carrying capacity of Earth? Would this figure remain the same from year to year, or even day to day?
8. Did the amount of land needed to sustain an individual for a year vary among people in your class? Explain the source of this variation.

Average amount of energy contained in human food that could be produced per square metre of land per year

Type of food	Yield (kJ/m^2/yr)	Type of food	Yield (kJ/m^2/yr)
bread	2720	cane sugar	14 640
oranges and grapefruit	4200	corn	1050
peanut butter	3850	milk	1760
rice or rice cereal	5230	cheese	170
potatoes	6700	eggs	840
carrots	3400	chicken	800
apples	6300	beef	540
pears or peaches	3750	fish	8
margarine	1250		

produced fertilizers and pesticides, and genetically modified crops. The **Green Revolution**, which began in the 1950s, consisted of an international effort to transfer the farming methods and crop varieties used so successfully in the more developed countries to less developed nations. Applying these highly mechanized, high-input methods in Africa, Asia, and Latin America dramatically increased yields. Worldwide rice production, for example, increased over 300 percent from 140 000 000 t in 1948 to 471 600 000 t in 1991, while the land being tilled increased only 71 percent. These increases have had costs, including environmental damage caused by chemicals and increased soil erosion. There has also been a loss of genetic diversity in the crop types used, which may make it more difficult to select new, insect- or drought-resistant strains in the future and may increase the susceptibility of a crop to disease. Therefore, it is probably unwise to assume that technology can solve all our problems without introducing new ones.

The Fewer Forks Approach

A second method of dealing with problems arising from human population growth has been called the *fewer forks approach*. This involves slowing or reducing population growth, most often by lowering fertility rates. Many methods of achieving this have been advocated and shown to have some success, including lowering infant mortality rates, increasing the economic well-being of families, improving the status and educational level of women, and promoting contraception. Most of these strategies move the population into a later stage of the demographic transition process, in which birth rates decline. Which of these will be the most successful or should be attempted in a particular area depends on many factors, including cultural values and the availability of economic opportunities.

The Better Manners Approach

Third, there is the *better manners approach* of dealing with the impact of population growth, which advocates improving how people interact. When individuals use the environment as a **source** (a supply house of specific materials) or as a **sink** (a substance or site that absorbs and holds on to materials, often for a long period), their actions have consequences for others. Ignoring these consequences has been referred to as the **tragedy of the commons**. This tragedy has been repeated over and over again all over the world, as groups — for example, nations, regional interests, or families — scramble to obtain a portion of a theoretically open resource, often preventing others from having their share. Without careful planning, such use may deplete or damage the resource. Therefore, this approach is not sustainable for the group that has access to the resource or for others who want access to it.

Many examples exist of resources becoming damaged. Overgrazing by cattle and logging without appropriate replanting on crown land (in Canada, this is land owned by the federal government) reduces the value of the land for other species as well as for humans. It also makes it more difficult for the land itself to recover and be available for the future. Overhunting of whales and overfishing by certain countries and/or segments of the fishing industry has depleted stocks of some species to the point where they may be unable to recover, as shown in Figure 15.14. In the mid-1950s, fishers in Atlantic Canada caught approximately 240 000 t of northern cod (*Gadus morhua*) per year. Shortly afterward, they began increasing their take of this species; in 1968, about 800 000 t were caught. After this, the amount fishers were able to catch declined steadily. By the 1970s, it was apparent that the species had been overfished and its numbers had declined. Many methods of fishing, such as using large nets dragged near the ocean bottom, have also damaged the ocean environment such that it is less able to support populations of different organisms.

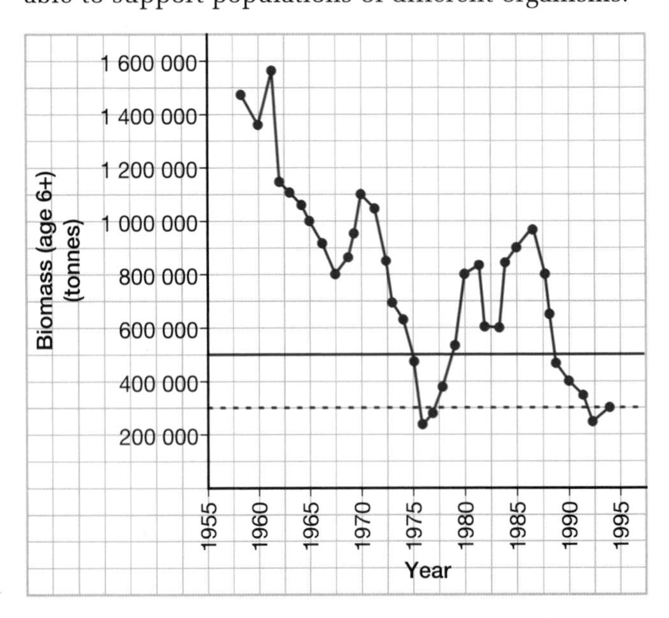

Figure 15.14 As shown in this graph, the biomass of fish aged 6 years or more decreased from a peak of 1 500 000 t to less than 300 000 t in 1975. At this time, Canada acted to limit foreign fishing in the area, and the fish stock rebounded for a time. But it declined again due to poor environmental conditions and overfishing. A complete moratorium on cod fishing was in effect from 1992 to 1999. (Data from the Canadian Department of Fisheries and Oceans.)

These and many other activities actually lower the carrying capacity of Earth for humans — and a variety of other living things. Our goal in attempting to estimate carrying capacity is to understand how big a human population Earth can sustain. To achieve this, it is as vital for us to consider our effect on the planet as it is to consider its effect on us. The two interact in determining how many of us can survive on Earth and under what conditions we must live. Since all of our actions have some impact, it is impossible to provide a complete review of these effects in a single chapter or even a single book. The final section of this chapter briefly examines some of the issues most commonly identified as being vitally important if we are to ensure the continued well-being of the human population.

COURSE CHALLENGE

It is extremely difficult for demographers to estimate Earth's carrying capacity for humans. Several factors affecting carrying capacity are described in the next section. Opinions vary regarding the impact that some factors have on Earth. Consider choosing one of these issues to analyze in detail. Write down your own opinion about the impact of this factor before you start your research. After completing your research, re-evaluate your position. Has it changed?

Canadians in Biology

Landscape Ecologist

Deer mice (*Peromyscus maniculatus*) are small, mainly rural rodents. Although they look innocuous, some deer mice carry hantavirus, which can cause respiratory illness and even death in humans who inhale airborne particles from the urine or droppings of infected deer mice. Humans are generally exposed to the virus when clearing or disturbing areas where deer mouse droppings have accumulated.

Dr. Lenore Fahrig and three colleagues did a cross-Canada study to determine how landscape structures influence the movement of deer mice.

Dr. Lenore Fahrig

They found that certain landscape variables can significantly affect virus incidence in deer mice. In descending order, the most important variables were: the presence of buildings (such as barns and granaries); amount of deer mouse habitat in the landscape surrounding a site; and fragmentation of the habitat (how broken up the patches of habitat were). Of lesser importance were variables such as temperature and seasonal change.

A professor of ecology at Carleton University, Dr. Fahrig specializes in the relatively new field of landscape ecology. "Landscape ecology," she explains, "is the study of how landscape structure affects the abundance and distribution of organisms." Traditional ecology has concentrated on variables such as availability of food and density of predators. Landscape ecology also investigates an area's spatial patterns and their effects on movement, reproduction, and mortality.

"The impact of landscape structure has been ignored, mostly because of the perceived difficulty of conducting such large-scale studies," says Dr. Fahrig. "This constraint is disappearing with the increasing availability of remotely sensed data, allowing much easier measurement of landscape structural variables."

Dr. Fahrig points out that "landscape structure is currently determined largely by human activities, such as forestry, agriculture, and development." Since this is the case, the results of landscape studies are relevant to land-use decisions. For example, Dr. Fahrig has studied deforestation and fragmentation of native Chilean forests, recommending more sustainable regional management to prevent the loss of species.

In a Canadian study, Dr. Fahrig and some colleagues investigated the effects of traffic volume on traffic-related deaths of amphibian species, such as leopard and green frogs. Dr. Fahrig is particularly interested in the effects of roads on distribution and persistence of populations, which species are most vulnerable to roads, and which road patterns are least damaging to wildlife populations.

Dr. Fahrig received her M.Sc. from Carleton University and her Ph.D. from the University of Toronto. She has a special interest in encouraging young women who are considering a career in science.

SECTION REVIEW

1. K/U What is the range (smallest to biggest value) of estimates of the carrying capacity of Earth for humans?

2. K/U Describe four methods that have been used to estimate carrying capacity.

3. K/U Give three reasons why it is difficult to obtain an accurate estimate of human carrying capacity.

4. I **(a)** How might you determine the carrying capacity of a particular region? That is, what would you need to measure and record?

   (b) Would the value you determine be permanent, or might it change over time?

5. K/U **(a)** What was the Green Revolution?

   (b) How has it changed food production on the planet?

6. MC In a sense, the introduction of genetically modified crops is a continuation of the Green Revolution. Do research to discover what advantages some of these crops provide and what disadvantages may be associated with their use. What problems do you think might occur as a result of introducing such crops into less developed nations?

7. K/U What is the difference between a source and a sink?

8. MC Do you think some parts of the world are used as sources more than sinks, or vice versa? Explain your answer and provide examples.

9. K/U Describe what is meant by the term "tragedy of the commons," and give examples.

10. MC How do the activities of people in the United States affect the carrying capacity of Canada? Give some specific examples.

11. MC Do you think the current trend toward globalization of industries and businesses will increase the carrying capacity of the world or decrease it? Might the effect be different in different countries or regions? Explain your answers.

12. I Where does your community get its water? Are limits placed on the amount of water or treated water available; that is, does water limit the carrying capacity of your community for humans? Design a study that allows you to obtain the answers to these questions. What information would you need and what would you have to calculate?

13. MC As you have read in this section, humans have overused resources in some cases to the extent of depleting them. Are there examples of resource overuse in Ontario? Do research that allows you to describe whether a particular resource of your choice (for example, a mineral; a particular species of tree, fish, or other animal; a particular species of plant; or a group of species) has been overused. Carefully record the source(s) of your information. Do you think it (they) might be biased in any way? Is there a diversity of opinion about whether this resource has been overused?

14. MC **(a)** Nets like the otter trawl shown here are used to catch fish and other organisms living on or near the ocean bottom. Dragged behind ships, part or all of this equipment disturbs the ocean floor. Are there beneficial effects of this method of fishing? If so, list them.

    (b) The dragging may scrape organisms (such as corals, which live attached to bedrock) off the sea floor, thereby killing them. Although this may not be an issue in environments that have been disturbed for many years by fishing activity, it can be detrimental in pristine environments. What detrimental effects might this activity have on organisms living on the ocean bottom?

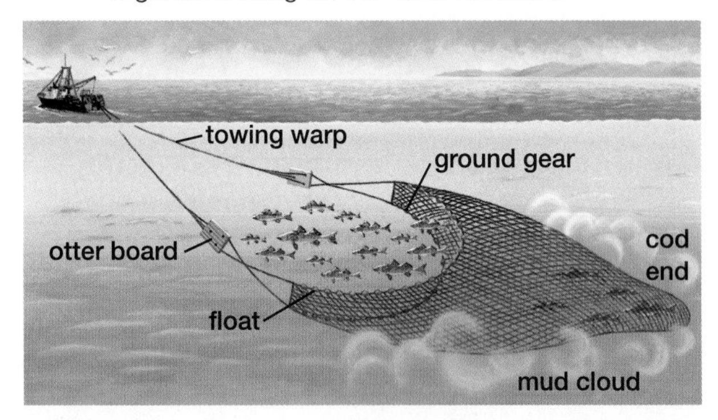

UNIT ISSUE PREP

If you have already decided on the environmental problem you will report on for your Unit 5 Issue Analysis, watch for additional information in the next section that you may be able to use. Be aware that opinions may vary on the seriousness and extent of the problem you have selected, and that you should be prepared to defend your own position with facts.

15.3 Negative Effects on Carrying Capacity

EXPECTATIONS

- **Distinguish among perpetual, renewable, and non-renewal resources.**
- **Describe why clean air, drinkable water, fertile soil, and biodiversity are vital renewable resources.**
- **Describe the effects of human population growth on the environment and quality of life, including the effects of growth on future generations.**
- **Describe differences in rates of consumption that exist among rich and poor segments of the world's population.**

To survive and achieve a condition of well-being, people consume manufactured goods and natural resources (such as water and food). Of course, the manufacture of goods also involves the use of natural resources. These resources provide the material from which the items are made or the energy that drives the manufacturing and distribution processes. Consumption in itself is not a bad thing. The problem arises when the rate of consumption is more than can be sustained by limited resources.

Some of the things we use as resources are not actually limited, at least from the perspective of the human time scale. Called **perpetual resources** by many, these include energy from the Sun, wind, and tides, as illustrated in Figure 15.15. The supply of these is essentially inexhaustible.

Other resources (including air, water, and soil) are referred to as **renewable resources**. The major difference between perpetual and renewable resources is that human influence cannot deplete perpetual resources. Biodiversity can also be considered a renewable resource.

The concept of **biodiversity** encompasses species diversity (the variety of plants, animals, and other organisms on Earth), the genetic diversity that exists within each of these species, and the diversity of ecosystems (ecosystem diversity) of which these species are a part. A diversity of ecosystems is integral to the functioning of the biosphere. For example, marsh ecosystems filter and purify water, while ocean and forest ecosystems act as sinks for excess carbon dioxide in the atmosphere. Of course, each of these ecosystems — and each of the other types as well — plays many roles, providing a home and services for a diversity of species, including humans.

Having a diversity of species within communities tends to stabilize ecosystems and make them more resilient to degradation. Genetic variability within a species also has many advantages. For example, it provides the potential for evolutionary change. If alleles exist that confer an advantage in a changed environment, the species may adapt as a result of natural selection, rather than going extinct.

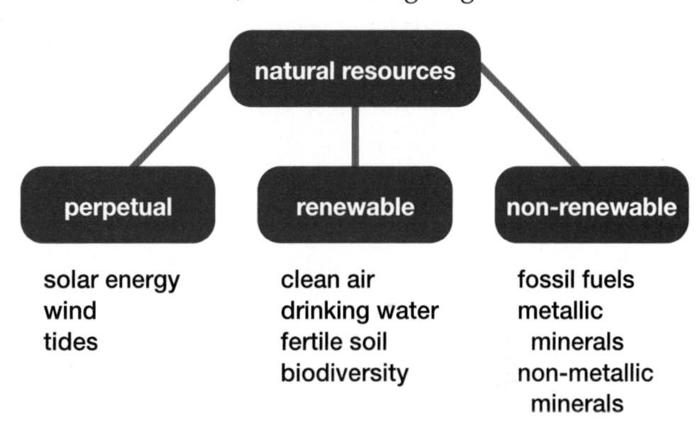

Figure 15.15 Natural resources can be grouped into three main categories when considered on a human time scale: perpetual, renewable, and non-renewable. Given that fossil fuels are derived from the bodies of dead plants and animals, why do you think they are classified as non-renewable?

Just because these resources are described as renewable does not mean that supplies are unlimited and everlasting. Air and water can be polluted, soil can become depleted, and populations can get so small that species go extinct. For the most part, however, these resources can be renewed or replenished by Earth's natural processes — as long as they are not used up too quickly. Some of these processes are part of biogeochemical cycles, such as the decomposition and carbon fixation that are part of the carbon cycle. These cycles tend to cleanse and recycle air, water, and soil. Other processes that are part of biological cycles, including meiosis and fertilization, replenish the size and genetic diversity of populations. Disturbance and succession are among the natural processes that maintain ecosystem diversity.

Finally, there are **non-renewable resources**. This group includes resources that cannot be replenished by Earth's natural processes, as well as those that are replaced by processes that require very long periods of time (often thousands or millions of years). These include metallic minerals (such as copper, iron, and aluminum), non-metallic minerals (phosphates), and fossil fuels (including oil and coal).

Population growth and increased per capita consumption has placed considerable strain on all types of resources. But according to the 1998 United Nations Human Development Report, the problems with non-renewable resources are not as serious as some researchers predicted several years ago. Human use of non-renewable resources has slowed in recent years, and fears that supplies of materials like oil and minerals might run out have not materialized. New reserves have been discovered, efficiency has improved, and recycling has been introduced. As well, per capita use of materials such as copper and steel has stabilized in many countries and declined in others.

Instead, the most urgent problem today is the deterioration of renewable resources as a result of overuse or because of environmental degradation, which reduces the ability of Earth's natural processes to replenish supplies. This deterioration happens not only as a result of the more industrialized countries overconsuming resources and producing excessive waste (much of it toxic), but also as a result of the poverty experienced in less developed nations. As many have noted, people who have difficulty finding enough food to feed their families are unlikely to have the time or desire to learn about environmental concerns. Although ecologists realize that cutting down rain-forest trees simply to supply charcoal for cooking fires is detrimental to ecosystems, the practice continues because this is the only way for people to obtain the money they need to survive. What is needed is a better balance between economic requirements and ecological concerns — in all parts of the world.

Such a balance is the goal of proponents of sustainable development. As defined by the World Commission on Environment and Development (also known as the Bruntland Commission) in 1987, **sustainable development** is development that meets the needs of the present without compromising the ability of future generations to meet their own needs. This seems like common sense. However, there are many examples of rapid and excessive exploitation of resources that have provided only short-term gain and have had serious environmental costs, as you can see in Figure 15.16. In contrast, sustainable development — which provides long-term gain and does not harm the environment — requires making difficult economic and political

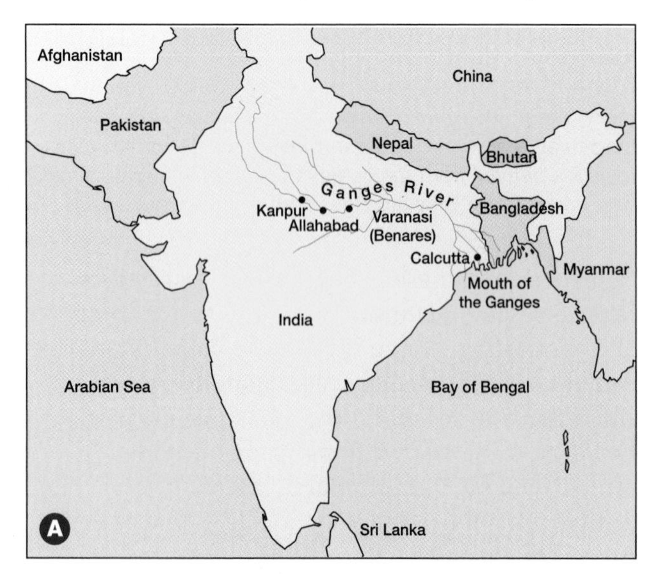

Figure 15.16 (A) Much of Bangladesh is traversed by tributaries of the Ganges River, downstream of India.
(B) Long ago, half of India was forested. Cutting the trees for timber and firewood, and clearing the land for agricultural purposes, has reduced the forests to 14 percent of the country's area. As a result, the water holding capacity of the soil is significantly reduced and soil erosion is a serious problem. Much of this soil, along with high levels of pollutants (including chemicals from leather tanning and large amounts of raw sewage), ends up in the Ganges River. This pollution, coupled with the natural tendency of this river to flood, has disastrous consequences for Bangladesh.

decisions. Such decisions require information about the costs and benefits of human practices, as well as about the details of how ecosystems function. This is a big undertaking. Looking at each of the renewable resources shown in Figure 15.15 on page 523 is a good place to begin.

BIO FACT

The Population Reference Bureau estimated in 1998 that it would cost $6 billion a year to provide basic education for all humans in the developing world and $9 billion a year for clean water and basic sanitation. The amount spent annually on cosmetics in the United States is $8 billion; approximately $11 billion is spent on ice cream in Europe every year.

Clean Air

Earth's atmosphere is made up of a mixture of gases — normally about 79 percent nitrogen, 21 percent oxygen, and 0.03 percent carbon dioxide. It also contains varying amounts of **pollutants** — thousands of chemicals that adversely affect living things. Our production of these pollutants currently exceeds the ability of Earth's sinks to absorb them or convert them into harmless forms. Four of the most important categories of air pollution today include excess greenhouse gases, acid precipitation, photochemical smog, and the production of chemicals that deplete the ozone layer. As Figure 15.17 shows, there may be several types of pollutants contributing to each of these problems, which often makes it difficult to determine who is responsible and what steps can be taken to remedy the situation.

Greenhouse Gases and Global Warming

Several of the gases in the atmosphere produce what is called the **greenhouse effect**; that is, they allow sunlight to pass through the atmosphere and hit Earth. As illustrated in Figure 15.18 on the following page, they then absorb and re-radiate most of the heat that bounces back. Some of the most important gases involved in this process include carbon dioxide, methane (CH_4), and nitrous oxide (NO_x). The trapping of heat is actually natural and normal. Without it, most of the Sun's heat would radiate away from Earth and average temperatures would be below freezing. Most of the carbon dioxide component of greenhouse gases has accumulated over millions of years as a result of the cellular respiration of living things; nitrous oxide and methane are also naturally produced, chiefly by the process of digestion in animals.

In recent decades, however, human activities have added new gases to the list of those that produce the greenhouse effect. These include nitrous oxide from a variety of sources (such as burning coal and nitrogen fertilizer), chlorofluorocarbons (commonly referred to as CFCs and used primarily as refrigerants), and halons (a group of gases used in fire extinguishers).

Even more significantly, these activities have dramatically increased the amounts of normal gases being produced, particularly carbon dioxide. In other words, humans have added **anthropogenic**

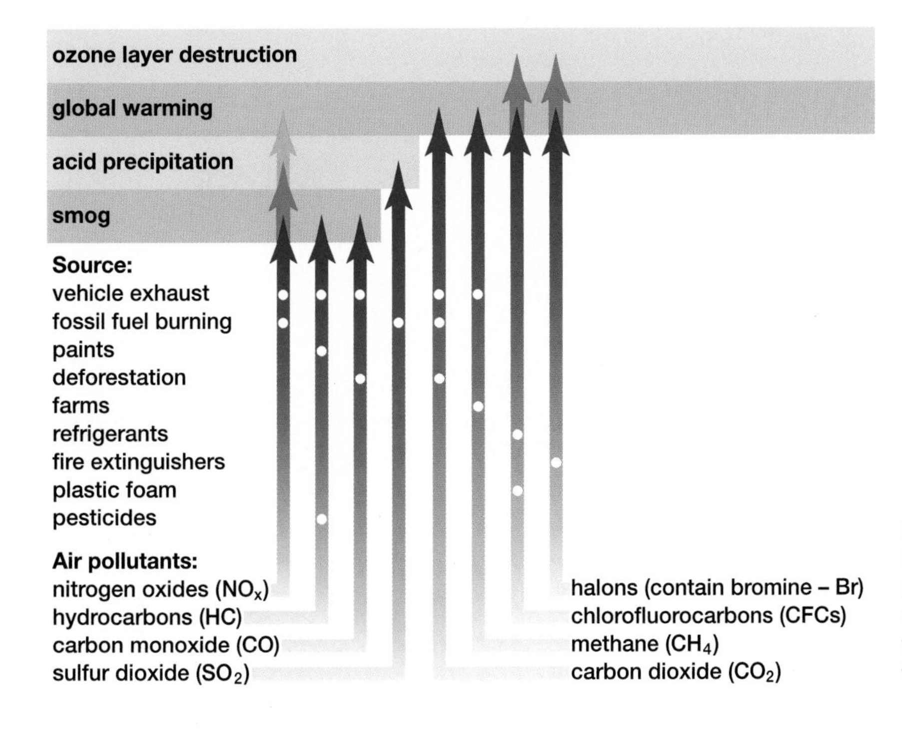

Figure 15.17 Each human activity may produce more than one type of air pollution, and each type of air pollution may have more than one effect. Which of the human activities shown contributes to the most types of air pollution?

(of human origin) sources of carbon dioxide to natural ones. The burning of fossil fuels and **deforestation** (the cutting down of forests) have been major contributors of carbon dioxide all over the world. Since living trees take up carbon dioxide for photosynthesis, they act as sinks that reduce the level of this gas in the atmosphere. As a result, forest fires and the burning of felled trees are doubly damaging, because they destroy the sinks and release large amounts of carbon dioxide.

Most scientists believe that excessive production of greenhouse gases could enhance the greenhouse effect and thus raise the average temperature on Earth, producing what is known as **global warming**. The effects of this are still being debated, but they may include major climate alterations, rising sea levels (enough to destroy coastal ecosystems and flood coastal cities), increased drought in some areas, and changed conditions for farming (possibly accompanied by a drop in production) in many regions. The Thinking Lab on page 527 will help you understand some of the potential problems related to global warming.

ELECTRONIC LEARNING PARTNER

Your Electronic Learning Partner has animation clips that will enhance your understanding of the greenhouse effect on Earth.

BIO FACT

Natural sources of carbon dioxide include cellular respiration (which produces about 60 billion tonnes per year) and the ocean (in which roughly 90 billion tonnes diffuse annually). Anthropogenic sources make up about seven billion tonnes per year. Of this latter amount, about three billion tonnes remain in the atmosphere and roughly two billion tonnes are absorbed into the oceans (which act as a sink for much more than this). Scientists think that much of the remaining two billion tonnes is absorbed by land vegetation. Recent studies conducted in Canada have indicated that the trees and soil of the boreal forest, for example, act as a sink for about 0.6 billion tonnes of carbon dioxide per year.

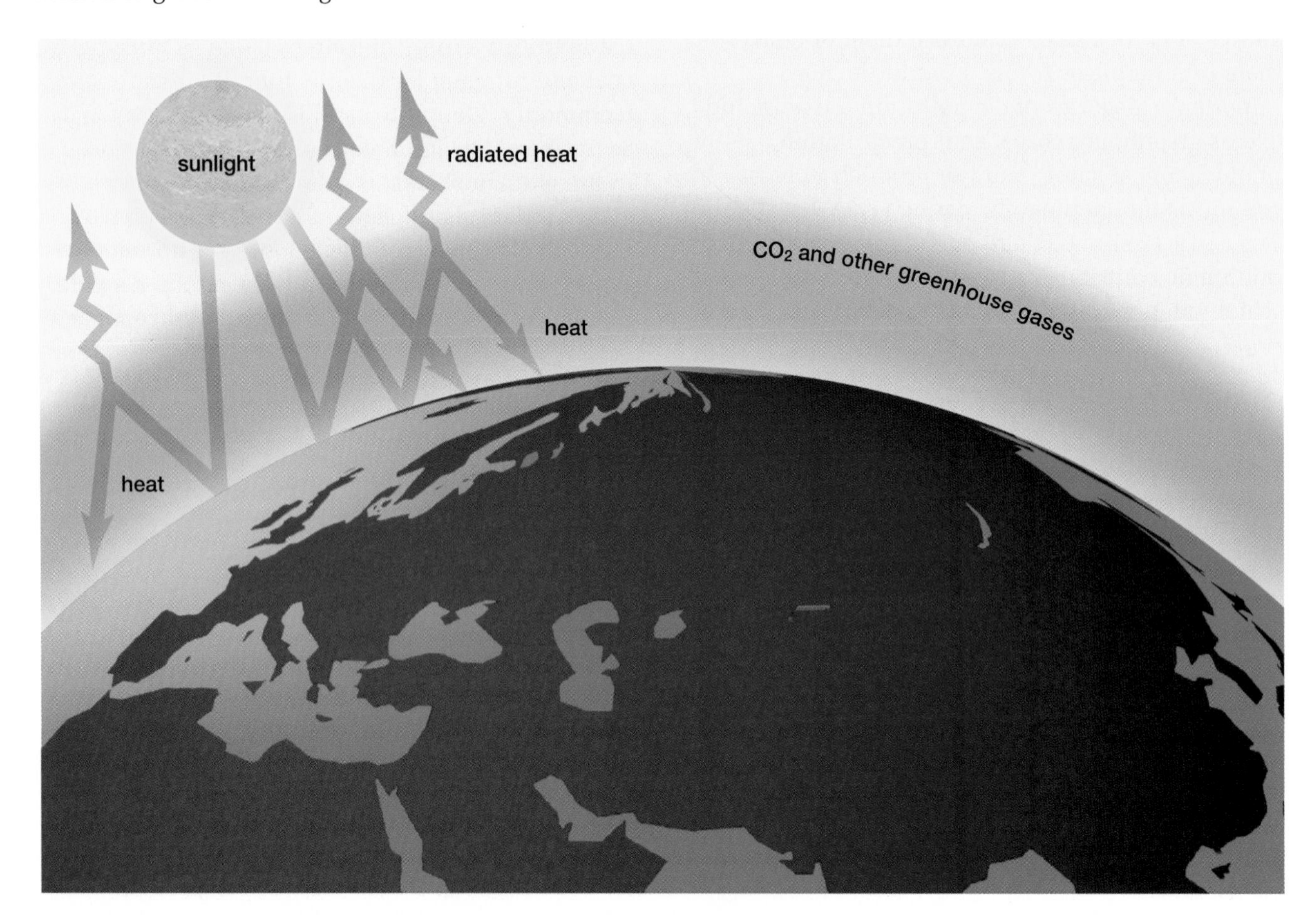

Figure 15.18 How do greenhouse gases affect the temperature of Earth?

WEB LINK

www.mcgrawhill.ca/links/biology12

Although caution must always be used with respect to the accuracy of information obtained from the World Wide Web, there are some very good sites that describe the greenhouse effect and global warming. Many of these sites provide actual data. Go to the web site above, and click on **Web Links**.

Ozone Depletion

In the atmosphere, ozone (O_3) is found both near Earth's surface and far from Earth in the upper limits of the atmosphere. Close to the surface, ozone is formed by the reaction of oxygen (O_2) with various substances found in car exhaust and industrial pollution, and it is a toxin that causes damage to plants and breathing difficulties in humans and other animals. High in the atmosphere, however, ozone is beneficial. It forms a thin layer that helps shield Earth from the harmful effects of the Sun's ultraviolet (UV) radiation. This type of radiation can cause skin cancer in humans, reduce the effectiveness of the immune system, and produce eye damage. Although research is ongoing, it is currently accepted that many other types of organisms, especially amphibians (which have thin skins) and the animals and plants inhabiting the top layer of the ocean, may also be harmed by UV radiation.

Recently, it has been discovered that the ozone layer has been significantly damaged. It has become

THINKING LAB

Looking at Global Warming

Background

Some researchers are still uncertain about whether Earth's temperatures are really increasing and whether this is related to anthropogenic sources of greenhouse gases — especially carbon dioxide. Could various carbon sinks (including forests and oceans) take any extra amount that is emitted such that the carbon cycle would continue to be balanced? This activity will give you an opportunity to see whether the amount of carbon emitted each year is correlated with the concentration of carbon dioxide in the atmosphere. You will also examine whether these variables are related to temperature.

You Try It

1. Graph the data in this table to discover if there is a relationship between year, carbon dioxide production, carbon dioxide concentration in the atmosphere, and temperature. This graph can be done using spreadsheet software or drawn by hand. Plotting all the variables on the same graph will help you see if there is a relationship.
2. Describe the pattern of change over time in the carbon dioxide and temperature variables.
3. Does a correlation between two variables necessarily indicate that one variable *causes* change in the other? What other explanations might there be for such a pattern?
4. Use a library or the Internet to research the following concepts and areas of controversy related to global warming, with a focus on answering the associated questions:
   - The concentration of carbon dioxide has varied widely over the history of Earth, and has been much higher than it is now. Has the rate of change ever been similar to or greater than the rate between 1960 and 2000?
   - Gases other than carbon dioxide also contribute to the greenhouse effect. Are all greenhouse gases comparable in their ability to trap heat? Might one or more of them be a more important cause of global warming?
   - The data in the table show that there has been only a small increase in temperature (less than the difference between winter and summer temperatures in Ontario) over the time period shown. Can a difference of this magnitude have any effect on the biosphere?

Relationship between year, carbon dioxide production, carbon dioxide concentration in the atmosphere, and temperature (from The World Watch Institute, 2001)

Year	Carbon dioxide emissions (million tonnes of C)	Carbon dioxide concentration in the atmosphere (ppm)	Global average temperature (°C)
1950	1612	not available	13.83
1955	2013	not available	13.91
1960	2535	316.7	13.96
1965	3087	319.9	13.89
1970	3997	325.5	14.03
1975	4518	331.0	13.94
1980	5155	338.5	14.18
1985	5271	345.7	14.10
1990	5931	354.0	14.41
1995	6190	360.9	14.39
2000	6299	369.4	14.36

thinner, particularly in certain parts of the world. This damage, or **ozone depletion**, is thought to be chiefly caused by chlorofluorocarbons (CFCs). When CFCs are released into the lower atmosphere, they move slowly upward and decompose, producing chlorine atoms that react with the ozone to produce oxygen. In 1987, even before current levels of damage appeared, an international agreement was signed promoting the phasing out of CFC production in most industrialized countries. Public awareness leading to the decreased use of aerosol sprays also had a significant impact on chlorofluorocarbon use. Since it takes 10 to 20 years for these pollutants to reach the upper atmosphere, there will be a delay before any benefits of reducing CFC production are realized. Figure 15.19 illustrates how ozone depletion occurs.

ELECTRONIC LEARNING PARTNER

Your Electronic Learning Partner has animation clips that will enhance your understanding of ozone depletion and its effect on Earth.

Acid Precipitation

The deleterious effects of acid precipitation were convincingly demonstrated first by Canadian scientists, including Dr. David Schindler and Harold Harvey. Dr. Schindler's group performed experiments in the Experimental Lakes Area, a study site located in northwestern Ontario. The major sources of atmospheric acid are sulfur dioxide (SO_2), released into the air as a result of burning high-sulfur coal and oil, and nitrogen oxides (NO_x) contained in automobile exhaust. These gases combine with water vapour in the air and dissolve in raindrops, which then fall as **acid precipitation** (rain or snow). As its name suggests, this precipitation has a pH of 4 to 5, much more acidic than that of normal rain (about 5.6). Acid precipitation corrodes metal and stone and harms trees. Because it makes lakes acidic, it alters aquatic ecosystems, killing some organisms and harming many others. Acid precipitation can also harm water supplies by causing contaminants to leach from surrounding soils and by dissolving metals, such as copper and lead, in water pipes.

The effects of acid precipitation are most obvious in the northeastern United States, southeastern Canada, and Europe. Recently, sulfur dioxide emissions have been reduced by the use of low-sulfur fuels and the installation of scrubbers

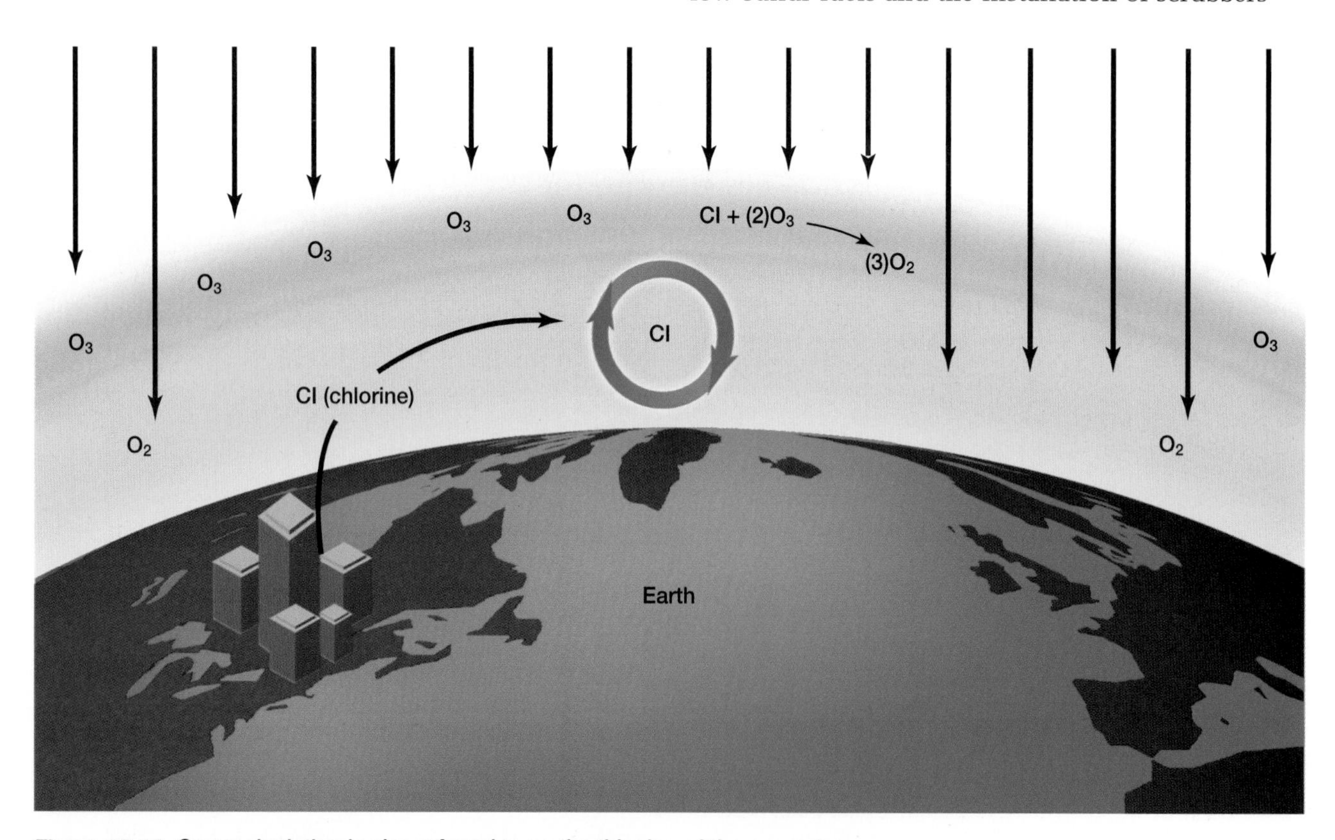

Figure 15.19 Ozone depletion is also referred to as the thinning of the ozone layer.

on smokestacks. Nitrogen oxide emissions have been reduced by design improvements in cars. Acid precipitation is still occurring, however, and is having damaging effects on ecosystems.

Photochemical Smog

Nitrogen oxides are also involved in other types of air pollution. In the presence of sunlight and water vapour, they react with another group of gases called hydrocarbons to produce low-atmosphere ozone and PANs (peroxyacyl nitrates). These are the major constituents of **photochemical smog** — the brown or grey haze that tends to hang over the area where it was produced. See Figure 15.20 for an example. Both of these constituent substances irritate eyes and lungs and may lead to long-term respiratory illnesses, such as emphysema. Smog may also contain particles of ash, asbestos, lead, dust, and other materials. It is more common in areas that have undergone rapid growth and industrialization. Many cities have now reached the stage where clean-up measures, often involving the use of cleaner-burning fuels in automobiles, have reduced the frequency and severity of smog. However, the problem is still increasing in some centres.

Figure 15.20 Have you ever experienced photochemical smog like that shown in this photo?

Drinking Water

There are two main issues relating to human impact on water supply. The first is that the supply is not limitless. Although water is a renewable resource, it is possible to use up the supply faster than it can be replenished. Second, human activities negatively affect the supply that does exist, reducing the useable portion still more.

Over 97 percent of the water on Earth is ocean water (which is salty and therefore undrinkable) and approximately two percent is frozen in glaciers and polar ice caps. This leaves less than one percent as fresh water, contained in bodies of surface water (lakes, rivers) and in **aquifers** (reservoirs of ground water held below Earth's surface). In addition, this relatively scarce supply of water is unequally distributed. While some parts of the world (for example, Canada) have a lot, desert and semiarid regions — which make up large portions of Africa, for example — have little.

The history of human population has, in large part, been dependent on water supply or on how we have manipulated it. Irrigation of crops made the first human cities possible. One of the most amazing features of Rome was its system of elevated aqueducts — channels that carried water great distances to supply the city. During the Industrial Revolution and the period of rapid population growth that followed it, water demand increased dramatically. The total amount of water taken out of rivers, lakes, and aquifers in 2000 was nine times greater than the amount withdrawn in 1900. This is mostly due to population growth, as annual per capita usage has only doubled since that time and has actually declined slightly in recent years.

Population Growth and the Water Supply

Despite this increase in efficiency, many experts worry that we will still soon be faced with a water shortage. Projections combining future population size with water availability suggest that by 2025, 40 percent of the world's population may experience serious difficulties in agriculture, industry, or human health if they have to rely on natural water sources. Twenty countries were already suffering from water stress in 1998, having less than 1000 cubic metres of water per person per year. (Per capita usage in Canada was 1622.7 cubic metres in 2000.) In addition, when humans divert water for their own purposes it is often at the cost of supplies for other species. In portions of the western United States, for example, choices already have to be made between irrigating crops or saving water for rivers used by salmon for spawning.

Human activities often reduce supplies of fresh water further as a result of pollution. Non-gaseous waste materials including untreated sewage, solid and liquid waste from factories, and run-off from agricultural fields and city streets (containing pesticides, fertilizers, oil, and other chemicals) must go somewhere. If they do not decompose naturally, they end up in the water or soil.

When organic materials from sewage and food processing plants, paper mills, and other sources enter the water, they serve as nutrients for the bacteria that degrade them. A substantial influx of these materials can cause rapid growth of the bacterial population. The cellular respiration of this large number of bacteria can deplete oxygen from the water and harm other aquatic organisms.

Many inorganic nutrients (such as nitrates and phosphates from fertilizers and sulfates from laundry detergents) also end up in water. This nutrient abundance can lead to excessive growth of algae and/or plants (called an algal bloom), which similarly rapidly depletes the available oxygen in the water (due to the high demand for oxygen to support cellular respiration). This then causes an increase in the death rate of the algae and plants. The large amounts of dead material are then decomposed by bacteria, which further deplete the oxygen supply, leading to the death of some aquatic animals. Figure 15.21 on page 531 shows one such situation. Some pollutants, such as oil, gasoline, heavy metals, and pesticides, are toxic to many forms of life. They cannot be degraded by bacteria easily and remain in the environment for long periods of time. These are the types of substances that are involved in biological magnification, as described in Chapter 13. Heat is another pollutant, creating the thermal pollution that affects many rivers and lakes. Produced by power plants and other industrial processes, this added heat reduces the amount of oxygen that can be dissolved in water and may eventually cause harm to aquatic organisms.

Many of the pollutants described above can have serious and immediate effects on human

Investigation 15 • B

SKILL FOCUS

- Predicting
- Hypothesizing
- Performing and recording
- Analyzing and interpreting
- Communicating results

How Can the Tragedy of the Commons Be Avoided?

When a resource is overused, it often becomes damaged in such a way that it becomes less available (or completely unavailable) to members of a community. To use resources in a *sustainable* way, they must be managed carefully so they are not degraded and they remain useable by future generations.

Pre-lab Questions

- What is meant by the phrase "tragedy of the commons"?
- What does it mean when an activity is said to be sustainable?

Problem

How can the members of a community prevent destruction or degradation of a shared resource?

Prediction

Predict how shared resources will be affected by community members over a specific time period, and suggest methods for sustaining that resource.

Materials

large number of white pebbles, marbles, beans, or other small objects
large number of red pebbles, marbles, beans, or other objects (they need not be red, but must be different in colour and similar in size to the white objects)
opaque bag (or box) for each community

Procedure

1. Divide the class into groups of four. Each group will represent a community.
2. Place 16 white pebbles in each community bag. The white pebbles represent one parcel of farmland.
3. Give each community member a large handful of red pebbles. Each red pebble represents the use of chemical fertilizers and pesticides.
4. For each white pebble drawn from the bag by a community member, a red pebble will be placed in the bag.
5. Each member of the community will, one by one and in his or her turn, extract one or more pebbles without looking in the bag. At least one white pebble must be drawn if the person is to survive; if a member fails to get a white pebble, he or she has lost his/her parcel of land and cannot continue play. Each member may take as many pebbles

health, especially if they affect sources of drinking water. This is also true of disease-causing organisms, which can be considered another form of pollution. These can include the bacteria, protists, or viruses that cause a diverse array of illnesses such as typhoid fever, cholera, and hepatitis. For example, *Escherichia coli* (various forms of which cause diverse illnesses), *Giardia*, and *Cryptosporidum* are organisms that commonly pollute water in areas where sewage treatment is inadequate or animal wastes can infect water supplies.

The following investigation will give you a chance to learn how resources can be adversely affected if not managed in a sustainable way.

Figure 15.21 Although there is a lot of biological activity going on in this pond, it is not a healthy place for animals.

as desired. The number of red pebbles drawn does not matter.

6. After all members of the community have drawn once, the number of white pebbles remaining in the bag are counted. Add the same number of white pebbles to the existing pebbles and place all of them back into the bag (that is, double the number of white pebbles in the bag). Leave all existing red pebbles in the bag.
7. Repeat steps 1 through 6. The completion of two rounds represents the first generation.
8. Repeat steps 1 through 6 for two more rounds. During these rounds, three red pebbles must be placed in the bag for each white pebble drawn. Rounds 3 and 4 represent the next generation — the children of those community members who started sharing the original land (white pebble) parcels.
9. Repeat steps 1 through 6 for two more rounds. Place three red pebbles in the bag for each white one drawn during this round as well. Rounds 5 and 6 represent the third generation — the grandchildren of the original community members.

Post-lab Questions

1. How many members of the community were forced out of the game during the first generation? in the second and third generations?
2. Why did each community member take as many pebbles as he or she did? (Each member should explain his/her reasoning to the other community members.)
3. Did any member of the community have an advantage over others? Explain your answer.

Conclude and Apply

4. How did the actions of the first generation affect the third generation? Do you consider this fair? Explain your answer.
5. Why did players increase the number of red pebbles added during Rounds 2 and 3? Is this a reasonable representation of what might take place in nature?

Exploring Further

6. Play the game again, this time without using the bags so that members of the community can monitor the condition of the shared resource and the amount of pollution being added. Does this change the length of survival time for all members of the community? Was it possible for any community to sustain the resource so the pollution levels were the same in the third generation as in the first? How much communication among community members was necessary to sustain the resource?

BIO FACT

In 2001, a group of pre-eminent scientists, writing in the journal *Science*, predicted that a rapid expansion of agriculture would occur over the next 50 years, driven by the demands of a wealthier and larger human population. This expansion would involve the conversion of 10 billion hectares of natural ecosystems to agriculture, and would increase the amount of nitrogen and phosphorus added to water by 2.4 and 2.7 times respectively. They predicted that this expansion, and the additional pesticide use that would occur, could cause massive damage to aquatic ecosystems and result in the extinction of many species.

Fertile Soil

Humans alter the landscape for a variety of purposes. We build dams in river valleys to produce electrical power, strip mountain tops or dig enormous holes to extract minerals or coal, cover the land with buildings and pavement, and cut down forests for lumber or to provide growing space for crops. All of these activities have an impact on the soil required to produce our food supply, as does agriculture itself. In 1998, United Nations experts estimated that one sixth of the world's land area (nearly two billion hectares) is now degraded as a result of overgrazing and poor farming practices. **Desertification**, the transformation of marginal dry lands into near-deserts, which are unsuitable for agriculture, is a serious problem in many parts of the world, as shown in Figure 15.22. This is particularly true in Africa where the Sahara desert is expanding, but also in parts of the southwestern United States (for example, Mojave desert areas) and India (along the Great Indian desert). Desertification can be reversed if it has not gone too far, but only slowly, at great expense, and with great effort.

Deforestation also has disastrous effects on soil quality, particularly in tropical regions, but also in North America, including all of the Canadian provinces. Some experts estimate that at least 40 percent of tropical rain forests have already been lost and that they are currently disappearing at the rate of roughly 400 km^2 per day. These forests are sometimes logged to supply lumber, but are often cleared simply to get at the mineral deposits beneath them. Commonly, they are cut down to obtain land for agriculture. In many cases, the fields that result are only fertile for 10 to 20 years or less. This is partly because, as you learned in Chapter 13, the soil in tropical rain forests is poor in nutrients. In addition, abundant rain in these areas causes erosion in cleared areas (particularly on steep slopes), and mudslides sweep crops and soil into the ocean. In Madagascar and Haiti, for example, soil erosion following deforestation is so severe that thousands of tonnes of soil wash into the sea each year.

Figure 15.22 What do you think would be required to restore this land to a useable state?

Biodiversity

All of the human impacts on renewable resources already described have serious consequences for other species as well. In other words, air and water pollution and soil erosion threaten all types of organisms. In some cases, the effect of humans has been very direct: over-exploitation of many fish, mammal, and bird populations for food, clothing, and other uses has put many species at risk. This has caused the **extinction** (complete disappearance from Earth of all members of a species) of some, and the **extirpation** (disappearance of a species from areas that were once part of its range) of others. In most cases, however, human impact on other species has occurred mainly as a result of habitat alteration, which reduces the ability of organisms adapted to that habitat to survive and/or reproduce.

As you saw in Chapter 12, extinction is a natural process and is constantly occurring. At various times in the history of Earth, there have been mass extinctions, in which large numbers of species died out. The period when most of the dinosaurs and many other species, especially marine forms, disappeared is one of these. Many researchers, however, believe that humans have played a significant role in causing the disappearance of other species during the past 50 000 years. They point to the fact that during this geologically recent time period it has been mainly conspicuous species of mammals, reptiles, and flightless birds that have disappeared. This suggests that human hunters have played a critical role in their demise. Other circumstantial evidence also points to the involvement of humans. For example, in places where the time of human arrival can be determined, most extinctions occurred shortly thereafter and among species humans would likely have hunted, as Figure 15.23 illustrates.

The list of endangered species (those at risk of extinction) is long and growing. The International Council for the Preservation of Birds, for example, estimates that almost 2000 species of birds have become extinct over the past 2000 years, most as a result of human activity. Comparable numbers for smaller, less conspicuous organisms are not available; experts predict that many species of insects went (and are going) extinct even before we discovered and named them, particularly in tropical rain forests.

Besides the ethical and moral reasons to conserve species, we constantly discover new drugs and other beneficial chemicals produced by species previously unknown to us. In many parts of the world, scientists are racing to learn all they can about species before they disappear. More importantly, the health of most ecosystems depends on the existence of a diversity of species, each contributing something to the myriad of interactions that sustain the biosphere.

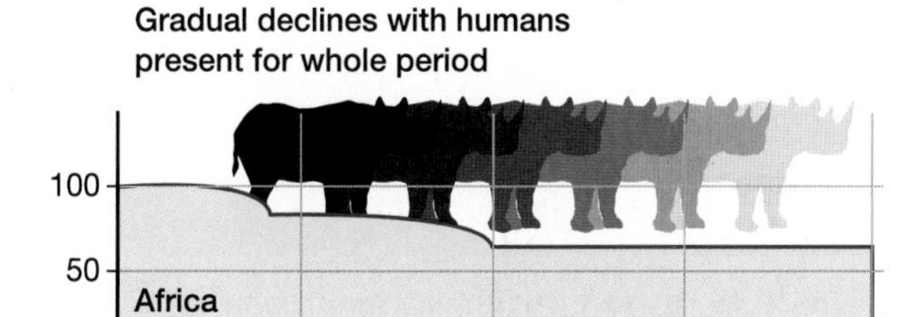

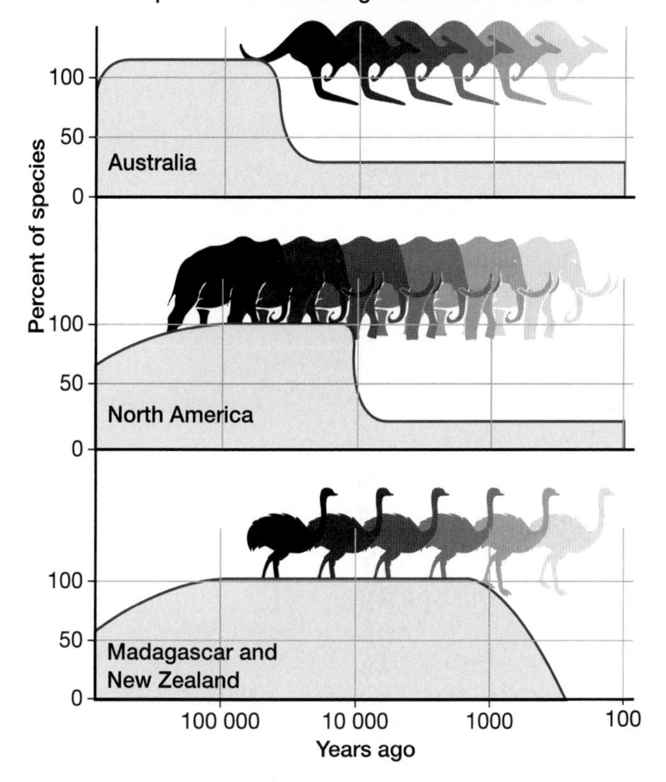

Figure 15.23 In many locations, the extinction rate of many species increased dramatically after the arrival of humans. Why do you think the rate of extinction was more or less constant in Africa over this same time period?

The Problem: Is It Population Growth or Consumption?

The global human population has increased dramatically during the last 100 years. So too has the disparity in wealth among different segments of the population. In industrialized countries, per capita consumption of all types of goods and resources (as calculated by United Nations staff) has increased steadily at a rate of about 2.3 percent annually. While the increase has occurred more slowly in less developed nations, per capita consumption has actually declined in some: the average household in some parts of Africa, for example, consumes 20 percent less today than it did 25 years ago. Nearly 60 percent of the 4.4 billion people in developing countries lack basic sanitation and about 30 percent do not have access to clean water. The inequalities in consumption among different parts of the world (and even within countries or regions) are greater than the differences in fertility rates. The richest 20 percent of the human population lives in the highest-income countries and consumes 45 percent of all meat and fish, 58 percent of the world's total energy supply (see Figure 15.24 on page 534), and 84 percent of all paper, and owns 87 percent of the world's cars. By comparison, the poorest 20 percent of the global population eats five percent of all meat and fish, uses less than four percent of the total energy, consumes 1.1 percent of all paper, and owns less than one percent of all cars in the world.

Figure 15.24 **Modern farming can produce results quickly, but it consumes a large amount of energy.**

On the waste side of the coin, the richest 20 percent of the population produces 53 percent of carbon dioxide emissions, while the poorest 20 percent accounts for three percent. Similar figures exist for garbage production. In Canada, landfills are overflowing and large cities are forced to transport their garbage to distant locations. In developing countries, much less material goes to waste as people find a way to use or re-use almost everything — and own much less excess to begin with (as shown in Figure 15.25). The result of this is that a typical child born in the industrial world adds 30 to 50 times more to consumption and pollution over his or her lifetime than a child born in a developing country. Which has the bigger impact on Earth: high fertility in developing countries or high consumption in industrial nations? Which nations are making more progress toward reducing their impact? Difficult questions perhaps, requiring difficult answers.

Figure 15.25 **In many parts of the world, farmers still use manual methods for harvesting. Waste from a harvest is minimal because all parts of the crop are used.**

SECTION REVIEW

1. K/U What are perpetual resources? Give two examples of this type of resource.
2. K/U Is oil defined as a renewable or non-renewable resource? Why?
3. K/U What is a pollutant? Give three examples of pollutants that affect the atmosphere.
4. I Design a study that would allow you to determine the amount of greenhouse gases contributed by an industry of your choice. What would you need to know before starting your study and how would you collect your data?
5. K/U How does deforestation affect global warming?
6. K/U What causes ozone depletion and why is this a problem?
7. MC What is the current status of the ban on chlorofluorocarbons, and which countries are complying? What more environmentally friendly substances are used in place of chlorofluorocarbons?
8. K/U What is acid precipitation? What causes it?
9. I How would you determine whether acid rain has had an effect on a certain lake? That is, what variables would you measure to decide this?
10. K/U What causes photochemical smog? What effects does it have on human health?
11. MC Have government programs like *Drive Clean* and *Smog Patrol* had an effect on the amount of smog or acid precipitation produced in Ontario? Explain your answer.
12. K/U What is desertification? Give two examples of areas where this is a problem.
13. K/U What is the difference between extirpation and extinction?
14. MC Illustrate differences in consumption rates between more and less developed countries. Provide examples.
15. I How might you compare the quality of life of an individual in a developing nation with that of someone in a developed country in order to see if the former was improving? What aspects of lifestyle would you examine and how would you measure them?
16. MC List 10 ways your level of resource consumption could be decreased. Do any of these also reduce your output of pollution? Explain how.

Summary of Expectations

Briefly explain each of the following points.

- Since the human population was small and controlled by density dependent factors for most of its history, humans have many traits typical of a *K*-selected species. (15.1)
- The Agricultural Revolution caused a gradual but consistent increase in the growth rate of the human population, which eventually began to grow exponentially. (15.1)
- As populations pass through the stages of a demographic transition, their birth and death rates change from a state of balance; through two stages in which the birth rate is higher than the death rate and the population grows; to the last stage in which they are again balanced and the population size remains stable (although larger than before). (15.1)
- Nations identified as more industrialized have completed their demographic transition, whereas those that are termed less developed tend to be in earlier stages and thus still growing. (15.1)
- The rate of growth of the human population appears to be slowing. (15.1)
- The age structure of a population (which can be illustrated in an age pyramid) affects its rate of growth. (15.1)
- The human population is rapidly becoming urbanized, a trend that has both costs and benefits. (15.1)
- Estimates of the carrying capacity of Earth for humans vary widely, and the variation among the estimates has increased over time. (15.2)
- Many approaches have been used in attempts to estimate the carrying capacity for humans; all make many assumptions that may or may not be true. (15.2)
- Earth's carrying capacity for humans is probably not static. It likely varies with many factors, including the diet of humans, weather, and environmental conditions. (15.2)
- The human population has a negative impact on Earth in ways that reduce its carrying capacity for humans. Three approaches to deal with this include what have been called the bigger pie theory, the fewer forks approach, and the better manners approach. (15.2)
- The natural resources that humans consume can be divided into three categories: perpetual, renewable, and non-renewable. (15.3)
- Humans have their biggest negative impact on renewable resources, which are being used up faster than they can be replenished. (15.3)
- Four renewable resources that are seriously affected by human overconsumption and environmental degradation are clean air, drinkable water, fertile soil, and biodiversity. (15.3)
- The goal of sustainable development is to balance economic and environmental concerns, thereby ensuring that all humans can maintain a good quality of life without causing irreparable damage to the environment. (15.3)

Language of Biology

State the biological significance of each of the following terms. Create concept maps to show your understanding of how they are related. Be prepared to explain your rationale.

- Agricultural Revolution
- demographic transition
- Industrial Revolution
- zero population growth
- more industrialized nation
- less industrialized nation
- fertility rate
- replacement fertility rate
- urban sprawl
- Green Revolution
- source
- sink
- tragedy of the commons
- perpetual resource
- renewable resource
- biodiversity
- non-renewable resource
- sustainable development
- pollutant
- greenhouse effect
- anthropogenic
- deforestation
- global warming
- ozone depletion
- acid precipitation
- photochemical smog
- aquifer
- desertification
- extinction
- extirpation

UNDERSTANDING CONCEPTS

1. (a) Compare the effects of the Agricultural and Industrial Revolutions on the growth rate of the human population.
   (b) What was the Green Revolution? How did it contribute to the growth of the human population?
2. Compare a country in Stage 2 of the demographic transition process with one in Stage 4 with respect to the following: expected growth rate, age structure of the population, fertility rate.

3. What is meant by the term "tragedy of the commons"? Give an example of this problem in Ontario.
4. Describe the range of environmental problems associated with the use of cars.
5. How do pollutants adversely affect living things? Give examples.
6. Irrigation can increase crop yields dramatically. What problems can it cause?
7. Do humans have a positive as well as a negative impact on their carrying capacity? Explain your answer.
8. In what places in Canada is smog likely to be the biggest problem? At what time of year is it worst?
9. Why do you think amphibians would be more affected by ozone depletion than organisms like reptiles? Amphibians have also been seriously affected by acid precipitation. Based on their biology, why do you think this might be the case?

INQUIRY

10. The data in the following table are United Nations Population Division statistics for 1995 and 2000, as well as projections (calculated in 2001 and based on a scenario of reaching an average worldwide fertility rate of 2.1 in 2050) for a number of subsequent years. Construct an age pyramid for each of the years. (Males and females are not separated in these data, so you will not have separate pyramids for the sexes.)
    (a) What do you notice about the proportion of individuals in the oldest age class? When members of your cohort are in this age class, what percentage of the population will consist of individuals of about your age?
    (b) What effect do you think the change in proportion of older individuals will have on the growth rate of the population? What about on the ability of working people to provide for the well-being of retired people and of those who are too young to work?

Human population for Earth (United Nations Population Division)

Population by age group (millions)					
Year	Total	0–14	15–59	60+	80+
1995	5666	1768	3354	544	62
2000	6055	1800	3650	605	70
2025	7824	1836	4807	1180	149
2050	8909	1747	5193	1970	370
2075	9319	1724	5166	2428	572
2100	9459	1713	5092	2654	717
2125	9573	1707	5078	2788	824
2150	9746	1706	5075	2964	956

11. In 1999, Canada was one of the top 10 emitters of carbon from fossil fuels in the world, producing very high levels compared with other countries (as shown in the table below). Why do you think this is so? Do research to discover what aspects of the Canadian lifestyle contribute to our high level of carbon emission. What steps can and should be taken to reduce these emissions?

Per capita emissions for selected countries (from the World Resources Institute)

Country	Total emissions (millions of tonnes)	Per capita emissions (tonnes)
United States	1520	5.6
United Kingdom	152	2.6
Canada	151	4.9
India	243	0.2
Australia	94	5.0
Mexico	101	1.0
France	109	1.8
Germany	230	2.8
Japan	307	2.4

12. Try to find out the quantity of waste deposited annually by your city, town, or region in its landfill, and find the quantity for one or two similar populations of your choice. Calculate the per capita amount of waste deposited in each landfill for each area. Does it differ? What factors might explain the difference? Suggest ways to reduce the amount of material sent to the landfill each year.

COMMUNICATING

13. Although China contains only about seven percent of Earth's arable land, its population was approximately 1.25 billion in 2001 — roughly 20 percent of the total human population. Early in the 1970s, the Chinese government realized that the growth rate of their country's population was unsustainable and took measures to reduce it. They instituted a policy that limited the vast majority of families to only one child — and strictly enforced it. Although the country has not yet stopped growing because of its momentum for growth, its fertility rate has actually been below replacement level for over a decade. What do you think are the benefits and drawbacks of the one-child policy? Have groups in your class take opposing sides and debate this topic. Be sure to consider all sides of the issue, including moral, ethical, and practical aspects.
14. Complete a T-chart or Venn diagram contrasting renewable and non-renewable natural resources. Make sure your chart addresses both similarities and differences, and give examples for each point.
15. Produce a poster or pamphlet that would help to educate the public about an environmental problem of your choice. You should briefly describe the effects of the problem and describe the steps that an individual or family could take to reduce the problem.
16. Imagine that a local industry has just closed and many people in your city are out of work. Another industry is interested in building a factory in your town, but it might create a considerable amount of air pollution. There will be a town meeting in which citizens can meet with representatives of the company and discuss the pros and cons of letting this industry set up shop in the area. Prepare a list of questions you want to ask and points you might make at the meeting.

MAKING CONNECTIONS

17. Which do you think would contribute more toward reducing ozone depletion: slowing population growth or changing consumption patterns? Explain your answer.
18. What shifts in survivorship curves might you expect to see if:
    (a) environmental problems worsen and pollution-related diseases increase?
    (b) cutbacks to prenatal and infant care are enacted?
19. What do you think is the difference between the *preservation* of a natural resource and its *conservation*?
20. Do research and prepare a short summary of the findings and conclusions of the Earth Summit. Other United Nations sponsored meetings have occurred since 1992 to address the issues raised at the Earth Summit. Has progress been made since then?
21. In the early 1970s, British scientist James Lovelock formulated the Gaia hypothesis. What is the Gaia hypothesis and what are its major strengths and weaknesses?
22. During the Green Revolution, many crop plants in less developed nations were replaced by high yield crop varieties exported from more developed nations. Although this had benefits, it also incurred costs. What are some of these costs? What effect might this have had on the biodiversity in the less industrialized nations, and why would this matter? What other costs might have been incurred by these crops?
23. Find out if Canada is involved in helping Bangladesh solve its flooding problems. If so, what is Canada doing? Evaluate whatever activities are being conducted in terms of what you feel their success rate might be. Consider whether these activities are dealing with the roots of the problem (for example, deforestation in India) or just the symptoms. A good place to start your search might be the web site of the Canadian International Development Agency (CIDA) or the International Development Research Council (IRDC), another Canadian organization that works on development issues in less industrialized nations. If you do not have access to the Internet, both of these organizations publish many reports on their activities. These reports can be obtained by writing or phoning their head offices.
24. Is deforestation a problem in Canada? in Ontario? Do you think Canada should be investing money in helping other countries with this problem before it is solved here? Explain your answer.

Sustainable Development: Population and the Environment

Background

In 1992, the largest-ever meeting of world leaders took place in Rio de Janeiro, Brazil at the United Nations Conference on Environment and Development. Commonly referred to as the Earth Summit, this two-week event was presided over by Canadian Maurice Strong (the conference's secretary-general). The goal of the Earth Summit was to promote the idea of sustainable development; that is, allowing present human generations to improve or maintain their quality of life without compromising that of future generations. Delegates recognized that the relationship between overpopulation and environmental degradation is not simple, since the conditions and practices associated with poverty often place as much stress on the environment as the excessive consumption of richer populations.

One of the basic issues related to quality of life is the availability of food; while some people have an abundance, others are starving. Although more equitable distribution of available food might help solve this issue, some researchers believe our current supply of food cannot keep up with the demands of the growing global population. And the more the population grows — the greater the potential for environmental damage. As described in Chapter 15, such damage reduces the carrying capacity of Earth for humans, making it difficult to produce sufficient food.

The 1992 Earth Summit Conference

ASSESSMENT

After you complete this issue analysis,

- **assess how clearly your report and presentation conveyed the information;**
- **assess your presentation according to the responses of your classmates;**
- **assess your research skills during the development of your report. Did your skills improve?**

A major outcome of the Earth Summit was the development of *Agenda 21*, a plan for achieving sustainable development in the twenty-first century. Various international agreements were negotiated, including the Framework Convention on Climate Change and the Convention on Biological Diversity. To monitor the implementation of these agreements, the UN Commission on Sustainable Development (CSD) was established in December of 1992. The CSD's annual meeting provides a forum for exchanging information and building partnerships among nations, businesses, industries, and other concerned groups.

In this activity, you will have the chance to make a presentation to the CSD as part of a group representing a specific country. You will describe your country's most pressing environmental issue and the state of its population. You will then propose solutions to some of the identified challenges.

Plan and Present

1. Working in groups of three or four, select a country. Half of the groups within your class will represent less industrialized nations (for example, Bangladesh, Botswana, Venezuela, Vietnam, or Iraq), while the other half will represent more industrialized nations (such as Japan, the United States, Germany, Australia, or the Netherlands). Your group will analyze and write a report on the status of your country's population and environment. You will then summarize this report in the form of a presentation to a meeting of the CSD. Your analysis and presentation should address the following four general questions.
   - What is the most pressing environmental issue affecting your country? Consider whether the following factors have an impact: thinning of the ozone layer, global warming, deforestation, desertification, air pollution, depletion of fresh-water resources, ocean pollution, loss of

biodiversity, soil erosion, accumulation of hazardous wastes.

- What are the main causes of this problem? Are they generated from within your own borders by your own citizens, or are they the result of activities going on in other countries? Is the problem due to overpopulation (too many people attempting to live on insufficient resources), overconsumption (use of excessive resources), or both?
- What is the state of your country's population in terms of its growth rate and quality of life for its citizens? Consider such factors as food availability, average income, income distribution (is wealth concentrated by a few individuals or evenly distributed), average life span, infant mortality rate, and access to health care. Is the future population of your country predicted to grow, stabilize, or decline?
- What measures might solve the problem or problems identified? Specifically, examine existing or potential partnerships with other countries. If you are representing a less industrialized nation, what sort of foreign aid might be useful? Does Canada participate in projects in your country, perhaps with funding from the Canadian International Development Agency (CIDA), the International Development Research Centre (IDRC), or other bodies? If so, are these projects successful in improving your country's quality of life? If you are representing a more industrialized nation, does your nation invest in foreign aid? Does your country participate jointly with Canada in any projects? If so, analyze the success of those projects.

2. When doing research on the major environmental issue facing your country, address the following points.
   - What is the science underlying the environmental issue? Provide data on the issue's causes and effects.
   - Does population density contribute to the issue? If so, in what way?
   - Does this particular environmental problem reduce the carrying capacity of your country for humans? If so, how?
3. Consider the resources available to you. For example, the Internet is an excellent source of current information, and many Canadian organizations (such as Pollution Probe) have information available on certain topics. Reports on CIDA and IDRC projects are often available from libraries or obtained from these organizations directly.
4. As a group, write a detailed report that will serve as a background document for your presentation. Each group member should focus on one of the four questions listed in step 1, do the research, and write about that topic. The entire group should discuss each topic before writing begins and provide feedback during the writing process.
5. Prepare a presentation to summarize your detailed report. The format of the presentation can be in the form of a speech, video, or drama.
6. Make your presentation. Be prepared to field questions from your classmates posing as representatives from other countries at the CSD meeting.

Evaluate the Results

1. Compare the content of your presentation with that presented by others. Do the environmental issues described as being important for other countries contribute to the difficulties in your country? Do opinions differ as to the cause of, or effective solution for, certain issues? If so, explain the reasons for these differences.
2. Consider the issue of bias in your work, as well as in the work of other groups. How many sources of information were used? Did any of them seem biased?
3. Did any of the presentations make you think about revising your own report or presentation? Describe the changes you might make and why you would make them.
4. Did your work on this issue analysis help you understand the information you learned in this unit? Might you integrate your learning into your lifestyle? Explain your answers.

UNIT 5 REVIEW

UNDERSTANDING CONCEPTS

Multiple Choice

In your notebook, write the letter of the best answer for each of the following questions.

1. The role played by a particular population in a community is referred to as its
   (a) habitat
   (b) biome
   (c) ecological niche
   (d) range
   (e) trophic structure
2. In any community, a species that is a herbivore is probably a
   (a) primary producer
   (b) primary consumer
   (c) secondary consumer
   (d) tertiary consumer
   (e) terrestrial organism
3. The number of individuals at each trophic level in an ecosystem can be represented by a
   (a) pyramid of numbers
   (b) food web
   (c) biomass pyramid
   (d) pyramid of productivity
   (e) biological magnification
4. In biogeochemical cycles, the processes of photosynthesis and respiration move materials
   (a) out of an abiotic reservoir in which elements are unavailable to organisms
   (b) between abiotic and biotic reservoirs in which nutrients are accessible to living things
   (c) between a biotic reservoir in which elements are unavailable to organisms to another biotic reservoir in which elements are accessible
   (d) between different abiotic reservoirs
   (e) between different biotic reservoirs
5. For a population that is not growing but is stable in size, the following is true.
   (a) the replacement rate (R) is two
   (b) the growth rate (r) of the population is one
   (c) the population is in Stage 2 or 3 of the demographic transition
   (d) the replacement rate (R) is one
   (e) the growth rate (r) is positive
6. Logistic population growth
   (a) occurs in unlimited environments
   (b) produces a J-shaped curve when graphed
   (c) is slow to begin with, but then speeds up and remains high until some abiotic factor reduces population size
   (d) is regulated by density-dependent factors, such as competition between individuals for limited resources
   (e) is not affected by the carrying capacity of the environment
7. Species with equilibrial (K-selected) life histories typically
   (a) have low juvenile mortality
   (b) have several reproductive events during a lifetime
   (c) produce single or a few offspring at each reproductive event
   (d) have relatively long life spans
   (e) all of the above
8. After a country has passed through a demographic transition (that is, completed Stage 4 of the transition), its population will likely
   (a) be smaller than it was before it started the transition
   (b) have a birth rate that is higher than its death rate
   (c) be the same size as it was before it started the transition
   (d) be poorer than it was before it started the transition
   (e) be larger than it was before it started the transition; and have birth and death rates that are about equal
9. Human activity
   (a) can decrease the carrying capacity of Earth for humans
   (b) has an effect on the carrying capacity of Earth for non-human organisms
   (c) can increase the carrying capacity of Earth for humans
   (d) both (a) and (b) are true
   (e) (a), (b), and (c) are true
10. Gases produced by the combustion of gasoline and released in the exhaust of vehicles contribute to
    (a) the production of photochemical smog
    (b) depletion of the ozone layer
    (c) the greenhouse effect
    (d) acid precipitation
    (e) (a), (c), and (d)

Short Answers

In your notebook, write a sentence or a short paragraph to answer each of the following questions.

11. Describe some features of the ecological niche of the panda. Is the panda a generalist or a specialist? Explain your answer.
12. Describe the phenomenon of biological magnification. Give an example of a substance that seems to have had an effect on other organisms as a result of this phenomenon.
13. Define the term "demography."
14. What four processes affect the size of populations?
15. A random sample of 100 mice from an island population are captured, marked, and released. Two weeks later, a second random sample of 150 mice is captured, of which 25 are recaptured marked individuals. Estimate the size of the mouse population on this island.
16. The birth rate for a population of 100 mice is 2.5 individuals per individual per month, and the death rate for this same population is 1.3 individuals per individual per month. What will be the size of this population one year from now?
17. The replacement rate for a population of annual plants is four individuals per individual per year. If we start studying this population when it consists of 100 plants, how big will it be five years from now (assuming environmental conditions remain constant)?
18. Examine the graph shown here and describe what is happening in the portions of the curve labelled (a), (b), and (c). Explain why growth is slow or fast in each of these regions of the curve.

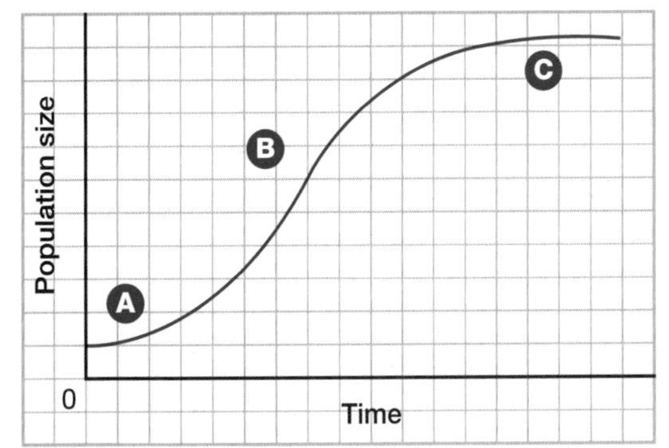

19. The carrying capacity for deer in a particular environment is 500. What does this mean?
20. Why do introduced (exotic) species, such as the zebra mussel, tend to display exponential growth? Give two other examples of introduced species.
21. Describe how density-independent and density-dependent factors might interact to limit the size of a moose population. Give examples of each type of factor.
22. Many populations, such as those of lynx and snowshoe hare, have cyclical population densities. What does this mean? What are some of the suggested explanations for the cycling of snowshoe hare populations?
23. Differentiate between interspecific and intraspecific competition, and describe the effects of each on populations.
24. Using the following three terms in a short paragraph, show the relationships among them: biotic potential, environmental resistance, carrying capacity.
25. What is survivorship? Describe three general patterns of survivorship and give an example of an organism that tends to fit each of these patterns.
26. Distinguish between opportunistic (*r*-selected) and equilibrial (*K*-selected) life histories with respect to patterns of age-specific fecundity and age-specific mortality. Give two examples of organisms with each type of life history.
27. What type of environment tends to favour an opportunistic (*r*-selected) life history? Why?
28. Describe the four stages of a demographic transition in terms of how the birth rate, death rate, and size of a population tend to change as it passes through each stage.
29. Are all human populations in which fertility rates are at or below replacement level characterized by zero population growth? Explain your answer.
30. Distinguish between perpetual, non-renewable, and renewable resources, giving an example of each.
31. Define the term "pollutant." Give five examples of pollutants.
32. In what way might deforestation enhance global warming?
33. What is ozone? Describe how it benefits humans and other organisms in some circumstances and harms them in others.

INQUIRY

34. A particularly severe outbreak of bubonic plague killed millions of humans in Europe and Asia between 1348 and 1400. The deaths of so many people had long-term social and economic effects, and some historians argue that it changed the world in many ways. Did the plague have a significant impact on the eventual size of the human population? The table on the right shows estimates of world population at the beginning of most centuries from 1200 to 2000, as well as in 1348. How might you determine the effect the plague has had on the current size of the human population? One approach might be to calculate what the size of the current population might have been had the plague not occurred, and then compare it with the actual size of today's population. Outline a procedure to do this using the equation $N_t = N_0e^{rt}$ (described in section 14.2). You may wish to perform the appropriate calculations and then make the comparison. (Hint: $\ln (N_t/N_0) = rt$) How large is the difference between what the population of the world might have been and what it actually is today?

Year	Population
1200	3.5×10^8
1300	3.8×10^8
1348	4.7×10^8
1400	3.7×10^8
1500	4.5×10^8
1600	4.9×10^8
1700	(no reliable estimate available)
1800	9.1×10^8
1900	1.6×10^9
2000	6.1×10^9

35. Outline the sampling technique you might use to determine the number of different tree species and the density of each species in a 25-hectare tract of forest.

36. Design an experiment that would help you determine whether a certain level of resource use is sustainable. You might decide to study potential limits on the use of water, a fossil fuel, wood, or some other resource.

37. How many people drive cars to your school? Is this enough to have an impact on the environment or are there too few cars to matter? How would you determine the answer to this question? What would you measure and how would you weigh the advantages and disadvantages of driving to school? Perform some research and analysis sufficient to answer this question.

38. Habitat restoration is a relatively new field of study in which ecologists are attempting to determine if it is possible to return degraded habitats to their original state and, if so, how this can be achieved. Consider a degraded habitat in or near your city or town — perhaps near your school. How would you determine what this habitat was like originally — what species it contained and the condition of the various abiotic components? How might you restore this habitat (as accurately as possible) to its original state? How would you evaluate your level of success? Work in teams or as a class to design a plan for the restoration of your chosen habitat. Then determine what needs to be done, carry out the plan, and assess the results.

COMMUNICATING

39. Agriculture increases the carrying capacity of Earth for humans but, as some ecologists have pointed out, it can also have negative effects that lower the carrying capacity. Do you think we should use all available forms of agricultural technology to support a growing human population? Or are there alternative methods of securing sufficient food for the people of the world? Hold a class debate to address this issue.

40. Think about or do research on the topic of urban sprawl for a city of your choice. Does this city have an overall plan for growth? If so, do you think this plan combats sprawl or makes it worse? Explain your answer. Write a letter to the mayor or city council commending them on their plan or make suggestions about how their plan might be improved.

41. Design your own city. How would you ensure that its residents have sufficient resources to maintain a good quality of life, while minimizing negative environmental impacts? You might

make a three-dimensional model of your city, or draw maps and diagrams to illustrate your plan. This should be accompanied by a report describing how your city will achieve sustainable resource use.

42. Research the impact of coffee growing on a country that exports this crop to Canada. Does the country achieve a net profit or a net loss as a result of growing coffee? Are environmental costs included in these calculations? Does this country produce any shade-grown coffee? What are the advantages and disadvantages of using this method to grow coffee beans? Design a pamphlet or poster that might serve as part of a publicity campaign to educate people about the environmental impact of coffee plantations. Alternatively, write a letter to the CEO or president of a company that imports or sells coffee, expressing your point of view.

43. Some introduced species are useful to humans (for example, most crops) and are relatively harmless to the environment. But others, like the zebra mussel, are referred to as *invasive*. Invasive species tend to grow over-abundantly, taking over ecosystems and disrupting normal ecological processes. Identify 20 to 50 species of plants, animals, or other organisms growing in the area of your school. Each member of the class should become an expert on one of the different species. Determine whether your species is native or introduced, and learn as much as you can about its biology. If your species is native, find out whether it is affected by (or at risk of being affected by) introduced species. If your species is introduced, find out when and how it was introduced and whether it is harmless or invasive. What impact has it had on other species? What methods, if any, are being used to control the spread of this species? Are they successful? Hold a classroom conference on the effects of introduced species. Each student should make a brief presentation on his or her species, describing what is known about the species and suggesting areas for further study.

MAKING CONNECTIONS

44. Many ecologists point out that in addition to the need to conserve ecosystems because they contain resources or *goods* (food, water, and various mineral elements, for example), we also need to recognize that ecosystems perform indispensable *services*. What do you think this means? Describe some of these services. Are there other equally effective ways to obtain these services? Or do you think this is another good reason to conserve a diversity of ecosystems? Give reasons for your answers.

45. Do you think the AIDS epidemic will have as much impact on the world as the bubonic plague epidemic that struck in the fourteenth century? Why or why not? The genetic structure of the human immunodeficiency virus (HIV) that causes AIDS changes frequently. What impact might this have on the virus's ability to affect human population growth?

46. As habitats around the world are lost, many species go extinct before we have even discovered them. What advantages are provided by high levels of species diversity, genetic diversity, and ecosystem diversity? In other words, what is the value of biodiversity? How might ecologists, or the citizens of countries with high levels of diversity that are rapidly being lost (such as countries in Central America, for example), use this "value" to preserve habitats and species? Answer this question as an essay, perhaps focussing on one country or region of the world.

47. Farmers in West Africa first began to cut and burn forests to increase the availability of arable land about 3000 years ago. Recently, scientists have suggested that the parasite that causes the disease malaria first evolved in West Africa at around the same time. If this is true, what have been the long-term effects of this farming practice? To answer this, find out more about malaria, including its cause and effect on individuals. In what areas of the world does malaria currently occur, and how widespread is it? What methods are used to treat and control it, and what are the environmental effects of these methods?

COURSE CHALLENGE

As you continue to prepare for the Biology Course Challenge, consider the environmental issues presented in Chapter 15 and their effects on the carrying capacity of Earth. Consider selecting one of these issues, and its effects on different species and their populations, as the topic for your Challenge.

Organize a Biology Symposium

In this culminating Course Challenge, you will have an opportunity to prepare a paper or a poster on a topic that integrates key knowledge and skills you have acquired throughout your Biology 12 course. With guidance from your teacher, you and your classmates will organize a simulated Biology Symposium. A symposium is a forum or conference during which participants share research results and ideas, usually through the formal presentation of a paper or through a less formal poster display. In fulfilling this Course Challenge, you can base your research on a wide variety of sources, and you can examine your topic from various perspectives. However, your symposium presentation should highlight a relevant STSE (Science/Technology/Society/ Environment) issue or issues associated with your chosen topic. As well, your work should aim to demonstrate knowledge and understanding, inquiry, and communication skills.

A symposium provides an opportunity for experts in a field of research to share findings and exchange information.

Challenge

Investigate a topic of interest in order to present a research paper or a poster display at the Biology Symposium. (A sample poster display is shown on page 546.) For example, you might decide to research a hereditary metabolic disorder such as porphyria. You could approach this topic from several perspectives, for instance:

- explaining how this disorder affects metabolic pathways in cells to demonstrate your understanding of those pathways
- describing cellular functions in the body of an individual affected by this condition and strategies required to diagnose and treat the condition
- exploring genetic expression and control of proteins
- researching the development of medical technologies related to treatment of disorders such as porphyria
- analyzing how porphyria affects the well-being of an individual diagnosed with this disorder
- examining how porphyria is distributed among the population and how this distribution affects societal decisions related to medical research and patient support

In researching and creating your paper or poster, do your best to make connections that link the content to the various units you have studied in your grade 12 Biology course. The topic of porphyria, for example, is related to content in Unit 1, Metabolic Processes; Unit 2, Homeostasis; Unit 3, Molecular Genetics; Unit 4, Evolution; and Unit 5, Population Dynamics. Time permitting, you might highlight an STSE relationship by conducting an interview with a scientific researcher, a medical professional such as a genetic counsellor, or, if appropriate, a person suffering from porphyria.

Possible Topics for the Course Challenge

In addition to the topic of porphyria, you might choose one of the following topics, or another topic approved by your teacher:

- The conservation of biodiversity at a local and/or global level has an impact on the human population. How effective are strategies to conserve species and what justifications exist for these efforts? Other questions to explore include: How can we better understand metabolic pathways in cells, and how would this understanding apply to species conservation? How do pharmacological and other business-based technologies benefit from biodiversity conservation? How does species diversity promote better understanding of genetic expression? How might a smaller gene pool for a species or reduced biodiversity degrade the health of an ecosystem? What role does biodiversity play in evolution?

The conservation of species diversity offers a potentially rich topic to explore in this Course Challenge.

- Human population movement is documented from the earliest days of human existence. Within the last few centuries, populations have shifted dramatically. Explore evidence of historical population movement, and explain how such research is done and the impact of population mergers.
- Aging and death are part of the life process. Explore how these processes occur at the level of cellular functions. Explore as well the impact of aging-related cellular changes on an individual and on populations.
- "Living things are remarkably effective at using energy and matter efficiently." Explore this statement at the levels of biological activity you have studied in this course, from molecular processes within the cell to populations within ecosystems.
- Lysosomal storage disorders such as Tay-Sachs are genetic problems that interrupt the normal metabolic pathways within the cell. Explore one of these disorders. Explain the basis of the disorder, how the cell's pathways are altered by the disorder, and how this alteration affects a patient's body. Examine as well the effects of the disorder on a population and society.

ASSESSMENT

Before beginning this performance task, work with your class to design a rubric or rubrics for assessing this Course Challenge. (Alternatively, your teacher may provide you with rubrics.) After you complete the Course Challenge, you will be assessed on

- **your ability to transfer the concepts you learned in this course to the new contexts of this Course Challenge;**
- **the quality of your research;**
- **the accuracy and depth of your understanding;**
- **your symposium presentation; and**
- **other criteria you decide on as a class.**

Materials

- Support your presentation by a portfolio of research findings, and a complete bibliography of references used. If you decide to produce a research paper, you might wish to prepare accompanying visuals, graphs, or lab results as a PowerPoint™ display. You might also tape-record an interview with a knowledgeable expert, or, alternatively, conduct an interview with an expert via email. If you wish, attach print versions of interviews, first-person accounts, or newspaper or journal articles as appendices to your paper. (See the sample interview with two genetic counsellors at the end of this Course Challenge.)
- To prepare a poster of your research findings, you will need pushpins, and a large sheet of cardboard or paper about 1.2 m × 1.2 m or 1.2 m × 1.6 m. Your poster display can feature a title, some photographs or diagrams, some tables and/or graphs, a literature summary, and an abstract (a brief summary or synopsis of your findings).

WEB LINK

www.mcgrawhill.ca/links/biology12
When you have chosen your topic, the McGraw-Hill Ryerson web site is a good place to begin your research.

Design Criteria

A. Collect your rough notes in a research portfolio.

B. Generally, poster displays are created in one of two ways: you can use a matte board (the coloured cardboard used in picture framing) in one piece or cut up into smaller sections; or you can prepare a one-piece poster on glossy paper.

C. Check with your teacher to determine the best way to format your information so that it can be displayed at the Biology Symposium. There is no specific format required for your poster layout. Check with your teacher for advice and suggestions. Examine the posters shown below. These were produced and presented by Dr. M. I. Sheppard and Dr. S.C. Sheppard at a conference in France. These scientists specialize in pollutant ecology and soil science. Your poster does not have to match this level of detail and complexity. However, it will contain similar design elements. Posters are used by students and professionals in many fields of biology. As you go through university, you will be required to produce posters and papers that can be easily understood by participants at meetings and conferences.

Action Plan

1. Along with your teacher and classmates, study the Achievement Chart for Science published in the Ontario Curriculum document shown on the following page. Discuss the categories and the criteria that you will strive to meet in completing this Course Challenge. Different types of background information — research essays, scientific papers, newspaper articles, interviews, lab reports, data analyses, etc. — may be better suited to fulfilling the criteria for different categories. Criteria for evaluating both the research paper and your presentation or the poster display can be established by the class, under the guidance of your teacher. These criteria can then be compiled in appropriate rubrics. Brainstorm criteria that could be used to evaluate the final product, as listed on the following page, for example:

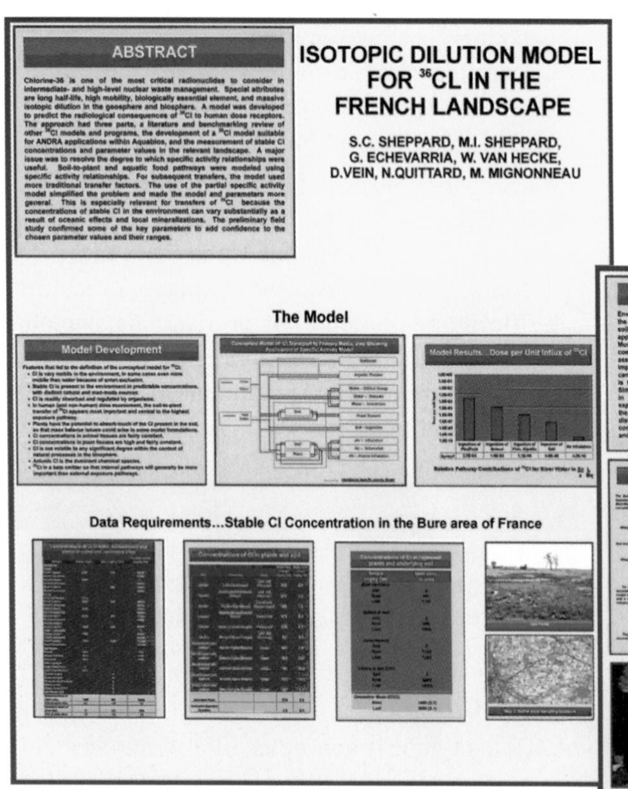

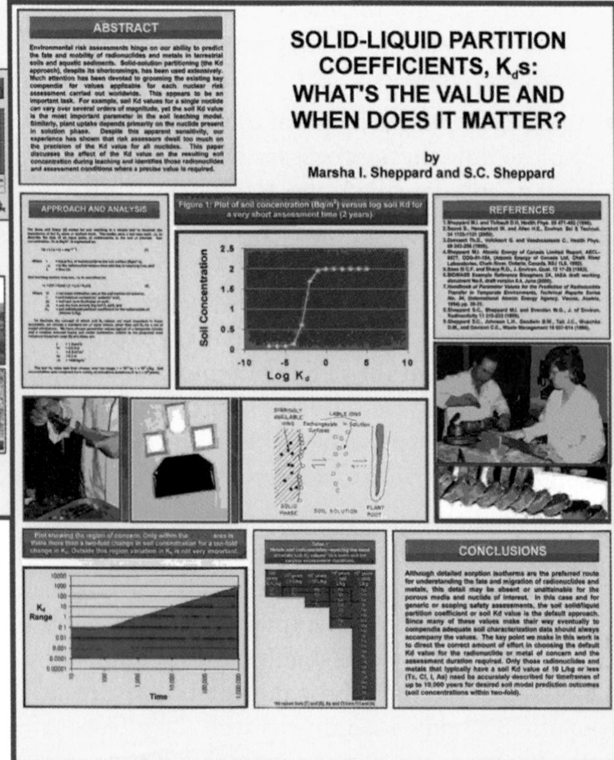

Posters are one means that biologists use to communicate their research information to their peers. Posters usually contain the following elements: a title; the authors' names; an abstract; some basic results, usually in the form of tables or graphs; a conclusion; and some interesting illustrations.

Achievement Chart

Knowledge and Understanding	Inquiry	Communication	Making Connections
■ Understanding of concepts, principles, laws, and theories ■ Knowledge of facts and terms ■ Transfer of concepts to new contexts ■ Understanding of relationships between concepts	■ Application of skills and strategies of scientific inquiry ■ Application of technical skills and procedures ■ Use of tools, equipment, and materials	■ Communication of information and ideas ■ Use of scientific terminology, symbols, conventions, and standard (SI) units ■ Communication for different audiences and purposes ■ Use of various forms of communication ■ Use of information technology for scientific purposes	■ Understanding of connections among science, technology, society, and the environment ■ Analysis of social and economic issues involving science and technology ■ Assessment of impacts of science and technology on the environment ■ Proposing courses of practical action in relation to science-and-technology-based problems

- originality of the topic
- quality of references used
- audio-visual aids used in the symposium presentation
- depth and breadth of topic coverage
- tie-ins to STSE issues

You may wish to assess specific components leading up to the final symposium presentation, as well as the presentation itself.

2. Consult with your teacher regarding a timeline for the completion of the Challenge so you can plan your time carefully.

3. Prepare an abstract and an outline for your proposed paper or poster.

(a) Your abstract should define the topic and briefly summarize your research results and a discussion of those results.

(b) In your outline, describe the scope of your paper or poster. You might include a possible strategy for researching your topic. Have your teacher review your potential sources of information to ensure they are appropriate.

4. Prepare a matrix based on the model (below) to show the relationships among your chosen topic, the Achievement Chart categories, and the five curriculum strands (units of study). Your symposium paper or poster must contain at least three different types of information sources, which clearly meet the expectations of the Achievement Chart. In your matrix, make links between your topic and the five units of study, as well as each category of the Achievement Chart.

Note: Some strands or categories may be addressed more than once.

- Knowledge and Understanding will be demonstrated in the content of your paper or poster.
- Inquiry skills can be addressed, for example, in terms of how you interpret experimental data related to your topic.
- Your actual symposium product (paper/poster plus presentation) will meet expectations related to Communication.

Sample Matrix

Achievement Chart Categories	Curriculum Strands				
	Unit 1: Metabolic Processes	Unit 2: Homeostasis	Unit 3: Molecular Genetics	Unit 4: Evolution	Unit 5: Population Dynamics
Knowledge/ Understanding					
Inquiry					
Communication					
Making Connections					

- How your topic connects science concepts to social, economic, environmental, and/or technological issues will reflect Making Connections.

5. Develop a plan to collect information for your presentation. Organize your information in a research portfolio.

6. Do an information search of your topic using key words or phrases. In a preliminary Works Cited, list any primary sources (lab notes taken by researchers in the course of their work, for example), and secondary sources (anything published or presented to an audience). The Works Cited list could take the form of an annotated bibliography.

7. Your abstract and outline must be submitted for approval to your teacher at this point.

8. Carry out your plan and make modifications as necessary. Search out your information sources, conduct interviews (if appropriate), and analyze any data that you find. Write your paper or create your poster, and prepare your final symposium presentation. Present ideas to classmates as a work-in-progress, sharing concerns and encouragement to help motivate each other.

The Future of Genetics

Guest speakers Dr. G. Simmons, Dr. P. Stanley, and Dr. T. Wu

Pronk Convention Centre
February 8 to 10
9:00 a.m. to 4:00 p.m. daily

A three-day symposium
featuring poster sessions
1:00 p.m. to 4:00 p.m. daily

A program for a conference or symposium lists presentations, names of presenters, and exhibit and display times.

9. Working in a group, you might like to prepare a program for the Biology Symposium, time permitting. You can invite school staff members, students in other classes, family members, and perhaps members of the public to attend.

10. During the symposium, review each presentation against the assessment criteria that you decided on as a class. Keep in mind that each student's research process and documentation are as important as the final product.

Evaluate Your Challenge

1. Using the assessment criteria (rubrics) that you have prepared, evaluate your work and your presentation. How would you evaluate your contribution to the symposium?

2. Evaluate your classmates' symposium presentations.

3. After analyzing the presentations of your classmates, what changes would you make to your own work if you had the opportunity to complete the Challenge again? Provide reasons for your proposed changes.

4. Did the process required to complete this challenge help you to think about what you have learned in this course? If so, how?

5. After the symposium, submit your paper or poster, and associated reference materials, to your teacher for a final evaluation. Again, the criteria for evaluation will be based on the Course Challenge rubrics.

Wrapping Up

In order to prepare a high-quality, in-depth Biology Symposium presentation, you will need to limit the amount of information that you attempt to present, focussing on the key points. Attempt to support your ideas with primary and secondary sources, and references to accepted scientific models, as appropriate. Give your symposium presentation added relevance by relating your topic to key societal issues, such as political, economic, or health-related considerations.

Use your Course Challenge presentation to assist your learning by drawing together topics from each unit of study, if possible. The quality of your final product will improve when knowledgeable links are made between topics. Remember as well to draw upon ingenuity and creativity in interpreting your research findings.

Interview: Genetic Counselling

Cheryl Shuman is Director of Genetic Counselling in the Division of Clinical and Metabolic Genetics at the Hospital for Sick Children and the Program Director of the M.Sc. Program in Genetic Counselling at the University of Toronto. Stacy Hewson is the Genetic Counsellor for the Genetic Metabolic Program at the hospital and is also a faculty member of the M.Sc. Program.

Q: What is the function of a genetic counsellor?

A: Genetic counsellors are health professionals who provide information and support to families whose members have birth defects or genetic disorders. We also help families who may be at risk for a variety of inherited conditions. For individuals or families seeking genetic counselling, we investigate the problem present in the family. We interpret information about the disorder, analyze inheritance patterns and risks of recurrence, and review available options with the family. We also provide supportive, non-directive counselling to individuals and to families. Genetic counsellors may also be involved in education and research activities.

Q: What type of training does a genetic counsellor require?

A: The educational background required to enter the field of genetic counselling is a four-year B.Sc. or B.A. with specific prerequisite courses. Then, graduate work, an M.Sc. degree, is done in the specific field of genetic counselling. This is a relatively young profession – about 30 years old. However, there is a need for more genetic counsellors, especially given the new developments in genetic research and testing that can be applied to patient care. Families need help gathering and interpreting data. They also need help assessing their personal risk in various situations. In addition, as the genetic basis for many of the serious or chronic disorders of adulthood becomes known, such as coronary artery disease or cancer, the need for genetic counsellors will continue to grow.

Genetic counsellors Cheryl Shuman (left) and Stacy Hewson (right) work in a team that helps individuals and families cope with the effects of genetically transmitted disorders.

Q: Have you ever been involved in the care of a patient with porphyria?

A: Porphyria is a very rare condition. I have been working in the field for over 20 years, and I have only dealt with a few families who have this condition.

Q: When a family comes to you for help and advice, what is their main concern?

A: With regard to porphyria, we see individuals for two main reasons. One is to help make a diagnosis of porphyria in a particular person if it has been suspected. The other is to help individuals concerned about their family history of this disorder who may wish information regarding their own risk for having porphyria or their risk for having an affected child. If a parent has an autosomal dominant form of porphyria, each of their children will have a 50 percent chance of inheriting the gene that causes this condition. Porphyria is quite variable, however, and not all individuals who inherit the gene will experience significant problems.

Q: How do you go about helping people in these cases?

A: We meet with individuals or families and take a detailed family history to construct a pedigree. This pedigree, along with medical documentation of the disorder, helps us to establish the mode of inheritance and estimate the risk of porphyria for that individual or for a couple. This information may also be helpful for a couple's children or if they are planning a pregnancy and are interested in prenatal testing options. The choices to pursue information and potential options in testing are up to each individual or couple.

Q: What other resources do you tap into in your job?

A: We work as a team with clinical geneticists. These specialists are physicians who assess patients directly for evidence of genetic disorders or syndromes. We also make use of laboratory resources. In the case of porphyria, lab resources would include tests for abnormal metabolites, for example.

Q: What advice can you give to students who may be interested in a job as a genetic counsellor?

A: A genetic counsellor needs a sound scientific background along with strong interpersonal and communication skills. Genetic counsellors must be adept at working both independently and as a team member. Other qualities would include an empathic nature, open-mindedness, and a drive for ongoing learning. Students interested in pursuing this profession should access genetic counselling resources on the Internet. Also, they might contact local genetic counsellors to obtain personal perspectives. Finally, they can volunteer for positions that involve training in communication skills.

WEB LINK

www.mcgrawhill.ca/links/biology12

To find out more about porphyria, visit the web site above, and click on **Web Links** to access information from the Canadian Porphyria Foundation.

APPENDIX 1

Safety Symbols and WHMIS Symbols

Safety Symbols

The following safety symbols are used in this *Biology 12* textbook to alert you to possible dangers. Make sure that you understand each symbol in a lab or investigation before you begin.

	Disposal Alert This symbol appears when care must be taken to dispose of materials properly.
	Biological Hazard This symbol appears when there is danger involving bacteria, fungi, or protists.
	Thermal Safety This symbol appears as a reminder to be careful when handling hot objects.
	Sharp Object Safety This symbol appears when there is danger of cuts or punctures caused by the use of sharp objects.
	Fume Safety This symbol appears when chemicals or chemical reactions could cause dangerous fumes.
	Electrical Safety This symbol appears as a reminder to be careful when using electrical equipment.
	Skin Protection Safety This symbol appears when the use of caustic chemicals might irritate the skin or when contact with micro-organisms might transmit infection.
	Clothing Protection Safety A lab apron should be worn when this symbol appears.
	Fire Safety This symbol appears as a reminder to be careful around open flames.
	Eye Safety This symbol appears when there is danger to the eyes and safety glasses should be worn.
	Poison Safety This symbol appears when poisonous substances are used.
	Chemical Safety This symbol appears when chemicals could cause burns or are poisonous if absorbed through the skin.
	Animal Safety This symbol appears when live animals are studied and the safety of the animals and the students must be ensured.

WHMIS Symbols

Look carefully at the WHMIS (Workplace Hazardous Materials Information System) safety symbols shown below. These symbols are used throughout Canada to identify dangerous materials in all workplaces, including schools. Make sure that you understand what these symbols mean. When you see these symbols on containers in your classroom, at home, or in a workplace, use safety precautions.

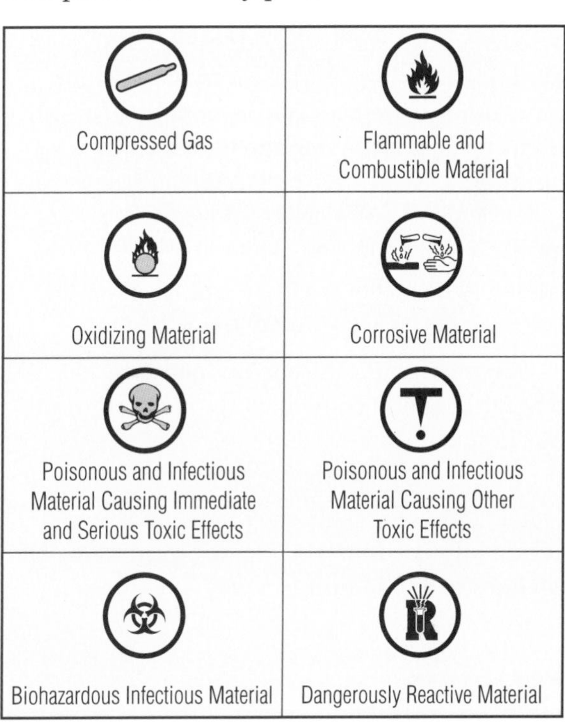

APPENDIX 2

Units of Measurement

When you take measurements in science, you use the International System of Measurement (commonly known as SI, from the French *Système international d'unités*). SI includes the metric system and other standard units. It is accepted as the standard for measurement throughout most of the world. This appendix provides you with a review of SI metric symbols and prefixes.

In SI, the base units include the metre, the kilogram, and the second. The size of any particular unit can be determined by the prefix used with the base unit. Larger and smaller units of measurement can be obtained by either dividing or multiplying the base unit by a multiple of 10.

For example, the prefix *kilo-* means multiplied by 1000. So, one kilogram is equivalent to 1000 grams:

$$1 \text{ kg} = 1000 \text{ g}$$

The prefix *milli-* means divided by 1000. So, one milligram is equivalent to one thousandth of a gram:

$$1 \text{ mg} = \frac{1}{1000} \text{ g}$$

The following table shows the most commonly used metric prefixes. Adding metric prefixes to a base unit is a way of expressing powers of 10.

Metric Prefixes

Prefix	Symbol	Relationship to the base unit
giga-	G	10^9 = 1 000 000 000
mega-	M	10^6 = 1 000 000
kilo-	k	10^3 = 1 000
hecto-	h	10^2 = 100
deca-	da	10^1 = 10
–	–	10^0 = 1
deci-	d	10^{-1} = 0.1
centi-	c	10^{-2} = 0.01
milli-	m	10^{-3} = 0.001
micro-	µ	10^{-6} = 0.000 001
nano-	n	10^{-9} = 0.000 000 001

Commonly Used Metric Quantities, Units, and Symbols

Quantity	Unit	Symbol
length	nanometre micrometre millimetre centimetre metre kilometre	nm µm mm cm m km
mass	gram kilogram tonne	g kg t
area	square metre square centimetre hectare	m^2 cm^2 ha (10 000 m^2)
volume	cubic centimetre cubic metre millilitre litre	cm^3 m^3 mL L
time	second	s
temperature	degree Celsius	°C
force	newton	N
energy	joule kilojoule *	J kJ
pressure	pascal kilopascal **	Pa kPa
electric current	ampere	A
quantity of electric charge	coulomb	C
frequency	hertz	Hz
power	watt	W

* Many dieticians in North America continue to measure nutritional energy in Calories, also known as kilocalories or dietetic Calories. In SI units, 1 Calorie = 4.186 kJ.

** In current North American medical practice, blood pressure is measured in millimetres of mercury, symbol mm Hg. In SI units, 1 mm Hg = 0.133 kPa.

APPENDIX 3

Significant Digits

In order to conduct experiments in biology and other sciences, you need to make measurements and manipulate numbers. However, there is always some degree of uncertainty in each measured value that you record. Scientists around the world have agreed that a measurement should include the number of digits that reasonably indicate its certainty. *In general, a measurement includes all digits that are certain, and one final digit that is estimated (uncertain).* Together, these digits are called *significant digits.*

For example, suppose you take a temperature reading using two different thermometers, as shown in Figure A3.1. The thermometer on the left has gradations of 0.1°C. Reading the thermometer carefully, you can be certain that the mercury is above 32.3°C. To report your measurement correctly, you must estimate the final digit. You might estimate 32.33°C, as shown in the diagram. Another person might estimate 32.35°C. The final digit is uncertain.

On the other hand, the thermometer on the right has gradations of 1°C. In this case, you can be certain that the mercury is above 32°C, but again you must estimate the final digit to report your measurement correctly. The diagram shows an estimate of 32.3°C.

A measurement of 32.33°C is more precise than a measurement of 32.3°C. The first measurement has four significant digits, while the second measurement has only three.

Which Digits Are Significant?

In the previous example, the number of significant digits was equal to the total number of digits. Notice that the measurements did not contain any zeros. *When a value does not contain any zeros, the number of significant digits is equal to the total number of digits in the value.* When a value does contain one or more zeros, however, determining the number of significant digits requires more thought. Zeros may or may not be significant depending on where they are located. The rules used to determine whether zeros in a quantity are significant are as follows.

1. Zeros to the right of a quantity that includes a decimal point are significant. Therefore, 7.50 mg has three significant digits.
2. Zeros located between two non-zero numbers are significant. Therefore, 909 g has three significant digits.
3. Zeros that are located to the left of a value are not significant. Therefore, 0.0001 mm has only one significant digit.
4. Zeros at the end of a whole number are ambiguous. For example, $1000 would be considered to be exact; however, 5 000 000 bacteria is probably not exact.

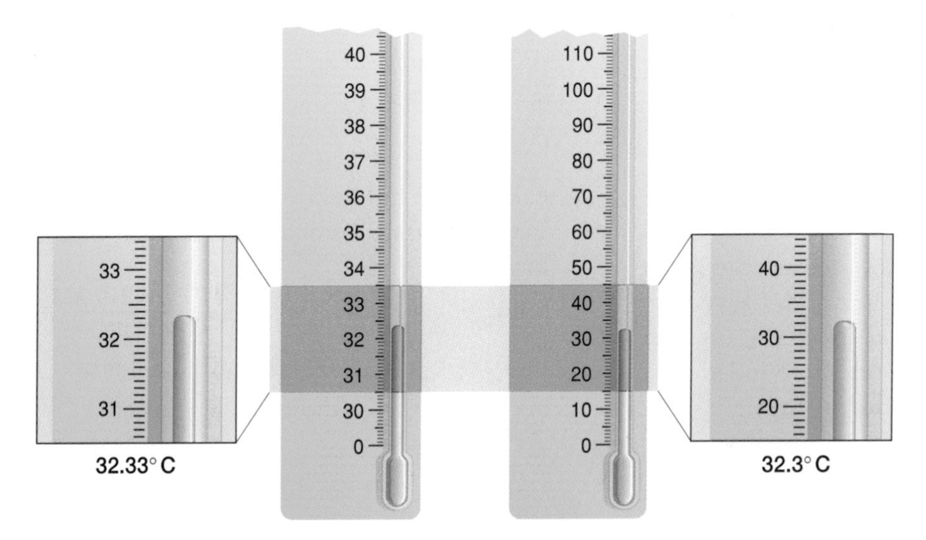

Figure A3.1 A temperature measurement made using the thermometer on the left will have more significant digits, hence more certainty, than a measurement made using the thermometer on the right.

One way to avoid ambiguity is to express numbers using scientific notation. Thus, in 5×10^6 bacteria, it is clear that there is one significant digit.

Mathematical Operations

You will sometimes need to perform calculations with measured quantities. It is important to keep track of which digits in your calculations are significant. Your results should not express more certainty than your measured quantities justify. Be particularly careful when using calculators — they usually report results with far more figures than your data warrant.

The following three rules will help you keep track of significant digits in calculations and results.

Rule 1: Multiplying and Dividing

The value with the *fewest number of significant digits*, going into the calculation, determines the number of significant digits that you should report in your answer.
For example, $25.00\text{ cm} \times 3.00\text{ cm} = 75.0\text{ cm}^2$.

Rule 2: Adding and Subtracting

The value with the *fewest number of decimal places*, going into the calculation, determines the number of decimal places that you should report in your answer.
For example, 34.50 mL + 23.1 mL = 57.6 mL

Rule 3: Rounding

To get the appropriate number of significant digits (rule 1) or decimal places (rule 2), you may need to round your answer.

- If your answer ends in a number that is greater than 5, increase the preceding digit by 1. For example, 2.346 can be rounded to 2.35.
- If your answer ends with a number that is less than 5, leave the preceding number unchanged. For example, 5.73 can be rounded to 5.7.
- If your answer ends with 5, increase the preceding number by 1 if it is odd. Leave the preceding number unchanged if it is even. For example, 18.35 can be rounded to 18.4, but 18.25 is rounded to 18.2.

Operations and Counted Quantities

If you see four bacteria in a microscope field, the value "4" is an exact number. If you say someone has two eyes, you mean exactly two, not 2.1 or 2.3. Counted quantities have no uncertainty. In other words, they do not limit the number of significant digits in a calculation.

When you multiply using counted quantities, you should treat the operation as an addition. For example, suppose you have four beakers, each containing 25.0 mL of solution. What is the total volume of solution?
$4 \times 25.0\text{ mL} = 100.0\text{ mL}$. Why are there four significant digits in the answer instead of three? The operation $4 \times 25.0\text{ mL}$ can be represented as an addition operation:
25.0 mL + 25.0 mL + 25.0 mL + 25.0 mL = 100.0 mL. Thus, the number of decimal places in the measured quantity being multiplied determines the number of decimal places in the answer. This is only the case when a measured quantity is multiplied by an exact, counted quantity.

Instant Practice

1. How many significant digits are there in the following measurements?
   (a) 2004 mm (b) 0.007 V
   (c) 26.30 cm (d) 21.6°C
2. Determine the number of significant digits in the following measurements. If you are unable to tell, briefly explain why.
   (a) 8.923×10^{-7}mg (b) 7.000 cm
   (c) 30 mL
   (d) $5.000\ 201 \times 10^2$ s
3. Calculate the following and round to the correct number of significant digits:
   (a) 705 mL + 0.40 mL (b) 13.07 s − 2.1 s
   (c) $206\text{ mm} \times 0.83\text{ mm}$ (d) 4.25 g/3 L
4. Perform the following calculations. Include the appropriate number of significant digits in your answer.
   (a) 55.682 g + 43.3 g (b) $\dfrac{1.732 \times 10^2\text{ g}}{15.5\text{ cm}^3}$
   (c) 0.0050 mL − 0.000 350 mL + 1.0 mL
   (d) 5×37.0 mL, where 5 is a counted quantity

APPENDIX 4

Cell Division

The cells in a multicellular organism fit into one of two categories: somatic cells and germ cells. **Somatic cells** divide by mitosis to make up the specialized cells of the body that the organism relies on for life processes. Somatic cells also include stem cells that retain the ability to develop into required specialized forms. The mature body has several different kinds of stem cells, each kind capable of forming a limited number of specialized cells. In addition to these cells, most multicellular organisms also have **germ cells** — those cells that are set aside to produce the next generation of organisms. Germ cells include the specialized gametes (eggs and sperm) formed by meiosis and the unspecialized cells that produce them.

Part A: Somatic Cells

Cells follow a cell cycle of growth and division. Unspecialized cells, like those found in the early embryo, can divide rapidly. As cells begin to specialize, some will specialize to the point that they are terminally differentiated, and are no longer able to divide. Terminally differentiated cells are fully specialized and unable to undergo further change. All of these cells follow the cell cycle as outlined in Figure A4.1. The cell cycle can be divided into the two main phases of mitosis (division) and interphase (growth and metabolism).

Interphase can be further divided into three discrete phases: G_1 (Gap 1), S (Synthesis) and G_2 (Gap 2). Cell activity varies through these different phases. Cells rely on a system of timed interactions among cell structures, and proteins that control these interactions control the sequence of events that lead towards cell division. Some cells leave the cell cycle for G_0 (Gap 0) as quiescent postmitotic cells that are metabolically active but do not grow or continue along the cell cycle towards mitosis. During the G_1 phase, the cell grows and there is an increased level of protein synthesis and DNA repair. Certain protein molecules act as inhibitory factors that stop further progress and keep a cell in G_1. Mutations associated with some of these inhibitory factors can lead to cancerous growth. Growth factors can induce postmitotic or quiescent cells to move into G_1 to continue in the cell cycle. This is important to stimulate tissue cells required to repair a wound or activate immune system cells that are required for an immune response. During the S phase, DNA replicates to double the genome of the cell. The centrosome, where microtubules are created and co-ordinated, also duplicates in preparation for mitosis. During the G_2 phase, DNA replication stops and various protein factors prepare the cell for mitosis, the M phase.

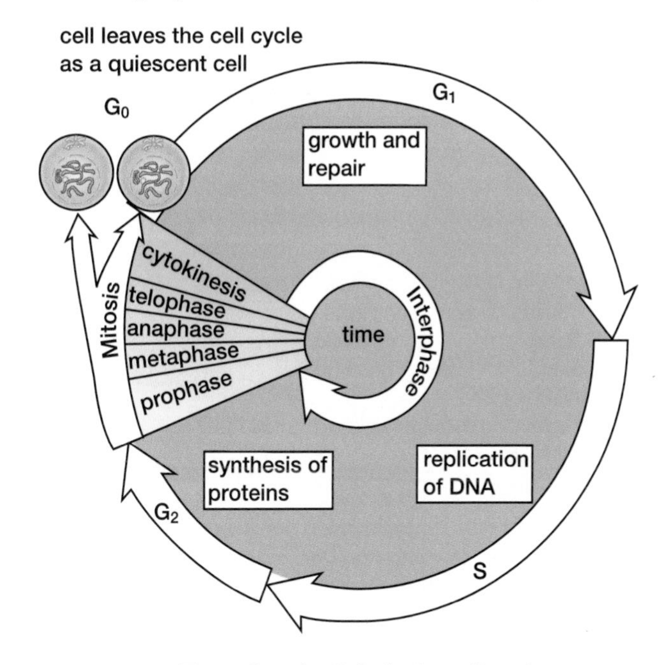

Figure A4.1 The cell cycle. Cells in the cell cycle go through G_1 (Gap 1), S (Synthesis), G_2 (Gap 2), and M (Mitosis). A specialized cell may leave the cell cycle for G_0 (Gap 0).

During the M phase, the cell goes through visibly dramatic activity, as illustrated in Figure A4.2 on page 556. During prophase, the replicated DNA in the nucleus coils and condenses to form distinct chromosomes. Initially the chromosomes appear as single threads, but later they appear as double threads showing two chromatids. The centrosomes separate to move away from each other and form poles that will define the daughter cells.

They also produce microtubule spindle fibres that connect to the centromeres of the chromosomes. The nuclear membrane disintegrates as nuclear lamins in the membrane become soluble. During metaphase, chromosomes with clear chromatids line up at the equatorial plane as spindle fibres pull each chromosome (paired chromatids) toward opposite centrosomes. During anaphase, the chromatids separate and slide along the spindle fibres to the centrosomes. Once the cell has successfully divided the genome into two identical daughter cells, telophase starts. The nuclear membrane re-forms, first around each new chromosome and later surrounding all of the chromosomes together. This process ends with cytokinesis when the cell cytoplasm divides and the cell membrane pinches along the equatorial plane to form separate cells.

Part B: Germ Cells

Unlike the cell cycle described above, meiosis is a linear process that produces terminally differentiated gametes. Gametes do not divide to produce other cells. During development as an embryo, some cells, called germ cells, were separated from the others. These germ cells migrated by amoeboid movement through abdominal cells to the gonads and divided to form cells that could undergo meiosis. There are two types of meiotic cells: spermatogonia (which will produce sperm) and oogonia (which will produce eggs). Germ cells also go through G_1, S, and G_2 before entering the M

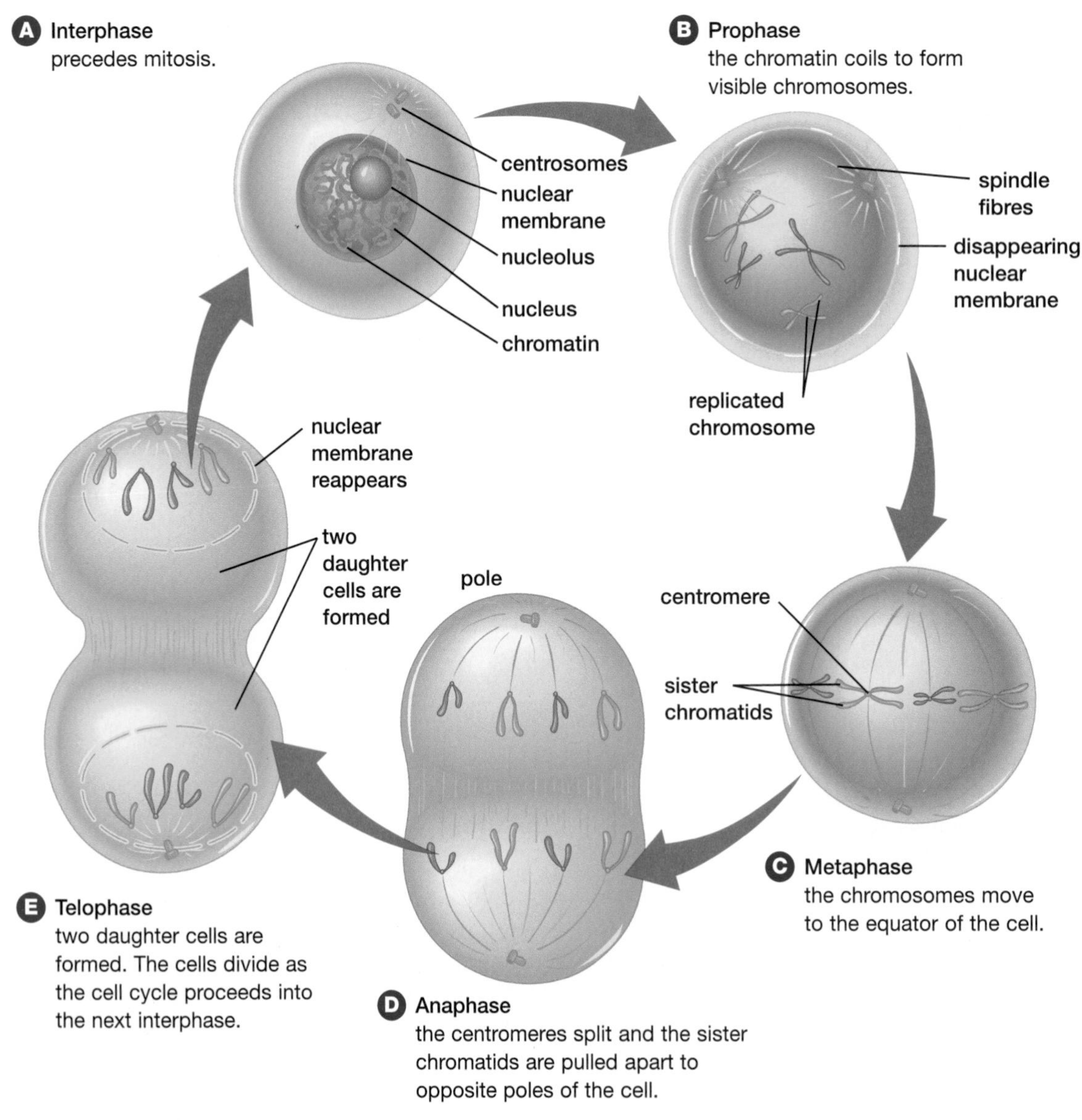

Figure A4.2 Mitosis in animal cells

phase, but here meiosis involves two division steps. The first step, meiosis I, reduces the number of chromosomes without separating the chromatids. During the second division, meiosis II, the chromatids are separated to form one or more haploid gamete cells.

Spermatogenesis

Spermatogonia are diploid germ cells that produce haploid sperm cells. In mammals and many other organisms, spermatogonia divide by mitosis to produce a dormant cell (that replaces the parent cell) and some cells that will actively divide by meiosis. This process is shown in Figure A4.3.

During meiotic prophase, homologous chromosomes come together and overlap in synapsis. At this time, enzyme complexes (recombination nodules) promote crossing over, the exchange of genetic material between homologous chromosomes. These synaptic pairs remain together until they are separated in anaphase I. Synapsis reduces the chances of an unequal distribution of chromosomes, which is also called nondisjunction.

At metaphase I, spindle fibres line the paired chromosomes at the equatorial plane and separate the chromosomes to opposite poles. The daughter cells then proceed through the steps of mitosis to form four haploid gametes, which then mature to form sperm cells.

Oogenesis

The egg cell, or oocyte, has a different function from the sperm cell. The egg cell provides the energy and materials necessary to support the growth and division of the embryo until additional materials are available. As a result, egg cells are relatively large and meiosis is modified to help this accumulation of cytoplasm by producing one haploid egg and two or three small cells called polar bodies. This process is illustrated in Figure A4.4 on the next page.

In most mammals, all the oogonia begin oogenesis before the organism is mature — even before birth. There are no reserve germ cells, and the oocytes are arrested during prophase in meiosis I. In human females, the hormones LH and FSH (discussed in Chapter 6) trigger the completion of the process by way of the ovarian cycle.

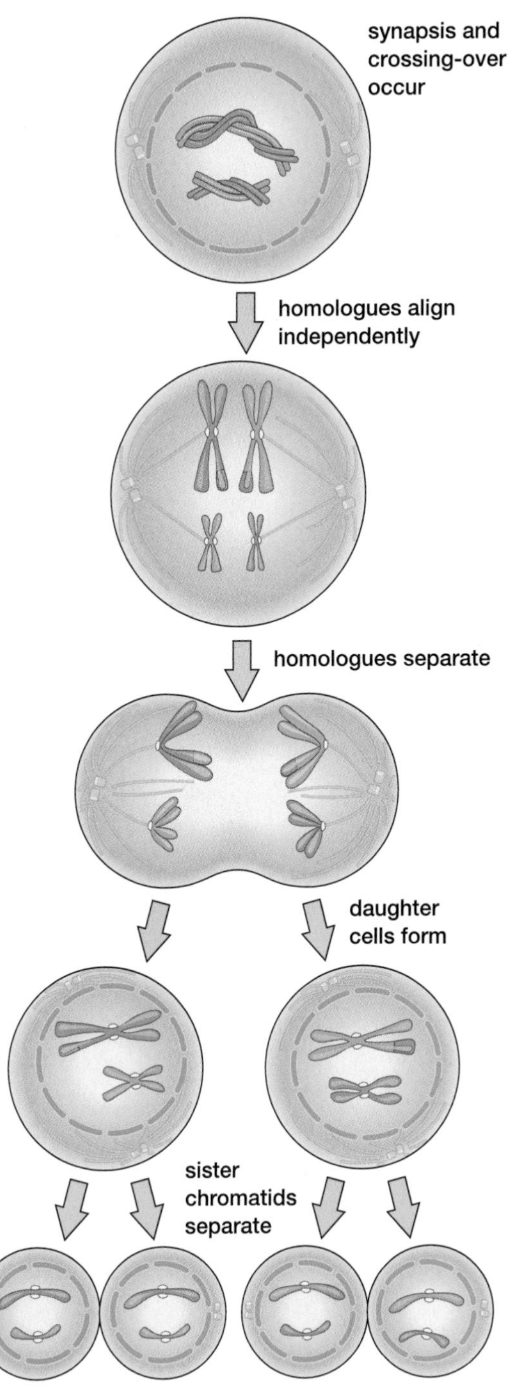

Figure A4.3 A brief overview of meiosis. Synapsis in prophase I allows the exchange of genetic information between homologous chromosomes.

During this arrested period, which may last up to 50 years for humans, some DNA sequences are transcribed to produce mRNA and proteins that support several mitotic divisions of the early embryo. Other proteins are produced that help the early embryo orient and develop as it grows. Once stimulated, meiosis I continues and one small polar body is produced. This polar body contains one set of duplicated chromosomes and very little cytoplasm. The oocyte continues through meiosis II and stops at metaphase II until it is fertilized. Fertilization by a sperm cell, and the arrival of the sperm pronucleus, lead to an increase of calcium inside the oocyte. This stimulates the oocyte to complete meiosis, producing one haploid pronucleus and a second polar body at the other centrosome pole.

The zygote has two pronuclei that move directly to the S phase. Since the egg stored quantities of mRNA and key proteins, the early embryo skips phases G_1 and G_2 and alternates between the S phase and M phase during mitotic cleavage. As the two pronuclei prepare for their first M stage, their membranes disintegrate and the chromosomes merge to form the nucleus of the new individual.

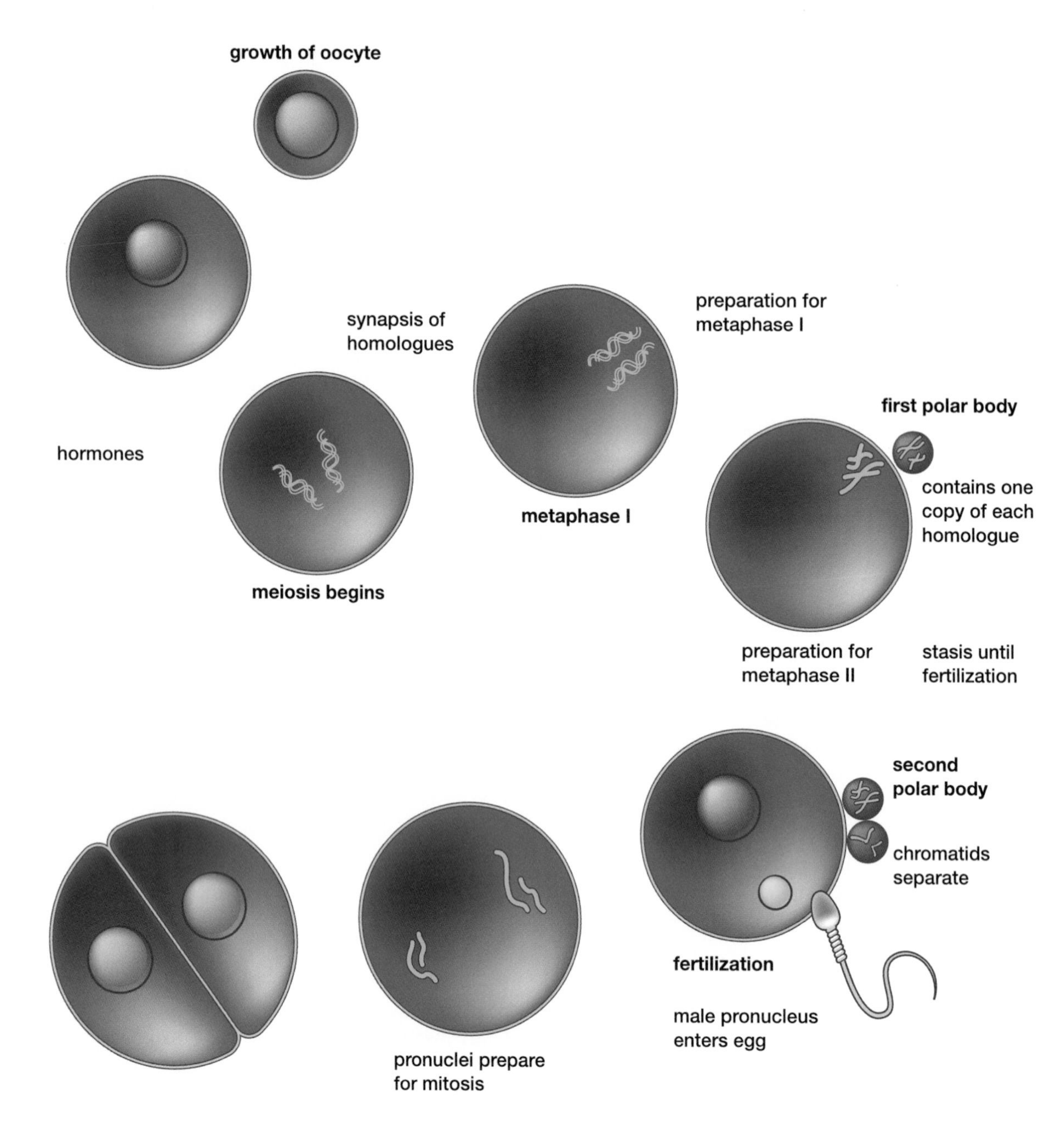

Figure A4.4 Oogenesis is a specialized form of meiosis. Cytoplasm is not evenly distributed. Hormones initiate meiosis, which arrests in metaphase II. Meiosis is completed after fertilization.

APPENDIX 5

Shapes of Selected Macromolecules

Part A: Protein Shape

The function of a protein is directly related to the specific shape of the protein. Ultimately, the shape of a protein depends on the sequence of amino acids that make up the protein.

Amino acids are biologically significant molecules, essential to life. As shown in Figure A5.1, an amino acid contains two reactive groups of atoms: an amine group, and a carboxyl group, joined by a central carbon atom. The central carbon atom is also bonded to a side chain, usually represented by the letter *R*. There are twenty common amino acids, each with a different side chain, or *R*-group.

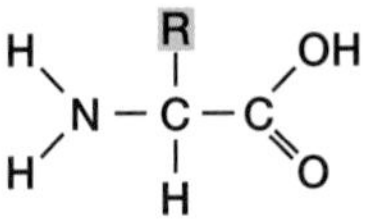

Figure A5.1 Amino acids have a standard form. An *R*-group is bonded to a carbon atom that is connected to a reactive amine group on one side and a reactive carboxyl group on the other side.

Primary Structure

The amine group on one amino acid can react with the carboxyl group on another amino acid to form a *peptide bond*. In this way, many amino acids can be joined by peptide bonds to form a long, linear chain of amino acids called a *polypeptide*, as shown in Figure A5.2. A *protein* is made of one or more polypeptides. The specific sequence of amino acids in a polypeptide is known as the **primary structure**. Part A of Figure A5.3, on the next page, illustrates the primary structure of a polypeptide.

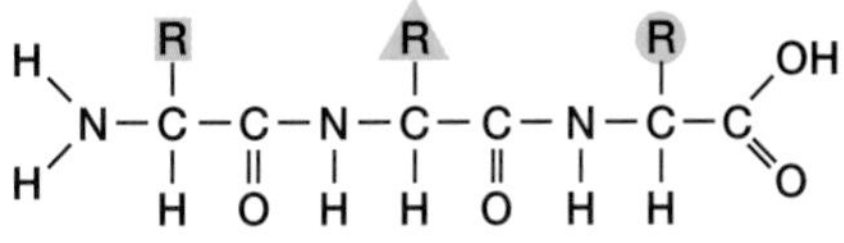

Figure A5.2 A short polypeptide chain with three amino acids, each with a different *R*-group

Secondary Structure

As you can see in Figure A5.2, the backbone of a polypeptide chain contains both N–H groups and C=O groups. These two groups can form hydrogen bonds with each other. This interaction causes polypeptide chains to fold into specific shapes, with hydrogen bonds holding the shape in place. These shapes are known as the **secondary structure**. Part B of Figure A5.3, on the next page, illustrates the secondary structure of a polypeptide.

Linear polypeptides spontaneously form one of two structures: the alpha-helix, and the beta-sheet, or pleated sheet. The alpha-helix is a right-handed spiral structure. In an alpha-helix, the hydrogen bonds are between the hydrogen and oxygen atoms of amino acids within the same polypeptide. The beta-sheet is formed by hydrogen bonding of amino acids between different polypeptides. These two structures are shown in Figure A5.4 on page 561. When a beta-sheet is formed from polypeptides that run in the same direction, the resultant shape is called a parallel beta-sheet. If the polypeptides run in opposite directions, the beta-sheet is antiparallel. This antiparallel structure is often the result of a chain that loops back on itself.

Tertiary Structure

When amino acids are joined together in a polypeptide, the *R*-group of each amino acid sticks out from the chain. These *R*-groups interact with each other and with the surroundings. Each *R*-group can be categorized as either hydrophilic ("water-loving") or hydrophobic ("water-hating"). The hydrophobic *R*-groups are non-polar, so they orient inward, away from water. The hydrophilic *R*-groups are attracted and oriented to water. As a result, when a protein is placed in water, it arranges itself so that the hydrophobic *R*-groups are protected within the protein, while the hydrophilic *R*-groups are on the outside of the structure. This organization makes proteins soluble in water.

As mentioned earlier, *R*-groups can interact with each other, as shown in Figure A5.4 on page 561. Some *R*-groups containing nitrogen and oxygen atoms can form hydrogen bonds. Other *R*-groups containing sulfur atoms can form disulphide bonds between different areas of a protein. As shown in Figure A5.4, these

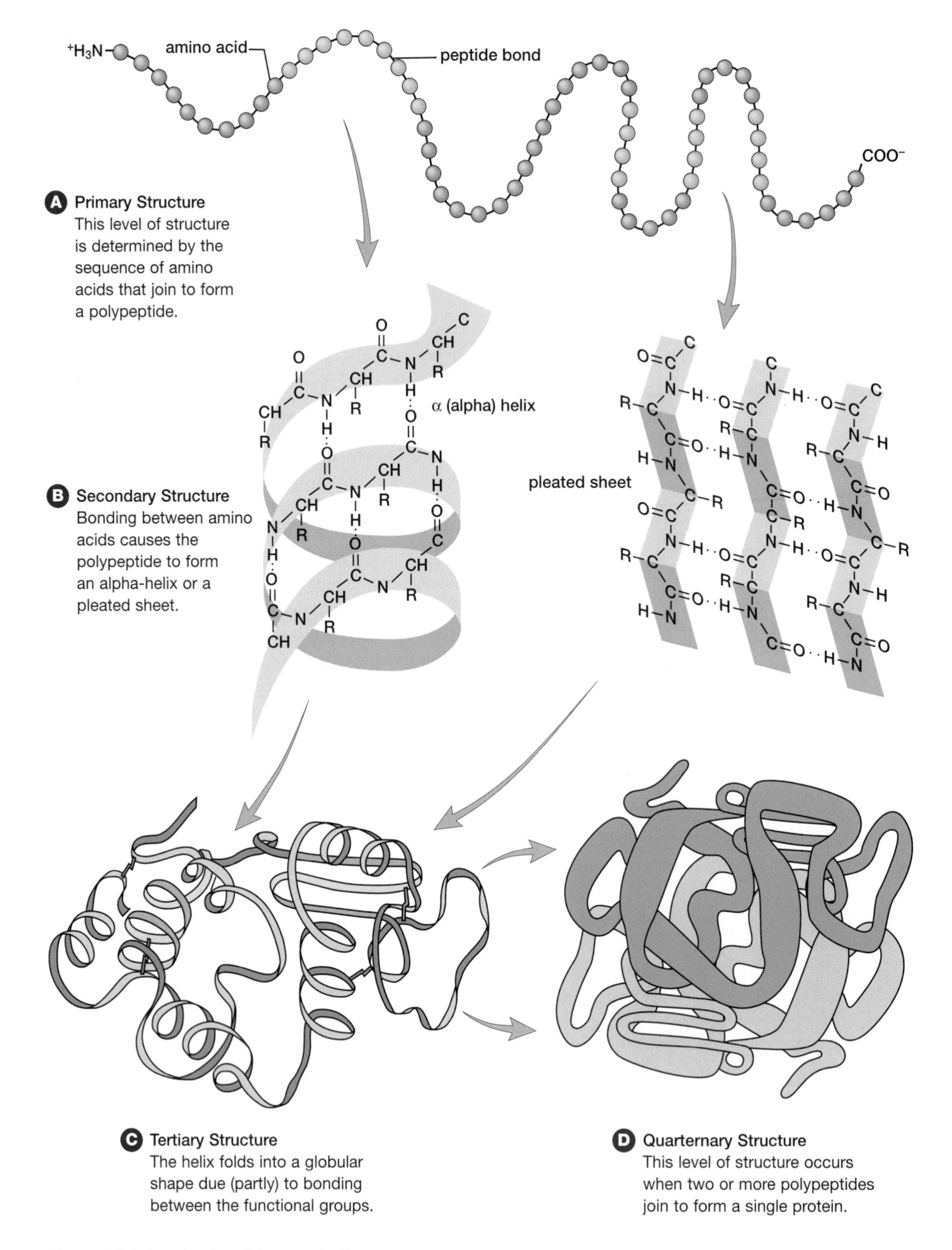

Figure A5.3 Levels of protein organization

and other interactions between *R*-groups in different parts of a protein can cause the protein to remain fixed in a complicated three-dimensional shape, called a **tertiary structure**.

Note that this three-dimensional shape is *in addition to* the alpha-helices and beta-sheets already formed. Part C of Figure A5.3 shows the tertiary structure of a polypeptide.

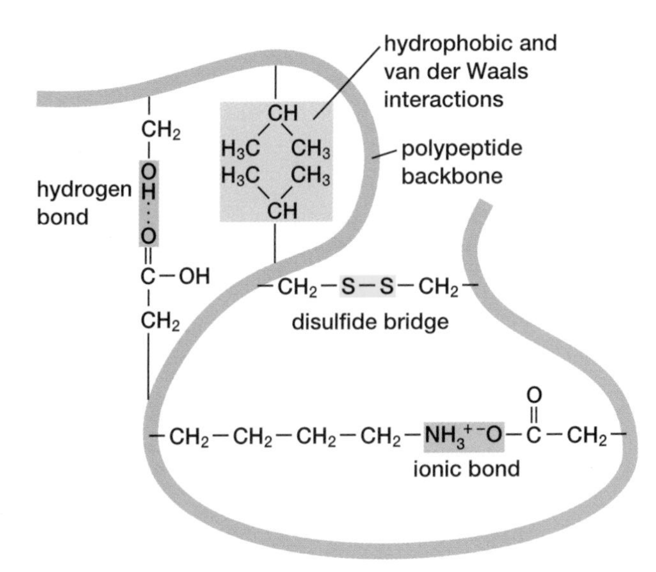

Figure A5.4 Interaction of *R*-groups

Quaternary Structure

When two or more polypeptides join and intertwine to form a protein, the resulting complex structure is called a **quaternary structure**. Part D of Figure A5.3 shows a quaternary structure made up of two polypeptides.

Why do proteins fold into these complicated structures? In fact, a folded protein is in a stable position of lowest energy. Hydrophobic groups are near other hydrophobic groups, while polar and ionic groups are interacting with other polar groups, or with polar water molecules. Because the structure of a protein is stable, the protein retains its specific shape unless the bonds between amino acids and *R*-groups are broken.

This specific shape allows a protein to perform complex functions. When a protein is unfolded, by heating or by some other process, the protein can no longer perform its function. If a protein is mutated, by substituting one amino acid in the chain of the protein for another, the shape and function of the protein is also disrupted. For example, sickle cell hemoglobin has a single amino acid substitution (valine replacing glutamic acid) near the outer edge of the protein. The placement of a non-polar amino acid on the surface of the hemoglobin protein creates a region that enables similarly mutated proteins to stick together, forming long chains that are unable to move through cell membranes.

Part B: The Cell Membrane and Passive Transport of Molecules

All cells are surrounded by a structure, the cell membrane, composed largely of a framework of lipid molecules, with proteins embedded in it. The cell membrane is so thin that it would take about 10 000 of these membranes, stacked on top of one another, to equal the thickness of a sheet of paper. Nevertheless, the fine, delicate structure of the cell membrane is essential to the life of the cell. Not only does it provide a structure for defining the shape and contents of the cell, but also it serves as a selective barrier, permitting only certain molecules and ions to enter and leave the cell.

The cell membrane consists largely of phospholipid molecules, each of which is made up of a phosphate functional group and two fatty acid chains that are bonded to a glycerol molecule by means of a condensation synthesis reaction. The resulting molecule has a non-polar, hydrophobic "tail" (the fatty acids) and a polar, hydrophilic "head" (the glycerol and phosphate group). Due to natural forces of repulsion and attraction with polar water molecules, the tail of the phospholipid molecule faces inward, away from water molecules, and the head faces outward, towards the water molecules. When large numbers of phospholipids interact with water, they spontaneously form a double-layered structure — a bilayer — with a non-polar interior.

The phospholipid bilayer is not a solid structure. Rather, its texture is more like that of thick olive oil; the phospholipid molecules are in constant motion, moving and undulating with the movement of the surrounding water molecules.

One method by which small molecules may pass through the cell membrane is by diffusion — the net movement of molecules from a region in which they are more concentrated to one in which they are less concentrated. The diffusion of a solvent, such as water, across a membrane that separates two solutions is called osmosis. The difference in concentrations of any molecules separated by the cell membrane is called a concentration gradient. When molecules are distributed equally on both sides of the membrane, there is no concentration gradient.

Small, non-polar molecules such as oxygen and carbon dioxide easily cross the cell membrane through diffusion. Because they are small, water molecules and other small polar molecules may also cross the membrane through minute imperfections between individual phospholipid molecules. Large polar and non-polar molecules and particles with charges are unable to cross the membrane on their own. The cell membrane, therefore, is selectively permeable, and certain essential substances (such as glucose) require assistance to cross it. In Chapter 1, you reviewed one method for transporting certain molecules, such as glucose, across the membrane, with the help of a carrier protein. Since carrier proteins accept only non-charged molecules with a specific shape, a different membrane protein, called a channel protein, is used to facilitate the movement of charged particles (ions) into and out of cells. The diagram on the right shows how a channel protein accomplishes this.

Keep in mind that channel proteins and carrier proteins involve the movement of substances along their concentration gradient. In other words, the cell provides no energy to assist in the transport of these substances. To move substances against their concentration gradient, the cell must expend energy through the process of active transport.

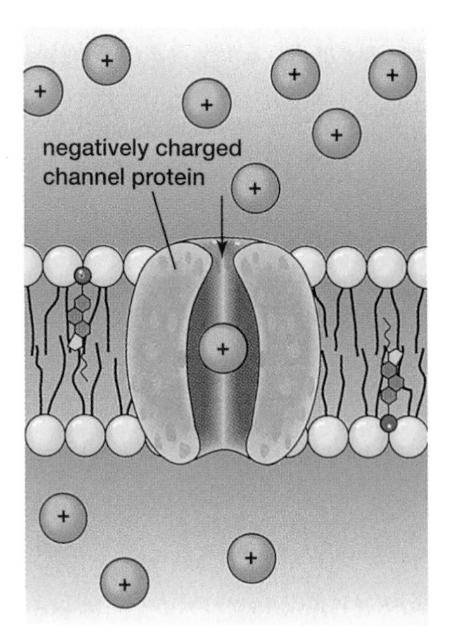

Figure A5.5 Dissolved ions (charged particles) cross the cell membrane by means of water-filled passages supplied by channel proteins.

Electronegativity:
A Quick Reference for the Biology Student

Electronegativity and the Periodic Table of the Elements

Chemist Linus Pauling, winner of the Nobel prize, first defined the concept of electronegativity in the 1930s. The electronegativity of an element is a measure of its ability to attract electrons in a bond. This chemical property is relevant only to bonding between atoms, not to individual atoms. Pauling created a quantitative scale of the electronegativity values of the elements listed in the periodic table. In general, electronegativity increases from left to right along each row (period) of the periodic table. Electronegativity also tends to increase from the bottom to the top of each column (group). Fluorine has the highest electronegativity value (4.0), so fluorine is said to be the most electronegative element. Francium and cesium each have the lowest electronegativity value (0.7), so they are the least electronegative elements.

The following table lists the electronegativity values of some biologically important *elements*.

Table A6.1
Electronegativity values of some biologically important elements

Element	Symbol	Electronegativity
potassium	K	0.8
sodium	Na	0.9
calcium	Ca	1.0
iron	Fe	1.8
hydrogen	H	2.1
phosphorus	P	2.2
carbon	C	2.5
sulfur	S	2.6
chlorine	Cl	3.0
nitrogen	N	3.0
oxygen	O	3.5

Electronegativity and Chemical Bonds

Two atoms of the same element have the same electronegativity, or the same ability to attract electrons in a bond. Therefore, if two atoms of the same element form a bond, as in a molecule of hydrogen (H_2), oxygen (O_2), or nitrogen (N_2), the bonding electrons are equally shared between the atoms. A bond in which electrons are shared between atoms is called a *covalent bond*. If the electrons are equally shared, the bond is said to be *non-polar covalent*.

If a covalent bond is formed between atoms of two different elements, such as hydrogen and chlorine in hydrogen chloride (HCl), the electronegativities of the elements determine how the bonding electrons are distributed. Because the electronegativities are different, the bonding electrons are shared unequally. A covalent bond in which electrons are shared unequally is called a *polar covalent bond*. As shown in Table A6.1, chlorine is more electronegative than hydrogen, so chlorine has the stronger attraction for the bonding electrons in a hydrogen chloride molecule. For this reason, the chlorine atom becomes slightly negative, and the hydrogen atom becomes slightly positive. The slightly charged ends of the bond are known as *poles*. Because two poles are present, the bond is said to have a *dipole*. The slightly positive and negative ends of the polar covalent bond can be represented using the symbols δ+ and δ–, as shown in Figure A6.1.

$$\overset{\delta+}{H}-\overset{\delta-}{Cl}$$

Figure A6.1 The bond between hydrogen and chlorine in a hydrogen chloride molecule is polar covalent.

If two elements form a covalent bond, the difference between their electronegativities determines how polar the bond is. Table A6.1 shows that the electronegativity difference between carbon and oxygen is 1.0, whereas the electronegativity difference between carbon and hydrogen is 0.4. Therefore, the sharing of electrons is more unequal between carbon and oxygen than between carbon and hydrogen. In other words, a carbon–oxygen bond is more polar than a carbon–hydrogen bond.

When two elements have very different electronegativities, the bonds the elements form can be described in terms of electron transfers, rather than electron sharing. For example, sodium and chlorine have an electronegativity difference of 2.1. When these elements react, electrons are transferred from sodium atoms to chlorine atoms to form positively charged sodium ions (Na^+) and negatively charged chloride ions (Cl^-). These ions attract each other. Bonds created by the transfer of electrons from one type of atom to another are called *ionic bonds*.

There is no sharp distinction between polar covalent bonds and ionic bonds. The greater the difference in the electronegativities of two bonded atoms, the more polar is the bond between them. When sharing is very unequal, chemists describe the electrons as being transferred from one atom to the other. For convenience, chemists have chosen an approximate value of the electronegativity difference that distinguishes polar covalent bonds from ionic bonds. In general, if the electronegativity difference equals or exceeds 1.7, the elements will tend to form ionic bonds. If the electronegativity difference is less than 1.7, the elements will tend to form covalent bonds. If the electronegativity difference is greater than 0 but less than 1.7, the bonds will be polar covalent. Strictly speaking, bonds are non-polar covalent only if the electronegativity difference is exactly 0, that is, if two atoms of the same element form a bond. However, when the electronegativity difference is less than 0.5 (for example, 0.4 for C and H), the electrons are shared almost equally, and the bond dipole is weak.

The electronegativity of carbon is 2.5, which is very close to the middle of the range of electronegativity values, from 0.7 to 4.0. Carbon can, therefore, form covalent bonds with elements that are more electronegative, such as oxygen, and with elements that are less electronegative, such as hydrogen. Carbon atoms are also able to form covalent bonds with other carbon atoms to produce a wide range of compounds that contain chains or rings of carbon atoms.

Molecular Polarity

You learned earlier that the bond in a hydrogen chloride molecule (HCl) is polar covalent. Because the bond is polar, and there is only one bond in the molecule, the whole molecule is also polar. However, most covalent molecules contain more than two atoms and more than one bond. In such cases, the polarity of a molecule depends on the polarity of the individual bonds and on the overall shape of the molecule. Two examples that show the importance of molecular shapes are carbon dioxide (CO_2) and water (H_2O).

In a carbon dioxide molecule, the two highly electronegative oxygen atoms form double bonds to a less electronegative carbon atom, as shown in Figure A6.2. The carbon–oxygen bonds are polar covalent. The shape of the molecule is straight or linear. As a result, the two dipoles counterbalance each other, and the molecule has no net dipole. Thus, even though a carbon dioxide molecule contains polar bonds, the molecule is non-polar.

$$\overset{\delta-}{O}\overset{\delta+\ \delta+}{=C=}\overset{\delta-}{O}$$

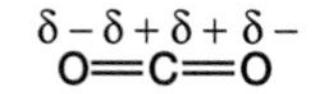

Figure A6.2 A linear carbon dioxide molecule contains polar covalent bonds, but the molecule is non-polar.

In a water molecule, one highly electronegative oxygen atom forms single bonds to two less electronegative hydrogen atoms, as shown in Figure A6.3. The hydrogen–oxygen bonds are polar covalent. In contrast to the linear carbon dioxide molecule, a water molecule is V-shaped or bent. As a result, the two dipoles do not counterbalance each other, and the molecule has a net dipole. Thus, a water molecule contains polar bonds, and the molecule is polar.

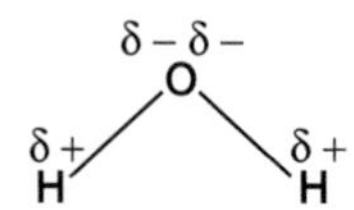

Figure A6.3 Because a water molecule is bent and contains polar covalent bonds, the molecule is polar.

Hydrocarbons, which are composed of carbon and hydrogen, are an important class of compounds in organic chemistry and biology. Hydrocarbon molecules, some of which contain large numbers of atoms, include carbon–hydrogen bonds that are slightly polar. However, because of the shapes of hydrocarbon molecules, these molecules are either non-polar or very low in polarity.

Some Biological Applications

The presence or absence of dipoles can have a great effect on the physical and chemical properties of covalent molecules. Therefore, an understanding of dipoles is very important in explaining many biological processes. Some examples of how electronegativity is important in biological processes are given below.

Functional Groups

Some important functional groups are listed in Figure 1.24 in section 1.4. All of the functional groups listed contain covalent bonds between elements with different electronegativities, so the functional groups contain polar covalent bonds. The presence of these polar bonds in biological molecules influences the physical interactions of the molecules and their chemical properties.

A hydroxyl group (–OH) contains a single covalent bond between a highly electronegative oxygen atom and a less electronegative hydrogen atom. As in a water molecule, the oxygen atom is at the slightly negative end of the dipole, and hydrogen is at the slightly positive end. Hydroxyl groups can attract some ions and polar molecules in a type of interaction called a hydrogen bond, which will be described below.

In a carbonyl group, highly electronegative oxygen forms a double bond with less electronegative carbon. Because the carbon atom is at the more positive end of the dipole, the carbon atom tends to react with substances that are electron donors. The oxygen atom at the more negative end of the dipole tends to react with substances that are electron acceptors.

A carboxyl group contains both a carbonyl group and a hydroxyl group connected together. Because a carboxyl group contains two highly electronegative oxygen atoms, the positive charge on the hydrogen atom is larger than in a hydroxyl group alone. In fact, carboxyl groups partially dissociate to form hydrogen ions (H^+). Therefore, carboxyl groups are present in organic acids, such as amino acids. An amino acid molecule also contains an amino group ($-NH_2$). In this functional group, the nitrogen atom is at the slightly negative end of each nitrogen–hydrogen dipole. The nitrogen atom can act as an electron donor, for example to a hydrogen ion. Thus, an amino acid contains a functional group (a carboxyl group) that can release hydrogen ions and another functional group (an amino group) that can react with them.

Hydrogen Bonds

When hydrogen atoms are bonded to small, highly electronegative atoms, such as oxygen, a weak force of attraction can exist between molecules. This force is known as a *hydrogen bond.* A hydrogen bond can be formed when a negative ion (such as Cl^-), or a slightly negative atom in a dipole (such as the oxygen in an –OH group) attracts a slightly positive hydrogen atom in another dipole. Therefore, water molecules form hydrogen bonds with each other, as shown in Figure A6.4. Hydrogen bonds are only about one-twentieth as strong as covalent bonds, but hydrogen bonds have important effects on the physical and chemical behaviour of many molecules in biological systems.

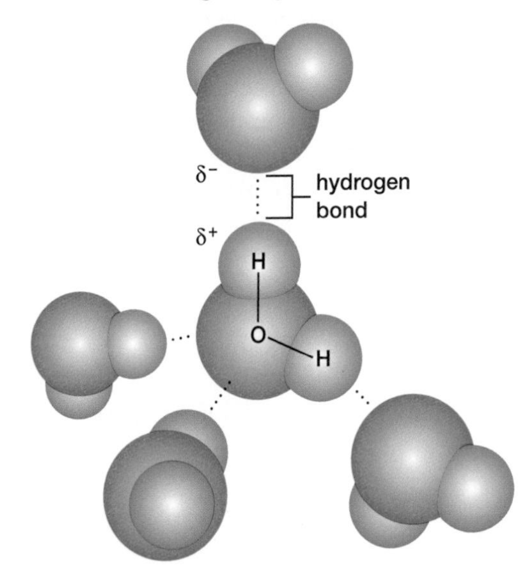

Figure A6.4 Hydrogen bonds exist between water molecules and account for many of the physical and chemical properties of water.

The hydrogen bonds between water molecules make water an ideal medium for life processes on Earth. Scientists believe that if there were no hydrogen bonds present, then water would be a gas, not a liquid, under the conditions in which we live. The polarity of water molecules, and their ability to form hydrogen bonds, explains the fact that water can act as a solvent for so many substances in living and non-living systems. When ionic and polar substances dissolve in water, they may form hydrogen bonds to water molecules. Hydrogen bonds can also bind the components of complex biological substances, such as the chains of nucleic acids that make up a DNA molecule. These two long intertwined chains are held to each other by thousands of hydrogen bonds. Proteins, such as enzymes, also require hydrogen bonding to maintain their shapes and their functions.

Aerobic Cellular Respiration

Aerobic cellular respiration involves the breakdown of glucose molecules (food) to form energy-rich molecules called ATP, adenosine triphosphate. Molecules of ATP are used to fuel many other chemical reactions in cells. ATP molecules are formed in several systems in aerobic cellular respiration. The system that produces the most ATP is called the electron transport chain (see Figure A6.5). In the electron transport chain, electrons (e^-) flow along the chain from one protein molecule to another. As electrons move along the chain, they release energy. This energy is used to make ATP molecules. Once electrons reach the end of the chain, however, they need to be removed from the system. Removal of the electrons is necessary in order to allow new electrons to enter at the start of the chain. The element that accepts the electrons at the end of the electron transport chain is oxygen. As you have learned, oxygen is one of the most electronegative elements in the periodic table. Oxygen accepts the electrons and combines with hydrogen ions (H^+), obtained from the breakdown of glucose, to form water molecules. Without oxygen to accept the electrons, the electron transport chain could not function and, as a result, could not produce ATP molecules. This is why oxygen is so important in cellular processes.

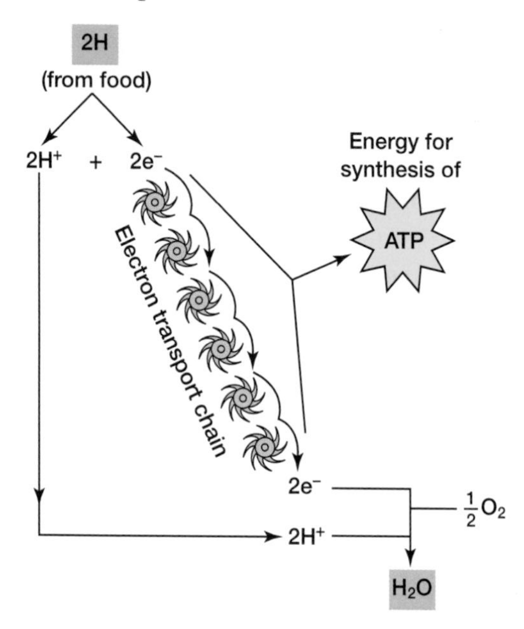

Figure A6.5 The electron transport chain

APPENDIX 7

Periodic Table and List of Elements

Metals
Metalloids
Nonmetals

atomic number
atomic mass

MAIN–GROUP ELEMENTS (1A–2A); TRANSITION ELEMENTS (3B–2B); MAIN–GROUP ELEMENTS (3A–8A)

Period	1A (1)	2A (2)	3B (3)	4B (4)	5B (5)	6B (6)	7B (7)	8B (8)	8B (9)	8B (10)	1B (11)	2B (12)	3A (13)	4A (14)	5A (15)	6A (16)	7A (17)	8A (18)
1	1 **H** 1.008																	2 **He** 4.003
2	3 **Li** 6.941	4 **Be** 9.012											5 **B** 10.81	6 **C** 12.01	7 **N** 14.01	8 **O** 16.00	9 **F** 19.00	10 **Ne** 20.18
3	11 **Na** 22.99	12 **Mg** 24.31											13 **Al** 26.98	14 **Si** 28.09	15 **P** 30.97	16 **S** 32.07	17 **Cl** 35.45	18 **Ar** 39.95
4	19 **K** 39.10	20 **Ca** 40.08	21 **Sc** 44.96	22 **Ti** 47.88	23 **V** 50.94	24 **Cr** 52.00	25 **Mn** 54.94	26 **Fe** 55.85	27 **Co** 58.93	28 **Ni** 58.69	29 **Cu** 63.55	30 **Zn** 65.39	31 **Ga** 69.72	32 **Ge** 72.61	33 **As** 74.92	34 **Se** 78.96	35 **Br** 79.90	36 **Kr** 83.80
5	37 **Rb** 85.47	38 **Sr** 87.62	39 **Y** 88.91	40 **Zr** 91.22	41 **Nb** 92.91	42 **Mo** 95.94	43 **Tc** (98)	44 **Ru** 101.1	45 **Rh** 102.9	46 **Pd** 106.4	47 **Ag** 107.9	48 **Cd** 112.4	49 **In** 114.8	50 **Sn** 118.7	51 **Sb** 121.8	52 **Te** 127.6	53 **I** 126.9	54 **Xe** 131.3
6	55 **Cs** 132.9	56 **Ba** 137.3	57 **La** 138.9	72 **Hf** 178.5	73 **Ta** 180.9	74 **W** 183.9	75 **Re** 186.2	76 **Os** 190.2	77 **Ir** 192.2	78 **Pt** 195.1	79 **Au** 197.0	80 **Hg** 200.6	81 **Tl** 204.4	82 **Pb** 207.2	83 **Bi** 209.0	84 **Po** (209)	85 **At** (210)	86 **Rn** (222)
7	87 **Fr** (223)	88 **Ra** (226)	89 **Ac** (227)	104 **Rf** (261)	105 **Db** (262)	106 **Sg** (266)	107 **Bh** (264)	108 **Hs** (269)	109 **Mt** (268)	110 **Uun** (271)	111 **Uuu** (272)	112 **Uub** (277)		114 **Uuq** (285)		116 **Uuh** (289)		

INNER TRANSITION ELEMENTS

Period	Series														
6	Lanthanoids	58 **Ce** 140.1	59 **Pr** 140.9	60 **Nd** 144.2	61 **Pm** (145)	62 **Sm** 150.4	63 **Eu** 152.0	64 **Gd** 157.3	65 **Tb** 158.9	66 **Dy** 162.5	67 **Ho** 164.9	68 **Er** 167.3	69 **Tm** 168.9	70 **Yb** 173.0	71 **Lu** 175.0
7	Actinoids	90 **Th** 232.0	91 **Pa** 231.0	92 **U** 238.0	93 **Np** 237.1	94 **Pu** (244)	95 **Am** (243)	96 **Cm** (247)	97 **Bk** (247)	98 **Cf** (251)	99 **Es** (252)	100 **Fm** (257)	101 **Md** (258)	102 **No** (259)	103 **Lr** (262)

List of Elements

Element	Symbol	Atomic Number	Atomic Mass*
Actinium	Ac	89	(227)
Aluminum	Al	13	26.98
Americium	Am	95	(243)
Antimony	Sb	51	121.8
Argon	Ar	18	39.95
Arsenic	As	33	74.92
Astatine	At	85	(210)
Barium	Ba	56	137.3
Berkelium	Bk	97	(247)
Beryllium	Be	4	9.012
Bismuth	Bi	83	209.0
Bohrium	Bh	107	(264)
Boron	B	5	10.81
Bromine	Br	35	79.90
Cadmium	Cd	48	112.4
Calcium	Ca	20	40.08
Californium	Cf	98	(249)
Carbon	C	6	12.01
Cerium	Ce	58	140.1
Cesium	Cs	55	132.9
Chlorine	Cl	17	35.45
Chromium	Cr	24	52.00
Cobalt	Co	27	58.93
Copper	Cu	29	63.55
Curium	Cm	96	(247)
Dubnium	Db	105	(262)
Dysprosium	Dy	66	162.5
Einsteinium	Es	99	(254)
Erbium	Er	68	167.3
Europium	Eu	63	152.0
Fermium	Fm	100	(253)
Fluorine	F	9	19.00
Francium	Fr	87	(223)
Gadolinium	Gd	64	157.3
Gallium	Ga	31	69.72
Germanium	Ge	32	72.61
Gold	Au	79	197.0
Hafnium	Hf	72	178.5
Hassium	Hs	108	(269)
Helium	He	2	4.003
Holmium	Ho	67	164.9
Hydrogen	H	1	1.008
Indium	In	49	114.8
Iodine	I	53	126.9
Iridium	Ir	77	192.2
Iron	Fe	26	55.85
Krypton	Kr	36	83.80
Lanthanum	La	57	138.9
Lawrencium	Lr	103	(262)
Lead	Pb	82	207.2
Lithium	Li	3	6.941
Lutetium	Lu	71	175.0
Magnesium	Mg	12	24.31
Manganese	Mn	25	54.94
Meitnerium	Mt	109	(268)
Mendelevium	Md	101	(256)

Element	Symbol	Atomic Number	Atomic Mass*
Mercury	Hg	80	200.6
Molybdenum	Mo	42	95.94
Neodymium	Nd	60	144.2
Neon	Ne	10	20.18
Neptunium	Np	93	237.1
Nickel	Ni	28	58.70
Niobium	Nb	41	92.91
Nitrogen	N	7	14.01
Nobelium	No	102	(253)
Osmium	Os	76	190.2
Oxygen	O	8	16.00
Palladium	Pd	46	106.4
Phosphorus	P	15	30.97
Platinum	Pt	78	195.1
Plutonium	Pu	94	(244)
Polonium	Po	84	(209)
Potassium	K	19	39.10
Praseodymium	Pr	59	140.9
Promethium	Pm	61	(145)
Protactinium	Pa	91	231.0
Radium	Ra	88	(226)
Radon	Rn	86	(222)
Rhenium	Re	75	186.2
Rhodium	Rh	45	102.9
Rubidium	Rb	37	85.47
Ruthenium	Ru	44	101.1
Rutherfordium	Rf	104	(261)
Samarium	Sm	62	150.4
Scandium	Sc	21	44.96
Seaborgium	Sg	106	(266)
Selenium	Se	34	78.96
Silicon	Si	14	28.09
Silver	Ag	47	107.9
Sodium	Na	11	22.99
Strontium	Sr	38	87.62
Sulfur	S	16	32.07
Tantalum	Ta	73	180.9
Technetium	Tc	43	(98)
Tellurium	Te	52	127.6
Terbium	Tb	65	158.9
Thallium	Tl	81	204.4
Thorium	Th	90	232.0
Thulium	Tm	69	168.9
Tin	Sn	50	118.7
Titanium	Ti	22	47.88
Tungsten	W	74	183.9
Ununbium	Uub	112	(277)
Ununhexium	Uuh	116	(289)
Ununnilium	Uun	110	(271)
Ununquadium	Uuq	114	(285)
Unununium	Uuu	111	(272)
Uranium	U	92	238.0
Vanadium	V	23	50.94
Xenon	Xe	54	131.3
Ytterbium	Yb	70	173.0
Yttrium	Y	39	88.91
Zinc	Zn	30	65.39
Zirconium	Zr	40	91.22

*All atomic masses are given to four significant figures. Values in parentheses represent the mass number of the most stable form of the atom.

Glossary

abiotic describes all the non-living physical and chemical factors that influence an organism's survival, such as temperature, light, water, and nutrients. (13.1)

accommodation adjustment the ciliary body makes to the shape of the lens in order to focus on objects at varying distances. (5.3)

acetylcholine the primary neurotransmitter of both the somatic nervous system and the parasympathetic nervous system. (5.2)

acetyl-CoA molecule produced during the transition reaction for use in the Krebs cycle. (3.1)

acid substance that donates H^+ ions when it dissolves or dissociates in water. (1.3)

acid precipitation rain or snow that has a pH of about 4 to 5, as compared to normal precipitation, which has a pH of about 6.5. (15.3)

acromegaly condition in which excess HGH production during adulthood causes thickening of bone tissue, leading to spinal deformities and abnormal growth of the head, hands, and feet. The condition is often caused by a tumour in the pituitary gland. (6.1)

action potential in an axon, the change in charge that occurs when the gates of the K^+ channels close and the gates of the Na^+ channels open after a wave of depolarization is triggered. (5.2)

activating enzyme enzyme that binds the correct amino acid to the tRNA molecule having the correct anticodon. Each activating enzyme has two binding sites: one that binds to the anticodon on the tRNA molecule, and one that binds to the amino acid corresponding to that anticodon. (8.3)

activation energy input of energy required to start a chemical reaction. (2.1)

activator molecule that promotes the action of enzymes by binding to the allosteric site. (2.2). Also a catabolite activator protein or CAP, that stimulates gene expression in *E. coli* by binding first to a cAMP molecule and then to a site near the P*lac* promoter. This action makes it easier for RNA polymerase to bind to the promoter, which in turn speeds up the rate at which *lac* genes are transcribed. (8.4)

active site site on an enzyme where substrates fit and where catalysis occurs. (2.2)

active transport process that moves particles across the cell membrane against a concentration gradient. The process requires the help of special proteins (often called pumps) and energy. (2.3)

adaptation the process by which a receptor responds to changes in the environment. If a stimulus continues for a long time, the brain becomes unaware of it. (5.3) Also, in evolution, a particular structure, physiology, or behaviour that helps an organism survive and reproduce in a particular environment. (10.1, 12.1)

adaptive radiation diversification of a common ancestral species into a variety of species, all of which are differently adapted. (12.3)

adenine (A) one of four nitrogen-containing bases in a nucleotide that make up DNA and RNA. (7.1)

adrenal cortex outer part of adrenal gland, which produces two types of steroid hormones: glucocorticoids (cortisol) and mineralcorticoids (aldosterone). (6.2)

adrenal gland organ composed of two layers: an outer cortex and an inner medulla. Each layer produces different hormones and functions as an independent organ. (6.2)

adrenal medulla inner part of adrenal gland, which secretes adrenaline and noradrenaline. (6.2)

adrenocorticotropic hormone (ACTH) hormone that regulates the production of cortisol and aldosterone. (6.2)

aerobic in an organism, environment, or cellular process, requiring oxygen. (3.1)

aerobic cellular respiration process composed of metabolic pathways that contribute to the production of ATP in the presence of oxygen. (3.1)

age pyramid age structure of a population represented in a way that shows the proportion of individuals in the population in each age class. (14.4)

age structure distribution of individuals among the age classes. (14.4)

Agricultural Revolution time period during which humans started to plant and harvest crops and domesticate animals, resulting in a stable food supply and increasing human population growth. (15.1)

alleles alternative forms of a gene. (11.1)

allopatric speciation formation of a new species that occurs when a population is split into two or more isolated groups by a geographical barrier. (12.3)

all-or-none principle principle that governs the response of an axon to a stimulus. If a neuron is stimulated sufficiently, an impulse will travel the length of an axon, if the stimulus is not sufficient, no impulse will travel down the axon.

The strength of the response is fixed and not dependent on the strength of the stimulus. (5.2)

allosteric regulation regulation of enzyme activity by inhibitors and activators. (2.2)

allosteric site location on an enzyme at which an inhibitor molecule binds. (2.2)

Ames test simple test that measures the potential for a chemical to be mutagenic. (9.1)

amino acid attachment site site at the 3′ end of the tRNA strand across from the anticodon loop that binds a particular amino acid, specified by the mRNA codon, to a tRNA molecule. (8.3)

amino-acetyl tRNA (aa-tRNA) tRNA molecule bound to a specific amino acid by an amino acid attachment site. (8.3)

anaerobic describes an organism, environment, or cellular process that does not require oxygen. (3.1)

anaerobic cellular respiration the breakdown of glucose and formation of ATP molecules in the absence of oxygen (e.g., during fermentation). (3.2)

analogous structures body parts in different species that have a similar function, but evolved separately (e.g., insect and bird wings). (10.3)

anion negatively charged ion (has more electrons than protons). (1.2)

antagonistic hormones hormones that have opposing physiological properties, but that work together. (4.3, 6.1)

antenna pigments chlorophyll molecules in a photosystem that collect and channel light energy. (3.3)

anthropogenic of human origin. (15.3)

antibody immunity defence against previously encountered diseases that is performed by B cells. (4.4)

anticodon specialized base triplet located on one lobe of a tRNA molecule that recognizes its complementary codon on an mRNA molecule. (8.3)

anti-diuretic hormone (ADH) hormone that regulates sodium levels in the bloodstream. The hormone is produced in the hypothalamus. (6.1)

antigen protein or other large molecule on the surface of a non-self cell (attacking or unfamiliar cell) that helps the body recognize the cell as non-self. (4.4)

antiparallel describes the property by which the 5′ to 3′ phosphate bridges run in opposite directions on each DNA strand. (7.2)

anti-sense strand strand of nucleotides that is complementary to a sense strand. (8.2)

apoptosis during an immune response, programmed cell death of plasma cells after the danger has passed. (4.4)

aqueous humour the liquid filling the area between the cornea and the lens of the eye. (5.3)

aquifer natural reservoir of groundwater held beneath the surface of Earth. (15.3)

artificial selection human selection of particular traits (e.g., faster horses, disease-resistant plants) by breeding. (10.1)

assortative mating type of non-random mating in which individuals choose partners that have similar phenotypes (e.g., size). (11.3)

astigmatism abnormality in the shape of the cornea or lens that results in uneven focus. (5.3)

atom the smallest unit of matter involved in chemical reactions. (1.1)

atomic mass the average mass of all the naturally occurring isotopes of an element (1.2)

atomic number the number of protons in the nucleus of one atom of an element. Atomic number is different for every element. (1.2)

ATP adenosine triphosphate. It is the universal energy unit in cells, and contains adenine (an amino acid) bound to ribose, which is bound to three phosphate groups. (2.3)

ATP synthase complex the multienzyme complex that forms ATP molecules through the movement of hydrogen ions across a membrane. (3.2)

autonomic nervous system the part of the nervous system that relays information to internal organs that are not under the conscious control of the individual. The system is made up of the sympathetic and parasympathetic nervous systems. (5.1)

autotroph a primary producer that gets energy from the Sun and builds organic molecules from water, carbon dioxide, and nutrients from soil. (13.2)

axon a long, cylindrical extension of a neuron's cell body that can range from 1 mm to 1 m in length. It transmits impulses along its length to the next neuron. (5.1)

BAC-to-BAC sequencing sequencing step in which the bacterial artificial chromosomes (BACs) are broken into smaller fragments that can then be sequenced using the chain termination reaction. (9.2)

bacterial artificial chromosome (BAC) bacterial vector used for DNA cloning. (9.2)

base compound that decreases the concentration of H^+ ions in solution, usually by either attracting H^+ ions or by releasing OH^- ions. (1.3)

behavioural adaptations adaptations associated with the ways in which species respond to their environment (e.g., migration, foraging behaviour). (12.1)

***beta*-oxidation** the removal of acetate units from fatty acids. (3.2)

biodiversity the variety of living organisms that inhabit Earth. (15.3)

biogeochemical cycle the route that chemical nutrients take through all biotic and abiotic components of an ecosystem. (13.4)

biogeography the study of the geographical distribution of species. (10.2)

biological barriers factors that keep species reproductively isolated even when they exist in the same region. (12.2)

biological species population that is reproductively compatible. Members of the population can interbreed and produce viable, fertile offspring. (12.2)

biomass amount of vegetation added to an ecosystem per area per year (g/m^2a). (13.3)

biomass pyramid type of graphical representation showing the pattern of mass changes through a trophic structure. (13.3)

biomes identifiable ecosystems found in specific regions on Earth that have particular combinations of biotic and abiotic factors. (13.1)

bioremediation use of living cells to perform environmental clean-up tasks, such as cleaning up PCBs using bacteria that naturally contain genes that degrade PCBs into harmless compounds. (9.3)

biosphere all ecosystems and their interactions on Earth. (13.1)

biotic describes the components of an organism's environment that include living things with which the organism interacts, such as predators, prey, symbionts, and competition. (13.1)

biotic potential the highest possible per capita growth rate possible for a given population. (10.2, 14.2)

births the number of individuals born during a given time period; in population dynamics, a factor that increases the population. (14.2)

bottleneck effect effect that occurs when a population is greatly reduced by events such as natural disaster or overhunting, resulting in certain alleles being overrepresented and other alleles being underrepresented or absent in the population, due to chance. The effect leads to a genetic drift and decreased genetic variation in the population. (11.3)

Bowman's capsule the receiving end of a renal tubule at which water and small solutes from the blood enter the proximal tubule from the glomerulus. (4.2)

buffers chemicals or combinations of chemicals that resist changes in pH by taking up extra hydrogen ions or hydroxide ions in solution. (1.3, 4.2)

calcitonin a hormone produced in the thyroid gland, which regulates calcium levels in the blood. Works in conjunction with the parathyroid hormone. (6.1)

Calvin cycle process in which green plants fix carbon from atomospheric CO_2 to produce carbohydrates. This process is the second of two major stages of photosynthesis (following the photo reactions). (3.3)

carbon fixation the initial incorporation of carbon from atmospheric CO_2 by green plants into organic molecules. (3.3)

carcinogenic describes a factor, such as a chemical mutagen, that is associated with one or more forms of cancer. (9.1)

cardiac centre part of the medulla oblongata that controls heart rate and the force of the heart's contraction. (5.1)

cardiolipin the inner membrane of the mitochondrion. (3.2)

carrier proteins proteins in a cell membrane that are able to move and change shape to assist large molecules to enter a cell by passive transport. (1.3)

carrying capacity (*K*) the maximum population size that can be sustained in a given environment over a long period of time. (14.2)

catalysis the acceleration of a chemical reaction by a substance that is regenerated, unchanged, at the end of a chemical reaction. (2.2)

catalytic cycle process in which substrates form an enzyme-substrate complex and are then are released from the complex as a product or products, freeing the enzyme for further reactions. (2.2)

cataract cloudy or opaque areas on the lens of the eye that increase in size over time and eventually cause blindness. (5.3)

catastrophism the idea that catastrophes such as floods, diseases, or droughts periodically destroyed species living in a particular region, allowing species from neighbouring regions to repopulate the area. (10.2)

cation positively charged ion (has more protons than electrons). (1.2)

cell body the main part of a neuron, containing the nucleus and other organelles. (5.1)

cellular immunity defense against a previously encountered disease; primarily a function of T cells. After a T cell recognizes a certain antigen, it multiplies through clonal expansion to produce several other, different T cells with specific functions. (4.4)

census complete count of all the members of a given population. (14.1)

central nervous system (CNS) the brain and spinal cord. (5.1)

cerebellum the part of the brain that controls muscle co-ordination. (5.1)

cerebral cortex the thin layer of grey matter that covers each hemisphere of the brain, enabling the individual to experience sensation, perform voluntary movements, and think. (5.1)

cerebrum the largest part of the brain, in which all the information from the senses is sorted and interpreted, voluntary muscles are stimulated, memories are stored, and decisions are made. (5.1)

chain termination sequencing process used to sequence DNA. The process relies on a modified form of the polymerase chain reaction. (9.2)

Chargaff's rule constant relationship in which in any sample of DNA, the amount of adenine is always equal to the amount of thymine, and the amount of cytosine is always equal to the amount of guanine. The relationship was discovered in the late 1940s by Erwin Chargaff. (7.1)

chemical bonds the forces that hold atoms to each other within molecules and in ionic compounds. (1.1)

chemical energy potential energy stored in the bonds of a molecule. (2.1)

chemical mutagen molecule that can enter the cell nucleus and induce a permanent change in the genetic material of the cell. (9.1)

chemiosmosis formation of ATP molecules through the movement of hydrogen ions across an electrochemical gradient. (3.2)

chlorophylls photosynthetic pigments within chloroplasts that absorb light energy. (3.3)

chloroplast organelle in which photosynthesis occurs. (3.3)

cholinesterase enzyme that breaks down the neurotransmitter acetylcholine. The enzyme is released from the pre-synaptic neuron, soon after the release of acetylcholine. (5.2)

choroid layer the middle layer of the eye, which absorbs light and prevents internal reflection. The layer forms the iris at the front of the eye. (5.3)

chromatin fibres of material that make up chromosomes. The fibres are composed of about 60 percent protein, 35 percent DNA, and 5 percent RNA. (7.2)

chromosome strand-like complex of nucleic acids and protein tightly bound together. Chromosomes contain the hereditary units, or genes. (7.1)

ciliary body thickened choroid muscle behind the iris that controls the shape of the lens of the eye. (5.3)

circadian rhythms daily biological cycles or regular patterns. (6.1)

climax community self-perpetuating (secure, self-maintaining) community with stable populations that interact with each other and the environment in a stable manner. In communities that are at this stage, succession has ended. (13.1)

cloning vector a molecule that replicates foreign DNA within a cell (e.g., a plasmid). (9.2)

closed system system that is isolated from its surroundings. Thermal energy and atoms cannot leave or enter the system. (2.1)

co-dominant describes a situation in which two alleles may be expressed equally. The situation occurs when two different alleles for a trait are both dominant. (11.1)

codon the basic unit, or "word," of the genetic code. It is a set of three adjacent nucleotides in DNA or mRNA that codes for amino acid placement on polypeptides. (8.1)

coenzymes organic cofactors. Vitamins are a part of many coenzymes. (2.2)

coevolution evolutionary process in which two species of organisms that are tightly linked (e.g., predator and prey) evolve together, each population responding to changes in the other population. (12.3)

cofactors inorganic ions and organic non-protein molecules that help some enzymes function as catalysts. (2.2)

cohort large group of individuals all born at the same time. (14.4)

collecting duct the location where all the kidney's filtrate gathers to be passed to the renal pelvis. (4.2)

colonizers the first organisms of a species to establish themselves in a habitat. (13.1)

commensalism relationship that exists between two or more organisms in which a symbiont benefits from a symbiotic relationship, but the host is not harmed. (14.3)

community all of the organisms in all the interacting populations in a given area. (13.1)

competitive exclusion principle the theory stating that species with niches that are exactly the same cannot co-exist. Species with sufficiently different niches can co-exist (possibly in limited numbers). (14.3)

competitive inhibition chemical compounds that bind to the active site of the enzyme and inhibit enzymatic reactions. (2.2)

complementary pairings of bases between nucleic acid strands. The two strands of a DNA double helix are complementary with each other — each purine base pairs with a pyrimidine base on the opposite strand (A pairs with T or U, and C pairs with G). (7.2)

condensation type of reaction in which two molecules react to form water and another molecule. This type of reaction is also known as dehydration synthesis. (1.3)

cones colour receptors in the eye. (5.3)

conjunctiva the thin, transparent membrane that covers the cornea and is kept moist by tears. (5.3)

conservative theory the theory stating that, in the replication of DNA, the two new strands form one molecule, while the two original strands re-form the original molecule. (7.3)

convergent evolution evolutionary process in which similar traits arise in two or more species because each species has independently adapted to similar environmental conditions, not because they share a common ancestor. (12.3)

copy DNA (cDNA) double-stranded molecule of DNA containing only the coding portions of the eukaryotic gene. (9.3)

co-repressor molecule that combines with a repressor protein, causing the repressor protein to have an increased affinity for the operator. When the repressor protein is bound to the operator, transcription is blocked. (8.4)

cornea the clear part of the sclera at the front of the eye. (5.3)

corpus callosum layer of white matter made up of axons that joins the two hemispheres of the cerebral cortex in the brain. (5.1)

corpus luteum follicle that has been emptied of its egg. The empty follicle produces progesterone. (6.3)

cortex (renal) the outer part of the kidney, where the glomerulus, Bowman's capsule, proximal tubule, and distal tubule of the nephrons are located. (4.2)

coupled reaction combinations of reactions in which the energy released by an exothermic reaction can be used to drive an endothermic reaction. (2.3)

covalent bond chemical bond formed by the sharing of electrons between two atoms. (1.1)

Cowper's gland gland in human males that secretes an alkaline fluid that neutralizes the acidity of the female reproductive tract during sexual intercourse. (6.3)

cryptic coloration structural adaptation characterized by camouflaging shape and coloring, enabling species to hide from predators. (12.1)

cyclic photophosphorylation addition of a phosphate group to ADP to form ATP. The process uses energy from an electron transport chain in which electrons are recycled in the photosystem. (3.3)

cytochrome c mitochondrial protein involved in respiration. The amino acid sequence of this protein is used to indicate relatedness among species. (10.3)

cytochromes electron carriers in the electron transport chain. (3.2)

cytosine (C) one of four nitrogen-containing bases in a nucleotide that make up DNA and RNA. (7.1)

cytosol fluid medium within a cell. (1.3)

cytotoxic T cell lymphocyte that binds to an infected cell and then destroys it by puncturing its membrane. (4.4)

deamination the removal of an amino group. (3.2)

death in population dynamics, a factor that decreases a population. (14.2)

decomposer organism that obtains its nutrition from dead organisms and animal waste. The organism breaks down larger organic molecules into smaller nutrients and returns them to the abiotic environment. (13.2)

deforestation the large-scale cutting down of forests by humans. (15.3)

demographic transition period of change in the growth rate (r) of a population. (15.1)

demography the study of populations, especially population size, density, age structure, and growth. (14.1)

dendrite the primary site on a nerve cell (neuron) for receiving signals from other neurons. There may be anywhere from one to several thousand dendrites on each neuron. (5.1)

density the number of individuals per unit of area or volume in a population. (14.1)

density-dependent factor variable, such as competition for resources, that affects the growth of a population when there is an increased number of individuals in a given area. (14.3)

density-independent factor variable, such as climate, that affects the growth of a population regardless of the number of individuals in the population in a given area. (14.3)

deoxyribonucleic acid DNA, also called deoxyribose nucleic acid. A nucleic acid consisting of long chains of individual nucleotides, each of which is composed of the five-carbon sugar deoxyribose, a phosphate group, and the nitrogenous base adenine, guanine, cytosine, or thymine. (7.1)

deoxyribose nucleic acid Phoebus Levene's name for deoxyribonucleic acid (DNA). (7.1)

descent with modification Darwin's theory that natural selection does not demonstrate progress (or evolution), but merely results from a species' ability to survive local conditions at a specific time. (10.2)

desertification the transformation of marginal dry lands into near-deserts that are unsuitable for agriculture. (15.3)

dideoxynucleotides variants of each of the four DNA nucleotides that resemble regular DNA nucleotides, but lack the 3′ hydroxyl group. (9.2)

differentiation the process by which certain portions of a genome are activated or silenced to enable a cell to take on the specific structure and function of a given tissue. (4.4, 9.4)

diffusion form of passive transport in which molecules pass through a membrane from a region of higher concentration to a region of lower concentration. (1.3)

diploid describes cells that contain two copies of every chromosome, characteristic of cells in organisms in which one set of chromosomes is inherited from each parent. (11.1)

directional selection selection that favours the phenotypes at one extreme over another, resulting in the distribution curve of phenotypes shifting in the direction of that extreme. (11.3)

directionality the specific orientation of each strand of DNA, read in the 5′ to 3′ direction. The orientation of one strand in DNA is opposite to the orientation of the other strand in a double helix. (7.2)

direct repair immediate recognition and correction of incorrectly paired nucleotides in DNA undergoing replication, accomplished by enzymes such as DNA polymerase. (9.1)

disaccharide sugar that is formed from two monosaccharides. (1.3)

dispersion distribution of individuals of a population within its geographical boundaries. (14.1)

dispersive theory the theory stating that in the replication of DNA, the parental DNA molecules are broken into fragments and that both strands of DNA in each of the daughter molecules are made up of an assortment of parental and new DNA. (7.3)

disruptive (diversifying) selection selection that favours the extremes of a range of phenotypes rather than intermediate phenotypes. This type of selection can result in the elimination of intermediate phenotypes. (11.3)

dissociation of water molecules the formation of H^+ and OH^- from water molecules. (1.3)

distal tubule the tubule that connects the loop of Henle to the collecting duct in the kidney. (4.2)

divergence species-forming pathway in which one or more new species arises from a parent species that continues to exist. (12.2)

divergent evolution process by which species that were once similar to an ancestral species become increasingly different (e.g., speciation, adaptive radiation). (12.3)

DNA *see* deoxyribonucleic acid. (7.1)

DNA amplification the process of generating a large sample of a target. (9.2)

DNA fingerprint pattern of bands formed by using gel electrophoresis on DNA fragments. (9.2)

DNA ligase enzyme that splices together Okazaki fragments during DNA replication on the lagging strand. DNA ligase catalyzes the formation of phosphate bonds between nucleotides. (7.3)

DNA particle gun in recombinant DNA technology, a device that fires DNA-coated microscopic metal particles directly into plant cells and their nuclei. It is used as a means of bringing DNA from the cytoplasm into the cell nucleus. This device was developed in 1988 by American researcher John Sanford. (9.3)

DNA polymerase during DNA replication, an enzyme that slips into the space between two strands, uses the parent strands as a template, and adds nucleotides to make complementary strands. (7.3)

DNA sequence from a single gene or DNA fragment. (9.2)

dominant allele in a heterozygous pairing, the allele that is fully expressed in an organism's phenotype. (11.1)

dopamine neurotransmitter that elevates mood and controls skeletal muscles. (5.2)

dynamic equilibrium state of balance achieved within an environment. Internal control mechanisms maintain the balance by continuously opposing outside forces that tend to change that environment. (4.1)

ecological niche the role that members of a population play in a community. The role includes the resources members need and how the members interact with other members of the population and the community. (13.1)

ecological succession process consisting of consecutive changes in species composition in a given area. (13.1)

ecology the study of the patterns of distribution and abundance of plants, animals, and other types of organisms on Earth. (13.1)

ecosystem community of living organisms, together with the biotic and abiotic factors that surround and affect the community. (13.1)

effectors structures that receive information from the homeostasis integrator, and make changes to the body's internal conditions (e.g., a gland, an organ, or a muscle). (4.1)

electron negatively charged subatomic particle, found outside the nucleus of an atom in an energy level. (1.1)

electron acceptor molecule that accepts an electron from the chlorophyll *a* reaction centre in a photosystem. (3.3)

electronegativity measure of the relative abilities of atoms to attract electrons. (1.1)

electron transport chain series of electron carrier molecules that move electrons during the redox reactions that form ATP. Also called electron transport system. (3.1)

electron transport system *see* electron transport chain. (3.3)

electrophoresis technique for analyzing DNA, used to examine the variability of genes in a population. Samples of DNA from individuals are placed in a gel that is in turn placed in a solution and connected to an electrical circuit. The DNA fragments move through the gel at varying speeds and a DNA "fingerprint" of bands is then stained and analyzed. (11.1)

element substance that cannot be broken down into simpler substances by chemical means. (1.1)

elongation the second step in transcription, in which the correct number of nucleotides from a DNA template is copied onto a messenger RNA molecule. (8.2) Also, in translation, the name of the second step, in which a polypeptide is synthesized. (8.3)

embryology study of embryos (10.2)

emigration movement of individuals away from a region; in population dynamics, a factor that decreases the population. (14.2)

endemic regarding species, unique to a certain region such as an island. (10.3)

endocrine glands ductless glands that secrete hormones directly into the bloodstream. (6.1)

endocrine system body system that works in parallel with the nervous system to maintain homeostasis by releasing chemical hormones from various glands. The system is composed of the hormone-producing glands and tissues of the body. (6.1)

endometrium the lining of the uterus. (6.3)

endothermic reaction reaction that requires an input of energy in order to proceed. (2.1)

energy the capacity for doing work. (2.1)

entropy disorder in a chemical reaction or the universe. (2.1)

environmental resistance situation in which biotic and abiotic factors in an environment limit a population's growth. (14.3)

enzymes specialized proteins that make cellular work possible in all cells by helping chemical reactions to occur. (2.2)

enzyme-substrate complex the combination of the substrate induced-fitted into the active site of an enzyme and the enzyme itself. (2.2)

epididymis in human males, the tube within the scrotum in which sperm mature and become motile. (6.3)

equilibrial life history reproductive strategy in which organisms produce a few large offspring and care for them to ensure their survival is favoured. This strategy is usually seen in populations that are at their carrying capacity. The strategy is also known as *K*-selected reproduction. (14.4)

evolution the relative change in the characteristics of populations that occurs over successive generations. (10.1)

exaptation process by which an adaptation that evolved for one purpose is co-opted for another use. (12.1)

excision repair replacement by the cell of a damaged section of DNA with a newly synthesized correct section of DNA. (9.1)

excitatory response process in which the neurotransmitter reaches the dendrites of a postsynaptic neuron, and a wave of depolarization is generated by the resultant opening of sodium gates (5.2)

exocrine glands glands that have ducts and secrete substances such as sweat, milk, or digestive enzymes. (6.1)

exon the coding region of a eukaryote gene. Each gene is composed of one or more exons. (7.4)

exothermic reaction reaction that is accompanied by a release of energy. (2.1)

exponential growth the growth of a population that occurs in an environment with unlimited resources. (14.2)

expression vector plasmid that contains a prokaryotic promoter sequence just ahead of a restriction enzyme target site. (9.3)

extinction complete disappearance from Earth of all members of a species. (15.3)

extirpation disappearance of a species from areas that were once part of its range. (15.3)

***ex vivo* therapy** in gene therapy, culturing some of the target cells from a patient's body and re-implanting genetically modified cells. (9.4)

fallopian tubes in a human female, two tubes that transfer eggs from the ovaries to the uterus. (6.3)

fecundity the average number of young produced by a female over her lifetime in a population. (14.4)

feedback inhibition type of non-competitive inhibition in which the end product of the pathway binds at an allosteric site on the first enzyme of the pathway. (2.2)

fermentation metabolic process that makes ATP from glucose in the absence of oxygen. The process does not use the electron transport chain; it produces either lactic acid or ethanol and CO_2. (3.1)

fertility rate the average number of children born alive to each woman in a population during her lifetime. (15.1)

fimbriae finger-like projections that sweep eggs from where they exit the ovaries into the fallopian tubes. (6.3)

fitness the suitability of an organism to its environmental conditions. (10.1)

5′ cap modified form of the G nucleotide added to the 5′ end of a pre-mRNA molecule. (8.2)

fixed describes a population in which all members are homozygous for the same allele. (11.1)

follicle one of many like cells held in the ovaries. Each one of these cells contains an ovum that will develop. (6.3)

follicular stage the first stage of the menstrual cycle, during which increased levels of FSH stimulate the follicles to release increased quantities of estrogen into the bloodstream. The endometrium thickens and the hypothalamus releases LH. (6.3)

food chain the transfer of nutrients from one trophic level to the next through a trophic structure. (13.2)

food web an interconnection of several possible food chains in an ecosystem. (13.2)

fossil fuels oil, gas, and other hydrocarbon fuels that were produced over millions of years, during which time large amounts of organic detritus built up, was buried, and was compressed. (13.4)

fossil record remains and traces of past life found in sedimentary rock, which has layers that correspond to time periods. The fossil record reveals the history of life on earth and the kinds of organisms that were alive in the past. (10.3)

founder effect cause of genetic drift due to a small group of individuals colonizing a new area; the small group probably will not contain all the genes represented in the parent population. (11.3)

fovea centralis concentration of cones on the retina located directly behind the centre of the lens. (5.3)

frameshift mutation permanent change in the genetic material of a cell caused by the insertion or deletion of one or two nucleotides within a sequence of codons. (9.1)

frequency the number of occurrences of a particular allele in a population divided by the total number of alleles in the population. (11.1)

frontal lobe the part of the brain that is involved in the control of motor area muscles and integration of information to help reasoning processes. (5.1)

functional groups specific groups of bonded atoms attached to a molecule such as a protein. (1.2)

GABA gamma aminobutyric acid; the most common inhibitory neurotransmitter of the brain. (5.2)

gel electrophoresis method in which molecules travel through a gel subjected to an electrical current. It is used to separate molecules according to mass and charge, and enables fragments of DNA to be separated for analysis. (9.2)

gene specific sequence of DNA that has the potential to be expressed; a discrete unit of hereditary information. (7.4)

gene flow movement of new genes into a gene pool. This movement can reduce differences between populations that were caused by isolation and genetic drift. (11.3)

gene pool the total of all the genes in a population at any one time. (10.1)

generalist organism that is flexible in its requirements and can be found in a wide variety of habitats. (13.1)

gene therapy the process of changing the function of genes in order to treat genetic disorders. (9.4)

genetic drift the change of frequencies of particular alleles in a small population, caused by chance alone. The change can be drastic. (11.3)

genetic structure the frequencies of all the alleles and genotypes in a population. (11.1)

genome the sum of all of the DNA carried in an organism's cells. (7.4)

genotype genetic make-up; remains constant throughout an individual's life. (11.1)

geographical barriers features such as rivers or mountains that prohibit interbreeding by keeping populations physically separated. (12.2)

germ cell reproductive cell (egg or sperm) produced through meiosis. (Appendix 4)

germ cell mutation permanent change in genetic material in a reproductive cell of an organism. (9.1)

germ-line therapy gene therapy that alters the genetic information contained in egg and sperm cells. In theory, this kind of therapy could eliminate inherited genetic disorders. Research into this kind of therapy is currently banned in Canada and in many other countries. (9.4)

gigantism excess of HGH during childhood causes elongation of the long bones, resulting in a very tall stature. This excess is often caused by a tumour in the pituitary gland. (6.1)

glaucoma build-up of the aqueous humour in the eye that irreversibly damages the nerve fibres responsible for peripheral vision. (5.3)

global warming rise in average temperatures on Earth. (15.3)

glomerular filtration the movement of water, salts, nutrients, and waste molecules from the blood to the kidney. (4.2)

glomerulus the blood vessel inside the Bowman's capsule, from which the water, salts, nutrient molecules, and waste molecules leave the blood to be filtered by the kidney. (4.2)

glucagon hormone produced in the islets of Langerhans in the pancreas that raises blood glucose by stimulating the breakdown of glycogen. (4.3)

glutamate neurotransmitter in the cerebral cortex that accounts for 75 percent of all excitatory transmissions in the brain. (5.2)

glycolysis metabolic pathway in which glucose molecules are split into pyruvate molecules. (3.1)

goiter swelling of the thyroid gland caused by insufficient dietary iodine. (6.1)

gradualism the theory that geographical change occurs through slow but steady processes. (10.2) Also, in evolution, the theory that change occurs slowly and steadily within a lineage, before and after a divergence. Large changes occur by means of the accumulation of many small changes. (12.3)

Grave's disease *see* hyperthyroidism. (6.1)

greenhouse effect natural phenomenon that traps heat in the atmosphere near the surface of Earth. (15.3)

Green Revolution international effort, started in the 1950s, to transfer the farming methods and crop varieties used successfully in the more developed countries to less developed nations. (15.2)

growth factors protein molecules that prolong the effects of HGH on bone and cartilage tissues. (6.1)

growth rate (*r*) the number of births minus the number of deaths over a given period of time. (14.2)

guanine (G) one of four nitrogen-containing bases in a nucleotide that make up DNA and RNA. Its name derives from the fact that it occurs in high concentration in the animal excrement guano. (7.1)

habitat place or area with a specific set of abiotic and biotic characteristics. Organisms have adaptations to the habitats where they typically live. (13.1)

Hardy-Weinberg equilibrium condition of a population in which genotypes of members maintain the same proportions through several generations. (11.2)

Hardy-Weinberg principle theory stating that in the absence of forces that change the proportions of the alleles at a given locus, the original genotype proportions will remain constant from one generation to the next, in a large population with random mating. (11.2)

helicases set of enzymes that cleave and unravel short segments of DNA just ahead of the replicating fork during DNA replication. (7.3)

helper T cell lymphocyte that gives off chemicals that stimulate other macrophages, B cells, and other T cells during an immune response. (4.4)

heterozygous describes an individual with two different alleles at a locus. (11.1)

histamines chemicals that increase the permeability of blood vessels, causing reddening and swelling of an area. (4.4)

histone complex of small, very basic polypeptides that form the core of nucleosomes, around which DNA is wrapped. (7.2)

homeostasis the body's maintenance of a relatively stable internal physiological environment. (4.1)

homologous structures body parts in different species that have the same evolutionary origin, but that have different structures and functions (e.g., wing and arm bones). (10.3)

homozygous describes an individual with two alleles at one locus that are identical. (11.1)

hormone chemical signal that is sent to many parts of the body; examples are adrenaline and noradrenaline from the neurons of the adrenal gland. (5.2, 6.1)

hormone replacement therapy administration of low levels of estrogen and/or progesterone to alleviate the symptoms of menopause. (6.3)

host the larger and more independent member in a symbiotic relationship. (14.3)

Human Genome Project (HGP) joint effort of thousands of researchers from laboratories worldwide that determined the sequence of the three billion base pairs that make up the human genome. A complete draft of the human genome was first published in February 2001. Among the HGP's immediate findings was that the DNA of all humans is more than 99.9 percent identical. (9.2)

human growth hormone (HGH) small protein molecule that spurs body growth by increasing intestinal absorption of calcium, increasing cell division and development, and stimulating protein synthesis and lipid metabolism. The protein's action prompts the body to use lipids for energy instead of proteins and carbohydrates. (6.1)

hydrogen bonds weak attractions between polar molecules that contain hydrogen atoms bonded to the more electronegative atoms oxygen, nitrogen, or fluorine. (1.1)

hydrolysis type of reaction in a molecule is split, or lysed, by the addition of water. (1.3)

hydrolytic enzymes enzymes that catalyze the addition of water in reactions. (2.2)

hydrophilic "water-loving"; describes molecules that interact with water. (1.1)

hydrophobic "water-hating"; describes molecules that do not interact with water. (1.1)

hydrothermal vent fissures in Earth's crust that release hot water and gases. (2.2)

hyperopia far-sightedness or difficulty seeing things that are nearby. The condition is caused by a too-short eyeball or weak ciliary muscles. (5.3)

hyperthermia heat exhaustion and unusually high body temperature. The condition is caused by the body's inability to release enough excess heat by vasodilation; can be fatal. (4.1)

hyperthyroidism (Grave's disease) an autoimmune disorder caused by an excess of thyroxine. Antibodies attach to TSH receptors on thyroid cells, causing the thyroid gland to produce too much thyroxine. Typically, the production of excess thyroxine results in increased metabolism, excessive heat production, dilation of blood vessels, and weight loss despite increased appetite. (6.1)

hypothalamus the part of the brain that acts as the main control centre for the autonomic nervous system, re-establishes homeostasis, and controls the endocrine hormone system. (5.1)

hypothermia condition in which the body no longer has the energy to shiver, extremities freeze, and blood flow to the brain slows, resulting in impaired judgment, sleepiness, eventual loss of consciousness, and even death. (4.1)

hypothyroidism deficiency in thyroxine (caused by iodine deficiency or an autoimmune disease) causes continual production of TSH and cell division in thyroid tissue. Typically, this deficiency results in reduced metabolism, reduced tolerance of cold, decreased heart rate, and weight gain despite decreased appetite. The condition is also known as myxedema. (6.1)

immigration movement of individuals into a region from elsewhere; a factor that increases a population. (14.2)

immunity the ability to resist a disease after having been exposed to it in the past. (4.4)

inbreeding mating between closely related partners. This type of mating results in increased numbers of homozygous members of the population. (11.3)

incomplete dominance blending of the traits of two different alleles at one locus that occurs when neither allele is dominant. (11.1)

induced in mutations, describes a permanent change in the genetic material of a cell caused by a mutagen outside the cell. (9.1)

induced fit change in shape of the active site of an enzyme to accommodate the substrate and to facilitate a chemical reaction. (2.2)

inducer substance, such as a lactose molecule, that stops the action of a repressor. RNA polymerase can then bind to the promoter sequence and begin transcription. (8.4)

Industrial Revolution period of transition in lifestyle during the eighteenth century during which people shifted from working in traditional agricultural jobs and making goods by hand to working in factories and making mass-produced goods. (15.1)

inheritance of acquired characteristics Lamarck's theory that characteristics acquired during an organism's lifetime could be passed to its offspring. (10.2)

inhibin hormone that acts on the hypothalamus to stimulate the production of the releasing factors that trigger release of FSH. The interaction of FSH and inhibin controls the rate of formation of sperm. (6.3)

inhibitors chemicals that bind to specific enzymes. (2.2)

inhibitory response process during which an inhibitory neurotransmitter is sent through the synapse from the presynaptic neuron to the postsynaptic neuron. This transmission raises the threshold of the all-or-none response for that postsynaptic neuron, making it less likely that the nerve impulse will be transmitted. The response is achieved by making the postsynaptic neuron more negatively charged by opening chloride channels. (5.2)

initiation the first step in transcription, in which the position along the original DNA strand where transcription is to begin is determined. (8.2) Also, in translation, the name given to the first step, in which a special initiator tRNA molecule binds to a ribosome-mRNA complex to begin translation. (8.3)

inorganic term generally used to describe substances that do not contain carbon, such as water and sodium chloride. (1.2)

***in situ* therapy** in gene therapy, injecting the vector directly into the tissue containing target cells. For example, a vector containing a gene that combats cancer can be injected directly into a tumour. (9.4)

insulin hormone that is produced in the islets of Langerhans in the pancreas and that lowers blood sugar levels. (4.3)

integrator in the nervous system, describes the function of the hypothalamus in the brain that receives messages from receptors and sends instructions to effectors. (4.1)

interspecific competition competition between or among members of different species for resources, such as food. (14.3)

interstitial cells cells surrounding the seminiferous tubules in the testes. (6.3)

interstitial fluid fluid in the pores between cells. (4.1)

intraspecific competition competition between or among members of the same species for resources, such as food. (14.3)

intron intervening non-coding sequence in a eukaryote gene. (7.4)

***in vivo* therapy** in gene therapy, a process that relies on a vector that can be taken up by the body through injection or inhalation. Once inside the body, the vector targets specific cells and inserts new genes directly into these cells. So far, these treatments have been unsuccessful. (9.4)

ion charged particle formed when an atom loses or gains electrons, so that the atom does not have an equal number of protons or electrons. (1.1)

ionic bond chemical bond formed by the transfer of electrons between atoms and the attraction between the oppositely charged ions that result. (1.1)

iris the muscle that adjusts the pupil to regulate the amount of light that enters the eye. (5.3)

islets of Langerhans groups of cells in the pancreas that produce insulin and glucagon. (4.3)

isomers two or more molecules with the same molecular formula but different structures. (1.2)

isotonic two solutions (usually, on either side of a membrane) that have the same concentrations of solutes and water. (4.2)

isotopes atoms of the same element that have different numbers of neutrons in their nuclei. (1.1)

J-shaped curve the shape of a line describing exponential growth of a population. (14.2)

kidney organ that filters blood to remove nitrogenous wastes and adjust the concentrations of salts in the blood. (4.2)

kinetic energy the energy associated with movement. (2.1)

***K*-selection** the type of selection at work in populations at their carrying capacities, in which individuals have few, large offspring and care for them. (14.4)

Krebs cycle metabolic pathway consisting of a series that completes the breakdown of glucose in the mitochondrion. (3.1)

lagging strand in DNA replication, the strand that is replicated by splicing together Okazaki fragments in the 5′ to 3′ direction (in the direction opposite to the movement of the replication fork). (7.3)

laws of thermodynamics set of laws governing the relationship between heat and other forms of energy. For example, the first law states that energy can be neither created nor destroyed, but can be transformed from one form of energy to another. The second law states that energy cannot be transformed from one form to another without a loss of energy. In other words, every energy transformation increases the total entropy of the universe. (2.1)

leading strand in DNA replication, the strand that is replicated continuously in the 5′ to 3′ direction (in the same direction as the movement of the replication fork). (7.3)

lens the clear, flexible part of the eye that adjusts its shape according to focal distance. (5.3)

less industrialized nation country with relatively few industries and a relatively low standard of living, compared to more industrialized nations. (15.1)

life history schedule of certain important events that occur during the life of an organism (e.g., age of sexual maturity). (14.4)

life table a summary of the relationship of survivorship, mortality, fecundity, and age, for a given population. (14.4)

lipids fats and phospholipids, such as those found in the cellular membrane. (1.3)

locus the location of a gene on a chromosome. The plural form is loci. (11.1)

logistic growth the type of population change that occurs in an environment where resources are limited. (14.2)

loop of Henle the long loop of the nephron that extends into the medulla of the kidney between the proximal tubule and the distal tubule. (4.2)

luteal stage the stage of the menstrual cycle that begins with ovulation. (6.3)

lymphocyte specialized white blood cells (B cells and T cells) formed in the bone marrow. B cells mature in the bone marrow, while T cells mature in the thymus gland. (4.4)

macro-evolution evolution on a large scale, such as the evolution of new species from a common ancestor. (11.1)

macromolecules large molecules that are often made of distinct smaller units. (1.2)

macrophage large cell that engulfs bacteria and detritus through phagocytosis as an immune response. (4.4)

mass extinction event in which many species become extinct at once. (12.3)

mass number the total number of protons and neutrons in the nucleus of one atom of an element. (1.1)

mechanoreceptor in the ear, a sensory receptor that translates the movement of air in the ear into a series of nerve impulses that the brain interprets as sound. (5.3)

medulla (renal) the inner part of the kidney, where the loop of Henle and the collecting ducts of the nephrons are. (4.2)

medulla oblongata structure attached to the spinal cord at the base of the brain; has cardiac, vasomotor, respiratory, vomiting, coughing, hiccupping, and swallowing functions. Damage to this part of the brain is usually fatal. (5.1)

meiosis type of cell division that occurs only in reproductive organs producing reproductive cells called gametes. (6.3)

memory T cell lymphocytes that remain in the bloodstream to promote a faster response if the same antigen appears in the body again. (4.4)

menopause period in a woman's life during which a decrease in estrogen and progesterone results in an end of menstrual cycles. The period usually occurs during middle age. (6.3)

menstrual cycle in a human female, period of 20–45 days during which hormones stimulate the development of the uterine lining, and an egg is developed and released from an ovary. If the egg is not fertilized, the uterine lining is shed, and the cycle begins again. (6.3)

menstruation the last stage of the menstrual cycle, in which the endometrium disintegrates and is expelled from the uterus. (6.3)

messenger RNA (mRNA) strand of RNA that carries genetic information from DNA to the protein synthesis machinery of the cell during transcription. (8.2)

metabolic pathway series of linked chemical reactions that take place in the cell. (2.2)

metabolism the collective name for all the chemical reactions that occur inside the cell. (1.1, 2.1)

micro-evolution the change in gene frequencies in a population over time. Evolution within a species, or evolution on a small scale. (11.1)

mimicry structural adaptation that allows one species to resemble another species or part of another species. (12.1)

mis-sense mutation permanent change in the genetic material of a cell that results in slightly altered but still functional proteins. (9.1)

modern synthesis the modern combination of Mendel's and Darwin's theories on evolution. (11.1)

monocyte white blood cell from which neutrophils and macrophages are derived. Neutrophils and macrophages assist in immune responses by phagocytosis. (4.4)

monomers small repeating units that join together to make up polymers. (1.2)

monosaccharide a carbohydrate containing three to seven carbon atoms per molecule. (1.3)

more industrialized nation country with relatively many industries requiring skilled workers, and with a relatively high standard of living. (15.1)

morphological species concept classification of organisms into species by measurable physical characteristics. (12.2)

mortality the proportion of individuals in a population that typically die at a given age over a given period of time. (14.4)

motor nerves nerves that transmit commands from the central nervous system to the muscles. (5.1)

multi-gene family in a eukaryotic cell, from a few hundred to hundreds of thousands of copies of the same or very similar genes. Members of these families may be clustered together on the same chromosome, or they may be distributed among a number of different chromosomes. (7.4)

mutagen substance or event that increases the rate of mutation in an organism. Mutagens can be physical or chemical. (9.1)

mutation permanent change in the genetic material of an organism. Mutations provide the genetic variation within a population. (9.1, 11.1)

mutualism type of symbiotic relationship in which both species benefit from the relationship and growth in one population. (14.3)

myelin sheath the fatty layer around the axon of a nerve cell, composed of Schwann cells. (5.2)

myopia near-sightedness or difficulty seeing things that are far away. It is caused by a too-long eyeball or too-strong ciliary muscles. (5.3)

NADH dehydrogenase complex a multienzyme complex that oxidizes the coenzymes NADH and $FADH_2$. (3.2)

natural killer cells white blood cells that devour any of the body's own cells that have become cancerous or infected. (4.4)

natural selection process whereby the characteristics of a population of organisms change because individuals with certain heritable traits survive specific local environmental conditions. (10.1)

negative feedback loop homeostatic mechanism that detects and reverses deviations from normal homeostasis levels. The mechanism consists of a receptor, an integrator, and an effector. (4.1)

negative gene regulation the process of shutting off transcription by the use of a regulatory site. A protein molecule interacts directly with the genome to turn off gene expression. (8.4)

nephric filtrate urine. About twenty percent of the blood plasma that enters the kidney becomes nephric filtrate. (4.2)

nephron there are about 1 million of these filters in each kidney, each filter consisting of a Bowman's capsule, the glomerulus, a proximal tubule, the loop of Henle, the distal tubule, and a collecting duct. (4.2)

nerve message pathway of the nervous system, made up of many neurons connected together. (5.1)

neuron the nerve cell that is the structural and functional unit of the nervous system, consisting of a nucleus, a cell body, dendrites, axons, and a myelin sheath. (5.1)

neurotransmitters chemicals that are secreted by neurons to stimulate motor neurons and central nervous system neurons. (5.2)

neutral atoms atoms that have the same number of protons and electrons, and that therefore have no overall charge. (1.2)

neutralization chemical change in which one compound in solution (a base) acquires a hydrogen cation from another compound (an acid). (1.3)

neutrons uncharged subatomic particles, found in the nuclei of most atoms. The particles have about the same mass as protons. (1.1)

neutrophils small white blood cells that engulf bacteria through phagocytosis as an immediate immune response. (4.4)

nitrogenous containing nitrogen. (7.1)

node of Ranvier the gap between Schwann cells around the axon of a nerve cell. The membrane of the axon is exposed and nerve impulses jump from one node of Ranvier to the next. (5.2)

non-competitive inhibition binding of an inhibitor molecule an enzyme at the allosteric site. (2.2)

non-cyclic photophosphorylation addition of a phosphate group to ADP to form ATP. This process uses energy from an electron transport chain in which the electrons are not recycled in the photosystem. (3.3)

non-overlapping generations populations in which parents and offspring are never present in the population at the same time. (14.2)

non-polar covalent bond chemical bond in which electrons are shared equally between two bonded atoms with the same electronegativity. (1.1)

non-random mating any situation in which individuals do not choose mates on a random basis, such as mating based on proximity, relatedness, or similarity of phenotype. (11.3)

non-renewable resource resource, such as petroleum, that cannot be replenished once it is used up. (15.3)

nonsense mutation permanent change in the genetic material of a cell that renders a gene unable to code for any functional polypeptide product. (9.1)

non-specific defences the ability to resist a disease, even if there has been no prior exposure to it. (4.4)

non-steroid hormones fat-insoluble hormones composed proteins, peptides, or amino acids. (6.1)

noradrenaline *see* norepinephrine. (5.2)

norepinephrine the primary neurotransmitter of the sympathetic nervous system. Also known as noradrenaline. (5.2)

nucleic acid a polymer of nucleotides, such as DNA and RNA. (1.3)

nucleic acid hybridization construction of a probe made of RNA or single-stranded DNA that is possible when at least part of the nucleic acid sequence of a gene is known. The probe consists of a nucleic acid sequence complementary to the known gene sequence, along with a radioactive or fluorescent tag. (9.3)

nucleoid specific region in the nuclear zone of the cell with an arrangement of DNA that prokaryotic cells use to pack their genetic material tightly. (7.2)

nucleosome the bead-like structural unit of chromosomes, composed of a short segment of DNA (about 200 base pairs) wrapped twice around a cluster of eight histone molecules. (7.2)

nucleotides the monomers of nucleic acids. Each nucleotide is composed of a five-carbon sugar, a phosphate group, and one of four nitrogen-containing bases. The bases found in DNA nucleotides are adenine (A), guanine (G), cytosine (C), and thymine (T). In RNA, uracil (U) is found instead of thymine. (7.1)

nucleus the small, dense core composed of protons and neutrons that is found at the centre of an atom. (1.1)

occipital lobe the part of the brain that receives optical information. (5.1)

Okazaki fragments short nucleotide fragments used during DNA replication of the lagging strand. They are made by DNA polymerase working in the direction opposite to the direction of movement of the replication fork, after which they are spliced together. (7.3)

open system system in which substances and/or energy can be exchanged with the surroundings. (2.1)

operator DNA sequence located between the promoter sequences and gene sequences that governs whether RNA polymerase can bind to the promotor sequences to begin transcription of the genes. (8.4)

operon stretch of DNA, common in prokaryotes, that contains one or more genes involved in a metabolic pathway along with a regulatory sequence called an operator. (8.4)

opportunistic life history *r*-selected reproduction in which individuals quickly produce many offspring when environmental conditions are favourable, but may quickly die off if environmental conditions become unfavourable. (14.4)

organic term generally used to describe substances made up of carbon-based molecules, such as carbohydrates and proteins. (1.2)

organic acids carbon-based molecules that are acidic, such as acetic acid and amino acids. (1.3)

organic bases carbon-based molecules that are basic, such as purines and pyrimidines. (1.3)

ovaries two glands that are suspended in the abdominal cavity and produce eggs (ova). (6.3)

overlapping generations populations in which parents do not die when they reproduce, so that two or more generations may be alive and reproducing in a population at the same time. (14.2)

ovulation the releasing of an egg from an ovary. (6.3)

oxidation the loss of electrons in a reaction. (1.3)

oxidative decarboxylation the process of removing a carbon dioxide from pyruvate to form an acetyl group that then combines with coenzyme A. (3.1)

oxidative enzyme enzymes that catalyze redox reactions. (2.2)

oxidative phosphorylation the formation of ATP using energy obtained from redox reactions in the electron transport chain. (3.2)

oxytocin hormone that triggers muscle contractions during childbirth and the release of milk from the breasts. (6.1)

ozone depletion reduction of the ozone layer. (15.3)

paleontology the study of fossils. (10.2)

pancreas a small gland near the small intestine, which is made of two types of tissues that independently function as exocrine and endocrine glands. As an exocrine organ, the pancreas secretes digestive enzymes into the duodenum. As an endocrine organ, the pancreas produces insulin and glucagons. (6.1)

parasitism type of symbiotic relationship in which a symbiont lives off and harms the host. (14.3)

parasympathetic nervous system counteracts the sympathetic nervous system to slow heart and breathing rates and relax muscles. (5.1)

parathyroid hormone (PTH) a hormone produced in the parathyroid glands, which regulates the calcium levels in the blood. Works in conjunction with the hormone calcitonin. (6.1)

parietal lobe the part of the brain that receives sensory information from the tastebuds, skin, and skeletal muscles. (5.1)

pathogen disease-causing agent (e.g., a bacterium or virus). (4.4)

pelvis (renal) part of the kidney where urine accumulates before moving through the ureters. (4.2)

PEP phosphenolpyruvate. (3.1)

peptide bonds the bonds that link amino acids together in protein molecules. (1.3)

peripheral nervous system (PNS) the nerves that enter and leave the brain and spinal cord (the CNS). It consists of the autonomic nervous system and the somatic nervous system. (5.1)

perpetual resource unlimited resource such as sunlight, wind, or tides. (15.3)

PGA 3-phosophoglycerate. (3.1)

PGAL glyceraldehyde-3-phosphate. (3.1)

PGAP 1-3-bisphosphoglycerate. (3.1, 3.3)

phagocytosis process in which a cell swallows and devours another cell, bacterium, or detritus. (4.4)

phenotype the physical and physiological traits of an organism. (11.1)

phospholipid bilayer the double layer of phospholipid molecules that line up tail to tail to create a hydrophilic cell membrane. (1.3)

phosphorylation the addition of one or more phosphate groups to a molecule. (3.2)

photochemical smog brown or grey haze of ozone and various nitrates formed by the reaction of nitrogen oxides and hydrocarbons in the atmosphere in the presence of sunlight and water vapour. (15.3)

photolysis the splitting of water molecules by a Z enzyme to release electrons and hydrogen ions (H^+) that are then used in a photosystem. Oxygen is produced in this process. (3.3)

photophosphorylation the process of forming ATP from ADP and an inorganic phosphate group by means of chemiosmosis through the thylakoid membrane as part of the light-dependent reaction of photosynthesis. (3.3)

photorespiration the oxidation of RuBP (ribulose bisphosphate) by RuBP carboxylase. (3.3)

photosynthesis the use of energy from light by cells to produce carbohydrates (food). (3.3)

photosynthetic unit collection of several hundred antenna pigment molecules together with the reaction centre. (3.3)

photosystem network of chlorophyll molecules that absorb light energy. (3.3)

phylogenetic tree pattern of descent. (10.3)

physical mutagen agent (e.g., X rays) that can forcibly break a nucleotide sequence, causing random changes in one or both strands of a DNA molecule. (9.1)

physiological adaptations adaptations associated with functions in organisms (e.g., enzymes for blood clotting, chemical defenses of plants). (12.1)

pineal gland a small gland located in the brain, which produces melatonin. (6.1)

pioneer species organisms, such as lichens, that are the very first to arrive in a barren landscape, and that change the landscape dramatically by their actions. (13.1)

pituitary dwarfism insufficient HGH production during childhood results in abnormally short stature. (6.1)

pituitary gland sometimes called the master gland. The gland is attached to the hypothalamus in the brain and produces hormones (e.g., TSH) that control many of the endocrine glands. (5.1)

plasmids small, circular double-stranded DNA molecules floating free in the cytoplasm of a prokaryotic cell. (7.2)

point mutation chemical change that affects one or just a few nucleotides. (9.1)

polar covalent bond chemical bond in which electrons are shared unequally between two bonded atoms with different electronegativities. (1.1)

polar molecule molecule that has an unequal distribution of charge and a net dipole, as a result of the polar bonds within the molecule and the shape of the molecule. (1.1)

pollutant chemical that adversely affects living things. (15.3)

poly-A tail long series of A nucleotides added to the 3′ end of a pre-mRNA molecule. (8.2)

polymerase chain reaction (PCR) almost entirely automated method of replicating DNA that allows researchers to target and amplify a very specific sequence within a DNA sample. The method is used to copy and analyze minute amounts of DNA from mummies or fossils. (9.2, 11.1)

polymers large molecules made of many repeating smaller molecules called monomers. (1.2)

polymorphic genetic loci that have more than one allele. (11.1)

polypeptides long chains of amino acids (proteins). (1.3)

polyploidy mutant condition in which errors in cell division result in extra sets of chromosomes. (12.3)

polyribosome a cluster of ribosomes bound to the same mRNA molecule. Each ribosome produces a copy of the same polypeptide. (8.3)

polysaccharide storage carbohydrate polymer made of a long chain of monosaccharides. (1.3)

population group of individuals of the same species living in the same geographical area. (11.1, 13.1)

porin the outer membrane of the mitochondrion. (3.2)

positive gene regulation the process of turning on transcription by the use of a regulatory site. The direct interaction of a protein molecule with the genome increases the rate of gene expression. (8.4)

postsynaptic neuron the state of a neuron after a synapse, i.e., after the neuron receives and transmits a stimulus (5.2)

post-zygotic barrier a factor that prevents hybrid zygotes from developing into normal, fertile individuals. Includes hybrid inviability, hybrid sterility, and hybrid breakdown. (12.2)

potential energy the energy stored by matter as a result of its position or arrangement in space. (2.1)

precursor mRNA (pre-mRNA) messenger RNA that is released when transcription ends. The RNA undergoes several changes before being transported out of the nucleus. (8.2)

presynaptic neuron the state of a neuron before a synapse, i.e., before the neuron carries a wave of polarization to a synapse leading to another nerve cell. (5.2)

pre-zygotic barrier factor that prevents mating between species, or that prevents fertilization if individuals from different species do attempt to mate. Pre-zygotic barriers include behavioral isolation, habitat isolation, temporal isolation, mechanical isolation, and gametic isolation. (12.2)

primary consumer herbivore, or, an organism that eats primary producers. This organism occupies the second trophic level. (13.2)

primary producer autotroph, or, an organism that is at the first trophic level. They produce organic molecules and the energy required for all the other trophic levels. (13.2)

primary productivity the amount of light energy that autotrophs in an ecosystem convert to chemical energy during a specific period of time. The amount is usually measured in terms of energy per area per year ($J/m^2/a$). (13.3)

primary structure the specific amino acid sequence of a protein. (Appendix 5)

primary succession the development of a new community in a previously barren area, such as after a volcanic eruption, or on a bare rock mountaintop. (13.1)

primase in DNA replication, an enzyme that forms a small strand of RNA that is complementary to a DNA template (a primer). The primer is necessary to begin the replication. (7.3)

primer short strand of RNA that works as a starting point for the attachment of new nucleotides during DNA replication. (7.3)

processing the final changes made to an mRNA molecule during transcription in a eukaryotic cell before the mRNA is transported from the nucleus to the cytoplasm. (8.2)

prodrug inactive drug that converts into its active form in the body only by metabolic activity. (2.2)

progesterone the female hormone that is released after an egg is released from the ovary. (6.3)

prolactin non-steroid protein hormone produced by the anterior pituitary that stimulates mammary gland tissue growth and milk production. (6.1)

promoter sequence particular nucleotide sequence on a DNA molecule that provides a binding site for RNA polymerase. (8.2)

prostaglandin E2 (PGE2) compound that increases the sensitivity of nerves to pain. (2.2)

prostate gland in human males, a gland that secretes an alkaline fluid that neutralizes the acidity of the female reproductive tract. (6.3)

proton positively charged subatomic particle, found in the nuclei of an atom. The number of protons in an atom's nucleus determines the identity of the atom. (1.1)

proton pumps multienzyme complexes that move hydrogen ions (H^+) from one side of a membrane to the other. (3.2)

proximal tubule in a kidney, the tube between the Bowman's capsule and the loop of Henle. (4.2)

pseudogenes in a multi-gene family, genes that are nearly identical to functional genes, but that have mutated until they are no longer functional. Pseudogenes are never expressed during the life cycle of the cell. (7.4)

puberty the stage of life during which reproductive hormones begin to be formed. Reproductive development occurs during this stage. (6.3)

punctuated equilibrium model that suggests that evolutionary history consists of long periods of stasis (stable equilibrium), punctuated by periods of divergence. (12.3)

pupil the aperture in the middle of the iris of the eye. The size of the aperture can be adjusted to control the amount of light. (5.3)

pupillary reflex the reflex contraction of the pupil of the eye that occurs when the eye is exposed to bright light. (5.3)

purine nitrogenous compound that has a double ring structure. The nucleotide bases adenine and guanine are derived from purines and always bond with pyrimidines in DNA. (7.2)

pyramid of numbers graphical representation of the number of organisms at each trophic level based on the decreasing amount of available energy through the trophic structure. (13.3)

pyramid of productivity graphical representation of the pattern of energy loss through a trophic structure. (13.3)

pyrimidine nitrogenous compound that has a single ring structure. The nucleotide bases thymine, cytosine, and uracil are derived from pyrimidines and always bond with purines in DNA. (7.2)

pyruvate three-carbon compound formed from the breakdown of glucose during glycolysis. (3.1)

pyruvate dehydrogenase complex the multienzyme complex responsible for the oxidative decarboxylation of pyruvate in the transition reaction. (3.2)

Q

quadrat area of determined size that is marked out for the purpose of sampling a population. (14.1)

quaternary structure protein shape formed by the intertwining of two or more tertiary-structure proteins. (Appendix 5)

R

radioactive isotopes unstable isotopes that break down (decay). (1.2)

range the geographical area where a population or species is found. The limits of the area are usually defined by habitat requirements. (13.1)

re-absorption in the kidneys, movement of water and nutrients back into the bloodstream from the tubes of the nephrons. (4.2)

reaction centre specific chlorophyll a molecule in a photosystem that transfers an electron to an electron acceptor. (3.3)

reaction rate the rate at which a reaction occurs, measured in terms of reactant used or product formed per unit time. (2.2)

reading frame collectively, codons within a nucleotide sequence that are translated or transcribed. (8.1)

recessive allele allele that is not expressed in an organism's phenotype in a heterozygous pairing with a dominant allele. (11.1)

recombinant DNA segments of DNA from two different species that are joined in the laboratory to form a single molecule of DNA. (9.2)

recombination repair use of a homologous portion of a sister chromatid by enzymes in the cell as a template to reconstruct damaged DNA. (9.1)

redox reaction reaction in which both reduction and oxidation occur. (1.3)

reduction the gaining of electrons by one substance in a redox reaction. (1.3)

reflex a very quick, involuntary nerve and muscle reaction to an outside stimulus. (5.1)

reflex arc the nerve path that leads from a stimulus to a reflex action. (5.1)

refractory period the brief time between the triggering of an impulse along an axon and the axon's readiness for the next impulse. During this brief time, the axon cannot transmit an impulse. (5.2)

regulatory sequence a strand of DNA that helps to determine when various genetic processes are activated. (7.4)

release factor a protein that binds to the A site in a ribosome and cleaves the completed polypeptide from the tRNA in the P site. This process causes translation to terminate. (8.3)

renewable resource natural resource, such as trees, that can be replaced when depleted. (15.3)

repetitive sequence DNA in eukaryotic cells, regions containing short sequences of nucleotides repeated thousands or millions of times. (7.4)

replacement fertility rate the fertility rate required to achieve long-term zero population growth. (15.1)

replacement rate (*R*) the growth rate of a population that has non-overlapping generations. (14.2)

replication forks during DNA replication, points at which the DNA helix unwind and new strands develop. (7.3)

replication machine complex involving dozens of different enzymes and other proteins working closely together in the process of DNA replication. (7.3)

replication origin specific nucleotide sequence of 100–200 base pairs in bacterial DNA. It is recognized by a group of enzymes that bind to the sequence and separate the two DNA strands to open a replication bubble. (7.3)

repressor protein that binds to the operator to make it impossible for RNA polymerase to bind to the promoter, thus preventing the *lac* operon genes from being transcribed. (8.4)

reproductive cloning the development of a cloned human embryo for the purpose of creating a cloned human being. (9.4)

respiratory centre the part of the medulla oblongata that controls the rate and depth of breathing. (5.1)

resting potential the difference in charge from the inside to the outside of a cell at rest. It is approximately –70 mV. (5.2)

restriction endonucleases family of enzymes made by prokaryotic organisms; these enzymes recognize a specific short sequence of nucleotides on a strand of DNA and cut the strand at a particular point within that sequence. (9.2)

restriction enzymes enzymes that cut the DNA strand during the DNA fingerprinting process. (2.2)

restriction fragment small segment of DNA cut from the DNA molecule by a restriction endonuclease enzyme. (9.2)

restriction site specific location on a strand of DNA where a restriction endonuclease will cut the strand of DNA. (9.2)

retina the innermost layer of the eye. (5.3)

reverse mutation change in a the genetic material of a cell; the change restores the effect of a previous change in the genetic material of the cell. (9.1)

reverse transcriptase enzyme that creates a DNA strand complementary to the RNA strand. (9.3)

rhodopsin purple pigment in the optical receptors of the eye; it enables the rod and cone cells to function. (5.3)

ribonucleic acid (RNA) nucleic acid consisting of long chains of individual nucleotides, each of which is composed of the five-carbon sugar ribose, a phosphate group, and the nitrogenous base adenine, guanine, cytosine, or uracil. It assists DNA in controlling protein synthesis in cells. Originally called "ribose nucleic acid" by its discoverer, Phoebus Levene. (7.1)

ribose nucleic acid Phoebus Levene's name for ribonucleic acid (RNA). (7.1)

ribosomal RNA (rRNA) most common class of RNA molecules. During polypeptide synthesis, these RNA molecules supply the site on the ribosome where the polypeptide is assembled. (8.3)

ribosome the cell structure that brings together the mRNA strand, the aa-tRNAs, and the enzymes involved in building polypeptides in order to enable protein synthesis to occur within the cytoplasm. (8.3)

RNA *see* ribonucleic acid. (7.1)

RNA polymerase main enzyme that catalyzes the formation of RNA from a DNA template. (8.2)

rods photoreceptors in the eye; more sensitive to light than cones, but unable to distinguish colour. (5.3)

***r*-selection** natural selection that produces opportunistic life histories. (14.4)

RuBP ribulose bisphosphate. (3.3)

RuBP carboxylase enzyme in green plants that catalyzes the fixation of carbon into organic molecules. (3.3)

sample small portion of an entire population that is counted or estimated. (14.1)

Schwann cells insulating cells around the axon of a nerve cell. (5.2)

sclera the thick white outer layer that gives the eye its shape. (5.3)

secondary consumer carnivore or omnivore that eats mainly herbivores and occupies the third trophic level. (13.2)

secondary structure in protein, the shape resulting from hydrogen bonding between the N–H and C=O groups of the primary structure. (Appendix 5)

secondary succession the redevelopment of a previously existing community after a disturbance. (13.1)

selective pressure environmental condition or conditions that select for certain characteristics of individuals, and select against others. (10.1)

semen fluid consisting of sperm and fluid from the prostate and Cowper's glands. The fluid is ejaculated from the male penis into the female vagina during sexual intercourse. (6.3)

semi-conservative theory the theory that in the replication of DNA, the daughter DNA molecules are each made up of one parental strand and one new strand. This was the model suggested by James Watson and Francis Crick, who proposed the double helix for the molecular structure of DNA. (7.3)

seminal vesicles glands that produce a mucus-like fluid containing fructose as energy for sperm. (6.3)

seminiferous tubules tubes inside the testes. (6.3)

sense strand strand of nucleotides containing the instructions that direct protein synthesis. It is located within a stretch of DNA that includes a gene. (8.2)

sensory nerves nerves that carry impulses from the body's sense organs to the CNS. (5.1)

sensory receptors scattered throughout the body, cells or groups of cells that work continually to receive information about the body's internal conditions, such as temperature, pH, glucose levels, and blood pressure. (4.1)

seratonin organic compound formed from tryptophan and found in tissue, especially the brain, blood serum, and gastric mucous membranes. The compound is active in vasoconstriction, muscle stimulation, transmission of impulses between nerve cells, and regulation of cyclic processes. (5.2)

Sertoli cells cells in the seminiferous tubules that support, regulate, and nourish developing sperm. (6.3)

sexual dimorphism the difference between phenotypes of males and females of the same species. (11.3)

sexual selection selection for mating based, in general, on competition between males and choices made by females. (11.3)

silent mutation permanent change in the genetic material of a cell that has no effect on the metabolism of the cell. (9.1)

sink a substance or site that absorbs and holds materials, often for a long period. (15.2)

small nuclear RNA (snRNA) the type of RNA that is paired with proteins to make a spliceosome. (8.2)

sodium-potassium pump a transport structure made up of special carrier proteins. The "pump" maintains an imbalance of sodium and potassium in cells, particularly nerve and muscle cells. (2.3)

somatic cell any body cell except germ cells. (Appendix 4)

somatic cell mutation permanent change in the genetic material of a body cell, not including germs cells, during the lifetime of the organism. (9.1)

somatic gene therapy therapy aimed at correcting genetic disorders in somatic cells. (9.4)

somatic nervous system relays information to and from skin and skeletal muscles under conscious control. (5.1)

somatotropin growth hormone produced by all mammals. (9.3)

source a supply of specific materials. (15.2)

specialist organism that has very particular climate or habitat requirements. These organisms generally have low tolerance of change. (13.1)

speciation the formation of species. (12.2)

specific immune system wide variety of cells, developed through exposure to disease, that recognize certain foreign substances and act to neutralize or destroy them. (4.4)

spermatogenesis the formation of sperm. (6.3)

sphincters two rings of muscle that control the exit of the bladder. One is voluntarily controlled, and the other is involuntarily controlled by the brain. (4.2)

spliceosome molecule made of proteins and snRNA that removes long stretches of nucleotides before polypeptides are constructed. The spliceosome cleaves the pre-mRNA at the ends of each intron and then splices the remaining exons. (8.2)

spontaneous mutation permanent change in the genetic material of a cell as a result of the molecular interactions that take place naturally within the cell. (9.1)

S-shaped curve the shape of a line describing the growth of a population as the population reaches its carrying capacity. (14.2)

stabilizing selection natural selection that favours intermediate phenotypes and acts against extreme variants. (11.3)

stereoisomers two molecules that have the same types of bonds, but different arrangements in space. Geometrical isomers can have very different physical properties but the same chemical properties; optical isomers (enantiomers) are mirror images of each other that have the same chemical and physical properties, but enzymes and proteins on the cell membrane can distinguish between them. (1.2)

steroid hormones hormones produced from cholesterol by the endocrine system. (6.1)

sticky ends short sequence of unpaired nucleotides remaining on a single strand of DNA at each end of a restriction fragment, after an endonuclease makes a staggered cut at the restriction site. (9.2)

structural adaptations adaptations that affect the appearance, shape, or arrangement of particular physical features, include mimicry and cryptic coloration. (12.1)

structural isomers two or more compounds with the same atoms bonded differently. (1.2)

subatomic particles tiny particles, such as protons, neutrons, and electrons, that make up an atom. (1.1)

substrate a reactant in an enzymatic reaction. (2.2)

substrate-level phosphorylation the formation of ATP by transferring a phosphate group to ADP from a substrate. (3.1)

supercoiled a strand of DNA in prokaryotic cells coils into a closed loop, then twists in on itself like a necklace that is coiled into a series of small loops. Such a structure is said to be supercoiled. This arrangement helps the prokaryote keep its entire DNA in the nucleoid region. (7.2)

suppressor T cell lymphocyte that slows and stops the cellular immune response when the danger has passed. (4.4)

survivorship the proportion of individuals in a population that typically live to a given age over a given period of time. (14.4)

suspensory ligament one of the ligaments that attach the ciliary muscles to the lens of the eye. (5.3)

sustainable development the use of renewable resources in a way that meets the needs of the present without compromising the ability of future generations to meet their own needs. (15.3)

symbiont organism that depends on close interaction with another organism (a host) for survival. (14.3)

symbiotic relationship close interaction of two species in a way that is beneficial to both. (14.3)

sympathetic nervous system the network of nerves that controls involuntary muscle reactions and organs in times of stress. (5.1)

sympatric speciation the differentiation of populations within the same geographic area into species. (12.3)

synapse junction between a neuron and another neuron or muscle cell. (5.2)

synaptic vesicle specialized vacuole in the bulb-like ends of the axons of a nerve cell containing neurotransmitters that are released into the synapse when a nerve impulse is received. (5.2)

system in science, the name for the components of a process under study. (2.1)

telomere specialized nontranscribed structure, typically rich in G nucleotides, at the end of each chromosome. The erosion of telomeres protects against the loss of other genetic material during cell division. (7.3)

temporal lobe the part of the brain that receives auditory information. (5.1)

termination in transcription of DNA, step that ends of the copying process. (8.2)

tertiary consumer carnivore that eats mainly other carnivores and occupies the fourth trophic level. (13.2)

tertiary structure the folded, three-dimensional shape of a protein molecule caused by hydrophobic interactions with water and covalent and ionic bonding between different amino acids side chains. (Appendix 5)

testosterone steroid hormone that is responsible for the development of male secondary sexual characteristics. (6.3)

thalamus sensory relay centre of the brain that governs the flow of information from all other parts of the nervous system. (5.1)

therapeutic cloning the culturing of human cells for use in treating medical disorders. (9.4)

thermodynamics the study of transformations of energy. (2.1)

thylakoids photosynthetic membranes within chloroplasts. (3.3)

thymine (T) one of four nitrogenous bases in a nucleotide that make up DNA. Thymine is not found in RNA. (7.1)

thymosin hormone produced in the thymus gland that stimulates the production and maturation of lymphocytes into T cells. (6.1)

thymus gland lymphatic tissue located between the lobes of the lungs in the upper chest cavity; the gland produces thymosin and usually disappears after puberty. The gland secretes several hormones that function in the regulation of the immune system and provides a location for the production of T cell lymphocytes. (6.1)

thyroid gland gland located below the larynx, which produces the hormone thyroxine. (6.1)

Ti plasmid tumour-inducing bacterial plasmid that infects a host plant by incorporating a segment of its DNA into the DNA of the plant. (9.3)

tragedy of the commons the destruction of a shared resource by individual greed. (15.2)

transcription the first stage of gene expression in which a strand of mRNA is produced that is complementary to a segment of DNA. (8.1)

transcription factors regulator proteins necessary for transcription in eukaryotes. (8.4)

transect a very long, relatively narrow rectangular area marked out in a study area for the purpose of sampling a population. (14.1)

transfer RNA (tRNA) RNA molecules that serve to link each codon along an mRNA strand with its corresponding amino acid. (8.3)

transformation in evolution, the process by which one species becomes a different species, as a result of accumulated changes over long periods of time. (12.2)

transgenic organism produced by moving DNA from one organism to another in order to create a new combination. (9.3)

transition reaction the oxidation of pyruvate to an acetyl group, also called oxidative decarboxylation. (3.1)

transitional fossil fossil that shows intermediary links between groups of organisms, and shares characteristics common to both groups. (10.3)

translation the second stage of gene expression, in which a ribosome produces a polypeptide, using the coded mRNA instructions. (8.1)

translocation the third stage in transcription, in which the ribosome moves a distance of three nucleotides along the mRNA molecule, resulting in the ribosome assembly moving a distance of exactly one mRNA codon in the 5′–3′ direction. (8.3)

transposons segments of DNA that move randomly throughout the cell genome. Also called "jumping genes." (7.4)

triplet hypothesis the notion that the genetic code is composed of codons of three nucleotides each. (8.1)

trophic level the position of a species in a trophic structure for a community, according to what the species eats. The first trophic level is occupied by primary producers, the second trophic level is occupied by primary consumers, the third trophic level is occupied by secondary consumers, and the fourth trophic level is occupied by tertiary consumers. (13.2)

trophic structure feeding relationships among members of a community or ecosystem. (13.2)

tubular secretion a process that uses active transport to draw substances, such as hydrogen ions, creatinine, and drugs (e.g., penicillin) out of the blood and into the filtrate of the kidney. (4.2)

U

uniformitarianism Charles Lyell's theory that geological processes repeat themselves. The key to understanding geological processes of the past lies in understanding geological processes of the present. (10.2)

uracil (U) one of four nitrogenous bases in a nucleotide that make up RNA. Uracil is not found in DNA. (7.1)

urban sprawl a pattern of scattered, uncontrolled development in which working, living, and recreational areas are widely separated. (15.1)

ureters tubes that carry waste fluid (urine) from the kidneys to the urinary bladder for temporary storage. (4.2)

urethra the tube through which urine exits the bladder (4.2)

valence electrons the electrons in the outermost occupied shell, or energy level, of an atom in its lowest energy state; valence electrons dictate the chemical properties of the element. (1.1)

valence level the outermost energy level of an atom. (1.1)

valence shell the outermost occupied shell or energy level of an atom in its lowest energy state. (1.1)

vasoconstriction decrease in the diameter of blood vessels near the skin to conserve body heat. Blood circulation is concentrated in the core of the body to keep the major organs functioning. (4.1)

vasodilation expansion of the diameter of blood vessels near the skin to bring more blood to the surface to help reduce body temperature. (4.1)

vasomotor centre part of the medulla oblongata that adjusts blood pressure by controlling the diameter of blood vessels. (5.1)

vestigial structures body parts that were functional in a species' ancestors, yet appear to have no current function (e.g., pelvic bones in baleen whales). (10.3)

vitreous humour the gel filling the compartment behind the lens in the eye. The gel helps maintain the shape of the eyeball. (5.3)

wave of depolarization nervous impulse transmitted along the axon of a nerve cell (neuron). (5.1)

whole genome shotgun sequencing computer program designed to sequence a large genome, developed in 1996 by Craig Venter. The program skips the genome mapping stage and breaks the entire genome into random fragments of first about 2000 and then about 1000 base pairs. The program then sequences and analyzes these fragments, and assembles nucleotide sequences corresponding to chromosomes. (9.2)

zero population growth situation in which the population is not increasing or decreasing because birth and death rates are approximately equal and immigration and emigration rates are approximately equal. (15.1)

Index

The page numbers in **boldface** type indicate the pages where the terms are defined. Terms that occur in Investigations (*inv*), Thinking Labs (*TL*), and MiniLabs (*ML*) are also indicated.

Photo Credits

xii (bottom), From *Inquiry Into Life 9th ed.* © 2000, 1997 by McGraw-Hill Companies Inc.; **xiii** (bottom), *Biology 11* © 2001 McGraw-Hill Ryerson; **iv** (bottom left), K.R. Porter/Science Source/Photo Researchers, Inc.; **vii** (centre left), Artbase Inc.; **2** (top centre), Artbase Inc.; **2** (bottom centre), Phototake; **3** (background), Bettmann/CORBIS/MAGMA; **4** (bottom right), SIU/Visuals Unlimited, Inc.; **4** (top left), Artbase Inc.; **5** (centre), © Alfred Pasieka/Science Photo Library; **6** (centre right), From *Inquiry Into Life 9th ed.* © 2000, 1997 by McGraw-Hill Companies Inc.; **6** (bottom left), © Rick Doucet/ Jennifer Martin/Fisheries and Oceans Canada; **7** (top centre), From Raven and Johnson, *Understanding Biology* © 1995 Wm. C. Brown Communications Inc.; **9** (centre), From *Inquiry Into Life 9th ed.* © 2000, 1997 by McGraw-Hill Companies Inc.; **9** (bottom centre), Charles Falco/Science Source/Photo Researchers, Inc.; **10** (bottom centre), From *Inquiry Into Life 9th ed.* © 2000, 1997 by McGraw-Hill Companies Inc.; **11** (top right), From *Inquiry Into Life 9th ed.* © 2000, 1997 by McGraw-Hill Companies Inc.; **12** (centre left), From *Inquiry Into Life 9th ed.* © 2000, 1997 by McGraw-Hill Companies Inc.; **12** (top right), From *Inquiry Into Life 9th ed.* © 2000, 1997 by McGraw-Hill Companies Inc.; **13** (top left), From *Inquiry Into Life 9th ed.* © 2000, 1997 by McGraw-Hill Companies Inc.; **14** (bottom left), S. Maslowski/Visuals Unlimited, Inc.; **15** (centre right), From *Chemistry: The Molecular Nature of Matter and Change* © 2000, The McGraw-Hill Companies, Inc.; **19** (top right), Quest/Science Photo Library/Photo Researchers, Inc.; **21** (bottom right), From *Chemistry 11* © 2001 McGraw-Hill Ryerson Limited; **23** (bottom left), From *Chemistry 11* © 2001 McGraw-Hill Ryerson Limited; **25** (top right), From Raven and Johnson, *Understanding Biology* © 1995 Wm. C. Brown Communications Inc.; **26** (bottom centre), From *Inquiry Into Life 9th ed.* © 2000, 1997 by McGraw-Hill Companies Inc.; **27** (top right), From *Inquiry Into Life 9th ed.* © 2000, 1997 by McGraw-Hill Companies Inc.; **27** (bottom centre), From *Inquiry Into Life 9th ed.* © 2000, 1997 by McGraw-Hill Companies Inc.; **28** (centre), From *Inquiry Into Life 9th ed.* © 2000, 1997 by McGraw-Hill Companies Inc.; **29** (centre), From *Inquiry Into Life 9th ed.* © 2000, 1997 by McGraw-Hill Companies Inc.; **29** (top centre), From *Inquiry Into Life 9th ed.* © 2000, 1997 by McGraw-Hill Companies Inc.; **30** (top right), From *Biology 11* © 2001 McGraw-Hill Ryerson; **32** (top left), From *Inquiry Into Life 9th ed.* © 2000, 1997 by McGraw-Hill Companies Inc.; **33** (top right), From Raven and Johnson, *Understanding Biology* © 1995 Wm. C. Brown Communications Inc.; **34** (bottom right), © Don W. Fawcett/Visuals Unlimited, Inc.; **35** (background), Ted Mathieu/Firstlight.ca; **36** (bottom left), Roy Ooms/ Masterfile; **40** (centre right), From Raven and Johnson, *Understanding Biology* © 1995 Wm. C. Brown Communications Inc.; **41** (bottom centre), From Raven and Johnson, *Understanding Biology* © 1995 Wm. C. Brown Communications Inc.; **43** (top centre), From Raven and Johnson, *Understanding Biology* © 1995 Wm. C. Brown Communications Inc.; **43** (bottom centre), From Raven and Johnson, *Understanding Biology* © 1995 Wm. C. Brown Communications Inc.; **45** (centre left), Artbase Inc.; **48** (top left), © Cheryl A. Ertelt/Visuals Unlimited, Inc.; **48** (bottom left), From Raven and Johnson, *Understanding Biology* © 1995 Wm. C. Brown Communications Inc.; **48** (top right), From *Inquiry Into Life 9th ed.* © 2000, 1997 by McGraw-Hill Companies Inc.; **49** (bottom left), Abbott Laboratories; **51** (top left), Elaine de Rooy; **52** (bottom left), Courtesy of the Nobel Prize Foundation; **53** (top left), Raymond Kao Photography Inc. Vancouver; **54** (bottom left), From *Inquiry Into Life 9th ed.* © 2000, 1997 by McGraw-Hill Companies Inc.; **55** (top right), © Norbert Rosing/Firstlight.ca; **55** (bottom left), From *Inquiry Into Life 9th ed.* © 2000, 1997 by McGraw-Hill Companies Inc.; **56** (top left), From Raven and Johnson, *Understanding Biology* © 1995 Wm. C. Brown Communications Inc.; **56** (top right), From Raven and Johnson, *Understanding Biology* © 1995 Wm. C. Brown Communications Inc.; **57** (centre left), From *Inquiry Into Life 9th ed.* © 2000, 1997 by McGraw-Hill Companies Inc.; **58** (top left), Artbase Inc.; **62** (bottom right), © George B. Chapman/Visuals Unlimited, Inc.; **63** (background), Allsport UK/Allsport; **64** (centre left), From *Inquiry Into Life 9th ed.* © 2000, 1997 by McGraw-Hill Companies Inc.; **64** (centre right), © David M. Phillips/Visuals Unlimited, Inc.; **65** (centre left), From *Inquiry Into Life 9th ed.* © 2000, 1997 by McGraw-Hill Companies Inc.; **66** (top right), From Raven and Johnson, *Understanding Biology* © 1995 Wm. C. Brown Communications Inc.; **67** (centre left), From *Inquiry Into Life 9th ed.* © 2000, 1997 by McGraw-Hill Companies Inc.; **69** (bottom right), K.R. Porter/Science Source/Photo Researchers, Inc.; **69** (bottom left), From *Inquiry Into Life 9th ed.* © 2000, 1997 by McGraw-Hill Companies Inc.; **71** (centre), From Raven and Johnson, *Understanding Biology* © 1995 Wm. C. Brown Communications Inc.; **73** (centre left), From *SciencePower 9* © 1999, McGraw-Hill Ryerson Limited, a subsidiary of The McGraw-Hill Companies.; **74** (centre left), From *Inquiry Into Life 9th ed.* © 2000, 1997 by McGraw-Hill Companies Inc.; **75** (bottom centre), © Tom Kennedy/Romark Illustrations; **76** (bottom right), From *Inquiry Into Life 9th ed.* © 2000, 1997 by McGraw-Hill Companies Inc.; **77** (bottom right), Duomo/CORBIS/MAGMA; **77** (top right), Phototake; **78** (bottom right), Artbase Inc.; **80** (centre left), From *Inquiry Into Life 9th ed.* © 2000, 1997 by McGraw-Hill Companies Inc.; **80** (bottom left), © David M. Phillips/Visuals Unlimited, Inc.; **83** (centre right), John Foster/Masterfile; **83** (bottom right), © James W. Richardson/ Visuals Unlimited, Inc.; **84** (top left), From Raven and Johnson, *Understanding Biology* © 1995 Wm. C. Brown Communications Inc.; **84** (centre right), From Raven and Johnson, *Understanding Biology* © 1995 Wm. C. Brown Communications Inc.; **85** (bottom right), © D. Newman/Visuals Unlimited, Inc.; **85**, From Oram, *Biology: Living Systems* 1998; **86** (top left), From Raven and Johnson, *Understanding Biology* © 1995 Wm. C. Brown Communications Inc.; **86** (centre left), From *Inquiry Into Life 9th ed.* © 2000, 1997 by McGraw-Hill Companies Inc.; **87** (bottom centre), © Beth Davidow/Visuals Unlimited; **88** (bottom left), From *Inquiry Into Life 9th ed.* © 2000, 1997 by McGraw-Hill Companies Inc.; **89** (bottom centre), From *Inquiry Into Life 9th ed.* © 2000, 1997 by McGraw-Hill Companies Inc.; **91** (centre right), Artbase Inc.; **91** (bottom right), Artbase Inc.; **91** (top right), From *Inquiry Into Life 9th ed.* © 2000, 1997 by McGraw-Hill Companies Inc.; **92** (top left), From *Inquiry Into Life 9th ed.* © 2000, 1997 by McGraw-Hill Companies Inc.; **92** (bottom right), © Jeremy Burgess/SPL/Photo Researchers, Inc.; **93** (bottom centre), From *Inquiry Into Life 9th ed.* © 2000, 1997 by McGraw-Hill Companies Inc.; **99** (top left), From *Chemistry: Concepts and Applications* © Glencoe/The McGraw-Hill

Companies, Inc.; **99** (top right), From *Chemistry: Concepts and Applications* © Glencoe/The McGraw-Hill Companies, Inc.; **104** (top centre), Artbase Inc.; **104** (bottom centre), Artbase Inc.; **105** (background), CORBIS/Warren Morgan/Firstlight.ca; **106** (top left), Artbase Inc.; **107** (background), Hans Blohm/Masterfile; **108** (bottom left), From *Inquiry Into Life 9th ed.* © 2000, 1997 by McGraw-Hill Companies Inc.; **110** (bottom centre), From Raven and Johnson, *Understanding Biology* © 1995 Wm. C. Brown Communications Inc.; **111** (centre), From *Anatomy & Physiology The Unity of Form and Function* Second Edition; **112** (bottom left), From Oram, *Biology: Living Systems*, 1998; **113** (centre left), AP/Wide World Photos; **114** (bottom left), From *Inquiry Into Life 9th ed.* © 2000, 1997 by McGraw-Hill Companies Inc.; **114** (bottom left), From Oram, *Biology: Living Systems*, 1998, Glencoe; **115** (top right), From *Inquiry Into Life 9th ed.* © 2000, 1997 by McGraw-Hill Companies Inc.; **116** (centre), From *Inquiry Into Life 9th ed.* © 2000, 1997 by McGraw-Hill Companies Inc.; **117** (bottom left), From Glencoe, *Biology: The Dynamics of Life* © The McGraw-Hill Companies Inc.; **119** (bottom right), Artbase Inc.; **119** (top right), Richard Megna/Fundamental Photos; **121** (centre right), © David M. Phillips/Visuals Unlimited, Inc.; **122** (centre left), From *Inquiry Into Life 9th ed.* © 2000, 1997 by McGraw-Hill Companies Inc.; **125** (centre right), © Kwangshin Kim/Photo Researchers, Inc.; **126** (top right), From Glencoe, *Biology: The Dynamics of Life* © The McGraw-Hill Companies Inc.; **127** (centre left), From *Inquiry Into Life 9th ed.* © 2000, 1997 by McGraw-Hill Companies Inc.; **128** (top right), From Glencoe, *Biology: The Dynamics of Life* © The McGraw-Hill Companies Inc.; **128** (top right), From Glencoe, *Biology: The Dynamics of Life* © The McGraw-Hill Companies Inc.; **128** (top right), From Glencoe, *Biology: The Dynamics of Life* © The McGraw-Hill Companies Inc.; **128** (bottom right), Martha Cooper/Peter Arnold, Inc.; **129** (top centre), From Glencoe, *Biology: The Dynamics of Life* © The McGraw-Hill Companies Inc.; **130** (top left), St. Louis Studio & Camera Inc.; **131** (bottom right), Ian Crysler; **136** (bottom right), Colin Chumbley/Science Source/Photo Researchers, Inc.; **137** (background), David & Peter Turnley/CORBIS/MAGMA; **138** (bottom left), CORBIS/Michael Pole/Firstlight.ca; **139** (centre), From Raven and Johnson, *Understanding Biology* © 1995 Wm. C. Brown Communications Inc.; **140** (bottom centre), Oram, *Biology: Living Systems*, 1998, Glencoe; **142** (top left), Melanie Brown/Photo Edit Inc.; **142** (top centre), From *Anatomy & Physiology The Unity of Form and Function* Second Edition; **142** (bottom left), American Philosophical Society; **142** (bottom right), Courtesy of Washington University Magazine 58(2), Summer, 1988; **143** (bottom centre), From *Inquiry Into Life 9th ed.* © 2000, 1997 by McGraw-Hill Companies Inc.; **145** (centre), From *Inquiry Into Life 9th ed.* © 2000, 1997 by McGraw-Hill Companies Inc.; **148** (centre left), From Glencoe, *Biology: The Dynamics of Life* © The McGraw-Hill Companies Inc.; **149** (bottom right), From *Inquiry Into Life 9th ed.* © 2000, 1997 by McGraw-Hill Companies Inc.; **149** (bottom right), AFP/CORBIS/MAGMA; **150** (bottom centre), From *Inquiry Into Life 9th ed.* © 2000, 1997 by McGraw-Hill Companies Inc.; **153** (centre), From *Inquiry Into Life 9th ed.* © 2000, 1997 by McGraw-Hill Companies Inc.; **154** (bottom right), Ian Crysler; **155** (top right), From *Inquiry Into Life 9th ed.* © 2000, 1997 by McGraw-Hill Companies Inc.; **156** (centre), From *Inquiry Into Life 9th ed.* © 2000, 1997 by McGraw-Hill Companies Inc.; **157** (top centre), From *Inquiry Into Life 9th ed.* © 2000, 1997 by McGraw-Hill Companies Inc.; **159** (bottom centre), From *Inquiry Into Life 9th ed.* © 2000, 1997 by McGraw-Hill Companies Inc.; **159** (top centre), From *Inquiry Into Life 9th ed.* © 2000, 1997 by McGraw-Hill Companies Inc.; **160** (top), From Glencoe, *Biology: The Dynamics of Life* © The McGraw-Hill Companies Inc.; **160** (bottom left), Photo Researchers Inc.; **164** (centre right), From *Inquiry Into Life 9th ed.* © 2000, 1997 by McGraw-Hill Companies Inc.; **166** (bottom right), Tim Beddow/Science Photo Library/Photo Researchers, Inc.; **167** (background), Jon Feingersch/The Stock Market/Firstlight.ca; **168** (bottom right), From *Inquiry Into Life 9th ed.* © 2000, 1997 by McGraw-Hill Companies Inc.; **171** (bottom), From *Inquiry Into Life 9th ed.* © 2000, 1997 by McGraw-Hill Companies Inc.; **172** (bottom right), Gary Carlson/Photo Researchers, Inc.; **173** (centre), From *Inquiry Into Life 9th ed.* © 2000, 1997 by McGraw-Hill Companies Inc.; **174** (centre right), © Ewing Galloway, Inc.; **175** (bottom right), Inga Spence/Visuals Unlimited, Inc.; **175** (top left), Courtesy Department of Illustrations, Washington University School of Medicine; **175** (top centre), Courtesy Department of Illustrations, Washington University School of Medicine; **175** (top right), Courtesy Department of Illustrations, Washington University School of Medicine; **177** (bottom right), K.R. Porter/Photo Researchers Inc.; **177** (top left), Lawrence Migdale/Photo Researchers, Inc.; **178** (top right), BSIP Bajande/Photo Researchers, Inc.; **179** (centre left), © Biophoto Associates/Photo Researchers, Inc.; **179** (centre right), David Young-Wolff/PhotoEdit; **180** (top centre), From *Inquiry Into Life 9th ed.* © 2000, 1997 by McGraw-Hill Companies Inc.; **182** (bottom right), Willie Hill, Jr./The Image Works; **184** (centre), From *Inquiry Into Life 9th ed.* © 2000, 1997 by McGraw-Hill Companies Inc.; **186** (bottom right), From *Atlas of Pediatric Physical Diagnosis*, Second Edition by Zitelli & Davis, 1992. Mosby-Wolfe Europe Limited, London, UK.; **186** (bottom right), From *Atlas of Pediatric Physical Diagnosis*, Second Edition by Zitelli & Davis, 1992. Mosby-Wolfe Europe Limited, London, UK.; **188** (top right), Amy Etra/PhotoEdit; **189** (bottom left), Ian Crysler; **190** (bottom left), AP/Wide World Photos; **192** (bottom left), Dandy Zipper/Getty Images/The Image Bank; **193** (bottom centre), From *Inquiry Into Life 9th ed.* © 2000, 1997 by McGraw-Hill Companies Inc.; **194** (top centre), © Biophoto Associates/Photo Researchers, Inc.; **195** (top right), From *Inquiry Into Life 9th ed.* © 2000, 1997 by McGraw-Hill Companies Inc.; **196** (centre), Paul A. Shouders/CORBIS/MAGMA; **196** (centre right), A. Remey/PhotoEdit; **198** (bottom centre), From *Inquiry Into Life 9th ed.* © 2000, 1997 by McGraw-Hill Companies Inc.; **199** (centre left), From Raven and Johnson, *Understanding Biology* © 1995 Wm. C. Brown Communications Inc.; **201** (bottom right), Artbase Inc.; **208** (bottom left), Photo Researchers, Inc.; **211** (top left), From *Inquiry Into Life 9th ed.* © 2000, 1997 by McGraw-Hill Companies Inc.; **213** (top right), Artbase Inc.; **214** (bottom right), E. Kiseleva-D. Fawcett/Visuals Unlimited, Inc.; **215** (background), G.C. Kelley/Photo Researchers, Inc.; **216** (bottom right), K.G. Murti/Visuals Unlimited, Inc.; **217** (background), Artbase Inc.; **218** (centre right), This picture was kindly provided by M. Courvoisier (+41)(61)267-3140) from the "Portrait Sammlung" of the University of Basel; **219** (top left), From Raven and Johnson, *Understanding Biology* © 1995 Wm. C. Brown Communications Inc.; **219** (bottom left), Edgar Fahs Smith Collection University of Pennsylvania Library; **219** (top right), From *Inquiry Into Life 9th ed.* © 2000, 1997 by McGraw-Hill Companies Inc.; **220** (centre), From Raven and Johnson, *Understanding Biology* © 1995 Wm. C. Brown Communications Inc.; **221** (top right), American Philosophical Society; **221** (bottom right), Lee D. Simon/Photo Researchers, Inc.; **224** (centre right), © Biophysics Dept., King's College, London, U.K.; **224** (centre left), © College Archives, King's College, London, U.K.; **225** (top left), © A Barrington Brown/Science Source/Photo Researchers, Inc.; **225** (centre right), From Raven and Johnson, *Understanding Biology* © 1995 Wm. C. Brown Communications Inc.; **227** (top centre), From *Inquiry Into Life 9th ed.* © 2000, 1997 by McGraw-Hill Companies Inc.; **228** (centre right), Biology Media/Science Source/Photo Researchers, Inc.; **229** (bottom left), K.G. Murti/Visuals Unlimited, Inc.; **231** (top left), K.G. Murti/Visuals Unlimited, Inc.; **231** (top right), Biophoto Associates/Photo Researchers, Inc.; **232** (bottom left), © Petit Format/Nestle/Science Source/Photo Researchers, Inc.; **235** (top left), NIH/Kakefuda/Science Source/Photo Researchers, Inc.; **238** (bottom right), Ian Crysler; **241** (bottom left), David Hall/Photo Researchers, Inc.; **242** (top centre), From Raven and Johnson, *Understanding Biology* © 1995 Wm. C. Brown Communications Inc.; **244** (bottom left), Dr. Renee Martin; **249** (top right), NIH/Kakefuda/Science Source/Photo Researchers Inc.; **250** (bottom right), Bonnie Sue/Photo Researchers, Inc.; **251** (background), Gary Meszaros/Visuals Unlimited, Inc.; **252** (bottom left), Artbase Inc.; **253** (centre right), From Raven and Johnson, *Understanding Biology* © 1995 Wm. C. Brown Communications Inc.; **255** (centre right), K.R. Porter/Photo Researchers, Inc.; **257** (centre right), Artbase Inc.; **259** (top centre), From Raven and Johnson, *Understanding Biology* © 1995 Wm. C. Brown Communications Inc.; **259** (centre left), Oscar L. Miller/Science Photo Library/Photo Researchers, Inc.; **260** (bottom right), Photo courtesy of Bert O'Malley; **263** (centre left), John Henley/Firstlight.ca; **269** (bottom left), Kiseleva-Fawcett/Visuals Unlimited, Inc.; **270** (bottom left), Doell Photo; **271** (top left), Professor Oscar Miller/Science Photo Library/Photo Researchers, Inc.; **272** (bottom left), Dan Guravich/Photo Researchers, Inc.; **272** (centre right), Andrew J. Martinez/Photo Researchers, Inc.; **277** (centre left), From *Inquiry Into Life 9th ed.* © 2000, 1997 by McGraw-Hill Companies Inc.; **277** (top right), From *Inquiry Into Life 9th ed.* © 2000, 1997 by McGraw-Hill Companies Inc.; **278** (top right), Owen Franken/CORBIS/MAGMA; **279** (bottom left), O.L. Miller, B.R. Beatty & D.W. Fawcett/Visuals Unlimited, Inc.; **279** (bottom right), E. Kiseleva & D. Fawcett/Visuals Unlimited, Inc.; **282** (centre right), Oscar L. Miller, Jr.; **284** (bottom right), VO TRUNG DUNG/CORBIS SYGMA/MAGMA; **285** (background), Carmela Leszczynski/Animals Animals; **286** (bottom left), Minnesota Pollution Control Agency; **288** (top left), From *Inquiry Into Life 9th ed.* © 2000, 1997 by McGraw-Hill Companies Inc.; **288** (bottom right), Will & Deni McIntyre/Photo Researchers, Inc.; **289** (bottom centre), From Raven and Johnson, *Understanding Biology* © 1995 Wm. C. Brown Communications Inc.; **293** (bottom left), Bettmann/CORBIS/MAGMA; **295** (top centre), From Raven and Johnson, *Understanding Biology* © 1995 Wm. C. Brown Communications Inc.; **295** (bottom centre), From *Inquiry Into Life 9th ed.* © 2000, 1997 by McGraw-Hill Companies Inc.; **296** (bottom left), From Raven and Johnson, *Understanding Biology* © 1995 Wm. C. Brown Communications Inc.; **297** (bottom right), Jean-Claude Revy/ISM-Phototake; **298** (top centre), From Raven and Johnson, *Understanding Biology* © 1995 Wm. C. Brown Communications Inc.; **299** (centre right), AFP/CORBIS/MAGMA; **300** (bottom right), Jeffrey L. Rotman/CORBIS/MAGMA; **303** (bottom left), Galleria degali Uffizi, Florence, Italy/Bridgeman Library; **306** (bottom centre), Courtesy of Dr. Stanley N. Cohen; **307** (top left), © Dr. John Sanford/Cornell University; **307** (centre right), Artbase Inc.; **309** (bottom left), Papilio/CORBIS/MAGMA; **310** (bottom left), Kevin R. Morris/CORBIS/MAGMA; **312** (bottom left), George Bernard/Science Photo Library/Photo Researchers, Inc.; **312** (bottom right), Biophoto Associates/Photo Researchers, Inc.; **313** (centre right), MURDO MACLEOD/CORBIS/MAGMA; **317** (centre left), Mark Richards/PhotoEdit; **322** (centre left), Reuters New Media Inc./CORBIS/MAGMA; **327** (top left), Kiseleva-Fawcett/Visuals Unlimited, Inc.; **330** (bottom centre), Andrew MacRae/Smithsonian Institute; **331** (background), Jim Zipp/Photo Researchers, Inc.; **332** (bottom right), Trevor Bonderud/Firstlight.ca; **333** (background), Tom & Pat Leeson/Photo Researchers, Inc.; **334** (bottom left), Ron Planck/Photo Researchers, Inc.; **335** (top centre), From *Inquiry Into Life 9th ed.* © 2000, 1997 by McGraw-Hill Companies Inc.; **338** (bottom), Carolyn A. McKeone/Photo Researchers, Inc.; **342** (bottom centre), From Raven and Johnson, *Understanding Biology* © 1995 Wm. C. Brown Communications Inc.; **343** (top left), From Raven and Johnson, *Understanding Biology* © 1995 Wm. C. Brown Communications Inc.; **343** (top left), From Raven and Johnson, *Understanding Biology* © 1995 Wm. C. Brown Communications Inc.; **343** (bottom left), Artbase Inc.; **343** (bottom right), Hal Beral/Visuals Unlimited, Inc.; **344** (bottom centre), From *Inquiry Into Life 9th ed.* © 2000, 1997 by McGraw-Hill Companies Inc.; **346** (bottom centre), From *Life Science* by Lucy Daniel © 1997 Glencoe/McGraw-Hill; **349** (bottom), Marvin Mattelson/National Geographic Society; **350** (bottom centre), From Glencoe, *Biology: The Dynamics of Life* © The McGraw-Hill Companies Inc.; **350** (top right), Jon Nelson; **351** (bottom centre), From Raven and Johnson, *Understanding Biology* © 1995 Wm. C. Brown Communications Inc.; **351** (centre right), James L. Amos/Photo Researchers, Inc.; **352** (centre left), University of Cambridge; **353** (top left), Victor Boswell/National Geographic Society; **353** (bottom left), Minden Pictures/Firstlight.ca; **353** (bottom centre), Brian Enting/Photo Researchers, Inc.; **353** (bottom right), Tom McHugh/Photo Researchers, Inc.; **354** (bottom right), From Raven and Johnson, *Understanding Biology* © 1995 Wm. C. Brown Communications Inc.; **355** (bottom left), From Raven and Johnson, *Understanding Biology* © 1995 Wm. C. Brown Communications Inc.; **355** (centre right), From *Life Science* by Lucy Daniel © 1997 Glencoe/McGraw-Hill; **356** (bottom right), Dr. Wu Xiao-Chun; **358** (top left), From Raven and Johnson, *Understanding Biology* © 1995 Wm. C. Brown Communications Inc.; **360** (top left), From Glencoe, *Biology: Living Systems* © The McGraw-Hill Companies Inc.; **362** (bottom right), David M. Phillips/Visuals Unlimited, Inc.; **363** (background), Artbase Inc.; **364** (bottom centre), From Glencoe, *Biology: The Dynamics of Life* © The McGraw-Hill Companies Inc.; **365** (bottom left), From Raven and Johnson, *Understanding Biology* © 1995 Wm. C. Brown Communications Inc.; **366** (centre right), From Glencoe, *Biology: Living Systems* © The McGraw-Hill Companies Inc.; **367** (top left), From Glencoe, *Biology: Living Systems* © The McGraw-Hill Companies Inc.; **367** (bottom centre), © 1988 Chris Bjornberg/Photo Researchers, Inc.; **368** (top left), Barbara Magnuson/Visuals

Unlimited, Inc.; **368** (top right), Scott W. Smith/Animals Animals Earth Science; **372** (bottom centre), From Glencoe, *Biology: The Dynamics of Life* © The McGraw-Hill Companies Inc.; **377** (bottom left), Joe McDonald/Visuals Unlimited, Inc.; **378** (top right), Bernard Photo Productions/Animals Animals Earth Science; **380** (centre right), David B. Fleetham/ Visuals Unlimited, Inc.; **381** (bottom centre), From Glencoe, *Biology: The Dynamics of Life* © The McGraw-Hill Companies Inc.; **383** (bottom left), From Raven and Johnson, *Understanding Biology* © 1995 Wm. C. Brown Communications Inc.; **383** (bottom right), From *Inquiry Into Life 9th ed.* © 2000, 1997 by McGraw-Hill Companies Inc.; **384** (bottom left), From *Inquiry Into Life 9th ed.* © 2000, 1997 by McGraw-Hill Companies Inc.; **385** (top left), From *Inquiry Into Life 9th ed.* © 2000, 1997 by McGraw-Hill Companies Inc.; **385** (bottom centre), Rob Simpson/Visuals Unlimited, Inc.; **385** (bottom right), Gerard Fuehrer/Visuals Unlimited, Inc.; **388** (bottom left), © 1988 Chris Bjornberg/ Photo Researchers, Inc.; **390** (bottom right), Budd Titlow/Visuals Unlimited, Inc.; **391** (background), BSIP/Science Source/Photo Researchers, Inc.; **392** (bottom left), Johnny Jensen/Visuals Unlimited, Inc.; **393** (centre right), Kenneth Read/Photo Researchers, Inc.; **393** (bottom right), Kjell Sandved/Photo Researchers, Inc.; **395** (bottom left), Artbase Inc.; **395** (bottom right), Scott Camazine/Photo Researchers, Inc.; **396** (bottom left), Artbase Inc.; **398** (bottom left), Ron Austing/Photo Researchers, Inc.; **398** (bottom centre), McDonald Wildlife Photography/Animals Animals Earth Science; **399** (centre right), Joseph T. Collins/Photo Researchers, Inc.; **399** (bottom right), Wild Nature/Animals Animals Earth Science; **400** (bottom right), Gregory Ochocki/Photo Researchers, Inc.; **401** (bottom left), Larry West/Photo Researchers, Inc.; **401** (bottom centre), Suzanne L. Collins & Joseph T. Collins/Photo Researchers, Inc.; **402** (centre right), A. Blachford; **404** (bottom centre), From Oram, *Biology: Living Systems* 1998, Glencoe/McGraw-Hill Companies; **405** (centre right), Norbert Wu Productions; **406** (bottom right), From *Inquiry Into Life 9th ed.* © 2000, 1997 by McGraw-Hill Companies Inc.; **407** (bottom centre), From Raven and Johnson, *Understanding Biology* © 1995 Wm. C. Brown Communications Inc.; **409** (centre), From Glencoe, *Biology: The Dynamics of Life* © The McGraw-Hill Companies; **410** (bottom centre), From Oram, *Biology: Living Systems* 1998, Glencoe/McGraw-Hill Companies; **411** (bottom left), Maslowski/Photo Researchers, Inc.; **412** (bottom left), Kjell B. Sandved/Photo Researchers, Inc.; **413** (bottom centre), © Tom Kennedy/Romark Illustrations; **414** (centre), From *Inquiry Into Life 9th ed.* © 2000, 1997 by McGraw-Hill Companies Inc.; **415** (bottom centre), © Tom Gagliano; **416** (top centre), Oram, *Biology: Living Systems*, 1998, Glencoe; **419** (top right), From Raven and Johnson, *Understanding Biology* © 1995 Wm. C. Brown Communications Inc.; **424** (centre), © Michael and Patricia Fogden; **426** (bottom), Tim Hauf/Visuals Unlimited, Inc.; **427** (background), Artbase Inc.; **428** (bottom right), Biosphere 2 Centre, Inc.; **429** (background), Artbase Inc.; **430** (bottom), From *SciencePower 10* © 2000, McGraw-Hill Ryerson Limited, a subsidiary of The McGraw-Hill Companies Inc.; **431** (bottom), From *Inquiry Into Life 9th ed.* © 2000, 1997 by McGraw-Hill Companies Inc.; **431** (top right), Richard R. Hansen/Photo Researchers, Inc.; **432** (top right), Kent and Donna Dannen/Photo Researchers, Inc.; **432** (bottom left), D.C. Lowe/Firstlight.ca; **432** (bottom right), Peter Dunwiddic/Visuals Unlimited, Inc.; **433** (bottom left), Georg Gerster/Photo Researchers, Inc.; **433** (top right), Gerald & Buff Corsi/Visuals Unlimited, Inc.; **434** (top left), Division of the Environment, University of Toronto; **434** (top right), From *Inquiry Into Life 9th ed.* © 2000, 1997 by McGraw-Hill Companies Inc.; **435** (bottom), From *Inquiry Into Life 9th ed.* © 2000, 1997 by McGraw-Hill Companies Inc.; **440** (bottom left), Jerome Wexler/Photo Researchers, Inc.; **442** (top left), Whoi D. Foster/Visuals Unlimited, Inc.; **447** (bottom), From *SciencePower 10* © McGraw-Hill Ryerson Limited; **450** (top), From *SciencePower 10* © McGraw-Hill Ryerson Limited; **456** (bottom left), John D. Cunningham/Visuals Unlimited, Inc.; **456** (bottom right), John D. Cunningham/Visuals Unlimited, Inc.; **457** (bottom left), Daryl Benson/Masterfile; **464** (bottom right), Grant Heilman/Comstock; **465** (background), Scott W. Smith/ Animals Animals Earth Science; **466** (centre right), Frank Lane Picture Agency/CORBIS/MAGMA; **467** (centre right), From *SciencePower 10* © McGraw-Hill Ryerson Limited; **470** (top left), David M. Grossman/Photo Researchers, Inc.; **470** (top right), Artbase Inc.; **472** (centre left), Austin J. Stevens/Animals Animals Earth Science; **472** (centre right), Firstlight.ca; **475** (bottom), From Raven and Johnson, *Understanding Biology* © 1995 Wm. C. Brown Communications Inc.; **480** (top left), Tim Davis/Photo Researchers, Inc.; **481** (bottom right), Ivy Images; **483** (centre right), © Lee F. Snyder/Photo Researchers Inc.; **483** (bottom right), John Sohlden/Visuals Unlimited, Inc.; **484** (top right), From *SciencePower 10* © McGraw-Hill Ryerson Limited; **488** (bottom right), Jeff J. Daly/Visuals Unlimited, Inc.; **489** (bottom left), Dan Guravich/Photo Researchers, Inc.; **495** (centre left), Artbase Inc.; **502** (bottom right), B. Banaszewski/Visuals Unlimited, Inc.; **503** (background), Jeff Greenberg/Visuals Unlimited, Inc.; **508** (bottom right), Beth Davidson/Visuals Unlimited, Inc.; **514** (centre right), Alain Evrard/Photo Researchers, Inc.; **514** (bottom left), Dick Hemingway; **515** (top right), Nik Wheeler/CORBIS/MAGMA; **518** (top right), Artbase Inc.; **521** (bottom left), Lenore Fahrig; **524** (bottom right), AFP/CORBIS/MAGMA; **529** (bottom left), Didier Dorval/Masterfile; **531** (top right), T.C.L./Masterfile; **532** (top right), Artbase Inc.; **534** (top left), Artbase Inc.; **534** (centre right), Paul Kenward/Getty Images/Stone; **538** (bottom left), AP/Wide World Photos; **544** (centre left), Jeff Greenberg/Visuals Unlimited, Inc.; **545** (top left), Artbase Inc.; **548** (bottom left), Artbase Inc.; **549** (top right), Catherine Little; **551** (centre left), *Biology 11* © 2001 McGraw-Hill Ryerson; **551** (centre right), *Biology 11* © 2001 McGraw-Hill Ryerson; **552** (bottom left), *Biology 11* © 2001 McGraw-Hill Ryerson; **552** (top right), *Biology 11* © 2001 McGraw-Hill Ryerson; **553** (bottom), From *Chemistry: The Molecular Nature of Matter and Change* Second Edition © 2000 The McGraw-Hill Companies, Inc.; **555** (centre right), From *Inquiry Into Life 9th ed.* © 2000, 1997 by McGraw-Hill Companies Inc.; **556** (centre), From *Biology 11* © 2001 McGraw-Hill Ryerson; **557** (centre right), From *Inquiry Into Life 9th ed.* © 2000, 1997 by McGraw-Hill Companies Inc.; **560** (centre), *Biology 11* © 2001 McGraw-Hill Ryerson; **562** (top right), © Tom Kennedy/Romark Illustrations/ *Biology: The Dynamics of Life* © 2000, McGraw-Hill Companies, Inc.; **567** (centre), From *Chemistry: The Molecular Nature of Matter and Change* Second Edition © 2000 The McGraw-Hill Companies, Inc.

Answers to Practice Problems

Student Textbook p. 370

1. genotype frequency:
   homozygous dominant (0.55)
   heterozygous (0.27)
   homozygous recessive (0.18)

   allele frequency:
   dominant allele (0.69)
   recessive allele (0.31)

2.

phenotype	purple	purple	white
genotype	AA	Aa	aa
number	320	160	20
genotype frequencies	0.64	0.32	0.04
number of alleles in gene pool	A = 800		a = 200
allele frequencies	A = 0.80		a = 0.20

Student Textbook p. 376

1. allele frequency: A = 0.6; a = 0.4
   genotype frequencies: AA = 0.36;
   Aa = 0.48; aa = 0.16

2. 21%

3. allele frequency: T = 0.9; t = 0.1
   genotype frequencies: TT = 0.81;
   Tt = 0.18; tt = 0.01

List of Elements

Element	Symbol	Atomic Number	Atomic Mass*
Actinium	Ac	89	(227)
Aluminum	Al	13	26.98
Americium	Am	95	(243)
Antimony	Sb	51	121.8
Argon	Ar	18	39.95
Arsenic	As	33	74.92
Astatine	At	85	(210)
Barium	Ba	56	137.3
Berkelium	Bk	97	(247)
Beryllium	Be	4	9.012
Bismuth	Bi	83	209.0
Bohrium	Bh	107	(264)
Boron	B	5	10.81
Bromine	Br	35	79.90
Cadmium	Cd	48	112.4
Calcium	Ca	20	40.08
Californium	Cf	98	(249)
Carbon	C	6	12.01
Cerium	Ce	58	140.1
Cesium	Cs	55	132.9
Chlorine	Cl	17	35.45
Chromium	Cr	24	52.00
Cobalt	Co	27	58.93
Copper	Cu	29	63.55
Curium	Cm	96	(247)
Dubnium	Db	105	(262)
Dysprosium	Dy	66	162.5
Einsteinium	Es	99	(254)
Erbium	Er	68	167.3
Europium	Eu	63	152.0
Fermium	Fm	100	(253)
Fluorine	F	9	19.00
Francium	Fr	87	(223)
Gadolinium	Gd	64	157.3
Gallium	Ga	31	69.72
Germanium	Ge	32	72.61
Gold	Au	79	197.0
Hafnium	Hf	72	178.5
Hassium	Hs	108	(269)
Helium	He	2	4.003
Holmium	Ho	67	164.9
Hydrogen	H	1	1.008
Indium	In	49	114.8
Iodine	I	53	126.9
Iridium	Ir	77	192.2
Iron	Fe	26	55.85
Krypton	Kr	36	83.80
Lanthanum	La	57	138.9
Lawrencium	Lr	103	(262)
Lead	Pb	82	207.2
Lithium	Li	3	6.941
Lutetium	Lu	71	175.0
Magnesium	Mg	12	24.31
Manganese	Mn	25	54.94
Meitnerium	Mt	109	(268)
Mendelevium	Md	101	(256)

Element	Symbol	Atomic Number	Atomic Mass*
Mercury	Hg	80	200.6
Molybdenum	Mo	42	95.94
Neodymium	Nd	60	144.2
Neon	Ne	10	20.18
Neptunium	Np	93	237.1
Nickel	Ni	28	58.70
Niobium	Nb	41	92.91
Nitrogen	N	7	14.01
Nobelium	No	102	(253)
Osmium	Os	76	190.2
Oxygen	O	8	16.00
Palladium	Pd	46	106.4
Phosphorus	P	15	30.97
Platinum	Pt	78	195.1
Plutonium	Pu	94	(244)
Polonium	Po	84	(209)
Potassium	K	19	39.10
Praseodymium	Pr	59	140.9
Promethium	Pm	61	(145)
Protactinium	Pa	91	231.0
Radium	Ra	88	(226)
Radon	Rn	86	(222)
Rhenium	Re	75	186.2
Rhodium	Rh	45	102.9
Rubidium	Rb	37	85.47
Ruthenium	Ru	44	101.1
Rutherfordium	Rf	104	(261)
Samarium	Sm	62	150.4
Scandium	Sc	21	44.96
Seaborgium	Sg	106	(266)
Selenium	Se	34	78.96
Silicon	Si	14	28.09
Silver	Ag	47	107.9
Sodium	Na	11	22.99
Strontium	Sr	38	87.62
Sulfur	S	16	32.07
Tantalum	Ta	73	180.9
Technetium	Tc	43	(98)
Tellurium	Te	52	127.6
Terbium	Tb	65	158.9
Thallium	Tl	81	204.4
Thorium	Th	90	232.0
Thulium	Tm	69	168.9
Tin	Sn	50	118.7
Titanium	Ti	22	47.88
Tungsten	W	74	183.9
Ununbium	Uub	112	(277)
Ununhexium	Uuh	116	(289)
Ununnilium	Uun	110	(271)
Ununquadium	Uuq	114	(285)
Unununium	Uuu	111	(272)
Uranium	U	92	238.0
Vanadium	V	23	50.94
Xenon	Xe	54	131.3
Ytterbium	Yb	70	173.0
Yttrium	Y	39	88.91
Zinc	Zn	30	65.39
Zirconium	Zr	40	91.22

*All atomic masses are given to four significant figures. Values in parentheses represent the mass number of the most stable form of the atom.

Periodic Table and List of Elements

Legend: Metals · Metalloids · Nonmetals

Key: atomic number / symbol / atomic mass

MAIN-GROUP ELEMENTS: groups 1A–2A and 3A–8A. TRANSITION ELEMENTS: groups 3B–2B.

Period	1A (1)	2A (2)	3B (3)	4B (4)	5B (5)	6B (6)	7B (7)	8B (8)	8B (9)	8B (10)	1B (11)	2B (12)	3A (13)	4A (14)	5A (15)	6A (16)	7A (17)	8A (18)
1	1 **H** 1.008																	2 **He** 4.003
2	3 **Li** 6.941	4 **Be** 9.012											5 **B** 10.81	6 **C** 12.01	7 **N** 14.01	8 **O** 16.00	9 **F** 19.00	10 **Ne** 20.18
3	11 **Na** 22.99	12 **Mg** 24.31											13 **Al** 26.98	14 **Si** 28.09	15 **P** 30.97	16 **S** 32.07	17 **Cl** 35.45	18 **Ar** 39.95
4	19 **K** 39.10	20 **Ca** 40.08	21 **Sc** 44.96	22 **Ti** 47.88	23 **V** 50.94	24 **Cr** 52.00	25 **Mn** 54.94	26 **Fe** 55.85	27 **Co** 58.93	28 **Ni** 58.69	29 **Cu** 63.55	30 **Zn** 65.39	31 **Ga** 69.72	32 **Ge** 72.61	33 **As** 74.92	34 **Se** 78.96	35 **Br** 79.90	36 **Kr** 83.80
5	37 **Rb** 85.47	38 **Sr** 87.62	39 **Y** 88.91	40 **Zr** 91.22	41 **Nb** 92.91	42 **Mo** 95.94	43 **Tc** (98)	44 **Ru** 101.1	45 **Rh** 102.9	46 **Pd** 106.4	47 **Ag** 107.9	48 **Cd** 112.4	49 **In** 114.8	50 **Sn** 118.7	51 **Sb** 121.8	52 **Te** 127.6	53 **I** 126.9	54 **Xe** 131.3
6	55 **Cs** 132.9	56 **Ba** 137.3	57 **La** 138.9	72 **Hf** 178.5	73 **Ta** 180.9	74 **W** 183.9	75 **Re** 186.2	76 **Os** 190.2	77 **Ir** 192.2	78 **Pt** 195.1	79 **Au** 197.0	80 **Hg** 200.6	81 **Tl** 204.4	82 **Pb** 207.2	83 **Bi** 209.0	84 **Po** (209)	85 **At** (210)	86 **Rn** (222)
7	87 **Fr** (223)	88 **Ra** (226)	89 **Ac** (227)	104 **Rf** (261)	105 **Db** (262)	106 **Sg** (266)	107 **Bh** (264)	108 **Hs** (269)	109 **Mt** (268)	110 **Uun** (271)	111 **Uuu** (272)	112 **Uub** (277)		114 **Uuq** (285)		116 **Uuh** (289)		

INNER TRANSITION ELEMENTS

Period															
6	Lanthanoids	58 **Ce** 140.1	59 **Pr** 140.9	60 **Nd** 144.2	61 **Pm** (145)	62 **Sm** 150.4	63 **Eu** 152.0	64 **Gd** 157.3	65 **Tb** 158.9	66 **Dy** 162.5	67 **Ho** 164.9	68 **Er** 167.3	69 **Tm** 168.9	70 **Yb** 173.0	71 **Lu** 175.0
7	Actinoids	90 **Th** 232.0	91 **Pa** 231.0	92 **U** 238.0	93 **Np** 237.1	94 **Pu** (244)	95 **Am** (243)	96 **Cm** (247)	97 **Bk** (247)	98 **Cf** (251)	99 **Es** (252)	100 **Fm** (257)	101 **Md** (258)	102 **No** (259)	103 **Lr** (262)